Moral Markets

Moral Markets

THE CRITICAL ROLE OF VALUES IN THE ECONOMY

Edited by

Paul J. Zak

With a foreword by Michael C. Jensen

PRINCETON UNIVERSITY PRESS • PRINCETON AND OXFORD

Library of Congress Cataloging-in-Publication Data

Moral markets : the critical role of values in the economy / edited by Paul J. Zak ; with a foreword by Michael C. Jensen.
    p. cm.
  Includes bibliographical references and index.
  ISBN 978-0-691-13522-9 (hardcover : alk. paper)—ISBN 978-0-691-13523-6 (pbk. : alk. paper)
  1. Economics—Moral and ethical aspects.   I. Zak, Paul J.
HB72.M558 2008
174—dc22   2007038797

British Library Cataloging-in-Publication Data is available

This book is dedicated to my big girl,

Elke Liesl Zak

The contributors to this book dedicate it to the memory of our friend and collaborator, Dr. Robert Solomon. Bob was an essential part of this research team, an enthusiastic mentor to all, and invariably dispensed his good humor, good spirits, and good sense in liberal quantities. We feel honored to have known and worked with him.

Contents

Foreword

Michael C. Jensen

My congratulations and admiration to the Gruter Institute for Law and Behavioral Research and the scholars who have evidenced the courage to tackle a subject as difficult and opaque as the Gruter Institute Project on Values and Free Enterprise. With the support of the John Templeton Foundation, the UCLA-Sloan Research Program on Business Organizations, and the Ann and Gordon Getty Foundation, the volume they have produced, *Moral Markets: The Critical Role of Values in the Economy*, promises to be a landmark effort in bringing scientific inquiry and analysis to the normative world of values. Economics, having traditionally focused on the positive analysis of alternative institutional structures, has far too long ignored the normative world.

By the term "positive analysis," I mean, of course, the analysis of the way the world is, how it behaves, independent of any normative value judgments about its desirability or undesirability. Such analysis leads to empirically testable propositions that are falsifiable in the sense of Karl Popper's revolutionary work "The Logic of Scientific Discovery."[1] By "normative," I mean establishing, relating to, or deriving from a standard or norm that specifies desirable or undesirable conduct or behavior, that is, what ought to be.[2]

The positive analysis of normative values may sound like a non sequitur, yet, in my opinion, it is among the major issues the world faces today. It does not take much reflection or study of history to begin to see the import of different judgments about normative values on the tensions and conflicts between human beings. There are, of course, many causes underlying the conflicts between humans in addition to those involving different value systems, including, for example, power, wealth, and so on. Nevertheless, the power of values, those deeply held personal beliefs about good versus bad behavior, desirable versus undesirable actions, right versus wrong, which cause, or at least encourage, human beings to commit the most horrific crimes against their neighbors is startling. Witness the multitudes of religious and civil conflicts ranging from the Crusades, the Holocaust, two world wars, and ethnic cleansing in Somalia to the brutal bombings, torture, and televised beheadings associated with the current conflicts involving radical Islam.

In the last two decades, we witnessed the end of three-quarters of a century of conflict and violence between the normative views of socialist and communist philosophy and those of capitalist philosophy over how the means of production in a society should be owned, directed, and coordinated, and how the output of that production should be distributed. Currently we are seeing the beginnings of what will likely be the next three-quarters of a century of conflict roiling civilization, in this case between radical Islam and most of the rest of the world over whether the very roots of modernity, the separation of church and state, will be overturned.

A potentially valuable source of knowledge in learning the source of these tensions, and therefore in learning how to manage and reduce them, is a much deeper understanding of the role of values in shaping and guiding human action and interaction. Clearly, this calls for a positive analysis of values: how they arise, how they change, how they interact across cultures, how they can be changed, and so on. Understanding how values are reflected in markets, and how markets either reduce or increase their positive or negative effects on human welfare, is obviously important. The work of the scholars represented in this volume contributes significantly to these issues and adds substantially to the foundations of analysis that will eventually lead to a richer and more complete understanding of these issues.

I look forward to seeing the creation of an entirely new field of inquiry in economics, and in its sister social sciences, focused deeply on the positive analysis of the role of values in elevating the possible outcomes of human interaction. This means understanding how values create, as well as facilitate, the management of conflicts between human beings. This does not require a discussion or debate about which values are better or worse but rather calls on us to examine the effects of differing values and how intentions diverge from actual results. A great start lies in creating a rich body of knowledge of how the values reflected in moral, ethical, and legal codes for standards of good versus bad behavior affect human interaction in families, groups, organizations, social cultures, and nations. The role of values is a purely positive question for economics and the other social sciences, and this volume begins that journey toward creating such a science.

Michael C. Jensen
Sarasota, Florida
February 2007

Notes

1. Karl Popper, *The Logic of Scientific Discovery* (New York: Basic Books, 1959).
2. See John Neville Keynes, *The Scope and Method of Political Economy*, 4th ed. (New York: Macmillan, 1917 [1891], pp. 34–35, 46; and Milton Friedman, *Essays in Positive Economics* (Chicago: University of Chicago Press, 1966), p. 3.

Introduction

Paul J. Zak

Just as nature is "red in tooth and claw," so is market exchange. All market participants—from business owners and managers to their employees to the grandmother shopping at her local grocery story—must eke out every possible efficiency or be crushed by the capitalist machine . . . or perhaps not. This book hopes to convince readers that both Alfred Tennyson's characterization of competition in nature, quoted above, and an uncritical reading about self-ish competition in markets in Adam Smith's *An Inquiry into the Nature and Causes of the Wealth of Nations* are not so much wrong as incomplete and, in the view of the contributors to this volume, woefully so.

The intertwining of biology and behavior is the leitmotif of this book. A biologically based approach both constrains and informs the analysis. The constraint restricts analyses that connect morals and markets to be consistent with evolution and behaviors in closely related species such as apes. Conclusions violating this constraint are fundamentally suspect. The biological approach is novel, as it elucidates variation along the continuum of behaviors, permitting one to extract that which is both relevant and amenable to policy. The rationale for this approach is to provide *convergent evidence* that modern market exchange is inconceivable without moral values. Although this idea is not new—scholars from Aristotle to Adam Smith have made related arguments—the revival of this idea and the modern evidence for it is controversial.

The controversy arises from the Marxist residue that invades our daily thought and language, as seen in the oft-repeated lines "Big business runs the economy" or "The little guy can't make a decent living." This economic victimization, however, is belied by the failure of large established corporations such as AT&T and the flourishing, world-changing success of businesses, such as the semiconductor-based computer industry, that began in a garage. How, one might ask, is any of this "moral," and, in fact, do moral values even exist?

The book begins by addressing the fundamental issue of whether moral values are identifiable, consistent, and real. Part 1 shows that they are. Once we have established that values can indeed be studied, we trace their emergence in certain species of monkeys and apes. We then look at the evolution of cultural norms and market institutions that reinforce virtuous propensities. This leads us to examine how current laws and public policy promote or inhibit our innate sense of values. Not everyone is virtuous in every circumstance, so exchange must occur in the shadow of enforcement. The book culminates by using these insights to directly examine how markets function when values are present and when they are in short supply.

The encompassing theme of *Moral Markets* is that human beings are a highly social species and gauge our own and others' behaviors against social expectations that manifest as values. Adam Smith, in his *Theory of Moral Sentiments* (1759), called this having "sympathy for others"; a more apt term today is "empathy," responding emotionally to another's needs. Emotional responses, in Smith's view, are the social glue that binds us together as a species and are the basis for moral behavior. Modern neuroscience research supports Smith's intuition.

Research on Morality

Understanding the basis for moral behavior is a topic of heightened academic and popular interest based on the spate of recent scientific articles and trade books on the topic.[1] Why do we care what makes others behave in good or evil ways? At one level, we cannot help it. We are hypersocial apes. We live and work among those to whom we are not closely related, strangers constantly in our midst. We are inveterate consumers and producers of gossip about others. We simply need to know the whats and whys.

Aristotle, Plato, Jesus, Buddha, and Mohammed all recognized that values are the foundation for happiness, and their wisdom has stood the test of time.[2] The modern tools of psychology and neuroscience are providing new, often compelling insights into moral values. In the dawn of the twenty-first century, every field of inquiry seems to have a "neuro" attached to it, and neuroethics has recently blossomed. Although the classical philosophers provided logical answers to ethical dilemmas, the question of why something was moral seemed forever in dispute, depending on exquisite (or excruciating) philosophical reasoning. By examining people's brains, an understanding of why we are moral (or not) is emerging.

Although an extensive survey of neuroethics is beyond the scope of this introduction, a brief overview will help contextualize many of the chapters in this book.[3] An important, consistent finding from neuroethics in the past ten years is that the Kantian notion of morality as being learned by rational deduction is generally incorrect. When viewing immoral acts, nearly all human beings have a visceral, emotional, and rapid neural response. In many cases, even children have similar reactions. In other words, one does not need to learn what is right and wrong, most of us know this, or perhaps better, sense this, immediately.

These moral emotions have been localized to evolutionarily old areas of the human brain. Yet, moral emotions were largely dismissed in the post-Cartesian world, where rationality was seen as mankind's crowning achievement and was thought to be in conflict with the emotions. Although distinct areas of the brain process information differently, which can provoke neural conflict, most often the parts of the brain associated with cognition and those associated with emotion work in tandem to guide our choices. This was elegantly demonstrated by Antonio Damasio's lab when studying patients with focal brain lesions in areas that process emotional stimuli. Damasio and colleagues found that these patients consistently make poor choices and have trouble adapting their behaviors to changing conditions.[4]

Many moral decisions have both cognitive and emotional components, including market decisions. For example, if your paycheck this month had an extra $10,000, you would likely feel joy ("extra money!"), and then perhaps fear ("I know they'll find me!"), and then maybe a cost-benefit calculation ("if I cash the check and play dumb, I can always return the money if they find out"). This might be followed by an emotionally weighted decision ("I didn't work for this money so I'm getting this unfairly, and therefore I should return it"). Not all of us would go through all these stages, but this example is meant to show that emotions and higher cognition are integrated and evolved to help us solve complex problems, including moral dilemmas.

Research on Markets

I recently attended a conference at University of California, Los Angeles. While walking from the parking structure two counties away to the faculty club, I spoke to some undergraduates who had set up a table with a sign that read "Students for Marxism." They were apparently trying to recruit members. I asked them why Marxism was the best way to organize a society. They repeated the usual egalitarian rhetoric, but could not answer my questions about how to spur innovation, and why the Soviet Union and other communist states had spectacularly failed; their best answer was "U.S. world domination," but this itself suggests that a democratic market system is a superior organizational form.

Why market economies have produced astonishing increases in wealth while socialist dictatorships have had disastrous economic performances has been the topic of extensive research.[5] The short answer is simple—we are not bees. Bees toil for the good of the hive because of the high degree of genetic relatedness between members and the near equality of productive skills between members (there is a simple division of labor in beehives between queens, workers, and drones). Human beings, on the other hand, have substantial, often identifiable, genetic variation. Our biological inclination to protect and nurture those who share our genes means that we favor the success of our family members over those to whom we are more distantly related.

Human beings have a wide variety of skills that they can offer to various businesses or use to start their own enterprises. We counsel our youth to find their niche in life, and indeed that is key to economic success. A market system generates a large number of organizations that seek people to provide a variety of services to them. Organizational types range from traditional, profit-oriented businesses to nonprofits to government to public and private universities. This plethora of opportunities and (often) objective markers of individual and organizational success allow the diversity of individual skills to find a remunerative employment niche.

In market economies, this niche-seeking behavior occurs organically as individuals and firms engage in bilateral search. In socialist economies, the assignment of individuals to jobs was typically done using a top-down approach. I may be able to work as an accountant, but it is not my passion, and I will have little incentive to improve myself as an accountant other than earning a larger paycheck. In socialist countries, even the latter incentive was removed. Because of genetic heterogeneity, the claimed equal sharing of resources triggered a race-to-the-bottom in work effort—my genes are little, if at all, helped by extra effort, so why bother? The design of market economies, then, is congruent with our evolved genetic predispositions.

Another extraordinary and initially unexpected value of market economies is using the "wisdom of crowds." A socialist planner needed to set the price of a loaf of bread, but the bread lines in the Soviet Union, East Germany, Cuba, North Korea, and so on, indicated that the price was often unable to satisfy the demand for bread. In market economies, thousands of individual bakeries and grocery stories set their own prices daily depending on market conditions. They have a powerful incentive to get prices right: too high and they will drive customers away, too low and they will lose money and their business will eventually fail. Individuals voting with their purses and wallets determine from whom they prefer to purchase bread, the kinds of bread they favor, and the preferred location and service. Even a large team of socialist planners could not match this minute-by-minute evolving wisdom when setting production goals and prices. Thus, seeking to safeguard our genes because of competition for profits leads to the broad satisfaction of others' desires.

A fascinating example of market wisdom is reported in a recent paper by Michael T. Maloney and J. Harold Mulherin. They examined the impact of the tragic 1986 space shuttle *Challenger* explosion on the valuation of the four companies that were the major contractors for the National Aeronautics and Space Administration (NASA).[6] All four companies, Rockwell International, McDonnell Douglas, Morton Thiokol, and Martin Marietta, were publicly traded. All four refused to comment on the *Challenger* disaster the day it crashed eight minutes after take-off, killing seven crewmembers. After a nearly five-month investigation by a presidential commission, the culprit for the disaster was found to be an O-ring produced by Morton Thiokol. Astonishingly, this conclusion was already known *three hours* after the *Challenger* exploded. Morton Thiokol stock fell 12 percent, while the other companies' stocks were off just 2.5 percent. Similar information revelation through individual choices in markets happens less dramatically every day. How this happens is not well understood, but investors have an incentive to use all available information and adapt to new circumstances as quickly as possible. This incentivized flexibility was, and is, completely absent in socialist economies.

In market economies, not only is exchange adaptation (varying prices and quantities bought and sold) important, but so is institutional flexibility (modifying the rules of exchange). Think of institutions as a bridge and exchange as cars driving over the bridge. Drivers adjust their speeds and lane choices to optimize the flow of cars over the bridge, and most drivers have an incentive to do so—time is limited, and, for most, driving is not a leisure activity. The bridge itself must also flex so that the pressure from the cars, wind, and other external forces do not cause it to rupture and collapse. Bridge design adheres to a "goldilocks principle": too much flexibility, as well as too little, can cause failure. In market economies, institutions that evolve have been shown to outperform those that do not. Flexible institutions are typically based on English common law: rules evolve case by case. Western inflexible institutions are based on the Napoleonic civil codes that fix violations and associated remedies.[7] A similar inflexibility cripples most Islamic economies.[8]

Finally, market exchange itself may lead to a society where individuals have stronger character values. The clearest evidence for this is the studies of fairness in small-scale societies conducted by Henrich and colleagues.[9] They showed that the likelihood of making fair offers to a stranger in one's society is more strongly predicted by the extent of trade in markets than any other factor they have found. Exchange is inherently other-regarding—both you and I must benefit if exchange is to occur. In this sense, exchange in markets is virtuous: one must consider not only one's own needs but also the needs of another.

Socialist economies, on the other hand, provide innumerable incentives to be non-virtuous. This lack of values is understandable when one seeks to survive, rather than thrive. Purges, gulags, and innumerable spies do not make a virtuous society. The feedback loops between social organization and behavior in markets is an issue that arises throughout this book.

The Reason for This Book

This volume is the culmination of more than two years of research by a transdisciplinary group of scholars, most of whom began the project with little knowledge of its subject. Although we believed that values might play an important role in supporting economic exchange, we did not know if this was supported by solid evidence or, in fact, what evidence to examine. We wondered if markets and market participants might be inherently immoral; we considered whether markets devalue human beings and degrade their dignity. Or, could the opposite be true?

This book calls into question the commonly held view that the economy is dominated by greed and selfishness. The cartoon of "greed is good" is nearly ubiquitous in Western countries, and often is even more strongly accepted in less developed nations. Yet, this pervasive belief is inconsistent with the scrutiny of scholars both ancient and modern. The desire to correct this misunderstanding of market exchange in a rigorous multidisciplinary way is the reason for this volume.

Those involved in this research program spent six working meetings together over two and a half years investigating moral markets. During the first several meetings, a common language and frames of analysis were developed so that philosophers could be understood by lawyers, and primatologists and economists could converse. Most of us were not even sure what values were, if anything. Normally academics refrain from admitting their own ignorance, but it was the very realization of not knowing that led us, with open minds, on an exploratory journey into morals and markets.

We came into this endeavor humbled by the breadth of the topic and sought to learn from one another. We began our first meetings by asking participants *not* to write anything—hard for academics! In our fourth meeting, participants wrote five to ten pages, a chapter draft by meeting 5, with revisions following. This approach produced a high degree of cross-fertilization across both individual contributors and fields. Each chapter evolved through the process of presentation, critique, and rewriting, which the book reflects: chapters extensively cross-reference findings in other chapters, as well as in the broader literature.

Contributors to this volume identify their affiliations as anthropology, biology, business, cognitive science, economics, law, neuroscience, philosophy, political science, and primatology, but this does not even touch upon the truly spectacular set of skills each contributor brought to this endeavor. Frans de Waal, a primatologist, has identified the sources of human compassion by studying our closest living ancestors. Peter Richerson, an anthropologist, uses evolutionary theory to understand how cultures change and transmit information over generations. The economist Robert Frank has pioneered the study of moral emotions when making economic decisions . . . and the list goes on.

Our explorations of moral markets also relied on experts who did not write chapters themselves but came to our meetings to inform those who did. Their influence is evident in many of the chapters, though their names do not always appear in the references. Our organizing principle was to attract the best people to the project, encourage them to talk to and learn from one another, identify new avenues for research, and provide them with an impetus to go down these paths.

Implications of Moral Markets

Our research revealed that most economic exchange, whether with strangers or known individuals, relies on character values such as honesty, trust, reliability, and fairness. Reliance on these values, we argue, arise in the normal course of human interactions, without overt enforcement—lawyers, judges, or the police are present in a paucity of economic transactions. Indeed, we show that legal regulations may perversely lead to an increase in immoral behaviors by crowding out our innate sense of fair play. Nevertheless, cheating does occur, and the institutional rules of exchange and their enforcement are a critical reinforcement of values. Civil laws can be understood, a fortiori, as an institutionalized approach to punish immoral behaviors in markets. Why, for example, as a society, do we spend millions of dollars to prosecute minor economic violations and punish the convicted with incarceration? Economic efficiency would argue for a simple payment to the aggrieved party, but such fines do not assuage our desire to punish moral violators.

We discovered that the very freedom to exchange in markets celebrates individual dignity and choice but also allows for transgressions. The exceptional increases in wealth in many areas of the world during the last two hundred years would not have been possible without the specialization of tasks and economies of scale. Yet, along with specialization came the decline of the sole proprietor craftsman and self-directed economic life. Although loss of autonomy is associated with increased wages, it has created a conflict in our economic lives. The burgeoning number of entrepreneurs, who often take pay cuts and bear large risks to start their own businesses, is evidence that autonomy itself is valued. There may be a middle ground, however, as my late colleague, Peter Drucker, has argued. Granting autonomy to employees within a business organization not only recognizes and restores the human need for self-direction but can also raise productivity. Yet, permitting employee autonomy runs counter to the standard principal-agent model of business organization, where unmonitored employees are assumed to shirk. This book seeks to reconcile these views.

Markets are moral in two senses. Moral behavior is necessary for exchange in moderately regulated markets, for example, to reduce cheating without exorbitant transactions costs. Market exchange itself can also lead to an understanding of fair exchange and in this way build social capital in non-market settings. Research has shown that the values that create social capital are a potent stimulus for economic development.[10] As a result, understanding moral markets is especially important in less developed countries. Nations from Iraq and Afghanistan to Mozambique and Tanzania are presently concentrating on developing their economies. For them, getting beyond the cartoon of evil capitalism is critical. Tanzanian capitalism needs to reflect the values of Tanzanian society. Nevertheless, as this book illustrates, many values are universally shared and it is these values that lead to moral markets.

Overview of the Book

The five sections of this book provide units of related thought, flowing from the most fundamental organizing principles to current implications of moral markets. We begin with the philosophical basis for establishing what values are and how they affect our behavior. Next, we offer evidence for the presence of values among our ancestors, and describe how values and societies co-evolved. We then examine how values influence and are influenced by laws. Finally, we conclude by examining insights presented throughout the volume to illuminate how values directly underpin economic exchange and business practices. Readers may choose, of course, to read the book in a different order. Indeed, each chapter draws on findings in other chapters so eclectic readers, after choosing their own starting point, are given a roadmap suggesting how one might proceed after reading a single chapter.

Part 1, "The Philosophical Foundations of Values," contains the first three chapters. Chapter 1, "The Stories Markets Tell" by William D. Casebeer, analyzes why market exchange has gotten a bad rap. Casebeer's conclusion is that it sells; simply put, it is more compelling to tell a tale of bad market behavior than the quotidian normalcy of purchase and delivery. Using the Aristotelian model of ethos, logos, and pathos, he shows why stories of abominable market behaviors persist, and then argues, using a virtue-theoretic lens, that participation in markets per se can promote moral development.

In chapter 2, "Free Enterprise, Sympathy, and Virtue," Robert C. Solomon identifies a person's virtues as essential to a good life, which follows from Aristotelian philosophy. Solomon argues that the inherent value of a virtuous life extends directly to ethical behavior in our business lives. The causal mechanism producing virtuous behavior, according to Adam Smith, is that we feel sympathy for our fellow human beings. This leads us to behave in socially responsible ways.

Robert H. Frank contends in chapter 3, "The Status of Moral Emotions in Consequentialist Moral Reasoning," that our moral intuitions are valuable guides to behavior, even when they lead to suboptimal outcomes. These intuitions often draw upon the emotion felt when observing (or imagining) a particular behavior. Moral emotions, Frank proposes, are commitment devices that keep us from breaking our word. We are endowed with these moral emotions because we are social creatures, dependent on others for survival.

Part 2, "Non-human Origins of Values," which includes chapters 4 and 5, deepens the foundation of how values affect exchange by studying the behavior of nonhuman primates. In chapter 4, "How Selfish an Animal?" Frans B. M. de Waal presents evidence showing that various nonhuman species behave altruistically (incurring a cost to help another), even without the cognitive machinery to fully understand the consequences of their actions. De Waal uses this framework to discuss how his studies of monkeys and apes reveal their strategies to enforce fairness and punish free riders. Based on this evidence, he proposes that the motivation in humans to cooperate is ancient and therefore deeply embedded in human nature.

Sarah F. Brosnan's contribution, "Fairness and Other-Regarding Preferences in Non-Human Primates," chapter 5, examines several types of altruistic behaviors in our closest nonhuman relatives. She begins with an evolutionary analysis of behavior in which altruism was valuable in a number of environments. The circumstances of the social interaction, Brosnan argues, are essential to understanding altruistic behaviors. Nonhuman primates have a wide range of contingent behaviors that lead to nuanced responses to requests for, and violations of, cooperation. For example, the gender and rank of the animals matter. Brosnan concludes that these evolutionary roots provide the basis for the human psychology of cooperative exchange in markets.

Part 3, "The Evolution of Values and Society," which includes chapters 6 and 7, examines how values filter through, and change, societies. In chapter 6, "The Evolution of Free-Enterprise Values," Peter J. Richerson and Robert Boyd provide evidence that market exchange systems co-evolved with the values of reciprocation and honesty. They also examine why it took biologically modern humans around forty thousand years to establish systems of market exchange. More succinctly, the authors examine the evolution of social institutions governing interpersonal trade. They contend that shared values, which they call the "moral hidden hand," are essential to understand the evolution of market institutions.

In chapter 7, "Building Trust by Wasting Time," Carl Bergstrom, Ben Kerr, and Michael Lachmann take a different approach from that of Richerson and Boyd and build a model showing that markets may create the values that allow them to function effectively. Their model shows that market exchange can generate other-regarding preferences from purely self-interested individuals by altering the institutions of exchange. This theory shows that "wasting time" is a commitment device in building a relationship with a trading partner. It does this by reducing the payoff from exploitation. Although Bergstrom and collaborators do not address proximate mechanisms, Robert Frank's discussion of moral emotions facilitating commitment in chapter 3 nicely complements this chapter's results.

Part 4, "Values, Law, and Public Policy," which includes chapters 8 to 11, examines the design of moral markets in modern societies. The authors of these chapters show that formal institutions governing exchange in the United States were intended to be value-free, even though the previous chapters have shown that the institutions of exchange depend critically on values. Lynn A. Stout's contribution in chapter 8, "Why We Overlook Conscience," analyzes why the legal academy (following economics) has taught students that people are narrowly self-interested maximizers. She argues that this is partly owing to the ordinariness of other-regarding behaviors, similar to

Casebeer's analysis in chapter 1. Stout provides six reasons why we overlook conscience as a source of moral guidance. She concludes that many laws seek to promote unselfish behaviors, and the law's ignorance of the role of conscience makes it less effective than it could be. Specifically, laws that permit the exercise of moral emotions such as shame may be more effective than explicitly amoral laws.

Erin O'Hara's contribution in chapter 9, "Trustworthiness and Contract," examines the role of good behavior in contract law. Her primary finding is that the threat of paying direct damages to a breach of contract reduces violations and therefore motivates trustworthiness, though imperfectly. Further, O'Hara suggests that penalty damages do not lead to sufficient trustworthiness because of the low likelihood of their imposition. She concludes that the flexibility of American contract law leads to near-optimal levels of contract fulfillment.

In chapter 10, David Schwab and Elinor Ostrom examine the role of values in the design and implementation of public policy. Their chapter, "The Vital Role of Norms and Rules in Maintaining Open Public and Private Economies," examines the role of institutions in providing information about one's reputation for trustworthiness. Institutions can do this effectively by marshaling economies of scale, making reputational information available at a low cost. An example is credit-reporting agencies. Schwab and Ostrom then analyze how institutions can be made immune to manipulation by political actors.

Chapter 11, "Values, Mechanism Design, and Fairness" by Oliver Goodenough, examines whether a value for fairness can be embedded in economic transactions. He concludes that it can, and often is, through the laws governing exchange. Laws provide a method to resolve welfare losses that may occur without fairness, something lacking, for example, in prisoner's dilemma games. He provides additional evidence that fairness reduces conflict and transactions costs and thus promotes social welfare and enhanced living standards. I offer a similar conclusion on economic regulation in chapter 12.

Part 5, "Values and the Economy," which includes the final chapters 12 to 15, directly analyzes the impact of values on personal and impersonal exchange, as well as on business organization. In chapter 12, "Value and Values: Moral Economics," I ask why corporate scandals such as Enron occur if values are essential in proscribing behaviors. I argue that the particular corporate culture and likely physiologic state of employees conspired to overcome their innate sense of moral values, and then use this example to analyze the neurophysiology of moral sentiments. I then distinguish environments likely to promote or inhibit the activation of the neural mechanisms supporting moral decisions. This chapter focuses on the role of the neuroactive peptide oxytocin in guiding moral decisions in markets by inducing a sense of empathy for others—a proximate mechanism supporting Adam Smith's intuition. As discussed in chapters 9 and 10, institutions play an important role in creating environments where values can flourish.

Chapter 13, "Building a Market: From Personal to Impersonal Exchange" by Erik Kimbrough, Vernon L Smith, and Bart J. Wilson, presents the results of a fascinating experiment on endogenous market formation. Kimbrough and colleagues set up a laboratory environment where one can earn money by selling what one produces and purchasing what others make. But, at what price do market participants exchange? And what are the rules of exchange? Their report of the conversations about rules between participants is insightful and often funny. For example, two trading partners exchange this message through a computer-mediated chat room when seeking to consummate a trade: "when i eat everyone [sic] eat." The authors find that subjects quickly work out the rules of exchange and find a trading partner. In large groups, however, nearly all exchange is bilateral, that is, it is personal. The move to impersonal exchange appears to require the imposition of formal institutions.

Chapter 14, "Corporate Honesty and Business Education," by Herbert Gintis and Rakesh Khurana, asks, as does Lynn Stout in chapter 8, whether moral violations in markets are exacerbated by the way business schools teach. The authors document the pervasive use in business education of "agency theory" and its underlying assumption that market participants are greedy and narrowly self-interested. Yet, this runs counter to most findings by psychologists and behavioral economists, as well as by casual observation. Gintis and Khurana propose a modification of the standard economic model to include character virtues. They also advocate teaching students this more accurate model as well as building a virtuous business community to re-professionalize business education.

The book concludes with a provocative essay by Charles Handy, "What's a Business For?" Like Gintis and Khurana in the preceding chapter, Handy criticizes an overemphasis on share price as the metric for corporate success. He proposes that the purpose of a business "is not to make a profit, full stop" but rather to produce a profit so that people who work for the business can do what pleases them, support their families, and sustain relationships with others. Handy argues that companies should be managed as communities of employees organized to serve communities of customers. By doing so, and using moral values, the system will retain internal and external integrity.

Just as most human beings have an intact and active moral compass, so, too, does economic exchange, the child of human minds, have a vigorous moral dimension. Markets may, in fact, promote and deepen our moral sense of honesty, trust, fairness, and reciprocity. The American patriot, and president, Thomas Jefferson, in 1814, eloquently captured the essence of the argument that we are social creatures and must consider the rights and feelings of others:

> These good acts give pleasure, but how happens it that they give us pleasure? Because nature hath implanted in our breasts a love of others, a sense of duty to them, a moral instinct, in short, which prompts us irresistibly to feel and to succor their distresses.

The standard model of *Homo economicus*, although a useful approximation, is ill designed to understand the subtleties of market exchange. The contributors to this volume, like Jefferson himself, seek to engage in "hostility against every form of tyranny over the mind of man." Our collective view is that the characterization of market actors as greedy and selfish is farcical and egregiously needs to be remedied by the verity of moral markets. By weaving a vibrant analytical tapestry designed around moral values and markets, we hope to demonstrate both the source and powerful impact of moral markets.

Notes

1. Good general audience books include, for example, Marc Hauser, *Moral Minds: How Nature Designed Our Universal Sense of Right and Wrong* (New York: Ecco/HarperCollins, 2006); Matt Ridley *The Origins of Virtue* (London: Viking, 1996); and Michael Shermer *The Science of Good and Evil: Why People Cheat, Gossip, Care, Share, and Follow the Golden Rule* (New York: Holt, 2004).

2. See, for example, Jonathan Haidt, *The Happiness Hypothesis: Finding Ancient Truth in Modern Wisdom* (New York: Basic Books, 2006).

3. For a broad survey of the findings in neuroethics, see the books cited in notes 1 and 2 above, as well as Joshua Greene and Jonathan Haidt, "How (and Where) Does Moral Judgment Work?" *Trends in Cognitive Science* 6 (2002): 517–523; and William D. Casebeer, "Moral Cognition and Its Neural Constituents," *Nature Reviews Neuroscience* 4 (2003): 840–847.

4. Antonio R. Damasio, *Descartes' Error: Emotion, Reason, and the Human Brain* (New York: Grosset/Putnam, 1994).

5. See, for example, David S. Landes, *The Wealth and Poverty of Nations: Why Some Are So Rich and Some So Poor* (New York: Norton, 1998).

6. Michael T. Maloney and J. Harold Mulherin, "The Complexity of Price Discovery in an Efficient Market: The Stock Market Reaction to the Challenger Crash," *Journal of Corporate Finance* 9, no. 4 (2003): 453–479.

7. See Paul G. Mahoney "The Common Law and Economic Growth: Hayek Might be Right," *Journal of Legal Studies* 30, no. 2 (2001): 503–525.

8. See Timur Kuran, *Islam and Mammon: The Economic Predicaments of Islamism*, Princeton, N.J.: Princeton University Press, 2004).

9. J. Henrich et al., *Foundations of Human Sociality; Economic Experiments and Ethnographic Evidence from Fifteen Small-Scale Societies* (Oxford: Oxford University Press, 2004).

10. See Paul J. Zak and Stephen Knack, "Trust and Growth," *The Economic Journal* 111 (2001): 295–321.

Preface

Is Free Enterprise Values in Action?

Oliver Goodenough and Monika Gruter Cheney

Economic systems based on private market exchange have brought remarkable increases in prosperity and liberty. Although such systems unquestionably create inherent challenges and problems, theory predicts and experience confirms that they are reliable engines of growth and social liberalization. All too frequently, however, private enterprise has been equated with unconstrained selfishness and viewed with suspicion in popular perceptions. Ironically, the caricature of free enterprise as the pinnacle of amoral self-interest is often advanced by the friends of private enterprise, as well as by detractors. This caricature is not just off-putting—it is also distorted and incomplete.

In fact, our open, self-organizing economic system, which some of us describe in shorthand as "free enterprise," is effective only because most of the time most of its participants abide by internally motivated "positive" values, such as trustworthiness, fairness and honesty. The common presence of these values is so *anticipated* that it goes largely unnoticed (see chapter 8, this volume), but it is this context within which reliable cooperation can occur without huge transaction costs for self-protection and third-party policing.

Exchange in both personal and impersonal markets requires solving fundamental problems of cooperation and reliability. Of course, our external institutions, such as the law, often do intervene when individuals and firms lapse from a given standard, providing reinforcement of value-based expectations. The law, and the constant threat of its enforcement, is critical in providing the predictability and stability needed for markets to work. Yet the law very often can be seen as a means of reinforcing values which are internally motivated. Moreover, most of us abide by certain internally motivated values even in situations where we are not certain what the law actually requires, or where we may know that there is no chance we will be caught disobeying the law. In short, more often than not, external enforcement is not required. Thus, while the law often reinforces values, law and values are not the same phenomena.

Viewed in this light, values are not merely well-intentioned but unreflective, aspirations. Rather, they are ubiquitous behavioral realities that play a critical role in facilitating the trustworthiness, fairness, and honesty that promote cooperation between individuals, firms, and institutions, and within society as a whole.

The purpose of this volume and of the larger project to which it belongs is to provide a more accurate understanding of what makes market systems work. This better understanding is essential to improving the institutions that support markets, not only in the U.S. and other developed economies but also in the many nations in the world where the quality of life could be dramatically improved through economic growth.

The caricature of an amoral system of free enterprise sometimes leads policy makers in developed economies to shape their institutions in ways that do not maximize the benefits of values. Values-based approaches, where they work, provide private, internal institutions that come at a far lower monetary cost than governmentally enforced laws and other externally enforced institutions. Further, over-reliance on external institutions may in some cases actually crowd out the spontaneous workings of values. The caricature also misinforms countries in the earlier stages of fostering a market-based economy, where the distasteful cartoon often provides ammunition to those who protest against policies of economic freedom.

In focusing on values in free enterprise, we do not argue that other forms of economic organization lack values. In fact, to the extent that they actually work in their own way to promote voluntary cooperative interactions, they, too, are likely to be drawing on the role values play in keeping us all within predictable behavioral bounds. We simply make the point that the underrated factor is critical in a self-organizing market system as well.

We also fully acknowledge the obvious role of self-interest as a critical ingredient in what makes markets work. Values are the forgotten ingredient but not the only one. We do not envision some sort of disinterested utopia, free of any self-interested behavior. Nor, frankly, is it clear that such a society would even be ideal, for who would get up in the morning and go to work if self-interest did not spur us to do so? There is, of course, no dearth of economic scholarship focusing on the importance of self-interest. The gap is on the other ingredients, and in this volume we seek to redress the imbalance and to show that pure, unbridled self-interest does not a market make. Institutions work to constrain and channel this self-interest in ways that lead to mutually productive outcomes among multiple players. And because a policeman does not lurk on every corner, our economic system cannot be—and is not—maintained solely by coercive enforcement. Simply put, self-interest is not the whole story. This volume demonstrates, we believe, that values, along with self-interest, are needed in the analysis.

The Project

This volume grows out of a project that was organized by the Gruter Institute for Law and Behavioral Research for the purpose of bringing the examination of values back into the story of economics. This project brought together scholars from disciplines including zoology, law, economics, evolutionary biology, political science, philosophy, neuroscience, and business to explore the fundamental role that values such as honesty, trust, reliability, and fairness *must* play at the core of successful economic systems, including those based in free market exchange. The combined perspectives of these participants have allowed us to tell an unusually complete story about the role of values; a narrative starting with the strategic underpinnings of behavior, continuing through the biology and psychology of human decision making, on to cultural and institutional forces such as custom and law, and culminating in economics, business, and business education.

Our project was made possible by the commitment of a number of talented academics and business practitioners, as well as the generous support of the John Templeton Foundation, the UCLA–Sloan Research Program on Business Organizations, the Ann and Gordon Getty Foundation, and our conference hosts, the UCLA School of Law, Georgetown University Law Center, the University of Cambridge, and the Harvard Business School. The work was structured as a targeted, interdisciplinary conversation. The conversation was not a simple one; it involved a total of more than forty contributors who came together in various combinations at six workshops and conferences held over a two-year period.

Although our project was driven by an idea, it was not, we hope, driven by a political ideology. Both the project and this volume present a breadth of views. The participants in our program espouse political opinions from Left to Right, at least within the range of current mainstream discourse, and the values considered include redistributive fairness. We do not argue that systems based on free enterprise are uniformly appropriate as a basis for organizing economic life, nor do we seek to deny that free enterprise creates challenges and problems where it is adopted. We understand that people in such a system do, at times, lie, cheat, and steal. Our goal, however, is to understand why they do not do so more often, and we believe we have shed new light on the question.

In the course of the planning and execution of our research program, three questions have arisen: What do we mean by values? Where do values come from? Why is the caricature of free enterprise so prevalent? Each in its way is a threshold question to the program as a whole, and our discussion here may help frame the book more clearly.

What Do We Mean by Values?

The concept of "values" has been in use for millennia, without attracting a fully accepted definition. Values, and related ideas such as virtues and character, are difficult to define in the abstract. Philosophers—and, for that matter, most of the rest of humanity (e.g. Phillips 2004)—have been trying to arrive at a consensus, without success. We have not focused on these debates in our conversations, nor do we pursue all possible approaches to the concept of values. In an approach shared in ancient times with Aristotle (Slote 2000) and in contemporary philosophy with some "virtue ethics" theorists (Slote 2000; Solomon 1999), we have generally viewed values in people as a critical link between principles and actions. Put more concretely, many of us have considered values as non-situational commitments to particular principles of character and action, commitments that may require sacrifice and self-denial, and that, particularly when shared, can reframe strategic interactions, providing solutions to dilemmas of positive interaction.

Values are generally more easily identified by example, in broadly admired traits such as honesty, reliability, friendship, fairness, and trustworthiness, or, by counterexample, in negative traits such as a failure to keep one's word or a callous disregard for others. But this, too, can lead to complications. The late philosopher Robert Solomon, a dearly missed contributor to our project and to this volume, suggested a catalog of business virtues (Solomon 1999), enumerating three he saw as "most basic"—honesty, fairness, and trustworthiness—together with a list of forty-five specific virtues, ranging alphabetically from ability to zeal. We believe that a discussion of the role of values can be useful without achieving a level of definitional closure on the term that has eluded philosophy for centuries.

Where Do Values Come From?

A further perennial question is—Where do values come from?—need not be settled here in order to discuss them in our context.

A traditional view is that values have a divine origin. Most religions make claims about the origins or sanctions for the basic rules of life, an approach some philosophers call the "divine command theory" (Quinn 2000). In the Judeo-Christian Bible, God presents the Ten Commandments to Moses. Mohammed receives divine guidance on moral questions which he relates to the wider world in the Koran. In the New Testament, the words of Jesus and the letters of his early followers contain many directives on the values that should form the basis of a moral life, many of which have economic consequences (e.g., Luke 16:10–12). The divine origin of values may not be susceptible to deductive proof (Quinn 2000), but it is deeply rooted in the faith of millions. Such commonly mentioned, religiously based virtues as honesty, dependability, respect, and concern for others represent commitments by which a good person should live, and these commitments map dependably onto the kinds of mechanisms that can reframe economic life to make cooperation a dominant strategy (see chapter 11).

More secular approaches look to rationality for the source of values (see, e.g., Kant 1797; Rawls 1971) or to history and culture (see, e.g., Wong 1991) or to the evolved biological inheritance of a social species (see, e.g., the intro-duction to this volume and chapter 4; and Hauser 2006). Certainly the ap-proaches followed in this volume involve a secular, scientific examination of values rooted in biology and cognitive neuroscience, but this need not be in contravention to the religious viewpoint. A disagreement over the relative role of the divine hand or the invisible hand of evolution in creating the hu-man eye will not greatly impact a discussion of the role and workings of the eye's lens, and the same is true with values and economic life.

In pursuing this project, we do not consider values as some kind of root-less abstraction, but rather as deeply seated commitments that guide the ac-tions of individuals. In this light, values have the potential not only to be good for their own sake but also to add value both to the individual and soci-ety (e.g., Gintis et al. 2003). They constitute a commitment to ways of acting that both are true in themselves and can lead to positive outcomes. There is a critical place for these values in keeping us on cooperative paths, by supple-menting external coercive institutions and by making trade possible even in the complete absence of external coercive institutions. In short, values oc-cupy a critical place in free enterprise.

Why Is the Caricature So Prevalent?

Why did the unbridled greed cartoon so eclipse the role of values in our busi-ness and economic discourse? Vernon Smith (1998) and William Casebeer (chapter 1) explain a number of the contributing factors. We supplement their insights with a few of our own.

One reason is the pendulum swings of social and intellectual history. The rhetorical necessity in one age often sows the seeds of misunderstanding in the next. For much of the history of human commerce, the major obstacle to the adoption and success of a free enterprise approach was the condemnation of self-interest. Hanging over the work of Adam Smith and his contemporaries in the 1700s, was a history of well-meaning but essentially ignorant economics. Revered figures such as Aristotle, St. Paul, St. Augustine, St. Thomas Aquinas, and Martin Luther all shared in these errors (Solomon 1997; see also Luke 16:13). In keeping with many other ecclesiastical and moral thinkers before the Age of Reason, Luther vilified profit seeking:

> When once the rogue's eye and greedy belly of a merchant find that people must have his wares, or that the buyer is poor and needs them, he takes advantage of him and raises the price. He considers not the value of the goods, or what his own ef-forts and risk have deserved, but only the other man's want and need. He notes it not that he may relieve it but that he may use it to his own advantage by raising the price of his goods, which he would not have raised had it not been for his neighbor's need. Because of his avarice, therefore, the goods must be priced as much higher as

the greater need of the other fellow will allow, so that the neighbor's need becomes as it were the measure of the good's worth and value. Tell me, isn't that an un-Christian and inhuman thing to do? (Luther 1520:182–183)

This condemnation of how the mechanisms of supply and demand operate illustrates the need for Adam Smith in *The Wealth of Nations* to focus on the positive role of self interest in free enterprise:

It is not from the benevolence of the butcher, the brewer, or the baker, that we expect our dinner, but from their regard for their own interest. We address ourselves, not to their humanity but their self-love, and never talk to them of our own necessities but of their advantages. (A. Smith 1776:20)

The pendulum has operated again in more recent times. For much of the twentieth century, the mainstream currents of social thought condemned self-interest. Reacting to the problems created by the liberal economics of the nineteenth century, and taking for granted its advances, many social critics advocated the need to reassert more communalist approaches. This kind of thought informed not only communism and socialism but also, to a surprising degree, the national socialist economics of fascism. The response to the failures of these approaches was, once again, to emphasize selfishness.

Milton Friedman's famous 1970 *New York Times* article, "The Social Responsibility of Business Is to Increase Profits," provides a paradigmatic example of this round in the rhetorical assertion of self-interest. Friedman explained, with ruthless clarity, the logic for asking corporate executives, in their institutional capacity, to seek to maximize earnings as their contribution to the common good (see also, e.g., Posner 1986; Kaplow and Shavell 2002). Friedman's article, at the time of its writing, was met with widespread criticism in a social climate that talked of different stakeholders in a company and gave somewhat higher value to corporate social action (see, generally, chapter 14 of this volume). Three decades later, a strong swing of the pendulum made Friedman's challenge the new orthodoxy, both in business and politics (see, generally, chapters 11 and 14); four decades later, we need to seek a better balance. We hope not to go all the way to the other pole but rather to create a complete description that reconciles values and self-interest.

Another explanation for the overemphasis on self-interest in our business and economic discourse involves the intense scrutiny of the workings of financial markets and our tendency to think that successful strategies in highly structured financial markets are also dominant in more personal, socialized settings. A short-sighted view of these highly structured, intensely regulated institutions suggests that they operate in a kind of amoral paradise. First of all, the undisguised self-interested behavior of the stock market trading floor only works because the necessary stability is provided by routine interactions conducted within tightly defined, highly supervised institutional structures. These structures can be seen as either embodying or substituting for the constraints values provide. In this sense, these markets are intensely moral.

Furthermore, the preoccupation with financial markets has led to the excessive application of their behavioral model to essentially non-market interactions. Economists have conspired in this because financial markets best fit their behavioral creation: the amoral *Homo economicus*. This self-interested rational actor is easy to formalize and, in the right context, gives useful results. In the context of highly structured markets (e.g., trading floors), microeconomic modeling shows us that short-term choices to buy and sell can be essentially rational and predatory. The institutions of the highly structured market have the necessary values embedded in them so that intensely selfish behavior around the individual trade can be a successful strategy.

Although our economy is often called a free market system, a large portion of the transactions that create the gains of specialization, scale, and trade occur in personally mediated settings. This is true of most interactions within firms, and for a significant portion of interactions between unaffiliated buyers and sellers. Personal exchange is deeply rooted in a very personal projection of values as a means of ensuring cooperative play. The application of the morals of the trading floor to such transactions often destabilizes them. Considering the intensely social atmosphere of a workplace as if it were an asocial circle of mutually predatory agents of pure selfishness will either lead to a poor description or a poor result for the company, if not both.

Conclusion: Putting Values Back into Free Enterprise

In their efforts to make self-interest respectable, proponents of free enterprise ever since Adam Smith have too often left out of their descriptions of free enterprise the structures of cooperation and restraint that must be coupled with self-interest to create the full recipe for economic success. Opponents of free enterprise, in their turn, seize on this emphasis in their public appeals for a more restricted economy. As a result, many people believe that business is valueless and amoral.

Our project, by contrast, explores the endless tensions between self-interest and other-regarding interest in human experience, and it does not posit one as taking rightful priority over the other across human experience. Free enterprise, along with any other functioning economic system, must consist of a balance of cooperative and selfish forces. In the relatively unregulated world of free exchange, personal values are not less important than they are in other systems; rather, they form internal institutions—a key psychological element in this balance. In a very real way, free enterprise must be regarded as values in action.

References

Friedman, Milton (1970). "The Social Responsibility of Business Is to Increase Profits." *New York Times*, September 13, 1970.

Gintis, Herbert, Bowles, Samuel, Boyd, Robert, and Fehr, Ernst (Eds.) (2003). *Moral Sentiments and Material Interests: On the Foundations of Cooperation in Economic Life.* Cambridge, Mass.: MIT Press.

Hauser, Marc (2006). *Moral Minds.* New York: Ecco/HarperCollins.

Kant, Immanuel (1797). *The Metaphysics of Morals.* Variously reprinted, including Cambridge: Cambridge University Press, 1996.

Kaplow, Louis, and Shavell, Steven (2002). *Fairness versus Welfare.* Cambridge Mass.: Harvard University Press.

Luther, Martin (1520). *Martin Luther Sermon on Trade and Usury.* Variously republished, including in, *Luther Works*, Vol. 45, ed. Jaroslav Pelikan. St. Louis, Mo.: Concordia, 1955.

Phillips, Christopher (2004). *Six Questions of Socrates: A Modern-Day Journey of Discovery through World Philosophy.* New York: Norton.

Posner, Richard (1986). *Economic Analysis of Law.* 3d ed. London: Little, Brown.

Quinn, Philip L. (2000). "Divine Command Theory." In *The Blackwell Guide to Ethical Theory*, ed. Hugh LaFollette. Oxford: Blackwell.

Rawls, John (1971). *A Theory of Justice.* Cambridge, Mass.: Harvard University Press.

Slote, Michael (2000). *Morals from Motives.* Oxford: Oxford University Press.

Smith, Adam (1776). *An Inquiry into the Nature and Causes of the Wealth of Nations.* Variously reprinted, including New York: Knopf, 1991.

Smith, Vernon (1998). "The Two Faces of Adam Smith." *Southern Economic Journal* 65:1–19.

Solomon, Robert (1997). *It's Good Business: Ethics and Free Enterprise for the New Millennium*, Lanham, Maryland, Rowman & Littlefield.

———. (1999). *The Joy of Philosophy.* Oxford: Oxford University Press.

Wong, David (1991). "Relativism." In *A Companion to Ethics*, ed. Peter Singer. Oxford: Blackwell.

Acknowledgments

I extend my sincere thanks to the John Templeton Foundation for its critical support of this research program. Barnaby Marsh, director of Venture Philanthropy Strategy and New Programs Development at the Foundation, repeatedly advised the contributors to this volume and shepherded the project over the course of more than two years to help us obtain the necessary resources. My co-investigator on this grant, which was administered by the Gruter Institute for Law and Behavioral Research by its executive director Monika Gruter Cheney and administration director Jeanne Giaccia, was the brilliant and assiduous Oliver Goodenough. Gruter Cheney, Giaccia, and Goodenough all labored intensively over three years to help make this research program a success. The project was conceived and developed by the late founder of the Gruter Institute, Dr. Margaret Gruter, based on her insight that values are critical to the functioning of markets. I also acknowledge the support of the Ann and Gordon Getty Foundation, the UCLA/Sloan Research Program on Business Organizations (through Lynn Stout), the Harvard Business School Leadership Program (through Rakesh Khurana), Emmanuel College at the University of Cambridge (through Simon Deakin), and Georgetown University Law Center (though John Mikhail). Thanks also to Tim Sullivan at Princeton University Press, who was unfailingly insightful, kind, and patient in my efforts to put this book together.

Any research program is only as good as its members, and the team of researchers, sponsors, and administrators for this project have been uniformly outstanding. I am deeply grateful to all of them for allowing me to be a part of it.

Contributors

Carl Bergstrom is an evolutionary biologist at the University of Washington studying the role of information in social and biological systems at scales from intracellular control of gene expression to population-wide linguistic communication. Working in close collaboration with empirical and experimental researchers, Dr. Bergstrom's group approaches these problems using mathematical models and computer simulations. Dr. Bergstrom's recent projects include contributions to the game theory of communication and deception, work on how immune systems avoid subversion by pathogens, and a number of more applied studies in disease evolution, including analysis of antibiotic resistant bacteria in hospital settings and models of the interaction between ecology and evolution in novel emerging pathogens such as SARS. Dr. Bergstrom received his Ph.D. in theoretical population genetics from Stanford University in 1998. After a two-year postdoctoral fellowship at Emory University, where he studied the ecology and evolution of infectious diseases, Dr. Bergstrom joined the faculty in the Department of Biology at the University of Washington in 2001.

Robert Boyd is a professor in the Department of Anthropology at the University of California, Los Angeles. His work is based on the premise that, unlike other organisms, humans acquire a rich body of information from others by teaching, imitation, and other forms of social learning, and this culturally transmitted information strongly influences human behavior. Culture is an essential part of the human adaptation, and as much a part of human biology as bipedal locomotion or thick enamel on our molars. Professor Boyd's research is focused on the evolutionary psychology of the mechanisms that give rise to, and shape, human culture, and how these mechanisms interact with population dynamic processes to shape human cultural variation. Much of this work has been done in collaboration with Peter J. Richerson at the University of California at Davis.

Sarah F. Brosnan is currently a postdoctoral fellow in the Department of Anthropology at Emory University. Her research focuses on the evolution of behavior at the intersection of complex social behavior and cognition in non-human primates, with an interest in both multi-species comparisons and cross-disciplinary approaches. Specifically, she is interested in the proximate mechanisms underlying such behaviors, particularly how animals perceive the value of different commodities and how the social environment affects the acquisition and use of this concept of value. Some projects include understanding how nonhuman primates perceive equity, whether chimpanzees have pro-social tendencies, and how these compare to those of humans, and primate economic behavior, including barter and the endowment effect. Ultimately, she hopes that these various projects in both nonhuman primates (primarily chimpanzees and capuchin monkeys) and humans will elucidate the evolution of cooperative and economic behavior across the primate lineage. She can be reached at http://userwww.service.emory.edu/~sbrosna.

Lt. Col. William D. Casebeer is an intelligence officer in the U.S. Air Force and most recently a Program Fellow at the Carr Center for Human Rights Policy at Harvard University. He holds degrees in political science from the U.S. Air Force Academy (B.S.), philosophy from the University of Arizona (M.A.), national security studies from the Naval Postgraduate School (MA), and cognitive science and philosophy from the University of California at San Diego (Ph.D.), where his dissertation received the campus-wide outstanding thesis award. Lt. Col. Casebeer's research interests include military ethics, interdisciplinary approaches to non-state political violence and terrorism, and the neural mechanisms of moral judgment and narrative processing. He is author of *Natural Ethical Facts* (MIT Press, 2003), and coauthor of *Warlords Rising* (Lexington Books, 2005). He has published on topics ranging from the morality of torture interrogation to the neural correlates of moral judgment (in venues such as *Nature Reviews Neuroscience*, *Biology and Philosophy*, and *International Studies*), and has experience as a Middle East affairs analyst. Formerly an associate professor of philosophy at the Air Force Academy, he is currently the chief of the Eurasian intelligence analysis division at NATO's military headquarters in Belgium. Lt. Col. Casebeer is a term member of the Council on Foreign Relations and an associate of the Institute for National Security Studies. He can be reached at drenbill@earth link.net.

Monika Gruter Cheney is Executive Director and President, Board of Directors, of the Gruter Institute for Law and Behavioral Research. Prior to her appointment as executive director, she served as the Institute's associate director. Before joining the Gruter Institute, Ms. Cheney practiced in the intellectual property and technology transactions group at Wilson, Sonsini, Goodrich, Rosati in Palo Alto, California. Representative transactions included licensing, development, service, distribution, manufacturing, electronic commerce and corporate partnering and merger agreements for clients ranging from Fortune 500 companies to early-stage technology and biotech startups. Ms. Cheney earned her J.D. in 1998 from Georgetown University Law Center, and her B.A. in Philosophy, Politics, and Economics, and International Relations from Claremont McKenna College in 1995.

Frans B. M. de Waal received a Ph.D. in biology from the University of Utrecht, in 1977. He is the C. H. Candler Professor of Primate Behavior at Emory University, and director of the Living Links Center at the Yerkes National Primate Research Center. He has authored numerous books including *Chimpanzee Politics* and *Our Inner Ape*. His research centers on primate social behavior, including conflict resolution, cooperation, inequity aversion, and food sharing, as well as the origins of morality and justice in human society. In 1993, he was elected to the Royal Dutch Academy of Sciences, and in 2004 to the U.S. National Academy of Sciences.

Robert H. Frank is the H. J. Louis Professor of Economics at Cornell University's Johnson School of Management, and also a monthly contributor to the "Economic Scene" column in the *New York Times*. Until 2001, he was the Goldwin Smith Professor of Economics, Ethics, and Public Policy in Cornell's College of Arts and Sciences. He has also served as a Peace Corps volunteer in rural Nepal, chief economist for the Civil Aeronautics Board, fellow at the Center for Advanced Study in the Behavioral Sciences, and was professor of American civilization at l'Ecole des Hautes Etudes en Sciences Sociales in Paris. His books include *Choosing the Right Pond*, *Passions within Reason*, *Microeconomics and Behavior*, *Principles of Economics* (coauthored with Ben Bernanke), *Luxury Fever*, *Falling Behind*, *The Economic Naturalist*, and *What Price the Moral High Ground?* The *Winner-Take-All Society* (coauthored with Philip Cook) was named a Notable Book of the Year by the *New York Times*, and was included in *Business Week*'s list of the ten best books for 1995.

Herbert Gintis, who received his Ph.D. in economics at Harvard University in 1969, is External Professor, Santa Fe Institute, and Professor of Economics, Central European University. He heads a multidisciplinary research project, the MacArthur Network on the Nature and Evolution of Norms and Preferences, which studies empathy, reciprocity, insider/outsider bias, vengefulness, and other human behaviors absent from the traditional economic model of the self-regarding agent. Recent books include *Game Theory Evolving* (Princeton University Press, 2000), and his coauthored volumes *Foundations of Human Sociality* (Oxford University Press, 2004) and *Moral Sentiments and Material Interests* (MIT Press, 2005). He is currently completing *A Cooperative Species* and *The Bounds of Reason*.

Oliver R. Goodenough has written on a far-ranging subjects relating to law, business, and cognitive and behavioral science. With Semir Zeki, he edited the 2004 special issue of the *Philosophical Transactions of the Royal Society* devoted to law and the brain, recently reprinted under that title by Oxford University Press. He is coauthor of *This Business of Television*, now in its third edition. Currently a professor at the Vermont Law School, Goodenough is also a research fellow of the Gruter Institute for Law and Behavioral Research and an adjunct professor at Dartmouth's Thayer School of Engineering. He has been a visiting research fellow at the Department of Zoology at the University of Cambridge, a lecturer in law at the University of Pennsylvania Law School, and a visiting professor at Humboldt University in Berlin. In the 2007/2008 academic year, he will be a fellow at the Berkman Center for Internet and Society at Harvard Law School.

Charles Handy, social philosopher, management scholar, best-selling author, and radio commentator, is an influential voice worldwide. One of the first to predict the massive downsizing of organizations and the emergence of self-employed professionals, Handy has a gift for looking twenty years ahead at ways in which society and its institutions are changing. Handy has written some of the most influential articles and books of the past decade, including *The Elephant and the Flea*, *The Age of Unreason*, and *The Age of Paradox*. After working for Shell International as a marketing executive, economist, and management educator, Handy helped start the London Business School in 1967. He has worked closely with leaders of business, nonprofit, and government organizations.

Michael C. Jensen, Jesse Isidor Straus Professor of Business Administration, *Emeritus*, joined the faculty of the Harvard Business School in 1985. Professor Jensen earned his Ph.D. in economics, finance, and accounting and his MBA in finance from the University of Chicago and an AB degree from Macalester College. He is the author of more than ninety scientific papers, as well as the books *Foundations of Organizational Strategy*, *Theory of the Firm: Governance, Residual Claims, and Organizational Forms*, and *CEO Pay and What to Do About It: Restoring Integrity to Both Executive Compensation and Capital-Market Relations*. In 1973, Professor Jensen founded the *Journal of Financial Economics*, one of the top three scientific journals in financial economics. In 2004, Jensen received the Tjalling C. Koopmans EFACT Conference Award, for "extraordinary contributions to the economic sciences, and to have reached the highest standards of quality of research."

Ben Kerr received his Ph.D. in biological sciences from Stanford University in 2002. While at Stanford, he worked with Marcus Feldman on modeling the evolution of flammability in resprouting plants, the evolution of animal learning, and the evolution of altruism. He also worked with Brendan Bohannan on evolution within microbial systems, and with Peter Godfrey-Smith on philosophical issues arising in the levels of selection controversy. Kerr spent three years as a postdoctoral research associate at the University of Minnesota, where he worked with David Stephens on modeling impulsive behavior in blue jays, with Tony Dean on the evolution of cooperation within a microbial host-pathogen system, and with Claudia Neuhauser on spatial dynamics within model population genetic systems. He joined the faculty at the University of Washington in 2005.

Rakesh Khurana is an associate professor of business administration in the Department of Organizational Behavior at the Harvard Business School. He is the author of *Searching for a Corporate Savior* (Princeton University Press, 2002), which examined the CEO labor market. His forthcoming book, *From Higher Aims to Hired Hands* (Princeton University Press, 2007), is a sociological and intellectual history of American business schools. Professor Khurana received his B.S. from Cornell University and his A.M. (sociology) and Ph.D. (organizational behavior) from Harvard University. Prior to attending graduate school, he worked as a founding member of Cambridge Technology Partners in Sales and Marketing.

Michael Lachmann, a researcher at the Max Planck Institute for Evolutionary Anthropology (MPI EVA), is an evolutionary biologist studying mainly the process of evolution itself, especially in relation to information. During his undergraduate studies, he worked on the origins of the evolutionary process, and on mechanisms of information transfer across generations other than via the DNA sequence itself (such as DNA methylation) and how such processes affect evolution. Then, during his Ph.D. work, he explored the evolution of its eusociality, and on informational aspects associated with its evolution—the control of differentiation and task allocation through signals, signaling between related individuals and its cost, and the benefits of sharing information in a group of related individuals. During a postdoctoral fellowship at the Santa Fe Institute, he focused mainly on the evolution of differentiation. Dr. Lachmann joined MPI EVA in 2003, and is using data obtained from large-scale gene expression experiments on closely related species, such as different species of apes or different species of mice, to understand the mechanisms of the evolution of tissues and the differences between tissues.

Erin Ann O'Hara, who received her B.A. at the University of Rochester (1987) and her J. D. at Georgetown University (1990), joined the faculty of Vanderbilt University Law School in 2001. Her most recent work includes a series of important articles on choice of law. Professor O'Hara taught at George Mason University School of Law from 1995 until 2001. Recently, Professor O'Hara was visiting associate professor at Georgetown University Law Center, where she taught criminal law, conflict of laws, and a law and economics seminar. Prior to joining the George Mason faculty, Professor O'Hara clerked for Chief Judge Dolores K. Sloviter, U.S. Court of Appeals for the Third Circuit, served as a Bigelow Teaching Fellow and lecturer in law at the University of Chicago Law School, and as a visiting assistant professor in the Departments of Legal Studies, Economics, and Finance at Clemson University. She can be reached at http://law.vanderbilt.edu/faculty/ohara.html.

Elinor Ostrom is Arthur F. Bentley professor of political science and co-director of the Workshop in Political Theory and Policy Analysis at Indiana University, Bloomington, and founding director of the Center for the Study of Institutional Diversity at Arizona State University, Tempe. She was elected to the National Academy of Sciences in 2001, is a member of the American Academy of Arts and Sciences, and a recipient of the Frank E. Seidman Prize in Political Economy, the Johan Skytte Prize in Political Science, the Atlas Economic Research Foundation's Lifetime Achievement Award, and the John J. Carty Award for the Advancement of Science. Her books include *Governing the Commons* (1990) and others she coauthored, including *Rules, Games, and Common-Pool Resources* (1994); *Local Commons and Global Interdependence* (1995); *Trust and Reciprocity* (2003); *The Commons in the New Millennium* (2003); *The Samaritan's Dilemma* (2005); and *Understanding Institutional Diversity* (2005). She is available at http://www.indiana.edu/~workshop or http://www.indiana.edu/~cipec.

David Schwab earned his B.A. in political science from Indiana University–Purdue University, Indianapolis, in 2002, and his M.A. from Indiana University in 2005. He is currently writing his Ph.D. dissertation on whether the "ideal speech situation" developed by Jurgen Habermas helps or hinders users of a common-pool resource to maximize their return. His additional projects include a genealogy of Habermas's "ideal speech situation," and a game-theoretic model of the effect of social influence on common-pool resource allocation. David's research interests are deliberative democracy, political communication, critical theory, game theory, and social dilemmas. He is particularly interested in integrating normative and positive theories of political conflict and determining which aspects of communication best support their resolution.

Peter J. Richerson, professor of environmental science and policy at the University of California at Davis, is applying theoretical and conceptual principles, along with methods of analysis of evolution mainly developed by evolutionary biologists, to study the processes of cultural evolution. His research models the evolutionary properties of human culture and animal social learning, and the processes of gene-culture co-evolution. His recent publications used theoretical models to try to understand some of the main events in human evolution, such as the evolution of the advanced capacity for imitation (and hence cumulative cultural evolution) in humans, the origins of tribal and larger-scale cooperation, and the origins of agriculture. He heads a research group funded by the National Science Foundation devoted to the study of cultural evolution in laboratory-scale microsocieties.

Vernon L. Smith, the 2002 Nobel Prize winner in Economic Sciences, is currently professor of economics and law at George Mason University, a research scholar in the Interdisciplinary Center for Economic Science, and a fellow of the Mercatus Center, all in Arlington, Virginia. He received his B.A. in electrical engineering from the California Institute of Technology, and his Ph.D. in economics from Harvard University. He has authored or coauthored more than two hundred articles and books on capital theory, finance, natural resource economics, and experimental economics. Cambridge University Press published his *Papers in Experimental Economics* in 1991, and a second collection of more recent papers, *Bargaining and Market Behavior*, in 2000. He has received numerous awards and distinctions, including Ford Foundation Fellow, Fellow of the Center for Advanced Study in the Behavioral Sciences, the 1995 Adam Smith Award conferred by the Association for Private Enterprise Education, and a Sherman Fairchild Distinguished Scholar at the California Institute of Technology. He was elected a member of the National Academy of Sciences in 1995, and received Cal Tech's distinguished alumni award in 1996. He has served as a consultant on the privatization of electric power in Australia and New Zealand, and participated in numerous private and public discussions of energy deregulation in the United States.

Robert C. Solomon was the Quincy Lee Centennial Professor of Business and Philosophy at the University of Texas at Austin, where he taught for more than thirty years. Professor Solomon passed away unexpectedly on January 2, 2007. He received his undergraduate degree in molecular biology from the University of Pennsylvania and his master's and doctoral degrees in philosophy and psychology from the University of Michigan. He has had visiting appointments at the University of Pennsylvania; the University of Auckland, New Zealand; University of California at Los Angeles; Princeton University; and Mount Holyoke College. Professor Solomon won many teaching honors, including the Standard Oil Outstanding Teaching Award; the President's Associates Teaching Award (twice); and the Chad Oliver Plan II Teaching Award. He was also a member of the Academy of Distinguished Teachers at the University of Texas, which is devoted to providing leadership in improving the quality and depth of undergraduate instruction. Professor Solomon wrote or edited more than 45 books including *The Passions* (1976); *About Love* (1988); *Ethics and Excellence* (1992); *A Short History of Philosophy*, with Kathleen Higgins (1996); *A Better Way to Think about Business* (1999); *The Joy of Philosophy* (1999); *Spirituality for the Skeptic* (2002); *Not Passion's Slave* (2003); and *In Defense of Sentimentality* (2004). His more than 150 articles have appeared in many of the leading philosophy journals and in numerous books.

Lynn A. Stout is the Paul Hastings Professor of Corporate and Securities Law at the University of California at Los Angeles, where she specializes in corporate governance, securities regulation, and law and economics. Professor Stout, a national figure in these fields, publishes extensively and lectures widely. She is the principal investigator for the UCLA–Sloan Foundation Research Program on Business Organizations and sits on the Board of Trustees for the Eaton Vance family of mutual funds and the Board of Directors of the American Law and Economics Association. She is past chair of the American Association of Law Schools (AALS) Section on Law and Economics and past chair of the AALS Section on Business Associations. Professor Stout has authored a casebook series on law and economics, as well as numerous articles on corporate governance, the theory of the corporation, stock markets, finance theory, and economic and behavioral analysis of law. Before joining UCLA, Professor Stout was professor of law at the Georgetown University Law Center and director of the Georgetown-Sloan Project on Business Institutions. She has also taught at Harvard Law School, New York University Law School, and the George Washington University National Law Center, and has served as a Guest Scholar at the Brookings Institution in Washington, D.C.

Bart J. Wilson is part of the Interdisciplinary Center for Economic Science (ICES) at George Mason University. He has an appointment in the Department of Economics as an associate professor, with affiliations in the Schools of Management and Law. Professor Wilson's broad fields of specialty are industrial organization, experimental economics, and econometrics. He is currently pursuing research on the foundations of exchange and specialization. His other research programs apply the experimental method to topics in e-commerce, electric power deregulation, and antitrust. He has published papers in the *Economic Journal*, *RAND Journal of Economics*, and the *Journal of Economic Theory*. The National Science Foundation, the Federal Trade Commission, and the International Foundation for Research in Experimental Economics have supported his research. Wilson uses experimental economics extensively in teaching undergraduate and graduate classes.

Paul J. Zak is the founding director of the Center for Neuroeconomics Studies (http://www.neuroeconomicstudies.org), a laboratory that integrates neuroscience and economics, at Claremont Graduate University in Claremont, California. Zak is also a professor of economics at Claremont, and professor of neurology at Loma Linda University Medical Center. He has degrees in mathematics and economics from San Diego State University, a Ph.D. in economics from the University of Pennsylvania, and postdoctoral training in neuroimaging from Harvard University. Zak is credited with the first published use of the term "neuroeconomics." His research, in 2004, was the first to identify the role of the mammalian neuropeptide oxytocin in mediating trusting behaviors between unacquainted humans. His recent publications include "Oxytocin Increases Trust In Humans" (*Nature*, 2005); "Oxytocin Is Associated with Human Trustworthiness (*Hormones and Behavior*, 2005); "The Neuroeconomics of Distrust" (*American Economic Review Papers and Proceedings*, 2005); and "Neuroeconomics" (*Philosophical Transactions of the Royal Society B*, 2004). He has been interviewed numerous times in print and on TV and radio programs, and gives many public lectures on the relationship between brain and behavior to groups ranging from professional money managers to elementary school students.

PART I

Philosophical Foundations of Values

One

The Stories Markets Tell
Affordances for Ethical Behavior in Free Exchange

William D. Casebeer

Opinions about free exchange have fallen on hard times. Major news maga-
zines such as the *Atlantic Monthly*[1] and the *New York Times* have devoted
column-inches to the anti-market zeitgeist, and self-organized protests plague
World Trade Organization meetings with regularity.[2] Given the palpable benefits of markets for all involved in them, why is the term *free market* as likely to
call to mind images of selfish and insensitive robber barons as it is to evoke
scrupulous and other-oriented small business owners and neighbors? This is
partly because cartoon versions of the nature and outcome of free exchange,
for multiple reasons, have carried the day in the court of public opinion.
These cartoons are typified by images of selfish capitalists exploiting labor
with glee for the good of no one but themselves. These cartoons, when true,
are only *partially* true, and present but one facet of the costs and benefits of
free exchange. They do make for colorful stories, however. Sinclair Lewis, for
example, would not have made his name writing about the friendly aspects of
garden-variety market operations.

These cartoons, however, leave off the considerable moral presupposi-
tions and ethical benefits that free exchange assumes and enables. Stories
widely held to be true are not necessarily informed by the results of our best
sciences, nor is colorful nature in storytelling tied to the likelihood of resembling reality in any law-like way. Cartoons that declaim free exchange as "selfish" or "exploitative" or "harmful" do have a point in some contexts. A fourth
story, however, less cartoon-like and more consilient with what we know
about the cognitive mechanisms that allow exchange to take place and which
are affected by it, is a *better* story, more true, with more fealty to the phenomenon it professes to be about.

The Freytag triangle is a theory of story that will pave the way for a quick
analysis of three cartoons illustrating that "exchange is bad": the Gordon
Gekko greed-is-good, the Karl Marx all-trade-exploits, and the Joseph Stiglitz
exchange-is-bad-all-round neo-Luddite narratives. When these stories score
points in the popular imagination, it is often because they leverage some aspect of the *ethos* (credibility), *logos* (logic), and *pathos* (emotional) domains,

which the venerable Greek philosopher Aristotle discusses in his *Rhetoric*.[3] A counter-narrative that points out the relationship between free exchange and the development and exercise of moral virtue might best allow us to add nuance to the cartoons and revivify the Adam Smith of *The Moral Sentiments* in the public mind.[4] Market mechanisms can often act as "moral affordances" as well as moral hazards, and this is something we should keep in mind as we explore the costs and benefits of free exchange.

What's in a Story?

Discussion of stories and narratives is hampered by the fact that there is no widely accepted definition regarding just what a story is. Indeed, an entire school of thought in literary criticism (postmodernism) is predicated on the idea that there are no necessary and sufficient conditions which a piece of text must meet in order to be a story, whether that text is verbal, written, or merely exists in the thoughts of a target audience. We can agree with the postmodernists that defining a "story" is difficult without thinking, however, that the concept plays *no* useful purpose. In that sense, the concept "story" is like the concept "game"—a game also does not require necessary and sufficient conditions for it to be a game, but that does not mean the concept is bankrupt nor that there cannot be "family resemblances" between games that would be useful to consider.

A good first hack, then, at a theory of stories comes from the nineteenth-century German writer Gustav Freytag; this is an admittedly Western notion of the structure of stories, and, though it cannot claim to be comprehensive, it is nonetheless an excellent place to begin. Freytag believed that narratives followed a general pattern: the story begins, a problem arises that leads to a climax, and the problem is resolved in the end. A coherent and unified story could thus be as short as three sentences providing the setup, climax, and resolution, for example: "John was hungry. He went to the store and bought a sandwich. It was delicious." Of course, this *particular* story is neither interesting nor compelling, but still it *is* a coherent narrative. This "Freytag Triangle," depicted in figure 1.1, captures the general structure of a story.[5]

The contemporary literary theorist Patrick Hogan amplifies the basic Freytag structure, pointing out that most plots involve an *agent* (normally, a *hero* or protagonist) striving to achieve some *goal* (usually despite the machinations of an antagonist, or *villain*)—there is a person (or group of persons) and a series of events driven by their attempts to achieve some objective. This familiar analysis is supported by the study of mythology (recall Joseph Campbell's analysis of the structure of most famous legends from antiquity), and by consideration of many forms of storytelling, whether they are oral, traditional, or contemporary.[6]

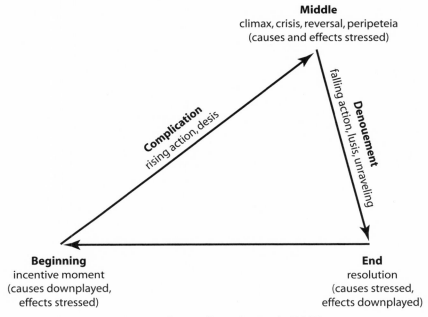

Figure 1.1. Gustave Freitag's triangle (1863)

Why Stories Are So Important

This working "theory of story" will enable us to gain insight into why stories are so important for structuring human thought. First, note that stories often are rich in metaphors and analogies; metaphors, in turn, affect our most basic attitudes toward the world. Suppose, for example, that I think of Islamic fundamentalism as a disease; a simple narrative about fundamentalist Islam might then be the following: "We want world communities to respect human rights. Fundamentalist Muslims disrespect some of those rights. We can prevent them from doing more harm by taking action now." This implies a series of actions I ought to do in reaction to fundamentalism: combat its spread; focus on this "public health problem" by inoculating people against it; consider those who try to spread it as evil agents up to no good—or, at the very least, modern day "Typhoid Marys"; and so on.[7]

A research program by Mark Johnson, George Lakoff, Giles Fauconnier, and Mark Turner have explored reasoning by metaphor and analogy in a rich research program; they and others conclude that our most complex mental tasks are usually carried out not by the "classical mechanics" of rational actor theory (where stories really have no place in the details) but rather by a set of abilities that enables us to make analogies and map metaphors, which forms the core of human cognition.[8] Exploration into the "storytelling mind" is a research program that combines metaphor and analogy into an examination of the powerful grip narrative has on human cognition; narratives can restructure our mental spaces in ways that profoundly affect our reasoning ability and, ultimately, what we make of the world. Think of the grip that the "Jihad versus McWorld"[9] narrative has on the terrorist organization Al Qaeda and how this affects the way the group thinks about the future; or consider conspiracy theorists, springloaded to see the world of markets entirely in terms of "us" (downtrodden consumers) versus "them" (corporate robber barons).[10] As Mark Turner noted, "Story is a basic principle of mind. Most of our experience, our knowledge, and our thinking is organized as stories."[11]

Even if making stories foundational to thought seems a stretch, however, there is *ample* evidence that stories influence our ability to recall events, motivate people to act, modulate our emotional reactions to events, cue certain heuristics and biases, structure our problem-solving capabilities, and ultimately, perhaps, even constitute our very identity.[12] Elucidating each of these points in detail, of course, is beyond the scope of this chapter, but it should be obvious here that the stories we tell ourselves about markets matter.

Some Stories about Free Exchange

At least three archetypical narratives exist about exchange: (1) it is selfish; (2) it is exploitative; and (3) it is, on balance, bad and ought to be rolled back in a neo-Luddite fashion.

The first narrative emphasizes the cutthroat nature of exchange, pointing out that it encourages selfish behavior in those who engage in it. This is the Gordon Gekko "selfish" cartoon. The fictitious Gordon Gekko, portrayed by Michael Douglas in the 1987 movie *Wall Street*, is a ruthless, heartless stock-market trader. In his Gucci suits and slicked-back hair, Gekko's single-minded pursuit of the almighty dollar wrecks lives and subordinates all other values to the selfish lining of one's nest with green bills. As Gekko says in the movie, "Greed is good." The Freytag structure of this narrative is obvious: Gekko becomes involved in market exchange; his soul is corrupted and he becomes an insensitive and immoral boor, causing misery to those around him; he dies a lonely and corrupted man.[13]

This story encourages us to view the world in an "us versus them" mode, where class differences define the in-group and out-group. It also emphasizes the profit motive, at the expense of the other motives that move people to engage in exchange. It goes hand-in-glove with a view of human nature that sees us as natural competitors rather than natural cooperators. It is also loaded with empirical claims about the effect on one's character of doing business. It is a superb example of the structural effects a narrative can have on just about everything, from how we reason about a situation even to how we think of our own identity.

The second cartoon is closely related. It emphasizes how free exchange, because it often involves the acquisition and accumulation of capital, necessarily exploits those involved, especially the line worker. As wealth accumulates, class distinctions are enhanced, and the owners of capital become distanced in multiple respects from those who labor on their behalf. Workers eventually become aware of these distinctions and, aside from a few under the spell of false consciousness, recognize their alienation from the products of their labor, and eventually rise up to reclaim the means of production on behalf of the people. This is the Karl Marx "exploitation" cartoon, and it exhibits a Freytag structure of "exploitation, rebellion, and reformation."

Many have criticized the Karl Marx cartoon, roundly and justifiably. Among this narrative's other cognitive effects, it encourages us to view even consensual exchange suspiciously, as masking exploitative relationships. It emphasizes a certain kind of identity, encouraging solidarity based upon class interest and one's position in the exchange relationship. It comes with a Manichean worldview, and provides a powerful "workers will rise" punch line, which in turn provides a pleasing justice-related climax to the setup.

The third cartoon is perhaps more moderate than the first two. It emphasizes the negative effects, on balance, of allowing free exchange to occur. Although it may reject the extreme Marxist view, and may have a more sanguine take on how the involvement in markets colors one's character, it nonetheless emphasizes the harmful effects of the operation of market mechanisms. If we allow markets to continue to operate, free exchange will lead to such dreaded occurrences as the destruction of the environment, a growing rift between the developed and undeveloped worlds, and an over-reliance on technology. This is the neo-Luddite "unhelpful" cartoon, and, at its most extreme, it would have us roll back the clock to a pre-Adam Smith state of affairs.

The neo-Luddite cartoon has powerful appeal. In its moderate form, it points out some of the faults that inflict capitalist manifestations of the free exchange ethos. Negative externalities exist; what are we to do about them? Sensible development is difficult; how does it progress? Joseph Stiglitz and others make neo-Luddite appeals in their writings. This narrative structure encourages us to be suspicious of progress, to look for hidden agendas behind trade and exchange, and to worry—perhaps to a fault—about the effect of globalization on myriad issues ranging from cultural authenticity to the gap between rich and poor.

To prevent misinterpretation, I do not contend that every aspect of these three cartoons is false. A fair examination of the benefits and burdens of exchange will probably include truthful elements from all three critiques (markets sometimes can corrupt; even exchange that is not coerced sometimes does dehumanize; unhampered, poorly thought out development can produce harmful consequences). Generally, however, these pictures of what free exchange does to us and to others are probably not fair, nor are they borne out by empirical examination. Most especially, all three, in their extreme forms, rely on unrealistic expectations regarding what "plot climaxes" will bring relief to the tension that drives their Freytag triangle–like plot.

Two explanatory burdens present themselves at this point. We need to understand why these narratives capture the public imagination, and to move toward an understanding of how we can redress the cartoons so they become more true and hence probably more useful.

An Aristotelian Evaluative Model: Ethos, Logos, and Pathos

In practice, an effective "counter-narrative strategy" will require understanding the components and content of the story being told so we can predict how it will influence the action of a target audience. In other words, we need a sophisticated understanding of strategic rhetoric. This is difficult to come by. Nonetheless, even well worn and simple models of this process, such as that offered by the ancient Greek philosopher Aristotle in his *Rhetoric*, can be useful for structuring our thinking. Aristotle would have us evaluate three components of a narrative relative to a target audience:

1. What is the *ethos* of the speaker or deliverer?
2. What is the *logos* of the message being delivered?
3. Does the message contain appropriate appeals to *pathos*?

Consideration of ethos would emphasize the need for us to establish credible channels of communication, fronted by actors who have the character and reputation required to ensure receipt and belief of the message. "You have bad ethos" is merely another way of saying, "You won't be believed by the target audience because they don't think *you* are *believable*." Consideration of logos involves the rational elements of the narrative. Is it logical? Is it consistent enough to be believed? Does it contain, from the target's perspective, non sequiturs and forms of reasoning not normally used day to day? Finally, pathos deals with the emotional content of the story. Does the story cue appropriate affective and emotive systems in the human brain? Does it appeal to emotion in a way that engages the whole person and that increases the chances the story will actually motivate action?

Thomas Coakley has summarized the Aristotelian model:

> **Ethos:** these are appeals the speaker makes to the audience to establish credibility. Essentially, ethos is what a speaker uses—implicitly or explicitly—to ensure that the audience can trust him or her. An example in advertising is an athlete endorsing an athletic product.

> **Pathos:** these are appeals the speaker makes to the audience's emotions. An example of this would be an advertisement for tires that emphasizes safety by portraying an infant cradled within the circle of the tire.

Logos: these are appeals to facts. More doctors recommend toothpaste X than any other brand.[14]

   Some of these Aristotelian considerations will be affected by *structural* elements of the story. Is the story coherent? Is it simple enough to be processed? Can it be remembered? Is it easy to transmit? If believed, will it motivate appropriate action? These aspects of stories will determine whether they can serve as effective mechanisms by which we change payoff matrices in exchange games (see chapter 11);[15] others will be affected by *content*. Does the narrative resonate with target audiences? Is the protagonist of the story a member of the target audience's in-group? Is the antagonist of the story a member of a hated out-group?[16] It is not the primary purpose of this chapter to articulate a comprehensive strategy for ensuring that the cartoons are burnished so that they become more true, as well as better, stories. They are effective stories currently, or they would not be popular (largely, I suspect, because of excellent appeals to pathos, good ethos with certain target audiences, and a logos that is at least internally consistent even if it is not true to the actual nature of free exchange). Rather, this detour through Aristotle serves to motivate the next section of the chapter: the perceived *logos* of the story will be influenced by the percolation of results from behavioral economics, neuroeconomics, moral philosophy, cognitive science, primatology, developmental economics, business ethics, and the like, into public awareness—hence the need for the present volume.
   The final two sections of this chapter add bulk to one line of (I hope) effective response to these three archetypical cartoons. Situations of free exchange offer multiple affordances for the development and exercise of traits and motivations praiseworthy from any reasonable moral perspective. Findings from the cognitive and social sciences are demonstrating that not only are human beings inclined by nature toward cooperation in many exchange situations (we are, as Aristotle pointed out, rational social animals) but also that the environment of exchange can influence which traits we exhibit. J. J. Gibson's concept of affordances is useful here, as is an understanding of the fundamental attribution error, a well-confirmed finding from social psychology. An explication of these two concepts and their application to the problem of cartoon narratives about exchange follows.

Affordances and the Importance of the Environment: Gibson and Contemporary Social Psychology

Environmental conditions can be considered *affordances*, to use the term coined by the psychologist J. J. Gibson.[17] This term denotes the capacities offered to us by things present in our environments; for instance, in an office environment, a chair "affords" sitting for people. In a garage, a hammer affords hitting nails. Critical to an understanding of affordances is the *relationship* between the organism and the thing in the organism's environment: some things will be affordances for us but not for other kinds of creatures (for instance, a telephone affords communication for an adult human, but it affords nothing—besides very bad chewing!—for a dog). Cartoon critics of exchange are keen to emphasize the negative aspects of environments where free exchange occurs, but ample positive aspects also exist. These "moral affordances" deserve just as much discussion as the others (more on this below, but see, too, chapter 13, where Kimbrough, Smith, and Wilson essentially discuss institutional affordances for cooperative behavior in their experimental economics work). This finding is counter-intuitive, admittedly, as we tend to "upload" causality into our mind/brain: the truly important causal structures are all in the head. This is too simple, of course, as the *interaction* of our mind/brain with our environment is what produces human behavior.

In a related vein, social psychologists often face an uphill battle when their results run contrary to folk wisdom. Convincing people that a major cause of their behavior is not their personalities but the situation they find themselves in can be truly challenging. Yet, the power of the situation is perhaps the foundation of much social psychological theory and some of the classic studies in the field.[18] When people search for causes of behavior, they tend to ignore the situation and blame (or credit) that which they *do* notice, the person. This is especially the case in individualistic cultures, such as those of the United States or Western Europe:

> Members of these cultures typically overestimate the impact and predictive power of observed or assumed individual differences in traits such as charitableness . . . they are apt to rely heavily on overly broad and simplistic notions of good or bad 'character,' both in their attempts to understand past behavior and in their efforts to predict future behavior.[19, 20]

This bias, attributing causes of behavior to actors (i.e., internal, dispositional factors) rather than the situation (i.e., external, environmental factors) is called the *fundamental attribution error* by social psychologists. In other words, whereas Gibson was primarily concerned about the role of extra-cognitive factors in enabling us to accomplish certain cognitive acts, contemporary social psychology has borne out a corollary to Gibson's concept: the environment is a critically important determinant of our behavior, moral or otherwise.

Bringing It All Together: How Free Exchange Can Enable Moral Growth

With this theoretical background, the final task is to demonstrate that free exchange is as much an *enabler* of moral growth and development as it is a negative influence upon it. Others in this volume have accomplished this much more thoroughly than I can in the limited space that remains (see, especially, chapters 2 and 15). Nonetheless, consider the following points:

1. *Exchange environments can themselves evolve creatures that are prone to cooperation as much as competition* (see chapters 5 and 14). Free exchange offers ecological niches that can be exploited by cooperative behavior as well as defection-laden instrumentally selfish behavior (see chapters 4, 5, and 7). Markets can provide ecological space for cooperative activity to continue to exist. Gekko may have been successful in the short run, but narrowly instrumental selfish agents will not flourish in large-scale exchange situations. Our basic neurobiology provides us with cooperative impulses, which themselves serve as a precursor for the trust and affiliative inclinations necessary to bootstrap exchange activity into existence in the first place (see chapters 3, 6, 11, and 12).[21]

2. *Exchange environments afford critical opportunities for cultivation of the classic virtues.*[22] Exchange situations oft-times lack an explicit enforcement mechanism; in some cases, at least, the motivation to follow through on exchange-based commitments must be purely internal (see chapters 2, 9, and 15). As character educators emphasize, practice is necessary for the virtuous both to become virtuous and to maintain possession of praiseworthy character traits. Trust and a handshake are as much a part of typical exchange as are punishment-based law and enforcement via the implicit threat of force or sanction. If we are to make the transition from Lawrence Kohlberg–style "stage 1" egoistic orientations to stage 2 (social convention) and ultimately stage 3 orientations (those of principled moral reasoners), we must be allowed an opportunity to observe others practicing the virtues, and be given the opportunity to practice them ourselves.[23] We also must know what factors in the environment might cause us to act in ways we would not be proud of in our quieter moments. These are familiar themes from the work of the classic virtue theorist Aristotle. To develop virtue, one must have the chance to practice being virtuous. This means cultivating a milieu where the environmental factors that influence human behavior make virtuous behavior the norm rather than the exception. Attention to how you react in these environments is important. People should be aware of the personality/environment interaction, for only then can they modulate their behavior accordingly so that taking virtuous action becomes more likely. As Aristotle states in his *Nichomachean Ethics* (in a passage that refers to his doctrine of the "Golden Mean"):

> We must also examine what we ourselves drift into easily. For different people have different natural tendencies toward different goals, and we shall come to know our own tendencies from the pleasure or pain that arises in us. We must drag ourselves off in the contrary direction; for if we pull far away from error, as they do in straightening bent wood, we shall reach the intermediate condition.[24]

3. *Exchange environments afford incentives to develop and maintain moral standards.* People fall prey to impulses to defect from the institutions that enable cooperative activity, of course—the cartoons are excellent at pointing this out. Markets, however, do not then stand in tension with morality; instead, they provide us incentive to cultivate fellow feeling and a sense of concern for the common good. Given the imperfection of instrumentally oriented enforcement mechanisms, markets actually motivate morality and the development of moral institutions that rely on internal enforcement of norms via a conscience or similar mental capacity. Because regulatory institutions fail, morality becomes a necessity. This is a point familiar to business ethicists (see chapters 2 and 15; for related thoughts, see chapters 8 and 10).

There are myriad other fashions in which parts of the cartoon stories about exchange examined in this chapter begin to fall apart. The aim here has not been to redress the shortcomings of the cartoons—for an admirable and successful attempt to do that, see Paul Zak's introduction to this volume—but to address threads common to all three cartoons and provide a framework for marshaling evidence in these other areas so as to give free exchange a more comprehensive consideration in, appropriately, the marketplace of ideas.

Conclusion

Economic systems anchored in free exchange have not played well in the popular imagination at the end of the twentieth century. Cartoon versions of "immoral markets" prosper because, in part, they make good stories, but recent advances in our understanding of the cognitive mechanisms that make exchange possible give credence to a narrative about free markets that emphasizes the opportunities they offer to exercise our moral potential. Markets tell many stories; a significant one emphasizes the affordances for ethical behavior that free exchange enables; simply put, markets are important character development mechanisms and provide environmental structure that can anchor moral behavior. The basic theory of story—the Freytag triangle—allowed us to leverage the Aristotelian model of ethos, logos, and pathos to analyze why the "selfish," "exploitative," and "unhelpful" cartoons examined in this chapter flourish in the story competition. We can be much more sanguine than these cartoon critics are about the importance of free exchange for cultivating and sustaining moral behavior. Using the notion of affordances and pointing out the fundamental attribution error allows us to articulate an alternate virtue-theoretic story.

Ideally, this narrative, which is more true to the consilient findings of the sciences discussed in this volume, will help shape public discourse in the twenty-first century: it is the *Aristotelian* alternative, which emphasizes *moral markets for ethical exercise.*

Notes

1. See Clive Crook, "Capitalism: The Movie—Why Americans Don't Value Markets Enough and Why That Matters," *Atlantic Monthly*, March 2006, vol. 297, no. 2., pp. 46–8.

2. Consider, especially, the November 1999 protests against the WTO in Seattle, Washington, which shocked trade delegates, involved tens of thousands of people, and received worldwide press coverage.

3. More correctly, in his *Rhetorica*, multiple translations of which are available. For discussion and references, see http://rhetorica.net/argument.htm.

4. Smith is usually associated with *The Wealth of Nations*, but his equally important work, *The Moral Sentiments*, highlights the role that fellow-feeling plays in human life, and can be considered a corrective to the simplistic version of Smith one gets by sampling the most popular quotations from *Nations*.

5. For more detail about Freytag, see Barbara McManus, a teacher of literary theory, at http://www.cnr.edu/home/bmcmanus/freytag.html. I am indebted to Dr. McManus for allowing me to use her Freytag Triangle graphic.

6. See Patrick Colm Hogan, *The Mind and Its Stories: Narrative Universals and Human Emotion* (Cambridge: Cambridge University Press, 2003).

7. This is an illustration only; this characterization of fundamentalism is flawed in multiple respects, and yet, in other respects, it is, in fact, illustrative.

8. Classic works here include George Lakoff and Mark Johnson, *Metaphors We Live By* (Chicago: University of Chicago Press, 1980); Dedre Gentner, Keith Holyoak, and Boicho Kokinov, *The Analogical Mind: Perspectives from Cognitive Science* (Cambridge, Mass.: MIT Press, 2001); and Gilles Fauconnier and Mark Turner, *The Way We Think: Conceptual Blending and the Mind's Hidden Complexities* (New York: Basic Books, 2002).

9. This is the structuring metaphor of Benjamin Barber's "clash of the world views" book *Jihad vs. McWorld: How Globalism and Tribalism Are Reshaping the World* (New York: Ballantine Books, 1996). See also Samuel P. Huntington's (in?)famous *Clash of Civilizations and the Remaking of World Order* (New York: Simon and Schuster, 1998).

10. See Fauconnier and Turner, *The Way We think*; or Mark Turner, *The Literary Mind* (New York: Oxford University Press, 1998).

11. Turner, *The Literary Mind*. See also the 2002 groundbreaking work by Anthony Patton, "The World as Story," unpublished manuscript.

12. See, e.g., Alicia Juarero, *Dynamics in Action* (Cambridge, Mass.: MIT Press, 1999); Troy S. Thomas, Stephen D. Kiser, and William D. Casebeer, *Warlords Rising: Confronting Violent Non-State Actors* (Lanham, Md.: Lexington Books, 2005), or Troy S. Thomas and William D. Casebeer, *Violent Systems,* Institute for National Security Studies Occasional Paper (2003); A. C. Graesser & G. V. Nakamura, "The Impact of a Schema on Comprehension and Memory," in H. Bower, ed., *The Psychology of Learning and Motivation* (New York: Academic Press, 1990), 16, 59–109; Maybel Chau-Ping Wong, "The Effects of Story Schemata on Narrative Recall," available https://repository.ust.hk/dspace/handle/1783.1/1337 (2004); and Daniel Dennett, "The Self as Center of Narrative Gravity," available http://ase.tufts.edu/cogstud/papers/selfctr.htm (1992).

13. It is not stated explicitly but only implied in the movie's critical structure that Gekko dies a lonely and unfulfilled human. In some respects, the Gekko character is secondary to the primary dramatic element in the movie involving the union boss and his Wall Street son. I thank Oliver Goodenough and Timothy Kane (of the Heritage Foundation) for reminding me of this point, as well as for other useful critiques and advice on the content and structure of this chapter.

14. Thomas Coakley, "The Argument against Terror: The Peruvian Experience, Globalization, and US Policy," Institute for National Security Studies (INSS) research paper, 2003; available from Casebeer at drenbill@earthlink.net or from INSS at james.smith@usafa.af.mil.

15. Indeed, some structure and content of stories may cause narratives to act as primary reinforcers—that is, just like food, drugs, or sex. A fascinating neurobiological exploration of this process of successful "cultural messaging" is being carried out by Casebeer and by neuroscientists such as Read Montague, the director of the Baylor College of Medicine's Human Neuroimaging Laboratory. Innovative new techniques such as "hyperscanning" allow the study of social cognition in vivo at the neural level using functional Magnetic Resonance Imaging.

16. It may well be that some aspects of narrative are evaluated in exactly the same way that theories in the sciences are evaluated: according to their simplicity, output power, explanatory power, justificatory power, coherence, breadth, clarity, and psychological plausibility.

17. For a more complete discussion of affordances, see "Affordance, Convention and Design," *Interactions*, May 1999, 38–43. Also available at http://www.jnd.org/dn .mss/affordances-interactions.html. Most standard psychology or cognitive science textbooks, especially those that stress embodied cognition, may also be consulted.

18. For an in-depth examination, see L. Ross & R. Nisbett, *The Person and the Situation: Perspectives of Social Psychology* (New York: McGraw-Hill, 1991).

19. L. Ross & D. Shestowsky, "Contemporary Psychology's Challenges to Legal Theory and Practice," *Northwestern University Law Review* 97, 1081–1114, 2003, p. 1093.

20. I am indebted to Professor Steve Samuels (Department of Behavioral Sciences and Leadership, U.S. Air Force Academy) for discussion about the fundamental attribution error, and for the substance of the preceding paragraph and the one that follows. For a more in-depth discussion of our views on how the fundamental attribution error influences character development practice, see Steven Samuels and William Casebeer, "A Social Psychological View of Morality: Why Knowledge of Situational Influences on Behavior Can Improve Character Development Practices," *Journal of Moral Education* 34, no. 1 (2005): 73–88.

21. See also William D. Casebeer and Patricia S. Churchland, "The Neural Mechanisms of Moral Cognition: A Multiple Aspect Approach to Moral Judgment and Decision-making," *Biology and Philosophy* 18 (2003): 169–194; William D. Casebeer, "Neurobiology Supports Virtue Theory on the Role of Heuristics in Moral Cognition," *Behavioral and Brain Sciences* 28, no. 4 (2005): 547–548; and Joshua Greene and Jonathan Haidt, "How (and Where) Does Moral Judgment Work?" *Trends in Cognitive Sciences* 6, no. 12 (2003): 512–523.

22. See William D. Casebeer, *Natural Ethical Facts: Evolution, Connectionism, and Moral Cognition* (Cambridge, Mass.: MIT Press, 2003).

23. Kohlberg's six-stage, three-level (ultimately schema-based) system of moral development is well studied. For a summary and critique, see W. C. Crain, *Theories of Development* (New York: Prentice-Hall, 1985), pp. 118–136.

24. Aristotle, *Nicomachean Ethics*, trans. with notes and glossary by Terence Irwin (384–322 BC/1999) (Indianapolis: Hackett, 1985), p. 29.

Two

Free Enterprise, Sympathy, and Virtue

Robert C. Solomon

> No man is devoid of a heart sensitive to the sufferings of others . . . whoever is devoid of the heart of compassion is not human.
> —Mencius, 4th century BC

> How selfish soever man may be supposed, there are evidently some principles in his nature, which interest him in the fortune of others, and render their happiness necessary to him, though he derives nothing from it except the pleasure of seeing it. Of this kind is pity or compassion, the emotion which we feel for the misery of others. . . . The greatest ruffian, the most hardened violator of the laws of society, is not altogether without it.
> —Adam Smith, *Theory of the Moral Sentiments*, 1759

It was common wisdom in the nineteenth century that business had become the great civilizing influence in Europe. Some writers may have rightly feared for a loss of culture, but in place of sectarian massacres in the name of religion and hundred-year feuds in the pursuit of a throne or some parcel of land, the new world of business offered compromise, mutual gain, innovation, and widespread wealth on a scale never before imagined. Napoleon was replaced by thousands of entrepreneurs, merchants and bankers. A decade of war (1805–1815) morphed quietly into a century of peace and prosperity. Thus, the modern history of business itself runs counter to the "cartoon"-like stories of commerce as all selfish greed and self-maximization (see chapters 1, 8, 9, and 14 of this volume).

As we enter the new millennium, an overriding question faces global corporate free enterprise: Can corporations that now or will control and effect so much of the planet's humanity and resources demonstrate not only their profitability but their *integrity*? The old quasi-theological debates still

Portions of this chapter were adapted from my books *In Defense of Sentimentality* (1999) and *A Better Way to Think about Business* (2004), both Oxford University Press, and my essay "Sympathie für Adam Smith: Einige aktuelle philosophische und psychologische Überlegungen," in *Adam Smith als Moralphilosoph* (German only), Verlag Walter de Gruyter GmbH, 2005.

persist: Do multinational corporations, and capitalism in general, best serve humanity? Are corporations and capitalism good or evil or, at best, amoral? Can corporations have a conscience, and can we expect them to act responsibly? In the year 2007, however, all this is merely academic. Whether or not this is "the end of history," the free enterprise mentality—along with its complements, consumerism, and corporatism—now rule the world. The fate of the world, for at least the first part of this critical century, lies in corporate hands. How these corporations conduct their business and how they conceive of their own identity is no longer simply "a business question." But free enterprise, according to its eloquent defender Adam Smith, is itself an activity based on virtue and integrity, or, more broadly, on a conception of human nature quite at odds with the self-interested model that is so often defended in his name.

This chapter explores in some detail Adam Smith's claim that *sympathy* or "fellow-feeling" is a "natural" moral sentiment. At least two large philosophical-psychological-biological questions are relevant here. First, what exactly does Smith mean by "sympathy," a term his writings by no means make clear? Second, is there sufficient evidence today for Smith's claim that sympathy is natural, or, after Darwin and Mendel, is sympathy something of a genetic predisposition or capacity and a product of evolution rather than socialization, education, or environment? This chapter primarily tries to answer the first question, which is more of a textual-philosophical-psychological investigation and appropriate to my own work on emotions and ethics. I also offer a few suggestions regarding the biological-evolutionary question, as this is obviously a primary focus of this volume, although such an investigation is far more in the domain of the other contributors.

If Smith is right about sympathy as a natural moral sentiment, then the economic models often presented in his name that presume self-interested behavior as an initial assumption are inappropriate. There are, of course, several varieties and spin-offs of these models, for instance, in contemporary decision theory and rational choice theory (for which Thomas Schelling won a Nobel Prize), which do not (or at least need not) make any such presumption. (See, especially, chapter 14 in this volume) But the argument that people are naturally and purely self-interested, so popular and shocking in the early eighteenth century, will not find its champion in Adam Smith. Like his good friend David Hume, Smith defended a rather broad practical notion of utility, though I would argue, also like Hume (and like his successor John Stuart Mill, a self-declared Utilitarian), that Adam Smith can best be understood as a follower of Aristotle, whose philosophy was also based on claims about what is "natural" to human beings but who clearly saw self-interest (or, at its extreme, *pleonexia*, "grasping" self-interest) as too common a perversion (that is, as not natural) but not itself definitive of human nature. What defined human nature, according to Aristotle, was the capacity for virtue, a desire for excellence, specified (as in Smith's own celebration on the specialization of labor) according to one's place in society. Putting all this together, I would argue that Adam Smith is an Aristotelian philosopher who (unlike his illustrious predecessor) also had deep insights into the workings and benefits of a market society. The basis of such a society, however, is not, as so often declared, self-interest. It is fellow-feeling, a sense of community, of "being in this together," and genuine discomfort at the very idea of causing harm to one's fellow citizens—in other words, a creature precisely the opposite of the pathological *Homo economicus* described in chapter 8. And if the market society in its current incarnation tends to block that vision, that is a cause for serious worry. (Aristotle precociously thought that it did block that vision, and that is why he condemned it.)

Thus, the many questions about business, ethics, and the fate of humanity in the new millennium all come together in one great concern: whether and how the virtues and integrity of the people who make up our corporations and the increasingly international business world can be implemented in that world and in those corporations. (I see this as a further extension of the story weaved in chapter 13 on how the impersonal market grew out of personal interactions and exchanges.) I do not doubt that most managers and executives take their own virtues and sense of integrity very seriously. They want the best from as well as for themselves, and they want the best for as well as from their organizations. They know it is not a law of nature or of economics that a business must pursue its profits *no matter what*. They also know that a corporation has obligations to its loyal employees and to the community that helped it flourish. And now, of course, that community is no longer local but global, so that what happens in one community has consequences and implications for all.

On the Sad Fate of Sympathy

> It was Midge Decter who declared that "compassion" is a term she cannot abide. The pages of *The Public Interest* are full of articles by young technocrats who may well be able to abide the emotion but who insist that it would cost too much.
> —Norman Birnbaum *The Nation*, April 23, 1988

It may have been the Kantian turn in philosophy that was responsible for the exile of the kindly sentiments and the "inclinations" in general from moral philosophy. In his *Groundwork of the Metaphysics of Morals* (*Grundlegung zur Metaphysik der Sitten* [1785]), the great German moral philosopher Immanuel Kant mentions "melting passion" with a kind of dripping sarcasm, referring to his illustrious predecessors (particularly David Hume). In a rightfully infamous biblical passage on love (that is, love of one's neighbor), Kant writes: "Love out of inclination cannot be commanded; but kindness done from duty—although no inclination impels us, and even although natural and un-conquerable disinclination stands in our way—is Practical, and not Patholog-ical love, residing in the will and not of melting compassion (*schmelzender Theilnehmung*)."[1] Worth mentioning, perhaps, is that Kant in his earlier years had insisted that compassion was "beautiful" but it nevertheless had no "moral worth." His many followers continue to give compassion short shrift, adding, as often as not, a slight sneer of condescension. Sympathy, by the end of the nineteenth century, had joined the much-abused ranks of the other tender sentiments under the rebuke of "sentimentality."

In contemporary politics, of course, compassion remains an endangered, often ridiculed sentiment despite the oxymoronic and hypocritical catch-phrase "compassionate conservatism." Kant, we should note, has been reno-vated in recent commentary, and, if he has not become a champion of the moral sentiments, at least he is no longer viewed as their nemesis.[2] Never-theless, a great many "Kantians," who control a good deal of contemporary moral philosophy (along with the various Utilitarians who are equally dubi-ous about the tender sentiments and prefer the hard acts of hedonism) are. One might object that the most influential moral philosopher of recent de-cades, the late John Rawls, though in obvious ways a follower of Kant, does not dismiss the sentiments. (It is often argued, convincingly and by Rawls's own admission, that he was, like Kant, heavily influenced by Jean-Jacques Rousseau.) Indeed, Rawls may emphasize the objectivity of his principles, but he also mentions the sentiments and our "sense" of justice in the inter-stices of the protracted "deduction" that makes up the bulk of his *Theory of Justice*. But such sentiments all too often appear to be no more than sponta-neous instantiations of abstract rational principles or casuistic evidence for them, despite the celebrated importance of our "intuitions" (sentimental feelings?) in Rawls's celebrated "reflective equilibrium." Rousseau, however, like Hume and Smith, thought that the moral sentiments were much more than this and undergirded (rather than simply instantiated) any morality of principles.

Against the Kantian and neo-conservative reactions, Adam Smith, de-spite his reputation among businessmen as the father of compassionless free market economics, should be viewed as a defender of moral sensitivity, even "sentimentality." This not only is compatible with free market thinking but is essential to it. (Several chapters in this volume thus emphasize the impor-tance of trust as the foundation of the market, notably chapters 7, 9, 12, and 14). In light of the Kantian tradition, it is always worth remembering how closely Kant (and Rawls) claimed to follow on the heels of Rousseau, one of the great defenders of the natural inclinations. In light of neo-conservative callousness, we should appreciate just how important it is that Smith de-fended the centrality of the natural sentiment of *sympathy* in morals, distin-guishing between sympathy and justice (which David Hume declared not to be "natural" at all). There is now a temptation to see Smith as anticipating Kant (by emphasizing the importance of the "impartial spectator" and the ideal of "perfect virtue," for example), but I think this is a mistake. As suggested above, Smith, in my opinion, can better be read (as can Hume) as a modern proponent of virtue ethics, with a powerful emphasis on what he called the "inclinations" of which sympathy is the most central. Indeed, I want to move Smith a bit farther away from Kant by suggesting that his concept of sympathy is even more "natural" than he and most of his commentators often suggest.

Moral sentiment theory began with the view that the basis of morality and (for Smith) justice is to be found in our natural disposition to have certain other-directed emotions. This did not mean that morality is not, in part, a function of reason. However, as Hume rather polemically insisted (and I do not think Smith ever disagreed with him on this), "reason is and ought to be the slave of the passions" or, more specifically, the moral sentiments. Reason certainly plays an important "instrumental" role in ensuring that the right things get done (and clearly the right things usually do not get done for the wrong reasons). But surely among the "right reasons" are the sympathetic emotions. Contrary to at least one dominant strand of the Kantian tradition, to be moral and to be just does not mean that one must act *on principle*. And counter to one leading conception of "economic rationality" (what I have referred to elsewhere as "smart selfishness"), rationality does not mean— whatever else it may mean—acting in one's own self-interest. (Kant, notably, made "making an exception of oneself the most egregious violation of morality. So, too, did the classic Utilitarians.) Thus, the "right reason" for doing something often involves the immediate awareness that someone else is in trouble.

A good reason for helping another person in need, accordingly, is not just because it is a Kantian duty but because one "feels sorry for him." Indeed, one is hard pressed to think of any other sort of reason so impervious to argument (which is not to say it is indefeasible). But reasoning here, and sympathy in particular, as Hume and Smith insisted, is subservient to the right sentiments. Yet, reading through the philosophy of much of the past two centuries, a nonphilosophical reader would be shocked at how often such reasons as "I feel sorry for him" and "because I care about her" are all but ignored as legitimate moral grounds.

What Is Sympathy?

Traditionally, moral sentiment theory has been concerned with a family of "natural" emotions, including benevolence, sympathy, compassion, and pity. (Care and caring, instructively, were not part of the standard list.) These were often lumped together and not infrequently treated as identical (by Hume, for example, at least sometimes). In the standard account of moral sentiment thinking, Francis Hutcheson's "moral sense" theory is usually included as a moral sentiment theory, even though Hutcheson explicitly denied any special moral role to the sentiments. Jean-Jacques Rousseau is usually not included as a moral sentiment theorist, although he was obviously one with them. With his Scottish colleagues he attacked the "selfishness" theories of Hobbes and Mandeville, and argued for the naturalness of pity, a kindred emotion to compassion. But as developed by Hume and Smith, in particular, the exemplary moral sentiment was *sympathy*. (In German, "*Sympathie*," but perhaps also "*Mitleid*" or "*Mitgefühl*"—though none of these translations is sufficiently precise to capture the nuances and ambiguities of the English word.) The sentiment of sympathy, I suggest, is an awkward amalgam of several emotional states and dispositions, including both compassion and a certain amount of what we have come to call "care." It also includes, especially for Smith, our contemporary notion of empathy.

The meaning of "sympathy" bears considerable confusion, both in the writings of the moral sentiment theorists and in our current conversations. In common parlance, *sympathy* means "feeling sorry for" someone, and, for many philosophers (notably Hume), it is conflated with benevolence. (Smith tries to keep these distinct.) "Feeling sorry for" can be a sign of caring, but surely a minimal one, as we can feel sorry for strangers, even our enemies. Benevolence has much in common with the more activist concept of "caring for," but benevolence has much greater scope than sympathy as such. We can feel benevolence in the abstract (without any particular object) and even for those whose feelings are utterly malicious or indifferent to us (e.g., in being merciful to a condemned and still hateful wrongdoer, perhaps as an expression of our own largesse but out of benevolence nevertheless). We often use the word "sympathy" or the verb "to sympathize" to register agreement or approval, although none of these qualify as an adequate philosophical conception of the term. Smith explicitly rejects "feeling sorry for" as an account of his term, but I do not think (and will try to argue) that he does not do away with this as an essential part of the meaning.

What Smith most often means by "sympathy" is fellow-feeling, feeling *with*. Technically, sympathy (literally, "feeling with," like "compassion") is the sharing of feeling, or, as a disposition, the ability to share the feelings of others. Or, if one wants to insist that the emotions can be individuated only according to the persons who have them and thus cannot be shared, one might say that sympathy is an "agreement of feelings" (*Random House Dictionary*), in the sense of "having the same [type of] emotion." One need not "agree with" in the sense of "approve of " the feeling in question, of course, any more than one must always enjoy, like, or approve of one's own emotions. The feelings may agree, but we need not; sharing a feeling is one thing, but accepting or approving of the feeling is something quite different. In grade B movies, we may well share the offended hero's rather fascist sense of revenge while berating ourselves for just that feeling. We might find ourselves sympathizing with someone envious or hateful and nevertheless criticize ourselves for doing so, thus sharing but not at all accepting or approving of the envy or the hatred.

Adam Smith uses the term in this technical way, as "agreement of emotion," but he does not thereby imply the agreement of any particular emotion or kind of emotion. Here is the serious ambiguity between sympathy as a specific sentiment and sympathy as a disposition to share sentiments (*whatever* sentiments) with others ("a fellow-feeling with any passion whatever" (*Theory of the Moral Sentiments* [*TMS*], I.i.5). Sympathy conceived as a disposition to share sentiments with others is not itself a sentiment at all but rather a *vehicle* for understanding other people's sentiments. One can sympathize with (that is, share) any number of feelings in another person, not only the kindly and social moral sentiments but also such unsocial sentiments as envy, anger, and resentment Smith then goes on to qualify this, however, by insisting that sympathy is *not* an actual sharing of sentiments (in the sense of "having the same feeling"). Rather, it is an act of imagination by which one can appreciate the feelings of another person by "putting oneself in his place," "a principle which interests him in the fortunes of others" (*TMS*, I.i.I.2). This gives Smith a way to account for how it can be that people are not essentially selfish or self-interested but are basically social creatures who can act on behalf of others whose feelings they do not (and logically cannot) actually share. But this raises a question: Can sympathy be the sort of motivating factor in our behavior which moral sentiment theory seeks to defend?

Moral sentiment theory depends on the idea that sympathy motivates moral behavior. In retrospect, this claim about motivation is the most dramatic difference between Kant and the Scots, namely, whether practical reason can itself be motivating (Kant thinks so; Hume, famously, thinks not) or whether the motivation must come from the inclinations. But on Smith's "fellow-feeling" interpretation of sympathy, it is not clear either that sympathy is an inclination and (therefore) can be motivating, nor is it clear that it can act in effective opposition to the equally natural (and often more powerful) sentiment of self-interest.[3] Emphasizing the impartial spectator, as current commentators do (for instance, Stephen Darwall), only seems to aggravate the problem. Smith covers himself here by insisting that the impartial spectator does have the sentiment of sympathy, in contrast to the wholly selfless and also passionless dictates of reason. ("The approbation of moral qualities most certainly is not deriv'd from reason, or any comparison of ideas; but proceeds entirely from a moral taste, and from certain sentiments of pleasure and disgust, which arise upon the contemplation and view of particular qualities or characters" [*TMS*]). But if sympathy is just fellow-feeling and not an actual sentiment (inclination) itself, then it does not seem to pack the requisite motivational punch to carry Smith's pre-Kantian, anti-Kantian argument.

Sympathy, according to Smith's definition as "fellow-feeling," seems to be more concerned with comprehension than feeling as such, and comprehension is too close to the "comparison of ideas" to provide the "sentiments of pleasure and disgust" and play the role that sympathy is called to play in morals.[4] ("As we have no immediate experience of what other men feel, we can form no idea of the manner in which they are affected, but by conceiving what we ourselves should feel in the like situation" [*TMS* I.i.I 2].) Sympathy cannot mean merely "comprehension," yet, on the other hand, sympathy as truly shared feeling (my suffering or envy is the mirror image of your suffering or envy) is too implausible.[5] Bringing in the (then as now) common sense understanding of sympathy as "feeling sorry for" provides the bridge between mere comprehension and true fellow-feeling. Feeling sorry for another presumes, at least in some minimal sense, understanding the other's plight. But it also requires, in some minimal sense, feeling on one's own part, though certainly much less than the mirror image of the other's suffering.

I do not mean to limit sympathy to an emotional reaction to another person's suffering, although this is (I think) how the term is most often employed. One might also share another's joy or pride, and I think that we do sometimes use the word "sympathize" (if not "empathize") in such cases. I argue, however, that Smith's use of the term is healthily complex, not quite to say inconsistent. On the one hand, he wants a mechanism for "fellow-feeling," on the other, a motive for morals. As fellow-feeling, it is not at all clear that he can have both, but if we assume that Smith does not mean what he says when he casts out the "feeling sorry for" interpretation and rather take him to be saying (correctly) that this alone does not even begin to capture the rich meaning of sympathy, then he can maintain both sympathy's fellow-feeling and its motivating power.[6]

My thesis here, put simply, is that sympathy, as Smith understands it, includes both the ability to feel *with* as well as *for* others, and it lies at the very foundation of our emotional lives and is the basis (though not the sole basis) of ethics.

Hume versus Smith on Justice, and All against Hobbes

Smith and, before him, David Hume and the other moral sentiment theorists, whatever their differences, made an all-important point against the then reigning cynicism of Hobbes and Mandeville, the idea that people are essentially self-interested, if not selfish, and no other motives could explain human behavior. Whether it was moral sense or moral sentiments, or merely primal man's "inner goodness," the case was put forward that human behavior had springs other than self-interest. Accordingly, sympathy is sometimes discussed as if it were no more than a generalized sense of altruism, a concern for others with no thought of benefit to oneself. But altruism, like benevolence, does not involve any sharing of feelings, as sympathy does, and altruism does not yet capture the central thesis of Smith's virtue ethics, which is that the very distinction between self-interest and altruism breaks down.

This point goes back to Socrates, though Aristotle probably explained it better. Socrates suggested that his virtuous behavior was in fact quite self-interested, because it was "for the good of his soul," and what could be more self-interested than that? Aristotle explained that a criterion for virtue was the enjoyment of its exercise, that a virtue should be "second nature." In other words, the post-pagan model of the battle between temptation and righteousness was not considered moral and was by no means the mark of the virtuous man. It is noteworthy that Kant needs to add the notions of the "holy will" and the "supererogatory" to explain the harmony of inclinations and rational will in special cases (e.g., the cases of Jesus and saints), but clearly such cases are the exception, not the rule. In virtue ethics the right course of action is motivating just because it is pleasant and self-fulfilling to the virtuous person.

Hume, if not Smith, quite proudly referred to himself as a pagan. His own post-pagan model of ethics harked back to Aristotle and the Ancients, and his ethics was a virtue ethics in this sense in particular, that the sign of "good character" was not inner struggle but inner harmony. For Smith, too, though he was too much a Christian to pride himself a pagan, ethics was, first of all, virtue ethics and a concern for good character and inner harmony. His claim that sympathy was a natural human sentiment was not so much a plea for altruism as a denial of total selfishness and the insistence that virtuous human behavior need not be one or the other. Smith's particularly brilliant emphasis on the natural need for others' approval has just this outcome: most of even our most "selfish" motives have as their objects the desire for camaraderie and communal agreement. What I want for myself is your approval, and to get it I will most likely do what you think I *should* do. No basic tension between what I want for myself and what I should want to do is much in evidence, and the cynic will have a hard time getting his cynicism in here.

Hume's earlier theory of sympathy and justice, which greatly influenced Smith, is somewhat different than this, as Smith sharply insists, and, to make matters more difficult, clearly Hume changed his mind between the writing of his early masterpiece, *A Treatise of Human Nature*, and his later *Inquiry Concerning the Principles of Morals*. In the early work, Hume treats sympathy rather casually, commenting that it is usually a weak emotion compared to most self-interest motives. In the later work, Hume defends sympathy as a universal sentiment sufficiently powerful to overcome self-interest in a great many cases. In the *Inquiry*, in particular, Hume takes sympathy to be a form of benevolence, a feeling for one's fellow citizens and a concern for their well-being. But for Hume, as for Smith, sympathy is too often countered and overwhelmed by selfishness, and, for this reason, a sense of justice is required. However, whereas Smith takes the sense of justice to be a natural revulsion at the harming of one's fellows, Hume takes justice to be an "artificial" virtue, constructed by reason for our mutual well-being. (It is important not to read "artificial" as "arbitrary." Artificial virtues can be precise and readily justifiable.) Justice, for Hume, is an advantageous conventional "scheme" rather than a natural sentiment as such. Thus, for Hume, sympathy is a genuine moral sentiment and justice is not. Even so, he admitted that justice was so beneficial that it became inseparably associated with the moral sentiments, for what could be more basic to these sentiments than our sense of the general good for everyone, "a feeling for the happiness of mankind and a resentment of their misery" (235). He writes,

> No virtue is more esteemed than justice, and no vice more detested than injustice; nor are there any qualities, which go farther to the fixing of character, either as amiable or odious. Now justice is a moral virtue, merely because it has that tendency to the good of mankind; and, indeed, is nothing but an artificial invention to that purpose.
>
> The whole scheme, however, of law and justice is advantageous to the society; and 'twas with a view to this advantage, that men, by their voluntary conventions, establish'd it. . . . Once established, it is *naturally* attended with a strong sentiment of morals. (Hume, *Treatise*, 577, 579)

Hume does not go so far as to say that justice itself is a matter of sentiment, but he insists that the moral sentiments in general, and sympathy for others, in particular, are so essential to morals that there can be no ethics without them. Both Hume and Smith are dead set against the Hobbesian view that people are motivated only by their own selfish interests, and both advocate the importance of distinctive, natural "social passions." Indeed, the core of their argument is, in Smith's terms, that "nature, when she formed man for society, endowed him with an original desire to please, and an original aversion to offend his brethren" (TMS III.2.6). Moreover, "nature endowed him not only with a desire for being approved of, but with a desire of being what ought to be approved of, or of being what he himself approves of in other men" (TMS III.2.7). It is not just sympathy but a whole complex of mutually perceiving and reciprocal passions that tie us together. Thus, it does not take too much tinkering with Scottish moral sentiment theory to incorporate justice along with sympathy under its auspices and take the whole as a welcome alternative to both the "man is essentially selfish" thesis and the overly intellectual "morality is rationality" view of Kant and most current justice theorists.

Part of the problem, for both Smith and Hume, is that there is an essential tension between sympathy and justice, but each resolves this in rather different ways. Hume, as just mentioned, distinguishes between sympathy as a natural emotion and variable according to proximity, and justice as an artificial virtue dependent on trans-personal conventions and institutions. Smith, instead, depends on his so-called impartial spectator who thereby transcends the particular preferences of personal proximity. I find both devices plausible, but in tandem and mutually complimentary. Justice is, as Rawls argues, the primary virtue of institutions, but it is also, as in Plato and Aristotle, a *personal* virtue, and the complex relationship between the institutional and the personal is partially captured but also greatly oversimplified in Hume's distinction. It is the personal that becomes institutionalized, not top-down from Rawlsian-type theories but bottom-up from the personal to the communal to the institutional. So, too, with Smith's imaginary impartial spectator, whom I imagine as a metaphor for a gradual and learned expansion from one's immediate experience with friends and family to the only dimly imagined probable experience of others at a distance. Thus, I find the hypostatization of the impartial spectator, treating him as a supposed person to whom one might "appeal" a terrible misunderstanding of Smith's conception, which rightly urges us to transcend our own perspectives and biases but makes no sense cut off from all perspectives. To be sure, Smith himself admits this and so supplements his still somewhat partial impartial spectator with a truly impartial spectator, namely, God. But here I would join Hume and Nietzsche and urge that we reject the idea of any such perspective-less perspective or God's-eye view, a "view from nowhere." Justice, like sympathy, is always a view from somewhere.

What both Hume and Smith are concerned to point out, against Hobbes, is that we genuinely and "naturally" *care* about other people, and we are capable of feeling *with* others as well as for ourselves. This is true of both intimate relations and the very distant considerations of justice. There are disagreements between them, of course, but they both deny the Hobbesian portrait of humanity as essentially selfish. (It is a mistake or at least unfair, I think, to attack Hume on the grounds that he is really an unreconstructed Hobbesian individualist who brings in "sympathy" only as a desperate measure to explain the non-coercive validity of morals.[7] Whatever their ambiguities and differences, both Smith and Hume provide united fronts against Hobbes and the egoists and the Kantian hordes to come.

Are Sympathy and Empathy Natural?

Smith and Hume celebrate sympathy as a "natural" emotion or sentiment and as the basis for all morality, although their notion of "sympathy" is ambiguous and, perhaps, over-ambitious. So, too, is their concept of "natural" unclear, at least by modern standards. Today we see as natural that which is biologically based, so we need to ask whether there is biological (neurological) or evolutionary evidence for this. In fact, contemporary philosophy and psychology has both clarified and confirmed much of what Smith and Hume had to say, and it is worth reviewing some of that research (as several of our colleagues have done elsewhere in this volume) to see if their conception of sympathy (and their attack on Hobbes and selfishness) stands up to scrutiny in light of current psychology and biology. The short answer is yes, although current thinking on the subject suggests a much more complicated answer than Hume and Smith anticipated.

First, however, we need to acknowledge the general problem in Smith and Hume that haunts empiricist accounts of emotion from Locke to (famously) William James. It is the idea that an emotion is some sort of "inner" perception, something akin to a sensation, the perception of something going on in the mind (and, in James, especially, the body). Hume's elaborate theory of the passions in the *Treatise* makes this quite clear. He defines a passion as a "secondary or reflective impression," that is, an impression of an impression, which trips him up enormously when he goes about actually trying to analyze particular passions. In his treatment of pride, for instance, he founders between the standard empiricist account of the emotion as a pleasant sensation and the much more illuminating idea that pride, like all emotions, is *intentional*. It is about the world, not (just) about the body or the mind. In the *Treatise*, Hume tries to reconcile his empiricist interpretation of pride with intentionality by suggesting a "monstrous heap of ideas"—about the self and so forth, but the result is (by his own admission) an incoherent mess. Neither Smith nor Hume subject sympathy to this kind of extensive analysis, but clearly this general framework looms in the background, and, I would argue, no emotion or sentiment—especially sympathy—is intelligible in its terms. Sympathy is not just a pleasant feeling (in the context, presumably, of another person). Sympathy is *about* the other person, a role that no mere sensation can play.[8]

I have also suggested that Smith has an ambiguous understanding of sympathy, one that he needs in order for the sentiment to do all the work he wants it to do for him. To express my argument in a more modern, less problematic way, I have suggested that Smith's sense of "sympathy" is importantly ambiguous between sympathy and *empathy*, between being upset about or having kindly feelings toward a fellow creature (usually in pain) and in some sense actually "sharing" the suffering or the emotions of others. Thus, sympathy can mean either feeling sorry *for* someone or feeling *with* them, and these are not the same. (Again, I want to leave room for sympathizing with the "positive" emotions of others, but I will focus—as is customary—on sympathizing with suffering.) The word "empathy" is more recent in origin and so was not available to the Scottish moral sentiment theorists (much less to Aristotle), but the current distinction between sympathy and empathy is of considerable importance and is a great help to us in our interpretation.

One can argue—as I suggested above—that empathy (Smith's own favored use of "sympathy") is not an emotion as such but rather refers to the *sharing* of emotion (any emotion) or sometimes the *capacity* to share emotions with others. There is some current research into what are appropriately called "mirror neurons," which seem to provide a neurological basis for our immediate non-cognitive ability to share others' feelings, what many psychologists, following Elaine Hatfield, call "emotional contagion." Paul Zak, in chapter 12 of this volume, provides a somewhat detailed account of this primitive neurological mechanism by which such "contagion" may take place. In all its various versions, however, empathy involves some sort of mirroring or sharing of others' emotions without it being, in itself, an actual emotion. Sympathy, by contrast, *is* a distinct emotion, a quite particular though rather diffuse and contextually defined emotion. In other words, it is defined by way of its intentionality, what it is about, the particular person in the particular situation, and this makes anything like a specific neurological basis unlikely (as opposed to a diffuse, complicated, and highly cognitive [cortical] activity). It is sympathy that does the motivational work Smith requires, and sympathy in turn requires empathy, the capacity to "read" and to some extent share other people's emotions. Thus, Smith is right, in my view, to house the two notions under one roof, so to speak, for although not all empathy involves sympathy, all sympathy does presuppose empathy. He is also right to suggest that empathy is the very foundation of our being human, and it is through sympathy that this basic human capacity is played out in ethics. What contemporary biology has to add is that there is good reason to think that empathy, at least, is a basic biological human function.

Empathy, in contemporary psychology, is defined by Nancy Eisenberg as "an affective response that stems from the apprehension or comprehension of another's emotional state or condition, and that is identical or very similar to what the other person is feeling or would be expected to feel."[9]

Eisenberg gives the following example: "If a woman sees or hears about a person who is sad and feels sad in response she is experiencing empathy."[10] She traces the term back to the early twentieth century, noting that early use of the term was heavily cognitive and involved imaginatively "taking the role of the other," very much like the German concept of *Verstehen*. Note, however, that Eisenberg emphasizes the "affective" aspect of empathy and points out that it need not involve thinking (or imagination) at all. Nevertheless, it is important to distinguish empathy from simply being upset or "personally distressed," and it requires, minimally, the "cognitive" separation of self and other. Infants, accordingly, cannot feel empathy, though they can get upset when their mothers are upset. Getting upset because someone else is upset may be wholly self-involved and aversive (for instance, growing uncomfortable when someone regales you with their tale of woe and just wanting to leave the room). This is not empathy. Nor is empathy merely "picking up" others' emotion by way of "emotional contagion." Thus, it is not just affective. Empathy is cognitive, and, in that regard, as noted above, is like the German *Verstehen*. That is not to say, however, that it must be as thoroughly entangled with the active imagination as Smith, in particular, supposes.

Much can be said about this distinction between the affective and the cognitive, but suffice it to say here that I find the distinction overdone.[11] Thus, the idea that empathy (and sympathy) are affective responses is in no way incompatible with their also being cognitive, involving concepts and ways of construing the world.[12] This is not to say that empathy (or sympathy) needs to involve thinking or imagining oneself "in the other person's shoes," and, in this sense, neither emotion is the same as *Verstehen*. Understanding may be minimal and merely tacit, not articulate, and it need not involve projection at all. A good distinction can be made following Nell Noddings in her book, *Caring*.[13] In writing about what she calls "the one-caring," she notes:

> We might want to call this relationship "empathy," but we should think about what we mean by this term. *The Oxford English Dictionary* defines *empathy* as "The power of projecting one's personality into, and so fully understanding, the object of contemplation." This is, perhaps, a peculiarly rational, Western, masculine way of looking at "feeling with."

Much as there is to disagree with in both the Oxford dictionary definition and Noddings's rather global dismissal of it, we should take very seriously her juxtaposition of being *receptive* to the feelings of others as opposed to projection and contemplation ("putting myself in the other's shoes"). Her example is a mother with a wet, crying baby. Noddings rightly comments that a mother does not project herself into her infant, and ask, "How would I feel if I were wet to the ribs?" She adds, "We do this only when the natural impulse fails." So without overdoing the distinction between the affective and the cognitive, or falling back to the notion of "emotional contagion," we might insist that empathy is primarily a pre-contemplative (not "pre-cognitive") feeling. The sense in which it is a *shared* feeling remains to be more carefully analyzed, for it is not as if the mother of the wet infant feels herself "wet to the ribs." Nevertheless, she "naturally" shares her infant's distress, and this means that she, too, is distressed, although obviously in a very different way via a much more sophisticated cognitive apparatus. Empathy is *not* needing to ask, "How would I feel in such a situation?"

In other words, we (and, no doubt, animals as well) have a far more primal capacity for empathy than our rather sophisticated ability to imagine ourselves in situations other than our own. It is too often assumed that empathy is just one emotional phenomenon rather than several (or many), but, depending on the amount of sophistication and imagination involved, it would seem to range from the mere fellow-feeling of "emotional contagion"—simply "picking up" another person's distress (or, perhaps, joy) by virtue of mere proximity—to the "higher cognitive" sharing of emotion through imaginatively and quite self-consciously "putting oneself in the other's place." (In chapter 12, Zak distinguishes three different "levels" of empathy, from the primitive mechanism of mirror neurons to what he calls "mentalizing" to affective representation (say, upon seeing an upsetting photograph or watching another person's facial expression), which he considers full-fledged empathy, "having an internal representation of others' emotions." I disagree with him about this, as I find the greatly overused notion of "internal representations" quite problematic; I argue, instead, that what Zak calls "mentalizing" or having a "theory of mind" divides up into several subsequent levels of sophistication. (Further, as I have also long argued, that the limbic regions of the brain are activated in emotional responses is an insufficient reason to say that such activities are *responsible* for the emotions in question.)

At the "low" end of the range, "emotional contagion" involves only minimal intentionality, some vague sense of shared discomfort or merriment, and it is by no means necessary, or even likely, that the subject of such an emotion could say clearly what it is "about," much less what it portends or what it means. (This is one place where I would argue that the language of "internal representation" leads us astray.) Even where such minimal awareness of causality is present, the question "Why are you feeling this?" would get only a causal answer, for instance, "because I'm with her, and she's upset," nothing like an account or an expression of the emotion felt. Considerably more sophisticated is Nell Noddings's "receptivity." Intentionality may still be minimal, but whereas "emotional contagion" happens to one by virtue of mere proximity, one must "open oneself up" to be receptive, one "lets it happen" (often hidden behind the observation that such emotional empathy is "spontaneous"). This may well involve a decision of sorts (although the crucial decision may have been made many years ago, for instance, to have a child or to foster this friendship). But the question "To what extent do we control (or "regulate") our emotions?" is an important "existential" question that is certainly not irrelevant to the fact that we expect people to be empathetic with others.

More cognitively complex still are those experiences of empathy that involve the imagination, one aspect of what Zak calls "mentalizing." (I think Noddings is right not to invoke the imagination in her account of receptivity. Also, it is evident at this stage that what goes on by way of empathy is something much less than having a "theory of [the other's] mind.") Imagination takes many forms, however. There is that basal form of imagination that Kant, for example, invokes as basic to all perception, which requires no effort or self-consciousness but is activated automatically as part of our awareness of the world. One might argue, too, that some basal form of imagination exists that is basic to our awareness of other people, which also requires no effort or self-consciousness. But whether or not this is so, the more obvious employment of the imagination is in the "comprehension" noted by Adam Smith, which does require some effort of the understanding. Again, the range of cognitive sophistication is considerable. I sit watching an early James Bond movie (*Goldfinger*) in a theater. The bad guy's bodyguard grasps Bond in a death grip, and Bond responds by kicking him in the groin. There was a collective "ungh!" from the men in the audience. (There were no such responses correlative to any of the other punching, gouging, slamming, etc.) I consider this is a straightforward if crude example of minimal imaginative empathy. To state the obvious, no one in the audience suffered at that moment from an actual pain in the groin, and yet the response (including the predictable self-protective gesture) was quite remarkable. There is nothing mysterious about the intentionality here, although it does make good sense, in this instance, to talk about an "internal representation" of another person's feelings (hardly an emotion, however).

Moving up in gentility, as well as cognitive sophistication, I remember sitting through any number of romantic films in which I "felt for" the hero or the heroine. (I have not discerned in myself any ultimate preference or discrimination, based on gender alone.) I find myself quite freely "identifying with" the character as I watch him or her on the screen.[14] It is worth noting the quite dramatic difference between empathizing with characters on the screen and doing so with characters in novels (and, of course, in nonfiction biographies and the like). It is often said that part of the effort of reading is "filling in" the details, using one's imagination to flesh out the skeleton provided by the verbal descriptions, adding colors (whether or not they are described in the book) and all sorts of details. Less often noted is the extent to which one can "identify" with the characters in a novel or biography by way of adding one's own details, including one's own emotional reactions (again, whether or not they are so described in the book). This is much more than mere imagination, although it is still much less than having anything like a "theory of mind," even in the case of reading a novel. A theory of mind requires a conception of *mind*, which is much more than even the best reader usually requires.

What might indeed require a "theory of mind," however, is at the top of the empathetic range, namely, the conscientious effort to "put oneself in the other's shoes," by consciously imagining the particular circumstances in which the other person finds him- or herself. This may require actual research, for instance, re-creating the historical circumstances in which a character or an individual lived. It may also require extravagant imaginative effort. I try to imagine, for instance, Joan d'Arc's feelings as she first approached Robert Baudricourt to offer her services to Charles VII. I am not a young woman—in fact, I am no longer even young; I am not religious, and, to my knowledge, neither God nor any other ethereal spirit has ever spoken to me; and I do not live in the Middle Ages (and know much less about that era than other periods of Western history). Yet, with considerable effort, I can empathize with young Joan's feelings by comparing them, with considerable trepidation, to my own experiences (of youthful enthusiasm, self-righteousness, intimidation by authorities, and aggressive ambition).

Compare that effort to something closer to home. I just finished writing a book about Nietzsche, a favorite philosopher of mine. One of the challenges of the book (and of understanding Nietzsche) was to "get inside his head," to understand why he said many of the rude and blasphemous things that he said. Part of the explanation, I concluded, was his personal resentment against his Christian upbringing, and against his lonely and sickly life. Another part was the mental condition that today we would call "bipolar." Well, I am also a philosopher (as opposed to an illiterate medieval girl saint). I know something of the nineteenth century, and I have visited the sites where Nietzsche often stayed. I have been lonely and occasionally sickly myself. I have suffered mild depressions and periods of great enthusiasm. I have found more than enough in my life to be resentful about (although, like Nietzsche, I despise my own resentment, giving rise to a tangle of conflicting emotions). And so, with minimal effort, I can "understand" Nietzsche, that is, I empathize with him. Nevertheless, it is an extravagant act of imagination, spanning some thirty years of reading, imagination, and research. It is empathy, to be sure, but it is anything but "spontaneous."

Is Sympathy a Basic Emotion?

Sympathy, in contrast to empathy, is defined by Eisenberg as "an affective response that consists of feeling sorrow or concern for the distressed or needy other (rather than the same emotion as the other person)." This is what Smith dismisses but what we have insisted upon as part of the meaning of "sympathy," "feeling sorry for." Philosophically, it is a much less interesting sentiment than empathy. Indeed, one might question whether empathy is even a sentiment or rather a capacity to have any number of emotions depending on the emotions of others. One would not ask that question regarding sympathy, however. Sympathy is a straightforward emotion in the sense that it is a distinctive emotion, dependent on the emotions of others, perhaps, but only as a response to others' emotions, not in any sense an imitation or reproduction of them. Indeed, one might sympathize with another person and get his or her emotional situation completely wrong. Empathy, by contrast, seems to require that one does indeed (in some complex sense) mirror the emotions of the other person.

Sympathy is a "basic" emotion, Peter Goldie argues, because it typically involves a distinctive emotional experience as well as characteristic emotional expression (though this can usually be read as such only in context), and it tends to motivate action (namely, helping or at least nurturing behavior).[15] I think that this position is quite right and a good defense of Smith (although Goldie is writing in defense of Hume and speaks of "compassion" rather than "empathy"). Goldie worries at length about sympathy's status as an "affect program," the current scientific conception of emotion. He argues, convincingly, that there is no good reason not to include sympathy [compassion] on the list of "basic emotions" along with such emotions as fear and anger. His worry about facial expression is particularly significant. Affect program theorists (Paul Ekman, Paul Griffiths) tend to eschew context. (Ekman's research consists largely of showing photographs of facial expressions, devoid of context or narrative.) Expressions of sympathy would be unintelligible with this restriction, indiscernible from mere sadness or distress. It is worth noting, however, that Goldie also distinguishes his view from Hume's view of sympathy, because he takes Hume to be emphasizing "imaginative identification" (as does Smith in the usual interpretation). I take it that Zak (see chapter 12) has a similar account of empathy in mind, though he improves on Goldie in acknowledging the various levels of empathy, some requiring imagination, others not. But on the question of whether empathy is a basic emotion, it would seem that Goldie and Zak agree, and, moreover, Zak supplies a fuller account of what the neurological process involves. In particular, it involves the neurochemical oxytocin, which, importantly, is also the chemical several other contributors to this volume cite in their accounts of trust. One must be careful not to be overly reductionist in this, but it seems clear that both trust and empathy are facilitated by oxytocin, which also raises all sorts of interesting questions about how the two emotional phenomena (neither one seems to be an emotion as such) are related.

I think we err when we turn either sympathy or empathy into something essentially projective and contemplative, a product of thought rather than shared feeling in a shared relationship. (Of course, the nature of both the sharing and the relationship may be uni-directional, as in a mother's empathy for her infant's distress.) I take this to be of enormous importance in understanding our emotional lives, this "natural" ability to tune into the emotions of others without the intermediary of a rich empathetic imagination. Our emotional lives are largely imitative (learned in what Ronnie De Sousa calls "paradigm scenarios"). We learn about the emotions of others long before we are capable of thinking ourselves into their shoes. We learn from them when an emotion is appropriate. We learn from them when one ought to (and ought not to) have an emotion. We learn from them what it is like to have an emotion. Children display signs of empathy from eighteen months to two years of age, no doubt by way of imitating their elders, although it is only with age and through learning that they actually come to understand something of the nature of other people's true feelings.

The "Golden Rule" is a powerful and arguably universal moral guideline, but following such a rule is by no means the basis of morality. That basis is empathy, and empathy is far more basic than the sophisticated thought process "how would I like it if . . . ?" or "what would that be like if . . . ?" To be sure, we do learn to think that way, and we can come to a very sophisticated understanding of another's feelings. But before we can understand "why?" we have to recognize that other people have feelings, and then we learn *what* emotion the other is having. First and foremost, perhaps, we have to learn that the other is "upset," and this, I suggest, is not something one can learn by reasoning. (Autistic people do this, perhaps, but that is all the more reason to see it as highly unusual.)

The distinctions between self and other, and between self-interest and the interest of others, are too often taken, by philosophers, as virtual absolutes. But Smith's insistence on the prevalence of empathy and his virtue ethics more generally suggests that this picture is wrong in its depiction of the complex emotional relationship between self and other. We do not just have our own interests; we *share* interests. And we do not simply have our own motives, desires, emotions, and moods; we *share* all these. This does not only consist, moreover, of having the same interest, emotion, and so on, at the same time. It means, quite literally, having the same interest, emotion, and so on (without violating the Cartesian rule that each person has his or her own emotion). Empathy is neither altruistic nor self-interested. Rather, it demonstrates the inadequacy of over-individualizing human nature. We are both social and emotional creatures, for whom mutual understanding—that is, the mutual understanding of one another's emotions—is essential. Without that understanding, it is not just as if each of us would be a closed emotional system, rich within ourselves but oblivious to the needs and interests of others. We would be thoroughly autistic, empty, devoid of even the most basic emotions, subject to fits and frustration but with none of the drama and complexity that comes with living in a shared emotional world.

This, I think, is the lasting importance of Smith's analysis. (It is less clear in Hume.) Against some of the monstrous theories of human nature he confronted—Hobbes, Mandeville, and so on—he argued a much more humane and admirable picture of human nature, one that greatly qualified and restricted the selfish aspects of humanity (for which he is mainly known, given a few famous comments in *Wealth of Nations*) and rendered human nature neither a "war of all against all" nor a buzzing beehive (Mandeville's biologically indefensible image) but a sympathetic community defined by fellow-feeling and an abhorrence of seeing one's fellow citizens in pain or suffering. This, I think, is also the basis of the free market, not "the profit motive" and other made-up euphemisms for self-interested behavior but the sensitivity to be concerned about other people, what they need and what they want. Other-interest, not self-interest, is the engine of a healthy free market system.

Forward to Aristotle

Over the years, I defended a theoretical framework which I call "an Aristotelian approach to business." It begins, as the above treatment of Adam

Smith would suggest, with a rejection of the idea that people are essentially selfish or self-interested. But whereas Smith posits a countervailing source of unselfish motivation, sympathy, Aristotle develops a view of human nature in which the very idea of selfish motivation becomes something exceptional, not the norm. It is not that Aristotle lacked either the concept or the experience of selfish behavior, particularly in business (which he viewed as a parasitic and "unnatural" activity). Such behavior, however, had little to do with his conception of human nature, rationality, or the Good and necessarily involved deviant motives and emotions.

Aristotle's central ethical concept, and what defined our natural tendencies, was a unified, all-embracing notion of "happiness" (or, more accurately, *eudaimonia*, perhaps better translated as "flourishing" or "doing well"). He spent a good portion of his *Nicomachean Ethics* analyzing what this means and what its ingredients are. To what extent are success and pleasure, for instance, necessary for happiness? But the truly essential ingredient in a happy life is *virtue*, which Aristotle defines as "the mean between the extremes" but describes in considerable detail (by way of descriptions of the particular virtues) in much less abstract terms. The central feature of all the virtues, however, is that they aim at solidifying our social relationships and our aspirations to *excellence*, whether this is in a profession, in one's social behavior, or in one's personal life. (The Greek word *aretē* translates both as "virtue" and as "excellence.") Decidedly, however, excellence does not mean "to one's own benefit or advantage," although certainly being excellent at whatever endeavor will, in fact, probably work to one's benefit and advantage. But this is not what motivates one's behavior. One is not courageous in order to win the battle or be thought well of, but both those results are likely to accrue from an act of courage. One is courageous because one wants to behave bravely and be a proper member of a warrior community (or a full member of a frank political community, although we have to go back in history to remember what that would be). So, too, one is generous not to win acclaim but to help another person. The motivation to be virtuous, in other words, is the felt need to be that kind of person, a virtuous person. Such motivation is neither selfish nor unselfish. It only requires the desire to be a good person.

There is no room in this picture for the false antagonism between "selfishness," on the one hand, and what is called "altruism" or "selflessness," on the other. For the properly constituted social self, the distinction between self-interest and social-mindedness is all but unintelligible, and what we call selfishness is guaranteed to be self-destructive as well. "Altruism" is too easily turned into self-sacrifice, for instance, by that self-appointed champion of selfishness Ayn Rand. But altruism is not self-sacrifice; it is just a more reasonable conception of self, as tied up intimately with community, with friends and family, who may indeed count (even to us) more than we do. What my Aristotelian approach to business ethics demands is not self-sacrifice or submerging oneself to the interests of the corporation, much less voluntary unhappiness. What it does say is that the distinctions and oppositions between self and society that give rise to these wrong-headed conceptions are themselves the problem, and the cause of so much unhappiness and dissatisfaction.

Similarly, the most serious single problem in business ethics is the false antagonism between profits and social responsibility. There is an academic prejudice, for example, against clever profit-making solutions—the obviously optimal solution to "social responsibility"-type problems. It is as if moralists have a vested interest in the nobility of self-sacrifice (that is, self-sacrifice by others). This is the same problem raised by Thomas Hobbes and his fellow egoists. According to their views, an action is either selfish or selfless, and the Aristotelian synthesis of self-interested, noble behavior is eliminated from view. Once one introduces such artificial oppositions between self-interest and shared interest, between profits and social responsibility, the debate becomes a "lose-lose" proposition, either wealth and irresponsibility or integrity and failure. Yet, I do not want to say that the Aristotelian approach offers us a "win-win" proposition, since that popular formulation already assumes a self-interested (albeit mutual self-interest) game theoretical situation. The truth is closer to this: by working together, *we* are better off (and woe to the corporation or society that keeps all the rewards at the top of the pyramid for too long, allowing the benefits only to "trickle down" to most of those in the category of "we"). In doing our jobs and doing them well, we are acting neither selfishly nor unselfishly but trusting in a communal system in which virtuous behavior is, by the way, rewarded.

The point is to view one's life as a whole and not separate the personal and the public or professional, or duty and pleasure. The point is also that doing what one ought to do, doing one's duty, fulfilling one's responsibilities and obligations is not counter but conducive to the good life, to becoming the sort of person one wants to become. Conversely, becoming the sort of person one wants to become—which presumably includes to a very large extent what one does "for a living"—is what happiness is all about. Happiness is "flourishing," and this means fitting into a world of other people and sharing the good life, including "a good job," with them. A good job, accordingly, is not just one that pays well or is relatively easy but one that means something, one that has (more or less) tangible and clearly beneficial results, one that (despite inevitable periods of frustration) one enjoys doing. Happiness (for us, as well as for Aristotle) is an all-inclusive, holistic concept. Ultimately one's character, one's integrity, determines happiness, not the bottom line. This is just as true, I insist, of giant corporations as it is of the individuals who work for them.

A word of caution, however. Aristotle is famous as an enemy of business. As I mentioned earlier, he accused the businessmen of his day of "parasitic" and "unnatural" behavior. But Aristotle was also the first economist, in many ways Adam Smith's most illustrious predecessor. He had much to say about the ethics of exchange and so might well be called the first (known) business ethicist as well. Aristotle distinguished two different senses of what we call economics, one of them "*oecinomicus*," or household trading, which he approved of and thought essential to the working of any even modestly complex society, and "*chrematisike*," which is trade for profit. The latter activity is the one Aristotle declared to be wholly devoid of virtue. Aristotle despised the financial community and, more generally, all of what we would call profit-seeking. He argued that goods should be exchanged for their "real value," their costs, including a "fair wage" for those who produced them, but he then concluded, mistakenly, that any profit (that is, over and above costs) required some sort of theft (for where else would that "surplus value" come from.) Consequently, he had special disdain for moneylenders and the illicit, unproductive practice of usury, which until only a few centuries ago was still a crime. ("Usury" did not originally mean excessive interest; it referred, instead, to any charge over and above cost.) Only outsiders at the fringe of society, not respectable citizens, engaged in such practices. (Shakespeare's Shylock, in *The Merchant of Venice*, was such an outsider and a usurer, though his idea of a forfeit was a bit unusual.)

All trade, Aristotle believed, was a kind of exploitation. Such was his view of what we call "business." Aristotle's greatest medieval disciple, St. Thomas Aquinas, shared "the Philosopher's" disdain for commerce, even while he struggled to permit limited usury (never by that term, of course) among his business patrons. (A charge for "lost use" of loaned funds was not the same as charging interest, he argued.) Even Martin Luther, at the door to modern times, insisted that usury was a sin and that a profitable business was (at best) suspicious. Aristotle's influence on business, one could argue, has been long-lasting—and nothing less than disastrous. Looking back, one might contend that Aristotle had too little sense of the importance of production and based his views wholly on the aristocratically proper urge for acquisition, thus introducing an unwarranted zero-sum thinking into his economics. Of course, it can be charged that Aristotle, like his teacher Plato, was a spokesman for the aristocratic class and quite unsympathetic to the commerce and livelihoods of foreigners and commoners. It is Aristotle who initiates so much of the history of business ethics as the wholesale attack on business and its practices. Aristotelian prejudices still underlie much of business criticism and the contempt for finance that preoccupies so much of Christian ethics even to this day, avaricious evangelicals notwithstanding.

The Aristotelian approach to business ethics, instead, begins with the two-pronged idea that individual virtue and integrity count but that good corporate and social policy encourage and nourish individual virtue and integrity. It is this picture of community, with reference to business and the corporation, that I want to explore. Here, one might speak of "communitarianism," but it is not at all evident that one must at the same time give up a robust sense of individuality (which is not the same as self-interested individualism).

Community and virtue form the core of the thesis I defend here. To call the approach "Aristotelian" is to emphasize the importance of community, the business community as such (I consider corporations as, first of all, communities) but also the larger community, even all of humanity and, perhaps, much of nature, too. This emphasis on community, however, should not be taken to eclipse the importance of the individual and individual responsibility. In fact, the contrary is true; only within the context of community is individuality developed and defined, and our all-important sense of individual integrity is dependent upon and not opposed to the community in which integrity derives both its meaning and its chance to prove itself. Thus, only within the context of a civilized community can a free market emerge, which gives an ethical turn to Kimbrough, Smith, and Wilson's account, in chapter 13, of how the "impersonal" market nevertheless develops from our more personal interactions with other people. What they term "mind reading" is no more than a fanciful rendition of what Smith called "sympathy" and what we call "empathy." In any case, they are right to place it in a position of great importance in our understanding of markets and how they work.

Aristotelian ethics also presupposes an ideal, an ultimate purpose, and, accordingly, had Aristotle been more suitably sympathetic, so does the ethics of the market. But the ideal or *telos* of business in general is not, as my undergraduates so smartly insist, "to make money." It is to serve society's demands and the public good, and to be rewarded for doing so. This in turn defines the mission of the corporation, and provides the criteria according to which corporations and everyone in business can be praised or criticized. "Better living through chemistry," "Quality at a good price," "Productivity through people," "Progress is our most important product"—these are not mere advertising slogans but reasons for working and for living. Without a mission, a company is just a bunch of people organized to make money while making up something to do (e.g., "beating the competition"). Such activities may or may not be innocent and legal, and they may, unintentionally, contribute to the public good. But Adam Smith's "invisible hand" metaphor never was a very reliable social strategy. (It is an image Smith used only once in the whole of *Wealth of Nations*, something you would never know if you just followed the legend that has grown up around that phrase.) The difference between intending to do good and doing good unintentionally is not just the rewarding sense of satisfaction that comes from the former. Contrary to the utterly irresponsible and obviously implausible argument that those do-gooders who try to do good in fact do more harm than good, the simple, self-evident truth is that most of the good in this world comes about through good and not selfish intentions, by way of the moral sentiments and not in the unconstrained exercise of the so-called profit motive. As Aristotle taught us twenty-five hundred years ago, meaningful human activity is that which intends the good rather than stumbling over it on the way to merely competitive or selfish goals, and the predictable outcome of such behavior is not the mysterious result of an invisible hand but of our own good intentions, amply rewarded.

Notes

1. Kant. *The Groundwork of the Metaphysics of Morals*, trans. H. J. Paton (New York: Harper and Row, 1964), p. 67 (p. 13 of the standard German edition), and W. Hamilton (London: Penguin, 1951), p. 43. Kant's phrase, "*schmelzender Theilnehmung*" (*Grundlegung, Werke*, Band IV p. 399), in *Grounding of the Metaphysics of Morals*, [trans.] James W. Ellington (Indianapolis: Hackett, 1981). The phrase is translated as "melting compassion" by H. J. Paton (New York: Harper and Row, 1964), p. 67; and as "tender sympathy" by Lewis White Beck (Indianapolis: Bobbs-Merrill, 1959), and Ellington, *Grounding of the Metaphysics of Morals*, p. 12, from whom the rest of the quotation is borrowed. Neither translation adequately captures what I take to be Kant's demeaning irony. "Melting" is much better than "tender" for "*schmelzender*," but neither "compassion" nor "sympathy" will do for "*Theilnehmung*," which is more like "participation" (and less like "*Mitleid*," usually translated as "compassion" or "pity").

2. See, for instance, Stephen Engstrom and Jennifer Whiting, eds., *Aristotle, Kant, and the Stoics: Rethinking Happiness and Duty* (Cambridge: Cambridge University Press, 1998); and Immanuel Kant, *Groundwork of the Metaphysics of Morals* (1785), ed. Mary Gregor, introduction by Christine M. Korsgaard (Cambridge: Cambridge University Press, 1998).

3. Some sentiments and emotions one might come to have via empathy might be motivating, of course. Empathizing with someone else's anger motivates one to punish the perpetrator and avenge the victim. But empathy as such has no motivating power.

4. Patricia Werhane, *Adam Smith and His Legacy for Modern Capitalism* (New York: Oxford University Press, 1991).

5. Joseph Cropsey, *Polity and Economy* (Westport, CT: Greenwood, 1957), p. 12.

6. Feeling sorry for someone need not motivate action, of course, but neither is it *merely* a spectator emotion. Sympathy sometimes motivates action; at other times it is unable to do so (because of distance, incapacity, the victim's refusal to be helped, and so on). Still, I can at least sympathize with the argument that if someone can readily help but does not, then what he feels cannot truly be sympathy.

7. Alasdair C. MacIntyre, *After Virtue: A Study in Moral Theory* (South Bend, Ind.: University of Notre Dame Press, 1981), pp. 214–215. Milton Friedman (characteristically): "Smith regarded sympathy as a human characteristic, but one that was itself rare and required to be economised" ("Adam Smith's Relevance for 1976," in *Selected Papers of the University of Chicago Graduate School of Business*, no. 50 (Chicago: University of Chicago Graduate School of Business, 1977), p. 16.

8. Peter Goldie refers to the "borrowed intentionality" of feelings in this regard. It is a nice phrase that underscores both the idea that sensations as such do not have intentionality but, as part of the "package" of an emotion, do, by association, take on the attributes of "aboutness."

9. Nancy Eisenberg, "Empathy and Sympathy," in *Handbook of Emotions*, ed. Michael Lewis and Jeannette M. Haviland-Jones, 2nd ed. (New York: Guilford, 2000), pp. 677–691.

10. Ibid., *p.* 677.

11. I refer the reader to either of my two books, *The Passions: Emotions and the Meaning of Life* (Indianapolis, Ind.: Hackett, 1993); and *Not Passion's Slave: Emotions and Choice (The Passionate Life)* (New York: Oxford University Press, 2007).

12. Martha Nussbaum offers an extensive defense of compassion as a purely cognitive emotion (an evaluative judgment) in part 2 of her *Upheavals of Thought: The Intelligence of Emotions* (Cambridge: Cambridge University Press, 2001). She claims, which I would not (or, at least, no longer), that compassion (and other emotions) need not involve any distinct feelings or "affect."

13. Nell Noddings, *Caring: A Feminine Approach to Ethics and Moral Education* (Berkeley: University of California Press, 1984).

14. At the same time, I should protest that this idea of "identification" is probably the most over-used and under-explained concept in film criticism. I will not try to pursue this here.

15. Peter Goldie, *Emotions* (Oxford: Oxford University Press, 1999), and idem, "Compassion: A Natural, Moral Emotion," in *Die Moralitaet der Gefuehle*, Special Issue of *Deutsche Zeitschrift fur Philosophie* 4, ed. S. A. Doering and V. Mayer, pp. 199–211 (Berlin: Akademie, 2002).

Bibliography and Further Readings

Aristotle. *Nichomachean Ethics*. In *The Works of Aristotle*. Translated by T. Irwin. Indianapolis: Hackett, 1985.

Aristotle. *Politics*. Translated by B. Jowett. New York: Modern Library, 1943.

Arrow, Kenneth, *Social Choice and Individual Values*. New Haven: Yale University Press>1963.

Axelrod, Robert. *The Evolution of Cooperation*. New York: Basic Books, 1984.

Bowie, Norman. *Business Ethics*. Engelwood Cliffs, NJ: Prentice-Hall, 1982.

———. "The Profit Seeking Paradox." In N. Dale Wright, ed., *Ethics of Administration*. Provo: Brigham Young University Press, 1988.

Brandeis, Louis. "Competition." *American Legal News* 44 (January 1913).

Carr, Alfred, "Is Business Bluffing Ethical?" *Harvard Business Review*, January–February 1968.

Ciulla, Joanne. *This Working Life*. New York: Random House, 1999.

Ciulla, Joanne, with Clancy Martin and Robert C. Solomon. *Honest Work*. New York: Oxford University Press, 2006.

Coleman, Jules L. *Markets, Morals and the Law*. Cambridge: Cambridge University Press, 1988.

Collard, David. *Altruism and Economics*. Oxford: Oxford University Press, 1978.

Cropsey, Joseph. *Polity and Economy*. Westport CT: Greenwood, 1957.

Davidson, Greg, and Paul Davidson. *Economics for a Civilized Society*. New York: Norton, 1989.

de George, Richard. *Ethics, Free Enterprise and Public Policy*. New York: Oxford University Press, 1978.

Donaldson, Thomas. *International Business Theory*. New York: Oxford University Press, 1990.

Etzioni, Amitai. *The Moral Dimension: Toward a New Economics*. New York: Free Press, 1989.

French, Peter A. "The Corporation as a Moral Person." *American Philosophical Quarterly* 16:3 (1979).

Friedman, Milton. "Adam Smith's Relevance for 1976. In *Selected Papers of the University of Chicago Graduate School of Business*, no. 50. Chicago: University of Chicago Graduate School of Business, 1977.

————. "The Social Responsibility of Business Is to Increase Its Profits." *New York Times*, Sunday Magazine, September 13, 1970, p. 32.

McGregor, Douglas A. *The Human Side of Enterprise*. New York: McGraw-Hill, 1960.

Sen, Amartya. *On Ethics and Economics*. Oxford: Blackwell, 1989.

Smith, Adam. *An Inquiry into the Nature and Causes of the Wealth of Nations*. New York: Hafner, 1948.

Smith, Adam, *The Theory of Moral Sentiments* (London: George Bell, 1880).

Solomon, Robert C. *A Better Way to Think about Business*. Oxford: Oxford University Press, 1999.

————. *Ethics and Excellence: Cooperation and Integrity in Business*. Oxford: Oxford University Press, 1992.

Werhane, Patricia. *Ethics and Economics: The Legacy of Adam Smith for Modern Capitalism*. Oxford: Oxford University Press, 1991.

Three

The Status of Moral Emotions in Consequentialist Moral Reasoning

Robert H. Frank

The philosopher Bernard Williams describes a situation in which a botanist wanders into a village in the jungle where ten innocent people are about to be shot. He is told that nine of them will be spared if he himself will shoot the tenth. What should the botanist do? Although most people would prefer to see only one innocent person die rather than ten, Williams argues that it would be wrong as a matter of principle for the botanist to shoot the innocent villager.[1] Most people seem to agree.

The force of the example is its appeal to a widely shared moral intuition. Yet some philosophers counter that such examples call into question the presumed validity of moral intuitions themselves (Singer 2002). These *consequentialists* insist that whether an action is morally right depends only on its consequences. The right choice, they argue, is always the one that leads to the best consequences overall.

I argue that consequentialists make a persuasive case that it is best to ignore moral intuitions in at least some specific cases. Many consequentialists, however, appear to take the stronger position that moral intuitions should play no role in moral choice. I argue against that position on grounds that should appeal to their way of thinking. As this chapter attempts to explain, ignoring moral intuitions would lead to undesirable consequences. My broader aim is to expand the consequentialist framework to take explicit account of moral sentiments.

Consequentialist versus Deontological Moral Theories

Consequentialism differs from traditional, or *deontological*, moral theories, which hold that the right choice must follow from underlying moral principles. These principles may spring from religious tradition (for example, the Ten Commandments) but need not (for example, Kant's categorical imperative). Whatever their source, the moral force of the principles invoked by deontologists increases with the extent to which these principles accord with strongly held moral intuitions.

For many cases, perhaps even the overwhelming majority, consequential-ist and deontological moral theories yield the same prescriptions. Both camps, for example, hold that it was wrong for Enron executives to lie about their company's earnings, and wrong for David Berkowitz to murder six innocent people. Even in cases in which there might appear to be ample room for dis-agreement about what constitutes moral behavior, most practitioners from both camps often take the same side.

Consider, for example, a variant of the familiar trolley-car problem dis-cussed by philosophers. You are standing by a railroad track when you see an out-of-control trolley car about to strike a group of five people standing on the tracks ahead. You can throw a nearby switch, diverting the trolley onto a side track, which would result in the death of one person standing there. Failure to throw the switch will result in the death of all five persons on the main track.

Consequentialists are virtually unanimous in concluding that the morally correct choice is for you to throw the switch. Some deontologists equivocate, arguing that the active step of throwing the switch would make you guilty of killing the person on the side track, whereas you would not be guilty of killing the five on the main track if you failed to intervene. Yet even most deontolo-gists conclude that the distinction between act and omission is not morally relevant in this example, and that your best available choice is to throw the switch.

But even though the two moral frameworks exhibit broad agreement with respect to the ethical choices we confront in practice, many deontologists re-main deeply hostile to the consequentialist framework.

The Status of Moral Intuitions

Critics often attack consequentialist moral theories by constructing examples in which the choice that consequentialism seems to prescribe violates strongly held moral intuitions. In another version of the trolley-car problem, for exam-ple, the trolley is again about to kill five people, but this time you are not standing near the tracks but on a footbridge above them. There is no switch to throw to divert the train. But there is a large stranger standing next to you, and if you push him off the bridge onto the tracks below, his body will derail the trolley, killing him in the process but sparing the lives of the five strangers. (Jumping down onto the tracks yourself will not work, because you are too small to derail the trolley.)

Consequentialism seems to prescribe pushing the large stranger from the bridge, since this would result in a net savings of four lives. Yet, most people, when asked what one should do in this situation, feel strongly that it would be wrong to push the stranger to his death. Those who share this intuition are naturally sympathetic to the deontologists' claim that the example somehow demonstrates a fundamental flaw in the consequentialist position. This version of the trolley problem thus elicits essentially the same moral judgment as Bernard Williams's example involving the botanist.

Many consequentialists, among them the Princeton philosopher Peter Singer, question the validity of the moral intuitions that such examples evoke (Singer 2002). To illustrate, Singer asks us to imagine another variant of the trolley problem, one that is identical to the first except for one detail. You can throw a switch that will divert the train not onto a side track but onto a loop that circles back onto the main track. Standing on the loop is a large stranger whose body would bring the trolley to a halt were it diverted onto the loop. Singer notes that this time most people say that diverting the trolley is the right choice, just as in the original example in which the switch diverted the trolley onto a side track rather than a loop. In both cases, throwing the switch caused the death of one stranger, but, in the process, spared the lives of five others on the main track.[2]

Singer's Princeton colleague Joshua Greene, a cognitive neuroscientist, has suggested that people's intuitions differ in these two examples not because the morally correct action differs but rather because the action that results in the large stranger's death is so much more vivid and personal in the footbridge case than in the looped-track case:

> Because people have a robust, negative emotional response to the personal viola-
> tion proposed in the footbridge case they immediately say that it's wrong . . . At the
> same time, people fail to have a strong negative emotional response to the rela-
> tively impersonal violation proposed in the original trolley case, and therefore re-
> vert to the most obvious moral principle, "minimize harm," which in turn leads
> them to say that the action in the original case is permissible. (Greene 2002, 178)

To test this explanation, Green used functional magnetic resonance imaging to examine activity patterns in the brains of subjects confronted with the two decisions. His prediction was that activity levels in brain regions associated with emotion would be higher when subjects considered pushing the stranger from a footbridge than when they considered diverting the trolley onto the looped track. He also reasoned that the minority of subjects who felt the right action was to push the stranger from the footbridge would reach that judgment only after overcoming their initial emotional reactions to the contrary. Thus, he also predicted that it would take longer for these subjects to make their decisions than for the majority who decided that it was wrong to push the stranger to his death, and also longer than it took for them to decide what to do in the looped-track example. Each of these predictions was confirmed.

Is it morally relevant that thinking about causing someone's death by pushing him from a footbridge elicits stronger emotions than thinking about causing his death by throwing a switch? Peter Singer argues that it is not—that the difference is a simple, non-normative consequence of our evolutionary past. Under the primitive, small-group conditions under which humans evolved, he argues, the act of harming others always entailed vivid personal contact at close quarters. One could not cause another's death by simply throwing a switch. So if it was adaptive to be emotionally reluctant to inflict harm on others—surely a plausible presumption—the relevant emotions ought to be much more likely to be triggered by vivid personal assaults than by abstract actions like throwing a switch.

A historical case in point helps highlight the distinction. Shortly after British intelligence officers had broken Nazi encryption schemes in World War II, Winston Churchill had an opportunity to spare the lives of British residents of Coventry by warning them of a pending bombing attack. To do so, however, would have revealed to the Nazis that their codes had been broken. In the belief that preserving the secret would ultimately save considerably more British lives, Churchill gave Coventry no warning, resulting in large numbers of preventable deaths.

It is difficult to imagine a more wrenching decision, and we celebrate Churchill's moral courage in making it. It is also easy to imagine, however, that Churchill would have chosen differently if he personally had to kill the Coventry residents at close quarters, rather than allowing their deaths merely by failing to warn them. The consequentialist claim is that, although this difference is a predictable consequence of the way that natural selection forged our emotions, it has no moral significance.

In sum, that consequentialist moral theories sometimes prescribe actions that conflict with moral intuitions cannot, by itself, be taken as evidence against these theories. Moral intuitions are contingent reactions shaped by the details of our evolutionary history. Often they will not be relevant for the moral choices we confront today.

Moral Sentiments as Commitment Devices

That sometimes it might be best to ignore moral emotions does not imply that ignoring them always, or even usually, is best. If we are to think clearly about the role of moral emotions in moral choice, we must consider the problems that natural selection intended these emotions to solve.

Most interesting moral questions concern actions the individual would prefer to take except for the possibility of causing undue harm to others. Unbridled pursuit of self-interest often results in worse outcomes for everyone. In such situations, an effective moral system curbs self-interest for the common good.

Moral systems, however, must not only identify which action is right, but they must also provide motives for taking that action (see chapters 2, 9, and 15). The difficulty of serving both goals at once is immediately apparent. Humans evolved in harsh environments in which the consequences of failure to pursue self-interest were often severe. Famines and other threats to survival were common. Polygyny was also the norm in early human societies, which, for men, meant that failure to achieve high rank ensured failure to marry. Under the circumstances, motivating individuals to forgo self-interest for the common good was obviously a challenge.

Yet, instances in which people forgo self-interest are actually quite common, even though the human nervous system is much the same today as it was tens of thousands of years ago. For example, although self-interest dictates leaving no tip in restaurants that you do not expect to visit again, most people tip at about the same rate at those restaurants that they do at local restaurants they expect to revisit.[3] Similarly, when sociologists performed the experiment of dropping wallets containing small amounts of cash on sidewalks in New York, about half the wallets were returned by mail with the cash intact.[4]

The Falklands War is another good example. The British could have bought the Falklanders out—giving each family, say, a castle in Scotland and a generous pension for life—for far less than the cost of sending their forces to confront the Argentines. Instead, they incurred considerable cost in treasure and lives. Yet, few in the United Kingdom opposed the decision to fight for the desolate South Atlantic islands. One could say that Margaret Thatcher gained politically by responding as she did, but this begs the question of why voters preferred retaliation to inaction. When pressed, most people speak of the nation's pride having been at stake.

People rescue others in distress even at great peril to themselves; they donate bone marrow to strangers. Such actions are in tension with the self-interest model that economists favor. These people seem to be motivated by moral sentiments, but where do these sentiments come from?

According to Adam Smith, we were endowed with moral sentiments by the creator for the good of society. Although it is true that society works better if moral sentiments motivate people to exercise restraint, selection pressures, as Darwin emphasized, are generally far weaker at the social level than at the level of the individual organism. Moral sentiments often motivate people to incur costs that they could avoid. On what basis might natural selection have favored these sentiments?

In *Passions Within Reason* (1988), I proposed a mechanism based on Tom Schelling's work on the difficulties people face when confronted with *commitment problems*.[5] Schelling illustrates the basic idea with an example of a kidnapper who seizes a victim and then gets cold feet. The kidnapper wants to set the victim free but knows that the victim, once freed, will reveal the kidnapper's identity to the police. Thus, the kidnapper reluctantly decides that he must kill the victim. In desperation, the victim promises not to go to the police. Here is the problem: both know that once the victim is out the door, his motive for keeping his promise will vanish.

Schelling suggests the following ingenious solution: if there is some evidence that the victim has committed a crime, he can share that evidence with the kidnapper, which then creates a bond to ensure his silence. The evidence of the victim's crime is a commitment device that makes an otherwise empty promise credible.

Schelling's basic insight can be extended to show why a trustworthy person might be able to prosper even in highly competitive market settings. Suppose you have a business that is doing so well that you know it would also thrive in a similar town three hundred miles away. Because you cannot monitor the manager who would run this business, however, he would be free to cheat you. Suppose a managerial candidate promises to manage honestly. You must then decide whether to open the branch outlet. If you do and your employee manages honestly, you each will profit very well, say, $1,000 each better than the status quo. If the manager cheats, however, you will lose $500 on the deal, and he will gain $1,500 relative to the status quo (see figure 3.1, for the relevant payoffs for the various options; note that these payoffs define a trust game like those discussed in chapters 7 and 10).

If you open the outlet, the manager finds himself on the top branch of the decision tree, where he faces a choice between cheating or not. If he cheats, his payoff is $1,500; if not, his payoff is only $1,000. Standard rational choice models assume that managers in these situations will be self-interested in the narrow sense. If that is your belief, you predict that the manager will cheat, which means your payoff will be -$500. Because that is worse than the

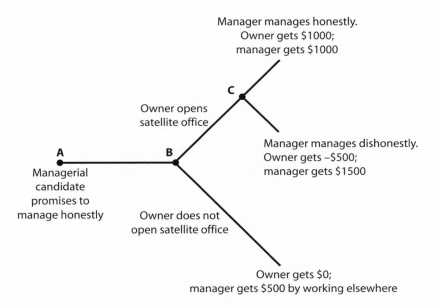

Figure 3.1. The Branch-Outlet Problem

payoff of zero you would get if you did not open the branch outlet, your best bet is not to open it. The pity is that this means a loss to both you and the manager relative to what you both could have achieved had you opened the branch outlet and the manager run it honestly.

Now suppose you can identify a managerial candidate who would be willing to pay $10,000 to avoid the guilt he would feel if he cheated you. Of course, using a financial penalty as a proxy for guilt feelings would be inappropriate in normative discourse. We would not say, for example, that it is all right to cheat as long as you gain enough to compensate for the resulting feelings of guilt. Nonetheless, the formulation does capture an important element of behavior. Incentives matter, and people are less likely to cheat when the penalties are higher.

It is clear, in any event, that this simple change transforms the outcome of the game. If the manager cheats, his payoff is not $1,500 but –$8,500 (after deducting the $10,000 psychological burden of cheating). Thus, if you open the branch outlet, the manager will choose to manage honestly, and both he and you come out ahead. If you could identify a trustworthy manager in this situation, he or she would not be at a disadvantage. On the contrary, both you and that manager would clearly profit (figure 3.2).

Note, however, that the managerial candidate won't be hired unless one can observe his taste for honesty. Thus, an honest candidate who is believed

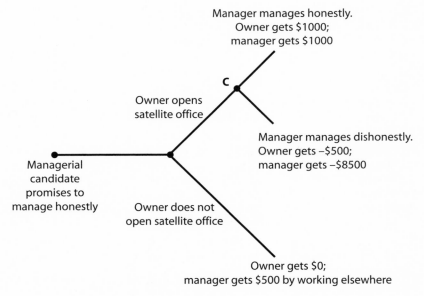

Figure 3.2. The Branch-Outlet Problem with an Honest Manager

to be dishonest fares worse than a dishonest candidate who is believed to be honest. The first doesn't even get hired. The second not only gets the job but also the fruits of cheating the owner.

Imagine a mutation that caused trustworthy people to be born with an identifying mark, such as a "C" on their foreheads (for "Cooperator"). The problem would then be solved. In such a world, the trustworthy types would drive the untrustworthy types to extinction. If the two types could be distinguished at no cost, the only equilibrium would be one with pure trustworthy types in the population.[6]

Generally, however, figuring out who is trustworthy is costly, and so people would not be vigilant in their choice of trading partners in an environment in which everyone was trustworthy. Being vigilant would not pay, just as buying an expensive security system for your apartment would not pay if you lived in a neighborhood where there had never been a burglary. Thus, a population consisting exclusively of trustworthy types could not be an evolutionarily stable equilibrium. Given reduced levels of vigilance, untrustworthy types could easily invade such a population. If character traits are costly to observe, therefore, the only sustainable equilibrium is one in which there is a mixed population consisting of both honest and dishonest individuals.

How might a signal of trustworthiness have emerged in the first place? Even if the first trustworthy person bore some observable marker, no one else would have had any idea what it meant. Nico Tinbergen argued that a signal of any trait must originate completely by happenstance.[7] In other words, if a trait is accompanied by an observable marker, then the link between the trait and the marker had to have originated by chance. For example, the dung beetle escapes predators by resembling the dung on which it feeds. How did it get to look like this? Unless it just happened to look enough like a fragment of dung to have fooled the most near-sighted predator, the first step toward a more dunglike appearance could not have been favored by natural selection. As Stephen Jay Gould asked, "Can there be any advantage in looking 5 percent like a turd?"[8] The problem is that no predator would be fooled. So how is the threshold level of resemblance reached? It must begin with a purely accidental link between appearance and surroundings. Once such a link exists, however, selection can begin to shape appearance systematically.

Similarly, we may ask, "How could a moral sentiment have emerged if no one initially knew the significance of its accompanying marker?" One hypothesis is suggested by the logic of the iterated prisoner's dilemma. There is no difficulty explaining why a self-interested person would cooperate in an iterated prisoner's dilemma.[9] If you are a tit-for-tat player, for example, and happen to pair with another such player on the first round, you and that other player will enjoy an endless string of mutual cooperation.[10] For this reason, even Attila the Hun, lacking any moral sentiments, would want to cooperate on the first move of an iterated prisoner's dilemma. The problem is that if you cooperate on the first move, you forgo some gain in the present moment, since defection on any iteration always yields a higher payoff than cooperation. It is well known that both humans and other animals tend to favor small immediate rewards over even much larger long-term rewards.[11] So, even though you expect to more than recoup the immediate sacrifice associated with cooperation, you may discount those future gains excessively. Successful cooperation, in short, requires self-control.

If you were endowed with a moral sentiment that made you feel bad when you cheated your partner, even if no one could see that you had that sentiment, this would make you better able to resist the temptation to cheat in the first round. That, in turn, would enable you to generate a reputation for being a cooperative person, which would clearly be to your advantage.

Moral emotions may thus have originated as impulse-control devices. This interpretation accords with observations made elsewhere in this volume by Bergstrom, Kerr, and Lachman (chapter 7), who argue, in effect, that a person's willingness to "waste" time in social relationships may serve as a commitment device. In their account, willingness to waste time would be a relatively costly step for defectors, who would be forced to seek other relationships anew if discovered cheating. My argument suggests a complementary interpretation: an inclination to spend seemingly unproductive time in social relationships may also be productive because it signals capacities to experience sympathy or empathy, which also make cheating more costly.

In any event, the activation of these emotions, like other forms of brain activation, may be accompanied by involuntary external symptoms that are observable. If so, the observable symptoms over time could have become associated in others' minds with the presence of these moral sentiments. Once that association was recognized, the moral emotions would be able to play a second role, namely, that of helping people solve one-shot prisoner's dilemmas and other commitment problems. The symptoms themselves can then be further refined by natural selection because of their capacity to help identify reliable partners in one-shot dilemmas.

How do you communicate something to another individual who has reason to be skeptical of what you are saying? Suppose, for example, that a toad meets a rival and both want the same mate. Among animals generally, the smaller of two rivals defers to the larger, thereby avoiding a costly fight that he would likely have lost anyway. Rival toads, however, often encounter each other at night, making visual assessment difficult. So they croak at each other, and the toad with the higher-pitched croak defers. Apparently, on average, the lower your croak, the bigger you are. Thus, it is prudent to defer to the lower croaker. This example illustrates the costly-to-fake principle: "I'll believe you not because you *say* you are a big toad, but rather because you are using a signal that is difficult to present unless you really *are* a big toad."[12]

Figure 3.3. Observable Correlates of Emotional States

It is the same when dogs face off: they seem to follow an algorithm of deferring to the larger dog. Consider the drawings in figure 3.3, taken from Charles Darwin's 1872 book, *The Expression of Emotion in Man and Animals*. The left panel portrays a dog confronting a rival. Darwin argued that we reliably infer what is going on emotionally in this dog's brain by observing that the numerous elements of its posture are so serviceable in the combat mode: the dog's hackles are raised, its fangs are bared, its ears are pricked, its eyes wide open and alert, and its body poised to spring forward. Darwin reasoned that any dog that had to go through a conscious checklist to manifest these postural elements one by one would be too slow on the draw to compete effectively against a rival in whom the entire process was activated autonomously by the relevant emotional arousal. That autonomous link, he concluded, provides a window into the dog's brain.

Darwin argued that humans display similar emotional symptoms.[13] Certain expressions, for example, spring unbidden to the human face in the presence of triggering emotions and yet are extremely difficult to present when those emotions are absent. People raised in different cultural traditions around the world can readily identify the schematic expression portrayed in figure 3.4 as one corresponding to emotions such as sadness or concern. As Paul Ekman and his colleagues have shown, most people are unable to reproduce this expression on command.[14] Various other emotions also have their characteristic signatures.

In the argument I am attempting to advance, emotions like sympathy and empathy play a central role. Other authors in this volume also stress a similar role played by these emotions. De Waal and Brosnan, for example, in chapters 4 and 5, respectively, report that precursors of these emotions are clearly visible in some primates and appear to motivate sharing. Solomon, in chapter 2, discusses the illustrious history of these emotions in moral discourse, beginning with David Hume and Adam Smith. Zak, in chapter 12, notes the relationship between the neuroactive hormone oxytocin levels and the experience

Figure 3.4. The Characteristic Expression of Sadness or Concern

of empathy and sympathy. Those of us who focus on these emotions stress their role in motivating individuals to forgo gain in deference to the interests of others. My particular concern is with the question of how a person might identify the presence of these emotions in others.

How, in other words, can you predict whether someone's sympathy for you will prevent him from cheating you, even when he has an opportunity to do so with no possibility of being punished? Clearly, it would not suffice for him merely to say that he was sympathetic to your interests. You need a more reliable signal. In the branch-outlet problem discussed earlier, for example, most people would feel more comfortable hiring an old friend than a perfect stranger.

I have been talking thus far as if there were good people and bad people, with some mixture of these two pure types comprising the entire population. Life is more complicated, of course: we have all done something decent, and we have all cheated at some point. The question is, under what conditions do we cheat? Evidence suggests that we are far less likely to cheat others with whom we enjoy strong sympathetic bonds.

Such bonds appear to form as a result of a complex dance that plays out among people over time.[15] When you have a commitment problem to solve, you pick somebody who you think cares about you. People are critical of George W. Bush for giving jobs to cronies. All leaders do that, however, and for good reason. Bush's particular problem is that many of his cronies are incompetent. The idea that you would pick someone well known to you is an intelligible, sensible thing to do. The question, in the end, is whether we can identify who will cheat and who won't.

Tom Gilovich, Dennis Regan, and I have done some experiments on this matter.[16] Our subjects had conversations in groups of three for thirty minutes, after which they played prisoner's dilemma games with each of their conversation partners. Subjects were sent to separate rooms to fill out forms on which they indicated, for each partner, whether they were going to cooperate or defect. They also recorded their predictions of what each partner would do when playing with them. Each subject's payoff was calculated as the sum of the payoffs from the relevant cells of the two games, plus a random term, so no one knew after the fact who had done what.

Almost 74 percent of the people cooperated in these pure one-shot prisoner's dilemmas. This finding is completely unpredicted by the standard self-interest model. Other empirical studies, however, have also found high cooperation rates in dilemmas when subjects were allowed to communicate.[17] Our particular concern was whether subjects could predict how each of their specific partners would play. When someone predicted that a partner would cooperate, there was an 81 percent likelihood of cooperation (as opposed to the 74 percent base rate). On the defection side, the base rate was just over 26 percent, but partners who were predicted to defect had a defection rate of almost 57 percent. This seems an astonishingly good prediction on the basis of just thirty minutes of informal conversation.

The following thought experiment also speaks to the question of whether people can accurately predict who will cheat them. Imagine that you have just returned from a crowded concert to discover that you have lost $1,000 in cash. The money, which was in an envelope with your name and address on it, apparently fell from your coat pocket while you were at the concert. Is there anyone you know, not related to you by blood or marriage, who you feel certain would return your money? Most people name such a person, but what makes them feel so confident?

It is extremely unlikely, of course, that they have experienced this situation before, but even if they had, they would not have known that the person they named had found their money and kept it. Under the circumstances, returning the money is a strict contradiction of the narrow self-interest model favored by economists.

Many people find it natural to say that the act of returning the money in such a situation must be motivated by some sort of moral emotion. Thus, people might predict that a friend would return their money because the prospect of keeping it would make the friend feel bad.

How did you pick the person you believe will return the money? Typically, it is someone with whom you have developed sympathetic bonds over an extended period. You feel you know enough about this person to say that if she found your money, she would not feel right about keeping it.

To say that trustworthiness could be an evolutionarily stable strategy does not mean that everyone is primed to cooperate all the time. Opportunism of the sort predicted by self-interest models is abundant. Yet, the prospects for sustaining cooperation are not as bleak as many economists seem to think. Many people are willing to set aside self-interest to promote the common good. Even if moral emotions are unobservable by others, they can still help you to be patient in repeated prisoner's dilemmas. If others recognize that you are a decent person, there are all sorts of ways in which you are valuable. If you are in business, your boss is likely to have a firm opinion about whether you would be the sort of person to return the lost $1,000 if you found it. You would like him to think that you'd return it. The best way to get him to think that, it appears, is actually to be the kind of person who would return it.

Do Our Models of Human Nature Matter?

As Gintis and Khaurana note in chapter 14, neoclassical economic models typically assume that people are self-interested in the narrow sense. Yet, abundant evidence suggests that motives other than self-interest are also important. An obvious consequence of inaccurate behavioral assumptions is that they often lead to inaccurate predictions. Casebeer and Stout note in chapters 1 and 8, respectively, another worrisome possibility, namely, that inaccurate portrayals of human nature may prove self-reinforcing.

The self-interest model of rational choice predicts that people will defect in one-shot prisoner's dilemmas. Does working with that model over the course of many years, as professional economists do, alter their expectations about what others will do in social dilemmas? And, if so, does this alter how economists themselves behave when confronted with such dilemmas? Tom Gilovich and Dennis Regan and I found that economics training—both its duration and content—affects the likelihood that undergraduate students will defect in prisoner's dilemma games.[18] In one version of our experiments, economics majors were almost twice as likely to defect as non-economics majors. This difference could stem in part from the fact that people who elect to study economics were different from others in the first place. But we also found at least weak evidence for the existence of a training effect. The differences in cooperation rates between economics majors and non-majors increased with the length of time that students had been enrolled in the major. We also found that, relative to a control group of freshmen astronomy students, students in an introductory microeconomics course were more likely in December than in September to expect opportunistic behavior on the part of others.

My point here is not that my fellow economists are wrong to stress the importance of self-interest but rather that those who insist it is the only important human motive are missing something important. Even more troubling, the narrow self-interest model, which encourages us to expect the worst in others, may bring out the worst in us as well.

Difficulties Confronting Emotion-Free Consequentialism

Consequentialist moral systems that ignore moral emotions face multiple challenges. It is one thing to say that we would all enjoy greater prosperity if we refrained from cheating one another; it is quite another, however, to persuade individuals not to cheat when cheating cannot be detected and punished.

Even for persons strongly motivated to do the right thing, consequentialist moral systems can sometimes make impossible demands on individuals. Imagine, for example, that five strangers are about to be killed by a runaway trolley, which at the flip of a switch you could divert onto a side track where it would kill four of your closest friends. Many consequentialists would argue that it is your moral duty to flip the switch, since it is better that only four die instead of five. A person capable of heeding such advice, however, is not likely to have had any close friends in the first place. Indeed, one can easily imagine that most people would become more reluctant to form close friendships if they believed it their duty to ignore the emotional bonds that such friendships inevitably entail.

The capacity to form deep bonds of sympathy and affection is important for solving various commitment problems. It is not a capacity easily abandoned, and even if we could abandon it, the emotional and material costs would be substantial.

Do Moral Emotions Define Right Conduct?

Since our current environment differs in many important ways from the environments in which our ancestors evolved, we should not be surprised that our intuitions sometimes mislead us about today's moral questions. Thus, we have not just Singer's example of an inhibition that is too strong (our reluctance to push the large man from the bridge to save the five strangers) but also many others in which our inhibitions are too weak (such as those against stealing from corporate employers, filing overstated insurance claims, or understating our income for tax purposes). In the latter examples, the weakness of inhibition is plausibly explained by the fact that in the environments in which we evolved, cheating always victimized specific persons rather than faceless institutions.

Our moral intuitions do not always mislead us, however. In the lost-envelope thought experiment, for example, my misgivings about keeping my friend's cash would push me to do what an independent consequentialist moral analysis says I ought to do under the circumstances, namely, return the cash. Indeed, our intuitions appear to provide sound moral guidance more often than not. For this reason, taking them at face value seems like a reasonable default option, provided we remain open to the possibility that they may be misleading in specific cases.

That said, I must emphasize that my argument about the moral emotions in *Passions Within Reason* was intended to serve one purpose only—to explain how people who evolved under the pressures of natural selection might nonetheless have inherited motivations to do the right thing under some circumstances in which such conduct entailed avoidable costs. I never claimed that our intuitions define right conduct.

As noted, the only equilibria sustainable in the evolutionary games I discuss entail populations containing at least some individuals who lack the usual moral inhibitions. We must incur costs to engage in character assessment, and it would make no sense to incur these costs if everybody were inclined to do the right thing all the time. But if people were never vigilant when choosing their trading partners, then mutant cheaters could invade an honest population at will. And since any population must therefore contain at least some cheaters in equilibrium, a given individual's moral intuitions simply cannot be used to define what constitutes right conduct. That assessment requires an independent analysis based on some sort of moral theory. Are our moral intuitions relevant to the choice of which moral theory to employ? In some cases, absolutely yes. In at least some others, however, I agree with Singer that we must be prepared to embrace a moral theory even though it might conflict with a specific moral intuition we hold dear.

If I read him correctly, however, Singer goes too far in claiming that moral intuitions should play no role at all in moral judgment—either in choosing among moral theories or in performing moral analysis within the framework of any given theory. Since our moral intuitions are in harmony with our moral theories most of the time, this claim seems strange on its face. (It is, of course, consistent with his contrarian nature!) Singer's point, though, is that the apparent harmony is less informative than it seems, because the authors of moral theories consciously strive to make them consistent with our intuitions. Fair enough, but that clearly does not imply that moral intuitions are generally irrelevant.

On the contrary, since they appear to provide useful guidance more often than not, we should be prepared to offer a coherent account for why a given intuition is misleading before proposing to dismiss it. That strategy works just fine in specific cases. For instance, it seems plausible to explain our relative lack of inhibition against using weapons that kill at great distances (and, by extension, our lack of inhibition about killing a stranger by flipping a trolley switch) by saying that killing in such remote ways simply was not possible in the ancestral environment.

But to say we should disregard moral emotions generally, one would have to offer a similar argument against each of them. And this, I believe, Singer cannot do. For this reason I find him unpersuasive when he insists that it is necessarily better to save two strangers than a single friend. His claim violates a strongly held intuition, but this time it is one that cannot easily be shown to be misleading.

Concluding Remarks

In brief, I have argued for a middle ground between Singer's position and that of John Rawls, who professed that progress in moral theory results from efforts to reconcile our moral theories with our moral intuitions.[19] Insofar as I believe that a moral theory is likely to be judged unacceptable if it systematically violates our moral intuitions, I am more or less on Rawls's side. With Singer, though, I am prepared to embrace a moral theory that violates a specific moral intuition if a plausible account can be given for why that intuition is misleading.

Notes

1. Williams 1973.
2. The looped-track example suggests that it was not the Kantian prohibition against using people merely as means that explains the earlier reluctance to push the stranger from the footbridge, since choosing to throw the switch in the looped-track example also entails using the stranger merely as a means to save the other five. In the original example, diverting the trolley onto the side track would have saved the others even had the stranger not been on the side track.
3. Bodvarsson and Gibson 1994. One can rationalize tipping in local restaurants as a self-interested activity: if you do not tip well, you might not get good service the next time. People resist the temptation to stiff the waiter because the shadow of the future is staring at them.
4. Hornstein 1976.
5. Schelling 1960.
6. Frank 1988, chap. 3.
7. Tinbergen 1952.
8. Gould 1977, 104.
9. Frank 1988, chap. 4.
10. Rapoport and Chammah 1965.
11. Ainslie 1992.
12. Frank 1988, chap. 6.
13. Darwin 1872.
14. Ekman 1985.

15. For a rich description, see Sally 2000.
16. Frank, Gilovich, and Regan 1993.
17. See Sally 1995.
18. Frank, Gilovich, and Regan (1993a), Marwell and Ames (1981), and Carter and Irons (1991) report similar findings.
19. Rawls 1971.

References

Ainslie, George. (1992). *Picoeconomics*. New York: Cambridge University Press.
Bodvarsson, O. B., and W. A. Gibson. (1994). Gratuities and Customer Appraisal of Service: Evidence from Minnesota Restaurants." *Journal of Socioeconomics* 23: 287–302.
Carter, John, and Michael Irons. (1991). "Are Economists Different, and, If So, Why?" *Journal of Economic Perspective* 5 (spring).
Darwin, Charles. (1965 [1872]). *The Expression of Emotions in Man and Animals*. Chicago: University of Chicago Press.
Ekman, Paul. (1985). *Telling Lies*. New York: Norton.
Frank, Robert H. (1988). *Passions Within Reason*. New York: Norton.
Frank, Robert H., Thomas Gilovich, and Dennis Regan. (1993a). "Does Studying Economics Inhibit Cooperation?" *Journal of Economic Perspectives* 7 (spring): 159–171.
———. (1993b). "The Evolution of One-Shot Cooperation." *Ethology and Sociobiology* 14 (July): 247–256.
Gould, Stephen Jay. (1977). *Ever since Darwin*. New York: Norton.
Greene, Joshua. (2002). "The Terrible, Horrible, No Good, Very Bad Truth about Morality, and What to Do about It." Ph.D. dissertation, Princeton University, Department of Philosophy.
Greene J. D., R. B. Sommerville, L. E. Nystrom, J. M. Darley, and J. D. Cohen. (2001). "An fMRI Investigation of Emotional Engagement in Moral Judgment." *Science* 293 (5537): 2105–2108.
Hornstein, Harvey. (1976). *Cruelty and Kindness*. Englewood Cliffs, N.J.: Prentice Hall.
Hume, David. (1978 [1740]). *A Treatise of Human Nature*. Oxford: Oxford University Press.
Marwell, Gerald, and Ruth Ames. (1981). "Economists Free Ride, Does Anyone Else?" *Journal of Public Economics* 15:295–310.
Rapoport, Anatol, and A. Chammah. (1965). *Prisoner's Dilemma*. Ann Arbor: University of Michigan Press.
Rawls, John. (1971). *A Theory of Justice*. Cambridge, Mass.: Harvard University Press, Belknap.
Sally, David. (1995). "Conversation and Cooperation in Social Dilemmas: A Meta-analysis of Experiments from 1958 to 1972." *Rationality and Society* 7:58–92.
———. (2000). A General Theory of Sympathy, Mind-Reading, and Social Interaction, with an Application to the Prisoners' Dilemma." *Social Science Information* 39 (4): 567–634.
Schelling, Thomas C. (1960). *The Strategy of Conflict*, New York: Oxford University Press.

Singer, Peter. (2002). "The Normative Significance of Our Growing Understanding of Ethics." Paper presented at the Ontology Conference, San Sebastian, Spain, October 3.

Smith, Adam. (1966 [1759]). *The Theory of Moral Sentiments*. New York: Kelley.

Tinbergen, Niko. (1952). Derived Activities: Their Causation, Biological Significance, and Emancipation during Evolution." *Quarterly Review of Biology* 27:1–32.

Williams, Bernard. (1973). "A Critique of Utilitarianism." In J.J.C. Smart and Bernard Williams, eds., *Utilitarianism: For and Against*. Cambridge: Cambridge University Press.

PART II

Nonhuman Origins of Values

Four

How Selfish an Animal?
The Case of Primate Cooperation

Frans B. M. de Waal

Biologists tend to classify behavior by its effects. If the actor benefits from a particular behavior, the behavior is called "selfish" regardless of motivation or intention. Thus, biologists will say that spiders build webs for "selfish" reasons even though spiders probably lack the ability to foresee what good a web will do. Similarly, "altruism" is defined as behavior that benefits a recipient at a cost to the actor itself regardless of whether the actor realizes this. A bee stinging an intruder is an "altruist" as she gives her life to protect the hive, even though her motivation is more likely aggressive than benign. As explained by Trivers (2002:6): "You begin with the effect of behavior on actors and recipients; you deal with the problem of internal motivation, which is a secondary problem, afterward. . . . If you start with motivation, you have given up the evolutionary analysis at the outset."

This means that if biologists claim that everything is selfishness, as they are wont to do, the psychologist should take this with a grain of salt. The biologist only means that all human and animal behavior is self-serving. It is not a statement about motivation or intention but about how and why behavior arose, that is, how it serves organisms that engage in the behavior. Even a vine overgrowing a tree is "selfish." This definition of "selfishness" is so broad that it also covers most altruism, because even the costliest act on behalf of another may—in the long run and often indirectly—benefit the actor itself, for example, if the recipient repays the favor. To again quote Trivers (1971:1): "Models that attempt to explain altruistic behavior in terms of natural selection are models designed to take the altruism out of altruism."

In the same way that evolutionary biologists willfully neglect motivational issues, I recommend doing the reverse when we turn our attention to motivation. Students of motivation should not occupy themselves with the fitness consequences of behavior. It is not for nothing that in biology we make a sharp distinction between "ultimate" and "proximate" domains of causation. The ultimate cause of a behavior refers to why it evolved over millions of years, in which case behavioral effects are paramount. The proximate cause of a behavior, on the other hand, refers to the immediate situation and stimuli that trigger it. Here, we wonder about the psychology and cognition behind it.

The ultimate and proximate domains can (and should) be treated independently: they represent different *levels* of analysis.

Let me illustrate this distinction with sexual behavior. Obviously, the ultimate reason for sex is reproduction, but does this mean that animals have sex with reproduction in mind? More than likely, they fail to connect the two mentally. They have sexual urges, and may seek sexual pleasure, but since reproduction is beyond their cognitive horizon, it does not figure in their motivation. When we say that animals have sex in order to reproduce, we are simply projecting our ultimate understanding onto their behavior, offering a shorthand that is incorrect when it comes to actual motivation.

The same applies to altruism. Human and animal altruism no doubt arose in the context of help to family members and help to those inclined to return the favor, hence in a self-serving context. Yet, altruistic behavior need not be tied to these specific contexts. Once in existence, it can float free from these constraints, at least to some degree. Kinship or reciprocity are not motivations; they are evolutionary causes. To believe that a chimpanzee, or any other animal, helps another with the goal of getting help back in the future is to assume a planning capacity for which there is no evidence. Most animals, we can be sure, perform altruism without payoffs in mind. In this sense, animal altruism is *more genuine* than most human altruism; we have an understanding of return benefits, and so we are capable of mixed motivations that take self-interest into account.

Apart from what motivates animals, we may ask whether their altruism is *intentional*. Do they realize their behavior benefits another? Do we? We show a host of behaviors for which we develop *post hoc* justifications. It is entirely possible, in my opinion, that we reach out and touch a grieving family member or lift up a fallen elderly person in the street before we fully realize the consequences of our actions. We say, "I felt I had to do something," but, in reality, our behavior was automatic and intuitive, following the common pattern that affect precedes cognition (Zajonc, 1980).

Similarly, it has been argued that much of our moral decision making is too rapid to be mediated by the cognition and self-reflection often assumed by moral philosophers (Haidt, 2001; Kahneman and Sunstein, 2005). Whereas we clearly have the *ability* of intentional altruism, we should be open to the possibility that much of the time we arrive at such behavior through quick-fire psychological processes similar to those of a chimpanzee reaching out to comfort another or sharing food with a beggar. Our vaunted rationality is partly illusory.

In sum, there are three kinds of altruism (table 4.1). *Evolutionary altruism* benefits the recipient at a cost to the actor regardless of the actor's motivations or intentions. This kind of altruism is common in a wide range of animals, from insects to mammals. *Psychological altruism*, on the other hand, occurs in reaction to the specific situation another individual finds itself in and is often motivated by other-directed emotions, such as nurturance or em-

TABLE 4.1

Altruistic behavior falls into three categories, dependent on whether or not it is socially motivated and whether or not the actor intends to benefit the other or not. The vast majority of altruism in the animal kingdom is only functionally altruistic, in that it takes place without an appreciation of how the behavior will impact the other and without any prediction of whether the other will return the service. Social mammals sometimes help others in response to distress or begging (socially motivated helping). Targeted helping may be limited to humans, apes, and a few other large-brained animals.

Evolutionary Altruism	Psychological Altruism	
Functionally altruistic: Cost to performer, benefit for recipient	Socially motivated: Response to the other's need, distress, or begging	Targeted helping: Intention is to benefit the other
Most animals		
	Social mammals and birds	
		Humans and a few large-brained animals

pathy. Such behavior may occur in the absence of any appreciation of how exactly the behavior will affect the other. An example would be a mother cat responding to the mews of her young by nursing them. It is unclear that the mother cat is aware that her behavior lessens hunger. Also note that, motivationally speaking, that her behavior serves genetic kin does not make it any less altruistic (to say so would be to confuse the evolutionary reason for maternal behavior with its proximate causation).

Sometimes, psychological altruism goes beyond other-directed emotions and motivations: it involves the explicit intention to help; that is, it is based on a prediction of how and why one's own behavior will benefit the other. An example from our chimpanzee colony is a young adult female who, when she noticed that an old, arthritic female struggled to get into a climbing structure, went behind her to push her up, placing both hands on her hips. Whereas psychological altruism is present in all animals that possess empathy (Preston & de Waal, 2002; de Waal, 2008), the subcategory of intentional altruism (also known as "targeted helping") requires knowledge of behavioral effects, and may be limited to a select few large-brained animals (de Waal, 1996, 2003).

Evolution sans Animals?

Human uniqueness is such a fundamental assumption in the social sciences that their current embrace of evolutionary theory risks being half-hearted.

Thus, some evolutionary psychology textbooks barely mention nonhuman animals at all. To keep evolution free from hairy creatures is a doomed enterprise, however, as it tries to follow the grand vision of Charles Darwin—a naturalist if ever there was one—without the inherent continuity among life forms that his vision implies.

An illustration of the uneasy marriage between human uniqueness claims and evolutionary theory is the current emphasis by behavioral economists on so-called strong reciprocity, according to which sanctioning of uncooperative behavior has created a uniquely human level of cooperation (e.g., Gintis, Bowles, Boyd & Fehr, 2003; Gürek, Irlenbusch, & Rockenbach, 2006). This may well be true, but sanctioning cannot possibly be the central issue in the evolution of cooperation. Without a tendency to form groups, to enter beneficial partnerships, and to strive for common goals, there would, of course, be no need for sanctioning. Cooperative tendencies came first. Strong reciprocity may well explain performance in cooperative games between strangers, yet this is not a likely context for the evolution of cooperation: cooperation probably originated among familiar individuals with reiterated encounters (Trivers, 2004). It seems logical, therefore, to first study the iceberg of cooperation and reciprocity before contemplating the little tip of "strong" reciprocity. Automatically, such a broader perspective will need to pay full attention to other animals.

Cooperation is widespread in the animal kingdom. Even the simple act of living together represents cooperation. In the absence of predators or enemies, animals do not need to stick together, and they might, in fact, be better off alone, away from conspecifics who eat the same foods and compete for the same mates (van Schaik, 1983). The first reason for group life is collective security.

In addition, many animals actively pursue common goals. By working together they attain benefits they could never attain alone. This means that each individual needs to monitor the division of spoils. Why would one lioness help another bring down a wildebeest if the other always claims the carcass for herself and her cubs? One cannot have sustained joint efforts without joint payoffs. With cooperation comes sensitivity to who gets what for how much effort (see chapter 5).

When we became cooperative animals, we abandoned the right-of-the-strongest principle and moved on to a right-of-the-contributor principle.

Reciprocity

Ever since Kropotkin (1902), the proposed solution to the evolution of cooperation among non-relatives has been that helping costs should be offset by return-benefits, either immediately or after a time interval. Formalized in modern evolutionary terms by Trivers (1971), this principle is known as *reciprocal altruism*.

Reciprocal altruism presupposes that (a) the exchanged acts are costly to the donor and beneficial to the recipient; (b) the roles of donor and recipient regularly reverse over time; (c) the average cost to the donor is less than the average benefit to the recipient; and (d) except for the first act, donation is contingent upon receipt. Although the initial work on cooperation (especially from the prisoner's dilemma perspective) focused primarily on the pay-off matrix to distinguish between reciprocity and mutualism, more recent efforts have included a significant time delay between given and received services as an additional requirement for reciprocal altruism (e.g., Taylor & McGuire, 1988).

Chimpanzees exchange multiple currencies, such as grooming, sex, support in fights, food, babysitting, and so on. This "marketplace of services," as I dubbed it in *Chimpanzee Politics* (de Waal, 1982), means that each individual needs to be on good terms with higher ups, foster grooming relations, and—if ambitious—strike deals with like-minded others. Chimpanzee males form coalitions to challenge the reigning ruler, a process fraught with risk. After an overthrow, the new ruler needs to keep his supporters content: an alpha male who tries to monopolize the privileges of power, such as access to females, is unlikely to keep his position for long—advice that goes back to Niccolò Machiavelli.

One of the commodities in the chimpanzee marketplace is food. Food-sharing lends itself uniquely to experimental research, because the quantity and type of food available, the initial possessor, and even the amount of food shared can be manipulated. *Active* food-sharing, a rare behavior, consists of one individual handing or giving food to another individual, whereas *passive* food-sharing—by far the more common type—consists of one individual obtaining food from another without the possessor's active help (figure 4.1).

We exploited the tendency of chimpanzees to share food by handing one of them a watermelon or some leafy branches. The food possessor would be at the center of a sharing cluster, soon to be followed by secondary clusters around individuals who had managed to obtain a major share, until all the food had trickled down to everyone. Claiming another's food by force is almost unheard of among chimpanzees—a phenomenon known as "respect of possession" (Kummer, 1991). Beggars hold out their hand, palm upward, much like human beggars in the street. They whimper and whine, but aggressive confrontations are rare. If these do occur, they are almost always initiated by the possessor to make someone leave the circle. She whacks them over the head with a sizable branch, or barks at them in a shrill voice until they leave her alone. Whatever their rank, possessors control the food flow (de Waal, 1989).

We analyzed nearly seven thousand approaches, comparing the possessor's tolerance of specific beggars with previously received services. We had detailed records of grooming on the mornings of days with planned food tests. If the top male, Socko, had groomed May, for example, his chances of obtaining

Figure 4.1. A cluster of food-sharing chimpanzees at the Yerkes Field Station. The female in the top-right corner is the possessor. The female in the lower left corner is tentatively reaching out for the first time, whether or not she can feed will depend on the possessor's reaction. Photograph by Frans de Waal.

a few branches from her in the afternoon were much improved. This relation between past and present behavior proved general. Ours was the first animal study to demonstrate a contingency between favors given and received. Moreover, these food-for-grooming deals were partner-specific, that is, May's tolerance benefited Socko, the one who had groomed her, but no one else (de Waal, 1997).

It was further found that grooming between individuals who rarely did so had a greater impact on subsequent food sharing than grooming between partners who commonly groomed. This finding may be interpreted in several ways. It could be that grooming from a partner who rarely grooms is more salient, leading to increased sharing by the food possessor. Chimpanzees may recognize unusual effort and reward accordingly. Second, individuals who groom frequently tend to be close associates, and favors may be less carefully tracked in these relationships. Reciprocity in close friendships may not have the high degree of conditionality found in more distant relationships. These explanations are not mutually exclusive: both will lead to a reduced level of conditionality with increased exchange within a relationship.

Of all existing examples of reciprocal altruism in nonhuman animals, the exchange of food for grooming in chimpanzees comes closest to demonstrating memory-based, partner-specific exchange. In our study, there was a significant time delay (i.e., a few hours) between favors given and received, and hence the favor was acted upon well after the previous positive interaction. Apart from memory of past events, for this to work we need to postulate that the memory of a received service, such as grooming, induces a positive attitude toward the same individual. In humans, this psychological mechanism is known as "gratitude," and there is no reason to call it anything else in chimpanzees (Bonnie & de Waal, 2004).

Capuchin Cooperation

Even though laboratory work on primate cooperation dates back to Crawford (1937), few experimental studies have been conducted since. Especially lacking is the experimental manipulation of "economic" variables, such as the relation between effort, reward allocation, and reciprocity. Recently, this situation has changed thanks to experiments on capuchin monkeys.

These monkeys show high levels of social tolerance around food and other attractive items, sharing them with a wide range of group members both in captivity and the field. This level of tolerance is unusual in nonhuman primates, and its evolution may well relate to cooperative hunting. Perry and Rose (1994) confirmed earlier reports that wild capuchins capture coati pups and share the meat. Since coati mothers defend their offspring, coordination among nest raiders may increase capture success. Rose (1997) proposed convergent evolution of food sharing in capuchins and chimpanzees, since both show group hunting (as do humans, for that matter). The precise level of cooperation is irrelevant for such evolution to occur: all that matters is that hunting success increases with the number of hunters. Under such circumstances, every hunter has an interest in the participation of others, something that can be promoted through subsequent sharing.

We mimicked this situation in the laboratory by having two capuchin monkeys work together to pull in a counterweighted tray at which point one or both of them would be rewarded (figure 4.2). This is similar to group hunts where many individuals surround the prey but only one will capture it. Our monkeys were placed in a test chamber separated from each other by a mesh partition. One monkey (the winner) of a pulling pair received a cup with apple pieces. Its partner (the laborer) had no food in front of it, and hence was pulling for the other's benefit. Food was placed in transparent bowls so that each monkey could see which one was about to receive the food.

We knew from previous tests that food possessors may bring food to the partition, where they permit their neighbor to reach for it through the mesh. On rare occasions, they push pieces to the other. We contrasted collective pulls

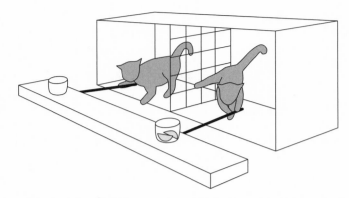

Figure 4.2. The test chamber used for the cooperative pulling task in capuchin monkeys inspired by Crawford's (1937) classical study. Two monkeys are situated in adjacent sections of the test chamber, separated by a mesh partition. The apparatus consists of a counterweighted tray with two pull bars, with each monkey having access to one bar. The bars can be removed. In the solo effort test, two monkeys were in the test chamber, but only one monkey had a pull bar and only this individual's food cup was baited. In the mutualism test, both monkeys were required to pull their respective pull bars, and both food cups were baited. In the cooperation test depicted here, both monkeys were required to pull, but only one individual's food cup was baited. Drawing by Sarah Brosnan.

with solo pulls. Under the latter condition, the partner lacked a pull bar, and the winner handled a lighter tray on its own. We counted more acts of food sharing after collective than solo pulls: winners were in effect compensating their partners for received assistance (figure 4.3; de Waal & Berger, 2000).

Furthermore, the partner pulled more frequently after successful trials. Since 90 percent of successful trials included food transfers to the helper, capuchins were assisting more frequently after having received food in a previous trial. The simplest interpretation of this result is that motivational persistence results in continued pulling after successful trials. A causal connection, however, is also possible, namely, that pulling after successful trials is a response to the obtained reward and the expectation of more.

Expectations about Reward Division

During the evolution of cooperation it may have become critical for parties to compare their own efforts and payoffs with those of others. Cooperative animals seem guided by a set of expectations about the outcome of cooperation and access to resources. De Waal (1991, p. 336) proposed a *sense of social regularity,* defined as: "A set of expectations about the way in which oneself (or others) should be treated and how resources should be divided. Whenever

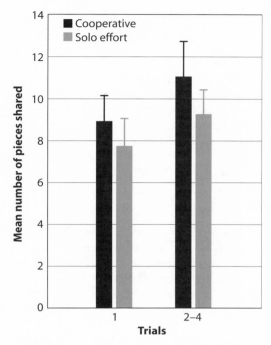

Figure 4.3. The amount of food sharing in successful cooperation tests (see figure 2) versus solitary controls. The mean (+SEM) number of times the partner collected food items through the mesh was significantly higher after cooperation than when food was obtained without help of the partner. From de Waal & Berger (2000).

reality deviates from these expectations to one's (or the other's) disadvantage, a negative reaction ensues, most commonly protest by subordinate individuals and punishment by dominant individuals."

The sense of how others should or should not behave is essentially egocentric, although the interests of individuals close to the actor, especially kin, may be taken into account (hence the parenthetical inclusion of others). Note that the expectations have not been specified: they are species-typical. Some primates are so hierarchical that the subordinate cannot expect anything from the dominant, whereas in other primates the dominants are prepared to share and, correspondingly, the subordinates have evolved all sorts of strategies (e.g., begging, whining) to extract food from possessors (de Waal, 1996). These animals negotiate about resources, as illustrated in the anecdote related below (videotaped) involving two female capuchin monkeys in the pulling task.

Cooperative pulling was done with two females, Bias and Sammy. In this case, both cups were baited. Sitting in separate sections of the test chamber, they successfully brought the food within reach. Sammy, however, was in such a hurry to collect her rewards that she released the tray before Bias had a chance to get hers. The tray bounced back, out of Bias's reach. While Sammy munched on her food, Bias threw a tantrum. She screamed her lungs out for half a minute until Sammy approached her pull-bar again. She then helped Bias bring in the tray a second time. Sammy did not do so for her own benefit, since by now her own cup was empty. Sammy's corrective response seemed the result of Bias's protest against the loss of an anticipated reward.

Against this background of cooperation, communication, and the fulfillment of an expectation (perhaps even an obligation), it makes sense to investigate primate reactions to reward division (see chapter 5).

The Free-Rider Problem

When capuchin monkeys perform cooperative pulling they seem to follow an "if—then" syntax according to which certain conditions need to be met before cooperative behavior will appear. If the partner is kin, it does not matter if food can easily be monopolized: cooperative tendencies are high in both partners. If the partner is unrelated and dominant, however, monopolizing food is an issue, and the cooperative tendency may dwindle if the partner fails to share (de Waal & Davis, 2003). The less tolerant the dominant partner, the less cooperation will take place. This means that the tendency to share payoffs of cooperative effort feeds back into the tendency to cooperate: sharing stabilizes cooperation. In fact, it is the dominant individuals who show the highest levels of tolerance in cooperative endeavors that benefit the majority (Brosnan, Freeman & de Waal, 2006).

This tool to stimulate cooperation may be explicitly wielded in chimpanzees. Boesch (1994) provides data suggesting that male chimpanzees who failed to contribute to the cooperative hunt were less successful than the hunters themselves in obtaining a share of meat. Perhaps this reflects recognition of effort. As Goodall (1971) noted, it is not unusual for the highest-ranking male to beg for (instead of physically claim) a share, which is why de Waal (1996) spoke of a "canceled" hierarchy when chimpanzees go into food-sharing mode. Reciprocity seems to take over from social dominance as regulator of access to resources.

Although active punishment is rare, it does occur. One anecdote from de Waal (1982) tells of a high-ranking female, named Puist, who took the trouble and risk to help her male friend, Luit, chase off a rival, Nikkie. However, Nikkie had a habit of singling out and cornering his rivals' allies after major confrontations to punish them for the part they played. This time, too, Nikkie displayed at Puist shortly after being attacked. Puist stretched out her hand to Luit for support, but Luit lifted not one finger to protect her. Immediately after Nikkie departed, Puist turned on Luit, barking furiously, chased him across the enclosure, even pummeled him. If Puist's fury was, in fact, the result of Luit's failure to reciprocate the help she had given him, the incident suggests that reciprocity in chimpanzees may be governed by similar "moralistic aggression" (cf. Trivers, 1971) as in humans.

Systematic evidence was obtained in the food-for-grooming study described above, where chimpanzees were more successful at obtaining food from individuals they had previously groomed. Food possessors showed aggressive resistance especially to approaching beggars who had failed to groom them. They were more than three times as likely to threaten such a beggar (figure 4.4; de Waal, 1997). This is not punishment per se but an aggressive reaction to those who try to get without giving, which, motivationally speaking, may not be far removed.

Chimpanzees also reciprocate in the negative sense: they know revenge. Revenge is the flip side of reciprocity. Data on several thousand aggressive interventions show a healthy correlation between how often individual A intervenes against B and how often B intervenes against A (de Waal & Luttrell, 1988). As a result, every choice has multiple consequences, both good and bad. The supported party in a conflict may repay the favor, whereas the slighted party may try to get even. It is obviously risky for a low-ranking individual to

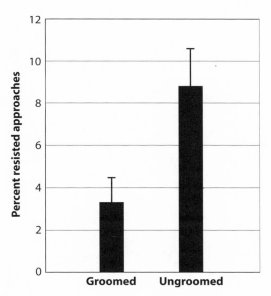

Figure 4.4. In the food-for-grooming experiment by de Waal (1997) it was found that the mean (+SEM) percentage of approaches meeting with active aggressive resistance from the food possessor depended on whether the possessor had been groomed by the approaching individual or not.

go against a high-ranking one, but if the latter is already under attack, there will be an opening to make that individual bleed. In chimpanzee society, payback is just a matter of time.

By far the most common tool to enforce cooperation, however, is partner choice. Unsatisfactory relationships can be abandoned and replaced by ones offering better benefits. With each individual shopping for the best partners and selling his or her own services, the framework becomes one of supply and demand as formalized in Noë and Hammerstein's (1994) Biological Market Theory. This theory applies whenever trading partners can choose whom to deal with. Market mechanisms are an effective way of sidelining profiteers.

It takes much energy, though, to keep a watchful eye on cheaters and the flow of favors. This is also why our own species relies on simpler forms of reciprocity much of the time. We form "buddy" relations with partners who have withstood the test of time. With spouses and good friends we relax the rules, and, in fact, consider keeping track of favors a sign of distrust. When it comes to distant relations, such as with colleagues and strangers, on the other hand, we do keep mental records and react strongly to imbalances, calling them "unfair" (Smaniotto, 2004). In chimpanzees, we found similar indications: the greater their familiarity, the longer the time frame over which chimpanzees seem to evaluate their relationships so that momentary imbalances matter less (see chapter 5).

All economic agents, whether human or animal, need to come to grips with the free-rider problem as well as the way rewards will be distributed following joint effort. Various mechanisms are available; "strong" reciprocity is merely one example, though perhaps the most relevant between human strangers. Like all social primates, we show emotional reactions to violated expectations. Other primates occasionally show aggressive reactions to unfulfilled expectations, whereas at other times they merely withdraw from cooperation. A truly evolutionary discipline of economics recognizes this shared psychology, and considers the possibility that the entire range of responses, from partner-switching to active sanctioning, are part of our background as cooperative animals.

Conclusion

If the term *selfish* refers to behavior that evolved to benefit its performer, then all human and animal behavior is ultimately selfish. Even behavior with obvious costs and no benefits for the actor—such as a risky rescue by a complete stranger—may still be an evolutionary product of life in small communities where helping tendencies served selfish interests.

This means, of course, that one needs to distinguish the evolutionary origin of behavior (the ultimate cause of its existence) from its actual motivation (the proximate cause). Once a tendency has come into existence, it is permitted to distance itself from its ultimate context. This happens all the time in both humans and other animals, with examples ranging from a bitch nursing kittens to a dolphin rescuing a drowning human. Bitches did not evolve to nurse kittens nor dolphins to rescue human swimmers, yet motivation becomes a force by itself.

It is best to learn from biologists, who long ago decided that a focus on the evolutionary level of explanation required them to ignore motivation. Similarly, anyone interested in the cognition and psychology behind altruistic behavior will do well to focus on the proximate level, and ignore fitness consequences. By so doing we keep the two levels of explanation apart. Ultimate causes simply do not exist in the minds of animals and rarely in those of humans. Thus, even if we explain a certain behavior as based on reciprocity, it is good to realize that reciprocity is not a motivation. Most of the time, the motivations guiding animal helping behavior are genuinely altruistic. Whether they are also intentionally altruistic is a separate issue, as this would require prediction of behavioral consequences. Only humans and very few other animals seem capable of such prediction.

References

Boesch, C. (1994). Cooperative hunting in wild chimpanzees. *Animal Behaviour* 48: 653–667.

Bonnie, K. E., & de Waal, F. B. M. (2004). Primate social reciprocity and the origin of gratitude. In *The Psychology of Gratitude,* ed. R. A. Emmons & M. E. McCullough, pp. 213–229. Oxford: Oxford University Press.

Brosnan, S. F., & de Waal, F. B. M. (2003). Monkeys reject unequal pay. *Nature* 425: 297–299.

Brosnan, S. F., Freeman, C., & de Waal, F. B. M. (2006). Partner's behavior, not reward distribution, determines success in an unequal cooperative task in capuchin monkeys. *American Journal of Primatology* 68:713–724.

Crawford, M. (1937). The cooperative solving of problems by young chimpanzees. *Comparative Psychology Monographs* 14:1–88.

de Waal, F. B. M. (1989). Food sharing and reciprocal obligations among chimpanzees. *Journal of Human Evolution* 18:433–459.

de Waal, F. B. M. (1991). The chimpanzee's sense of social regularity and its relation to the human sense of justice. *American Behavioral Scientist* 34: 335–349.

de Waal, F. B. M. (1996). *Good Natured: The Origins of Right and Wrong in Humans and Other Animals.* Cambridge, MA: Harvard University Press.

de Waal, F. B. M. (1997). The chimpanzee's service economy: Food for grooming. *Evolution and Human Behavior* 18: 375–386.

de Waal, F. B. M. (1998 [orig. 1982]). *Chimpanzee Politics: Power and Sex among Apes.* Baltimore, MD: Johns Hopkins University Press.

de Waal, F. B. M. (2003). On the possibility of animal empathy. In: *Feelings and Emotions: The Amsterdam Symposium,* T. Manstead, N. Frijda, & A. Fischer (Eds.), pp. 379–399. Cambridge: Cambridge University Press.

de Waal, F. B. M. (2008). Putting the altruism back into altruism: The evolution of empathy. *Annual Review of Psychology* 59: doi:10.1146/annurev.psych.59.103006.093625

de Waal, F. B. M., & Berger, M. L. (2000). Payment for labour in monkeys. *Nature* 404:563.

de Waal, F. B. M., & Davis, J. M. (2003). Capuchin cognitive ecology: Cooperation based on projected returns. *Neuropsychologia* 41:221–228.

de Waal, F. B. M., & Luttrell, L. M. (1988). Mechanisms of social reciprocity in three primate species: Symmetrical relationship characteristics or cognition? *Ethology and Sociobiology* 9:101–118.

Gintis, H., Bowles, S., Boyd, R., & Fehr, E. (2003). Explaining altruistic behavior in humans. *Evolution and Human Behavior* 24:153–172.

Goodall, J. (1971). *In the Shadow of Man.* Boston: Houghton Mifflin.

Gürek, Ö., Irlenbusch, B., & Rockenbach, B. (2006). The competitive advantage of sanctioning institutions. *Science* 312:108–111.

Haidt, J. (2001). The emotional dog and its rational tail: A social intuitionist approach to moral judgment. *Psychological Review* 108:814–834.

Kahneman, D., & Sunstein, C. R. (2005). Cognitive psychology and moral intuitions. In: *Neurobiology of Human Values.* J-P Changeux, A. R. Damasio, W. Singer, & Y. Christen (Eds.), pp. 91–105. Berlin: Springer.

Kropotkin, P. (1972 [orig. 1902]) *Mutual Aid: A Factor of Evolution.* New York: New York University Press.

Kummer, H. (1991). Evolutionary transformations of possessive behavior. *Social Behavior and Personality* 6:75–83.

Noë, R., & Hammerstein, P. (1994). Biological markets: Supply and demand determine the effect of partner choice in cooperation, mutualism, and mating. *Behavioral Ecology and Sociobiology* 35:1–11.

Perry, S., & Rose, L. (1994). Begging and transfer of coati meat by white-faced capuchin monkeys, *Cebus capucinus. Primates* 35:409–415.

Preston, S. D., & de Waal, F. B. M. (2002). Empathy: Its ultimate and proximate bases. *Behavioral and Brain Sciences* 25:1–72.

Rose, L. (1997). Vertebrate predation and food-sharing in Cebus and Pan. *International Journal of Primatology* 18:727–765.

Smaniotto, R. C. (2004). "You scratch my back and I scratch yours" versus "Love thy neighbour." Two proximate mechanisms of reciprocal altruism. Ph.D. thesis, University of Groningen.

Taylor, C. E., & McGuire, M. T. (1988). Reciprocal altruism: 15 years later. *Ethology and Sociobiology* 9:67–72.

Trivers, R. L. (1971). The evolution of reciprocal altruism. *Quarterly Review of Biology* 46:35–57.

Trivers, R. L. (2002). *Natural Selection and Social Theory.* Oxford: Oxford University Press.

Trivers, R. L. (2004). Mutual benefits at all levels of life. *Science* 304:964–965.

van Schaik, C. P. (1983). Why are diurnal primates living in groups? *Behaviour* 87:120–144.

Zajonc, R. B. (1980). Feeling and thinking: Preferences need no inferences. *American Psychologist* 35:151–75.

Five

Fairness and Other-Regarding Preferences in Nonhuman Primates

Sarah F. Brosnan

Perhaps studying nonhuman primates to learn something about our reactions to free enterprise and market economies seems like a most illogical place to begin. Arguably, however, it is precisely where we should look. Studying these mirrors of ourselves and our evolutionary past can tell us a great deal about what to expect from ourselves in various situations (Boehm, 1999; de Waal, 1996; Wrangham & Peterson, 1996). My work centers on studying inequity in nonhuman primates, with an eye toward the evolution of this response. This is of prime concern to many; one of the basic assumptions in the west is that being treated fairly is a human right.

A key question, however, is whether this is an ideal we have created and should strive to live up to where possible, or whether reacting to unfair situations increased the fitness of our ancestors to the degree that this is an evolved characteristic of humans and something we all expect. More than just an academic debate, this answer will help us understand the degree of flexibility in the system and how institutions can be aligned with our preexisting biases toward fairness and unselfish behavior. For instance, it is important to know if a given degree of inequity will never be tolerated, no matter what the justification, or if humans have the flexibility to change their perceptions of inequity, using, for instance, narrative attempts (see chapter 1). Our nonhuman primate cousins are not us, but, by their very closeness, they can help us unpack some understanding about ourselves.

The "cartoon" of free enterprise tends to promote the idea that the system is self-serving to the few at the top, unfair to the masses, and generally cutthroat. Why is the system perceived this way? Is it really an unfair system, or are certain elements simply perceived that way in differing cultures or presentations? And, if this is the case, what can we do to promote the "real" version of free enterprise?

This chapter addresses some of these questions in nonhuman primates to glean insights about the human reaction to fairness and the human propensity to have other-regarding preferences. Monkeys and apes, including our closest living ancestor, the chimpanzee, do respond negatively to inequity, or situations in which they receive less than another of their group mates for the same amount of work. This probably sounds familiar: imagine how you would feel if your boss suddenly announced that a coworker with no experience or skills beyond yours was to receive a 10 percent raise, while you received nothing. The response, however, is also finely tuned to the situation and the social partners available. What if the above scenario happened the year before, and it was you who received the 10 percent raise while your colleague received nothing? What if your boss had explained that only one raise was available per year, so he would alternate who got it? Monkeys are also sensitive to these types of contingencies. Capuchin monkeys who have to cooperate for unequal rewards will continue to do so, but only if their partner takes turns receiving the windfall. Thus, maybe a little inequity is okay, as long as everyone wins sometimes.

But this does not necessarily mean that we are all wholly other-regarding beings looking out for everyone else's interests. Several recent experiments show that chimpanzees are apparently disinterested in assisting their partner to receive a better outcome when there is nothing in it for them. These studies are new, and more research needs to be done, but perhaps, in some cases, cooperation that helps others (as well as the self) can be achieved primarily through self-interest. This, too, makes intuitive sense to us; designing systems in which everyone wins means that everyone has an incentive to participate.

These results help us understand our own predispositions for fairness. We clearly evolved to be interested in our own outcomes, particularly in regard to what others around us are receiving. We are willing to tolerate some inequity but only if there is some reciprocation of effort. Such evolutionary history needs to be considered when designing economic systems. It is crucial that individuals involved feel as if they are being treated equitably. Moreover, as this chapter will show, reactions formed against a system may keep individuals from cooperating even if it would be in their best interests. Monkeys who are cooperating with a partner who does not share the wealth generally quit participating in any situation, not just those in which they stand to be treated inequitably.

Why Study Nonhuman Primates

Behavioral Phylogeny

Nonhuman primates are a window into our evolutionary past. They provide a way of examining what traits may have been present in our closest living ancestor. This is typically done through a technique known as *behavioral phylogeny*. Like physical traits, behavioral traits can be compared to ascertain the relative likelihood of common descent (C. Boehm, 1999; Preuschoft & van Hooff, 1995; Wrangham & Peterson, 1996). This technique can be used for human traits by comparing our behaviors with those of great apes and monkeys;

if traits are shared across species, it is assumed that they are commonly derived from the shared ancestor. If traits vary, it is then assumed that the traits arose after the divergence of the species in question.

Of course, just because human and nonhuman animals have a trait in common does not necessarily mean that there is phylogenetic continuity. Similarity can be caused by common descent (*homology*) or by independent evolutionary events (*analogy*). For example, wasps, bluebirds, and Mexican free-tailed bats all have wings, but this is not the result of common evolutionary descent but the independent evolution of wings at least three times during the evolution of animals. (Note, though, that such information on analogous evolution is still useful, as it can help determine the situations under which it was beneficial for a trait, such as wings, to arise in the first place.) Careful assessment of various species can help us understand which traits are shared by common descent and which arose as a result of common ancestral environments and constraints.

What Do Nonhuman Primates Have to Offer?

Studying nonhuman primates (or any animal species) offers us a wider understanding of behaviors. Animals, unlike humans, are not constrained by complex cultural phenomena such as religions, schools, and governments. In humans, these cultural phenomena affect responses to various situations and lead us further from our instinctive responses. (Of course, that humans have complex culture is a phenomenon not to be ignored. It is clearly an evolutionary adaptation in its own right and may be sufficiently powerful to have affected human genetic evolution [Richerson & Boyd, 2005]). In nonhuman species, we get a trimmed-down version which allows us a glimpse into our past.

Furthermore, in nonhumans we are required to pull all our data from observation of behavior rather than from self-report analyses. Although not being able to ask for an animal's underlying motivations is surely a drawback, the science may be more rigorous as a result. Scientists studying animal behavior are forced to rely on purely behavioral measures, reporting on how we see our study organism act rather than why they tell us they act in a certain way. In fact, this can lead to more accurate results. First, it may be helpful in emotionally charged situations, where people may report that they would do one thing but in fact do another. Second, it makes it more difficult for our subjects to (intentionally or accidentally) deceive us.

A "Sense of Fairness"

Few would disagree that humans have a sense of fairness. We respond badly when treated unfairly; we give more than the minimum required in experimental games (Henrich, Boyd, Bowles, Camerer, Fehr, Gintis & McElreath,

2001) and we frequently punish in situations in which another individual behaves non-cooperatively (Fehr & Rockenbach, 2003; Kahneman, Knetsch & Thaler, 1986; Zizzo & Oswald, 2001). To varying degrees, these inequity averse responses are seen across a vast array of cultures and differ significantly depending upon the quality of the relationship between the individuals involved (Clark & Grote, 2003). They have recently been linked to emotional, as well as rational, processes (Frank, 1988, 2001; Sanfey, Rilling, Aronson, Nystrom & Cohen, 2003).

This has been studied widely in many fields. In behavioral economics, it is often referred to as *inequity aversion*, and the bulk of research relates to how humans respond to situations involving division of cash (see, e.g., Camerer, 2003). In the early social psychology literature, such research was carried out under the rubric of equity theory, and researchers were interested not only in how individuals restored equity but also in the mechanisms individuals use to do so, including compensation and, in lieu of direct equity restoration, psychological leveling mechanisms (Walster [Hatfield], Walster & Berscheid, 1978). Finally, current research in social psychology addresses the issue in relation to procedural justice, organizational justice, and distributive justice.[1]

This background, useful though it may be, must be re-envisioned somewhat to be relevant when discussing nonhuman species. Although understanding psychological mechanisms is vital, in social psychology, fairness or justice is based on people's subjective appreciation of the equity of the situation. Such subjective approaches are impossible in nonhumans, as we cannot ask them to relate their feelings on the matter or administer a survey in an attempt to assess their understanding of the situation. Thus, we must rely on the information provided by their behavioral reactions to inequitable situations, much as is done in current work on behavioral economics (in which money spent is the variable of interest), which makes it much more difficult to uncover psychological mechanisms.

However, scientists can assess nonhuman animals in this realm through behavioral observations of their reactions to different situations. Most basically, we can determine individuals' responses when they receive an outcome that is less desirable than that of a conspecific (a smaller bit of food or a less tasty morsel). By determining their behavioral response to distributional inequity, we learn something about their psychology. This can be extended to investigate differences in the relative level of effort or even responses to behaviors that affect another. We can potentially examine self-sacrifice in which individuals choose sub-optimal outcomes for themselves in order to maintain the conventions or norms societally considered to be "fair."

It is difficult, however, to refer to the presence of any of these behaviors, in isolation, as a "sense of fairness." Rather, each appears to be an element that would be required for a sense of fairness or, better yet, a reaction to an inequitable situation. This also does not address how individuals may come to violate fairness norms in the first place. Thus, to maintain consistency throughout the chapter, I use the term *inequity response* (Brosnan, 2006b) when referring to animals, and suggest its use in the future to cover the entire broad group of responses to inequity. This emphasizes that, in many cases involving nonhumans, it is a response, or behavior, to an inequitable situation that is being studied, and not necessarily fairness in the sense usually intended in the human social psychology literature.

In fact, in the human literature, one might argue, it is best to subdivide "fairness" into its different elements and investigate them all individually. There tends to be a laundry list of characteristics that fall under the heading "sense of fairness," some of them even confounded in the same studies. This helps no one, as many of these elements may vary in expression depending on the time, place, situation, and other individuals involved.

Thus, in this chapter, I address important aspects of the sense of fairness and rarely discuss "fairness" as a concept. This is, of course, partly because it is difficult at this point to say anything meaningful about the sense of fairness as a whole in animals, but it is also because I think it is important to recognize that there are different aspects to fairness, and *all* of these must be addressed separately when tackling a problem as complex as people's reactions to free enterprise.

Why Respond to Inequity

A strong reaction to inequity is thought to be beneficial in several areas. First, in the social arena, an aversion to inequity may promote beneficial cooperative interactions, because individuals who recognize that they are consistently getting less than a partner can look for another partner with whom to cooperate more successfully (Fehr & Schmidt, 1999). This should provide a more profitable environment for interaction for all cooperators.

Second, and related to the first area mentioned, having strong reactions to inequity may be one of Frank's proposed *commitment devices* (Frank, 1988, 2001, 2004), which are strong, emotional responses not under conscious control that cause people to respond in predictable ways to situations that might be against their best interest. Such devices may cause individuals to make an "irrational" choice in the short term, committing themselves to a path (such as revenge or romantic love) that leads to more propitious situations in the long term by convincing others of their high level of dedication to the situation. Similarly, reacting to an inequitable situation may cause a short-term loss if the individual sacrifices a potential (and inequitable) gain (e.g., refuses to accept advantageous inequity). This may have future benefits, however, as it may discourage others from acting inequitably toward them in the future, or it may encourage other equitably-minded individuals to seek them out for interactions. Thus, responding to inequity may act as a sort of insurance policy to ensure future beneficial cooperative interactions. Aversion to inequity could serve a similar role in nonhumans, who show high levels of cooperation and evidence of emotional responses to inequity.

What May Affect the Inequity Response—Lessons from Nonhumans

Individual Recognition

Nonhuman primates in many species are known to recognize one another. Capuchin monkeys are able to keep track of social relationships and dominance (Perry, Barrett & Manson, 2004). Vervet monkeys and rhesus macaques not only can identify one another, but they can identify the mothers of offspring that are using the mother's vocalizations (Cheney & Seyfarth, 1990; Gouzoules, Gouzoules & Marler, 1984).

Cooperation in primate species is also limited to individuals from the group. There is good evidence in humans that reciprocity and cooperation take place primarily among individuals with good relationships; in other words, these favors are not given indiscriminately, nor are there some individuals who are "cooperative" and others who are not. Instead, each individual varies in their cooperativeness depending on the identity of their partner (Frank, 1988, 2001). I suspect that nonhuman primates behave similarly. Chimpanzees are known to cooperate best with those individuals with whom they have a tolerant relationship (Melis, Hare & Tomasello, 2006a) and to actively recruit these individuals (Melis, Hare & Tomasello, 2006b). As I discuss in detail later on in this chapter, chimpanzees also seem to respond differently to inequity depending upon the identity of their partner (Brosnan, Schiff & de Waal, 2005) and capuchin monkeys are more interested in the behavior of their partner than the reward distribution in an inequitable cooperative task (Brosnan, Freeman & de Waal, 2006).

In fact, this is a potential problem with any sort of large-scale cooperative endeavor, like free enterprise. Individuals will behave differently with others depending on the partner's identity. Moreover, identifying individuals with whom one will cooperate requires face-to-face, or at least personal, interaction, which almost by definition is missing in many of these large-scale situations. This may be especially true for individuals in societies where almost nothing is done anonymously. It can be overcome, but it will require careful execution and understanding from those implementing any such large scale system.

Responses to Out-Groups

Adding to the problem of variation in response depending upon the individuals involved, our race has an evolutionary history of xenophobia. It is crucial to state here that this is *not* a justification for racism or xenophobia but is something that we must take into account when designing systems involving more than a person's in-group.

In primates, individuals never cooperate and are often quite violent in interactions with individuals from other groups. Chimpanzees are known to be murderous in their inter-group encounters, and they apparently intentionally set out on these forays. These same individuals, however, can be quite cooperative with conspecifics during coalition formation, group hunts, or even these deadly patrols (Mitani, 2006).[2] Much as we humans strive to avoid such biases, evidence indicates that some of the same sort of behavior exists. For instance, a study has shown that undergraduates dividing a "scholarship stipend" between a candidate from their school and another school are biased in their decision making toward the in-group candidate (Diekmann, Samuels, Ross & Bazerman, 1997).

There is a lesson for us here. Our closest ancestors are capable of great cooperation among friends, but outside the group it is quite different. Such behavior is known among humans and is likely to play just as great a role. We need to consider that our psychology is more attuned to equity among individuals we know (and, presumably, like) than individuals from an out-group, however it may be defined. Again, this is a surmountable obstacle, as any two people can meet and become friends, but that element may be required before easy and trustful cooperation can come about.

Tolerance and Social Relationships

The level of tolerance in the dominance hierarchy probably has a significant impact on reactions to inequity, and the level of tolerance differs dramatically between species. Although none of the species discussed in this chapter lacks a dominance hierarchy altogether, in some of them one's position in the hierarchy makes a much bigger difference in one's access to resources and general quality of life.

In relatively tolerant species, if the group finds a food bonanza, all individuals can expect to share at least somewhat and maintain possession of the food acquired. In relatively less tolerant species, individuals may not even maintain possession of food that they have stored in a cheek-pouch, and any transgression against a dominant will be met with quick violence. Moreover, a subordinate individual in a relatively tolerant species (e.g., a chimpanzee) can throw a temper tantrum in front of a dominant with little expectation of retribution, but a subordinate member of a relatively despotic species (e.g., a rhesus macaque) has no recourse against a more dominant individual (de Waal, 1991b). Of course, bear in mind that this relative level of tolerance does not mean that rank is nonexistent or that subordinates frequently protest against dominants, but it does mean that subordinates can protest without fear of retribution. This creates a baseline level of expectation of equity that makes individuals more likely to react to inequitable situations, and hence makes it more likely that reactions to inequity would evolve (de Waal, 1996).

Significant variation in responses may also result from social group identification or relationship quality. In humans, there is apparently considerable variation among different cultural groups in how their members respond to unfair situations (Henrich, Boyd, Bowles, Camerer, Fehr, Gintis & McElreath, 2001; Lind & Earley, 1992). Furthermore, there is a great deal of variation based upon individual relationships. Individuals in positive or neutral relationships tend to be much more oriented toward equity than those in negative relationships (Loewenstein, Thompson, & Bazerman, 1989), and individuals in close relationships are far more likely to possess a communal orientation than are those in more distant relationships, who show a contingent response (Clark & Grote, 2003). This may be so either because of a difference in perception or a difference in the time span used for assessment of equity.

Hierarchies and Dominance

Because of the hierarchical nature of animal societies and the often extreme imbalances in power, distributive fairness in animals will almost certainly not be about equality. Instead, fairness will probably center on equity, keeping the distributions between individuals even enough that all the players are willing to continue their participation in the joint effort. This is also true in humans, and the same sorts of constraints are likely to be factors in what humans consider to be fair or are willing to put up with.

In all animal societies, most of the variation between individuals is probably owing to the character of the relationship between dominants and subordinates (de Waal, 1991a, 1996). Individuals who are dominant in the group get more. They may or may not provide more (for instance, having a large alpha male may help in conflicts between groups of the same species, which benefits all group members), but dominants will collect most of the benefits in terms of food, mates, social partners, and breeding success. Subordinates, on the other hand, can choose to stay in the group and accept their lesser share, to diminish their participation to the lowest acceptable level (think here of the employee who does just enough to avoid being fired), or can leave the group and try their chances somewhere else.

However, no social animal is completely free to leave its group, so the ease with which the dominants can coerce subordinates will depend upon the ease with which subordinates can leave (Vehrencamp, 1983). This is a trade-off for both parties. Subordinates who emigrate leave offspring and other family as well as a chance to be the dominant breeder in the future (Kokko & Johnstone, 1999); they also face a steep cost when entering another group. All this makes them unlikely to leave the group without great cause and empowers the dominant to maintain the inequitable balance. All social animals, however, including dominants, rely on one another for food, mates, social interaction, and protection from both predators and other conspecifics (Krebs & Davies, 1993), so dominants who tip the distribution too far to the inequitable side may find themselves in the untenable situation of having no group mates. Much of the variation in dominance hierarchies may be the result of outside ecological pressures; despotic dominance hierarchies, in which the dominants have full coercive power, are much more likely in situations with high competition for food or serious outside pressure from other conspecific groups or predators (Van Schaik, 1989). Such a trade-off of alternatives also occurs in humans. For instance, in close relationships any individual who is less dependent on the other becomes dominant in decision making because of their greater ability to seek alternatives (Kelley, 1979).

Thus, equity in a group of animals may boil down to finding a balance in which dominants get the lion's share but make staying in the group attractive enough that their subordinates do not abandon them. These parameters obviously shift from species to species and from group to group, depending on the factors just discussed. Conventions regarding fair distributions depend a great deal on the particular species and the individual's place within the hierarchy of the group.[3]

Contingent Behavior

As with humans, animals are also capable of showing contingent behavior and can take into account the social environment and how it will affect expected payoffs. For instance, several species adjust their food calling depending on both the social environment and the quantity or divisibility of available food (Caine, Addington & Windfelder, 1995; Chapman & Lefebvre, 1990; Elgar, 1986; Evans & Marler, 1994; Hauser, Teixidor Fields, & Flaherty, 1993; van Krunkelsven, Dupain, van Elsacker & Verheyen, 1996). The findings of one study indicated that individuals suppressed calls for a small, unshareable amount of food but not for a larger amount of food (fifteen apples) that presumably could not have been consumed before the food possessor was discovered by the group (Brosnan & de Waal, 2003b). In another study, unrelated to food calling, capuchin monkeys adjusted their willingness to participate in a cooperative task depending on the food distribution. If the food was distributed in a way that each individual could easily obtain food upon participating, cooperation occurred frequently. However, if the position of the food was adjusted so that one individual could easily dominate food rewards, cooperation failed to occur even at the first trial, indicating that the subordinate monkeys did not need experience to recognize that this situation was unlikely to be rewarding (de Waal & Davis, 2002). Reactions to inequity are also likely to hinge on the current situation.

Variation in Sense of Value

Animals may have a different assessment of value depending upon other factors unrelated to the current interaction. The behavior of individuals may be affected by their gender, their rank, whether they are pregnant or lactating, whether they have recently eaten, been groomed, or mated, as well as their personal preferences for various commodities (Seyfarth & Cheney, 1988). Such variation makes observational research (in humans and nonhumans) challenging, and requires a careful tuning of rewards and efforts in experimental research.

In humans, we need to remember that not only previous experience but also personal tastes play a large role. This ties back in with the ways in which social partners influence decisions. Problems may arise simply because two individuals do not like each other. There is no easy solution to rectifying these variations based on experience and taste, but they should be kept in mind as a somewhat random variable that may have great effects on peoples' choices.

The Evolution of the Sense of Fairness

If the inequity response has evolved because it increases fitness, which I believe is the case, this response probably did not arrive *de novo* (Brosnan, 2006a, 2006b; Brosnan & de Waal, 2004b). The links between current behavior and the long-term payoffs of rejecting inferior (yet positive) rewards seem too tenuous to have given rise directly to the sort of complex inequity responses we see in humans. I believe that the inequity response arose through a series of steps that at each point increased the individual's relative fitness. Assuming that this evolution took place before the hominid lineage split from the rest of the primates, these evolutionary precursors to inequity aversion (IA) are likely to be seen in other species, particularly nonhuman primates. By outlining the elements that are presumably required, we can look for inequity-related behaviors in other species, learn something about the preconditions for the behavior, and perhaps extrapolate information concerning the evolution of the behavior.

Disadvantageous inequity aversion refers to the willingness to sacrifice potential gain to block another individual from receiving a superior reward (Fehr & Schmidt, 1999), yet its evolutionary precursors may be much simpler. A first step is the recognition that other individuals obtain rewards that are different from one's own. This may seem somewhat self-evident, but it is not at all clear that nonhumans pay such close attention to conspecifics' rewards. Such an ability is also necessary for social learning, which is essentially based on recognizing that another individual has a better outcome than oneself, and attempting to achieve that outcome. Thus, irrespective of the mechanism individuals use to socially learn (stimulus enhancement, emulation, imitation, etc.), an individual must initially notice that the other individual is obtaining some benefit they cannot obtain themselves. Such social learning abilities have been documented in a number of species, including chimpanzees (Whiten, 1998; Whiten, Custance, Gomez, Teixidor & Bard, 1996) and capuchin monkeys (Brosnan & de Waal, 2004c; Custance, 1999). Later, after understanding that rewards vary between conspecifics, individuals may be motivated to obtain these better rewards as well.

At the second level, the maligned individual reacts to this discrepancy, which requires that the individual feels strongly enough about the difference to alter his or her behavior, leading to a negative response to discrepancies against the self. In fact, this may not always lead to an absolutely better outcome for the individual but may lead instead to alternate behaviors which might, in the long run, minimize both inequity and the cost to achieve equity. The best example of this thinking is the work on commitment problems by Robert Frank, as discussed earlier (Frank, 1988, 2001). He argues that individuals may make short-term sacrifices that lead to long-term benefits. For instance, the willingness to punish irrationally can become a signal of one's commitment to good treatment and lead to future positive interactions with others without further cost accrued. Regarding the evolution of inequity, reacting negatively to conspecifics who are not equitable in their treatment of oneself may lead to finding new cooperation partners who will presumably show better treatment.

At the third level is the willingness to sacrifice in order to punish a lucky individual. This is known as costly punishment, a situation in which individuals are willing to pay to rectify inequity. This presumably leads to further cooperation (and more equity) as punished individuals toe the line in the future (Fehr & Gachter, 2002). In humans, individuals are also known to respond negatively to advantageous inequity, or inequity that is detrimental to another (Fehr & Schmidt, 1999; Hatfield, Walster & Berscheid, 1978; Loewenstein, Thompson & Bazerman, 1989). Either through attempts to rectify the situation or psychological leveling mechanisms, individuals work to bring equity to these situations as well. Thus, we reach the full range of inequity averse behavior present in humans.

Following the evolution of disadvantageous inequity aversion, advantageous inequity aversion may develop. It seems unlikely that one would notice or react to inequitable circumstances harming another individual before noticing one's own negative outcomes. Thus, prior to the evolution of a reaction when one is under-benefited, there was probably little to no selective pressure for advantageous inequity aversion. Such advantageous IA leads to the development of a proto-moral system whereby individuals attempt to rectify discrepancies relative to others. Advantageous IA is quite important when thinking about the free enterprise system and is discussed later in this chapter.

Evidence of "Fairness" Affecting Fitness

Given how new this field is, there is little evidence as yet directly relating any element of fairness to fitness. However, the exception comes from some excellent work exploring fair play among social canids (Bekoff, 2001, 2004). Bekoff, in his studies of social play, has found, in both wolves and domestic dogs, that individuals who do not engage in fair play seem to be at a disadvantage. In the play behavior of young canids, much stronger or older individuals typically self-handicap during play, and excessive aggression is not tolerated. This behavior apparently represents a group norm. Bekoff found that individuals who did not exhibit such leveling behavior were excluded from play sessions. Playmates failed to solicit these rough individuals for play, and even left the play group if these individuals joined in. The result is that individuals who violate group norms do not interact nearly as much with other pups their age. Play, apparently, is important for these social animals as an opportunity to learn how to cooperate and negotiate social agreements in a situation where mistakes or transgressions will be forgiven (Bekoff, 2004).

More startling is evidence that there is a direct fitness consequence to this behavior. When coyotes mature, Bekoff has found, individuals ostracized from play sessions are much more likely to leave the pack than those who are included. Furthermore, the mortality rate for individuals who stay in the pack is 20 percent compared to 55 percent for those who leave the pack, more than double that of their stay-at-home counterparts. Thus, a direct fitness link is apparent between playing fairly and early survival, rendering this behavior extremely susceptible to alteration through natural selection. To my knowledge, this is the first good evidence of a direct fitness effect of violating social norms or standards.

Responses to Distributional Inequity in Nonhuman Primates

In my own work, I have examined inequity aversion in two different primate species, capuchin monkeys (*Cebus apella*) and chimpanzees (*Pan troglodytes*) (Brosnan & de Waal, 2003a; Brosnan, Schiff & de Waal, 2005). These species were chosen because they demonstrate many of the characteristics that seem to be prerequisites for inequity aversion. Both primate species are highly tolerant, so that individuals can expect to maintain possession of items or receive some share of resources, even if they are subordinate. These species are also highly cooperative and, as inequity aversion is proposed to promote human cooperation, it seemed logical to first look at two cooperative species (Boesch, 1994; Brosnan & de Waal, 2002; de Waal & Berger, 2000; de Waal & Brosnan, 2006; de Waal & Davis, 2002; Mendres & de Waal, 2000).

An experiment was intended to elucidate only one aspect of a disadvantageous inequity aversion—whether the primates reacted when another individual received a superior reward—and one aspect of advantageous inequity aversion—whether individuals reacted when they were over-benefited (discussed later). Testing was done on unrelated pairs made up of five adult female brown capuchin monkeys from two social groups and twenty adult chimpanzees from two social groups and two pair-housed groups. Testing was based on an exchange, in which subjects were given a token (a small granite rock for capuchins and a white PVC pipe for chimpanzees) which they had to return to the experimenter to receive a food reward (either a highly preferred grape or a less preferred cucumber). Each test session consisted of a series of fifty exchanges alternating between the partner and the subject such that each individual received twenty-five trials per session and the partner always exchanged immediately prior to the subject.[4]

Each subject underwent four tests. The Equity Test (ET) was a baseline test in which both the subject and the partner exchanged for a low-value reward (cucumber or celery). For the Inequity Test (IT), which determined their response to an unequal reward distribution, the partner initially exchanged a token for a high-valued reward (grape) followed by the subject exchanging a token for a low-value reward. There were also several controls. The Effort Control (EC) Test allowed us to separate the amount of effort from the value of the rewards. In this, the partner was initially handed a high-valued reward without having to exchange any tokens for it (e.g., it was a gift), after which the subject had to exchange a token to receive the low-value reward. More important was the Food Control (FC) Test, which examined how subjects reacted when a higher-valued food was available but no other individual received it. This test separates what might colloquially be called "envy" (wanting what another has) from "greed" (as discussed in the above section). In this test, the higher-valued reward was present but not given to a conspecific.

For each individual, we measured the frequency of refusals to exchange, which was further divided into two categories, not returning the token and refusing the reward. Because subjects were not shown which reward they would receive prior to exchange, both responses represented an unusual reaction to the testing situation. These responses included passive refusals (refusing to return the token or accept the reward) as well as active ones (throwing the token or reward out of the cage). To be conservative, an exchange was only considered a refusal to accept the reward if it never came into the vicinity of the subject's mouth.

Capuchin and Chimpanzee Reactions to Inequity

Compared to their reactions when both received the same reward, capuchin monkeys and chimpanzees were both much less likely to be willing to complete the exchange or accept the reward when their partner received the preferable grape; *Cebus* $F_{3,16} = 25.78$, $p < 0.001$; *Pan* $F_{24,55} = 3.87$, $p < 0.0001$; Brosnan & de Waal, 2003a; Brosnan et al., 2005). Moreover, among capuchin monkeys, if the partner did not have to do the work to receive the better reward but was given it for "free," the subjects were even more likely to quit participating, indicating that they do pay attention to the effort as well as the value of the reward. Notably, the same was not true of chimpanzees, who showed no increased reaction for these no-effort trials.

This can be explained in several ways. First of all, chimpanzees, unlike capuchins, simply may not pay attention to the effort required to obtain a reward. This seems highly unlikely, however, particularly since both species attend to reward differences. More likely is that the amount of effort required to exchange the token was insignificant for the chimpanzees, and so they showed no response. For a capuchin monkey, returning the token typically required both hands and a full-body movement, whereas the chimpanzees could return the token merely by moving a single arm and remaining in their position. Further testing requiring a greater effort on the part of the chimpanzees is required to determine whether chimpanzees attend to effort.

The results of the critical fourth control, where a higher-valued food was visible but no other individual could receive it, indicated that while subjects clearly show some response to the presence of a higher-valued food, their reactions are also driven by social comparison. Chimpanzees, but not capuchins, generally showed a lower level of response in this Effort Control Test than in the Inequity Test, implying that they do notice and react differently when a partner does not receive the better reward. In both species, however, the most striking difference was in the level of reaction over the course of the twenty-five-trial session. Compared to sessions where their partner got the better reward (IT and EC Tests), subjects' reaction to the presence of this high-valued food decreased significantly over the course of the session, indicating that although higher-valued food items are clearly preferred, they do not elicit quite such dismay if no other primate receives them. In tests where their partner received the higher-valued food, the capuchin's frequency of refusal to participate increased over time, whereas the chimpanzee's frequency of refusal to participate remained uniformly high.

Rank had no effect on reactions in either species. We expected to find that more dominant individuals would be much more upset by inequitable treatment than their more subordinate counterparts, because even in these relatively tolerant species the dominants are accustomed to receiving more. One possible explanation is that the subjects perceived that the inequity was brought on not by their partner but by the human experimenter. Supporting this suggestion is that any negative behavioral reactions (e.g., threats) were directed at the experimenter, not at the partner, and, after the experiment, there was no retaliation against the primate who received the preferable reward.

Social Effects on the Inequity Response

The chimpanzees and capuchin monkeys exhibited one large difference (Brosnan, Schiff & de Waal, 2005). Individuals from multiple groups were tested in both species. The capuchins came from two different social groups, but the level of response did not differ between the two groups. Among the chimpanzees, however, one group had been housed together for more than thirty years (i.e., long-term social group), and all subjects but one were born and reared within the group; the exceptional subject was present at the group formation and was the alpha female at the time of testing. The other social group had been put together a mere eight years before the study (Seres, Aureli & de Waal, 2001) (i.e., short-term social group). Thus, no subject had been born in the group, and most individuals had been introduced as adults. These apes showed large differences among the three groups ($F_{2,13} = 4.84, p = 0.0269$). The four pair-housed subjects and six subjects from a social group that had been created 8 years previously (when the individuals were already adults) showed a strong response to inequity, similar to that of the capuchins. However, there was a vastly different response from ten subjects in another social group that had been housed together for more than thirty years and in which all but one of the subjects were born into the group. These individuals who had grown up together showed virtually no response to inequity, with individuals consistently accepting the inequitable distribution. Given that the other groups all reacted to inequity, it seems unlikely that this group does not notice when the situation is to their disadvantage. Instead, they apparently do not react to this negative situation.

More intriguing, these individuals who always complete the exchange interaction also take about half the time to do so than the chimpanzees, who frequently refuse to exchange ($t = 7.361$, df $= 4$, $p = 0.005$). This is true in all cases including the equity test, indicating that the speed of interaction is not related to the distribution of food rewards. From this test, however, we cannot determine causation, yet it is intriguing that those individuals who take more time, and thus may have time to evaluate the situation, are more likely to react to inequity.

This response to inequity also fits well with other data from these groups. Aside from their response to inequity, the long-term social group in our study shows high levels of reciprocity in food sharing and grooming, a behavior presumably associated with equity (de Waal, 1997a), extensive reconciliation after fights (Preuschoft, Wang, Aureli & de Waal, 2002), and a tendency to avoid confrontation (Hare, 2000), whereas those in the short-term group were still working out issues four years after their formation (Seres, Aureli & de Waal, 2001). Although one test is insufficient to understand all the contingencies of how relationships affect responses to inequity, it is noteworthy that individuals who had grown up together and had social interactions implying harmony showed communal orientation to the inequity test whereas those from the less stable situations showed more contingent rules. It is known, moreover, that chimpanzees will alter their behavior depending on the current social situation (Brosnan & de Waal, 2003b), their housing situation (Aureli, 1997; Baker, Seres, Aureli & de Waal, 2000), or their social group (Whiten, Goodall, McGrew, Nishida, Reynolds, Sugiyama, Tutin, Wrangham & Boesch, 1999). Because of the scarcity of chimpanzees, however, most behavioral testing utilizes individuals from only a single social group (or chimpanzees from pair- or single-housed situations).

A note of caution is in order, however: in this study, the chimpanzees' relationship and time of cohabitation was confounded in these two groups. The group that had lived together more than three times as long also showed more stable relationships. Thus, it is unclear which is more important (and, in fact, the two may be interrelated in that social relationships are important but become more stable over a longer period). Colleagues and I are currently investigating the role between certain social behaviors and social relationships within the group with six groups of chimpanzees who have been co-housed for the same period. This will shed more light on the role of social relationships in decisions about social interactions.

These data do fit well, however, with the present understanding of social relationships on behavior in humans. Current theory proposes that individuals in close or positive relationships follow communal rules that do not pay overt attention to fairness, whereas those in less stable relationships utilize contingent rule-based behavior such as equity or inequality (Clark & Grote, 2003; Loewenstein, Thompson & Bazerman, 1989). Such committed relationships are also highly correlated with a willingness to sacrifice (Van Lange et al., 1997). If our long-term group of chimpanzees has similarly close relationships, the inequity presented to them may be largely irrelevant within the context of their relationships. Further study of multiple social groups is certainly warranted, both in primates and other species, so that we may obtain a full picture of animal responses. Obviously, this result has profound implications for free enterprise, which forces individuals to interact with those outside their social group. This may lead to different reactions than they would normally have and should be considered when drawing up policy.

Distributional Inequity and Free Enterprise

This ability to recognize when you are receiving unfair treatment is a response we might also expect in humans. In fact, its presence in two nonhuman primates certainly implies a long evolutionary history, with the responses of the ape and monkey representing stages in the evolution of the complex responses to inequity exhibited by humans. This means that our responses are deeply ingrained and not just the result of social conditioning or our culture.

A striking aspect of the primates' responses is their seeming irrationality. The primates gain nothing by refusing to participate. Their partners continue to be rewarded, and, moreover, refusing the rewards means they end up with both absolutely and relatively less than their potential. Whether the decisions of nonhuman primates, like in humans, are driven by emotion is unknown. However, this irrationality, coupled with behaviors such as throwing tokens during testing (common among both species), seems to implicate some level of emotional involvement.

In humans, it is essential to keep this link in mind. Many responses to free enterprise are probably not rational but are a "gut response" to the situation. So, even if a system will make individuals more successful, both overall and relative to others, if it does not seem fair to the participants, they may reject the system. This violation of rationality goes against many of the tenets we assume about humans, especially in economics, and is vital to remember.

That this result does not vary based on social rank is also striking, as it implies, again, that reactions against distributional inequity are universal and do not depend entirely on circumstances. The one notable exception, however, is the apparent link between social relationships and inequity responses. That chimpanzees who grew up together are far less likely to respond to inequity than those who met as adults is reminiscent of human work, but also of other chimpanzee responses. Recent work with chimpanzees indicates that individuals who have highly tolerant relationships are much more likely to cooperate than those who do not (Melis, Hare & Tomasello, 2006a). Clearly, the relationship between individuals is important. Given the impersonal nature of most economic markets, individuals may have to overcome a fundamental difficulty in order to learn to function in them.

Does the Partner's Behavior Matter?

A recent study examining the role of choice in nonhumans' reactions to inequity provides more tangential support for the idea that social emotions may play a role (Brosnan, Freeman & de Waal, 2006). Here, monkeys were allocated a reward by the experimenter that differed from their partners' reward but was not controlled by their partners. Since the investigator did not decide the distributions, the monkeys were able to decide whether to cooperate and, if so, which monkey received which reward. Moreover, in this situation either monkey could affect the other's outcome by refusing to participate.

This was done using a mutualistic barpull task, which previously had been used extensively to test cooperation and altruism in capuchin monkeys (see chapter 4; also see Brosnan & de Waal, 2002; de Waal & Berger, 2000; de Waal & Davis, 2002; and Mendres & de Waal, 2000). In this study, a tray was weighted so that no capuchin alone could pull it in, but, working together, two individuals could. Rewards for the pullers were set on the tray, one on each side. Previous studies had shown that the monkey would receive the food that is in front of the bar they were pulling (de Waal & Davis, 2002). Thus, we could manipulate rewards so that the monkeys were pulling for the same or different rewards.

In this case, the tray was baited with either two low-value rewards (apples), two high-valued rewards (grapes), or an inequitable distribution of one low- and one high-valued food. Individuals, it turned out, did not vary their pulling based on the distribution of the reward; they were equally likely (or unlikely) to pull in all three conditions. We did notice an interesting discrepancy, however. Some pairs were very good at dividing the reward in the unequal situation. In these cases, both individuals received the preferred grape in about half the trials, indicating some sharing or reciprocity across trials. These pairs, which we dubbed "equitable pairs," were very successful with the barpull, regardless of condition. In the other pairs, one individual tended to dominate the preferable reward in the successful unequal trials—almost 80 percent of the time. In these "inequitable pairs," the success rate was quite low (half that of the equitable pairs) in all conditions.

These findings have two major implications. First, when the individuals interact, it is apparently not the reward distribution but the partner's behavior that is important in determining "fairness." This may have profound implications for group activities, such as cooperative hunts, where many individuals participate but the reward is not readily divisible. Second, this may shed light on the mechanism these monkeys are using to make these decisions. It is enlightening that, in the inequitable pairs, the success rate was low across all conditions, including those where there was no possibility for inequity. This may imply that the decision to cooperate is based on an emotional response to the partner rather than a calculated determination of one's potential rewards. In the latter case, we would expect individuals to participate regardless of their partner's behavior in the two conditions where the rewards are identical, and only refrain when the partner will claim more. Perhaps in neither monkeys nor humans can we expect responses to be maximizing at all times. Preferences for the partner can affect even interactions where an unfair outcome seems impossible.

Distributional Inequity versus Partner's Behavior

Research on capuchin monkeys tells us that we must consider two aspects of the reaction against inequity: distress over inequitable rewards and distress over the partner's inequitable behavior. Each is important in different situations.

First, in situations where individuals feel that neither they (nor another participating individual) have the power to alter the situation, the equitability of the distribution is the most important. This may seem irrelevant since someone can always change the rewards, but further research is needed to determine how distantly removed that individual must feel before their efforts are discounted by participants in the system. In situations such as these, individuals must feel that the distributions are equitable among those receiving the benefit.

Second, in situations where individuals clearly have some control, the distribution of the rewards may be less important than the behavior of the players. In this case, an unequal payoff is acceptable if all the individuals feel that they benefit at some point. Here, the actual inequity of the distribution is overlooked because of the "fair" behavior of the players involved. The flip side has darker implications; individuals may be willing to sacrifice absolute gain in order to punish those perceived as taking advantage of their position or, worse, simply refuse to participate in those cases, even if participation would actually be in their best interest. As a result, in situations where individuals feel cheated or taken advantage of, getting them to participate again voluntarily may be extremely difficult.

This distinction may be vital in setting up a viable economic system. Remember that multiple levels are involved and that both the distributions *and* the behavior of those involved must be perceived as fair by those participating in the system. Regaining people's trust in the system after they have evaluated it as "unfair" may be quite difficult.

Advantageous Inequity and Other-Regarding Preferences

Scholars disagree regarding the strength of the effects of advantageous IA in humans, reflecting, in many respects, the particular discipline looking at the evidence. In economics, for example, evidence indicates that people are more concerned about disadvantageous IA than with advantageous IA. Perhaps this reflects the discipline's focus on the exchange of money rather than psychological mechanisms. Whereas people prefer equity to any sort of inequity, of course, if inequity is present, advantageous inequity is typically preferred to disadvantageous inequity (Loewenstein, Thompson & Bazerman, 1989). Consider the following experiment where subjects had to choose between two distributions, giving cash to oneself or to another; if the subjects were given an option between having complete knowledge of both distributions or only knowing their own payoffs, subjects typically chose to ignore information that could have led to a more equal outcome at some cost to the themselves (Dana, Weber & Kuang, 2003).

In psychology, on the other hand, research indicates that advantageous IA can be just as important as disadvantageous IA. People tend to respond negatively to both. Furthermore, when actual equity cannot be restored, people often respond with psychological leveling mechanisms rather than material compensation. For example, a person may justify why he or she deserved a superior reward, and, in this way, level the playing field (Walster, Walster &

Berscheid, 1978). In experiments where individuals can either allocate hypothetical funds or judge existing hypothetical allocations, allocators typically
made equal allocations, even when they could have used available information to justify benefiting themselves more than others. Although judges
tended to see these hypothetical allocations that overly benefited themselves
as being almost as inequitable as those benefiting their partners, in one situation there was a strong bias favoring the in-group (individuals from their university), indicating that, in certain social situations, people believe that
equity is not the most desirable outcome (Diekmann, Samuels, Ross & Bazerman, 1997). Finally, behavior may be altered simply when one feels that he or
she has overly benefited; for instance, individuals who perceive they were favored as children experience altered relationships with parents as well as siblings (Boll, Ferring & Fillipp, Fillipp, 2005).

Responses to Advantageous Inequity Aversion in Nonhuman Primates

The results discussed here came out of the first study reported on distributional inequity using the exchange paradigm. Although the focus was on the
response of the disadvantaged subject, data were also gathered on the reaction
of the *advantaged* subject, who received a grape while their partner continued
to receive only a cucumber. This made it possible to look at how individuals
who get the grape (the advantaged ones) responded when their partner continued to receive the lower-value cucumber.

   Neither species showed advantageous inequity aversion. Among the capuchins, no partner ever shared a grape with a subject receiving a slice of
cucumber (Brosnan & de Waal, 2003a). In fact, in several situations where
subjects rejected the cucumber slice, partners would finish their grapes and
then reach through the mesh to take the subjects' cucumber slices and eat
that as well! Apparently, to monkeys, cucumbers taste better if you have already had a grape.

   Among the chimpanzees, in the IT and EC tests, where the partner received the superior reward but the subject received the lower-value reward,
not one of the twenty partners ever refused to exchange the token. There
were four instances of subjects potentially sharing the grape out of two thousand interactions where one individual received a grape and the other a cucumber, representing, at best, a sharing rate of 0.002 percent (Brosnan, Schiff
& de Waal, 2005), which is much lower than spontaneous sharing within a
group of chimpanzees (de Waal, 1989a). In several of these cases, however,
the partner seemed to have accidentally dropped the grape (the anatomy of a
chimpanzee's hands makes grasping such a small object difficult) and then allowed the partner to collect it. In no instance was there active sharing, where
the advantaged partner handed the disadvantaged subject their grape. Finally,
the partner's latency to exchange was no different if the subject received a

lesser-valued reward than if the subject received the same reward (comparing partner latency to exchange in the ET vs. IT: $t = -1.44$, $df = 19$, $p = 0.168$), indicating that they did not pause to investigate this inequitable situation.

Although it is noteworthy that the partner never acted to equalize his or her relatively advantaged state, whether this lack of reaction when an individual is overly benefited reflects an absence of societal norms for equity, and hence a lack of fairness in the human sense, is still debatable. This experiment provided no evidence that the lucky individual would make an effort to rectify the situation or display any sort of psychological distress, which might have been suggested through changes in the speed of the interaction.

However, this experiment was not designed to test explicitly for advantageous inequity aversion, and so this may have been missed. A better test would give the advantaged partners some mechanism to rectify the inequity without requiring them to actively sacrifice some of their own gain, and a recent experiment doing just that is discussed in the next section.

Other-Regarding Preferences in Chimpanzees

Silk and colleagues explicitly examined how chimpanzees respond in a situation where they might reward a partner at no cost to themselves (Silk, Brosnan, Vonk, Henrich, Povinelli, Richardson, Lambeth, Mascaro & Schapiro, 2005). Because humans exhibit other-regarding preferences, such as donating blood, tipping at restaurants to which they will never return, and assisting others in catastrophes, for example, after Hurricane Katrina or the 9/11 disaster, the findings here are particularly notable to determine whether this behavior is uniquely human or, as with other behaviors, such as negative reactions to inequity, exists in other species.

In this study, which was conducted at two different sites,[5] donor chimpanzees were given the choice of either giving themselves a food reward or giving both themselves and conspecific partners a food reward. All the rewards were the same, so the donors had no incentive to choose one over the other, except whether or not the option would also reward their partners. All the partners were from the donors' social groups, and so they were familiar. None of the partners were related to the donors and, among the chimps at the MD Anderson site, reciprocity was not possible as the recipient was never given a chance to choose. The critical measure in this test was how often the donor chose the option with a food reward for each chimpanzee when a chimpanzee was present to receive it versus when the donor was alone.

The study revealed that chimpanzees did not choose the option that brought food to both themselves and their partner any more often when the partner was present than absent. This was true at both study sites (at MD Anderson, chimpanzees chose this option 48 percent of the time whether the partner was present or absent; at New Iberia, chimpanzees chose this option 58 percent of the time when the partner was present and 56 percent of the time when the partner was absent). Thus, in this situation, the chimpanzees are not motivated to provide rewards to a familiar conspecific at no cost to themselves. The same results were found with a different group of chimpanzees in a similar study (Jensen, Hare, Call & Tomasello, 2006).

This result matches my previous results on advantageous inequity aversion quite well (see above). In these earlier studies, the chimpanzees who received the higher-valued food reward did not change their behavior when they received this superior reward compared to when they and their partner both received the same lesser-valued reward. Thus, chimpanzees may not be as motivated to assist conspecifics as humans apparently are.

More research is needed to investigate these behaviors in nonhuman species. Much evidence of reciprocity, for instance, involves social behaviors rather than food (see chapter 4, and also Brosnan & de Waal, 2002; de Waal & Brosnan 2006; and Mitani & Watts, 2001), and there is some evidence that other-regarding behavior may exist in non-food situations (Preston & de Waal, 2002; Warneken & Tomasello, 2006). Research is also needed to distinguish between the motivations of the individuals involved and fitness consequences that presumably drove the evolution of the behavior. Even among humans, not all altruistic behavior is necessarily intentional, and this may also be the case in nonhuman subjects (see chapter 4).

Human Correlates

In this area of study, humans seem to display quite different behaviors than nonhuman primates. Humans do react to receiving more—we prefer to receive the same as someone else to either receiving more or less (Loewenstein, Thompson & Bazerman, 1989)—and we will act to rectify inequity through either direct compensation or psychological leveling mechanisms (Walster, Walster & Berscheid, 1978). We also consider that helping others is morally correct and that ignoring or trampling on the needs of others is morally reprehensible. In fact, this is the basis for charitable giving, for example, and may even explain the discrepancies in salaries, for instance, between morally "correct" jobs (i.e., working for a nonprofit organization) and morally "suspect" jobs (i.e., testifying in favor of tobacco companies) (see chapter 3, and Frank, 2004).

Such other-regarding preferences clearly need to be taken into account when designing or implementing large-scale systems. A system can only be harmed if cast in the light of "help yourself and everyone else be damned." This aversion to being (or appearing) completely self-regarding probably explains much of the animosity toward the free enterprise cartoon, which sees us all as completely self-interested beings out only for our own good.

What Does All This Mean for Free Enterprise?

The most important take-home message from the data presented in this chapter is that fairness counts. Both human and nonhuman primates dislike being treated inequitably, whether as a result of unequal distribution or an unfair partner. These responses are not always rational, and may be based on social emotions rather than overt calculation. Moreover, myriad other social factors play into this, making it much more difficult to easily predict people's responses.

In sum, we know from work on nonhuman primates that both the distribution of the rewards *and* the behavior of the partner play a role in decisions. Perhaps, as with the monkeys, in situations where the interactant clearly has some control over the situation, we expect a certain equity in their behavior toward us. Otherwise, we will cease to participate, even in situations where it is not only to our advantage, but also fair, to continue to participate.

On the other hand, in a situation where the interactant clearly cannot control the outcome (as in our study in exchange and distributional inequity), the presence of an inequitable distribution is sufficient to elicit a negative response. It is important to note that this response occurs *even though* the partner had nothing to do with the inequity. This seems irrational but is nonetheless an important constraint to consider.

Moreover, as we all know, it is unlikely that any system will ever be perceived as fair by all of the people all of the time. As discussed earlier, there are myriad social and environmental effects on perceptions of fairness in nonhuman societies. For instance, one's ability to leave the group, one's status, and one's relative level of satiation all have effects on perceptions of fairness over differing timetables. These situations occur in both nonhuman and human primates, and it is essential to consider them. Previous studies have shown that people do not make their decisions without taking social factors into account, nor are people always rational. Assuming that either is the case will jeopardize the success of any venture.

Ultimately, we stand to gain a great deal by studying nonhuman species to help us understand our own reactions. Other species can be studied without the biases and quirks that mask our true intentions and allow us to distinguish behaviors that are genetically predetermined preferences from those that are culturally based. Such distinctions provide a better understanding of the best approach to take when casting a new program. Studying other species, furthermore, can help us elicit the evolutionary trajectory of behaviors, which would then allow us a fuller understanding of them. Humans are capable of great feats of cooperation, and, with some understanding of ourselves gained from our closest relatives, we contribute to the success of an economic system designed to benefit us all.

Notes

1. For a recent review, see *Personality and Social Psychology Review* 7, no 4 (2003), edited by Linda J. Skitka and Faye J. Crosby.

2. Chimpanzees can also be murderous within the group, particularly during fights for alpha status or in female elimination of a rival's offspring.

3. The ideas presented in this subsection draw heavily on a presentation given by Oliver Goodenough at a conference on "Law, Behavior, and the Brain," hosted by the Gruter Institute for Law and Behavior, in May 2004, in Squaw Valley, California. I am grateful to Oliver for bringing this aspect to my attention and for a long discussion that helped me formulate these ideas.

4. For more details on the experimental set-up, see Brosnan, 2006b; Brosnan and de Waal 2003a; Brosnan, Schiff, and de Waal 2005.

5. The study was conducted at the UT/MD Anderson Cancer Center in Bastrop, Texas, and at the New Iberia Primate Research Center in Lafayette, Louisiana.

References

Aureli, F., and de Waal, Frans B. M. (1997). Inhibition of social behavior in chimpanzees under high-density conditions. *American Journal of Primatology, 41*, 213–228.

Baker, K., Seres, M., Aureli, F., & de Waal, F.B.M. (2000). Injury risks among chimpanzees in three housing conditions. *American Journal of Primatology, 51*, 161–175.

Bekoff, M. (2001). Social play behavior: Cooperation, fairness, trust, and the evolution of morality. *Journal of Consciousness Studies, 8*(2), 81–90.

Bekoff, M. (2004). Wild justice, cooperation, and fair play: Minding manners, being nice, and feeling good. In R. Sussman & A. Chapman (Eds.), *The Origins and Nature of Sociality* (pp. 53–79). Chicago: Aldine.

Boehm, C. (1999). *Hierarchy in the forest: The evolution of egalitarian behavior.* Cambridge, Mass.: Harvard University Press.

Boesch, C. (1994). Cooperative hunting in wild chimpanzees. *Animal Behavior, 48*, 653–667.

Boll, T., Ferring, D., & Fillipp, S.-H. (2005). Effects of parental differential treatment on relationship quality with siblings and parents: Justice evaluations as mediators. *Journal of Social Justice, 18*(2), 155–182.

Brosnan, S. F. (2006a). At a crossroads of disciplines. *Social Justice Research, 19*, 218–227.

Brosnan, S. F. (2006b). Nonhuman species' reactions to inequity and their implications for fairness. *Social Justice Research, 19*, 153–185.

Brosnan, S. F., & de Waal, F.B.M. (2002). A proximate perspective on reciprocal altruism. *Human Nature, 13*(1), 129–152.

Brosnan, S. F., & de Waal, F.B.M. (2003a). Monkeys reject unequal pay. *Nature, 425*, 297–299.

Brosnan, S. F., & de Waal, F.B.M. (2003b). Regulation of vocal output by chimpanzees finding food in the presence or absence of an audience. *Evolution of Communication, 4*(2), 211–224.

Brosnan, S. F., & de Waal, F.B.M. (2004a). A concept of value during experimental exchange in brown capuchin monkeys. *Folia primatologica, 75*, 317–330.

Brosnan, S. F., & de Waal, F.B.M. (2004b). Reply to Henrich and Wynne. *Nature, 428,* 140.

Brosnan, S. F., & de Waal, F.B.M. (2004c). Socially learned preferences for differentially rewarded tokens in the brown capuchin monkey, *Cebus apella. Journal of Comparative Psychology, 118*(2), 133–139.

Brosnan, S. F., Freeman, C., & de Waal, F.B.M. (2006). Partner's behavior, not reward distribution, determines success in an unequal cooperative task in capuchin monkeys. *American Journal of Primatology, 68,* 713–724.

Brosnan, S. F., Schiff, H. C., & de Waal, F.B.M. (2005). Tolerance for inequity may increase with social closeness in chimpanzees. *Proceedings of the Royal Society of London,* Series B, *1560,* 253–258.

Caine, N. G., Addington, R. L., & Windfelder, T. L. (1995). Factors affecting the rates of food calls given by red-bellied tamarins. *Animal Behaviour, 50,* 53–60.

Camerer, C. (2003). *Behavioral Game Theory: Experiments in Strategic Interaction.* Princeton, N.J.: Princeton University Press.

Chapman, C. A., & Lefebvre, L. (1990). Manipulating foraging group size: Spider monkey food calls at fruiting trees. *Animal Behavior, 39,* 891–896.

Cheney, D. L., & Seyfarth, R. M. (1990). *How Monkeys See the World: Inside the Mind of Another Species.* Chicago: University of Chicago Press.

Clark, M. S., & Grote, N. K. (2003). Close relationships. In T. Millon & M. J. Lerner (Eds.), *Handbook of Psychology: Personality and Social Psychology* (Vol. 5, pp. 447–461). New York: Wiley.

Custance, D., Whiten, A., and Fredman, T. (1999). Social learning of an artificial fruit task in capuchin monkeys (*Cebus apella*). *Journal of Comparative Psychology, 113*(1), 13–23.

Dana, J. D., Weber, R. A., & Kuang, J. (2003). Exploiting moral wriggle room: Behavior inconsistent with a preference for fair outcomes. *Carnegie Mellon Behavioral Decision Research Working Paper No. 349,* http://ssrn.com/abstract=400900.

de Waal, F.B.M. (1989a). Food sharing and reciprocal obligations among chimpanzees. *Journal of Human Evolution, 18,* 433–459.

de Waal, F.B.M. (1989b). *Peacemaking among Primates.* Cambridge, Mass.: Harvard University Press.

de Waal, F.B.M. (1991a). The chimpanzee's sense of social regularity and its relation to the human sense of justice. *American Behavioral Scientist, 34*(3), 335–349.

de Waal, F.B.M. (1991b). Rank distance as a central feature of rhesus monkey social organization: A sociometric analysis. *Animal Behavior, 41,* 383–395.

de Waal, F.B.M. (1996). *Good Natured: The Origins of Right and Wrong in Humans and Other Animals.* Cambridge, Mass.: Harvard University Press.

de Waal, F.B.M. (1997a). The chimpanzee's service economy: Food for grooming. *Evolution and Human Behavior, 18,* 375–386.

de Waal, F.B.M. (1997b). Food transfers through mesh in brown capuchins. *Journal of Comparative Psychology, 111*(4), 370–378.

de Waal, F.B.M., & Berger, M. L. (2000). Payment for labour in monkeys. *Nature, 404,* 563.

de Waal, F.B.M., & Brosnan, S. F. (2006). Simple and complex reciprocity in primates. In P. Kapeller & C. P. van Schaik (Eds.), *Cooperation in Primates and Humans: Evolution and Mechanisms.* Berlin: Springer.

de Waal, F.B.M., & Davis, J. M. (2002). Capuchin cognitive ecology: Cooperation based on projected returns. *Neuropsychologia, 1492,* 1–8.

di Bitetti, M. S. (1997). Evidence for an important social role of allogrooming in a platyrrhine primate. *Animal Behaviour, 54,* 199–211.

Diekmann, K. A., Samuels, S. M., Ross, L., & Bazerman, M. H. (1997). Self-interest and fairness in problems of resource allocation: Allocators versus recipients. *Journal of Personality and Social Psychology, 72*(5), 1061–1074.

Elgar, M. A. (1986). House sparrows establish foraging flocks by giving chirrup calls if the resources are divisible. *Animal Behaviour, 34,* 169–174.

Evans, C. S., & Marler, P. (1994). Food calling and audience effects in male chickens, *Gallus gallus*: Their relationships to food availability, courtship, and social facilitation. *Animal Behaviour, 47,* 1159–1170.

Fehr, E., & Gachter, S. (2002). Altruistic punishment in humans. *Nature, 415,* 137–140.

Fehr, E., & Rockenbach, B. (2003). Detrimental effects of sanctions on human altruism. *Nature, 422,* 137–140.

Fehr, E., & Schmidt, K. M. (1999). A theory of fairness, competition, and cooperation. *Quarterly Journal of Economics, 114,* 817–868.

Frank, R. H. (1988). *Passions within Reason: The Strategic Role of the Emotions.* New York: Norton.

Frank, R. H. (2001). Cooperation through emotional commitment. In R. M. Nesse (Ed.), *Evolution and the Capacity for Commitment* (pp. 57–76). New York: Russell Sage Foundation.

Frank, R. H. (2004). *What Price the Moral High Ground? Ethical Dilemmas in Competitive Environments.* Princeton, N.J.: Princeton University Press.

Gouzoules, S., Gouzoules, H., & Marler, P. (1984). Rhesus monkey (*Macaca mulatta*) screams: Representational signalling in the recruitment of agonistic aid. *Animal Behaviour, 32,* 182–193.

Hare, B., Call, J., Agnetta, B., & Tomasello, M. (2000). Chimpanzees know what conspecifics do and do not see. *Animal Behaviour, 59*(4), 771–785.

Hatfield, E., Walster, G. W., & Berscheid, E. (1978). *Equity: Theory and Research.* Boston: Allyn and Bacon.

Hauser, M. D., Teixidor, P., Fields, L., & Flaherty, R. (1993). Food-elicited calls in chimpanzees: Effects of food quantity and divisibility. *Animal Behavior, 45,* 817–819.

Henrich, J., Boyd, R., Bowles, S., Camerer, C., Fehr, E., Gintis, H., & McElreath, R. (2001). In search of *Homo economicus*: Behavioral experiments in 15 small-scale societies. *American Economic Review, 91,* 73–78.

Jensen, K., Hare, B., Call, J., & Tomasello, M. (2006). Are chimpanzees spiteful or altruistic when sharing food? *Proc. R. Soc. Lond. B, 273,* 1013–1021.

Kahneman, D., Knetsch, J. L., & Thaler, R. (1986). Fairness as a constraint on profit seeking: Entitlements in the market. *American Economic Review, 76,* 728–741.

Kelley, H. H. (1979). *Personal Relationships: Their Structures and Processes.* Hillsdale, N.J.: Erlbaum.

Kokko, H., & Johnstone, R. A. (1999). Social queuing in animal societies: A dynamic model of reproductive skew. *Proc. R. Soc. Lond. B, 266,* 571–578.

Krebs, J. R., & Davies, N. B. (1993). *An Introduction to Behavioral Ecology.* Oxford: Blackwell.

Lind, E. A., & Earley, P. C. (1992). Procedural justice and culture. *International Journal of Psychology, 27*(2), 227–242.

Loewenstein, G. F., Thompson, L., & Bazerman, M. H. (1989). Social utility and decision making in interpersonal contexts. *Journal of Personality and Social Psychology, 57*(3), 426–441.

Melis, A. P., Hare, B., & Tomasello, M. (2006a). Engineering cooperation in chimpanzees: Tolerance constraints on cooperation. *Animal Behavior, 72,* 275–286.

Melis, A. P., Hare, B., & Tomasello, M. (2006b). Chimpanzees recruit the best collaborators. *Science, 311,* 1297–1300.

Mendres, K. A., & de Waal, F.B.M. (2000). Capuchins do cooperate: The advantage of an intuitive task. *Animal Behaviour, 60*(4), 523–529.

Mitani, J. C. (2006). Reciprocal exchange in chimpanzees and other primates. In P. Kapeller & C. P. van Schaik (Eds.), *Cooperation in Primates and Humans: Evolution and Mechanisms.* Berlin: Springer.

Mitani, J. C., & Watts, D. P. (2001). Why do chimpanzees hunt and share meat? *Animal Behaviour, 61*(5), 915–924.

Perry, S., Barrett, H. C., & Manson, J. H. (2004). White-faced capuchin monkeys show triadic awareness in their choice of allies. *Animal Behavior, 67,* 165–170.

Preston, S. D., & de Waal, F.B.M. (2002). Empathy: Its ultimate and proximate bases. *Behavioral and Brain Sciences, 25,* 1–72.

Preuschoft, S., & van Hooff, J.A.R.A.M. (1995). Homologizing primate facial displays: A critical review of methods. *Folia primatologica, 65*(3), 121–137.

Preuschoft, S., Wang, X., Aureli, F., & de Waal, F.B.M. (2002). Reconciliation in captive chimpanzees: A reevaluation with controlled methods. *International Journal of Primatology, 23,* 29–50.

Richerson, P. J., & Boyd, R. (2005). *Not by Genes Alone: How Culture Transformed Human Evolution.* Chicago: University of Chicago Press.

Sanfey, A. G., Rilling, J. K., Aronson, J. A., Nystrom, L. E., & Cohen, J. D. (2003). The neural basis of economic decision-making in the Ultimatum game. *Science, 300,* 1755–1758.

Seres, M., Aureli, F., & de Waal, F.B.M. (2001). Successful formation of a large chimpanzee group out of two preexisting subgroups. *Zoo Biology, 20,* 501–515.

Seyfarth, R. M., & Cheney, D. L. (1988). Empirical tests of reciprocity theory: Problems in assessment. *Ethology and Sociobiology, 9,* 181–187.

Silk, J. B., Brosnan, S. F., Vonk, J., Henrich, J., Povinelli, D. J., Richardson, A. S., Lambeth, S. P., Mascaro, J., & Schapiro, S. (2005). Chimpanzees are indifferent to the welfare of unrelated group members. *Nature, 437,* 1357–1359.

van Krunkelsven, E., Dupain, J., van Elsacker, L., & Verheyen, R. F. (1996). Food calling by captive bonobos (*Pan paniscus*): An experiment. *International Journal of Primatology, 17*(2), 207–217.

Van Lange, P. A. M., Drigotas, S. M., Rusbult, C. E., Arriaga, X. B., Witcher, B. S., & Cox, C. L. (1997). Willingness to sacrifice in close relationships. *Journal of Personality and Social Psychology, 72*(6), 1373–1395.

Van Schaik, C. P. (1989). The ecology of social relationships amongst female primates. In A. R. F. V. Standen (Ed.), *Comparative Socioecology: The Behavioral Ecology of Humans and other Mammals* (pp. 195–218). Oxford: Blackwell.

Vehrencamp, S. L. (1983). A model for the evolution of despotic versus egalitarian societies. *Animal Behavior, 31*(3), 667–682.

Walster, H. E., Walster, G. W., & Berscheid, E. (1978). *Equity: Theory and Research*. Boston: Allyn and Bacon.

Warneken, F., & Tomasello, M. (2006). Altruistic helping in human infants and young chimpanzees. *Science, 311*, 1301–1303.

Whiten, A. (1998). Imitation of the sequential structure of actions by chimpanzees (*Pan troglodytes*). *Journal of Comparative Psychology, 112*(3), 270–281.

Whiten, A., Custance, D. M., Gomez, J.-C., Teixidor, P., & Bard, K. A. (1996). Imitative learning of artificial fruit processing in children (*Homo sapiens*) and chimpanzees (*Pan troglodytes*). *Journal of Comparative Psychology, 110*(1), 3–14.

Whiten, A., Goodall, J., McGrew, W. C., Nishida, T., Reynolds, V., Sugiyama, Y., Tutin, C. E. G., Wrangham, R. W., & Boesch C. (1999). Cultures in chimpanzees. *Nature, 399*, 682–685.

Wrangham, R., & Peterson, D. (1996). *Demonic Males*. Boston: Houghton Mifflin.

Zizzo, D. J., & Oswald, A. (2001). Are people willing to pay to reduce others' incomes? *Annales d'Economie et de Statistique, 63–64*, 39–62.

PART III

The Evolution of Values and Society

Six

The Evolution of Free Enterprise Values

Peter J. Richerson and Robert Boyd

The free enterprise system that dominates the world economy today has deep evolutionary roots even though it has a shallow history. As Darwin (1874) argued cogently in the *Descent of Man*, long before geneticists showed that humans have unusually little genetic diversity, all human populations have essentially the same "mental and moral faculties."

> Although the existing races differ in many respects, as in color, hair, shape of the skull, proportions of the body, etc., yet, if their whole structure be taken into consideration, they are found to resemble each other closely on a multitude of points. Many of these are so unimportant or of so singular a nature that it is extremely improbable that they should have been independently acquired by aboriginally distinct species or races. The same remark holds good with equal or greater force with respect to the numerous points of mental similarity between the most distinct races of man. The American aborigines, Negroes, and Europeans are as different from each other in mind as any three races that can be named; yet I was constantly struck, while living with the Fuegians on board the "Beagle," with the many little traits of character showing how similar their minds were to ours; and so it was with a full-blooded Negro with whom I happened once to be intimate. (237)

He gave the emotion of sympathy a foundational role in generating our ethical systems. Adam Smith (1790) gave sympathy the same key role in his *Theory of Moral Sentiments*. Paul Zak's experiments show that we are some ways toward understanding the neurobiology of sympathy (see chapter 12). Darwin suggested that sympathy evolved in "primeval times" among the tribal ancestors of all living humans and proposed what we nowadays call a group selection hypothesis to explain how the moral faculties arose:

> It must not be forgotten that although a high standard of morality gives but a slight or no advantage to each individual man and his children over other men of the same tribe, yet that an increase in the number of well-endowed men and an advancement in the standard of morality will certainly give an immense advantage to one tribe over another. A tribe including many members who, from possessing in a high degree the spirit of patriotism, fidelity, obedience, courage, and sympathy, were always ready to aid one another, and to sacrifice themselves for the common good, would be victorious over most other tribes; and this would be natural selection. (178–179)[1]

This left Darwin with the problem of accounting for the rise of civilized nations from tribal ones. Since tribal societies still exist even today in many parts of the world, how does one explain why some societies have "advanced"[2] and others not? In the *Descent of Man* he speculates about the role of a temperate climate and the need for civilization to be based upon agricultural production, but concludes: "the problem, however, of the first advance of savages toward civilization is at present much too difficult to be solved" (179–180).

By now, anthropologists, archaeologists, and historians have given us a very rich picture of how and where complex societies have evolved from simpler ones (McNeill 1963; Diamond 1997; Klein 1999; Johnson and Earle 2000). Note that the civilization-barbarism-savagery terminology has dropped out of technical usage because of the invidious comparisons it implies; after the horrors that twentieth-century "civilizations" inflicted upon one another, the ethical superiority of the civilized nations came in for even more skepticism than Darwin expressed. Much of the puzzle remains, however. Anatomically, modern people arose some two hundred thousand years ago in Africa, and their behavior had a distinctly modern cast by, at the latest, fifty thousand years ago. Agriculture first arose in the Levant more than eleven thousand years ago. The first societies with a complex division of labor arose in Mesopotamia about five thousand years ago. Modern free enterprise economic systems began to emerge in Adam Smith's time in Britain and the U.S. and were still few when Darwin wrote. Yet, today, the democratic, free enterprise society, usually with a more or less generous welfare safety net, has outcompeted all challengers and is spreading rapidly across the globe.

The mystery is, why now and why so rapidly? The puzzle is made more acute by the fact that even the simplest human societies that anthropologists have had a chance to study are characterized by high levels of individual autonomy, a simple division of labor based on age and sex, and respectable amounts of intra- and inter-group trade. Modern social systems would seem to be a straightforward extrapolation from abilities and proclivities possessed by humans living tens if not hundreds of thousands of years ago. Indeed, in some key respects, modern free enterprise societies recapture some of the desirable features of tribal societies. Our ancestral tribes managed to combine a high degree of individual autonomy with considerable cooperation, including simple institutions designed to mitigate risk (Boehm 1999). Trade within and between tribes was often well developed, including the use of standardized media of exchange, such as shell bead money in California (Heizer 1978). The state-level societies that arose in the wake of agriculture were typically dominated by small elites that restricted individual autonomy and usually provided little social insurance for common people. Max Weber (1930) famously argued that the spirit of capitalism derived from events that combined the Calvinist concept of everyday business as a religious calling with the secular rationalism of Western philosophy. In figures like Benjamin Franklin, the asceticism derived from Calvinism and the rationality of the philosopher produced men devoted to growing their businesses through investment and technical acumen, and creating a political space in which business could thrive. The political values fostered by pragmatic free enterprise enthusiasts like Franklin, and enshrined in documents like the U.S. Declaration of Independence, led gradually, but seemingly inexorably, to the spread of political rights and ultimately to universal suffrage. Universal suffrage, in turn, favored the expansion of social insurance schemes and universal state-subsidized education (Lindert 2004). Free enterprise societies have managed to combine a considerable degree of individual autonomy (liberty) with a productive economy (wealth) and low risks of want, disease, and death (welfare). Why didn't Franklin-like innovators launch human societies on the path to free enterprise thousands or even tens thousands of years ago?

Evidently, the wheels of cultural evolution roll on the time scale of millennia, even though, when we look closely at any one society over short periods of time, change is often readily perceptible. Biological evolution, incidentally, exhibits the same pattern. Evolution from generation to generation in the wild is often fast enough to be measured, yet average rates of evolution as measured over long time spans are very slow (Gingerich 1983), so slow that they would be immeasurable in any scientist's lifetime. Most often, both cultural and genetic evolution seem to noodle about aimlessly in response to local events and forces, either moving only very gradually in any particular direction or making sudden, rare excursions (both patterns appear in the fossil record (Carroll 1997).

This chapter reviews what we have learned since Darwin about the processes of cultural evolution, particularly the evolution of social institutions, the rules of our social life. The scientific problems posed by cultural evolution are usefully divided into two sorts, *microevolutionary* and *macroevolutionary*. Microevolutionary processes are those that occur over time periods short enough that we can study them by direct observation and experiment. Macroevolutionary problems are the long-term trends and big events we observe indirectly by historical, archaeological, and paleontological reconstruction. Microevolutionary processes we assume to be the basic engine of macroevolution, borrowing the geologists' doctrine of uniformitarianism. The mathematical theory of evolution is built to allow us to extrapolate from the generation-to-generation microevolutionary time scale out to the time scales of long-term trends and major events. We want to understand events like the origin of agriculture and the development of free enterprise economies in terms of the day-to-day decisions imitators and teachers make about what cultural variants to adopt and the consequences of those decisions.

The key to understanding cultural evolution is the idea that it is a population-level phenomenon. In other words, values[3] are the individual-level motor of social institutions, but we cannot understand the evolution of institutions only in terms of the individual-level processes. Put another way, even if we have perfect information about the innate aspects of human behavior, we can only go part way toward understanding values and institutions. Cultural history matters over medium time scales, and genetic history over somewhat longer time scales. A complex concatenation of evolutionary processes affects the evolution of socially transmitted values over many generations, indeed as noted in the previous paragraph, over millennia. The origins of the institutions of free enterprise lie in the cumulative results of decisions and consequences of decisions going back many millennia, although a number of the most distinctive departures from traditional agrarian societies evolved during the last half-millennium. We argue in this chapter that the social instincts we inherit from our tribal past were shaped by gene-culture coevolution in which group selection on cultural variation played the leading role. This process seems to have been especially active between about 250,000 years ago and 50,000 years ago (McBrearty and Brooks 2000). *The macroevolutionary puzzle, therefore, is why, given the current adaptive success of free enterprise societies and the fact that humans were apparently completely capable 50,000 years ago, did cultural evolution proceed so slowly over the past fifty millennia?* We are still near the beginning of this explanatory endeavor, but we know roughly what dots we need to connect and have the means to draw some interesting hypothetical dashed lines between them. Douglass North (1994) remarked not so long ago that neoclassical economic theory needs to be supplemented by theory that has a richer appreciation of individual psychology and cumulative learning of societies in order to understand the history of economies. Cultural evolutionary theory is a key element of this enlarged theory (Bowles 2003).

The Theory of Cultural Evolution

Organic evolutionists began to use mathematical models to investigate the properties of evolution in the first quarter of the twentieth century. The aim of the effort was to take the micro-scale properties of individuals and genes, scale them up to a population of individuals, and deduce the long-term evolutionary consequences of the assumed micro-level processes. Empiricists have a handle on both the micro-scale processes and the long-term results but not on what happens over many generations in between. Moreover, human intuition is not so good at envisioning the behavior of populations over long spans of time. Hence, mathematics proved an invaluable aid.

Beginning with the pioneering work of Lucca Cavalli-Sforza and Marc Feldman (1981) in the early 1970s, these methods were adapted to study cultural evolution. The problem is similar to organic evolution. People acquire information from others by learning and teaching. Cultural transmission is imperfect, so the transmission is not always exact. People invent new cultural variants, making culture a system for the inheritance of acquired variation.

People also pick and choose the cultural variants they adopt and use, processes that are not possible in the genetic system (although, in the case of sexual selection, individuals may choose mates with the objective of getting good genes for their offspring). Social scientists know a fair amount about such things, enough to build reasonable mathematical representations of the micro-level processes of cultural evolution. The theory is of the form

$$p_{t+1} = p_t + effects \ of \ forces$$

where p measures something interesting about the culture of a population, for example, the fraction of employees who are earnest workers. Teaching and imitation, all else equal, tend to replicate culture. The fraction of workers in a culture who are earnest tends to remain similar from one generation to the next. Earnest workers model earnest behavior for others to imitate and try to teach earnestness to new employees—likewise slackers. Typically, several processes we call *forces* will act simultaneously to change culture over time. For example, management may find it difficult to discover and sanction slacking. Earnest workers may experiment with slacking and find that there are seldom any adverse consequences. Hence, some earnest employees may become slackers. New employees may observe that some people slack and some work hard. They may tend to prefer the easier path. At the same time, firms with a high frequency of slackers will tend to fail whereas those with many earnest workers may prosper. Prosperous firms will have the opportunity to socialize many more new workers than those that fail prematurely. The overall quality of the economy's workforce in the long run will be determined by the balance of forces favoring slacking versus those favoring earnestness. Theorists are interested in the abstract properties of such evolutionary models. Empiricists are interested in finding the models that best describe actual evolving systems. Real-world practitioners are interested in predicting the outcomes of policies that might improve or harm the quality of a firm's or an economy's workforce.

Our own interest (Boyd and Richerson 1985; Richerson and Boyd 2005) has been to use such models to answer a series of substantive questions. We have been interested in the adaptive costs and benefits of culture, rates of different kinds of cultural evolution, the evolution of symbolic systems, and the role of culture in the evolution of cooperation. Most relevant to this chapter, we have tried to ferret out the factors that retard the rate of cultural evolution and thus explain the fifty-thousand-year gap between the last major genetic changes in our lineage and our current extraordinarily successful societies (Boyd and Richerson 1992).

Correcting the Oversimplifications of Selfish Rationality

Challenging the emphasis of selfish rationality in conventional economic theory is the main theme of this book. Neither the neo-classical assumption of selfishness nor the assumption of rationality is an innocent simplification. Cultural evolutionary models are based upon a model of a human decision maker that exercises effort to select cultural variants in an attempt to increase his or her genetic fitness. If we make this decision maker omnisciently rational and selfish, we end up mating neoclassical economics to basic evolutionary theory, giving an in-principle complete theory of human behavior, as Jack Hirshleifer (1977) and Paul Samuelson (1985) have noted. However, if we introduce the simple realistic consideration that rationality is imperfect because information is costly to acquire (Simon 1959), we immediately spawn a great deal of *evolutionary* complexity. In organisms with little or no culture, variable amounts of investment in big brains and other devices for individual-level behavioral and anatomical flexibility evolve. But such devices only function at the margin; most organic adaptations are gene-based. Thus, we have millions of species with highly evolved, specialized innate abilities rather than a few species that flexibly occupy a wide range of ecological niches. Humans are the biggest exception to this rule. Even as hunter-gatherers we spread to almost all the presently habitable bits of the world, including harsh environments like the high arctic—quite an accomplishment for a tropical ape! In the course of doing so, we have made our living in an almost limitless number of ways. But not just any human can live in the high arctic. You have to have the requisite technology (kayaks, parkas, snow houses) and social organization (institutions underpinning cooperative hunting, emergency assistance, and inter-group trade) that you learn from your parents and others. The latest version of the High Arctic cultural adaptation (the Thule culture and the ethnographic Eskimo) spread from Alaska to Greenland only after AD 600 or so (Dumond 1984).

The Adaptive Advantage of Culture

The origins of human culture are a macroevolutionary puzzle of the first magnitude. Most "killer adaptations" evolved long ago, multiple times, and are retained by most of the lineages in which they evolved. Take eyes. Eyes are ancient structures, and they come in a wide variety of forms (Nilsson 1989). They often perform at close to optically perfect limits. Both compound eyes and camera-style eyes, the most sophisticated forms, evolved in more than one lineage. Paleozoic animals evolved all the basic types of eyes. Most animals living in lighted environments have eyes. Humans' advanced abilities to teach and imitate, and hence create cumulative culture, would also seem to be a killer adaptation on the evidence that it is our basis for becoming the earth's dominant organism. Yet, this fancy adaptation only evolved in the middle and late Pleistocene rather than in the Paleozoic and is largely, if not entirely, restricted to our species.

A considerable amount of evidence supports the hypothesis that recent increases in environmental variability are behind the evolution of culture in humans (Richerson et al. 2005). Theoretical models show that a costly system for social transmission of information out-competes the familiar pattern of genetic inheritance plus individual learning in highly variable environments. The earth's climate has been cooling and drying over the last sixty million years or so, and the Pleistocene Era has seen the cyclical growth and recession of ice sheets. The cyclical component of this variation is probably at too long a time scale (twenty thousand years and longer) to directly favor culture (but see Potts 1996). During at least the last couple of glacial cycles, glacial climates were extremely variable on millennial and submillennial time scales. Variation on these time scales is exactly what theoretical models suggest should favor culture. These time scales are too rapid for extensive adjustment by innate mechanisms, but cultural evolution, by adding decision making to the effects of natural selection, evolves faster than genes and can track millennial and submillennial scale variation. Low-cost decision rules that people are known to deploy (Gigerenzer and Goldstein 1996), like "satisficing," going along with the majority and copying successful individuals, are quite effective at moving culture right along on such time scales. Humans are not the only animals whose brains have gotten larger. Mammalian brains in many lineages have gotten larger as the climate has deteriorated (Jerison 1973), and the rate of evolution in many of them shot up during the Pleistocene. The amount of both social and individual learning is correlated with brain size in nonhuman primates (Reader and Laland 2002) and probably in many other lineages including birds, the living branch of the notoriously small-brained dinosaur lineage.

Even the most complex examples of nonhuman cultural traditions are quite simple compared to human culture. In particular, nonhuman culture shows little or no evidence for the cumulative evolution of complex artifacts and social systems in which many successive innovations have led to exceedingly complex, often highly adaptive, traditions (Boyd and Richerson 1996). Animals often make simple tools (e.g., Caledonian crows [Hunt 2000]), but even the most complex animal tools fall short of a stone-tipped spear, much less an Arctic hunter's kayak (Arima 1987). Even chimpanzees, a species quite good at social learning compared to most animals, imitate much less accurately than human children (Tomasello 1996). We are probably the only species with advanced culture because we got there first with the most. Bipedal locomotion evolved in our lineage long before our brain size increased relative to other apes and long before any evidence of sophisticated toolmaking. When toolmaking did come along, our hands were relatively free to evolve into technology-making organs. Apes are rather social as mammals go, and we, too, were pre-adapted in this way to evolve more complex social systems. Once humans acquired the capacity for complex culture, we radiated rapidly and spread to distant parts of the earth. Any creature tending to repeat this trajectory faces stiff competition from humans, who, we might imagine, occupy most of the ecological niches where advanced culture would be a large advantage.

Cultural Evolution and the Evolution of Tribal Social Instincts

The discoveries of the cultural evolutionists regarding human social systems have two important legs. First, we now have a much deeper insight into human nature than was possible in the absence of an understanding of cultural evolution. Humans have evolved a social psychology that mixes a strong element of cooperative dispositions, deriving from group selection on cultural variation, with an equally strong selfish element deriving from more ancient primate dispositions (Richerson and Boyd 1998). Evolutionary biologists have long argued that group selection of the sort Darwin envisioned for humans is not a very important force in nature (Williams 1966).[4] The problem in theory, and probably in practice, is that, short of the strong reproductive isolation of species, genetic variation between competing populations is hard to maintain. Migration is a powerful homogenizing force. Theoretical models suggest that cultural variation is much more susceptible to group selection than genetic variation is (Henrich 2004). For example, if people use conformity to the majority to bias their acquisition of culture, rare migrants will tend to be ignored, and the homogenizing effects of migration will be enhanced. Several other mechanisms have a similar effect. For instance, human groups are often symbolically marked by linguistic and other differences, and people tend to prefer to imitate similar others (McElreath et al. 2003). The upshot is that Darwin's picture of group selection among tribes is plausible for humans if we assume that the variation being selected is cultural, not genetic. At the same time, human genetic reproduction is mainly the business of individuals and families, so selection on genes still favors individual selfishness and kin-based nepotistic cooperation. Your family is liable to take more interest in your genetic reproduction than even your close friends do (Newson et al. 2005). This sets up a coevolutionary antagonism between genes and culture. Humans are much like a domestic animal—docile and inclined to conform to social conventions (Simon 1990). The flush of oxytocin-induced empathetic feelings that result from acts of trust provide a proximal mechanism for this domesticity (see chapter 12). Human social behavior differs from that of other primates (see chapter 5); a natural hypothesis is that the evolution of human sympathy was driven by increasing oxytocin production in response to empathetic acts of others and to our own empathetic acts. But, like domestic animals, we retain a healthy interest in our own comfort and opportunities to reproduce. We call this the "tribal social instincts hypothesis" in honor of Darwin's original formulation.

Thus, we are imperfect and often reluctant, though often very effective, cooperators. We are contingent cooperators. Few will continue cooperating when others do not. The effectiveness of our cooperation is *not* just a product of our social psychology; rather, our social psychology creates evolutionary forces that build *cultural systems of morality and convention* that, in turn, make possible sophisticated systems of cooperation such as businesses. Individuals are not really that rational. We depend on cultural evolution to generate values and social institutions over many generations that are more group-functional than individuals can hope to devise based directly on our social instincts.

Several contributions to this volume point to the importance of culturally evolved institutions in regulating human social life. Kimbrough et al.'s experiment, discussed in chapter 13, illustrates how much trouble individuals have in discovering efficient solutions to complex cooperative problems. Schwab and Ostrom's contribution, chapter 10, argues that humans use quite complex multidimensional institutions to manage common property resources. O'Hara argues, in chapter 9, that contract law provides an important but blunt tool for encouraging trustworthiness among business partners. We also understand something of how institutions recruit individual-level mechanisms like hormone release to do their work (Nisbett and Cohen 1996). The theory of gene-culture coevolution adds an account of where institutions come from and how a biology that can respond to institutions could have evolved.

Tribal Human Nature, Work-arounds, and Organizational Management

The understanding that human nature is fundamentally tribal is, we believe, an important insight, but it leaves unexplained the rise of supra-tribal social systems beginning about five thousand years ago. The organizations of complex societies are made possible, but not easy, by a tribal human nature that is conditionally cooperative. Given the right culturally transmitted rules and enough of our peers willing to honor them, most of us are also willing to honor them. The organizations of complex societies succeed when they manage to recruit the group favoring the tribal impulses that most of us have, but they also have to work against the fact that such organizations face a more constrained job than tribes do. Tribes worked only for their members' benefit, whereas businesses and other organizations within complex societies have a broad array of "stakeholders" to satisfy—customers, suppliers, owners, lenders, neighbors, and regulators. The great vice of tribes was inter-tribal anarchy. The small compass of tribal patriotism frequently led, particularly when population densities increased after the evolution of agricultural subsistence, to chronic military insecurity. Complex societies use grants of power and other devices as "work-arounds" to control inter-"tribal" anarchy in the interests of domestic tranquility and an efficient division of labor. A collection of tribes that owed obedience to a paramount chief who settled disputes among them could thus mitigate inter-tribal anarchy. A professional priestly establishment might invent dogma and design rituals by which a tribe might maintain a sense of common culture and social solidarity as it grew far larger than the egalitarian tribe of the ancestral hunters and gatherers. But such workarounds often lead to management problems, such as abuses of power for selfish ends

(Richerson and Boyd 1999). Successful management is thus substantially the art of using work-arounds to tap the tribal social instincts while at the same time minimizing their inherent vices.

The Moral Hidden Hand and the Functioning of Organizations

Conditional cooperation and the existence of social rules, to which we more or less readily conform, constitute a moral hidden hand. One can depend on most people, most of the time, to be spontaneously helpful and honest—even to strangers. Just as no corps of central planners needs to work out the details of how a market economy is to operate, no central authority needs to comprehensively supervise the day-to-day interactions of a human community to ensure that we all take account of one another's needs and behave decently and honestly. Democracies work, for example, because most voters vote even though rational selfish individuals would not bother. The evidence also suggests that people vote their principles rather than their pocketbooks (Sears and Funk 1991).

Thus, the moral hidden hand deriving from our tribal social instincts is one foundation upon which our immensely successful free enterprise systems rest. This system has a claim to have better developed work-arounds than competing styles for organizing complex societies. It refines work-arounds to reduce the conflict between the tribal social instincts and the functional demands of large-scale organizations. The vote in democratic polities and consumer and producer sovereignty in markets restore some of the individual and family autonomy that characterized egalitarian tribes. The rule of law subjects the exercise of power to constraints that favor its pro-social exercise. Common miscreants are punished, but the powerful are also constrained to act out roles that limit abuses of power. The trick is to get the balance right. In particular, attempts to control individual behavior by the use of power to set up incentives designed to appeal to selfish motives risk "crowding out" (Frey and Jegen 2001). When power holders use individualized incentives, such as salary bonuses, to motivate the desired behavior, these incentives may impair the functioning of the moral hidden hand. Workers are liable to take incentives as a lesson that management is trying to teach them; the smart, if not the right, thing to do is to respond only to incentives that would appeal to the rational selfish worker. If constructs resulting from the operation of the moral hidden hand, like professionalism and pride in honesty, are neglected in favor of responding to incentives, and if incentives are imperfect or game-able, incentives can easily damage, not improve, performance. As Stout tells us in chapter 8 of this volume, *Homo economicus* sounds suspiciously like a sociopath. The very idea that we have culture depends on us learning the lessons society teaches. If it teaches *Homo economicus* . . . Evidence for the operation of crowding-out includes experimental results and a good deal of evidence from the field. Some business management scholars believe that the influence of neoclassical economists with their rational-selfish models and intuitions has crowded out the moral hidden hand in the behavior of management school graduates (Ferraro, Pfeffer, and Sutton 2005; Ghoshal 2005; see also chapter 14, this volume). The use of draconian punishment systems to coerce correct behavior leads to a collapse of a society's morale, as in the East Germans' cynical saying, "they pretend to pay us, we pretend to work."

Contrariwise, legal systems attempt to "crowd in" virtuous behavior (Salter 1995; see chapter 9, this volume). A trial is a morality play in which offenders, jury members, and spectators are invited to see offending behavior as violating the standards of the community. First offenders, especially those that confess and show remorse, may be let off with a lecture. Elinor Ostrom (1990) describes the care with which village-scale commons management systems punish offenders, ramping up punishments for repeat offenders but treating first offenders as perhaps just not understanding the rules. Strong sanctions must ultimately be applied to the minority of sociopathic strategists who, in the lab and in the field, can cause organizations to collapse.

Nevertheless, legal systems often fail to crowd in virtue. For example, in highly stratified and ethnically divided societies, cultures of resistance may arise if sub-communities, perhaps with reason, feel that the exercise of power is illegitimate (Sidanius and Pratto 1999). In American underclass communities, tribal-scale street gangs may arise partly to serve social functions like protection not furnished by legitimate institutions, partly as symbolic cultures of resistance, and partly to exploit black market business opportunities. Since these cultures are viewed by the dominant culture as evil, they reinforce the dominants' view that what subordinate communities see as abuses of power are, in fact, a necessary and legitimate part of the legal system. A sick coevolutionary spiral sets in with a culture of increasing resistance driving a culture of increasing repression. Thus, we end up with the spectacle of the world's largest free nation offering economic opportunity to wave after wave of immigrants while at the same time incarcerating about the same number of its own citizens as the Soviet Union did. A number of inherent difficulties in improving and maintaining the functioning of complex societies led these authors to characterize them as "crude superorganisms" (Richerson and Boyd 1999).

Adam Smith and Charles Darwin both made empathy the cornerstone of their theories of virtue as we have seen. They observed that, without the other-regarding virtue of sympathy, the social life humans enjoy today would not be possible, much less reforms aimed at improving our social life. Market forces certainly do exert important hidden hand effects, but the effects of everyday virtues are equally pervasive and nearly as hidden in the sense that formal legal institutions and formal policies and procedures represent only a small part of their effect (Ellickson 1991). Informal rules and everyday virtues affect our behavior in a multitude of unforced, unplanned ways. Formal law is costly and cumbersome, and is most often invoked when custom and everyday virtue fail in some way.

Smith's and Darwin's old insights are buttressed by modern theoretical and empirical studies that show how far human behavior deviates from the neoclassical economist's selfish rational assumption. For example, an important component of the moral hidden hand is that many people will altruistically punish cheaters in social games (Fehr and Gachter 2002). Given such results, we should not be surprised that businesses attending to their social and environmental responsibilities make no less money than the average business and, in many cases, seem to make more money than ones that focus more ruthlessly on the bottom line (Orlitzky et al. 2003).

Businesses and other modern organizations are complex cooperative systems that function best when the moral hidden hand is operating most freely. A business full of high-morale cooperators will tend to earn the firm respectable profits and still have plenty of spare energy to help people and the environment. The firm that focuses excessively on the bottom line may find that it has inadvertently disadvantaged the moral hidden hand by encouraging employees to focus selfishly on their personal bottom lines, which might include diverting the firms resources for their own gain by concentrating on personal agendas, padding expense accounts, pilfering the supply cabinet, running up sales commissions by making expensive promises to customers, and by the many other ways that selfish employees can exploit the organization. Most economists are surprised by findings such as Orlitzky et al.'s (as they are by many of the cultural-evolutionary findings that underpin our analysis). Economists have been trained to expect a *trade-off* to exist between a firm's profitability and any *special* attention it pays to social or environmental concerns rather than the *synergy* between these goals predicted by cultural evolution (and supported by laboratory experiments). Economics students, incidentally, are quite resistant to the moral hidden hand in the laboratory compared to other students, and they have trouble making cooperation work (Marwell and Ames 1981)! Having imbibed the selfish rational assumption, they are at a disadvantage in running the model businesses we set up in the laboratory. Economics, we should add, is changing very rapidly because some of the most elegant support for the moral hidden hand has come from the studies of pioneering experimental economists brought up in the neoclassical tradition.

Explaining Fifty Millennia of Cultural History

The Moral Hidden Hand and the Evolution of Institutions

The moral hidden hand not only functions directly in the operation of organizations, it also certainly functions as an agent of cultural evolution. All else equal, individual decision makers should espouse norms that result in better-functioning institutions and norms. Collective decision making, as we have seen with the tendency of people to vote altruistically, ought also to favor prosocial norms and institutions. We have said that the main adaptive advantage of culture is its rapidity of adaptation. We know from personal experience that cultural evolution is indeed rather rapid, at least at some times and in some places. If our argument that our tribal social instincts and cognitive capacities were already modern fifty thousand years ago, this leaves us with a lot of historical noodling around to explain. Do we have a serious macroevolutionary flaw in the cultural evolutionary explanation? We think the answer is that we do have plenty of macroevolutionary puzzles to solve but that we also have plenty of candidate hypotheses to explain the fifty millennia.

External and Internal Explanations for Macroevolutionary Events and Trends

Two sorts of processes might be responsible for the patterns we see in history. First, events and processes external to the biological and cultural system that is evolving may be driving the evolutionary process. In invoking the onset of higher-amplitude, higher-frequency environmental variation in the Pleistocene to explain the evolution of large brains in many lineages and culture in the human line, we have constructed an externalist explanation. In imagining that the time period from 250,000 years ago to 50,000 years ago was the main period when gene-culture coevolution was building our tribal social instincts, we might be appealing to the idea that major genetic change is rather slow. At any rate, responses to selection alone would be slower than the response of culture to selection plus decision-making forces on culture. If the environment changes suddenly and sets evolution in motion in a steady direction toward new cultural and genetic adaptations, new cultural adaptations will evolve at a limiting rate set by the genes rather quickly. New cultural "environments" will then create selection on genes; genetic changes will, in turn, permit more "advanced" culture. Whether genetic evolution was sticky enough to account for the whole 200,000 years of coevolution in Africa is hard to say. Perhaps the evolution of our large brain was a sticky enough process to explain the whole 2-plus million years of brain expansion in our lineage. Recently raised cores from Lake Malawi promise to give us high-resolution climate data for Africa for the last million years. When these data become available, we will be able to estimate the lag between the onset of high-amplitude, millennial- and sub-millennial-scale variation and the gene-culture coevolutionary response. It is sufficient to say that, at short enough time scales, internal hypotheses are always important; so long as we are dealing with evolution at all, we will see some lag between even changed environments and cultural change. Similarly, at the time scale of the geological evolution of the earth we can be sure that external changes are important. In between, the debate will rage.

The Origins of Agriculture Experiment

Several independent trajectories of subsistence intensification, often leading to agriculture, began during the Holocene (Richerson et al. 2001). By intensification we mean a cycle of innovations in subsistence efficiency per unit of land leading to population growth that, in turn, leads to denser settlement per unit area of land. No plant-rich intensifications are known from the Pleistocene. Subsistence in the Pleistocene seems to have depended substantially on relatively high-quality animal and plant resources that held human populations to modest densities. Population growth is a rapid process on time scales shorter than a millennium. Cultural evolution is a rapid process on time scales of ten millennia. If agriculture had been possible in the Pleistocene, it should have appeared before the Pleistocene-Holocene transition. The high-amplitude variation that, as we argued above, favored the evolution of our capacity for culture would have made heavily plant-based subsistence strategies, especially agriculture, impossible during the last glacial. Climate and weather variation are a major difficulty for farmers today, and, with much higher amplitude variations, human populations were probably forced to forego specializing on a narrow spectrum of high-productivity resources such as proto-domesticates like wild wheat and barley, although we do know that they sometimes used them (Kislev et al. 1992). In addition, last-glacial environments were generally drier than in the Holocene, and the concentration of carbon dioxide in the atmosphere was significantly lower. The quite abrupt final amelioration of the climate at the onset of the Holocene 11,500 years B.P. was followed immediately by the beginnings of plant-intensive strategies in some areas, although the turn to plants was much later elsewhere. Almost all trajectories of subsistence intensification in the Holocene are progressive, and eventually agriculture became the dominant strategy in all but the most marginal environments. The Polynesian expansion of the last 1,500 years and the European expansion of the last 500 years pioneered agriculture in the Pacific Islands, Australia, and large parts of Western North America, the last substantial areas of the earth's surface favorable to it.

Thus, evolution of human subsistence systems during the career of anatomically modern humans seems to divide quite neatly into two regimes, a Pleistocene regime of hunting and gathering subsistence and low population density enforced by an external factor, climate, and a Holocene regime of increasingly agricultural subsistence and relatively high and rising population densities. Climate change was a minor factor in the Holocene, and most likely internal factors account for the slow growth of economic and political complexity over the last eleven millennia.

The dispersed resources and low mean density of populations in the Pleistocene meant that relatively few people could be aggregated together at any one time and place. The lack of domestic livestock meant that movement of goods on land would be limited to what humans could carry. No evidence of extensive use of boats to transport goods appears in the archaeological record of the late Pleistocene, although some significant water crossings were necessary for people to reach Australia and the larger islands of "near" Oceania, like New Zealand. Low-density, logistically limited human populations have small (but far from negligible) scope for exploiting returns to scale in cooperation, coordination, and division of labor, and their institutions remain comparatively simple.

Intensified subsistence and higher population densities multiply the number of people and volume of commodities that societies can mobilize for economic and political purposes. Expanded exchange allows societies to exploit a finer division of labor. Larger armies are possible to deal with external threats or to coerce neighbors. Expanding the number of people sharing a common language and customs will accelerate the spread of useful ideas. *Given appropriate institutions*, the denser societies made possible by agriculture can realize considerable returns to better exploitation of the potential of cooperation, coordination, and the division of labor. Corning (2005) elaborates the advantage of large-scale and more complex social organization along these lines under his synergism hypothesis. Thus, in the Holocene, the origins of agriculture and its rising productivity over succeeding millennia at least permit the evolution of more complex societies.

A Competitive Ratchet

Intra- and inter-society competition put a sharp point on the potential for more complex societies. Holding the sophistication of institutions constant, marginal increases in subsistence productivity per unit of land will lead to denser or richer populations that can outcompete societies with less intensive subsistence systems. Holding subsistence productivity constant, societies with marginally more sophisticated social organization will also outcompete rivals. Within groups, contending political interests with innovations that promise greater rewards for altered social organization can use either selfish or patriotic appeals to advance their cause. Successful reformers may entrench themselves in power for a considerable period. Malthusian growth will tend to convert increases in subsistence efficiency and security against depredations to greater population density, making losses of more complex institutions painful and further advance rewarding. Richerson and colleagues (2001) argue mathematically that the rate-limiting process for intensification trajectories must almost always be the rate of innovation of subsistence technology or subsistence-related social organization. At the observed rates of innovation, rates of population growth will always be rapid enough to sustain a high level of population pressure favoring further subsistence and social-organization innovations. Competition may be economic, political/military, or for the hearts and minds of people. Typically all three forms will operate simultaneously. In the Holocene, agriculture and complex social organization are, in the long run, compulsory. In the most dramatic episode of the expansion of social complexity, from the sixteenth through the nineteenth centuries, European populations settled many parts of the world and overwhelmed native populations with less efficient subsistence and less complex social organization. In regions such as Asia, where disparities of subsistence and social organization with the West were less striking, societies like China, Japan, and India retained or reclaimed their political independence at the cost of humiliating exposure to Western power and of borrowing many technical and social-organizational techniques from the West. Note that in areas where geography disadvantaged the armies of states, tribal-scale social institutions retained much of their vitality. The diseases of Sub-Saharan Africa and the mountainous topography and protection from the sea enjoyed by Switzerland and Afghanistan allowed much local autonomy. The power of states and empires is forever being contested by tribal-scale organizations (Garthwaite 1993).

The tendency of a population to grow rapidly, and for knowledge of advanced techniques to be retained somewhere, act as pawls on the competitive ratchet. Even during the European Dark Ages, when the pawls slipped several cogs on the ratchet, the slide backward was halted and eventually reversed in a few hundred years.[5]

Replications of the Experiment

Agricultural subsistence evolved independently at least seven times in the Holocene, and many more societies have acquired at least some key agricultural innovations by diffusion (Richerson, Boyd, and Bettinger 2001). Although none of these origins are earlier than the early Holocene, many are much later. The trajectory of institutional evolution is similar. To take one benchmark, the origin of the state level of political organization began in Mesopotamia around 5,500 B.P., but most are later, some much later (Service 1975; Feinman and Marcus 1998). For example, the Polynesian polities of Hawaii and Tonga-Samoa became complex chiefdoms on the cusp of the transition to states just before European contact (Kirch 1984). Pristine states evolved independently, perhaps ten or so times, in several parts of the world, and traditions of statecraft in various areas of the world evolved in substantial isolation for significant periods.

If our basic hypothesis is correct, the climate shift at the Pleistocene-Holocene transition removed a tight constraint on the evolution of human subsistence systems and hence on the institutional evolution. On the evidence of the competitive success of modern industrial societies, subsistence evolution has yet to exhaust the potential for more efficient subsistence inherent in agricultural production, and ongoing increases in the complexity of social institutions suggests that institutional evolution is still discovering more

synergistic potential in human cooperation, coordination, and division of labor. The out-of-equilibrium progressive trend in human evolution over the
last eleven millennia means that we can achieve a certain conceptual and
probably empirical simplification of the problem of the evolution of institutions in the Holocene. We can assume a strong, worldwide tendency, driven
by the competitive ratchet, toward societies at least as complex as current industrial societies. We can assume that changes in climate and similar nonso-
cial environmental factors play a small role in the Holocene. Granted these
assumptions, we are left with three internalist questions about subsistence and
institutional evolution:

1. Why are rates of change so rapid in some areas (Western Eurasia) and
   slow in others (Western North America)? The competitive ratchet
   seems to have been routinely cranked faster in some places than others. What are the factors that limit the rate of cultural evolution in
   some cases relative to others? We shall argue that several processes can
   retard the rate of cultural evolution sufficiently to account for the observed rates of change.
2. How do we explain the multilinear pattern of the evolution of institutional complexity? Although an upward trend of complexity characterizes most Holocene cultural traditions, the details of the trajectory
   vary considerably from case to case. The operation of the ratchet is
   very far from pulling all evolving social systems through the same
   stages; only relatively loose parallels exist between the cases.
3. Why does the ratchet sometimes slip some cogs? In no particular cultural tradition is progress even and steady. Episodes of temporary stag-
   nation and regression are commonplace.

What Regulates the Tempo and Mode of Institutional Evolution?

The overall pattern of subsistence intensification and increase in social complexity is clearly consistent with the hypothesis that agriculture, and hence
complex social institutions, were impossible in the Pleistocene but eventually
mandatory in the Holocene. The real test, however, is whether we can give a
satisfactory account of the variation in the rate and sequence of cultural evolution. Work on this project is in its infancy, and what follows is only a brief
sketch of the issues involved.

GEOGRAPHY MAY PLAY A BIG ROLE

Diamond (1997) argues that Eurasia has had the fastest rates of cultural
evolution in the Holocene because of its size and, to a lesser extent, its orientation. Plausibly, the number of innovations that occur in a population increases

with total population size and the flow of ideas between sub-populations. Since we know that the original centers of cultural innovation were relatively small compared to the areas where they later spread, most societies acquired most complex cultural forms by diffusion. Societies isolated by geography will have few opportunities to acquire innovations from other societies. Contact of isolated areas with the larger world can have big impacts. The most isolated agricultural region in the world, Highland New Guinea, underwent an economic and social revolution in the last few centuries with the advent of American sweet potatoes, a crop that thrives in the cooler highlands above the malaria belt of lowland New Guinea (Wiessner and Tumu 1998). The Americas, though quite respectable in size, are oriented with their major axis north-south. Consequently, innovations have to spread mainly across lines of latitude from the homeland environment to quite different ones, unlike in Eurasia where huge east-west expanses exist in each latitude belt. The pace of institutional change in Eurasian societies mirrors this region's early development of agriculture and the more rapid rate of subsistence intensification.

CLIMATE CHANGE MAY PLAY A SMALL ROLE

The Holocene climate is only invariant relative to the high-frequency, high-amplitude oscillations of the last glacial (Lamb 1977). For example, seasonality (the difference between summer and winter insolation) was at a maximum near the beginning of the Holocene and has since fallen. The so-called Climatic Optimum, a broad period of warmer temperatures during the middle Holocene, caused a wetter Sahara and the expansion of early pastoralism into what is now forbidding desert. The late medieval onset of the Little Ice Age caused the extinction of the Greenland Norse colony (Kleivan 1984). Agriculture at marginal altitudes in places like the Andes seems to respond to Holocene climatic fluctuation (Kent 1987). The fluctuating success of state-level political systems in the cool, arid Lake Titicaca region is plausibly caused by wetter episodes permitting economies that support states, whereas, during arid periods, these systems collapse or fade. Although we must always keep in mind the effect of Holocene climate fluctuations on regional sequences (e.g., Kennett 2005), the dominance of the underlying monotonic tendency to increase subsistence intensification and evolve more complex institutions is likely to be driven by other processes.

COEVOLUTIONARY PROCESSES PROBABLY PLAY A BIG ROLE

The full exploitation of a revolutionary new subsistence system like agriculture requires the evolution of domesticated strains of plants and animals. Human social institutions must undergo a revolution to cope with the increased population densities that follow from agricultural production. Human biology changes to cope with the novel dietary requirements of agricultural subsistence.

AGRICULTURE REQUIRES PRE-ADAPTED PLANTS AND ANIMALS

In each center of domestication, people domesticated only a handful of the wild plants that they formerly collected, and, of this handful, even fewer are widely adopted outside those centers. Zohary and Hopf (2001) have listed some of the desirable features in plant domesticates. California has so many climatic, topographic, and ecological parallels with the precocious Fertile Crescent that its very tardy development of plant-intensive subsistence systems is a considerable enigma. Diamond (1997), drawing on the work of Blumler (1992), notes that the Near Eastern region has a flora unusually rich in large-seeded grasses. California, by contrast, lacks large-seeded grasses; not a single species passed Blumler's criterion. Aside from obvious factors like large seed size, most Near Eastern domesticates had high rates of self-fertilization. This means that farmers can select desirable varieties and propagate them with little danger of gene flow from other varieties or from weedy relatives. Maize, by contrast, outcrosses at high rates. Perhaps the later and slower evolution of maize compared to Near Eastern domesticates is owing to the difficulty of generating responses to selection in the face of gene flow from unselected populations (Diamond 1997:137). Smith (1995) discusses the many constraints on potential animal domesticates.

Even in the most favorable cases, the evolution of new domesticates is not an instantaneous process. Blumler and Byrne (1991) identify the rate of evolution of domesticated characters like non-dehiscence as one of the major unsolved problems in archaeobotany. Coevolution theorists like Rindos (1984) imagine a long drawn out period of modification leading up to the first cultivation. Blumler and Byrne, on the other hand, conclude that the rate of evolution of domesticates *may* be rapid, but they stress the uncertainties deriving from our poor understanding of the genetics and population genetics of domestication. The simulations of Hillman and Davies (1990) indicate that the evolution of a tough rachis (the primary archaeological criterion of domestication) in inbreeding plants like wheat and barley could easily be so rapid as to be archaeologically invisible, as, indeed, it so far is. Their calculations also suggest that outcrossed plants, such as maize, will respond to cultivator selection pressures on the much longer time scales that Rindos and Diamond envision.

HUMANS HAVE TO ADAPT BIOLOGICALLY TO AGRICULTURAL ENVIRONMENTS

Although the transition from hunting and gathering to agriculture resulted in no genetic revolution in humans, a number of modest new biological adaptations were likely involved in becoming farmers. The best-documented case is the evolution of adult lactose absorption in human populations with long histories of dairying (Durham 1991), but one of the fruits of the Human Genome Project has been the detection of other genes that appear to be rapidly evolving in humans. Many genes related to diet and epidemic disease seem to have evolved since the origins of agriculture (Sabeti et al. 2006). To some extent, the relatively slow rate of human biological adaptation may act as a drag on the rate of cultural innovations leading to subsistence intensification and on institutional advances.

DISEASES LIMIT POPULATION EXPANSIONS, PROTECTING INTER-REGIONAL DIVERSITY

McNeill (1976) and Crosby (1986) draw our attention to the coevolution of people and diseases. The increases in population density that resulted from the intensification of subsistence invited the evolution of epidemic diseases that could not spread at lower population densities. One result of this process may be to slow population growth to limits imposed by the evolution of cultural or genetic adaptations to diseases. For example, a suite of hemoglobins have arisen in different parts of the world that confer partial protection against malarial parasitism, and these adaptations may have arisen only with the increases in human population densities associated with agriculture (Cavalli-Sforza, Menozzi, and Piazza 1994). Cavalli-Sforza and colleagues estimate that it would take about two thousand years for a new mutant hemoglobin variant to reach equilibrium in a population of fifty thousand or so individuals. Serious epidemics also have direct impacts on social institutions when they carry away large numbers of occupants fulfilling crucial roles at the height of their powers. In such epidemics, significant losses of institutional expertise could occur, directly setting back progressive evolution. (Or, alternatively, epidemics might sweep away the Old Guard and make a progressive change easier.) Regional suites of diseases disadvantage immigrants and travelers, thus tending to isolate societies from the full effects of cultural diffusion.

CULTURAL EVOLUTIONARY PROCESSES PLAY A DECISIVE ROLE

The processes of cultural evolution may generally be more rapid than biological evolution, but cultural change often takes appreciable time. Richerson and Boyd (2005) view cultural evolution as a Darwinian process of descent with modification. Evidence about characteristic rates of modification is important for understanding the relative significance of various processes in cultural evolution. In one limit, the conservative, blind, transmission of cultural variants from parents to offspring, the main adaptive force on cultural variants would be natural selection, and rates of cultural evolution would approximate those of genes. At the other extreme, humans may pick and choose among any of the cultural variants available in the community and may use cognitive strategies to generate novel behaviors directly in light of environmental contingencies. In the limit of economist's omniscient rational actors, evolutionary adjustments are modeled as if they are instantaneous. We believe that, for many cultural traits, human decisions have relatively weak effects in the short run and at the individual level, although they can be powerful when integrated over many people and appreciable spans of time. For example, the four streams of British migration to North America have led to regional differences that have persisted for centuries (Fischer 1989; Nisbett and Cohen 1996). Archaeological and historical data on the rates of change in different domains of culture will be some of the most important evidence to muster to understand the tempo and mode of cultural evolution. Much work remains to be done before we understand the regulation of rates of cultural evolution, but some preliminary speculation is possible.

NEW TECHNOLOGICAL COMPLEXES EVOLVE WITH DIFFICULTY

One problem that will tend to slow the rate of cultural (and organic) evolution is the sheer complexity of adaptive design problems. As engineers have learned when studying the design of complex functional systems, discovering optimal designs is quite difficult. Blind search algorithms often get stuck on local optima, and complex design problems often have many of these. Piecemeal improvements at the margin are not guaranteed to find globally optimal adaptations by myopic search. Yet, myopic searches are what Darwinian processes do (Boyd and Richerson 1992). Even modern engineering approaches to design, for all their sophistication, are more limited by myopic cut and try than engineers would like.

Parallel problems are probably rife in human subsistence systems. The shift to plant-rich diets is complicated because plant foods are typically deficient in essential nutrients, have toxic compounds to protect them from herbivore attack, and are labor-intensive to prepare. Finding a mix of plant and animal foods that provides adequate diet at a feasible labor cost is not a trivial problem. For example, New World farmers eventually discovered that boiling maize in wood ashes improved its nutritional value. The hot alkaline solution breaks down an otherwise indigestible seed coat protein that contains some lysine, an amino acid that is low in maize relative to human requirements (Katz, Hediger, and Valleroy 1974). Hominy and *masa harina*, the corn flour used to make Mexican tortillas, are forms of alkali-treated maize. The value of this practice could not have been obvious to its inventors or later adopters, yet most American populations that made heavy use of maize employed it. The dates of origin and spread of alkali cooking are not known. It has not been reinvented in Africa, even though many African populations have used maize as a staple for centuries.

NEW SOCIAL INSTITUTIONS EVOLVE WITH DIFFICULTY

An excellent case can be made that the rate of institutional innovation is more often limiting than the rate of technological innovation. As anthropologists and sociologists such as Julian Steward (1955) have long emphasized, human economies are social economies. Even in the simplest human societies, hunting and gathering is never a solitary occupation. Minimally, such societies have division of labor between men and women. Hunting is typically a cooperative venture. The unpredictable nature of hunting returns usually favors risk sharing at the level of bands composed of a few cooperating families, as most hunters are successful only every week or so (Winterhalder et al. 1999). Portions of kills are distributed widely, sometimes exactly equally, among band members.

The deployment of new technology requires changes in social institutions to make the best use of innovations, often at the expense of entrenched interests, as Marx argued. The increasing scale of social institutions associated with rising population densities during the Holocene have dramatically reshaped human social life. Richerson and Boyd (1998, 1999) discuss the complex problems involved in an evolutionary trajectory from small-scale, egalitarian societies to large-scale complex societies with stratification and hierarchical political systems. For example, even the first steps of intensification required significant social changes. Gathering was generally the province of women, and hunting that of men. Male prestige systems were often based on hunting success. A shift to plant resources required scheduling activities around women's work rather than men's pursuit of prestige. Using more plants conflicted with men's preferences as driven by a desire for hunting success; a certain degree of women's liberation was required to intensify subsistence. Since men generally dominated women in group decision making ("egalitarian," small-scale societies seldom granted women equal political rights), male chauvinism tended to limit intensification. Bettinger and Baumhoff (1982) argued that the spread of Numic speakers across the Great Basin a few hundred years ago was the result of the development of a plant-intensive subsistence system in the Owens Valley. Apparently, the groups that specialized in the hunt would not or could not shift to the more productive economy to defend themselves, perhaps because males clung to the outmoded, plant-poor subsistence. Winterhalder and Goland (1997) used optimal foraging analysis to argue that the shift from foraging to agriculture would have required a substantial shift in risk management institutions, from minimizing risk by intraband and interband sharing to reducing risk by field dispersal by individual families. Some ethnographically known Eastern Woodland societies that mixed farming and hunting—for example, the Huron—seemed not to have made this transition and to have suffered frequent catastrophic food shortages.

Institutional evolution no doubt involves complex design problems. For example, Blanton (1998) described some of the alternative sources of power in archaic states. He noted that archaic states differ widely in time and space as their evolution wanders about in a large space of alternative social institutions. Thus, the classical Greek system of small egalitarian city-states, with wide participation in governance, was a far different system from those of Egypt, with divine royal leaders from near its inception as a state, or the bureaucracies common in Western Asia. Philip, Alexander, and their successors substantially rebuilt the Greek state along Western Asian lines in order to conquer and administer empires in Asia. Much of the medium-term change in archaic and classical state institutions seems to involve wandering about in a large design space without discovering any decisively superior new institutional arrangements (Feinman and Marcus 1998).

The spread of complex social institutions by diffusion is arguably more difficult than the diffusion of technological innovations. Social institutions violate four of the conditions that tend to facilitate diffusion (Rogers 1983). Foreign social institutions are often incompatible with existing institutions, complex, difficult to observe, and difficult to try out on a small scale.

Thus, the evolution of social institutions rather than technology will tend to be the rate-limiting step of the intensification process. For example, North and Thomas (1973) argued that new and better systems of property rights set off the modern industrial revolution rather than the easier task of technical invention itself. A major revolution in property rights is likely also necessary for intensive hunting and gathering, and agriculture, to occur (Bettinger 1999). Slow diffusion also means that historical differences in social organization can be quite persistent, even though one form of organization is inferior. As a result, the comparative history of the social institutions of intensifying societies exhibits many examples of societies having a persistent competitive advantage over others in one dimension or another because they possess an institutional innovation that their competitors do not acquire. For example, the Chinese merit-based bureaucratic system of government was established at the expense of the landed aristocracy, beginning in the Han dynasty (2200 B.P.), and was completed in the Tang (1,400 B.P.) (Fairbank 1992). This system has become widespread only in the modern era and still operates quite imperfectly in many societies.

To the extent that games of coordination are important in social organization, changes from one coordination solution to another may by greatly inhibited. Games of coordination are those, like which side of the road to drive on, for which it matters a lot that everyone agrees on a single solution and less on which solution is chosen (Sugden 1986). Notoriously, armies with divided command are defeated. A poor general's plan formulated promptly and obeyed without question by all is usually superior to two good generals' plans needing long negotiations to reconcile or leaving subordinates with choices of whose plan to follow. We care less whether gold, silver, or paper money is legal tender than we care that we have a single standard. Many, if not most, social institutions probably have strong elements of coordination. Take marriage rules. Some societies allow successful men to marry multiple wives, and others forbid the practice. One system may or may not be intrinsically better, but everyone is better off playing from one set of rules. Since the strategies appropriate for one possibility are quite different from the other, marriage partners would like agreement on the ground rules of marriage up front to save costly negotiation or worse later on. Hence, many institutions are in the form of a socially policed norm or standard contract ("love, honor, cherish, and obey until death do us part"), solving what seems like it ought to be a private coordination problem. Except in pure cases, however, different coordination equilibria will also have different average payoffs and different distributions of payoffs than others. Even if most agree that a society can profitably shift from one simple pure coordination equilibrium to another (as when the Swedish switched from driving on the left to the right a couple of decades ago to conform to their neighbors' practices), the change is not simple to orchestrate. One of our universities voted recently not to switch from the quarter to the semester system, despite a widespread recognition that a mistake was made thirty years ago when the quarter system was instituted. Large, uncertain costs that many semester-friendly faculty reckoned would attend such a switch caused them to vote no. Larger-scale changes, such as the Russian attempt to move from a Soviet to a free enterprise economy, face huge problems that are plausibly the result of the need to renegotiate solutions to a large number of games of coordination as much as any other cause.

The design complexity, importance of coordination, slow evolution, limited diffusion, and difficulty of coordination shifts probably conspire to make the evolution of social institutions highly historically contingent. The multilinear pattern of evolution of social complexity could result from two causes. Societies might be evolving from diverse starting points toward a single common optimal state surrounded by a smooth "topography" toward whose summit optimizing evolutionary processes are climbing. Or, societies may be evolving up a complex topography with many local optima and many potential pathways toward higher peaks. In the latter case, even if societies start out at very similar initial points, they will tend to diverge with time. We believe that at least part of the historical contingency in cultural evolution is the result of slow evolution on complex topographies (Boyd and Richerson 1992).

IDEOLOGY MAY PLAY A ROLE.

Nonutilitarian processes may strongly influence the evolution of fads, fashion, and belief systems. Such forces are susceptible to feedback and runaway dynamics that defy common sense (Boyd and Richerson 1985, chap. 8). The links between belief systems and subsistence are nevertheless incontestably strong. To build a cathedral requires an economy that produces surpluses that can be devoted to grand gestures on the part of the faithful. The moral precepts inculcated by the clergy in the cathedral underpin the institutions that in turn regulate the economy. Arguably, ideological innovations often drive economic change. Recall Max Weber's classical argument about the role of Calvinism in the rise of capitalism.

COMPLEX SOCIAL SYSTEMS ARE VULNERABLE

We suggest that the fragility of institutions derives from compromises and trade-offs that are caused by conflicts between the functional demands of large-scale organizations and the trajectory of small-scale cultural evolution often driven by psychological forces rooted in the ancient and tribal social instincts. The evolution of work-arounds seldom results in perfect adaptations. Resistance to the pull of the ratchet can increase sharply when external pressures, such as competition from other societies, demographic catastrophes, or internal processes like the evolution of a new religion, put weak work-arounds in jeopardy. All complex societies may have weak work-arounds lurking among their institutions. As noted above, each of the major types of institutional workarounds has defects that lead to intra-societal conflict. Small-scale societies have appreciable crudities deriving at least partly from conflicts, both intra-psychic and political, between individual and kinship interests and the larger tribe (Edgerton 1992). If our argument is correct, larger-scale societies do not eliminate these conflicts but add to them manifold opportunities for conflict between different elements of the larger system. Even the best of such systems current at any one time are full of crudities, and the worst are often highly dysfunctional. A considerable vulnerability to crisis, change without progress, setback and collapse is inherent in an evolutionary system subject to strong evolutionary forces operating at different levels (Turchin 2003).

Maynard Smith and Szathmáry (1995) treat the rise of human ultra-sociality as analogous to other major evolutionary transformations in the history of life. As with our tribal social instincts and work-around proposals, the key feature of transitions from, say, cellular grade organisms to multicellular ones is the improbable and rare origin of a system in which group selection works at a larger scale to suppress conflict at the smaller, and eventually to perfect the larger-scale super-organisms. Actually, *perfect* is too strong a word; distinct traces of conflict remain in multicellular organisms and honeybee colonies, too. We suggest that human societies are recently evolved and remain rather crude super-organisms, heavily burdened by conflict between lower- and higher-level functions and not infrequently undone by them. Outside the realm of utopian speculation and science fiction, there does not appear to be an easy solution. Muddle along is the rule, pulled on the trajectory toward more social complexity by the competitive ratchet.

CHANGES IN THE RATE OF CULTURAL EVOLUTION AND THE SIZES OF CULTURAL REPERTOIRES

Rates of social and technical evolution appear to be rising toward the present. Modern individuals know more than their ancestors, and social complexity has increased. The cultural evolution in the Holocene began at a stately pace. Not for some six thousand years after the initial domestication of plants and animals in southwestern Asia did the first state-level societies finally decisively transcend tribal-scale roots of human sociality. Tribes, city-states, and empires competed to govern Eurasia for another forty-five hundred years while the first states emerged in the New World and Africa. The rise of the West over the last millennium has brought revolutions in subsistence and social organization, particularly during the last half-millennium. Today, globalization is spreading the culture of free enterprise societies. Even in Eurasia, the last pastoral and hunting tribes of the interior were only defeated by Chinese and Russian firearm armies a couple of centuries ago. Only for the last century or two has cultural evolution been sufficiently rapid so that almost everyone is aware of major changes within their lifetime. Malthus, writing around the turn of the nineteenth century, still regarded technical innovation as quite slow, on sound empirical grounds. Only a couple of decades after his death would cautious empiricists have good grounds to argue that the Industrial Revolution, coupled with free enterprise, was something new under the sun (Lindert 1985). The accelerating growth of the global population is a product of these changes, and the curve of population growth is one reasonable overall index of cultural change. Another is the increasing division of labor. Innovations on the subsistence side at first rather gradually, and then lately very rapidly, reduced the personnel devoted to agricultural production and shifted labor into an expanding list of mercantile, manufacturing, government, and service occupations.

The reasons for the accelerating rate of increase are likely several. First, the sheer increase in numbers of people must have some effect on the supply of innovations. Second, the invention of writing and mathematics provided tools for supplementing memories, aiding the application of rationality, and for the long distance communication of ideas. Scribes in small numbers first used their new skills to manage state supply depots, tax roles, and land mensuration. Only gradually did procedures in different fields become written and mathematics come to be used to solve an expanding array of problems. Third, books ultimately became a means of both conserving and communicating ideas, at first only to an educated elite. Fourth, quite recently, the mass of people in many societies became literate and numerate, allowing most people to take up occupations that depended on prolonged formal education, policy and procedural handbooks, technical manuals, reference books, and elaborate calculations. Fifth, the rise of cheap mass communication, beginning with the printed book, has given individuals access to ever larger stores of information. The Internet promises to give everyone able to operate a workstation access to all the public information in the world. Donald (1991) counted the spread of literacy and numeracy as a mental revolution on the same scale as the evolution of imitation and spoken language. Sixth, institutions dedicated to deliberately promoting technical and social change have grown much more sophisticated. Boehm (1996) argued that even acephalous societies usually have legitimate, customary institutions by which the society can reach a consensus on actions to take in emergencies, such as the threat of war or famine (see also Turner 1995:16–17).

Institutions organized as a matter of social policy to further change continue to increase perceptibly in scope and sophistication. Institutions, like patents that give innovators a socially regulated property right in their inventions, ushered in the Industrial Revolution. At about the turn of the twentieth century, private companies began to invest in new technology, under the eye of government regulators. Government bureaucracies have conducted useful research from the public purse beginning in a small way in the nineteenth century. Research universities recruit some of the best minds available, place them in an intellectual hothouse, and reward scholars for whatever new ideas they are prepared to pursue. Masses of young people are educated by such innovators and their students, especially during the last fifty years. Johann Murmann (2003) traced the development of the synthetic dye industry in the nineteenth century. German dye manufacturers cooperated to foster the development of research chemistry departments and came to dominate an industry pioneered in Great Britain; this occurred because university-trained chemists, working both in the universities and in industrial labs, were at the forefront of innovation in dyes. Development institutions, such as agricultural extension services and teaching hospitals, move innovations in some fields from the university to the farm or doctor's office at a smart pace. Think-tanks ponder public policy in the light of research; national academies of science craft White Papers based on elaborate searches for expert consensus; and legislatures hold hearings trying to match the desires of constituents with the findings of the experts in order to produce new policies and programs.

WE HAVE A SHADOWY OUTLINE OF WHY FREE ENTERPRISE VALUES EVOLVED WHEN AND WHERE THEY DID

The Pleistocene onset of high-amplitude, high-frequency climate variation probably drove the increases in brain size we see in many mammalian lineages. The evolution of humans capable of creating complex cultural institutions was only complete about fifty thousand years ago, perhaps driven by ongoing climate deterioration. The large, rapid change in environment at the Pleistocene-Holocene transition set off the trend of subsistence intensification and institutional complexity of which modern free enterprise societies are just the latest examples. If our hypothesis is correct, the reduction in climate variability, the increase in the carbon dioxide content of the atmosphere, and increases in rainfall rather abruptly changed the earth from a regime where agriculture was everywhere impossible to one where it was possible in many places. Since groups that utilize efficient, plant-rich subsistence systems and deploy the resulting larger population more effectively will normally outcompete groups that make less efficient use of land and people, the Holocene has been characterized by a persistent tendency toward subsistence intensification and growth in institutional sophistication and complexity. The diversity of trajectories taken by the various regional human sub-populations since ≈11,600 B.P. are natural experiments that will help us elucidate the factors controlling the tempo and mode of cultural evolution leading to more efficient subsistence systems and the more complex societies these systems support. A long list of processes interacted to regulate the trajectory of subsistence intensification, population growth, and institutional change that the world's societies have followed in the Holocene. Social scientists habitually treat these processes as mutually exclusive hypotheses. They seem to us to be competing but certainly not mutually exclusive. Many are not routinely given any attention in the historical social sciences. At the level of qualitative empiricism, tossing any one out entirely leaves puzzles that are hard to account for and produces a caricature of the actual record of change. If this conclusion is correct, the task for historically minded social scientists is to refine estimates of the rates of change that are attributable to the various evolutionary processes and to estimate how those rates change as a function of natural and socio-cultural circumstances. We lack a quantitative understanding of the burden of flawed work-arounds and other features of complexity that retard and locally reverse tendencies to greater complexity. We only incompletely understand the processes generating historical contingency.

The free enterprise societies' combination of individual autonomy, wealth, and welfare bear a strong resemblance to the preferences that are rooted in our ancient and tribal social instincts. The rational-selfish picture of free enterprise captures only the first of these. The societies of our ape ancestors, if they were anything like those of living apes, were closer to the rational selfish model than we are. Chimpanzees live in groups regulated by dominance hierarchies leavened with some kin-based altruism. They have very little of the cooperative economic enterprise that characterize humans. They are wild animals, as people who try to raise them as if they were dogs or children soon discover (Hayes 1951). Chimpanzees are good at autonomy but weak on wealth and welfare. Humans evolved to be the sort of species we are by adding cooperative wealth acquisition and mutual aid to our repertoire. When eighteenth-century social theorists like Adam Smith began to tinker with the ideas that would flower in European countries in the nineteenth and twentieth centuries, they spied a way to free the mass of people from economic and political oppression without risk of political anarchy. Smith and Darwin are often held up as one-dimensional ideologues espousing theories of competitive individualism, whereas both rested their theories of ethics on what we have called the moral hidden hand. Utopians, on the other hand, often think that they can educate or coerce the individual and the tribesperson out of us in the interests of some sort of social insect–like, hyper-cooperative system. If our diagnosis is correct, humans, subject to selection at both the individual and tribal level, are capable of neither sort of society. Humans exhibit the sympathy and patriotism that make our social life possible, but we know that we also have to look out for our own interests because no one else does, at least not perfectly and reliably. Darwin (1874:192), we think, put his finger on the motor that makes moral progress possible:

> With highly civilized nations continued progress depends in subordinate degree on natural selection; for such nations do supplant and exterminate one another as do savage tribes. . . . The more efficient causes of progress seem to consist of a good education during youth when the brain is impressible, and to a high standard of excellence, inculcated by the ablest and best men, embodied in the laws, customs and traditions of the nation, and enforced by public opinion.

In other words, if we let the moral hidden hand have the largest possible scope when we attend to cultural innovation and cultural transmission, we can make progress in spite of the selfish and chauvinistic aspects of our social psychology and the customs they have favored.

While we are quite happy to celebrate the accomplishments of free enterprise, we are not complacent. Not only are many societies still unfree, but even the freest have imperfections. We cannot help even willing recipients to build free societies easily or rapidly. Some contemporary evolutionary trends are disturbing. The present extremely high rates of technical and institutional evolution are a problem of immense applied importance. For example, our headlong quest for increased material prosperity that guides so much current calculated institutional change not only takes great risks of environmental deterioration and a hard landing on the path to sustainability (e.g., National Research Council 2002) but seems flawed when it comes to satisfying human needs and wants (Frank and Cook 1995; Easterlin 2001). The collapse of birth rates in the developed, and increasingly the developing, world is starting to replace fears of a population explosion with fears of an implosion (Bongaarts and Watkins 1996). The growth of international institutions in the twentieth century was impressive (Jones 2002), but much work needs to be done to bring the dangerous adventurism of states to heel even as the technical capability of states, and even stateless groups like Al Qaeda, to cause mischief grows.

Notes

1. Solomon and Stout, in chapters 2 and 8, respectively, note the importance of sympathy in the systems of several moral philosophers, notably Adam Smith.

2. The word *advanced* is in quotes here, because Darwin had complex opinions about the concept of progress, as many of us do today. Darwin did think that moral progress was possible, but he also had a nuanced critique of the moral behavior of peoples of different grades of civilization. He was prone to celebrating the moral accomplishments of "savages" and, particularly in his passionate critique of slavery in the

Voyage of the Beagle (Darwin 1845), he excoriated the civilized nations that tolerated slavery. The term *social Darwinism* and the oft-repeated idea that Darwin was a typical Victorian racist (Alland 1985) have led an unfortunately large number of people to have a highly erroneous concept of Darwin's views on races and progress. Historians note that Darwin's politics were somewhat to the left and that he subscribed fully to the doctrine of the psychic unity of humankind, as our quotes above illustrate (Richards 1987).

3. Most of the other authors in this volume allude to individual-level and institutional-level explanations of behavior. Evolutionists are driven to ask where values, emotions, norms, and institutions come from. We assume that some mixture of genetic and cultural evolution shapes the raw material out of which individuals and communities continually reconstruct their behavior.

4. For a contrary view, see Sober and Wilson 1998.

5. See Turchin 2003, for one explanation of the cycles of boom and bust that afflicted agrarian states.

References

Alland, A. 1985. *Human Nature, Darwin's View*. New York: Columbia University Press.

Arima, E. Y. 1987. *Inuit Kayaks in Canada: A Review of Historical Records and Construction*. Based mainly on the Canadian Museum of Civilization's collection. Canadian Museum of Civilization, Ottawa.

Bettinger, R. L. 1999. From traveler to processor: Regional trajectories of hunter-gatherer sedentism in the Inyo-Mono region, California. In B. R. Billman and G. M. Feinman, eds., *Settlement Pattern Studies in the Americas: Fifty Years since Virú*, 39–55. Washington D.C.: Smithsonian Institution Press.

Bettinger, R. L., and M. A. Baumhoff. 1982. The numic spread: Great Basin cultures in competition. *American Antiquity* 47:485–503.

Blanton, R. E. 1998. Beyond centralization: Steps toward a theory of egalitarian behavior in archaic states. In G. M. Feinman and J. Marcus, eds., *Archaic States*, 135–172. Santa Fe: School of American Research.

Blumler, M. A. 1992. Seed weight and environment in Mediterranean-type grasslands in California and Israel. Ph.D. diss., University of California, Berkeley.

Blumler, M. A., and R. Byrne. 1991. The ecological genetics of domestication and the origins of agriculture. *Current Anthropology* 32:23–53.

Boehm, C. 1996. Emergency decisions, cultural-selection mechanics, and group selection. *Current Anthropology* 37:763–793.

Boehm, C. 1999. *Hierarchy in the Forest: The Evolution of Egalitarian Behavior*. Cambridge, Mass.: Harvard University Press.

Bongaarts, J., and S. C. Watkins. 1996. Social interactions and contemporary fertility transitions. *Population and Development Review* 22:639–682.

Bowles, S. 2003. *Microeconomics: Behavior, Institutions and Evolution*. Princeton, N.J.: Princeton University Press.

Boyd, R., and P. J. Richerson. 1985. *Culture and the Evolutionary Process*. Chicago: University of Chicago Press.

Boyd, R., and P. J. Richerson. 1992. How microevolutionary processes give rise to history. In M. H. Nitecki and D. V. Nitecki, eds., *History and Evolution*, 179–209. Albany: State University of New York Press.

Boyd, R., and P. J. Richerson. 1996. Why culture is common but cultural evolution is rare. *Proceedings of the British Academy* 88:73–93.

Carroll, R. L. 1997. Patterns and Processes of Vertebrate Evolution. New York: Cambridge University Press.

Cavalli-Sforza, L. L., and M. W. Feldman. 1981. *Cultural Transmission and Evolution: A Quantitative Approach*. Princeton, N.J.: Princeton University Press.

Cavalli-Sforza, L. L., P. Menozzi, and A. Piazza. 1994. *The History and Geography of Human Genes*. Princeton, N.J.: Princeton University Press.

Corning, P. A. 2005. *Holistic Darwinism: Synergy, Cybernetics, and the Bioeconomics of Evolution*. Chicago: University of Chicago Press.

Crosby, A. W. 1986. *Ecological Imperialism: The Biological Expansion of Europe, 900–1900*. Cambridge: Cambridge University Press.

Darwin, C. 1845. Journal of researches into the natural history and geology of the countries visited during the voyage of H.M.S. Beagle round the world, under the command of Capt. Fitz Roy, R.N. London: John Murray.

Darwin, C. 1874. *The Descent of Man and Selection in Relation to Sex*. 2nd ed. New York: American Home Library.

Diamond, J. 1997. *Guns, Germs, and Steel: The Fates of Human Societies*. New York: Norton.

Donald, M. 1991. *Origins of the Modern Mind: Three Stages in the Evolution of Culture and Cognition*. Cambridge, Mass.: Harvard University Press.

Dumond, D. E. 1984. Prehistory: Summary. In D. Damas, ed., *Arctic*, 72–79. Washington, D.C.: Smithsonian Institution.

Durham, W. H. 1991. *Coevolution: Genes, Culture, and Human Diversity*. Stanford, Calif.: Stanford University Press.

Easterlin, R. A. 2001. Subjective well-being and economic analysis: A brief introduction. *Journal of Economic Behavior & Organization* 45:225–226.

Edgerton, R. B. 1992. *Sick Societies: Challenging the Myth of Primitive Harmony*. New York: Free Press.

Ellickson, R. C. 1991. *Order without Law: How Neighbors Settle Disputes*. Cambridge, Mass.: Harvard University Press.

Fairbank, J. K. 1992. *China: A New History*. Cambridge, Mass.: Harvard University Press.

Fehr, E., and S. Gachter. 2002. Altruistic punishment in humans. *Nature* 415:137–140.

Feinman, G. M., and J. Marcus. 1998. *Archaic States*. 1st ed. Santa Fe: School of American Research Press.

Ferraro, F., J. Pfeffer, and R. I. Sutton. 2005. Economics language and assumptions: How theories can become self-fulfilling. *Academy of Management Review* 30:8–24.

Fischer, D. H. 1989. *Albion's Seed: Four British Folkways in America*. New York: Oxford University Press.

Frank, R. H., and P. J. Cook. 1995. *The Winner-Take-All Society*. New York: Free Press.

Frey, B., and R. Jegen. 2001. Motivational crowding theory: A survey of empirical evidence. *Journal of Economic Surveys* 15:589–611.

Garthwaite, G. R. 1993. Reimagined internal frontiers: Tribes and nationalism—Bakhtiyari and Kurds. In D. F. Eickelman, ed., *Russia's Muslim Frontiers: New Directions in Cross-Cultural Analysis*, 130–148. Bloomington: Indiana University Press.

Ghoshal, S. 2005. Bad management theories are destroying good management practice. *Academy of Management Learning and Education* 4:75–91.

Gigerenzer, G., and D. G. Goldstein. 1996. Reasoning the fast and frugal way: Models of bounded rationality. *Psychological Review* 103:650–669.

Gingerich, P. D. 1983. Rates of evolution: Effects of time and temporal scaling. *Science* 222:159–161.

Hayes, C. 1951. *The Ape in Our House*. New York: Harper.

Heizer, R. F. 1978. Trade and trails. In R. F. Heizer, ed., *Handbook of North American Indians*, 8:690–693. Washington, D.C.: Smithsonian Institution.

Henrich, J. 2004. Cultural group selection, coevolutionary processes and large-scale cooperation. A target article with commentary. *Journal of Economic Behavior and Organization* 53:3–143.

Hillman, G. C., and M. S. Davies. 1990. Measured domestication rates in wild wheats and barley under primitive cultivation, and their archaeological implications. *Journal of World Prehistory* 4: 157–222

Hirshleifer, J. 1977. Economics from a biological viewpoint. *Journal of Law and Economics* 20:1–52.

Hunt, G. R. 2000. Human-like, population-level specialization in the manufacture of pandanus tools by New Caledonian crows *Corvus moneduloides*. *Proceedings of the Royal Society of London Series B–Biological Sciences* 267:403–413.

Jerison, H. J. 1973. *Evolution of the Brain and Intelligence*. New York: Academic Press.

Johnson, A. W., and T. K. Earle. 2000. *The Evolution of Human Societies: From Foraging Group to Agrarian State*. 2nd ed. Stanford, Calif.: Stanford University Press.

Jones, D. V. 2002. *Toward a Just World: The Critical Years in the Search for International Justice*. Chicago: University of Chicago Press.

Katz, S. H., M. L. Hediger, and L. A. Valleroy. 1974. Traditional maize processing techniques in the New World: Traditional alkali processing enhances the nutritional quality of maize. *Science* 184:765–773.

Kennett, D. J. 2005. *The Island Chumash: Behavioral Ecology of a Maritime Society*. Berkeley: University of California Press.

Kent, J. D. 1987. Periodic aridity and prehispanic Titicaca basin settlement patterns. In D. L. Browman, ed., *Arid Land Use Strategies and Risk Management in the Andes: A Regional Anthropological Perspective*. 297–314. Boulder, Colo.: Westview.

Kirch, P. V. 1984. *The Evolution of the Polynesian Chiefdoms*. Cambridge: Cambridge University Press.

Kislev, M. E., D. I. Nadel, and I. Carmi. 1992. Epipalaeolithic (19,000 BP) cereal and fruit diet at Ohalo II, Sea of Galilee, Israel. *Review of Palaeobotany and Palynology* 73:161–166.

Klein, R. G. 1999. *The Human Career: Human Biological and Cultural Origins* 2nd ed. Chicago: University of Chicago Press.

Kleivan, I. 1984. History of Norse Greenland. In D. Damas, ed., *Arctic*, 549–555. Washington, D.C.: Smithsonian Institution.

Lamb, H. H. 1977. *Climatic History and the Future*. Princeton, N.J.: Princeton University Press.

Lindert, P. H. 1985. English population, wages, and prices: 1541–1913. *Journal of Interdisciplinary History* 15:609–634.

Lindert, P. H. 2004. *Growing Public: Social Spending and Economic Growth since the 18th Century.* Cambridge: Cambridge University Press.

Marwell, G., and R. E. Ames. 1981. Economist free ride: Does anyone else? *Journal of Public Economics* 15:295–310.

Maynard Smith, J., and E. Szathmary. 1995. *The Major Transitions in Evolution.* Oxford: Freeman/Spectrum.

McBrearty, S., and A. S. Brooks. 2000. The revolution that wasn't: A new interpretation of the origin of modern human behavior. *Journal of Human Evolution* 39:453–563.

McElreath, R., R. Boyd, and P. J. Richerson. 2003. Shared norms and the evolution of ethnic markers. *Current Anthropology* 44:122–129.

McNeill, W. H. 1963. *The Rise of the West : A History of the Human Community.* New York: New American Library.

McNeill, W. H. 1976. Plagues and peoples. 1st ed. Garden City, N.Y.: Anchor.

Murmann, J. P. 2003. *Knowledge and Competitive Advantage: The Coevolution of Firms, Technology, and National Institutions.* Cambridge: Cambridge University Press.

National Research Council. 2002. *Abrupt Climate Change: Inevitable Surprises.* Washington, D.C.: National Academy Press.

Newson, L., T. Postmes, S. E. G. Lea, and P. Webley. 2005. Why are modern families small? Toward an evolutionary and cultural explanation for the demographic transition. *Personality and Social Psychology Review* 9:360–375.

Nilsson, D. E. 1989. Vision optics and evolution—nature's engineering has produced astonishing diversity in eye design. *Bioscience* 39:289–307.

Nisbett, R. E., and D. Cohen. 1996. *Culture of Honor: The Psychology of Violence in the South.* Boulder, Colo.: Westview.

North, D. C. 1994. Economic performance through time. *American Economic Review* 84:359–368.

North, D. C., and R. P. Thomas. 1973. The Rise of the Western World: A New Economic History. Cambridge University Press, Cambridge.

Orlitzky, M., James G., F. L. Schmidt, and S. L. Rynes. 2003. Corporate social and financial performance: A meta-analysis. *Organization Studies* 24:403–441.

Ostrom, E. 1990. *Governing the Commons: The Evolution of Institutions for Collective Action.* Cambridge: Cambridge University Press.

Potts, R. 1996. *Humanity's Descent: The Consequences of Ecological Instability.* New York: Avon.

Reader, S. M., and K. N. Laland. 2002. Social intelligence, innovation, and enhanced brain size in primates. *Proceedings of the National Academy of Sciences USA* 99: 4436–4441.

Richards, R. J. 1987. *Darwin and the Emergence of Evolutionary Theories of Mind and Behavior.* Chicago: University of Chicago Press.

Richerson, P. J., R. L. Bettinger, and R. Boyd. 2005. Evolution on a restless planet: Were environmental variability and environmental change major drivers of human evolution? In F. M. Wuketits and F. J. Ayala, eds., *Handbook of Evolution: Evolution of Living Systems (including Hominids)*, 223–242. Weinheim, Germany: Wiley-VCH.

Richerson, P. J., and R. Boyd. 1998. The evolution of human ultrasociality. In I. Eibl-Eibesfeldt and F. K. Salter, eds., *Indoctrinability, Ideology, and Warfare; Evolutionary Perspectives*, 71–95. New York: Berghahn Books.

Richerson, P. J., and R. Boyd. 1999. Complex societies: The evolutionary origins of a crude superorganism. *Human Nature—an Interdisciplinary Biosocial Perspective* 10:253–289.

Richerson, P. J., and R. Boyd. 2005. *Not by Genes Alone: How Culture Transformed Human Evolution*. Chicago: University of Chicago Press.

Richerson, P. J., R. Boyd, and R. L. Bettinger. 2001. Was agriculture impossible during the Pleistocene but mandatory during the Holocene? A climate change hypothesis. *American Antiquity* 66:387–411.

Rindos, D. 1984. *The Origins of Agriculture. An Evolutionary Perspective*. New York: Academic Press.

Rogers, E. M. 1983. *Diffusion of Innovations*, 3rd ed. New York: Free Press.

Sabeti, P. C., S. F. Schaffner, B. Fry, J. Lohmueller, P. Varilly, O. Shamovsky, A. Palma, T. S. Mikkelsen, D. Altshuler, and E. S. Lander. 2006. Positive natural selection in the human lineage. *Science* 312:1615–1620, and Supplementary Material online.

Salter, F. K. 1995. *Emotions in Command: A Naturalistic Study of Institutional Dominance*. Oxford: Oxford University Press.

Samuelson, P. A. 1985. Modes of thought in economics and biology. *American Economic Review* 75:166–172.

Sears, D. O., and C. L. Funk. 1991. The role of self-interest in social and political attitudes. *Advances in Experimental Social Psychology* 24:1–81.

Service, E. R. 1975. *Origins of the State and Civilization : The Process of Cultural Evolution*. 1st ed. New York: Norton.

Sidanius, J., and F. Pratto. 1999. *Social Dominance: An Intergroup Theory of Social Hierarchy and Oppression*. Cambridge: Cambridge University Press.

Simon, H. A. 1959. Theories of decision-making in economics and behavioral science. *American Economic Review* 49:253–283.

Simon, H. A. 1990. A mechanism for social selection and successful altruism. *Science* 250:1665–1668.

Smith, A. 1790. *The Theory of Moral Sentiments*. 6th ed. London: A. Millar.

Smith, B. D. 1995. *The Emergence of Agriculture*. New York: Scientific American Library.

Sober, E., and D. S. Wilson. 1998. *Unto Others: the Evolution and Psychology of Unselfish Behavior*. Cambridge, Mass.: Harvard University Press.

Steward, J. H. 1955. *Theory of Culture Change: the Methodology of Multilinear Evolution*. Urbana: University of Illinois Press.

Sugden, R. 1986. *The Economics of Rights, Co-operation, and Welfare*. Oxford: B. Blackwell.

Tomasello, M. 1996. Do apes ape? In C. M. Heyes and B. G. Galef, Jr., eds., *Social Learning in Animals: The Roots of Culture*. New York: Academic Press.

Turchin, P. 2003. *Historical Dynamics*. Princeton, N.J.: Princeton University Press.

Turner, J. H. 1995. *Macrodynamics: Toward a Theory on the Organization of Human Populations*. New Brunswick, N.J.: Rutgers University Press.

Weber, M. 1930. *The Protestant Ethic and the Spirit of Capitalism*. London: Unwin Hyman.

Wiessner, P. W., and A. Tumu. 1998. *Historical Vines: Enga Networks of Exchange, Ritual, and Warfare in Papua New Guinea*. Washington, D.C.: Smithsonian Institution Press.

Williams, G. C. 1966. *Adaptation and Natural Selection: A Critique of Some Current Evolutionary Thought*. Princeton, N.J.: Princeton University Press.

Winterhalder, B., and C. Goland. 1997. An evolutionary ecological perspective on diet choice, risk, and plant domestication. In K. J. Gremillion, ed., *People, Plants, and Landscapes: Studies in Paleoethnobotany*, 123–160. Tuscaloosa: University of Alabama Press.

Winterhalder, B., F. Lu, and B. Tucker. 1999. Risk-sensitive adaptive tactics: Models and evidence from subsistence studies in biology and anthropology. *Journal of Archaeological Research* 4:301–348.

Zohary, D., and M. Hopf. 2001. *Domestication of Plants in the Old World: The Origin and Spread of Cultivated Plants in West Asia, Europe, and the Nile Valley*. Oxford: Oxford University Press.

Seven

Building Trust by Wasting Time

Carl Bergstrom, Ben Kerr, and Michael Lachmann

> Time waste differs from material waste in that there can be no salvage.
> —Henry Ford, My *Life and Work*, 1922

Much of neoclassical economics is founded upon models in which wholly self-interested agents interact—but an overwhelming preponderance of empirical evidence suggests that human individuals are not purely self-interested. Rather, they express other-regarding preferences such as fairness, trustworthiness, or generosity that we collectively label *values*. Although these values surely facilitate social and economic exchange and, with them, the concomitant gains to communication and trade (see chapters 12 and 13, in this volume), we should not leap too hastily from this observation to the conclusion that these values are part of our genetic nature—and that, without them, modern commerce would be impossible.

Rather than try to justify the prevalence of other-regarding behavior in economic exchange by means of a model in which economic structure is jury-rigged on top of a system of values that was perhaps adaptive in small-group interpersonal interactions, we invert the causal chain. We argue that markets need not *rest upon* values that arose before them; instead, markets may *create* the values that allow them to function effectively. Markets are human constructions; in creating them, the participants engage in a process of mechanism design, selecting the rules of the strategic games in which they will be involved. These rule choices give rise to conventions of behavior—and where such conventions are granted normative force, they may appear to us as values.

To illustrate this process, we provide an example in which other-regarding values of fidelity and loyalty emerge within a system of primarily self-regarding individuals, as mechanisms to alter the structure of the games so as to facilitate and stabilize beneficial outcomes. These are important values in our society; when surveyed about attributes that are highly valued in a potential mate, men and women rank the values of loyalty and fidelty as among the most important [1, 2]. To Illustrate how these values can arise from conventions of behavior, we address an unlikely question: Why do we build trust by wasting time? We then look at how conventions of lengthy courtship and values of trustworthiness or faithfulness might arise as optimal strategies for purely self-interested agents.

Wasting Time

Animals court prospective mates for days or weeks, even when the breeding season is precariously narrow. Adolescents "hang out" with their friends, killing time in shopping malls and on street corners when they could be earning money or mingling with the opposite sex or developing their prowess in any number of academic or athletic pursuits. Firms invest heavily in lengthy and expensive processes of contract negotiation before initiating cooperative relations, even when these contracts may be largely insufficient to provide adequate legal recompense in the event of a unilateral defection. Economists (Spence 1973; Farmer and Horowitz, 2004) and biologists (Dawkins, 1976; Wachmeister and Enquist, 1999; Sozou and Seymour, 2005) commonly interpret these time costs either as attempts to acquire information by assessing the quality of potential partners, or as efforts to transmit information via costly signaling, with signal costs paid in a currency of time. Although these may be important components of time-wasting behavior, we propose an alternative hypothesis that may operate instead: we argue that wasting time functions to change the incentive structure in such a way that long-term cooperative behavior becomes strategically stable. This allows us to bring waste to bear upon another problem that faces those working in biological, economic, and legal fields alike: that of maintaining cooperation in the face of individual incentives for defection. While the stand-by explanations of kin selection or reciprocity may be adequate to explain some cases of cooperation, the conditions for these mechanisms seem too restrictive to explain most cooperative interactions (Hammerstein, 2003). Recent authors have started to replace kin-selection and reciprocity with explicit market analogies, affording a key role to partner choice, reputation, and mechanism design (Noë and Hammerstein, 1995; Hammerstein, 2003). The present proposal rests very much within that line of explanation.

A Simple Model of Courtship

We developed a simple evolutionary game as a model of courtship.[1] Because we want to know the cost of wasting time, we keep track both of payoffs Π from the game and the amount of time T taken to play the game, and we look at the metagame of maximizing $E[\text{payoff}]/E[\text{elapsed time}]$ as in forgaging theory (Stephens and Krebs, 1986; Stephens, Kerr, and Fernández-Juricic, 2004). This is equivalent to maximizing the weighted average E^* of payoff rates $E^*[\text{payoff}/\text{elapsed time}]$ where the rates are weighted by the amount of time spent at earning each rate.

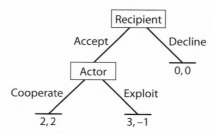

Figure 7.1. Trust game in extended form

The Basic Interaction

Two individuals engage in a basic Trust Game akin to that treated by Schwab and Ostrom in chapter 10. The intereaction begins with an opt-in/opt-out decision by the first player whom we call the "recipient." If the recipient opts out (declines), the game is over; if the recipient opts in (accepts), a simple Dictator Game follows with the second player, whom we call the "actor," in the role of dictator. The actor chooses whether to cooperate (C), yielding a payoff of 2 to each party, or to exploit (E), yielding a payoff of 3 to himself and a payoff of -1 to the recipient. Either way, the game takes 1 time unit. Figure 7.1 shows this game in extended form.

We treat this as a one-shot game with anonymous partners drawn from a large population. Individuals do not recognize other individuals, nor do they play again with the same partner once the partnership is terminated. This game has a single subgame-perfect equilibrium (Decline, Exploit), and under replicator dynamics, a population would spend almost all its time on a neutral component along which the recipient plays Decline always and the actor plays Exploit with at least $\frac{1}{3}$ probability (see, e.g., Samuelson, 1997).

Iterated Play and Payoff Rates

Now we extend the game to consider the possibility of repeated play between partners. As illustrated in figure 7.2, each round of play takes some duration of time t_b for the receiver to make a decision and an additional duration t_c for the actor to make a decision if granted the opportunity. If the recipient declines the partnership, each individual is randomly assigned a new partner and role, and this pair formation process takes some duration of time t_a. To keep the algebra maximally simple, we assume that play can continue indefinitely without temporal discounting or accidental disruption of the partnership. Although some of the details would change if we relaxed this assumption, the basic conclusions would continue to hold.

To further simplify the algebra, we proscribe all strategies that require the ability to count or more than one round of memory on the player's part. The

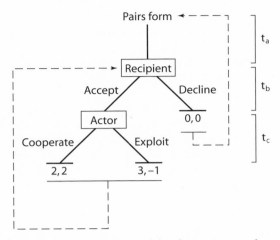

Figure 7.2. Repeated version of the choice game, with time

basic points that we are making here will survive the extension to a more complex set of memory-based strategies, but that extension comes at a heavy algebraic cost.

Actors cannot count rounds, and they only get to move if receivers opt in. Thus the actors' only strategy choice is to play a possibly mixed strategy in each round with a probability of $1-p$ of cooperating and p of exploiting. Recipients can recall only the most recent round, and we restrict them to pure strategies; thus their strategy choices are limited to DA, decline always; DE, decline if and only if exploited; or DN, decline never.[2] Defining the *time cost of re-pairing* as $t_1 = t_a + t_b$ and the *time cost of playing* as $t_2 = t_b + t_c$, the payoff rates are then as follows (see the appendix to this chapter):

$$\Pi(\text{DA}, \, p) = (0, 0)$$

$$\Pi(\text{DE}, \, p) = \left(\frac{2-3p}{pt_1 + t_2}, \, \frac{2+p}{pt_1 + t_2} \right)$$ (1)

$$\Pi(\text{DN}, \, p) = \left(\frac{2-3p}{t_2}, \, \frac{2+p}{t_2} \right)$$

Now we make a few observations. First, if the receiver plays DN, then the actor's payoff increases monotonically with p, and therefore the actors should always exploit, yielding a negative payoff to the receiver. Thus, the receiver does better with DA than with DN. Therefore, we will not expect to see receivers playing DN at equilibrium.

Next, what is the optimal strategy for the actor if the receiver plays DE? We differentiate the actor's payoff from $\Pi(\text{DE}, p)$ with respect to actor strategy p to get $(t_2 - 2t_1)/(pt_1 + t_2)^2$. When $t_1 < t_2/2$, this is strictly positive for p on

[0, 1] and thus we get an edge solution of $p=1$. When $t_1 > t_2/2$, this is strictly negative for p on [0, 1] and thus we get an edge solution of $p=0$.

As a result, when t_1 is small, actors' best response to DE is to exploit always, and receivers therefore do better playing DA than DE. When t_1 is large, actors' best response to DE is to exploit never, and in that case DE outperforms DA.

Therefore, we expect the following phenotypic equilibria of the replicator dynamics: When $t_1 < t_2/2$, we cannot maintain cooperation and, instead, receivers will decline to interact at all. When $t_1 > t_2/2$, receivers will accept interactions and actors will cooperate. Thus, when t_1 is sufficiently large, there is no problem maintaining cooperation. But what can the participants do if t_1 is too small?

Incorporating Courtship

To answer this, assume $t_1 < t_2/2$. We extend the game slightly, adding a convention where actors must first "court" recipients for a period of duration t_d before the recipient decides whether to accept or decline the interaction. This courtship period offers no direct payoff return to either player; it is simply "wasted time." This game is shown in figure 7.3.

Now we take what we learned above. By adding a courtship period, we have effectively extended the time cost of re-pairing from $t_1 = t_a + t_b$ to $t_{1*} = t_a + t_d + t_b$. When the time cost of re-pairing t_{1*} is less than $t_2/2$, actors will do best to exploit always (EA) and, in this setting, recipients can do no better than to decline always (DA) for payoff rates $\Pi = (0, 0)$. When the time cost of re-pairing exceeds $t_2/2$, actors will do best to cooperate always. With a courtship time t_d just beyond that needed to induce the actor to cooperate, i.e., $t_d = t_2/2 - t_1 + \varepsilon = (t_c - t_b)/2 - t_a + \varepsilon$, we will have an equilibrium in this game in the players court for time t_d, receivers agree to play, and actors cooperate always. This yields asymptotic payoffs $\Pi = (2/t_2, 2/t_2)$. By instituting the additional courtship stage of the game, which is purely wasteful in any direct sense given that it uses time and confers no direct payoff, we have stabilized the socially efficient outcome which otherwise would have been unobtainable. Thus the need to allow effective cooperative exchange has led to the development of a convention—wasteful courtship and subsequent cooperation—for the game.

Figure 7.4, panels A and B, provide a graphical illustration of the courtship duration necessary to stabilize cooperation by the actor, given that the receiver is playing DE. The dashed line indicates the average payoff rate should the actor exploit always ($p=1$); the solid line indicates the average payoff rate should the actor cooperate always ($p=0$). When the courtship period is short, exploiting and then finding a new partner provides a higher payoff rate. When the courtship period is long, repeated cooperation offers a higher payoff rate. This explanation closely parallels Charnov's marginal

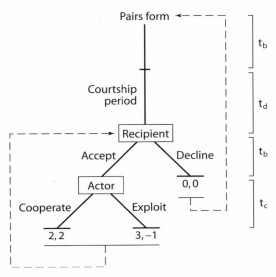

Figure 7.3. Extended version of the choice game, with time. The actor now courts the recipient for some time before the participants play the game treated previously.

value theorem used to compute the optimal duration of patch use (Charnov, 1976).

Other Kinds of Waste

In the present model, we are concerned with payoff rate defined as payoff over time; we have considered models where the participants initially reduce payoff rate by increasing the denominator, wasting time at the beginning of their interaction. Alternatively, individuals could reduce payoff rates by decreasing the numerator, and this will have a similar strategic influence on the game. For example, if payoff is accrued in monetary units, the players could waste money instead of time. Panel C of Figure 4 illustrates this graphically.

Economists have treated this latter case in considerable detail. Carmichael and MacLeod develop a mathematical model of gift exchange, in which the institution of gift exchange stabilizes cooperation among agents playing a prisoner's dilemma. Their logic is closely analogous to that presented here:

> Under this convention, an exchange of gifts at the beginning of a new match will break down mistrust and allow cooperation to start immediately. The reason is simple. A parasite in a gift-giving society will have to buy a succession of gifts, while

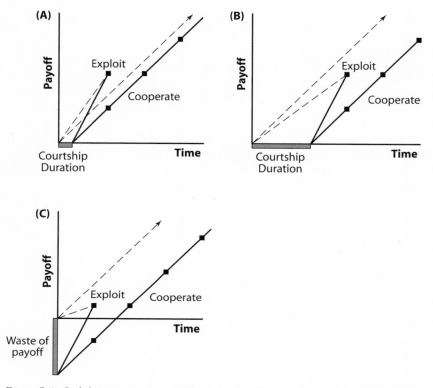

Figure 7.4. Stabilizing cooperation. The actor does best to choose the strategy which yields the highest payoff rate over time. In each figure payoff rate for exploiting and then finding a new partner is shown by the dashed line from the origin to the point labeled "exploit." The payoff rate for cooperating repeatedly is shown by the dashed line through the origin, drawn parallel to the slope of the cooperate trajectory.
(a) When courtship times are short, exploit offers the higher payoff rate to the actor.
(b) When courtship times are long, cooperate offers the highest payoff rate to the actor.
(c) Wasting payoff can stabilize cooperation as well. Given a payoff waste as shown, cooperation offers a higher payoff rate than exploitation.

> incumbent honest types need buy the gift only once. The gift-giving custom lowers
> the relative return to being a parasite. (486)

Bolle (2001) developed a model of gift giving even more closely related to the model presented here. He extended the Prisoner's Dilemma approach of Carmichael and MacLeod to the Trust Game treated here; this allowed him to deal with asymmetric games (such as pairing between males and females), where the two players have different preferences. Bolle also noted that the time costs of courtship can serve a gift-exchange function: "Evaluated by their respective wages, the time spent together [by a courting pair] is a large-scale exchange of gifts. Both may enjoy this time—but from an 'objective' point of view there is hardly any utility from this extensive being together" (1). The major differences between Bolle's model and ours result from our focus on time-wasting: to handle this in depth, we separate time from payoff, treating time explicitly within the model and maximizing the *payoff rate* rather than the expected payoff.

The theory of non-salvagable assets (Nelson, 1974; Klein and Leffler, 1981) puts forth a similar explanation of expenditures on advertising, custom furnishings, lavish showrooms, and other highly firm-specific investments that neither directly generate revenue nor are recoverable should the firm be dissolved. By this explanation, advertising is effective precisely because it is expensive. Advertisements convey no direct information about the quality of the product. Rather, through the expense of advertisement a firm signals that it has sunk appreciable cost into its reputation, and thus indirectly signals its intention to engage in long-term cooperative relationships rather than short-term exploitative ones. Although we treat investment in time here, we notice that time is, after all, an utterly non-salvagable asset.

Time differs from many other non-salvagable assets in that, unlike advertisements or custom furnishings, the costs of wasted time accrue directly to both parties. Similarly, time investment in courtship differs from unidirectional gift giving in its reciprocal nature. The *reciprocal exposure* inherent in wasting time thereby both protects the recipient from exploitation by the actor, and protects the actor from having a recipient opt out subsequent to the courtship period. Models suggest that gift-giving (whether as a signal or as a way of changing the game payoffs) may work most efficiently with gifts that are costly to produce but have low value to the recipient, because low-value gifts avoid the risk to the actor of a recipient who simply collects gifts and then opts out (Bolle, 2001; Sozou and Seymour, 2005). As Spence (1973) and later Bolle (2001) point out, time wasted (alone or, better yet, together) serves beautifully in this respect.

Discussion

The model presented here shows how a group of purely self-interested agents can establish a courtship convention and standards of subsequent behavior that resemble values such as generosity and trustworthiness, particularly should they come to be imbued with normative force. Here potential gains through exchange or trade, as modeled by a trust game, are realized not because the players come to the game with generous tendencies or placing value on faithfulness. Rather, this opportunity for gain through trade provides all participants with an incentive to expand the game itself, as in figure 7.3, so as to generate within the new game the incentives for non-exploitative behavior and extended partnership fidelity.

We believe that a vital step in the evolution of smoothly functioning social and financial institutions is the process by which the players themselves choose or modify the rules of the game (see chapter 11). By turning simple games, often with noncooperative solutions, into mechanism design problems, agents can expand the possibilities for stable pro-social interaction, thereby facilitating exchange and generating values.

Legal structure surrounding contract law serves a similar function; O'Hara, in chapter 9, describes a number of ways by which "the law's commitment to enforce contracts can, at the margin, provide added assurances that the risk to dealing with strangers is minimized." Notably, it would be inefficient for the legal system to offer too complete protection. As O'Hara points out, full legal compensation can generate a moral hazard problem in which interacting parties lack any incentive to gather relevant information about their partners' past performance and likely future intent.

Our model bears a close relationship to models of punishment. First of all, the duration of courtship necessary to stabilize cooperation (or the payoff wasted, if we are wasting payoff currency) is exactly equal to the amount that would have to be extracted as a punishment for exploitation, in order to deter exploitation by payoff-maximizing players. Once the initial courtship becomes an established custom, we can think of the initial courtship period as a sort of advance punishment in that if all recipients require courtship and all recipients break off a partnership after exploitation, the cost of exploitation will be the immediate need to go through a courtship period. In this way, the recipient's decision to break off the repeated interaction imposes a de facto cost equal to the duration of courtship, and this cost serves as sufficient punishment to deter exploitation.

A major problem with most models of punishment is that, if there is any cost to the act of punishing, then punishing itself becomes a sort of altruistic act. Wronged individuals typically do better to cut their losses and move on than to invest in costly post-hoc punishment. When the timing of the punishment shifts from after the fact (and conditioned on misbehavior) to before the interaction (and necessary to initiate any interaction), the individual incentives not to punish disappear. Indeed, all individuals have to go through the full courtship period so as to avoid providing the actor with an immediate incentive to exploit. Thus the act of imposing the punishment costs necessary for cooperation—which are the same whether they occur at the start of each partnership or after each transgression—shifts from an altruistic act to a self-interested one. This mechanism offers another solution to the second-order public goods problem associated with altruistic punishment (Boyd, Gintis, Bowles, and Richardson, 2003).

Appendix

To compute the payoff $\Pi(DE, p)$, we use first-step analysis [27], with the census point at the recipient's decision point immediately after pair formation or retention; notice that a recipient who plays DE always accepts upon reaching the census point.

Let T_n be the expected time elapsed after n plays of the subgame in which the actor actually reaches his decision point (i.e., after n plays in which the recipient accepts either because the partnership is a new one, or because the recipient has cooperated on the previous round). If the actor cooperates, then the players return to the census point after one decision interval t_b by the recipient and one t_c by the actor for a total of $t_b + t_c$. If the actor exploits, $t_b + t_c$ elapses as before, but now an additional t_b elapses during which the recipient chooses to decline and an additional t_a elapses during the formation of new pairs. This gives a total elapsed time of $t_a + 2t_b + t_c$. We can express T_n recursively:

$$T_n = p(T_{n-1} + t_a + 2t_b + t_c) + (1-p)(T_{n-1} + t_b + t_c)$$
$$= T_{n-1} + p(t_a + t_b) + t_b + t_c \qquad (2)$$

Similarly, the expected payoffs P_n and Q_n to the actor and recipient respectively are

$$P_n = 2 + p + P_{n-1}$$
$$Q_n = 2 - 3p + Q_{n-1} \qquad (3)$$

Our initial census point occurs after the first pairs are formed, before any points have been scored, so $T_0 = t_a$, $P_0 = 0$, and $Q_0 = 0$. Thus we can write the recursions (3) in explicit form:

$$T_n = n(p(t_a + t_b) + t_b + t_c) + t_a$$
$$P_n = n(2 + p) \qquad (4)$$
$$Q_n = n(2 - 3p)$$

The expected payoff rates for actor and recipient are then $\lim_{n \to \infty} P_n/T_n = (2+p)/(p(t_a + t_b) + t_b + t_c)$ and $\lim_{n \to \infty} Q_n/T_n = (2 - 3p)/(p(t_a + t_b) + t_b + t_c)$, respectively.

Notes

The authors thank Mark Grady for bringing our attention to this problem, Ted Bergstrom for sharing his unpublished manuscript on this problem, Oliver Goodenough for the epigraph, and the many participants of the Gruter Institute program "Free Enterprise: Values in Action" who provided us with numerous provocative discussions and helpful suggestions.

1. T. Bergstrom and Ponti independently developed a closely related model in an unpublished 1998 working paper [13]; the remarkable similarities between the present analysis and that previous one are enough to make one believe in genetics, vertical cultural inheritance, or both.

2. A fourth possibility, Decline if and only if not exploited, breaks up favorable partnerships while retaining unfavorable ones, and it is not be considered here.

References

Bergstrom, T. C., and G. Ponti. Long-term partnership and desertion. Working paper, November 1998.

Bolle, F. When to buy your darling flowers: On cooperation and exploitation. *Theory and Decision*, 50:1–28, 2001.

Boyd, R., H. Gintis, S. Bowles, and P. J. Richardson. The evolution of altruistic punishment. *Proceedings of the National Academy of Sciences USA*, 100:3531–3535, 2003.

Buss, D. M., and M. Barnes. Preferences in human mate selection. *Journal of Personality and Social Psychology*, 50:559–570, 1986.

Carmichael, H. L., and W. B. MacLeod. Gift giving and the evolution of cooperation. *International Economic Review*, 38:485–509, 1997.

Charnov, E. L. Optimal foraging: The marginal value theorem. *Theoretical Population Biology*, 9:129–136, 1976.

Dawkins, R. *The Selfish Gene*. Oxford University Press, 1976.

Farmer, A., and A. W. Horowitz. The engagement game. *Journal of Population Economics*, 17:627–644, 2004.

Hammerstein, P., ed. *Genetic and Cultural Evolution of Cooperation*. MIT Press, Cambridge, Mass., 2003.

Hammerstein, P. Understanding cooperation: An interdisciplinary challenge. In S. Bowles and P. Hammerstein, eds., *Dahlem Conference Report: Genetic and Cultural Evolution of Cooperation*. MIT Press, Cambridge, Mass., 2003.

Klein, B., and K. B. Leffler. The role of market forces in assuring contractual performance. *Journal of Political Economy*, 89:615–641, 1981.

Nelson, P. Advertising as information. *Journal of Political Economy*, 82:729–754, 1974.

Noë, R., and P. Hammerstein. Biological markets. *Trends in ecology and evolution*, 10:336–339, 1995.

Samuelson, L. *Evolutionary Games and Equilibrium Selection*. MIT Press, Cambridge, Mass., 1997.

Simpson, J. A., and S. W. Gangestad. Sociosexuality and romantic partner choice. *Journal of Personality*, 60:31–51, 1992.

Sozou, P. D., and R. M. Seymour. Costly but worthless gifts facilitate courtship. *Proceedings of the Royal Society of London Series B*, 272:1877–1884, 2005.

Spence, A. M. Time and communication in economic and social interaction. *Quarterly Journal of Economics*, 87:651–660, 1973.

Stephens, D. W., B. Kerr, and E. Fernández-Juricic. Impulsiveness without discounting: The ecological rationality hypothesis. *Proceedings of the Royal Society of London Series B*, 271:2459–2465, 2004.

Stephens, D. W., and J. R. Krebs. *Foraging Theory*. Princeton University Press, Princeton, N.J., 1986.

Taylor, H. M., and S. Karlin. *An Introduction to Stochastic Modeling*. 3rd ed. Academic Press, New York, 1998.

Wachmeister, C.-A., and M. Enquist. The evolution of female coyness: Trading time for information. *Ethology*, 105:983–992, 1999.

Williamson, O. E. Credible commitments: Using hostages to support exchange. *American Economic Review*, 73:519–540, 1983.

PART IV

Values and the Law

Eight

Taking Conscience Seriously

Lynn A. Stout

> We are not ready to suspect any person of being defective in selfishness.
> —Adam Smith, *The Theory of Moral Sentiments*, 1759

Imagine you are enjoying your morning coffee at a local café on a lovely spring day. Looking out the window, you see a dirty, unshaven man lying unconscious on the sidewalk. Beside the man rests a cardboard sign reading, "Homeless, Please Help." Next to the sign is a cup that holds some dollar bills and loose change. As your coffee cools, you watch dozens of people walk by the unconscious man. No one puts any money in the cup. What thoughts run through your mind?

You may speculate that the homeless man is an alcoholic or mentally ill. You may shake your head at a society that allows its citizens to fall into such a state. You may think to yourself, "People are so unkind." Here is one thought that almost surely will *not* cross your mind: "How remarkably unselfish people are! Person after person is walking by *without stealing any of the money in the cup.*"

Yet, you are witnessing repeated acts of unselfishness as you watch the pedestrians pass by without stealing from the homeless man's cup. The money is there for the taking. The homeless man is unconscious and unable to defend his property. The police are nowhere in sight. Anyone who walks by without taking money is committing an altruistic act—sacrificing an opportunity to make himself or herself better off, at least in material terms, by harming another.[1]

Civilized life in urban societies depends in large part on this sort of "passive" altruism. Newspapers are left in driveways when no one is about; brawny, young men wait peacefully in line behind frail senior citizens; people use ATM machines without hiring armed guards; stores stock their shelves with valuable goods watched over only by a few salesclerks. Each of these common arrangements requires that the vast majority of people behave altruistically, in the sense that they refrain from relentlessly taking advantage of every possible opportunity to make themselves better off. In the language of social science, these arrangements assume that most people, most of the time, act in an "other-regarding" rather than purely selfish fashion. In lay terms, they assume that most people act as if they have a conscience.

Enter *Homo economicus*

The claim that most people act as if they have a conscience, may appear so obvious to some readers that any suggestion to the contrary seems bizarre. Nevertheless, as Gintis and Khurana discuss in chapter 14 of this volume, this is not what modern professors teach their students in economics, law, and business courses. Rather, they teach students to assume that people are "rational maximizers" who behave like members of the mythical species *Homo economicus*. Economic Man does not worry about morality, ethics, or other people. Instead, Economic Man is cold and calculating, worries only about himself, and pursues whatever course brings him the greatest material advantage.

Students and professors both know, of course, that *Homo sapiens* do not always act like *Homo economicus*. Some people give to charity anonymously; some plunge into icy waters to rescue strangers. Nevertheless (the professors tell the students), these are exceptions that prove the rule. Most people, most of the time, act selfishly. Unselfish behavior is rare, unpredictable, and unworthy of serious study.

Although this cynical account of human nature has long been a staple of neoclassical economic theory (no wonder it's called the dismal science), today the *Homo economicus* approach has spread far beyond the study of anonymous markets. Led by such Nobel Prize winners as James Buchanan and Gary Becker, economists now apply "rational choice" to problems of crime, education, and corporate governance. Political scientists analyze lawmakers as self-interested actors. Public policy departments and business schools incorporate economics into the basic curriculums and serve up the *Homo economicus* account as standard student fare. Even in law schools—where the idea of conscience would seem of obvious significance—scholars embrace "law and economics," an approach that applies the tools of economic theory, including the assumption of selfish rationality, to legal rules and institutions. Are people cheating on their taxes? Increase the fine for tax evasion. Are students failing to learn to read? Tie teacher pay to test scores. Are corporate executives shirking rather than working for shareholders? "Incentivize" them with stock options.

Economic Man as Sociopath

This enthusiasm for applying the *Homo economicus* model to all types of social problems is a bit disturbing. Not to put too fine a point on it, *Homo economicus* is a sociopath. The hallmark of sociopathy is extreme selfishness as shown by a willingness "to lie, cheat, take advantage, [and] exploit" (Wolman 1987, 42). According to the American Psychiatric Association's *Diagnostic and Statistical Manual of Mental Disorders* (2000, 701–702), an individual suffers from Antisocial Personality Disorder (the formal psychiatric name for sociopathy) if he or she shows three of the following seven characteristics:

1. failure to conform to social norms with respect to lawful behaviors as indicated by repeatedly performing acts that are grounds for arrest;
2. deceitfulness, as indicated by repeated lying, use of aliases, or conning others for personal profit or pleasure;
3. impulsivity or failure to plan ahead;
4. irritability and aggressiveness, as indicated by repeated physical fights or assaults;
5. reckless disregard for safety of self or others
6. consistent irresponsibility, as indicated by repeated failure to sustain steady work or honor financial obligations;
7. lack of remorse, as indicated by being indifferent to or rationalizing having hurt, mistreated, or stolen from another.

Let us see how our friend *Homo economicus* stacks up against the list. Lack of remorse (item 7)? Obviously; why would *Homo economicus* feel bad just because he hurt or misled another, if he advanced his own material welfare? Irresponsibility and reckless disregard for the safety of others (items 5 and 6)? *Homo economicus* feels responsible for, and cares about, no one but himself. Deceitfulness (item 2)? *Homo economicus* is happy to lie any time it serves his interests. Failure to conform to social norms with respect to lawful behaviors (item 1)? Whenever and wherever the police aren't around describes *Homo economicus*.

Although *Homo economicus* is neither cranky nor impulsive—items 3 and 4—he has five of the seven characteristics on the list. Unburdened by pity or remorse, he will lie, cheat, steal, neglect duties, break promises—even murder—if a cold calculation of the likely consequences leads him to conclude that he will be better off. Like any sociopath, *Homo economicus* lacks a conscience.

Blind to Conscience

Luckily, very few people actually act like this. The American Psychiatric Association estimates that only a small percentage of the U.S. population suffers from Antisocial Personality Disorder, and many of these individuals are safely locked away in prison. Nevertheless, the *Homo economicus* model has become the dominant approach for modeling human behavior in today's economic, legal, policy, and business circles. Why? Why do so many authorities in so many fields, including not only economics but also business, political science, policy, and law, find *Homo economicus* so appealing?

One suspects his charming personality does not provide the answer. Instead, "rational choice" dominates modern legal, business, and policy discussions for two rather distinct reasons. The first is the relative simplicity and tractability of the *Homo economicus* account. Rational choice has flourished in part because academics find selfish behavior easy to predict and easy to model with a mathematical formula. This gives the rational choice approach an attractive patina of "scientific" rigor. In contrast, unselfish behavior seems quirky, unstable, and difficult to reduce to an algorithm that can be used to make hard predictions.

The second important reason why many experts enthusiastically embrace the *Homo economicus* account is the focus of this chapter, and it has to do with an odd phenomenon of perception. In brief, many people believe that rational selfishness is a more accurate and universal model of human behavior than it actually is because they *tend not to notice common forms of unselfish behavior.* As a result, they view the *Homo economicus* account as a good description of how most people behave most of the time.

This view is demonstrably false. Laboratory experiments confirm that altruistic behavior—sacrificing one's own material payoffs in order to benefit others—is, in fact, extremely common. This is most clearly demonstrated by the results of such well-known experimental treatments as the Ultimatum Game, the Dictator Game, the Trust Game, and the Social Dilemma, experiments designed to test what real people do when placed in positions where their self-interest conflicts with the interests of others. Hundreds of these sorts of experiments have been reported. With remarkable consistency, they confirm that most people routinely act as if they consider not only their own welfare but also the welfare of others.[2] Indeed, as Paul Zak describes in chapter 12, altruism can be seen not only in our behavior but in our biology, where it has been linked with the hormone oxytocin and with activity in certain brain structures.

To paraphrase a popular saying, "unselfishness happens." Many observers—including many intelligent and sophisticated observers—nevertheless resist the notion that unselfish behavior is either common or important. This tendency is illustrated by the thought experiment that began this chapter. Many people observing such a situation might be struck by the apparent heartlessness of the pedestrians walking by the homeless man without donating. Few, if any, would notice the pedestrians' altruism in refraining from stealing from the homeless man, even though refraining from taking advantage of another's weakness is an altruistic act.

Eight Reasons Why We Overlook Conscience

Why do we overlook such obvious evidence of the force of conscience? The remainder of this chapter is devoted to answering this question. It turns out that at least eight different factors conspire to make it difficult for us to recognize altruistic behavior, *even when it occurs under our noses*. Taken alone, any one of the eight can lead us to overlook the reality of widespread unselfishness. Taken together, they blind even the most sophisticated observers to our remarkable human capacity for decency.

1. Confusing Altruistic Actions with Altruistic Feelings

One reason why it is easy to overlook the reality of unselfish behavior is, oddly enough, the English language. This is not because English lacks words for altruism. Indeed, it has many, including *generous, benevolent, kind, charitable, compassionate, considerate, open-handed, warmhearted, giving, decent*, and so on. Unfortunately, none of these words distinguishes clearly between *altruistic behavior* and *altruistic emotion*. For example, if we describe someone as "charitable," it is not immediately clear whether we are describing a behavior (an actual gift of money) or an attitude (a desire to give money, if only one had any).

This lack of clarity makes it easy to confuse a claim that people often *act* unselfishly with a claim that they have unselfish *feelings*. The confusion is important, because, although altruistic action is common, truly altruistic feelings may be rare. When we show consideration for others, introspection often makes us aware of "selfish" emotions underlying our apparent decency. For example, one person might decline to steal from the homeless man's cup because he wants to avoid the painful pang of guilt. Another might yearn for the warm glow that goes with feeling good about herself. Still another might refrain from the theft out of fear of punishment in some afterlife purgatory. Because such subjective concerns "feel" selfish, it is easy to conclude that purely altruistic emotions are rare or nonexistent.

Yet even if this were true, it is essential to recognize that, however egoistic our motive, when we sacrifice our material welfare to help or to avoid hurting someone else, our action remains objectively unselfish. The individual who declines to steal from the unconscious homeless man when she could easily do so has passed up a chance to make herself wealthier than she was before. One does not need to believe in altruistic emotions in order to believe in—or value—this sort of altruistic action. Indeed, to a regulator or policymaker, it doesn't matter if "selfish" feelings prompt people to keep promises, follow rules, and help others. What matters is that they *do* keep promises, follow rules, and help others, even when they have little or no external incentive to do so.

2. The Misunderstood Nature of "Morality"

A second reason why many experts may tend to discount the force of conscience may be that, at least among intellectuals, the idea of conscience has become uncomfortably entwined with the idea of morality, and the idea of morality has acquired something of a bad reputation. This is not because intellectuals are particularly immoral (I suspect most compare favorably to, say, lobbyists or used car salesmen). Rather, many members of the intelligentsia tend to view the idea of "morality" with suspicion because they associate it with culturally idiosyncratic rules about dress, diet, and deportment, especially sexual deportment.

Diet, dress, and deportment rules vary from place to place and from time to time. Although most Westerners think it immoral to eat human flesh, Hindus reject not only human flesh but cow flesh as well, and many Buddhists shun all meat entirely. Some cultures teach that nakedness is evil; others find masturbation wicked; still others view swearing as a sin. Despite (indeed, perhaps because of) the fact that such rules vary widely among cultures, their proponents often speak of them in moral terms, a practice that conveniently relieves the speaker from the burden of having to explain or justify the rules. As a result, the idea of morality is so often and so easily misused that it becomes easy to suspect it has no content.

But universal moral rules do exist, and these rules have two notable elements in common. First, they are indeed universal, shared in some form by every orderly society. In his important book, *Human Universals*, anthropologist Donald Brown described hundreds of "universals" of human behavior and thought that have been identified by ethnographers in every culture. Moral universals include proscriptions against murder and rape; restrictions on some forms of violence; concepts of property and inheritance; the ideas of fairness and of taking turns; admiration for generosity; and a distinction between actions under self-control and those not under self-control (Brown 1991, 13–14). This is not to say that every member of every culture always complies with universal moral rules. Yet, around the globe, violations are condemned, except when the victim is deemed deserving of punishment or is not viewed as a member of the relevant community whose interests deserve consideration. (This last observation explains why some of the most enduring and acrimonious moral debates we see today, including debates over abortion, animal rights, and the fate of Palestine, center on who exactly qualifies as a member of the "community" whose welfare should be taken into account.)

Second, universal moral rules generally have to do with helping, or at least not harming, other people in one's in-group. In *The Origins of Virtue*, science writer Matt Ridley argues that, worldwide, murder, theft, rape, and fraud are viewed as "crimes of great importance" because they involve one person placing his or her own comfort and interests above the interests of another person—in Ridley's words, "because they are selfish or spiteful acts that are committed for the benefit of the actor and the detriment of the victim" (Ridley 1996, 142). Darwin reached a similar conclusion from his voyaging on the H.M.S. *Beagle*, observing that "to do good unto others—to do unto others as ye would they should do unto you—is the foundation-stone of morality" (Darwin 1871, 322).

That culturally idiosyncratic rules of diet, dress, and deportment are often described in terms of "morals" or "values" accordingly should not tempt us into becoming moral relativists, nor into dismissing the importance of conscience as a constraint on behavior. At its most basic level, conscience demands that in some circumstances we take account of the interests of others—and not only our own selfish desires—in making our decisions. Of course, even "values" that seem eccentric to outsiders—for example, the Muslim proscription against leaving the dead unburied for more than a day—are "other-regarding" in the sense that they reflect a concern for something other than one's own material interests. Universal moral values, however, reflect concern not for some abstract "other" like a divinity or religious text but direct concern for the concrete welfare of other living, breathing humans in one's community. At this level, morality is both an important concept and a widely shared one.

3. The Banality of Goodness

Once we understand that it is a mistake to confine the notion of morality to disputes over Janet Jackson's "wardrobe malfunction" at the 2004 Super Bowl, it becomes easier to discern a third factor that often keeps people from recognizing the importance of conscience. This third factor, ironically, is the sheer *ordinariness* of much conscience-driven behavior.

Many unselfish acts fail to capture our notice because they are—as the phrase "common decency" suggests—so commonplace. (No newspaper would bother to run stories with headlines like "Traveler Tips in Restaurant She'll Never Visit Again" or "Man Doesn't Steal, Even When No One's Looking!") This is especially true for what might be termed *passive altruism*, meaning instances where people refrain from taking advantage of others' vulnerability, as when passersby refrain from stealing from an unconscious homeless man.

Human beings tend to overlook such widespread evidence of conscience, because they perceive their environment with what psychologists term *selective attention*. The world is simply too busy and too complicated a place for us to monitor everything going on around us. Instead, we filter out most stimuli, taking conscious note only of the limited objects and events that capture our attention. A classic example of this is the "cocktail party" effect. When you are at a bustling party speaking to a friend, you are unlikely to take in any details from the hubbub of voices around you—until someone across the room speaks your name.

An important category of events that capture attention are *unusual* events. This is why extreme instances of *active altruism*, as when someone plunges into icy waters to save a drowning swimmer, attract attention. But no one is surprised by small acts of active altruism, as when a traveler leaves a tip, or a local resident stops to give directions to a lost stranger. Nor do we tend to notice the passive altruism of people who show others common courtesy and decline to violate the law in situations where a violation would not be detected or punished. In a society where considerate, law-abiding behavior is the norm, it is the opposite of considerate, law-abiding behavior—rudeness and cheating—that captures attention. This makes it easy for us to overestimate selfishness and underestimate unselfishness.

Writing on the war crimes trial of Otto Adolf Eichmann, philosopher Hannah Arendt famously described the bureaucrat who engineered the Nazi death camps as exemplifying "the banality of evil." (Arendt 1963) Yet goodness, too, is often banal—too ordinary and mundane to be of interest. The nation was shocked to see hundreds of looters running wild in the streets of New Orleans in the lawless aftermath of Hurricane Katrina. Few people stopped to marvel at the tens of thousands of New Orleans residents who were *not* looting. Conscience is so deeply intertwined into our daily lives that we simply do not see it—until it disappears.

4. The Peculiar Salience of Cheating

In addition to being likely to overlook routine forms of other-regarding behavior for the simple reason that they are routine, some evolutionary psychologists have argued that we are particularly likely to pay attention to selfish behavior that involves violations of commonly accepted social rules. In lay terms, we may have evolved a special sensitivity to "cheating."

Psychologist Leda Cosmides (1989) provided evidence for this thesis with the following experiment. Suppose a researcher shows you four cards laid face down on a table. The backs of the cards are marked D, F, 3, and 7. The researcher tells you that each card has a number on one side and a letter on the other. She also tells you a rule: if a card is marked "D" on one side, the other side should be marked with a "3." She then asks you to identify which of the four cards you would need to turn over in order to determine if the cards all follow this rule. What would you answer?

Most people's eyes promptly glaze over when they are asked to solve this problem, and fewer than 25 percent answer it correctly. (You would need to turn over the cards marked D and 7, to see if the other sides were marked with a 3 or a D, respectively.) So consider a different experiment. Suppose four people are drinking in a bar. One person is drinking beer; one person is drinking cola; one person looks forty years old; and one person looks fourteen. The bar owner tells you that no one should be drinking an alcoholic drink unless the person is at least twenty-one years old. Which of the four people's ID would you need to check to ensure that the rule is being followed?

A large majority of test subjects quickly figure out that they would need to check the ID of the person drinking beer (to see if she is twenty-one or older) and the person who looks fourteen (to see if he is drinking alcohol). This is curious, because the first problem and the second are *logically identical* versions of something called the "Wason selection task." Why does the second version seem so much easier to solve?

Cosmides asked experimental subjects to take on several different versions of the Wason task in order to find out why some versions seemed easier to solve than others. She concluded that people performed the Wason task best when the solution involved *detecting cheating*. Cosmides and anthropologist John Tooby went on to argue that human beings, as highly social animals, may have evolved an ability to quickly spot instances when others are not "following the rules." They hypothesized that the human brain has a specialized "anti-cheating module" that makes even unsophisticated individuals skilled at solving logic puzzles that involve cheating (Barkow, Cosmides, and Tooby 1992).

Cosmides' and Tooby's modular thesis is controversial. Their results may demonstrate not that the brain has evolved specialized systems to detect cheating, but rather that cheating sparks our interest, motivating us to apply greater brainpower to a particular problem than we are willing to apply to a more abstract and less compelling task. Whatever the reason for Cosmides' results, however, the results themselves indicate that human beings have a collective talent for spotting selfish misbehavior.

This "nose for cheating" may lead us to fixate our attention on the relatively rare instances where people behave badly (a fixation both evidenced, and reinforced, by the headlines of any major newspaper). It may also distract us away from, and make us relatively insensitive to, the many instances where people behave well. This can lead us to assume, often unrealistically, that other people are not as "nice" as we are—not as honest, responsible, considerate, or law-abiding. The result, once again, is a tendency to view the *Homo economicus* account as a more accurate and universal description of human nature than it really is.

5. Naïve Realism

Yet another psychological phenomenon that may tempt us to overestimate others' selfishness is a habit of thinking that psychologists have dubbed *naïve realism* (Ross and Ward 1995, 279). In lay terms, *naïve realism* means that we tend to assume that our own perceptions accurately reflect objective reality, and that other rational people must therefore share our perceptions and see the world as we do. One consequence of this belief is a tendency to assume that when others act in a way that seems to disagree with our views, their apparent disagreement is driven not by a different perception of the world but by some bias—especially the bias that flows from self-interest.

Suppose, for example, you are driving down the highway and suddenly realize you need to merge into the right-hand lane to take the next exit ramp off the highway. To you, it appears that there is a gap in the traffic in the right-hand lane that leaves you just enough room to merge in front of a blue Toyota. When you hit your right-turn signal, however, the Toyota speeds up and closes the gap, making it impossible for you to merge into the right-hand lane. What is your reaction? You may easily jump to the conclusion that the Toyota's driver shared your perception there was room to merge safely but sped up out of a selfish desire not to let another car get in front. You are likely to discount the alternative possibility that the Toyota driver was not motivated by selfishness but, instead, sincerely believed that the gap was too small to allow a safe merger.

Naïve realism can amplify our tendency to notice—indeed, fixate on—instances of others' apparently selfish behavior. The end result, again, is a tendency to believe that other people are more self-serving than they actually are. By natural extension, the *Homo economicus* account seems a more accurate and compelling portrait description of human nature than it actually is.

6. The Correlation of External and Internal Sanctions

Previous sections have explored psychological biases that can interfere with our ability to "see" common forms of unselfish behavior. This blindness can be seriously exacerbated by yet another important obstacle to recognizing widespread unselfishness: the fact that most well-functioning societies are structured so that actions that help others are rewarded at least to some extent, whereas actions that harm others are punished at least to some extent.

To take one obvious example, Western tort law uses the threat of liability to discourage people from negligently injuring others. The result is a correlation between internal sanctions (conscience) and external sanctions (legal damages) that makes it easy to quickly conclude that only the threat of legal sanctions, not the force of conscience, keeps people from negligently injuring others. Reputation is another kind of external force that can encourage other-regarding behavior. Many of the considerate acts we observe in daily life (waiting one's turn, donating blood) involve people who are acquainted with one another or who live in the same neighborhood or share the same workplace. This raises the possibility that apparently unselfish behavior is actually motivated by selfish concern for reputation.

Such external incentives for "good" behavior are common in orderly societies, and even a small possibility of an external incentive can make it difficult to convince a skeptic that conscience also drives behavior. Suppose, for example, that I walk down the street past several husky individuals, and none of them mugs me. (I have performed this experiment on many occasions with uniformly successful results.) I suspect the main reason I have not been mugged is that the vast majority of people unselfishly prefer not to hurt me just to get the contents of my wallet. Yet I cannot exclude the possibility that the only reason I have not been mugged is that would-be muggers fear being arrested and jailed. Nor can I prove they are not, instead, deterred by the fear that someone who knows them might observe the mugging and carry news of their misbehavior back to their community, thus harming their reputation.

The common correlation between internal and external sanctions can make it difficult to convince cynics that the incidence of unselfish behavior we observe in daily life is far greater than can be explained by legal and reputational incentives alone. (And, as we have already seen, we may be predisposed to be cynics.) The end result, as political scientist Jane Mansbridge has written, is that "we seriously underestimate the frequency of altruism when, having designed our lives to make self-interest and altruism coincide, we interpret such coincidences as demonstrating the pervasiveness of self-interest rather than altruism" (Mansbridge 1990, 141).

7. Darwinian Doubts About Altruists' Survival

The factors we have thus far considered—the conflation of altruistic action (which is common) with altruistic emotion (which may be rare); general distrust of the idea of moral constraints; psychological biases that lead us to fixate on selfish behavior and overlook unselfish actions; the common correlation between internal and external sanctions—all work to blind expert and lay person alike to the power of conscience. There is a yet another influence, however, that may discourage academics in particular from taking the idea of conscience seriously. This influence is evolutionary biology.

Someone acquainted with Darwin's theory of natural selection may easily conclude that unselfish behavior is likely to be rare because altruistic individuals must, of necessity, lose out to egoists in the Darwinian struggle for "survival of the fittest." The evolutionary argument against altruism is straightforward. Even if some people are capable of unselfish behavior, it is obviously true that many also are capable of selfishly pursuing their own welfare at others' expense. In a world where altruists and egoists interact (the argument goes), the latter enjoy a distinct advantage. Consider two primitive hunters, an altruist and an egoist. When the altruist snares a rabbit, he shares the meat; when the egoist snares a rabbit, he keeps it for himself. One can readily see which of the two will get thin in hard times. Similarly, when the altruist sees the egoist threatened by a leopard, he rushes to the egoist's defense; when the egoist sees the altruist threatened by a leopard, he uses the big cat's distraction as an opportunity to flee. This sort of thinking has lead many authorities to conclude that pop singer Billy Joel must have gotten it right when he sang *Only the Good Die Young*.

Yet the claim that Darwinian pressures should quickly snuff out altruists runs into a significant problem: the empirical evidence does not support it. Perhaps altruism among strangers should not be common, but it *is* common. This pattern is evidenced not only by everyday life but also, as noted earlier, by the results of hundreds of formal experiments.

The empirical reality of widespread altruism has prompted evolutionary theorists in recent years to revisit the old notion that altruism should be seen only among blood relatives (the "selfish gene" hypothesis popularized by Richard Dawkins [1976]). Although altruism within the family is certainly common, altruism outside the family is common as well. As touched on in chapters 6 and 7 of this volume, evolutionary biologists have proffered a number of plausible explanations for why this is so, including theories based on reciprocity (Trivers 1971); sexual selection (Miller 2000); commitment and partner selection (Frank 1988; see also chapter 7); group-level evolution (Sober and Wilson 1998); and gene-culture evolution (Richerson and Boyd 2005).

A full exploration of each of these theories lies beyond the scope of this discussion. Nevertheless, the bottom line remains clear: despite Darwinian pressures, altruistic behaviors can indeed evolve in social species.

8. *Who Studies* Homo economicus?

Finally, let us turn to one last factor that may help explain why so many experts have come to view the *Homo economicus* model of human behavior as largely accurate, despite the evidence of conscience all around them. That final factor is the nature and training of the experts themselves.

A number of experimental studies have suggested that beliefs about others' selfishness or unselfishness can be influenced by formal training, especially formal training in economics. Some evidence indicates, for example, that students who take courses in economics describe others as more selfish than students who have not taken economics courses do. (Frank, Gilovich, and Regan 1993). This finding likely reflects, at least in part, the results of direct instruction. Economic theory teaches students to assume that people are selfish "rational maximizers." (This pedagogical strategy not only encourages students to see more selfishness in others but actually glorifies selfish behavior, implying that unselfishness somehow is "irrational.") It is easy to see how individuals who receive formal instruction in economic analysis might therefore come to perceive people as more selfish than they really are. (A second disturbing possibility, which we must leave for another day, is that formal training in economics actually *makes* students more selfish [ibid.]).

A second source of biased perception that may be at least as important as overt indoctrination is that students and researchers in certain fields, including both economics and law, tend to devote the lion's share of their time and attention to studying situations where people, in fact, behave selfishly. Conventional economics, for example, analyzes anonymous markets, an arena where self-interested behavior is both common and socially acceptable. Studying law may produce a similar prejudice. Formal law is usually brought into play only when cooperation breaks down and individuals become embroiled in conflicts requiring outside resolution. This makes it easy for law students who spend hours studying case law—the documentary debris of crimes, horrible accidents, difficult divorces, and bitter contract disputes—to conclude that selfish behavior is the rule.

Thus experts who specialize in law or in economics, and others who receive economic or legal training (a category that includes not only economists and lawyers but also judges, politicians, businessmen, policy experts, and, indeed, most college graduates) may, through that training, come to perceive people as more selfish than they really are. The phenomenon is similar to that of the police officer who, after years of dealing mostly with criminals, cynically comes to expect the worst from his fellow citizens.

Nor is it inconceivable that many individuals who choose to specialize in economics, business, and possibly law tend to perceive others as relatively selfish because they are more inclined toward self-interest themselves. (One can see how someone who is relatively rational and selfish himself might find the *Homo economicus* model especially persuasive.) Although this possibility has received little formal study, a survey of Swiss graduate school students found that those who chose to specialize in business were significantly more selfish in their attitudes and dealings with others than students pursuing studies in psychology (Meier and Frey 2004). Similarly, in a *New Yorker* magazine profile of one of the most influential proponents of the law-and-economics approach, Judge Richard Posner, Posner described his own character as "exactly the same personality as my cat . . . cold, furtive, callous, snobbish, selfish, and playful, but with a streak of cruelty" (MacFarquhar 2001). Leave out the playfulness and the cruelty, and one has a wonderful description of *Homo economicus*.

Such observations are not intended to suggest that everyone who embraces the *Homo economicus* model is sociopathic, or even that one might not want to invite them to parties.[3] Rather, the point is that individuals who themselves are relatively self-regarding might be prone to assuming that others are selfish as well. This, combined with the biased perceptions that can result from

formal training in the *Homo economicus* approach or in areas where selfish behavior is the norm, might easily persuade experts in certain fields—including
law, business, and economics—to overestimate the accuracy of the *Homo economicus* model.

Conclusion

Selfishness happens. But unselfish behavior also happens. Indeed, unselfish
behavior is commonplace in our daily lives. Nevertheless, like gravity, the
force of conscience is so omnipresent we almost never take conscious note
of it.

This chapter has explored some of the reasons why. A surprising variety
of factors work together to encourage people in general—and experts trained
in economics, business, and law in particular—to overlook the role conscience can play in constraining human behavior. Some of these factors
spring from the limits of language, and others from the structure of incentives
in modern societies; still others reflect psychological quirks our species
evolved thousands or even millions of years ago, and there are others that
stem from the tastes and training of those who study law, economics, and policy. All conspire to make us greatly overestimate the incidence of selfish behavior, and to seriously underestimate the incidence of unselfishness.

Should this be cause for concern? For policymakers, scholars, reformers,
businesspeople, and leaders who care about accurately predicting and
changing human behavior, the obvious answer to this question is yes. Experts in a wide range of disciplines continue to rely routinely on the *Homo
economicus* model to analyze and devise solutions to problems, not only because they find this model tractable but also because they believe it to be a
reasonably accurate and universal description of human nature.[4] But what if
unselfishness is not rare? What if it is all around us, and we simply have difficulty seeing it?

This possibility should be of interest to anyone who deals with other human beings. If conscience is indeed an omnipresent and powerful force, and if
we can find some way to use it deliberately to change human behavior, it may
offer enormous leverage in any quest to promote a better, more just, and more
productive society. Before we can hope to employ the force of conscience,
however, we must first understand it. Before we can understand conscience,
we must first recognize it exists and take it seriously as a source of human behavior. This chapter offers some first steps in that direction.

Notes

1. Some skeptics might be tempted to argue that the passersby do not take the
money out of the cup not because they are altruists but because they selfishly fear that

if they did take money from the cup, a witness might call attention to the theft. This simply moves the problem back a step by raising the question of why any witness would make a fuss. Making a fuss to help a homeless man is an altruistic act; a selfish person would choose not to "get involved."

2. For surveys, see Blair and Stout 2001; Camerer and Thaler 1995; Dawes and Thaler 1988; Frank, Gilovich, and Regan 1993; Nowak, Page, and Sigmund 2000; and Sally 1995.

3. The author certainly hopes this is not the case, having once been a champion of *Homo economicus* herself (Barnes and Stout 1992).

4. This is not to suggest that people always act unselfishly. In some situations, the *Homo economicus* account seems a reasonable facsimile of actual human behavior. The interesting question is when it does not.

References

American Psychiatric Association. (2000). *Diagnostic and Statistical Manual of Mental Disorders*. 4th ed. Washington, D.C.: American Psychiatric Association.

Arendt, Hannah. (1963). *Eichmann in Jerusalem: A Report on the Banality of Evil.*

Barnes, David A., and Lynn A. Stout. (1992). *Law and Economics: Cases and Materials.*

Barkow, J., Leda Cosmides, and John Tooby. (1992). *The Adapted Mind.*

Blair, Margaret M., and Lynn A. Stout. (2001). Trust, trustworthiness, and the behavioral foundations of corporate law. 149 *University of Pennsylvania Law Review* 1735–1880.

Brown, Donald E. (1991). *Human Universals.*

Camerer, Colin, and Richard H. Thaler. (1995). Ultimatums, dictators and manners. 9 *Journal of Economic Perspectives.* 209.

Cosmides, Leda. (1989). The logic of social exchange: Has natural selection shaped how humans reason? Studies with the Wason Selection Task. 31 (3) *Cognition* 187–276.

Darwin, Charles. (1871). *The Descent of Man.*

Dawes, Robyn M., and Richard H. Thaler. (1988). Cooperation. 2 *Journal of Economic Perspectives.* 187.

Frank, Robert. (1988). *Passions within Reason.*

Frank, Robert, Thomas Gilovich, and Dennis T. Regan. (1993). Does studying economics inhibit cooperation? 7 *Journal of Economic Perspectives.* 159–171.

Mansbridge, Jane J. (1990). On the relation of altruism and self-Interest. In *Beyond Self-Interest* (Jane J. Mansbridge, Ed.).

MacFarquhar, Larissa. (2001). The bench burner: How did a judge with subversive ideas become a leading influence on American legal opinion? *New Yorker*, December 10, 78.

Meier, Stephan, and Bruno S. Frey. (2004). Do business students make good citizens? 11(2) *International Journal of Business.* 141–163.

Miller, Geoffrey. (2000). *The Mating Mind.*

Nowak, Martin A., Karen M. Page, and Karl Sigmund. (2000). Fairness versus reason in the ultimatum game. 289 *Science* 1773.

Ridley, Matt. (1996). *The Origins of Virtue: Human Instincts and the Evolution of Cooperation.*

Ross, Lee, and Andrew Ward. (1995). Psychological barriers to dispute resolution. 27 *Advances in Experimental Social Psychology.* 255–304.

Sally, David. (1995). Conversation and cooperation in social dilemmas: A meta-analysis of experiments from 1958 to 1992. 7 *Rationality & Society.* 58.

Sober, Elliott, and David Sloan Wilson. (1998). *Unto Others: The Evolution and Psychology of Unselfish Behavior.*

Trivers, R. L. (1971). The evolution of reciprocal altruism. 46 *Quarterly Review of Biology.* 35–57.

Wolman, Benjamin J. (1987). *The Sociopathic Personality.*

Nine

Trustworthiness and Contract

Erin Ann O'Hara

Open any business textbook these days and you are likely to find claims about the link between markets and efficiency. According to the currently predominant view, successful markets function efficiently, and efficiency is enhanced when market actors are motivated solely to increase profits.[1] In recent years, however, a growing number of prominent scholars are beginning to question these claims by focusing on the importance of interpersonal trust for the success of markets. For example, Francis Fukuyama studied the economic success of large-scale corporations and the necessity of nonkin trust to the creation and maintenance of these entities.[2] Others have focused on the fact that interpersonal trust can increase the gains from exchange by improving parties' performance and minimizing contract negotiation, performance bonding, and monitoring costs.[3] And, of course, whether A trusts B has much to do with whether B is thought to be trustworthy.

This link between success and trustworthiness is at some level intuitively obvious to most of us. Parents, teachers, scout groups, and religious leaders all recognize the importance of trustworthy behavior and the development of internal commitment norms associated with trustworthiness.[4] No doubt trustworthiness is a virtue in its own right, but it can also provide significant material benefits to those who practice it. One who performs one's promises even when performance is costly may lose in the short run. However, most people prefer to associate with trustworthy people,[5] so presumably they are provided with more opportunities over time than are the nontrustworthy.[6] Trust and trustworthiness work hand-in-hand in the sense that one is more likely to trust another who is trustworthy, and the two together can contribute to economic efficiency.

At some point, however, a potential tension between trust and market efficiency develops. Put differently, in a world where trading partners are chosen purely on the basis of interpersonal trust, people sort themselves into small trading groups consisting of family, friends, and acquaintances. This preference for known others can limit the potential gains from exchange.[7] If, instead, people were willing to transact with complete strangers, then goods and services could be allocated to those who value them more highly. The problem with strangers, however, is that the parties typically lack reliable information about each other's commitments to trustworthiness.

In well-functioning markets, middlemen often can ameliorate the diffi-culties posed by stranger transactions. Although one might be reluctant to deal with a total stranger, well-known middlemen can separately contract with people who each know them but would be strangers to one another. Think, for example, of the many buyers and sellers that WalMart brings to-gether every day. More recently, middlemen such as EBay have sprung up on the Internet to act as "trust intermediaries" between consumers and un-known vendors. In addition, rating agencies, bonding and insuring agencies, and others function, in part, to encourage people to trade with those they do not know. And third-party experts such as mechanics, architects, jewelers, and tradesmen can serve as inspectors of the quality of high-priced goods and services.

The parties themselves often can adopt strategies to encourage strangers to deal with them. For example, the supplier of goods and services can offer to cut her price in order to get others to incur the risk of trading with a stranger. Similarly, the supplier can throw in extra goods and services for free or offer a free sample. Some providers of goods and services invest large amounts of money in up-front advertisement as a type of bonding mechanism. If the money invested is sufficiently large so that it can only be recouped through sales over a long period, the provider has little incentive to perform poorly along the way.[8] Moreover, suppliers of goods and services can provide access to their facilities and business records to enable monitoring by wary pur-chasers. If, on the other hand, the supplier is wary of the buyer, then the buyer can offer to pay in full up front for goods and services.

These private mechanisms are crucial for promoting trade between strangers, and, when highly successful, they can work alone to ensure that mar-kets function smoothly. In virtually all societies where strangers routinely transact, however, more formal institutional mechanisms, including legal rules, evolve to help bolster the private mechanisms for promoting exchange. In the United States, for example, high rates of immigration, internal migration, and business enterprise instability during the late nineteenth and early twentieth centuries led to a disruption of informal interpersonal trust and to the neces-sity of providing alternative institutional mechanisms for promoting trade, in-cluding legislation and regulation.[9] This chapter focuses on a third legal institution for promoting trade between strangers: contract law and doctrine. Contract law provides external incentives that promote a party's ability to "trust that" the other will not behave opportunistically. Put differently, the law's commitment to enforce contracts can, at the margin, provide added as-surances that the risk to dealing with strangers is minimized.

To be sure, the external incentives that contract law provides may be a weak substitute for the personal knowledge of another's commitment to per-form well. At best, contract law can promote trust and trustworthiness only along a small margin. But in a large market consisting of millions of people and trillions of dollars in trade each week, small margins can translate into considerable wealth and well-being.

In at least three ways contract law helps to create a sense that one can trust that another will perform well.[10] First, if courts are willing to enforce parties' promises, then the parties know that at least a remedy for breach of promise may be possible. The availability of a remedy for breach of contract provides a type of safety net that helps to minimize the sense of vulnerability that makes trust assessments necessary in the first place. Moreover, because the breaching party can be held liable for damages he causes by breaching his contract, he has a greater incentive to keep his promise.

Second, the enormous flexibility built into contract law enables parties to structure their own transactions to help contain the amount and types of risk that each party undertakes. If a high-quality carpenter has a reputation for falling behind in his work, the parties might provide a bonus for on-time job completion. If an employee can harm her company by quitting, the employer can encourage her to stay by offering a generous pension plan or other generous employee benefits with a vesting date in the future. If the buyer of expensive goods is concerned that they may not be of sufficiently high quality that they justify the price, the seller can offer to allow the buyer to return the goods or can provide warranties to ensure that the goods will last. By enabling parties to replace off-the-shelf contract rules with commitments more carefully tailored to the parties' concerns, and by thereafter working to enforce those tailored commitments, contract law helps parties to more effectively address their own trust concerns.

Third, rules applied to contract interpretation and party conduct can be used to help support trustworthy commercial norms. For example, U.S. courts will apply a duty of good faith to the performance obligations of all contracting parties.[11] The tort of wrongful discharge limits the ability of an employer to fire at-will employees.[12] Further, fiduciary duties are imposed to set limits on professionals who routinely work with individuals who are specially vulnerable.[13] Though the law can help to promote these norms by serving an expressive function, if the norms proliferate successfully, parties will not need recourse to the courts to enforce these rules.

This chapter explores the relationship between trustworthiness and contract law, paying particular attention to how contract damages rules interact with interparty trust and trustworthiness. Its central claim is that the optimal contract damages rule provides substantial but not complete protection against breaches of contract. However, the optimal rule should be a default rule—one that the parties can contract around when special trust issues arise for the particular parties. At the same time, however, penalty damages provisions would promote suboptimal trust levels and should therefore not be enforced in courts of law. Although the trust framework does not enable a precise specification of optimal damages levels, the general damages rules proposed in this chapter as optimal look surprisingly similar to those that have developed in American contracts law.

Part 1 of this chapter briefly defines and functionally describes the concepts of trust and trustworthiness and their role in promoting economic opportunity and cooperation. Part 2 then describes the role that substantive contract doctrine plays in promoting trustworthiness, and therefore trust. Specifically, both the flexibility and the norm-shaping function of contract doctrine are highlighted in part 2. Part 3 then explores contract damages and discusses which damages measures are likely to produce optimal levels of trust and trustworthiness in society.

1. Trust and Trustworthiness

Trust and trustworthiness are amorphous and complicated concepts. The subjects fill many books and are still not fully understood, but a few points about these concepts are worth mentioning to facilitate the remaining discussion. To start, we need a working definition of the concept of trust. Although not all scholars agree on a single definition of trust, many at least agree that *trust* is a cognitive assessment tool that suggests to the assessor that it would not be unwise to make oneself vulnerable to another for the prospect of a potential gain.[14] Note that this definition limits trust to interpersonal assessments. The assessment entails a prediction about the likely behavior of another. Lay expressions intended to predict events, such as "I trust that it will rain tomorrow," do not count as trust under this definition. In addition, that assessment about the behavior of another must have some bearing on the well-being of the assessor. Predictions that do not implicate the vulnerability of the assessor also do not count as trust under this definition.

Moreover, although trust promotes cooperation, trust is neither necessary nor sufficient to induce cooperative behavior. Often one chooses not to cooperate or otherwise engage in a relationship with those one does not trust. At the same time, however, we are sometimes forced for other reasons to interact and even cooperate with individuals we do not trust. Conversely, though trusting another often does incline one to cooperate with that person, one can choose not to cooperate or engage with the other for reasons having nothing to do with the trust. Notwithstanding these caveats, trust, as a cognitive assessment tool, does incline the assessor toward interacting in potentially mutually productive ways with the target of the trust assessment.

Recently, scholars have also noted that trust and distrust are not mutually exclusive in relationships; trust and distrust can and often do coexist in the same relationship.[15] For example, a buyer of widgets may come to trust her supplier to deliver high-quality widgets at fair prices but to distrust the supplier's ability to deliver on time. This coexistence of trust and distrust is important for commercial relationships, because one contracting party can trust the other with regard to twenty-five facets of their relationship and yet be wary of the others' performance regarding a twenty-sixth facet. Legal rules might never be able to transform a fundamentally distrustful relationship into a trustful one, but they can help to convince a party to forge ahead with a transaction despite uncertainty about one facet (or a few) of the other party's performance.

In contrast to trust (a cognitive assessment tool), *trustworthiness* is a commitment to a pattern of behavior. Unlike trust, trustworthiness is a value. Put differently, it is neither socially nor individually useful for a person to have a commitment to trust others, despite the surrounding circumstances confronted by the assessor; in contrast, we tend to think that people should commit themselves to trustworthiness, even in the face of significant temptation to behave otherwise. Trustworthiness is considered to be a behavioral norm in its own right, but trustworthiness usually entails also complying with subsidiary norms of behavior. Specifically, trustworthy behavior typically is also cooperative, reciprocal, loyal, competent, diligent, and often honest.

Trust and trustworthiness are closely related in that most people respond to trustworthy behavior with trust and to untrustworthy behavior with distrust. Moreover, trustworthy people tend to trust others more than untrustworthy people do, and, perhaps as a consequence, high trust individuals are generally preferred to low trust individuals.[16]

The relationship between trust and trustworthiness is potentially complicated, however, once one takes into account the distinction between trust expressed as "trust in" and trust expressed as "trust that."[17] If A trusts in B, A predicts that B's behavior regarding A will be trustworthy based on an assessment of B's character. The more A observes B engaged in relevant trustworthy behavior,[18] the more A will trust in B. In contrast, "trust that" trust need not have anything to do with A's knowledge of B's character. A might trust that B will deliver what he has promised for a number of other reasons. Social norms and reputational sanctions, legal rules, and the hope for A's future business all can have the effect of creating incentives for B to behave well. One might argue that because "trust that" trust is driven by external incentives this category of trust should not count as trust at all.[19] However, "trust that" assessments often influence the determination that, despite possible vulnerability, one can cooperate with another for potential mutual gain. From a common sense intuitive perspective, then, these assessments appear to be about trust.

In the extreme, something akin to "trust that" trust could exist even in the absence of any information about the target's trustworthiness. For example, if a despot could credibly threaten to kill anyone who breaks a promise to him, he could fairly confidently predict that others would keep those promises. In fact, he might be so confident that he would not care to gather any information about the trustworthiness of those he deals with. Note that the despot has so much power over his subjects that he is no longer vulnerable to the risk that the subjects will fail to keep their promises. When all sense of vulnerability has been removed, it is likely that the prediction involving the behavior of others is no longer an assessment based on trust, given that trust was defined as inextricably linked to vulnerability. For those without perfect assurances that the structure or nature of the transaction/relationship will induce cooperation, however, some faith in the character of the other must accompany whatever "trust that" trust is present.

When contract law is successful, it helps to support an institutional structure that increases the likelihood of people keeping their promises, which in turn helps to minimize the "trust in" trust necessary for parties to be willing to experiment with transacting with one another. Smaller-scale transactions may be possible with little or no trust-relevant information about the other. If initial transactions are successful, the parties begin to rely on each other's trustworthiness, and larger-scale transactions (which entail greater vulnerability) are then made possible. Similarly, where parties start with more confined relationships, as they become confident that the other can be trusted regarding the relatively constrained dimensions of their relationship, they can allow their relationship to become richer, and to take on a more multifaceted nature. To the extent that contract (or other) law functions as a safety net for parties exploring the beginnings of a commercial or other relationship, those relationships and transactions are encouraged.

At the same time, however, trustworthy behavior is good for society for many reasons, and trustworthy behavior is encouraged only if people still have an incentive to gather and act on information about the trustworthiness of others. The despot cited above has no incentive to gather trustworthy information, and in turn his subjects have no reason to exhibit their trustworthy natures to the despot. Indeed, recent experimental studies indicate that externally imposed sanction systems can actually work to reduce trustworthy behavior by focusing subjects instead on the minimum they need to do to avoid imposition of the sanction.[20] The intrinsic motivation to do the right thing is apparently dampened with the imposition of the external sanction.

A government might shift to a focus on compensation to reduce vulnerability from contracting while avoiding the problem created by sanctions, but this strategy would create other problems. Consider, for example, what might happen if the government served as insurer against nonperformance in transactions. Whenever a contracting party suffered a breach, that party could apply to the government for full compensation for all losses. By entirely removing a party's vulnerability in entering into transactions, the government can certainly increase individuals' willingness to deal with strangers. Unfortunately, however, the compensation system would do little to encourage either trustworthy behavior (performing as promised) or the gathering of trust-relevant information (i.e., information about the other's reputation) prior to making a contracting decision. And, if the system exclusively relied on compensation rather than trustworthiness to drive transacting decisions, our tax bills would no doubt expand considerably.

The challenge of the law is to provide a system for handling breaches of responsibility so that people are encouraged to trade with strangers but not at the same time discouraged from gathering trust-relevant information about each other. That system, moreover, should be one that encourages trustworthy behavior by enabling trustworthy individuals to reap rewards for their virtue. To function optimally, then, contract law should create "trust that" trust that partially, but not completely, substitutes for "trust in" trust. The next two parts describe ways in which American contract doctrine serves these goals.

2. Contract and Trust

How, precisely, does American contract law facilitate exchange by providing "trust that" trust to parties? The following two subsections answer that question, first by focusing on the role of the contract doctrine's flexibility, and then by examining the norm-shaping functions of particular contract doctrines.

A. Flexible Doctrine

Contract doctrine is replete with rules and standards that apply to contracts as a matter of default. Put another way, the parties are able to contract around many, if not most, of American contract rules. These "default rules" take many forms. For example, if the parties fail to specify a price in their contract, a "reasonable price" is determined by the courts.[21] If no time for delivery is specified, a reasonable time is assumed.[22] If no place for delivery is specified, then the seller's place of business (or residence) is assumed.[23] If warranties are not specifically excluded or modified, then implied warranties of fitness and merchantability attach to the sale of goods.[24]

Contract law also builds constructive conditions into contracts to help the parties deal with potential performance disputes. For example, if a contract contemplates the exchange of multiple or continuing performance for payment, courts typically assume that the performances must be completed before the performer is entitled to demand payment.[25] Where a single performance is exchanged for a single payment, each party's duty to perform what he promised is conditioned on the other party being ready, willing, and able to perform what he promised in exchange.[26]

Notwithstanding these constructive conditions, sometimes a court will interpret a contract to be divisible—or broken up into several contracts—to entitle one party to payment before performance is complete. Whether the court is willing to interpret a contract as divisible turns on the equities of the case. Calamari and Perillo explain:

> If A and B agree that A will act as B's secretary for one year at a salary of $1000 per week, the contract is said to be divisible. Once A has worked for a week, A becomes entitled to $1000 irrespective of any subsequent events. Thus, even if A breaches the contract by wrongfully quitting, A is nonetheless entitled to $1000 less whatever damages were caused by the material breach. In effect, for the purpose of payment the contract is deemed to be divided into 52 exchanges of performances. However, if the secretary failed, without justification, to work for four days out of a particular week and the employer wished to discharge the secretary, . . . [the contract would be treated as a single, year-long contract.][27]

These constructive conditions are default rules in the sense that the parties remain free to specify that the contract is divisible, or that the employee is entitled to daily or weekly payment of wages for work completed, or both.[28]

Default rules also control other contract interpretations. If a court cannot tell whether the parties intended to enter into a bilateral or a unilateral contract, courts typically will assume that the parties intended to enter into a bilateral contract.[29] The preference for a bilateral contract turns on the fact that the "offeree" in a unilateral contract is vulnerable to the whimsical change of mind of the "offeror.". Suppose, for example, that A offers to pay B $50 for a key lime pie. B goes to the store, buys the ingredients, and begins to bake the pie. Before B actually delivers the pie, however, A changes his mind and decides to go on a low-carbohydrate diet. If the parties had entered into a unilateral contract, traditional contract law provided that there was no contract until B actually delivered the pie. This meant that A could revoke his offer at any point prior to delivery. Because of the possibility that B will make an investment but suffer a default of her right to return performance (here, payment for the pie), contract law assumes that the parties intended a bilateral contract unless they state otherwise. If A offered to enter into a bilateral contract, then the contract is formed as soon as B either promises to make the pie or begins her performance of making the pie.[30] Once the contract is formed, each side is entitled to rely on the other's promise, and each is liable to the other in the event that the promise is broken.

Similarly, if a court cannot tell whether a contract term is intended to be a condition or a promise, a court will presume that the parties intended the language to be promise.[31] If language in the contract is interpreted to be promise, then a failure to comply with the contract language is a breach of contract, but immaterial breaches by one party do not excuse performance by the other party.[32] If, instead, language is interpreted to be condition, no liability for breach of that condition follows, but the other party is excused from any contract obligations if the condition is not fully satisfied.[33] Suppose, for example, that A and B enter into an arrangement whereby B is to deliver widgets by 6:00 PM on Friday, and A is to pay $300 for the widgets. What happens if B delivers the widgets at 6:30 PM instead of 6:00 PM? If the language in the contract is tantamount to a promise to deliver by 6:00 PM, then B is liable in damages to A for delivering late, but A still has an obligation to take and pay for the widgets. If, instead, the language is condition, then A's obligation to take and pay is extinguished when B fails to satisfy the condition that he deliver by 6:00 PM. A can walk away from the deal entirely, but B is not liable for failing to satisfy the condition. By interpreting the language as promise rather than condition, B is protected from A's using a minor deviation to defeat B's right to any return performance. And, if A really does want the right to walk away from the deal at 6:01 PM, she can condition her obligation on timely delivery so long as she makes her condition clear when they form their deal.

In addition, the doctrines of impracticability, frustration of purpose, and mistake each defines circumstances where a party will be excused from performing a promise if circumstances either are not as he supposed or have changed between contracting and performance.[34] Under each of these doctrines, if a basic assumption of the contract turns out not to be true, courts can release the harmed party from the contract without liability.[35] For example, when parties contracted for the sale of a barren cow, the seller was permitted to rescind the contract when it was discovered that the cow was pregnant and therefore worth more than ten times the value of a barren cow.[36] For another example, when an individual contracted to pay premium prices to let an apartment in England for two days to watch the coronation procession for King Edward VII, he was excused from paying remaining amounts due when the parade was canceled, frustrating the entire purpose for which he was willing to pay for the lease (a fact known to the apartment owner at the time of contracting).[37] In these cases, courts consider the terms of the contract, its price, and the surrounding circumstances to glean a sense of the risks that each party likely was undertaking when entering into the contract.[38] These doctrines function as default rules in that the parties can obviate the courts' inquiry by themselves assigning risks regarding the nature of the matter contracted for or the circumstances surrounding the contract itself.[39] For instance, the cow contract can provide that the sale is final and that the cow is being sold "as is," or it can state that the seller assumes the risk that the cow might turn out to be fertile.

These default rules and others give the parties wide latitude to structure their rights and obligations under a contract. Parties can define payment, quantity, duration, delivery, warranties, terminations, specific obligations, and so on. Moreover, within broad limits,[40] parties can allocate or divide risks. For example, in the lease agreement discussed above, the contract could provide that in the event the coronation is canceled, the lessee is still obligated to pay half (or some other portion) of the agreed rental amount.[41]

Default rules serve several functions that are recognized in the contract law literature. These rules enable parties to economize on contract negotiation and drafting by providing a set of off-the-rack rules that will apply to a contract when the parties do not think it worthwhile to tailor a rule to suit their individual contract.[42] In contrast, default rules sometimes serve, instead, as penalties that force the parties to communicate with each other by specifying the terms they wish to govern their contracts.[43] In construction contracts, for example, courts will impose a default rule that a contractor is not entitled to any payment for his work until the job is completed.[44] Most contractors cannot afford to wait until the end of a job to receive payment, but courts are at a loss to figure out on their own what portion of payments should be due, and when.[45] The default rule forces the contractor to draft a payment schedule for the job. Default rules also disenable a party from relying on the indefiniteness of contract terms to defeat contract liability where the parties clearly acted as though they intended to form a contract.

I suggest here a third function of default rules: many seem intended to promote optimal levels of trust between contracting parties. How do default rules work to promote optimal trust levels? As indicated earlier, many default rules direct courts (or juries) to write in reasonable terms when the parties have failed to specify their own terms. In the event of a dispute, then, a neutral arbiter rather than one of the interested parties will fill in the missing term. A similar focus on neutrality (or at least a denial of biased influence) is reflected in the canon of construction that directs courts to construe a contract against its drafter.[46] The canon enables the contract to be construed in favor of the drafter where it is clear but in favor of the nondrafter where the terms are not clear. The canon thus serves to limit the ability of the drafter to use hidden devices to dictate the terms of the contract. Each of these default rules serves to protect a party against the unscrupulous or unreasonable tactics of the other.

The default rules promote roughly optimal trust in other ways as well by considering the equities of individual contracting problems. For example, the employment contract discussed earlier provides an example of how constructive conditions can enable one party to receive return performance when it is typically due while disenabling the other party from taking advantage of the technicalities of the conditions. The balance is a subtle one for contract law, because contract rules that protect one party from opportunistic behavior can be so strong that it induces the protected party to engage in opportunism himself. More specifically, a party generally should not be entitled to performance by the other unless she is planning to perform herself. However, where payment for part performance can be reasonably allocated, then the partial payment should be forthcoming. The default rules applied to constructive conditions seem to carefully calibrate parties' mutual trust concerns. Notably, however, because the rules are default rules, they can be reworked to deal with unilateral trust concerns and with more nuanced mutual trust issues. So, for example, if A as B's secretary is habitually late for work, B might offer A $2,000 as a bonus conditioned on her not being late for work more than five days in a given year. The explicit condition in this agreement enables B to address a unilateral trust problem—that of A's unreliability.

Parties' contracts routinely deal with trust concerns that are particular to the parties or to the type of transaction. Consider the purchaser of goods or services who is a bad credit risk. The seller may be unwilling to deal at all with the purchaser under the default rules which suggest that goods or services must be tendered before the seller has a right to payment.[47] The parties can strike an alternative agreement, however, whereby the purchaser agrees to pay in full before receiving the goods or services. By switching the default rule regarding order of performance, the specific trust concerns of the provider can be addressed.

Note also that the default rule is set in such a way that a party's effort to deviate from the default rule in negotiations signals a trust problem. The effects of this signaling are not fully understood at this time. Some experimental evidence indicates that people respond to distrust signals with untrustworthy behavior.[48] For example, consider what might happen if one party has discretion in performing services and the other indicates a clear lack of trust in the service provider's capacity or willingness to perform those services well. If the untrusting party attempts to fill the contract with duties intended to protect himself against the service provider, the service provider may well decide to shirk where his discretion remains. However, it is unclear to what extent these findings would be replicated in other transactional settings. Suppose, for example, that a seller has concerns about a buyer's financial reliability. To assuage his concerns, he might request that the contract include a term which states that the buyer defaults on his right to receive those goods and services in the event of nonpayment. In this situation, the buyer might decide not to deal with the seller, but if he chooses, instead, to enter into the contract, he likely has no incentive to behave in an untrustworthy fashion. Even if the buyer does respond to this distrust signal with untrustworthy behavior, the seller might be able to remove his own vulnerability by requiring prepayment, and more transactions may be facilitated on net than might be the case without the seller's protective device. Moreover, reliable prepayment can help a buyer restore his reputation for trustworthiness so that, in the future, prepayment may no longer be required. Note that the experimental games showing distrust responsiveness do not incorporate reputations about the parties' trustworthiness, nor do they make it possible for one or both players to choose not to play with each other. In those settings, threats of punishment are imposed on individuals without good reason, and players respond with distrustful behavior. Where untrusting signals are grounded in a party's poor reputation or prior unreliable behavior, however, the untrusted individual might well respond with trustworthy actions. In addition, contractual terms addressing modest trust concerns with reasonable mechanisms might not provide distrust responsiveness in the other party at all.

Parties can contract around other types of default rules to handle unilateral trust issues. Consider, for example, the "best efforts" requirement in exclusive dealing contracts. If a person agrees to be the exclusive agent for the sale of a particular type of product in a particular area, courts will assume that the agent promises to use best efforts to promote the sale of the goods and the manufacturer/seller obligates itself to use best efforts to supply the goods needed.[49] The rule enables each party to rely on the other to perform the contract diligently. At the same time, the rule is a default rather than a mandatory rule. A seller or buyer need not obligate himself to use best efforts if he does not wish to do so.

Why might a party not wish to commit himself to use best efforts in the context of an exclusive dealing contract? To answer that question, let us turn to the facts of a famous exclusive dealing case, *Wood v. Lucy, Lady Duff-Gordon*.[50] Here Wood and Lucy entered into a contract whereby Wood was to have the exclusive right to sell Lucy's name to endorse other products (she had become famous herself for high-end ladies' apparel design). Half the proceeds that Wood generated would be paid to Lucy. In the case itself, Lucy entered into a very sizable contract behind Wood's back, and Wood sued to enforce his right to half of the proceeds from the sale. To enforce Lucy's promise to pay Wood, Judge Benjamin Cardozo needed a reciprocal promise by Wood to Lucy, but it did not appear on the face of the contract as though Wood had promised to do anything. Judge Cardozo implied an obligation on the part of Wood to use reasonable efforts to promote Lucy's endorsement. It is worth pondering, however, whether Wood, in fact, would have wanted to undertake an obligation to use reasonable efforts or, using today's default rule, best efforts. Lucy apparently was quite flamboyant, and she was also a spendthrift. Given her impulsive and erratic ways, she was quite capable of quickly besmirching her own name such that it would no longer be a valuable commodity. If Lucy could not be trusted to protect her own reputation, would Wood wish to enter into a moderate-term contract where he was required to use any efforts to sell that reputation? The high commission rate Lucy promised to Wood alone gave the latter a clear incentive to sell her name wherever a profit could be made, so it is not clear that Lucy really needed the protection of an implied "best efforts" clause on top of the explicit compensation clause.[51] Perhaps the parties structured their deal to handle the special problem of Lucy's untrustworthiness while maintaining Wood's incentive to use efforts where it made sense.[52] If so, the case provides an illustration of why the best efforts requirement at most should be a default rather than a mandatory rule.

Note that, in some of these examples, the default rule was intended to address one type of common trust problem, and parties who are contracting around the default rule have different trust concerns that are exacerbated by the default rule. In the exclusive dealing case, for example, the best efforts requirement is intended to protect people from the opportunistically lazy behavior of the one person who has the exclusive rights to deal on behalf of the other. The one who grants the exclusive right must worry that the grant makes her vulnerable to the one who possesses the exclusive right. The best efforts requirement helps to alleviate some of these concerns. In the Wood case, however, Wood apparently had more reason to worry about Lucy than vice versa. If he wished to protect himself from Lucy's volatile behaviors today, he could eliminate his obligation to use best efforts in their contract.

Nuanced mutual trust problems can also be addressed by contracting around default rules. Consider again the constructive conditions that are written into construction contracts. Suppose, for example, that a contractor and a home owner enter into a contract for major home renovations and additions. The contractor might well worry that the home owner lacks the solvency necessary to pay for the work. Moreover, the contractor might worry that the home owner will be demanding and difficult to please in order to get out of paying for at least part of the work. The home owner typically worries that the contractor will attempt to cut corners with low-quality work and will delay completing the small details of the work in favor of more lucrative jobs. The home owner does not want to have to pay the contractor up front, and the contractor often lacks both the liquidity and the trust necessary to perform all the work before receiving payment from the home owner.[53] The parties often can successfully treat these trust concerns with specific contract terms. For example, construction contracts often are contingent on satisfactory credit checks. Moreover, the home owner might be required to pay a significant down payment, as well as specified periodic payments during the course of the contractor's performance. In addition, a portion of the final payment is typically held back until a neutral, third-party expert certifies that the work was completed according to the contract.[54] Finally, these contracts often provide specified penalties in the event that completion of construction is delayed.[55] By enabling the parties to contract around the off-the-shelf constructive conditions, the parties can much more creatively address their concerns.

In sum, the default rule nature of most contract law results in an enormously flexible body of contract doctrine that treats many common trust issues while simultaneously enabling contracting parties to circumvent the default rule to deal with nuanced or specialized unilateral or mutual trust problems unique to the particular parties or transaction. The next section shifts focus to discuss the mandatory contract rules found in American contract law. These mandatory rules also serve varied functions, but here we focus on their role of shaping the norms by which contracting parties deal with each other.

B. Norm-shaping Doctrine

Although American contract law is primarily made up of default rules, some contract law takes the form of *mandatory rules*, or those rules and obligations that the parties cannot circumvent in their contract. Mandatory rules can take two different forms: statutory regulation and common law contract doctrine. Regulations are laws that replace contract law with rules that contracting parties must comply with. For example, minimum wage laws, regulations of the Occupational Safety and Health Administration, and Social Security tax payment obligations all affect the terms as well as the performance of employment contracts. Most of these regulations result from legislative efforts to change the employment relationship previously defined by the common law of contracts. This section deals with the second type of mandatory rules—those that are part of the common law doctrine applied to contracts. These latter mandatory rules were developed alongside the default rules by judges who intended to limit the flexibility of contract law to impose certain basic standards of conduct on the parties.

The point of many mandatory rules is to help shape the norms by which contracting parties conduct themselves. Contract law serves a type of expressive function by communicating to tradesmen that certain standards of decency will be required of their conduct. The claim here is not that courts are needed to get contracting parties to behave decently—in most cases, the norms of decency likely developed prior to court involvement. However, the courts are needed to prevent opportunistic people from taking advantage of the decent behavior of others and to help strengthen the force of these values in the commercial arena. By helping to shape and strengthen the values according to which commercial agents operate, contract law works to reinforce the expectations of contracting parties in each other's performance. If values associated with trustworthiness are strongly embedded into the norms of the marketplace, then strangers are more likely to trust each other, at least on an experimental basis.

Three basic mandatory rules that operate on contracts are (1) the good faith obligation; (2) the prohibition against unconscionable contract terms; and (3) fiduciary obligations. The first two mandatory rules operate on all types of contracts, whereas the third, fiduciary obligations, are imposed only on contracts where one party is highly dependent on (and therefore unilaterally vulnerable to) the other.

First, consider the good faith obligation. Every contract includes an obligation by both parties to execute and perform the obligations under the contract in good faith. According to the Uniform Commercial Code, good faith means "honesty in fact" and, at least in the case of merchants, compliance with reasonable commercial standards of fair dealing.[56] Although it is not entirely clear what the good faith obligation entails, it has been described as excluding behavior inconsistent with common standards of decency, fairness, and reasonableness as well as behavior that is inconsistent with the parties' common purposes and justified expectations.[57] Parties are not permitted to use contract language to advantage themselves opportunistically. Thus, when a buyer promises to buy a house on the condition that he finds suitable financing, he cannot get out of the contract by not looking for financing. Notwithstanding the absence of specific contract language on the point, the buyer must seek financing in good faith.[58] Similarly, when a contract for the sale of potatoes required that the potatoes meet the buyer's satisfaction, the buyer could not claim that he was dissatisfied merely as an excuse to cancel a contract after the market price fell dramatically.[59] Finally, when a party has the right to terminate a contract at will, he cannot exercise his right to terminate opportunistically or for reasons that might violate public policy.[60]

Judge Richard Posner has described the good faith obligation as one designed to minimize parties' defensive behaviors and their accompanying costs.[61] Moreover, the duty of good faith can impose an obligation on the part of one contracting party to perform his obligations with the other party's interests in mind, although complete selflessness is not required.[62] Under American contract law, it is worth noting, the duty of good faith applies to contractual performance but does not apply to contract negotiations.[63] In effect, contracting parties are entitled to behave in a completely self-interested fashion when they are choosing contracting partners and negotiating contract terms. Once the relationship has been formed, however, the parties are expected to treat the contract as a kind of partnership—the relationship is supposed to benefit both parties, and performance or termination which deprives one of the parties of the substantial value of the contract is simply unacceptable. In short, the formation of a contractual relationship changes the parties' expectations to one of a certain degree of mutual trust:

> Before the contract is signed, the parties confront each other with a natural wariness. Neither expects the other to be particularly forthcoming, and therefore there is no deception when one is not. Afterwards the situation is different. The parties are now in a cooperative relationship the costs of which will be considerably reduced by a measure of trust. So each lowers his guard a little bit, and now silence is more apt to be deceptive.[64]

Unlike the contract doctrines mentioned earlier, the duty of good faith is a mandatory duty imposed on all contracting parties which cannot be eliminated by incorporating a contrary term into the document.[65] The good faith obligation does remain flexible in that its contours in an individual case very much depend upon the specific obligations the parties have undertaken in their contract. Nevertheless, the parties cannot shape their obligations under the contract such that the good faith obligation (or lack thereof) remaining has become manifestly unreasonable.[66]

Consider also the unconscionability prohibition as a second mandatory rule that can enhance trust between contracting parties. Whereas the good faith obligation polices the exercise of discretion provided by the silence in the terms of the contract, the unconscionability prohibition polices the specified terms of the contract. Parties are given broad control to craft the terms of their agreement, but the unconscionability doctrine remains available to courts to strike down contracts or individual clauses that they find fundamentally unfair. The unconscionability doctrine is not invoked frequently by courts. When it is invoked, typically the substantive unfairness of the terms of the contract is coupled with circumstances that lead the court to believe that one party did not understand or have the opportunity to read and assent to the terms of the contract. For example, when a poorly educated person with limited English-language skills signs an unfair contract, a court is more likely to strike fundamentally unfair terms as unconscionable.[67] When a welfare mother purchased furniture but was never given a copy of the sales and financial agreement that gave the company unfair authority to repossess the furniture, the court stepped in to scrutinize the repossession term.[68] Moreover, complicated boilerplate language presented in detailed contracts with inconspicuous placement of unfair terms also presents circumstances where courts will scrutinize the terms more carefully.[69] Finally, if consumers are unable to get better terms elsewhere, the courts are more likely to step in to police the bargain.[70] In each of these cases, courts prevented a contracting party from exploiting the other party's ignorance or vulnerability to capture an unfair advantage. By policing contract terms where they are least likely to be read or understood, courts minimize the extent to which consumers need to be wary of the other party's contract drafting.

Finally, the imposition of fiduciary obligations in some contract relationships further enhances trustworthiness. Fiduciary obligations are duties that rise above and beyond the good faith obligations imposed on ordinary contracting parties. They are imposed on one of the parties in certain categories of contracting relationships. These relationships are generally characterized by high levels of dependency or vulnerability for one or both parties. Examples of fiduciary relationships include the doctor-patient, attorney-client, parent-child, and corporate officer/director-shareholder relationships.[71] The specific content of the fiduciary obligations varies with the type of relationship,[72] of course, but often the fiduciary obligations include a duty of care and a duty of loyalty. The duty of loyalty requires that the fiduciary behave in ways that are honest and do not conflict with the interests of the vulnerable party. Common fiduciary obligations based on this duty of loyalty include a duty to segregate and earmark entrusted funds, a duty to avoid or at least disclose any potential conflicts of interest between the fiduciary and the vulnerable party, and a duty to provide the vulnerable party with relevant information and periodic accounting.[73] The duty of care requires that the fiduciary exercise reasonable care and skill in carrying out her duties. The duty of care often entails an obligation to gather pertinent information, carefully deliberate prior to acting, and exercise reasonable skill when acting.[74]

Together the duties require that the trusted party give extra efforts to the vulnerable party to perform her duties in a manner that is careful, reliable, honest, forthcoming, and loyal. Moreover, a certain selflessness is required of the fiduciary. As Judge Cardozo famously observed:

> Many forms of conduct permissible in a workaday world for those acting at arm's length, are forbidden to those bound by fiduciary ties. A trustee is held to something stricter than the morals of the market place. Not honesty alone, but the punctilio of an honor the most sensitive is then the standard of behavior. As to this there has developed a tradition that is unbending and inveterate.[75]

Fiduciary duties acknowledge, in part, that one of the parties to the relationship must place great trust in the other for the relationship to be effective. Attorneys cannot effectively serve their clients' needs, for example, if the client does not trust the lawyer with sensitive information. Beneficiaries of trusts must have faith in the honesty of the trustee. Patients must trust in the competence of the doctor to treat the patient's ailments. Regardless of whether the fiduciary duties actually work to enhance the trust of any individual vulnerable party, they certainly work to foster the trustworthy behaviors of the fiduciaries. To the extent that fiduciaries as a group behave in trustworthy fashion, those of us who are vulnerable to the fiduciaries take more comfort that these relationships will work to our benefit. In short, fiduciary duties help to strengthen norms of trustworthiness in relationships where high trust is both necessary and beneficial.

3. Contract Damages and Trustworthiness

A. Substantial but Incomplete Compensation Is Optimal

Courts can award compensation to those who suffer financial losses from another's breach of contract, and these damages awards serve two important functions related to trust. First, if a contracting party knows that he might be able to obtain a damages award if the other party breaches, then the amount of vulnerability the party perceives from entering the contract is reduced. Second, the possibility of liability for breach of contract provides an extra incentive for a would-be breacher to honor his contract. Here, too, the party making a trust assessment perceives that his vulnerability is smaller than it would be without contract law damages. At the margin, this legal protection might make the difference between entering into a contract with a stranger and not.

If the point of the law were simply to maximize the extent to which strangers will transact with each other, then we might expect contract damages to be fully compensatory. If each party knows that he or she will suffer no harm from dealing with a stranger, then trust assessments become irrelevant, and trust-relevant information is no longer a basis for separating out potential contracting parties.

In theory, contract damages are fully compensatory under American contract law. The most common measure of damages for breach of contract is expectations damages; the non-breaching party is entitled to the monetary equivalent of the benefit of his bargain.[76] Suppose, for example, that A contracts to buy a porcelain vase from B for $50. Later, B breaches that contract by calling A and telling him that he will not sell the vase to A after all. A has paid no money to B, but A claims that she has suffered a loss because A is unable to find a substitute vase and, had B actually delivered the vase, then A would have had a vase with an actual market value of $75. If A can prove the value of the vase, A would be entitled to recover $25 in contract damages from B. Here, the benefit of A's bargain was that she had contracted to get a $75 vase for only $50. She is entitled to the value she was deprived of, even if she in fact never paid over a penny of the agreed upon $50 to B. If A had paid the $50 to B before B announced his breach, then A would be entitled to $75 in damages from B. In that case, she would have performed her half of the bargain—paying over the $50 she promised—and now she would be entitled to the value of what was promised her—a vase worth $75 or, here, the $75.

Consider another example. Suppose that C promises to paint D's house for $1,000 and then breaches before D pays her anything. Suppose also that when D seeks a substitute painter, he discovers that it will cost him $1,500 to get someone else to paint the house. Here, D is entitled to the difference in cost between the contract price, $1,000, and the substitute price, $1,500, or $500. D is entitled to the benefit of his bargain, which is to get his house painted without taking more than $1,000 out of his own pocket. Note that in both these examples, in the event of a breach, the non-breaching party is fully compensated, not just for her losses but for her lost gains.

It turns out, however, that although in theory contracting parties are entitled to full compensation, in practice the damages rules almost never work to provide plaintiffs with full compensation after litigation. First, a substantial portion of the damages award typically is paid to the plaintiff's attorney to cover legal fees and court or arbitrator costs. Somewhere between 30 and 45 percent of a damages award typically is used to pay someone other than the plaintiff.[77] In addition, the plaintiff is not separately compensated for the time and effort that must be expended to gathering evidence, finding the defendant, and proving a right to compensation.

Second, damages are recoverable only to the extent that they are provable with reasonable certainty.[78] Often, especially in the case of lost profits, claimed damages are not recoverable because they are deemed speculative.[79] This reasonable certainty standard is not unique to contracts, but in other areas of law—torts for example—supplemental damages awards are available to help alleviate the non-provable elements of damages. In tort, the plaintiff can recover damages for pain and suffering, and, in some cases, plaintiffs are entitled to punitive damages, which are awarded above and beyond the compensatory damages. In contract actions, however, with rare exception, neither pain and suffering nor punitive damages are recoverable.[80]

Moreover, whereas in many civil law countries plaintiffs can be awarded specific performance as a matter of course, it is much more difficult to obtain specific performance in American courts.[81] Specific performance is a remedy whereby the breaching party is ordered to perform the contract instead of paying damages. In the painting contract above, if a court awarded specific performance, it would order C to paint the house. In the vase sales contract, it would order B to turn over the vase to A. American courts will not order a party to perform personal services in a breach of contract action,[82] so specific performance would be unavailable in the painting contract. In the vase sales contract, specific performance would only be available if B still had the vase and the vase was so rare that it was irreplaceable.[83] Otherwise, non-breaching parties are stuck asking for monetary damages even though the true damages, for example, subjective value, might be hard to prove.[84]

Finally, the largest contract damages award in the world does the plaintiff no good if the defendant or his assets cannot be found or attached to satisfy the award. The harder it is to actually satisfy a judgment, the greater the vulnerability that remains when contracting with strangers. To make contract damages more fully compensatory, the government could set up a fund that plaintiffs could use to recover unpaid damages claims in case the defendant disappears or is effectively judgment-proof. Yet rarely, if ever, are such funds put in place. Together, these factors leave the average plaintiff significantly undercompensated.

Some legal scholars have criticized the practical undercompensation of contract plaintiffs. To discourage inefficient breaches, and to encourage people to keep their promises, these scholars have argued that contractual remedies need to be expanded. Some call for increased availability for specific performance and injunctions whereas others have called for recovery of punitive damages and nonpecuniary damages, as well as relaxation of the proof requirements associated with recovering expectations damages.[85]

From a trust perspective, however, it is not optimal to provide the non-breaching party will full compensation. If we treat contract damages measures as a set of default rules that are at least to some extent modifiable by the parties, then to promote optimal trust contract damages measures should provide substantial but not complete compensation. Substantial compensation serves the purpose of encouraging strangers to be willing to experiment with doing business with each other on a small scale. Complete, guaranteed compensation provides greater assurances, of course, but complete compensation would come at a very high price to society. The societal goal of encouraging strangers to transact with each other conflicts at some point with the social goal of encouraging trustworthy behavior in its own right. Consider, first, the American damages rules that provide substantial but not complete compensation on average. With this protection, parties are willing to enter into small-scale exploratory transactions with each other even if the protection is not sufficient to induce strangers to enter into large-stakes transactions right away. By inducing the small to moderate transactions, strangers experiment with each other and each knows that the other may be willing to enter into a larger, more extensive business relationship if the exploratory transactions prove successful. Put differently, in a world with substantial but incomplete compensation, parties still require that the other party perform his side of the bargain in a trustworthy fashion as a precondition to a willingness to take more substantial steps.

By forcing the non-breaching party to incur a co-payment in the event of breach, trustworthiness is encouraged. Those who conduct themselves in a trustworthy fashion reap rewards, and those who are not trustworthy suffer economically. Trustworthiness provides individuals with significant personal benefits in that we all prefer to socialize with people who are trustworthy. If the law can work to bolster these personal benefits with financial rewards, then trustworthiness is further encouraged. In order for the trustworthy to reap financial rewards, compensation for breach of contract must remain incomplete. To understand this last point, consider a market where every person was guaranteed complete compensation for breach of contract. In that market, parties would have no incentive to discriminate between trustworthy and untrustworthy trading partners. With no costs to breach, people of commerce have little need for obtaining, processing, or acting on trust-relevant information. In other words, compensation can induce transactions that encourage trustworthiness between strangers, but perfect compensation induces transactions without regard to trustworthiness.

Experimenters have discovered that subjects behave in a less trustworthy fashion when their contractual partners have an incentive to enter into that contract even in the event of breach.[86] When the compensation for breach of contract is sufficiently high that parties no longer discriminate among possible trading partners, dishonest behavior can actually become more prevalent, at least to the extent that the compensation does not work to deprive the breacher of all benefits for breaching. One might then think of the relationship between trustworthiness and contract damages as a kind of Laffer curve. As compensation rises from zero, more trade and therefore more potential gains from trustworthy contract behavior are created. At some point, however, the compensation for breach of contract rises to the point where the gains from trustworthy contract behavior actually begin to fall. As a matter of contract default rules, then, there presumably is an optimal compensation percentage that is less than 100 percent of the damages. Trustworthiness, not trust, is maximized.

The preceding analysis assumes that trustworthiness rather than trust should be maximized. But why? Trustworthiness is generally beneficial for several reasons. First, as a value, it is important in its own right. We seem to have a preference for people who are trustworthy. As humans, we derive pleasure from seeing people act loyally, honestly, dependably, competently, and so on. Second, we feel more secure in communities where people tend to behave in a trustworthy fashion. The more commonly trustworthy the actions of those around us, the less often we find it necessary to engage in the unpleasant task of assessing the extent to which another can be trusted not to act against our interests. If people feel free to enter into contracts without being trustworthy, perhaps their more general commitment to trustworthy behavior would also be relaxed. If so, we would be more generally impoverished under a contracts regime that failed to encourage trustworthy behaviors. Third, trustworthiness provides efficiencies to a market at lower cost than other mechanisms for promoting trade. Consider a regime under which plaintiffs were guaranteed full compensation. This guarantee would require that plaintiffs be compensated by the government or a common pool produced through some form of product/ service taxation if the defendant was unreachable or unable to satisfy the full compensation judgment. Both these mechanisms for full compensation would produce some dead weight losses. More important, however, contracting parties would discover over time that their individual trustworthiness has little to do with their future business opportunities. If the gains to trustworthy behavior are eliminated, then in this otherwise competitive environment, people in business would no doubt be tempted to cut corners to eliminate the costs to trustworthy behavior as well. This will produce more breaches and more reliance on funds to compensate plaintiffs. The need for third-party funding of the common funds would grow.

B. Changing the Default Damages Rules

At best, this chapter has successfully argued that to promote trustworthiness, contract damages ordinarily should not be fully compensatory. Although American contract damages rules generally lead to substantial but not complete compensation, the impact of those rules do not fall evenly onto the shoulders of all contracting parties. For example, foreign and poor defendants are less likely to have assets available to satisfy judgments. Further, because some types of damages—for example, lost profits and subjective value—are significantly less likely to be recoverable than others, parties who are more likely to suffer these types of damages will probably face a greater gap between damages and compensation than others will. Although no contract plaintiffs receive full compensation for their losses, the proportion of damages ultimately recoverable will vary significantly from one party to another. Some plaintiffs' recovery will no doubt be insufficient to promote socially optimal levels of trust. If these parties know that they are likely to suffer undercompensation, then presumably they should be able to supplement the default damages rules to enhance the likelihood that the vulnerable party will be willing to contract.[87]

Even if the damages rules affected all contracting parties uniformly, the trust issues that parties confront in their transactions vary significantly from contract to contract. One party could be particularly vulnerable to post-contract opportunism, for example, if she is required to invest a great deal in order to make profitable use of a good or service to be supplied by the other party. Alternatively, one party to the contract might have a reputation for being less than ideally trustworthy. Consider, for example, a contractor who has a reputation for delaying completion of a project or for cutting corners in construction. The contractor could suffer a market penalty for such behavior. Suppose, on the other hand, that the contractor realizes that more trustworthy behavior is warranted and seeks to reform her errant ways. When one party is uniquely vulnerable or the other has a reputation for untrustworthy behavior, then the vulnerable party estimates that there is a greater likelihood that he will suffer a breach by the other party. Even if contract damages rules could provide all plaintiffs with a 70 percent recovery in the event of breach, those who estimate a higher probability of suffering a breach might choose not to enter into a contract. To offset this higher probability, the particularly vulnerable party would demand a higher recovery rate. Moreover, the repentant contractor might offer higher or guaranteed compensation as a mechanism for trying to recapture a trustworthy reputation with customers. Here, too, the parties should be able to contract around the default damages rules.

Note the signaling effects of contracting around the default rules. When parties specify a higher or more secure damage amount in their contract, one party is suggesting either that he is uniquely vulnerable to delay or that he does not trust the other party to perform the contract adequately without special incentives. Consider the latter situation by returning to the errant contractor. The customer who is concerned that the contractor will cut corners or delay her performance has a couple of tools available to treat these issues in the contract. Regarding delay, the customer can insist on a clause that specifies amounts of money that will be forfeited by the contractor for each day the project's completion is delayed. Further, regarding the quality of the work, the customer can demand that part of the payment for the construction be held back until an architect, engineer, or other expert certifies that the work complies with the standards of the industry or of those specified in the contract or both. In each case, the clause works to throw the onus on the contractor to seek payment for the work performed rather than placing the onus on the customer to seek damages from the contractor. The clause can also work in each case to enable the customer to specify a damage or escrow amount that will guarantee him more complete compensation in the event of breach. Again, in each case, the customer is signaling to the contractor (and to a later court, if necessary) that he perceived a significant trust issue that needed to be specially handled in the contract. To the extent that the construction contract includes more detailed or larger forfeiture provisions than other industry contracts, the clause makes salient that the contractor's untrustworthiness remains quite costly to her. In short, the signaling effect of specialized damages provisions helps to further encourage trustworthy behavior.

This last claim might appear to conflict with experimental findings that subjects tend to respond to sanction threats with less rather than more trustworthy behavior. For example, Falk and Kosfeld found that additional voluntary performance by agents shrunk considerably where principals restricted their agents' performance choices by imposing a minimum performance level for the agent.[88] Similarly, Fehr and List found that both CEO and student subjects behaved in a less trustworthy fashion when the person they were interacting with exercised her authority to build sanctions for noncooperative behavior into their transactions.[89]

In each of those experiments, however, subjects did not know each other, and each subject was required to play the trust game at issue. In the real world of transactions, however, parties do have the option to forgo dealing with each other. The customized clauses might cause problems in that they do signal distrust. When the distrust signal is a response to relevant information about the trustworthiness of another, the signal might be a necessary cost incident to providing the protection necessary for the parties to move forward. Moreover, the distrust signal places the burdened party on notice that future transactions may be jeopardized by her untrustworthy reputation. In these cases, the distrust signal might actually increase trustworthy behavior.

Alternatively, a party might wish to limit the damages recoverable in contract to an amount below the default standard in order to prevent the other party from relying on him too much. Consider the security alarm contracts mentioned earlier in this chapter. Security companies promise to quickly notify the police in the event that a customer's alarm goes off. The contract is intended to protect the customer against theft, but the security company worries that customers will over-rely on the security company's performance to his or her own detriment. To deter burglary, for example, home owners should also lock their doors, and arrange for mail and newspaper pickup when they are away. A timer on lights in the house will also provide added deterrence. Security companies fairly routinely limit the damages that can be recovered in the event of a break-in. One possible justification for this limitation is to prevent the customer from placing too much trust in the security company.

There are a few places in the law where the contracts default rule to undercompensate is replaced with a more fully compensatory damages rule. For example, in fiduciary relationships, discussed earlier, breaches of fiduciary duties entitle the injured party to tort damages.[90] In these cases, one party to the contract is specially vulnerable, and, as a matter of public policy, it seems important to send a strong message that special care is needed. Note also that, contrary to some scholars' claims, imposing these special duties in special cases is unlikely to signal that these people/companies are untrustable. Instead, the problem often is that the vulnerable individual is unable to carefully calibrate the degree to which trust is appropriate. Although trust and distrust routinely coexist in most relationships, for some the benefits of high trust are so large that trust assessments become more heuristic and less deliberative.[91] For example, medical patients seem incapable of separately assessing their doctor's loyalty, competence, and honesty. A competent doctor is assumed to be loyal and honest, and a doctor suspected of dishonesty is suddenly assumed to be incompetent and disloyal as well.[92] At the same time, most patients report unreasonably high rates of trust in their doctors.[93] Inaccurately high trust assessments might serve the vulnerable party's interests, however. The vulnerable party in fiduciary relationships seeks help because of that vulnerability, and often the patient's or client's interests are best served with high levels of trust. Clients who trust their lawyers, and patients who trust their doctors, are more likely to provide the fiduciary with sensitive information vital to the fiduciary's services. Patients are also more likely to seek evaluation and follow prescribed treatment if they have high trust in their doctor. Moreover, patients seem to report feeling better as soon as they know they have a doctor whom they trust to help them with their health problems.[94]

In any event, in the fiduciary context, carefully calibrated trust copayments are unlikely to produce carefully calibrated trust assessments by the vulnerable party, and, even if they could, the vulnerable party is more likely to fully cooperate, disclose, or invest if he or she can fully trust the fiduciary. In these cases, the Laffer curve notion of the relationship between trust and damages proffered earlier seems inapplicable. Here, the law seeks more complete compensation.

Returning to nonfiduciaries, although contracting parties do have some abilities to deal with their special trust issues by changing the default rules, the parties' abilities to carve out special treatment of their individual trust concerns are limited. Some problems are easy for the parties to address. If a customer is less likely than others to pay his debts on time, a company or salesperson can insist that the customer pay for goods or services before they are delivered or performed. Customers with good credit can purchase on credit, but customers with lousy credit histories must use the layaway method. If a contractor is likely to cut corners, the customer can hold back some portion of the payment until the job is completed. The greater the perceived problem, the larger the amount that can be held back. Other problems are more difficult to solve with specialized contract provisions. Suppose, for example, that an employer is considering whether to hire an employee to perform work at the employer's place of business. The employer knows that the employee is a hard worker, but the employer is worried that the employee will quit without notice at some future time. The employer can use contract terms to induce the employee to stick around. He can create benefit plans that vest in the future or give generous bonuses at the end of the year to employees who have stayed and worked hard. The employer is limited, however, in his abilities to try to guarantee that the employee does not leave without notice. The employee may be effectively judgment-proof against harm from leaving, and statutes might prevent the employer from withholding the employee's pay for quitting work. Even if a statute does not prevent this behavior, under contract law the employer cannot withhold a portion of the employee's pay as a type of penalty bond to ensure performance. Further, the employer cannot get a court to order the employee to work as promised because specific performance is unavailable for personal services.

The employment context might be particularly problematic given the personal services nature of the contract, the regulatory environment in which the contract sits, and the potentially limited assets of the employee. However, in other contexts parties also face limitations on their abilities to contract around the default damages rules. In particular, contract law provides that parties can specify the damages that will be paid in the event of breach only so long as the amount specified is deemed reasonable in light of anticipated or actual harm. Provisions that work to provide a penalty for breach rather than for compensation for harm will be struck down.[95] For some courts, moreover, liquidated damages provisions are only allowable in cases where the injury caused by the breach is uncertain or difficult to quantify.[96] Although the limitations on liquidated damages might be serving multiple goals, clearly one of these goals is to prevent a party from recovering more than the damages anticipated or actually suffered as a consequence of breach. Despite this limitation, however, a party can provide for much fuller compensation in a contract with the following provisions:

1. A liquidated damages provision that estimates the actual harm suffered as a consequence of breach
2. A provision that holds the breaching party liable for reasonable attorneys fees incurred by the other party
3. A provision that holds the breaching party liable for interest accruing from the time of breach until the time of compensation (this clause can be useful in jurisdictions that do not include prejudgment interest in awards or that include prejudgment interest at lower than market rates)

It appears, moreover, that a few courts are now allowing parties in some contracts to provide for specific performance as a remedy for breach. To strengthen the trend, the recent amendments to the Uniform Commercial

Code, which no state has yet adopted, provide that in non-consumer contracts for the sale of goods the parties can contract for specific performance.[97]

Specific performance and specified damages coupled with attorney fees and prejudgment interest payments work to ensure that the non-breaching party is more fully compensated in the event of breach. Although the law enables these provisions, it does not enable parties to contract for penalties or to include provisions that otherwise have the effect of guaranteeing their own performance (other than the specific performance provisions). Why might courts allow parties to contract for specific performance but not allow them to include a penalty provision so large that it effectively ensures the other's performance? The problem with the latter tool is that it opens the door too wide to opportunism on the part of the party not pledging the penalty provision. Put another way, when a party pledges an amount so high that it takes away all incentives for breach, then probably the non-pledging party would actually profit from the breach. If so, the non-pledging party might have an incentive to induce the pledging party to somehow breach the contract or to claim that the other party has breached the contract when that has not occurred. To prevent creating incentives toward untrustworthy behavior on the part of the non-breaching party, the contract damages rules enable parties to contract for full or nearly full, but never excessive, compensation.

4. Conclusion

Much of contract law operates on the assumption that some parties are aware of, and motivated to respond to, the incentives contract law produces. Consequently, proponents of law and economics often analyze contract principles, especially damages measures, from the perspective of efficiency. Recently, scholars have focused on the importance of trust to creating market efficiencies, and, following that lead, this chapter has explored the many ways in which contract doctrines and damages rules can be said to promote trust between contracting parties. Contract doctrines work to prevent parties from behaving in opportunistic fashion with each other while at the same time providing contracting parties with enormous flexibility to structure their transactions and relationships in order to address their individualized trust concerns.

An analysis of the contract damages rules indicates that they work to encourage strangers to transact with each other while also maximizing the extent to which trustworthiness still plays an important role in the choice of a contracting party. To encourage trustworthy behavior, the law provides substantial but not complete compensation in the event of breach. The contract damages rules are a blunt tool for calibrating party incentives in many individual cases, however. To effectively treat party heterogeneity, the contract damages rules serve as a type of semi-default rule. So long as the parties do not attempt to overcompensate in the event of breach, they can design their own remedies to ensure more complete compensation.

Notes

1. For a discussion of the predominance of neoclassical agency theory in business school and the importance of profit-maximization to that theory, see chapter 14 of this volume.

2. Francis Fukuyama, *Trust: The Social Virtues and the Creation of Prosperity* (1995).

3. See, e.g., Margaret M. Blair & Lynn A. Stout, "Trust, Trustworthiness and the Behavioral Foundations of Corporate Law," 149 *University of Pennsylvania Law Review* 1735 (2001); Lawrence E. Mitchell, "Fairness and Trust in Corporate Law," 43 *Duke Law Journal* 425 (1993).

4. For a discussion of internal norms related to personal commitments, see Oliver Goodenough, "Law and the Biology of Commitment," in *Evolution and the Capacity for Commitment* 269 (R.M. Nesse, ed. 2001); Michael P. Vandenbergh, "Beyond Elegance: A Testable Typology of Social Norms in Corporate Environmental Compliance," 22 *Stanford Environmental Law Journal* 55 (2003).

5. See, e.g., Anne M. Burr, "Ethics in Negotiation: Does Getting to Yes Require Candor?" 56 *Dispute Resolution Journal* 8 (July 2001).

6. See Russell Hardin, "Gaming Trust," in *Trust and Reciprocity: Interdisciplinary Lessons from Experimental Research* 80, 87 (Elinor Ostrom & James Walker, eds., 2003).

7. See chapter 13, this volume, for a discussion of games where subjects tend to get locked in to bilateral exchange relationships and, as a consequence, fail to find and exchange with their optimal trading partners.

8. See Benjamin Klein & Keith Leffler, "The Role of Market Forces in Assuring Contractual Performance," 89 *Journal of Political Economy* 615 (1981).

9. Lynne G. Zucker, "Production of Trust: Institutional Forces of Economic Structure, 1840–1920," 8 *Research in Organizational Behavior* 53 (1986).

10. One might legitimately question whether these external incentives for cooperation are really trust at all. See, for example, Oliver E. Williamson, "Calculativeness, Trust and Economic Organization," 36 *Journal of Law & Economics* 453, 485 (1993). Because laypeople often talk about their predictions of strangers' behavior as trust (as in "I trust that this unknown online vendor will, in fact, send the product it purports to sell"), I use the term here. However, nothing in the analysis that follows turns on these incentives in fact representing trust.

11. Restatement (Second) of Contracts § 205 (1981); Uniform Commercial Code § 2–306 (2001).

12. John D. Calamari & Joseph M. Perillo, *The Law of Contracts* 59–60 (4th ed., 1998) (describing situations where discharge is considered wrongful under state or federal law).

13. See *infra* notes 71–75 and accompanying text.

14. Harvey S. James, Jr., "The Trust Paradox: A Survey of Economic Inquiries into the Nature of Trust and Trustworthiness," 47 (3) *Journal of Economic Behavior & Organization* 291, 292 (2002).

15. Roy J. Lewicki et al., "Trust and Distrust: New Relationships and Realities, 23 *Academy of Management Review* 438 (1998).

16. Blair & Stout, *supra* note 3, at 1765–66.

17. For a more detailed discussion of these two different types of trust, see Claire Hill & Erin Ann O'Hara "A Cognitive Theory of Trust," forthcoming *Washington University Law Quarterly* (2007).

18. I use the qualifier "relevant" because not all of B's trustworthy behavior will be relevant to A's assessment. B's positive behavior toward his wife and children may be of limited usefulness in determining whether B will behave well when dealing with strangers. Moreover, B might unfailingly deliver high-quality products on time but might nevertheless consistently fail to pay back money that he borrows from others.

19. Williamson, *supra* note 10.

20. See, e.g., Ernst Fehr & Simon Gachter, "Do Incentive Contracts Crowd Out Voluntary Cooperation?" (draft manuscript available at http://papers.SSRN.com/abstract id=229047, accessed July 16, 2007); Ernst Fehr & John A. List, "The Hidden Costs and Returns of Incentives—Trust and Trustworthiness among CEOs" (draft manuscript available at http://www.ssrn.com, accessed July 16, 2007).

21. See, e.g., Uniform Commercial Code § 2–305 (1). This provision goes on to state that if the parties intended not to be bound unless they fixed a price, a failure to specify a price means that the parties have no enforceable contract. *Id.* at § 2–305(4).

22. UCC § 2–309(1).

23. UCC § 2–308(1).

24. UCC § 2–314 through 2–316.

25. Restatement (Second) of Contracts § 234(1).

26. See, e.g., UCC § 2–507.

27. Calamari & Perillo, *supra* note 12, at 431.

28. On the other hand, statutes now mandate that employees are entitled to some periodic payments of salary and wages. Restatement (Second) of Contracts § 240, comment a.

29. See, e.g., Restatement (Second) of Contracts §32 and comment b (1978) (suggesting it takes clear language to make an offer that can be accepted only by performance).

30. See, e.g., Restatement (Second) of Contracts § § 30, 50.

31. See, e.g., Restatement (Second) of Contracts § 227.

32. See, e.g., Restatement (Second) of Contracts § 237 (performance excused if other party's lack of performance constitutes "material failure" of duty).

33. See, e.g., *id.* at § 225.

34. See, generally, Calamari & Perillo on *Contracts*, chap. 13 & § 9.25–9.26 (5th ed. 2003).

35. See, e.g., Restatement of Contracts § § 152(1) (mistake must go to basic assumption of contract); *id.* at § 261 (impracticability requires event contrary to basic assumption of contract); *id.* at § 265 (frustration requires event contrary to basic assumption of contract).

36. *Sherwood v. Walker*, 33 N.W. 919 (Mich. 1887).

37. *Krell v. Henry*, 2 K.B. 740 (1903).

38. See Restatement of Contracts § 154 (mistake); Uniform Commercial Code § 2–615, Official Comment (excuse by failure of presupposed conditions).

39. See, e.g., Restatement (Second) of Contracts § § 152–54 (mistake); § § 261, 265, 266 (impossibility and frustration).

40. See discussion of unconscionability *infra.*

41. The textual claim is subject to the caveat, discussed *infra,* that like all contract claims this risk-splitting clause can nevertheless be scrutinized for fundamental unfairness under the unconscionability doctrine.

42. Schwab, "A Coasean Experiment on Contract Presumptions," 17 *Journal of Legal Studies* 237 (1988); Goetz & Scott, "The Mitigation Principle: Toward A General Theory of Contractual Obligation," 69 *Virginia Law Review* 767 (1983).

43. Ian Ayres & Robert Gertner, "Filling Gaps in Incomplete Contracts: An Economic Theory of Default Rules," 99 *Yale Law Journal* 87 (1989).

44. E. Allan Farnsworth, *Contracts* 546 (4th ed. 2004).

45. *Id.*

46. Restatement (Second) of Contracts § 206.

47. Uniform Commercial Code § 2–507.

48. See, e.g., Armin Falk & Michael Kosfeld, "Distrust—The Hidden Cost of Control" (draft, July 2004), available at http://ssrn.com/abstract=5606224, accessed July 16, 2007).

49. UCC § 2–306(2).

50. 118 N.E. 214 (N.Y. 1917).

51. Note also that Wood could not deal with Lucy's reputational issues by entering into a very brief contract period, because the contract period needed to be long enough to protect Wood from Lucy's efforts to delay a deal and keep all the profits.

52. I thank my colleague Bob Rasmussen for providing me with this insight into the Wood case.

53. The contractor would be required to complete all the work prior to payment under the contract default rule, Restatement (Second) of Contracts § 234(2), but standard-form construction contracts nearly universally contract around the default rule. Note that standard-form contracts enable one or both parties to deal with industry-wide trust issues without signaling particularized distrust in a contracting partner.

54. *See* Farnsworth, *supra* note 44, at 522–23 (discussing common requirement that architect certify satisfactory completion of work).

55. See, for example, *Graham v. Lieber,* 191 So. 2d 204 (La. App. 2d Cir., 1966).

56. Uniform Commercial Code § 1–201 (2001).

57. Robert S. Summers, "The General Duty of Good Faith—Its Recognition and Conceptualization," 67 *Cornell Law Review* 810, 820, 826 (1982).

58. *Billman v. Hensel,* 391 N.E. 2d 671 (Ct. App. Indiana, 3d Dist. 1979).

59. *Neumiller Farms, Inc. v. Cornett,* 368 So. 2d 272 (Ala. 1979).

60. See, generally, Edward J. Murphy et al. *Studies in Contract Law* 711–23 (6th ed., 2003).

61. *Market Street Associates Limited Partnership v. Frey,* 941 F. 2d 588 (7th Cir., 1991).

62. See *Fortune v. National Cash Register Co.,* 364 NE 2d 1251 (Mass. R77).

63. Murphy et al., *supra* note 60, at 667–69.

64. *Market Street Associates, supra* note 61, at 594.

65. Farnsworth, *supra* note 44, at 489.

66. Restatement (Second) of Contracts § 204 (1981).

67. See *Williams v. Walker-Thomas Furniture Co.*, 350 F. 2d 445 (D.C. Cir. 1965).

68. *Id.* at 450.

69. John D. Calamari & Joseph M. Perillo, *The Law of Contracts* 373–74 (4th ed. 1998).

70. *Henningsen v. Bloomfield Motors*, 161 A. 2d 69 (N.J. 1960).

71. See, e.g., D. Gordon Smith, "The Critical Resource Theory of Fiduciary Duty," 55 *Vanderbilt Law Review* 1399 (2002) (discussing various fiduciary duties); *Black's Law Dictionary* 564 (5th ed., 1979) (defining "fiduciary or confidential relationship").

72. P. D. Finn, *Fiduciary Obligations* 2 (1977).

73. Tamar Frankel, "Fiduciary Duties," in *The New Palgrave Dictionary of Law and Economics* 127, 129–30 (1998).

74. *Id.* at 130.

75. *Meinhard v. Salmon*, 164 N.E. 545, 546 (1928).

76. John P. Dawson et al., *Contracts: Cases and Comment* 3 (8th ed. 2003).

77. Herbert M. Kritzer & Joel B. Grossman, "The Cost of Ordinary Litigation," 31 *U.C.L.A. Law Review* 72, 91–92 (1983).

78. See Calamari & Perillo, *supra* note 12, at chap. 14.8.

79. Established businesses have an easier time recovering lost profits, but even in those cases recovery can be limited by the reasonable certainty rule. *Id.*

80. For a discussion of these damages categories and their limited recoverability, see Farnsworth, *supra* note 44, at 760–64 (punitive damages), and *id.* at 809–10 (emotional distress damages).

81. See, e,g., Farnsworth, *supra* note 44, at 741 (discussing differing attitudes of civil and common law systems toward specific performance).

82. Calamari & Perillo, *supra* note 12, at 617–18.

83. *Id.* at 611–12 (discussing requirement that legal remedy of money damages be inadequate).

84. The Uniform Commercial Code has promoted a more liberal attitude toward specific performance, however. The Code provides that "[s]pecific performance may be decreed where the goods are unique or in other proper circumstances." UCC § 2–716 (1) and Official Comment (1). Notwithstanding this provision, most courts remain reluctant to grant specific performance without being convinced that it would be impossible to obtain substitute goods. Calamari & Perillo *supra* note 12, at 616.

85. See, e.g., Frank A. Cavico, Jr., "Punitive Damages for Breach of Contract—A Principled Approach," 22 *St. Mary's Law Journal* 357, 370–74 (1990); John A. Sebert, Jr., "Punitive and Nonpecuniary Damages in Actions Based upon Contract: Toward Achieving the Objective of Full Compensation, 33 *U.C.L.A. Law Review* 1565 (1986); see also Daniel A. Farber, "Reassessing the Economic Efficiency of Compensatory Damages for Breach of Contract," 66 *Virginia Law Review* 1443 (1980).

86. Iris Bohnet, Bruno S. Frey, & Steffen Huck, "More Order With Less Law: On Contract Enforcement, Trust and Crowding" (draft manuscript available at http://papers.ssrn.com/paper.taf?abstract_id=236476, accessed July 16, 2007).

87. For some parties, contract damages do little to make the parties feel comfortable contracting. Consider, for example, alumina smelters. Smelters need bauxite, and, without it, the plants destroy themselves. If a smelter enters into a contract with a

bauxite producer, it might be that no contract damages award can make a smelter comfortable with the possibility that the plant could be destroyed. In these cases, the ownership structure of the bauxite producers is often altered to deal with the trust problems the smelters confront. First, some smelters own their own bauxite sources. Second, some bauxite sources are government-owned, and, because they are, political risk insurance becomes available to the smelters as a source of complete compensation. In both cases, the ownership of the bauxite sources is manipulated to reduce the smelter's ultimate vulnerability.

88. Armin Falk & Michael Kosfeld, "Distrust—The Hidden Cost of Control" (draft manuscript available at http://ssrn.com/abstract=560624).

89. Fehr & List, *supra* note 20.

90. Restatement (Second) of Torts § 874 (1979).

91. Claire Hill & Erin Ann O'Hara, "A Cognitive Theory of Trust," forthcoming, *Washington University Law Quarterly* (2007).

92. Mark A. Hall, "Law, Medicine and Trust," 55 *Stanford Law Review* 463, 477 (2002).

93. Hill & O'Hara, *supra*.

94. Hall, *supra*.

95. Dawson et al., *supra* note 76, at 132; Restatement (Second) of Contracts § 356.

96. Joseph M. Perillo, Calamari & Perillo on Contracts § 14.31 (5th ed. 2003).

97. Uniform Commercial Code § 716(1) (2003) (newly amended).

Ten

The Vital Role of Norms and Rules in Maintaining Open Public and Private Economies

David Schwab and Elinor Ostrom

After the fall of the Berlin Wall, the presumption of many policy analysts was that rapid development in the former Soviet Union would result from the creation of effective market institutions and the privatization of land and productive enterprises. Achieving the long-term goals of economic growth and prosperity for citizens has proved to be far more difficult than many policy analysts initially presumed (Kikeri and Nellis, 2004; Shivakumar, 2005). Building systems of law and order is more complex and takes substantially more time and effort than simple textbook examples illustrate, or analysts thought, when confronting post-colonial and post-Soviet political-economic systems (World Bank, 2001, 2002).

This chapter draws on systematic methods for analyzing strategies, norms, and rules developed by colleagues at the Workshop in Political Theory and Policy Analysis at Indiana University. These methods make it possible to dig under casual assumptions of "law and order" in a slow and methodological manner to illuminate the diverse challenges of establishing effective law and order, and trust, among citizens in any society. Establishing a high level of trust after massive failure of governance systems is even more difficult.

We begin with a simple model, the game of Snatch, which has been used in past work (Plott and Meyer, 1975) to illustrate that non-simultaneous exchange in the absence of "law and order" leads to inefficient outcomes. We do not present this game to illustrate all exchange settings that occur informally in many settings. Rather, we start with this simple model to illustrate the basic dilemma of non-simultaneous exchange prior to a more complex analysis. The two-person Prisoners' Dilemma is frequently used in this fashion. We then demonstrate that, although reciprocity norms can overcome the inefficient outcome of the game of Snatch, these norms depend on contextual factors such as community integration and stability.

The Game of Snatch in a "Hobbesian State of Nature"

Consider the following situation: you own a farm in an isolated area. Another farmer has recently arrived in the area. To keep things simple, assume that you only grow corn, and the other farmer only grows beans. Eating corn all the time has lost its appeal, and your neighbor is tired of bean soup, so you propose to trade him a bag of your corn for an equal amount of his beans. He agrees and suggests that since he passes your place to go to work on another plot this Friday, you leave a bag of corn out front for him, and he'll exchange it for a bag of beans. You think about this and realize that you will not be in the field when he stops by, so it is possible for him to take your corn and not leave anything in return. Now assume, for argument's sake, that no police are available to complain to, and that you and the other farmer are unlikely to see each other again. Do you leave the corn out or not?[1]

Classical game theory—predicated on purely self-interested, economically maximizing players—advises against leaving the corn for your neighbor: he gets more utility from taking the corn and leaving you nothing than taking the corn and leaving behind some beans.[2] Let us solve the game in figure 10.1 through backward induction. Clearly, the other farmer receives more utility from playing Snatch than playing Exchange. Yet, the form of the game is common knowledge, so you know the other farmer's payoffs and can predict that he will choose Snatch. Thus, you choose not to leave the corn out for him, leading to a Nash equilibrium of (Don't Offer, Snatch). In this way, you avoid being a sucker. Unfortunately, you also forego any gains from trade: both you and the other farmer would benefit by trading corn for beans, but because there is no guarantee that he won't steal your corn and leave you nothing, you decide not to take the risk.

While backward induction is an accepted solution concept for a finite extensive-form game, it is not without its problems. Like all static solution concepts, it assumes hyper-rational players who never make mistakes and

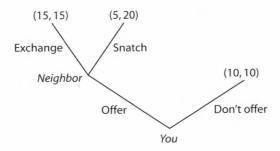

Figure 10.1. Game of Snatch

have perfect recall. Furthermore, experimental evidence from Centipede and Dictator games conflicts with the predictions of backward induction (Camerer, 2003). For this reason, it is valuable to explore a dynamic solution concept to the game of Snatch.

Replicator dynamics, which measure the fitness of a given population relative to the average fitness of its members, are commonly used to do this (Samuelson, 1997). We let x equal the percentage of Farmer 1's who play Don't Offer, and y equal the percentage of Farmer 2's who play Snatch. Then, the evolution of x and y over time t is given by the following replicator equations:[3]

$$dx/dt = 5x\,(1-x)(2y-1)$$

$$dy/dt = -5y\,(y-1)(1-x)$$

Because this game has asymmetrical payoffs, its behavior can be quite complex and counterintuitive. Space considerations prevent a full evolutionary analysis, but we can make the following observations of interest.

By examining the second replicator equation, we see that it is always increasing for $y<1$ and $x<1$. This means that, regardless of the strategy chosen by the initial population of Farmer 2's, this entire population will always end up playing Snatch. This is not surprising, as the strategy of Snatch strictly dominates the strategy of Exchange for Farmer 2.

However, while the population of Farmer 2's playing Snatch is always increasing toward 1, its rate of increase varies inversely with the size of the population of Farmer 1's playing Don't Offer. In other words, as more Farmer 1's play Don't Offer, the rate of increase of Farmer 2's playing Snatch slows down. Thus, while the population of Farmer 2's always converges to all playing Snatch, how long this convergence takes depends on the growth rate of the population of Farmer 1's playing Don't Offer.

What can we say about this growth rate? Examining the first replicator equation, we see that dx/dt is increasing provided $x>0$ and $y>\frac{1}{2}$. However, when $x>0$ and $y<\frac{1}{2}$, dx/dt is decreasing. Thus, the proportion of Farmer 2's playing Snatch has to reach a certain threshold ($y=\frac{1}{2}$) before the population of Farmer 1's begin to respond. So long as fewer than half the population of Farmer 2's are playing Snatch, they can get away with their malfeasance. However, since this population is increasing for all values of x and y, eventually the Farmer 1's begin to adapt and play Don't Offer instead of Offer.

In fact, since the population of Farmer 2's playing Snatch converges to 1, and the population of Farmer 1's playing Don't Offer is increasing for $y>\frac{1}{2}$, then eventually they also converge to 1. We say that the point $x=y=1$, where everyone is playing (Don't Offer, Snatch), is a fixed point of this system because $dx/dt = dy/dt = 0$. This means that once both strategy sets converge to this point, they will stay there provided everyone in both populations continues to play only these strategies. However, if players from either population make mistakes—that is, play Offer or Exchange—then the strategy sets may diverge from the fixed point $x = y = 1$. In particular, since the strategy (Don't Offer, Exchange) is only weakly dominated by (Don't Offer, Snatch), it cannot be eliminated from the strategy set of Farmer 1 by evolutionary pressure (Samuelson, 1997).

Does this mean that the strategy (Don't Offer, Snatch) is evolutionarily stable? The answer depends on one's definition of stability. According to Selten's theorem, it is not an evolutionarily stable strategy (ESS), because the weakly dominated strategy (Don't Offer, Exchange) is not completely eliminated by the evolutionary process. However, as Samuelson (1997) notes, the criteria for a strategy to be an ESS are quite strong and eliminate some strategies that are, for all intents and purposes, evolutionarily stable. The question we need to answer is whether the failure to completely eliminate the weakly dominated strategy of (Don't Offer, Exchange) affects the evolutionary dynamics of this game in an important way. At first glance it does not seem to, although a more complete evolutionary analysis is needed to fully answer this question.

We have seen that non-simultaneous exchange is not a Nash equilibrium of the game of Snatch. Furthermore, while the strategy (Don't Offer, Snatch) is not an ESS, it is fixed-point stable. Yet, we observe and participate in numerous exchanges of this sort every day, which leads us to believe that something more must be at work. Indeed, numerous explanations have been offered that make (Offer, Exchange) an equilibrium of this game. Below, we offer details of one of these explanations, that of trust-enhancing institutions. The point of the game of Snatch is that it is contrived: by abstracting from the shared strategies, norms, and rules that operate in the world we live in, we can more clearly see the necessity of having them in order for exchange to occur.

In this chapter, we argue that many trust-enhancing institutions may evolve through the efforts of participants in long-term, repeated market-exchange environments or when they are linked together as providers and consumers of public goods or common-pool resources. These trust-enhancing institutions make it easier to establish a reputation as a trustworthy participant as well as making exchange less costly, more stable, and more effective than would be possible without such institutions. Furthermore, efforts to design such institutions without understanding the context of relationships can sometimes crowd out trust rather than enhance it.

Trust in the "State of Nature": The Need for Institutions

The game of Snatch illustrates that, without any norms or rules related to the private exchange of goods, a Hobbesian state of nature exists. Hobbes, like modern game theorists, carefully laid out his assumptions. First, he set aside "the arts grounded upon words" (Hobbes [1651] 1960: 80). Second, he examined a setting where "men live without a common power to keep them all in awe" (ibid.). He then presumed that in such a state, where every man is against every other man,

> there is no place for industry; because the fruit thereof is uncertain: and consequently no culture of the earth; no navigation, nor use of the commodities that may be imported by sea; no commodious building; no instruments of moving, and removing, such things as require much force; no knowledge of the face of the earth; no account of time; no arts; no letters; no society; and which is worst of all, continual fear, and the danger of violent death; and the life of man, solitary, poor, nasty, brutish, and short. (ibid.)

In such a state of nature, a game theoretic analysis predicts that households will not engage in exchange owing to lack of trust. This contradicts not only our intuition derived from our own everyday experience in open, democratic societies but also disagrees with empirical evidence that people do trust one another in many social dilemma situations far more than is theoretically predicted (Camerer, 2003; E. Ostrom and Walker, 2003). Political theorists have also shown that a "common power that keeps them all in awe" is not the only way that "law and order" is achieved (V. Ostrom, 1987, 1997). Accounting for this contradiction will lay the groundwork for the remainder of this chapter.

In game theory, equilibria are mathematically determined. Yet, these equilibria are initially determined by the structure of the game assumed by the theorist. Alternative models often have different equilibria, and the question of which model best represents a particular puzzle or problem in field settings can be difficult to answer. Below, we consider alternative models of the game of Snatch that lead to different equilibria than the one determined by our initial model.

Trust in a World with Norms and Warm Glows

One possibility is that players adopt norms that lead them to derive utility by self-consciously refraining from snatching the goods. As Lynn Stout argues in chapter 8 of this volume, we often overestimate how selfish people are, because selfish behaviors stand out from everyday observations. Andreoni (1989) suggested that people receive a "warm glow" from performing good deeds and adhering to social norms. Following Crawford and Ostrom (2005), we can model this "warm glow" as a delta (δ) parameter that is added to the utility of the player who follows a norm to refrain from snatching goods, and thus exchanges rather than snatches goods. The game of Snatch with Warm Glow is illustrated in figure 10.2.

Examination shows that (Offer, Exchange) is now the Nash equilibrium, provided that the value of δ is large enough to offset the advantage gained from snatching the goods. Paul Zak and others offer evidence in this volume

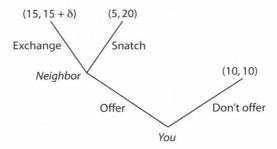

Figure 10.2. Game of Snatch with "Warm Glow"

that the δ may indeed be large enough in many cases (see chapter 12). This makes the "warm glow" hypothesis compelling; much research demonstrates, however, that social norms are needed to instantiate and maintain high enough δ's to sustain a stable free-market society (Arce, 1994; Axelrod, 1986; Barner-Barry, 1986; Basu, 1995; Brooks, 2001). Thus, even if most players refrain from snatching because of a "warm glow," social norms are necessary to cultivate and sustain this level of cooperation. The game of Snatch with Warm Glow could be an appropriate way to model close-knit settings with repeated interactions between members of a community, and where the individuals do have strong δ's. While there is no single external force to keep all participants "in awe," the setting allows for the establishment of norms so long as stability of the population and shared commitment to the norms are maintained.

Trust in a World with Law and Order

A second alternative is the establishment of "law and order" by creating a new position in the community and adding the power of sanctioning to that actor's authority to be used when goods owned by one member of the community are taken illegally. Now, if your neighbor snatched your goods, you would be motivated—so long as you could get your commodities returned—to go to the judge, ask for your commodities back, and request the judge to sanction your neighbor by taking his commodities away. Once a position of judge is created with authority to sanction someone who took someone else's goods, your neighbor would face a new set of alternatives if he snatched, or better phrased now, *stole* your goods.[4] Now, your neighbor knows that you would be motivated to call in the judge who would confiscate his goods and return yours to you. We call this the game of Snatch with Property Rights and Sanctions (see figure 10.3).

Examination shows that (Offer, Exchange) is the Nash equilibrium of this game, so long as the sanction involves a loss of utility from being sanctioned

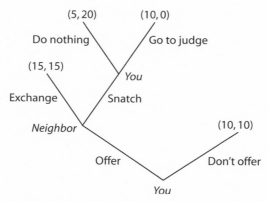

Figure 10.3. Game of Snatch with Property Rights and Sanctions
Source: E. Ostrom (2005).

that is greater than the expected gain from snatching the goods. Yet, for the sanction to be effective, you must be willing to go to the judge and the judge must punish infractions. In the game in figure 10.3, you have a real incentive to go to the judge, but the judge is represented as a player without a choice.

At a more general level, passing rules and enforcing them involves costs, which leads to a second-order, collective-action problem: what would be the incentives for players to monitor and sanction others when they receive no direct benefit from this action? Furthermore, the game in figure 10.3 assumes that monitoring and sanctioning are both done perfectly: the guilty are always caught and the innocent never punished. Our more general point is that the process of "assuming law and order" is not a simple one that can be imposed by some strong central authority. Those who participate in ongoing transactions must be committed to following the rules and incentives in place. If the judge wanted a large share of your goods as well as confiscating those of your neighbor, you might never go to the judge.

A third possibility is that players can develop reputations for trustworthiness through their behavior. Players can then choose whom they wish to trade with based on those players' reputations, even in a one-shot game (provided the game itself is repeated with a sufficient number of other players to generate reputations). Substantial evidence shows that people do rely on reputational information in determining whom they will trust (Colson, 1974; Sally, 2002; Wedekind and Milinski, 2000). We believe that reputation provides a plausible explanation for one of the key mechanisms enabling sustainable free trade in a society. However, reputations cannot lead to an increased level of trust without institutions that enhance knowledge about and reliability of reputations.

Trust-Enhancing Institutions and Reputational Management

There is substantial evidence, as noted above, that people rely on reputational information in deciding whom to trust. Yet, reputations only provide a partial solution to that dilemma, because learning about these reputations is costly in a world of imperfect information. If this cost is too high, even perfectly accurate information will not be disseminated at an effective level. Here we argue that one effect of trust-enhancing institutions is to help people manage reputations by disseminating information about who is trustworthy and who is not.

Strong evidence exists that people examine the reputations of others to decide if they are trustworthy. E. Ostrom (1998) and E. Ostrom and Walker (2003: chap. 2) developed a behavioral theory linking the interaction of trust, reputation, and reciprocity to success in solving collective-action problems. Their theory is founded on persuasive empirical support. For example, Fafchamps and Minten (1999) described how agricultural traders in Madagascar rely upon personal networks in order to procure goods and manage risk. Clay (1997) examined the question of how, in the absence of formal legal systems in California during the early 1800s, merchants were able to engage in active trading. Drawing on evidence from the merchant's business correspondence, Clay examined the coalitions merchants formed to mitigate the commitment problem involved when many actors handle goods, and the punishment merchants imposed on cheaters. She also looked at the expansion of the coalitions toward mid-century and their eventual collapse at the time of the Gold Rush, when coalitions were overwhelmed with many new participants without reputations.

Milgrom, North, and Weingast (1990) took this work a step further by analyzing the development of exchange markets in Champagne, France. Michele Fratianni (forthcoming) provides a fascinating history of the evolution of a formal association—San Giorgio—aimed at protecting creditors' rights and reducing the risk of the monarch repudiating debt. Reputation, norms, and rules are all involved in this historical evolution. Eric Feldman (2005) analyzed a dynamic market in modern Tokyo that had developed a highly formal legal system backed by extensive norms to cope with the contentious problems related to the quality of tuna and uncovering the party responsible for spoiled fish.

Formal models also demonstrate the importance of this interaction. Annen (2003) used an infinitely repeated Prisoner's Dilemma game to show how the inclusiveness, communication capacity, and complexity of an exchange setting affects an individual's ability to manage social capital. Bravo and Tamburino (2008) used an agent-based model to delimit the initial conditions that lead to the development of a "sense of trust" among a population. They speculate that:

> A system where trust, reputation and reciprocity are strictly linked in a positive feedback relationship is likely to assume only two states: one where the cooperation

level is high and one where it is low, while intermediate cooperation levels are less likely. If this idea is right, a relatively thin threshold should mark the transition between high and low cooperation equilibrium and even close initial configurations may therefore end in fully different states.

As demonstrated above, reputations can, under certain circumstances, resolve the trust dilemma that is at the heart of the game of Snatch. Yet, if Bravo and Tamburino are correct, it is essential to determine what these circumstances are.

As a first step, we considered the difference between perfect and imperfect information for developing a stable interaction between trust, reputation, and reciprocity. It is well known in game theory that, in an infinitely repeated trust game, cooperation can be maintained if players do not switch partners and have perfect information and recall about each other's past actions (Fudenberg and Tirole, 1991). Kandori (1992) showed that this result holds even if players switch partners every period, provided that the conditions of perfect information and recall are maintained. In this case, perfect information leads to indirect reciprocity. Further, players are willing to punish defectors even at a cost to themselves in order to maintain the stability of the system. In effect, perfect information resolves the second-order, collective-action problem of punishing defecting players. This leads to an equilibrium where snatching the goods is more costly than exchange, because no one will trade with you in the future if you have a reputation for snatching goods.

Fudenberg and Tirole (1991) and Kandori (1992) provided valuable insight into the conditions necessary for developing and maintaining a stable interaction between trust, reputation, and reciprocity. They do not go far enough, however, since, in the real world, acquiring and disseminating reliable information is costly for all involved. People can only directly observe the decisions of a small fraction of those they may have to trust; for the rest, they must rely on information from other sources, such as the reports of others or heuristics such as group membership or likeability.

Of these substitutes, our analysis focuses on the reports others provide about a given reputation, because this best illustrates the differences that arise in a world of imperfect information. Recall that perfect information can sustain trust in a society of one-shot trust exchanges, even when players switch their partner every turn. This is because all players know the reputation of their exchange partners prior to every exchange game. Imperfect information removes this certainty about the reputation of one's partner. There is now a temptation to cheat—to snatch the goods—because doing so improves one's utility and does not necessarily decrease one's future utility. If reputation is not common knowledge, it does not have the same deterrent effect.

One way to overcome this problem is for people to tell one another when they have been cheated. In this way, one can approximate a world of perfect information and exclude untrustworthy people from exchange relationships. This could lead to the creation of a "sense of trust" in a society, especially given recent evidence that many people are predisposed to trust one another (de Waal, 1996; Kurzban, 2003). Yet, there remains the problem that people may not have an incentive to report to others that they have been cheated unless they already exist in tight networks of relationships. Reporting that one has been cheated is a costly action. These costs grow as the society increases in size. This creates a second-order, collective-action problem: everyone would benefit if people did report that they had been cheated, but no one individual has an incentive to do so. For this reason, we must investigate the interaction of trust, reputation, and reciprocity under conditions of imperfect information. In doing so, we will see that one function of institutions is to help people manage reputations by disseminating information about who is trustworthy and who is not.

It is necessary, first, to clarify what we mean by "institution." We draw on the framework presented in E. Ostrom (2005), which is based on more than two decades of theoretical and empirical study. We define *institutions* as the prescriptions humans use to organize all forms of repetitive and structured interactions. Institutions are present in families, neighborhoods, markets, firms, sports leagues, churches, private associations, and government at all scales. They do not have to be written down or codified, and frequently they are not. A farmers' market, a rotation system developed on a farmer-managed irrigation system, and a legal code are all examples of institutions.

In studying institutions, it is useful to distinguish between three mechanisms that help players coordinate their behavior. Crawford and Ostrom (2005) developed an institutional grammar for this purpose. We use this grammar to describe the coordinative mechanisms of shared strategies, norms, and rules, and to delimit the similarities and differences between them.

In the institutional grammar, all three mechanisms are expressed in the following syntax. This syntax includes five components: ATTRIBUTES, DEONTIC, AIM, CONDITIONS, and OR ELSE. Each mechanism is composed of different groupings of these components, which is clarified in the following descriptions.

A *shared strategy* is defined as ATTRIBUTES of players with a specific AIM that may be performed under certain CONDITIONS. For example, in the game of trust, the players have the ATTRIBUTES of Trustor and Trustee. The AIM of the Trustor is to decide how much (if any) of his or her endowment to give to the Trustee, and the AIM of the Trustee is to decide how much (if any) of this money to return to the Trustor. The CONDITIONS for the Trustor delimit how he or she will behave based upon the actions of the Trustee. For example, the CONDITIONS for a "grim trigger" strategy would be to "cooperate in each round of the game unless the Trustee defects, in which case defect for the rest of the game." CONDITIONS for the Trustee follow a similar logic.

A shared strategy depends solely upon players' expectations about each other's future behavior and prudential decision making. It noticeably lacks a

DEONTIC component specifying what players must, may, or may not do. Instead, it is entirely the strategic interaction of each player's strategies that determines the equilibrium of a game.

A *norm* is defined as ATTRIBUTES of players who, according to certain DEONTIC conditions, must, may, or may not perform a specific AIM under certain CONDITIONS. Thus, a norm differs from a shared strategy in that it specifies that players must, may, or may not act in certain ways. In the Trust game with norms, the DEONTIC may state that the Trustee must return a certain portion of the money that he or she received from the Trustor.

A *rule* is defined as ATTRIBUTES of players who, according to certain DEONTIC conditions, must, may, or may not perform a specific AIM under certain CONDITIONS, OR ELSE. Thus, the difference between a *norm* and a *rule* is the presence of a sanction: the OR ELSE condition. In the trust game, this condition may define which sanctions the Trustee is subject to if he or she does not follow the condition specified in the DEONTIC. This necessarily is only a brief overview of the institutional grammar. Interested readers are referred to chapter 5 in E. Ostrom (2005) for a more complete treatment.

We now can use the above grammar to describe three interactions of trust, reputation, and reciprocity of increasing complexity. In doing so, we will see the problems that arise from simple solutions to trust generation; how more complex institutions arise to resolve these problems; and how these solutions lead in turn to different problems that then require a careful understanding of their complexity to mediate and resolve.

Three Examples of Trust-Enhancing Institutions

We begin with a game illustrating how shared strategies can lead to a general "sense of trust" in a society. In a forthcoming article, Bravo and Tamburino describe an agent-based model based on a repeated sequential Prisoner's Dilemma. Figure 10.4 illustrates this game. Examination shows that this game places the first mover in the role of Trustor, and the second mover in the role of Trustee. This is because if the first player chooses to Cooperate, this increases the possible payoffs of both players, but also requires the first player to trust the second player not to Defect and take a higher payoff. As with our game of Snatch, backward induction shows that the Trustor should choose to Opt-Out—that is, refuse to trust the Trustee. The Nash equilibrium of the one-shot game is therefore (Opt-Out, Defect), resulting in no gains from trust for either player.

How does adding reputation change the dynamics of this game? Bravo and Tamburino use an agent-based model to answer this question. In this model, each of one thousand agents plays the above game with other randomly selected agents for ten thousand generations. Each generation consists of twenty-four

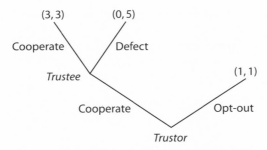

Figure 10.4. Sequential Prisoner's Dilemma with Opt-Out Option

periods. The role of each agent is randomly selected for each pairing; that is, sometimes the agent is the Trustor, and other times it is the Trustee.

Each agent is defined by two parameters: the probability (p) that the agent will choose to Cooperate when it is the Trustor, and the probability (q) that the agent will choose to Cooperate when it is the Trustee. Values of p and q are randomly assigned at the beginning of the first generation and evolve throughout the simulation. At the end of every generation, each agent generates a number of offspring proportional to the sum of the payoffs earned during the twenty-four periods forming the generation. Each offspring inherits the characteristics of its parent, with a 0.5 percent probability of a random mutation in one of its two parameters. At the end of the reproduction process, one thousand offspring are randomly selected to form the next generation.

Bravo and Tamburino model six different states. The ones of interest to us involve imperfect information. In these states, every time a Trustee moves, only a randomly selected number of agents are allowed to observe its behavior and use this information in subsequent periods. Each of these states was tested under two different conditions: stranger matching, where agents switch partners every round, and partner matching, where agents play six periods before being re-matched.

What do Bravo and Tamburino observe in these two conditions under imperfect information? First, they note that the results obtained in both conditions depend heavily on the proportion of agents who can observe a given exchange. In turn, these results are different in the stranger-matching condition than in the partner-matching condition: in the latter, high cooperation levels are sustainable with about half as many observers as are needed in the former. Bravo and Tamburino attribute this to the ability of agents in the partner-matching condition to rely on direct reciprocity, as well as indirect reciprocity. As agents must rely on indirect reciprocity in the stranger-matching condition, a higher proportion of observer agents are needed to maintain high cooperation levels.

Bravo and Tamburino's second finding is that, in a world of incomplete information, something akin to a general "sense of trust" emerges because agents cannot consistently base their decision on the reputation of their opponents since they lack full information regarding this. They summarize this "sense of trust" as follows:

> By emergence of a general trust, we mean that the micro-level effect of trust is independent from the corresponding effect of reputation. First-players are more likely to be trustful even if they do not know anything about their opponents' past behaviors. . . . Most agents will therefore start cooperatively any interaction with an unknown opponent and only subsequently adapt their moves as a function of the other's behavior, a clear example of direct reciprocity strategy. (2008)

Finally, Bravo and Tamburino note that in both conditions, the system under incomplete information seems to be able to reach only two states: one close to the noncooperative equilibrium and one close to the cooperative optimum. Through parameter analysis, they note that when the effect of reputation is weakened and the number of observers is proportionally low, the system rapidly closes in on the noncooperative equilibrium.

What lessons can we learn from this analysis? First, it is important to note that using the institutional grammar outlined above, the agents are relying on shared strategies rather than norms or rules. Bravo and Tamburino demonstrate that, under conditions of incomplete information, these shared strategies can generate a "sense of trust" to a degree independent of reputational effects. However, this "sense of trust" is highly dependent upon the initial parameters of the model, especially on the number of agents who are able to observe a given exchange. Furthermore, when these parameters are not sufficiently high, the model rapidly closes in on the noncooperative equilibrium. This demonstrates that shared strategies can lead to high levels of cooperation, but that achieving this result is highly dependent on the initial levels of trust and trustworthiness present in a population. It is unlikely that many large and mobile populations will be able to achieve high levels of cooperation based on shared strategies alone. Furthermore, even populations with the right initial conditions have a strong incentive to look for ways to bolster their initial "sense of trust" against internal and external shocks.

One way to do this is by switching from shared strategies to norms. Norms differ from shared strategies in that they have a DEONTIC component that specifies actions the player may, must, or must not take. E. Ostrom (2005) described them:

> Norms can be represented in formal analyses as a delta parameter that represents the intrinsic benefits or costs of obeying a normative prescription in a particular setting. The changes may occur as a result of intrinsic motivation such as pride when keeping a norm or guilt when breaking a norm. The delta parameter may also

occur as a result of the action being observed by others leading to esteem for following a norm or shame for breaking it. (121–122)

We observed an example of this type of modeling in figure 10.2 above, the Game of Snatch with Warm Glow. Here, the second player has an internal norm against snatching the goods from the first player. This norm is modeled as a delta (δ) parameter that is added to the payoff of the second player if he or she does not snatch the goods. If the value of δ is greater than the marginal gain from snatching the goods, then the second player gains more utility from not exchanging goods than from snatching them.

Do norms, in fact, influence behavior in this way? There is strong evidence that they do. Gouldner (1960) theorizes a norm of reciprocity present in most societies. This norm is present even in children, and likely learned by children from their parents (Harbaugh et al., 2003). Further evidence comes from Kerr and Kaufman-Gilliland's (1994) meta-analysis of social dilemma experiments. After examining competing explanations for how groups manage to resolve social dilemmas, the authors conclude that successful groups rely on explicit or implicit promises to abide by a group agreement that, if followed, will resolve the dilemma. In most cases, this agreement is made during face-to-face communication with other group members, and in almost all cases, defection cannot be punished by the group (this is certainly true when the players communicate only once). Yet, in a majority of cases, players in groups that used face-to-face communication to arrive at successful agreements did abide by them (Sally, 1995; see also Bochet, Page, and Putterman, 2006). Since each player in a social dilemma has an incentive to defect from the group agreement (particularly if the other group members abide by it), this is strong evidence that norms do alter individual behavior.

At a macro-social level, norms can stabilize the initial "sense of trust" generated by shared strategies. There is strong evidence that humans have both an innate propensity to learn norms and a keen ability to recognize violations of norms in others (Pinker, 1994; Manktelow and Over, 1991; Price, 2006). This could make untrustworthy behavior more costly, even under conditions of imperfect information.

Certainly, the norms that one learns are highly dependent upon the culture where one grows up. However, some norms are nearly universal, such as that of reciprocity (Gouldner, 1960). There is also evidence for a norm of strong reciprocity, where people are willing not only to reciprocate trustworthy behavior but to punish untrustworthy behavior even at a cost to themselves (see chapter 14, this volume). In this way, norms can resolve the second-order, collective-action problem associated with costly punishment.

One way to observe norms in action is to examine the behavior of nonhuman primates such as chimpanzees and capuchins. As Brosnan demonstrates in chapter 5, these nonhuman primates exhibit behavior consistent with the possession of internal and external norms of in-group reciprocity and inequity aversion. For example, Brosnan found that the willingness of capuchins to accept a food reward depended on whether their reward was equal to or less satisfying than the reward of another member of their in-group. Using the institutional grammar, we can define this norm as follows: a capuchin who receives a food reward that is less satisfying than the reward of another in-group member must not accept this food reward.

The advantage of this methodology is that one can separate behavior motivated solely by norms from behavior motivated by a combination of norms and rules or rules alone. While the social relations of nonhuman primates are not simple, nonhuman primates are less likely to be intentionally deceptive about their purposes than human primates. In other words, the social and dominance hierarchies that mark all primate societies, including those of humans, are typically more visible to the trained observer than those of human societies. Furthermore, the absence of a differentiated, abstract language capability lessens the complexity of analysis: where rules are present, the sanctions are immediately observable behaviors, which leads to strong predictions about the motivation for this behavior. In sum, we join with Brosnan and the other colleagues in this volume in calling for further study of nonhuman primates as a way to isolate the fundamental norms that affect prosocial behavior from the confounding effects of an abstract, differentiated social structure.

As the above analysis shows, norms can stabilize the initial "sense of trust" generated by shared strategies. Yet norms are not a complete solution. Because they lack an explicit sanction (the OR ELSE statement in our institutional grammar), societies that rely solely on norms are subject to invasion by purely self-interested actors. Moreover, norms must be transmitted within a cultural framework, making them vulnerable to the weakening of this framework. This makes norms ill-equipped to handle the expansion and specialization necessitated by economic growth. Thus, although norms can stabilize the "sense of trust" in a society, the need to impose specific, graduated sanctions eventually leads societies that experience this growth to develop institutions relying not only on norms but on rules as well.

One example of this development is the rise of the medieval Law Merchant described by Milgrom, North, and Weingast (1990). The Law Merchant was a set of legal codes governing commercial transactions and administered by private judges drawn from the commercial ranks. The purpose of these codes was to enforce contracts between merchants from different localities. A merchant who felt that he or she had been cheated by another merchant could file a grievance with the local private judge, who would then conduct a trial and, if the grievance was justified, enter a judgment on behalf of the aggrieved merchant. As Milgrom, North, and Weingast note, the development of the Law Merchant code was vital to reducing the uncertainty of trade associated with varying local practices and limiting the ability of localities to discriminate against alien merchants. This in turn led to increasing economic differentiation and specialization that helped make early modern Europe a center of economic exchange.

This development is surprising in that the judges had little or no power either to force parties to abide by the terms of their contract or to sanction those who did not comply. This lack of sanctioning power was a result of the prevalence of foreign merchants trading in Europe, against whom the judges had no legal jurisdiction. Thus, although judges could enter judgments on behalf of an aggrieved merchant, they lacked the legal and political power to enforce these judgments. This creates the following paradox, nicely summarized by Milgrom, North, and Weingast:

> It is not clear why such a system would be effective. What prevents a merchant from cheating by supplying lower quality goods than promised, and then leaving the Fairs before being detected? In these circumstances the cheated merchant might be able to get a judgment against his supplier, but what good would it do if the supplier never returned to the Fairs? Perhaps ostracism by the other merchants might be an effective way to enforce the payment of judgments. However, if that is so, why was a legal system needed at all? (1990:6)

Their resolution of this paradox ties in nicely with our previous discussion of the interaction between trust, reputation, and reciprocity. Recall that with perfect information, a high level of cooperation is sustainable even if players switch partners every turn. Bravo and Tamburino demonstrated that this level of cooperation was possible even given imperfect information, although achieving cooperation rather than the uncooperative equilibrium was heavily dependent upon how much information players had about each other's reputations. As our analysis demonstrates, one function of norms is to stabilize the system of cooperative exchange against internal and external shocks. Yet, norms become increasingly unworkable as the society becomes more economically differentiated. Societies must then switch from norms to rules such as the Law Merchant. In doing so, they do not switch from one set of problems to another; rather, as Milgrom, North, and Weingast demonstrated, the Law Merchant makes possible the system of increasingly anonymous exchange by providing accurate information about the reputation of merchants to other merchants at a low cost. Thus, the core problem of maintaining a reputational system under imperfect information still exists, and the function of the Law Merchant is not to harness state power to enforce contracts but rather to disseminate reputational information at a low cost to all merchants trading at a specific locale. Merchants then use this information to choose with whom to trade, and can effectively ostracize untrustworthy merchants, even without direct observation of their untrustworthy actions.[5]

Yet, for the Law Merchant to perform this function, it must provide an incentive for merchants to register grievances when they feel they have been cheated. As we have demonstrated, it is costly for merchants to register these grievances, and, given the judges' lack of formal enforcement powers, an individual merchant benefits little by doing so. In other words, the Law Merchant runs into the second-order collective action outlined above. The solution to this problem is ingenious: judges only allow merchants to use the system to resolve disputes if they make the appropriate queries prior to engaging in trade. This allows the judge to collect payments simply for providing information. In this way, both judges and merchants have an interest to use the Law Merchant courts, which in turn resolves the collective-action problem and leads to the dissemination of low-cost information. We can represent this solution by the following rules:

1. Merchants must make queries prior to engaging in trade, OR ELSE they cannot use the Law Merchant courts to register grievances.
2. Judges must provide accurate information to merchants in response to these queries, OR ELSE they cannot collect a fee for their services.

Milgrom, North, and Weingast (1990) showed that this system is sustainable provided the costs of making queries and adjudicating disputes are not too high relative to the frequency and profitability of trade.

The Law Merchant courts are one example of how rules can structure complex and differentiated exchange relationships. They also demonstrate one way that rules evolve from norms in order to fix problems that norms are not capable of resolving. However, rules are not a panacea; note that the Law Merchant system does not have to work exactly as described above. In particular, merchants who have bad reputations may bribe judges to conceal this fact from other merchants. Because judges have an information advantage in their relationship with the merchants, there is a real possibility for this type of corruption to occur. Not all judges will be honest, and, as Milgrom, North, and Weingast demonstrated, this can lead to perverse outcomes and ultimately to the system's collapse.

A first reaction to this problem is frequently to create more rules: perhaps a rule sanctioning judges if they are caught taking bribes. Yet this may lead to further problems. Rules are never costless. All rules are potentially subject to manipulation by political actors (Riker, 1982). Furthermore, each additional layer of rules creates not only additional governance costs but increased complexity in the system. This complexity can lead to perverse outcomes and make the detection of corruption more difficult, and both can cause the system to lose its legitimacy in the eyes of those who must rely on it.

Two Principles of Effective Institutional Design

The dilemma we have outlined is as old as politics: Who will guard the guardians? In this section, we argue that a slight rephrasing is in order: How should the guardians be guarded, and what are the consequences of guarding them incorrectly?

We begin with the second question, because the consequences of incorrect institutional design are so well known in contemporary society. In the private market, the Enron scandal demonstrates the potential for harm when powerful actors manipulate rules for their own gain at the expense of the weak. There are also devastating consequences in public economies when institutions fail, including environmental destruction and paralyzing government corruption. This leads not only to economic inefficiency but to a pervasive distrust of private and public institutions by the public. At the extreme, this distrust turns to cynicism and erodes the social capital necessary for both public and private economies to function efficiently.

It is not our purpose to speculate on the extent to which this is happening in the world today. Nor are we claiming that the Enron scandal or government corruption are solely the result of poor institutional design: both are complex phenomena, of which institutions compose only one part. Yet it is a vitally important part, and if we are to prevent future economic disasters, we must understand the basic principles of successful institutional design. There are many such principles. Here we want to focus on the following two.

First, it is vital to understand that the shared strategies, norms, and rules of an institutional arrangement are context-sensitive. There is no one perfect institutional solution to all problems, or even all problems of a certain type. Let us consider a well-known example involving common-pool resource management. For many years, it was thought that there were only two solutions to the "tragedy of the commons" outlined by Hardin (1968). These solutions were, on the one hand, instituting private property among resource users, and, alternatively, instituting a sanctioning apparatus to punish users who overuse the resource. Both solutions had their partisans, who advocated their solution to all common-pool resource dilemmas and explained contrary results as anomalies.

What went largely unnoticed prior to a National Research Council study in 1986, and empirical research that it stimulated (e.g., Berkes, 1989; E. Ostrom, 1990; Acheson, 2003), was that the common-pool resource dilemma is not a single problem but rather an abstract way of describing a large class of problems with certain similarities. These similarities provided a theoretical model with which to think about these problems. Yet, some scholars reified the dilemma as existing everywhere when multiple users harvest from the same resource, and they recommended either privatization or government ownership as a panacea. Using only one way of thinking about common-rule-resource problems obscured important distinctions between different common-pool resource situations. These variations are more likely to be recognized now; yet many precious resources were destroyed in learning this lesson.

Thus, our first principle of successful institutional design is to pay attention to context-specific differences between similar classes of institutional problems. Selling commodities in a farmers' market is quite different than selling options to buy or sell these commodities in a futures market. In the wake of the Enron scandal, it is tempting to lay blame at a single cause and seek a corresponding solution. Yet, this is precisely the wrong action to take. Institutional reform is needed in both private and public economies, but this reform must be closely attuned to the individual problem under consideration. Otherwise, we risk forcing problems to conform to institutions, rather than the other way around.

Second, we must understand that institutional solutions, if poorly implemented, can crowd out trust rather than enhance it (Frey, 1994). This problem is especially prevalent with solutions imposed upon a group of people by outside experts. Any functioning system has evolved some institutions for its own maintenance. These institutions evolve through the reciprocal adaptation between the people and the overall process in which they are engaged. They are rarely identical to the formal rules of the process, and often poorly understood by outsiders. As a result, imposed solutions often crowd out these adaptive institutions.

This is especially true where trust is concerned. Just as there is evidence for an innate tendency to trust others, there is also evidence that this tendency can be overridden if people are forced to trust others by sanctioning mechanisms or other means. For example, Cardenas, Stranlund, and Willis (2000) found that imposing rules on resource users without communication resulted in less efficient usage than allowing users the choice to adopt or reject rules during face-to-face communication. In a similar vein, Janssen et al. (2006) found that subjects in a foraging experiment who vote to instantiate property rights showed greater compliance with these rights than when they were imposed upon them with no vote. Clearly, in designing and repairing institutions that will not crowd out people's willingness to trust one another, we must be careful not to impose sanctioning mechanisms upon them.

Conclusion

We began this chapter with a simple game in order to make a simple point: exchange of goods and services does not occur between strangers in the absence of trust-enhancing institutions. By tracing the interaction of trust, reputation, and reciprocity through the evolution of a simple institution to a complex one, we argued that the problem of imperfect information makes it difficult for exchange partners to rely on each other's reputations, and that one reason institutions evolve is to make reputation management less costly, more stable, and more effective.

In concluding, we note three important points: first, although ours is not the only solution to the problems of corporate scandal and government corruption, we strongly believe that any solution that does not take seriously the effect institutions have on exchange relationships is likely to fail.

Second, although we have largely ignored questions of equity in order to focus on questions of efficiency, the two are not the same. Any political-economic system allows a great deal of leeway in how institutions are designed, and this flexibility can institutionalize privilege in the hands of the powerful. This is an important factor to keep in mind when deciding which reforms to implement and which to forego.

Finally, it is worth noting that institutions—even well-designed ones—will not lead to beneficial outcomes by themselves. Institutions are inseparable from the people who make use of them, and, as noted above, all rules are subject to manipulation by political actors. Thus, at some point, we must cease to rely upon institutional corrections and place our faith in a citizenry well educated in virtue.[6] Ultimately, we must guard the guardians. It is our hope that, by keeping our rules for institutional design in mind, we can guard them well.

Appendix

To arrive at the replicator equations, we use the replicator dynamics first defined by Maynard Smith (1982). A full description of the assumptions behind this method analysis can be found in Samuelson (1997).

Define players 1 and 2 with the strategy sets and payoffs represented by the following normal form game shown in table 10.1.

Let x equal the percentage of Player 1's who play Don't Offer, and $(1-x)$ equal the percentage of Player 1's who play Offer. Similarly, let y equal the percentage of Player 2's who play Snatch, and $(1-y)$ equal the percentage of

TABLE 10.1
Game of Snatch (Normal Form)

	Exchange	Snatch
Offer	15, 15	5, 20
Don't Offer	10, 10	10, 10

Player 2's who play Exchange. Player 1 then has the following utility functions for each strategy:

$$u_1(\text{Offer}) = 15(1-y) + 5y$$

$$u_1(\text{Don't Offer}) = 10(1-y) + 10y$$

Similarly, Player 2 has the utility functions:

$$u_2(\text{Exchange}) = 15(1-x) + 10x$$

$$u_2(\text{Snatch}) = 20(1-x) + 10x$$

Player 1's average utility is found by adding the expected utility of playing Offer and Don't Offer. This gives us:

$$u_{1\text{AVG}} = x[10(1-y) + 10y] + (1-x)\ [15(1-y) + 5y],$$

that simplifies to

$$u_{1\text{AVG}} = 10x + (1-x)(15 - 10y)$$

Player 2's average utility is found by adding the expected utility of player Exchange and Snatch, giving us:

$$u_{2\text{AVG}} = y[20(1-x) + 10x] + (1-y)[15(1-x) + 1\,0\,x]$$

that simplifies to:

$$u_{2\text{AVG}} = y(20 - 10x) + (1-y)(15 - 5x).$$

We want to find the rate of change of x and y with respect to time t as a function of the difference between each player's payoff from playing the Nash equilibrium (Don't Offer, Snatch) and the average payoff for each player. This gives us the following two replicator equations:

$$dx/dt = x[10(1-y) + 10y - (10x + (1-x)(15 - 10y))]$$

$$dy/dt = y[20(1-x) + 10x - (y(20 - 10x) + (1-y)(15 - 5x))]$$

that simplify to

$$dx/dt = 5x\ (1-x)(2y - 1) \text{ and } dy/dt = -5y\ (y-1)(1-x).$$

Notes

1. The game of Snatch has a similar structure to the Choice game developed by Carl Bergstrom in chapter 7 of this volume. Dating and market exchanges both rely on the development of reputations, norms, and sometimes rules as we discuss in more depth in this chapter.

2. This assumes that utility corresponds only with individual economic gain. Many economists now question this assumption: for examples, see Rabin 1993 and Cain 1998.

3. See the appendix to this chapter for the derivation of these equations.

4. Creating a system of "law and order," however, is just one change in the context and structure of a game. As E. Ostrom (2005:211–214) illustrates, at least four rules—position, boundary, choice, and aggregation—need to be changed to move from the "state of nature" game of figure 10.1 to a game with property rights and sanctions. These rules do not need to be changed by a single center of authority for an entire country, however, as Hobbes posited. In many countries, local communities have established their own rules related to diverse transactions taking place within their boundaries for centuries (Berman, 1983).

5. Merchants worldwide have faced similar problems to those described by Milgrom, North, and Weingast. Gracia Clark (2000, 2002, 2004) studied the evolution of market rules in Kumasi, Ghana, and the important role that Asante women traders and their leaders play in both resolving everyday disputes among traders, building reputations, and increasing the returns that traders can obtain from using this market as well as the rules and norms that have evolved in these arenas.

6. See chapters 2 and 14 in this volume.

References

Acheson, James M. 2003. *Capturing the Commons: Devising Institutions to Manage the Maine Lobster Industry.* New Haven, Conn.: University Press of New England.

Andreoni, James. 1989. "Giving with Impure Altruism: Applications to Charity and Ricardian Equivalence." *Journal of Political Economy* 97:1447–1458.

Annen, Kurt. 2003. "Social Capital, Inclusive Networks, and Economic Performance." *Journal of Economic Behavior and Organization* 50 (4): 449–463.

Arce, Daniel G. 1994. "Stability Criteria for Social Norms with Applications to the Prisoner's Dilemma." *Journal of Conflict Resolution* 38 (4): 749–765.

Axelrod, Robert. 1986. "An Evolutionary Approach to Norms." *American Political Science Review* 80:1095–1111.

Barner-Barry, Carol. 1986. "Rob: Children's Tacit Use of Peer Ostracism to Control Aggressive Behavior." *Ethology and Sociobiology* 7:281–293.

Basu, Kaushik. 1995. "Civil Institutions and Evolution: Concepts, Critique and Models." *Journal of Development Economics* 46 (1): 19–33.

Berkes, Fikret, ed. 1989. *Common Property Resources: Ecology and Community-Based Sustainable Development.* London: Belhaven.

Berman, Harold. 1983. *Law and Revolution: The Formation of the Western Legal Tradition.* Cambridge, Mass.: Harvard University Press.

Bochet, Olivier, Talbot Page, and Louis Putterman. 2006. "Communication and Punishment in Voluntary Contribution Experiments." *Journal of Economic Behavior and Organization* 60 (1): 11–16.

Bravo, Giangiacomo, and Lucia Tamburino. 2008. "The Evolution of Trust in Non-Simultaneous Exchange Situations." *Rationality and Society*, forthcoming.

Brooks, Nancy. 2001. "The Effects of Community Characteristics on Community Social Behavior." *Journal of Economic Behavior and Organization* 44 (3): 249–267.

Cain, Michael. 1998. "An Experimental Investigation of Motives and Information in the Prisoner's Dilemma Game." *Advances in Group Processes* 15:33–60.

Camerer, Colin. 2003. *Behavioral Game Theory*. New York: Russell Sage Foundation.

Cardenas, Juan Camilo, John Stranlund, and Cleve Willis. 2000. "Local Environmental Control and Institutional Crowding-Out." *World Development* 28:1719–1733.

Clark, Gracia. 2000. "Small-Scale Traders' Key Role in Stabilizing and Diversifying Ghana's Rural Communities and Livelihoods." In *Women Farmers and Commercial Ventures*, ed. Anita Spring, 253–270. Boulder, Colo.: Lynne Rienner.

———. 2002. "Market Association Leaders' Strategic Use of Language and Narrative in Market Disputes and Negotiations in Kumasi, Ghana." *Africa Today* 49 (1): 42–58.

———. 2004. "Managing Transitions and Continuities in Ghanian Trading Contexts." *African Economic History* 32:65–88.

Clay, Karen. 1997. "Trade without Law: Private-Order Institutions in Mexican California." *Journal of Law, Economics, and Organization* 13 (1): 202–231.

Colson, Elizabeth. 1974. *Tradition and Contract: The Problem of Order*. London: Heinemann.

Crawford, Sue, and Elinor Ostrom. 2005. "A Grammar of Institutions." In *Understanding Institutional Diversity*, ed. Elinor Ostrom, 137–174. Princeton, N.J.: Princeton University Press. Originally published in *American Political Science Review* 89 (3) (1995): 582–600.

de Waal, Frans. 1996. *Good Natured: The Origins of Right and Wrong in Humans and Other Animals*. Cambridge, Mass.: Harvard University Press.

Fafchamps, Marcel, and Bart Minten. 1999. "Relationships and Traders in Madagascar." *Journal of Development Studies* 35 (6): 1–35.

Feldman, Eric. 2005. "The Tuna Court: Law and Norms in the World's Premier Fish Market." Paper no. 63. Philadelphia: University of Pennsylvania Law School, Berkeley Electronic Press (bepress).

Fratianni, Michele. Forthcoming. "Government Debt, Reputation, and Creditors' Protections: The Tale of San Giogio." *Review of Finance*.

Frey, Bruno S. 1994. "How Intrinsic Motivation Is Crowded Out and In." *Rationality and Society* 6:334–352.

Fudenberg, Drew, and Jean Tirole. 1991. *Game Theory*. Cambridge, Mass.: MIT Press.

Gouldner, Alvin. 1960. "The Norm of Reciprocity: A Preliminary Statement." *American Sociological Review* 25:161–178.

Harbaugh, William T., Kate Krause, Steven G. Liday Jr., and Lise Vesterlund. 2003. "Trust in Children." In *Trust and Reciprocity*, ed. Elinor Ostrom and James Walker, 302–322. New York: Russell Sage Foundation.

Hardin, Garrett. 1968. "The Tragedy of the Commons." *Science* 162:1243–1248.

Hobbes, Thomas. [1651] 1960. *Leviathan, or the Matter, Form and Power of a Commonwealth Ecclesiastical and Civil*. Oxford: Basil Blackwell.

Janssen, Marco A., Robert L. Goldstone, Filippo Menczer, and Elinor Ostrom. 2006. "Effect of Rule Choice in Dynamic Interactive Commons." CIPEC Working Paper. Bloomington: Indiana University, Center for the Study of Institutions, Population, and Environmental Change.

Kandori, Michihiro. 1992. "Social Norms and Community Enforcement." *Review of Economic Studies* 59 (1): 63–80.

Kerr, Norbert L., and C. M. Kaufman-Gilliland. 1994. "Communication, Commitment, and Cooperation in Social Dilemmas." *Journal of Personality and Social Psychology* 66:513–529.

Kikeri, S., and J. Nellis. 2004. "An Assessment of Privatization." *World Bank Research Observer* 19 (1): 87–118.

Kurzban, Robert. 2003. "Biological Foundations of Reciprocity." In *Trust and Reciprocity,* ed. Elinor Ostrom and James Walker, 105–127. New York: Russell Sage Foundation.

Manktelow, Ken, and David Over. 1991. "Social Roles and Utilities in Reasoning with Deontic Conditionals." *Cognition* 39:85–105.

Milgrom, Paul, Douglass North, and Barry Weingast. 1990. "The Role of Institutions in the Revival of Trade: The Law Merchant, Private Judges, and the Champagne Fairs." *Economics and Politics* 2 (1): 1–25.

Ostrom, Elinor. 1990. *Governing the Commons: The Evolution of Institutions for Collective Action.* Cambridge, Mass.: Cambridge University Press.

———. 1998. "A Behavioral Approach to the Rational Choice Theory of Collective Action." *American Political Science Review* 92 (1): 1–22.

———. 2005. *Understanding Institutional Diversity.* Princeton, N.J.: Princeton University Press.

Ostrom, Elinor, and James Walker. 2003. *Trust and Reciprocity: Interdisciplinary Lessons from Experimental Research.* New York: Russell Sage Foundation.

Ostrom, Vincent. 1987. *The Political Theory of a Common Republic: Designing the American Experiment.* 2nd ed. San Francisco, Calif.: ICS Press.

———. 1997. *The Meaning of Democracy and the Vulnerability of Democracies: A Response to Tocqueville's Challenge.* Ann Arbor: University of Michigan Press.

Pinker, Steven. 1994. *The Language Instinct.* New York: Morrow.

Plott, Charles, and Robert A. Meyer. 1975. "The Technology of Public Goods, Externalities, and the Exclusion Principle." In *Economic Analysis of Environmental Problems,* ed. Edwin S. Mills, 65–94. New York: Columbia University Press.

Price, Michael. 2006. "Monitoring, Reputation, and 'Greenbeard' Reciprocity in a Shuar Work Team." *Journal of Organizational Behavior* 27 (2): 201–219.

Rabin, Matthew. 1993. "Incorporating Fairness in Game Theory and Economics." *American Economic Review* 83:1281–1302.

Riker, William H. 1982. *Liberalism against Populism: A Confrontation between the Theory of Democracy and the Theory of Social Choice.* San Fransisco, Calif.: W. H. Freeman.

Sally, David. 1995. "Conversation and Cooperation in Social Dilemmas: A Meta-Analysis of Experiments from 1958 to 1992." *Rationality and Society* 7:58–92.

———. 2002. "Two Economic Applications of Sympathy." *Journal of Law, Economics, and Organization* 18 (2): 455–487.

Samuelson, Larry. 1997. *Evolutionary Games and Equilibrium Selection.* Cambridge, Mass.: MIT Press.

Shivakuman, Sujai. 2005. *The Constitution of Development: Crafting Capabilities for Self-Governance.* New York: Palgrave MacMillan.

Smith, John Maynard. 1982. *Evolution and the Theory of Games.* Cambridge: Cambridge University Press.

Wedekind, C., and M. Milinski. 2000. "Cooperation through Image Scoring in Humans." *Science* 288 (5467): 850–852.

World Bank. 2001. *Private Sector Development (PSD): Findings and Lessons from Selected Countries.* Washington, D.C.: World Bank.

———. 2002. *World Development Report, 2002. Building Institutions for Markets.* Washington, D.C.: World Bank.

Eleven

Values, Mechanism Design, and Fairness

Oliver R. Goodenough

Making Values Part of Rigorous Economic Theory

Economic theorists have generally underestimated values as critical elements in human choice and behavior. Until recently, economists have found values hard to define, hard to measure, and seemingly at odds with the calculating rationality that was the starting point for traditional economic modeling (see chapter 8, this volume). These traditional economists have viewed values with a kind of puzzled detachment; they are not ready to fully dismiss them, but they tend to look at them merely as nice add-ons that probably contribute to efficiency in some vague way. In his widely recognized text, *Economic Analysis of Law*, Richard Posner allocated a few pages to morality, mainly to link it with efficiency: "Honesty, trustworthiness and love reduce the costs of transactions," he declared (1998, 284).

Good economists may understand the limits of the self-interested, rational actor as a proxy for human decision making (e.g. Smith 2002; Posner 1998), but in scholarly rhetoric and public conception this imaginary *Homo economicus* has morphed into a cartoon version of himself, a heartless sociopath dominated by the "anti-value" of selfishness (see chapters 1 and 8). Vernon Smith, in his Nobel Lecture (2002), calls such distorted reasoning a "vulgar impression." Accurate or not, this dismal, excessively simplified version of the rational actor approach has come to dominate the public vision of free economic systems. As a result, much of humanity views free enterprise[1] as a kind of Jurassic Park—a Hobbesian island of eat or be eaten, a world where Herbert Spencer and Social Darwinism still rule (Solomon 1993; chapter 1, this volume).

This vision is not just propagated by opponents of free enterprise. Its friends have all too often contributed to the picture. Adam Smith, in his *Wealth of Nations* (1776) famously stated:

> It is not from the benevolence of the butcher, the brewer, or the baker, that we expect our dinner, but from their regard for their own interest. We address ourselves, not to their humanity but their self-love, and never talk to them of our own necessities but of their advantages. (20)

As Vernon Smith (1999), and Paul Zak and Robert Solomon (in chapters 12 and 2, respectively) remind us, Adam Smith's rhetorical sally can be understood in context by reference to his other great work, *The Theory of Moral Sentiments* (1759). Nonetheless, the cartoon version is salient in the common understanding.

Bob McTeer is a representative contemporary voice in this strain. Currently Chancellor of the Texas A&M University System, he was president of the Federal Reserve Bank of Dallas and a member of the Federal Open Market Committee for almost fourteen years. It would be hard to find a more committed proponent of free enterprise. He is also, for better or worse, a happy purveyor of the cartoon. To give him his due, he prefaces his *Free Enterprise Primer* (McTeer 2006) by declaring his *intention* to oversimplify:

> This primer on free enterprise is my oversimplified and unapologetic description of how a market economy works. It is meant not only to be informative about the subject, but also to create an appreciation for a unique system that has given us something as remarkable as supermarkets.

In the body of his primer, whose substance reflects traditional economics, his rhetoric is all about competition and selfishness. In considering the effect of the focused pursuit of individual economic goals, he opines:

> You may ask, isn't that selfish? The answer is yes. But Adam Smith showed us more than 200 years ago in *The Wealth of Nations* that pursuit of self-interest in a competitive market economy is, as if by some "invisible hand," consistent with promoting the public interest.

Focusing on selfishness and ignoring values may make life easier for economic theorists of a particular viewpoint, but it is highly unsatisfactory to most other people. We have been raised, at least most of us, to consider values of prime importance. Don't lie. Be fair. Keep your promises. Our families, our friends, our schools, our religions, all promote values of some sort. They may be qualified by context, group identification, and other conditional factors, but they most definitely are part of the social and emotional landscape. Successful business practitioners know this (e.g., Templeton 1987; chapter 2, this volume), but traditional economic theorists often forget (as pointed out in chapters 3 and 14). Any behavioral theory that ignores this reality will seem fundamentally deficient to humans around the world, and the seeming absence of values in the cartoon presented by McTeer and others probably does more damage than good in the spread of free economic systems (see chapter 1). The better approach to looking at values and free enterprise is to assume that values exist for a reason, and not that reason somehow contradicts their existence. This kind of motivational realism in the context of fairness has been advocated by Kahneman, Knetsch, and Thaler (1986), who suggest we need to adopt more complex models for the psychology of economic actors.

Functionality Not at Odds with Subjective Experience

What would we think of a supposedly scientific psychology that marginalized love, friendship, or taste? It would seem desperately incomplete. Rather than dismissing such "irrational" elements of our psyches, we need to incorporate them into our theories. Viewed functionally, there are good, materialistic explanations, rooted in a rational logic, for why we should experience these critical elements of our inner life. Love helps us to make the necessary strong connections to support reproduction and child rearing in a species where high parental investment is a good strategy for success (e.g., Fisher 1992). Friendship can be seen as a good heuristic for long-term cooperative partnering (Hrushka and Henrich 2006; chapter 4, this volume; Smaniotto 2005). Our enjoyment of chocolate cake derives from the value of sugar and fat in an ancestral foraging landscape (Durrant and Ellis 2002). We intuitively worry that this kind of explanation somehow devalues the subjective experience, but this need not be the case (see, generally, Dawkins 1998; U. Goodenough 1998). A functional understanding from a scientific perspective does not contradict our reliance on, and celebration of, the psychological manifestations of love, friendship, and taste as a basis for successful human action.

For instance, understanding the reproductive explanation for love might lead some to a cynical abandonment of its constraints in the pursuit of sexual gratification. Such a syndrome has been anecdotally noted among some adherents of an evolutionary approach to human behavior, but it also occurs among plumbers and film stars. A person who takes an excessively short-sighted view of the rational actor model in the reproductive context, denying the importance of affiliative love, might be successful as a lothario but would probably be unsuccessful as a spouse or parent. And since being a relatively reliable spouse and parent (at least during critical years; see Fisher 1992) appears to be a good strategy for long-term human reproductive success, the full-blown lothario model is not generally the dominant behavioral mode for humans (as opposed to roosters or elephant seals, for example, see Judson 2002). Love is part of how humans do reproductive success.

A key step in adopting a scientific approach to these strong, subjective experiences is to honor their rich, deeply felt, psychological manifestations. Our inner life is a critical mechanism that helps us implement our shared, human strategies of nutrition, reproduction, and sociality. A rigorous understanding of the importance of these responses in our survival and success should coexist with their subjective experience and with their religious and moral components. These elements are not flaws; they are how the system works for most of us.

We should take a similar starting point for values in human economic life (Binmore 2005). Posner was partly right here—values do serve a function—but he was vastly mistaken in emphasis—their function is foundational. Free enterprise, when successful, is values in action. Instead of a theory that says values are not important, we need a theory that puts them front and center, in a balance with self-interest. We need a theory that rigorously describes the often critical role values play in a free economic system, so that Chancellor McTeer makes *Values* a major section heading in his free enterprise primer. And we need a theory that honors their subjective importance as well as their functional utility.

Goals

This chapter has two overall goals. The first is to delineate the critical role values play in a free economic system. This is, for the most part, not a ground-breaking exercise. Both in this volume and elsewhere, behavioral and institutional economists, game theorists, biologists, cognitive psychologists, sociologists, anthropologists, philosophers, and even legal and business scholars have crafted the pieces of the puzzle and, in large measure, put them into place. My hope is to tell the story clearly, succinctly, and in context. Working across so many fields and in a short space, I will inevitably omit detail and nuance. Through the references, I point to more complete tellings by others, where such shortcomings can be rectified.

The second goal is to analyze the role of a particular value: fairness. Fairness is a bit of an orphan in the free enterprise world. Some traditional approaches wonder what such a concept can add to welfare maximization (Kaplow and Shavell 2002), rejecting a consideration that seems to constrain efficiency and welfare. Once again, if we ask what it could be for, rather than theorizing about why it cannot be helpful, we can attempt to correct important shortcomings in the understandings of traditional economics.

The Classic Preoccupation with Output

Until recently, economic theory has been largely preoccupied with questions of maximizing output, typically through the efficient allocation of productive resources in ways that match output potentials to preferences and demand. These are important questions, but they are essentially *secondary*, by which I mean they can only be addressed after other issues are settled. The modeling and analysis of output generally presume the existence of stable regimes of ownership and exchange (e.g., Shavell 2004). The *primary* questions about how such regimes might come into existence and be sustained are left unaddressed. In addition to an economics of output, we have needed an economics of mechanisms, and, most important, stable mechanisms. As Binmore (2005) pointed out, stability is the first level of priority in understanding a social system.

But understanding mechanisms requires a theory of social interaction, and, as Posner admitted, "traditional economics generally assumed . . . that people made decisions without considering the reactions of other people" (1998, 17). This has led to the predominance of an asocial model of rationality in economic thinking, an impersonal, cost/benefit rationality that is perfectly appropriate in a context of foraging or, for that matter, of a highly structured, impersonal market, and largely inappropriate in a context of strategic interaction between potentially cooperative parties.

The development of institutional economics over the last three decades has allowed economics to take a leap forward, explicitly raising questions of mechanism and providing an increasing body of answers about how stable social and transactional mechanisms might work (North 1981; chapters 10 and 13, this volume). The synthesis of traditional and institutional economics allows us to understand that these two factors—output and stability—are both necessary problems for any economic system to solve. Their simultaneous solution will determine the productive potential of any society.

Obstacles and Pathways to Successful Cooperation: The Science of Games

Humans are members of a gregarious, social species, like chimpanzees, perhaps even pack animals, and wolves (Jones and Goldsmith 2005). Being part of interactive groups imposes costs for individuals but also creates significant opportunities. As we have understood from the days of Adam Smith, the gains of trade and cooperative interaction are the sources of most increases in our well-being and prosperity (A. Smith 1776). Economic cooperation permits specialization, increases of scale, tackling projects beyond the scope of individual effort, accumulation of capital, and many other opportunities for bigger and better payoffs.

The potential rewards may be high, but paths to successful cooperation are not always easy to find and follow. Game theory provides the clearest understanding of this very significant challenge, and it stands as a critical addition to classical economic approaches (e.g., Binmore 2005; Bowles 2004; Gintis 2000; von Neumann and Morgenstern 1944). The crux of the difficulty is that human economic interactions often contain opportunities to grab short-term advantage, "defections," that leave the other player in the lurch, often worse off for having tried to reach the cooperative gains.

The so-called Prisoners' Dilemma is a classic, widely familiar example of such a structure (e.g., Bowles 2004; Skyrms 2004; Gintis 2000; Badcock 1998, 2000). It consists of a simultaneous, two-player game, where there is an aggregate mutual upside to cooperation, but where each player, acting individually, will be better off by defecting, no matter what the other player does. Therefore, if played "rationally," both players will defect, achieving a lower mutual payoff. This outcome, although it does not maximize the joint payoff, is said to dominate the cooperative alternative.

Cooperative opportunities and defection pitfalls come in a number of shapes and flavors, beyond any hope of fully cataloging here. Many contain a variation of the "first-mover problem," where, unlike the Prisoners' Dilemma, the moves are made in a sequence, not simultaneously (Dixit and Skeath 2004). Here, to get on the pathway to a better mutual payoff, one of the actors must incur a cost, from which the other can take a short-term advantage. In some instances, the second actor directly appropriates the investment the other makes, without adding any value. In other cases, the value is added by the second actor, but that individual then gets to collect the benefit of mutuality, and can choose not to share it. The Snatch game, described by Ostrom and Schwab in chapter 10 in the context of trading farmers, is just such a structure.

The problem can also be illustrated by a variation of the hunting game.[2] Imagine a grouse hunt, where one hunter is the "beater," making noise to drive the grouse out of their cover and toward a second hunter with a shotgun, lying in wait where the grouse will emerge. When the second hunter makes a kill, there will be some time before the beater emerges, enough time for the second to abscond with the meat from the successful hunt. A variant of the first-mover problem gives spice to the so called trust games often used in experimental economics (e.g., McCabe and Smith 2000).

If vegetable bartering or grouse hunting seems a bit unrealistic in the modern world, consider the problem of a one-time trade between a buyer in the United States and a seller in China, an exchange from which both will significantly benefit. The buyer places an order and asks the seller to ship the goods, to be paid for on receipt. The seller is happy to comply but needs to receive the money first. The first one to move is vulnerable to being taken advantage of: the goods may be received but not paid for, or the money sent and the goods never shipped. Untrammeled rational self-interest suggests no reason why the second actor should not do exactly that (e.g., Binmore 2005). Absent some institutional intervention (see chapter 10), like enforceable contract law (see chapter 9), which in effect restructures the game, the exchange is unlikely to occur, and the potential gains of trade will be lost.[3]

Avenues for Restructuring Games

As the mention of institutions reminds us, competent strategic actors are not simply stuck with the game forms presented to them. If the hillside is full of grouse, a pair of clever, self-interested hunters will first hunt for ways to structure their relationship so as to make non-defection the dominant strategy set, giving a reliable expectation both that the second move will be made and that the benefits of success will be shared on a satisfactory basis between the players. A similar search for structure can be conducted by the potential vegetable (see chapter 10) or international trading partners. Such a restructuring is sometimes thought of as "stabilizing" cooperation, but it is really a case of changing the form of the game itself so that plus-sum interaction becomes the dominant expectation and not a forlorn hope.

Of course, not all avenues to cooperation are initially booby-trapped by defection-dominant outcomes; some are naturally cooperative (Skyrms 2004). The Prisoners' Dilemma is a worst case, not the universal case (Binmore 2005). For instance, in the "driving game," deciding which side of the road to drive on, the dominant solutions involve converging on a mutually observed convention; both right and left will work (Binmore 2005; see also Skyrms 2000). These coordination games can be relatively easy to bring into cooperative equilibrium (but fiendishly hard to change; see chapter 6). Nonetheless, many of the important opportunities in human economic life for plus-sum, mutually-enhanced outcomes come in defection-prone packages, and can only be accessed by some form of strategic restructuring (see chapter 10).

What do strategies for restructuring look like? Game theory, economics, and evolutionary biology have identified several, which I briefly summarize here. In application, they are not necessarily mutually exclusive; successful restructuring may involve combinations and layers of complimentary approaches.

Reciprocity. A classic avenue identified in biology and economics involves reciprocity. In a seminal 1971 paper, Robert Trivers explored how exchanges of benefits, staggered over time in repeat interactions, could create a self-sustaining climate of reliable cooperative moves, a mechanism he called "reciprocal altruism" (Trivers 1971; see also, e.g., chapter 4, this volume, and Badcock 2000). The challenge to maintaining productive reciprocity is essentially the first-mover problem; at any given point the other actor can take the current benefit and not reciprocate when the next need arises. In theory, a system of reciprocity with a known end point should unravel backward from the last move (Binmore 2005; Gintis 2000; Badcock 2000); however, experiments with humans suggest that it often sustains itself for much of the possible benefit period (Gintis 2000).

In a series of famous computer modeling "tournaments" run by Robert Axelrod in the early 1980s, a selective reciprocity strategy called "tit for tat" was identified as a good solution in a repeat Prisoners' Dilemma environment (Binmore 2005; Axelrod 1984). Even more robust was a forgiving version of tit for tat (Dixit and Skeath 2004), which allowed for getting reciprocity started once again with a similarly minded player if the defection was random, or some sort of mistake, although excessive forgiveness can open the process to defection once again (Boyd and Lorberbaum 1987).

Punishment and "Strong Reciprocity." Punishment for defection can be a very effective mechanism for making continued cooperation the dominant choice in many otherwise defection-prone games. The punishment for "cheating" can be as simple as a cessation of cooperation in return, as in tit for tat (Binmore 2005; chapter 14, this volume). Punishment may be even more effective if some kind of affirmative harm is inflicted either by the disappointed partner or by some bystander or designated guarantor of the process. The threat of pun-ishment must be followed through on, even though costly in the short run for the punisher, if this mechanism is to have long-term effectiveness (Fehr and Gächter 2002). Indeed, the cost of punishment to the punisher is part of what makes the strategy work. This approach is often termed "strong reciprocity" (Fehr, Fischbacher, and Gächter 2002; Bowles and Gintis 2000; Bowles 2004)

Reputation. Another mechanism that provides reliability is reputation. In a multiplayer context, reliable play can be observed by others and can form the basis for the expectation of reciprocal cooperation with a new player in turn (Engseld and Bergh 2005). A reputation for reliability can be an asset in persuading another to take the risk of a cooperative move (see chapter 10; Binmore 2005; O. Goodenough 2001), effectively turning one-shot games into an open-ended string. Its efficacy has been shown in modeling experi-ments (Nowak and Sigmund 1998) and in such human activities as law prac-tice (O. Goodenough 2001). A contemporary business application involves eBay, where reputation mechanisms are integral parts of maintaining trust and cooperation as a dominant strategy in the eBay marketplace, notwith-standing massive potential first-mover problems in what are individually one-shot interactions.

Commitment. Commitment mechanisms are very important devices in shaping the decisions of the other player in an interaction (Nesse 2001; Dixit and Skeath 2004). Commitments are often strategies that create some kind of counterbalancing loss, through a reliable link, if a defecting move is made (see chapter 3). In this context it can be viewed as a kind of self-inflicted punish-ment. A commitment can be made around reputation. For instance, the high-cost marble architecture of a traditional bank is a commitment that will be lost if the bank acquires a reputation for unreliability (O. Goodenough 2001). Mortgages with legal enforcement can be seen as a kind of commitment, as can contracts (see chapter 9; O. Goodenough 2001), and so can wasting time in courtship (see chapter 7).

Communication. When information can be passed back and forth between players in a game, it can help tip the choices toward cooperation (Skyrms 2004). A key element in effective communication is the reliability of the in-formation exchanged. One way to make information reliable is to make it costly to produce. This is widely accepted as the basis for the peacock's tail: it tells a story of health and vigor that a sickly peacock is in no position to fake (e.g., Zahavi and Zahavi 1997). Indeed, there is a whole class of circumstances where wasteful messages about quality gain reliability by their shear profligacy, a phenomenon that underpins excessive hospitality customs, Super Bowl advertising, and conspicuous consumption (see chapter 7). There are also cir-cumstances where "cheap talk" can be reliable (e.g., Skyrms 2004; Bergstrom et al. 2001). In one variation, the message is cheap if it is honest, and will only prove costly if it is false. One zoological example is the way the size of a black chest patch on the Harris's Sparrow serves as a signal of dominance potential. If the self-report is accurate, the bird leads a quiet life and there is little cost or consequence; if inaccurate, the bird incurs high costs in lost fights with more dominant birds challenging the presumptuous upstart (Rohwer 1982; Rohwer and Rohwer 1978).

Hierarchy/Following a Leader. Leadership/dominance can lead to reliable cooperation. A third party with the power to inflict harm can simply order two subordinates to cooperate; as in most of these mechanisms, the payoffs of defection are radically diminished, and the dominant solution to the game shifted toward the pro-social effort (Rubin 2000). Leaders can also help find better outcomes without threat or coercion if the problem is simply one of coordination or communication (Binmore 2005). There is a particular downside to hierarchy as the solution, however, namely, keeping the demands of the hierarch for an increasing share of the surplus under control. A "good" dominant individual will extract a relatively low rent for her share in making the cooperation happen; a "bad" one will take most of the benefit (see chapter 10). Where the relationship is one in which the transaction is between the dominant and the subordinate directly, the hierarch can exploit the position even further, at worst to the point of slavery.

The Dual-Key Lock Box. One way to solve the distribution of benefits problem illustrated by the grouse hunters is to establish a mechanism that requires the presence of both players to free up the cooperation surplus. This can be done through a purely physical mechanism, such as the dual-key lock box, or safe. This works best in a business with two partners, where all the receipts are put into a box that requires two keys, used simultaneously, to open. Each partner has a key, and the payoffs are reliably shared. Forms of this mechanism are widely incorporated in cooperative institutions, ranging from two-signature checking accounts through the unanimity requirements for certain important actions embodied in traditional American partnership law, and even into the launch controls of nuclear missiles; more on this below.

Contract. Contracting is a widely understood mechanism for successfully restructuring interactions to ensure a positive outcome (see chapter 9; O. Goodenough 2001). It is often just assumed to be available in traditional economic modeling. Indeed, in one definitional convention, games where a binding agreement can be made on all critical points are termed "cooperative games," and those where such agreement is not possible "noncooperative games" (Bowles 2004). As Dixit and Skeath (2004, 26) point out, however, this standard terminology "is somewhat unfortunate because it gives the impression that the former will produce cooperative outcomes and the latter will not." In fact, both kinds can lead to "a lot of mutual gain." Contract can be viewed as a composite of mechanisms, a kind of voluntary commitment about future performance, often, but not necessarily, linked to punishment through the intervention of a dominant third party, such as the state. Like the double key lock box, contracting mechanisms can be physically constructed and can also exist in such social and psychological mediums as religion and law.

Property. Property is another mechanism widely assumed as a starting point for classical economics. At heart, it is a convention by an actor to respect the possession of a resource by another, even if appropriation by the actor were possible, either by stealth or by force. Some have argued that property is necessarily a creation of the state (e.g., Hobbes 1660), a composite solution only available through punishment and hierarchical dictate. Maynard Smith, however, demonstrated that aspects of a property convention can take hold in a population purely as a matter of un-coerced evolution, a result he called the "bourgeois strategy" (Maynard Smith and Parker 1976; see, generally, Stake 2004). Property creates a number of positive possibilities, some of the transactional kind focused on here and in chapter 10, but also including benefits from investment, conservation, and capital formation.

The foregoing is at best only a partial list. Furthermore, since some of the mechanisms can be restated in terms of other mechanisms, there are undoubtedly linkages overlooked in even this partial catalog. Nonetheless, it serves as a stepping-off point for the next stages in the argument.

Mechanism Design

As already suggested, the game structures of our cooperative opportunities are subject to some choice and change. A player can often choose which kinds of opportunities to pursue, and with whom (Jackson and Watts 2005; Engsled and Bergh 2005). A player can also work to convert a potential opportunity into a preferable form. Jackson and Watts (2005) suggest the term "social games" for interactions "where players not only choose strategies but also choose with whom they play."

A branch of game theory called "mechanism design" deals with *intentional* versions of this game-structuring process (see chapter 7; Binmore 2005; Parkes 2001; Papadimitriou 2001). The field has focused particularly on applications to computer commerce and the design of auction procedures. Scholars of mechanism design typically view the designing and enforcement as done by a knowledgeable, powerful third party—a boss, a king, a government, a programmer, in short, the Leviathan. Papadimitriou (2001) described this approach succinctly: "Given desired goals (such as to maximize a society's total welfare), design a game (strategy sets and payoffs) in such a clever way that individual players, motivated solely by self-interest, end up achieving the designer's goals."

Doubtless, many human economic institutions have come into being in just such a fashion. The express rules of the New York Stock Exchange, for example, are largely the intentional creation of human intelligence, although probably through a committee process. Indeed, it is possible to view much of the law as an elaborate exercise in mechanism design, allowing the restructuring of games before the fact (transactions, contracts, firms, etc.) as well as after the fact (litigation, arbitration, etc.).

But there is a bootstrapping problem to all of this: the intentional design of formal and legal institutions/mechanisms takes place against the background of an existing level of successful human cooperation that came about far less intentionally. Trial-and-error learning at the individual level can provide part of the story of institutional design (e.g., Young 2001), but it still rests on a foundation of existing human sociality.

Traditional economic scholarship simply ignores the problem. It generally assumes the presence of contract and property in their strongest forms, and moves on to talk about output. For instance, Steven Shavell (2004, 293) specified that "contracts are assumed to be enforced by a court, which generally will be interpreted to be a state authorized court." This is a perfectly legitimate starting point, provided you understand that you are missing half the story and leaving out a lot of the hard questions. If property and contract did simply exist as the gift of a disinterested, all-enforcing designer, then focusing on output as the only important solution would be justified, and marginalizing values would indeed be correct; they would have little to add. These solutions, however, do not drop for free out of the sky whenever needed. Institutions, wherever embedded, are the structural piece of the puzzle that must be solved simultaneously with output. And the genesis of those institutions leads us back to values.

What Do Values Add?

We have explored the challenges involved in creating structures of interaction that will allow humans to capture the gains of sociality and cooperation, and we have looked at the key role institutions can play as devices for restructuring the nature of the games involved so as to make cooperative outcomes dominant solutions for the players. We now turn to the real question at hand: What is the role of values in our economic system? The functional role of values is to serve as a class of institutions that can accomplish this reshaping. Our working definition of values—non-situational commitments to particular principles of character and action, which may require sacrifice and self-denial—casts them as evolved psychological manifestations which give us access to many of the changing and choosing mechanisms described above. How is this accomplished?

The catalog of values and virtues with application to economic life is long and varied; Solomon, for instance, listed forty-five virtues, ranging from ability to zeal (Solomon 1999; see also the discussion in chapter 12 of this volume). In the same work however, he identifies three "most basic" business virtues—honesty, fairness, and trustworthiness—which I focus on here. These three map directly onto the mechanisms for strategic redesign set out above. Trustworthiness represents a kind of commitment (see chapter 9) and can help support reciprocity, in both its regular and strong flavors. It is a key target for tales about reputation (see chapter 10). Honesty is a core virtue of communication and signaling, and also helps to support reciprocity. Fairness is often thought of in two flavors as well: procedural and distributional. Procedural fairness goes to the integrity of the game itself. Is the promised structure available to all who play? Will those who set and police the rules respect them as well? Distributional fairness is also critical, serving functions addressed later in the chapter.

If you are dealing with someone who you believe, as a general matter, possesses these traits—trustworthiness, honesty, and fairness—you will probably be willing to rely somewhat on the power of these traits to constrain the choices this person will make in dealing with you. This reliance can reshape many otherwise defection-prone games of economic interaction into opportunities on which it is worth taking a chance. Values are thus the building blocks of personally constructed institutions within our own psychology. These personal, internal institutions may be supplemented by, and interact with, external institutions such as government or law. Still, their independence from such outside forces gives them a particular salience in a free enterprise system, which is dedicated to a relatively low level of governmental involvement. Some traditional economists may argue that the role of government should be limited to interventions to support contract and property (see the discussion in Hoffman 2004); values such as honesty, trustworthiness, and fairness provide the opportunity to capture at least some of the potential benefits of economic specialization and cooperation without even these supports. Values don't just lower transaction *costs*, as Posner suggests; in many contexts, values make transactions *possible*.

A key component in a successful economic system is the availability of strategic structures with dominant cooperative solutions for capturing at least a reasonable portion of the gains of specialization, coordination, and scale available to it. When free enterprise cultivates values, along with other critical institutions such as law and markets, it does pretty well on this task; when it forgets values, a lot of potential gain is left on the table (see chapter 14).

How Do Values Arise?

Where does the capacity for such internal commitments come from? Some look to religion, and indeed a divine, designing power would have good reason as a matter of mechanism design to put such a capacity into humans, a gift as essential to their eventual well-being as sight and locomotion. But such a divine gift is not the province of science; we rely on that wonderful mechanism for bootstrapping adaptive design: evolution.

The key question for this study, therefore, is whether values of this kind can evolve. My answer is yes, but that is currently more a matter for supported assertion than for clear proof. Evolutionary evidence and arguments of a traditional kind have been offered by scholars such as deWaal (chapter 4) and Frank (chapter 3). A full exploration of this question awaits the more complete development of the field of evolutionary mechanism design. Still, strides are being made in works such as *Moral Sentiments and Material Interests* (Gintis et al. 2003), which makes an explicit connection between evolutionary processes, both biological and cultural, our moral sentiments, and the solving of cooperation problems.

Biological evolution could produce a psychology of values. In the same way that it is individually advantageous for a male peacock, as a matter of commitment, communication, and sexual selection, to have the opportunity to grow a resplendent tail, so, too, could it be advantageous for a human to have the opportunity to develop and commit to character (although the communication of this commitment may be more complex). Smith and Bird (2005) suggest one process for such evolution, in a costly signaling context. Given the potential payoffs flowing from transactional gains made possible by an honest display of character, however, it may be a signal whose costs are highest when the message is false, like the chest plumage of the sparrow. Indeed, we often see significant punishment for the false assumption of prosocial values.

Additionally, once a psychology of values becomes available, with some reliability of honesty, a commitment to character becomes a "green beard." This term, coined by Dawkins (1976, 1982), describes an arbitrary, recognizable phenotypic signal linked to altruistic behavioral traits. Such a signal allows committed cooperators to seek each other out, and to safely open themselves to the dangers of defection in potentially profitable games with otherwise bad dominant structures (see chapter 3).

Science is beginning to work out the biology of values, particularly those associated with empathy, trust, and related aspects of attachment (Zak, forthcoming; see, too, chapter 12). Zak and his collaborators have identified oxytocin as a neurochemical highly correlated with trusting and trustworthy behavior (Zak, forthcoming; Kosfeld et al. 2005), and Smith, working with Wilson and Kimbrough (chapter 13) and others have been studying the structural neurology of trust.

Values also receive cultural support (see chapter 6). By making the possession, profession, and application of these principles such as honesty and trustworthiness a matter of general praise and approbation, culture helps to cement them in the "meta-game" that underlies the design of solutions in particular instances. And the replication dynamics of cultural evolution can support group-level selective effects (e.g., see chapter 6, this volume; Boyd and Richerson 1985; Laland and Brown 2002; O. Goodenough 1995), helping to offset individual payoffs from defection that benefits the individual at the expense of diminished payoffs to others in the group. In an explicitly economic context, Somanathan and Rubin (2004) suggest a model for the cultural co-evolution of honesty and capital.

Finding ways to capture the benefits of the evolution of cooperative solutions is not unique to human economic cooperation. As Maynard Smith and Szathmary (1995) have pointed out, many critical stages in the evolution of terrestrial life have involved biochemical solutions to similar problems in cell organization, multi-cellular organisms, sexual reproduction, and multi-organism sociality.

Locating Values in the Human Composite: The Example of the Lock Box

The dual-key lock box mechanism, mentioned above, exemplifies how a generalized game design mechanism can be instantiated in a number of different physical, cultural, legal, and psychological institutions. Recall that many cooperative opportunities face an absconding-with-the-benefits defection strategy. This is a particular challenge in business partnerships with a small number of members. In my own family lore, there is a story of how a brief moment of Goodenough family prosperity was undone early in the twentieth century when my great-grandfather's two partners in a Long Island real estate company took all the money and fled to happy exile in Mexico, leaving him, an intensely honorable man himself (E. Goodenough 1961), with a lifetime chore of paying off the debts.

The joint action mechanism which helps to restructure the game so as to limit this defection possibility can take a largely physical form—the lock box itself. A brief web exploration reveals commercially available models such as the Cobalt SDS-01K. It has a hinged slot at the top which drops deposits into a lower, protected chamber that requires two keys to open. The eBay description proclaims it to be "Perfect for Retail or Restaurant use . . . This is a great economic drop safe for any small business. . . . Take the extra step to securing your profits today with this easy to use low cost solution."

A formally identical approach can be put in place through a dual signature bank account, where the ability to access the partnership's money is again linked to joint participation; here the institution is both physical (the signature) and cultural (a mixture of law and the practices of the bank where the account is held). U.S. partnership law contains provisions that set unanimity among the partners as the default rule for actions outside the ordinary course of business (Uniform Partnership Act [1997] §401 [j]), and a partnership agreement could add to that list a partner's withdrawal of funds from the partnership. These culturally fixed rules constitute a somewhat more abstracted version of the same strategy.

Finally, the dual-key mechanism can exist in a set of value-based expectations about loyalty and trustworthiness among partners. An internal commitment to reliability and mutual benefit can create the psychological equivalent of the two-signature bank account and the dual-key lock box. Of course, as my ancestor's experience demonstrates, values, along with signatures and keys, can be faked (see also chapter 3). Some failure is inevitable. Nonetheless, a layering of physical, cultural, and values-based versions of this solution can provide sufficient expectation of a positive payoff to encourage many happier partnership experiences.

The opinion of Benjamin Cardozo in the well-known case of *Meinhard v. Salmon* provides a direct example of such a partnership concern. It also employs beautiful language to describe the value-based expectations of a partnership relationship and the role of the law in supporting these expectations. Meinhard and Salmon had been "joint adventurers" (effectively partners) in a twenty-year real estate venture at the corner of 42nd Street and Fifth Avenue in New York City. Salmon had been the manager of the business. When the lease for the property expired in 1922, the owner approached Salmon individually about a larger project that included, but was not limited to, the property that had been the subject of the partnership. When he heard about the new project, Meinhard sued to be included in the deal. The question here was not one of absconding with what was in the box, but rather whether this opportunity should have gone into the box to be exploited for joint benefit.

In holding that the opportunity was to be shared, Cardozo wrote an evocative description of the trustworthiness that should exist between partners:

> Joint adventurers, like copartners, owe to one another, while the enterprise continues, the duty of the finest loyalty. Many forms of conduct permissible in a workaday world for those acting at arm's length are forbidden to those bound by fiduciary ties. A trustee is held to something stricter than the morals of the market place. Not honesty alone, but the punctilio of an honor the most sensitive, is then the standard of behavior. As to this there has developed a tradition that is unbending and inveterate.

He continued by stressing that law should link itself to the most virtuous aspects of this vision, and not dilute it as it moves the institution from psychology to the law.

> Uncompromising rigidity has been the attitude of courts of equity when petitioned to undermine the rule of undivided loyalty by the "disintegrating erosion" of particular exceptions (*Wendt v. Fischer*, 243 N. Y. 439, 444). Only thus has the level of conduct for fiduciaries been kept at a level higher than that trodden by the crowd. It will not consciously be lowered by any judgment of this court.

Georgakipoulos (1999) calls the opinion a gem of rhetoric and morality. Noting that it has "superb" economic ramifications, he details a number of benefits that accrue from this approach. While he lauds it as a crucial step in the transition to *impersonal* management, it is the transmission of the existing values-based understanding of partnership out of psychology and into the more extensive and impersonal institution of law that makes this step a possibility.

This example helps to demonstrate how values—or their strategic equivalents—can be located in many places in the physical, cultural, and psychological composite that is humanity. Indeed, an interesting aspect of highly structured markets is that they incorporate the safeguards of values into the rules of the exchange, allowing a more overt selfishness in that limited context (see, in this volume, Goodenough and Cheney's précis "Is Free Enterprise Values in Action?"). Before turning to fairness, it is only fair to give Adam Smith a final word on the role of values in the mix. Although he applauded formal institutions, particularly law, as providing the security that trade and exchange required (Evensky 2005), he also recognized the centrality of values:

> What institution of government could tend so much to promote the happiness of mankind as the general prevalence of wisdom and virtue? All government is but an imperfect remedy for the deficiency of these. Whatever beauty, therefore, can belong to civil government upon account of its utility, must in a far superior degree belong to these. (Smith 1759, chap. 2)

Fairness: A Mechanism for Evaluating and Choosing

Fairness is also an essential value in free enterprise, although in a different context than transactional values such as honesty and trustworthiness. As with the other values, there is a long and rich history of thought about what fairness can mean (compare, e.g., Rawles 1971; Kaplow and Shavell 2002; and see Zajac 1995). It has been a somewhat slippery concept, representing for many a conclusion with application both to distributional results and to the process that produces them. In the context of the approach developed here, however, we can productively view fairness as a psychological mechanism for deciding what games we should take part in, both as a matter of initial choice or of staying in a game once it has begun. We can also use this aspect of fairness to help anticipate the conclusions of others in turn facing this choice.

The transactional values provide stability through making outcomes reliable; fairness provides a metric for analyzing the distributional aspects of the payoff structure which reliable play will deliver. "Unfair" distribution of the payoffs can be as effective an obstacle to having a transaction go forward as a high risk of defection. The considerations on which we make such fairness choices weigh the payoffs from staying in the game against the payoffs available for sitting on the sidelines, pursuing other games, or even preying on the results achieved by others.

This is not the only possible approach to distributional fairness; other, more egalitarian considerations also seem to motivate us. For example, a sweatshop worker with few alternatives could view the low wage offered as personally "fair" in this way, helping to make the social structure within which she works stable for the time being; that conclusion does not relieve the rest of us from considering the "fairness" of her position and pay in a larger sense. The approach set out here does, however, supply both a useful baseline that apparently captures a good portion of our *personal* sense of distributive fairness and explains fairness as a critical value necessary for a stable free enterprise system resting on the freely made choices of *all* its participants.

Fairness and Mechanism Design

With this approach to fairness, we change our focus from game properties that make certain payoffs reliable to the size and distribution of the payoffs themselves. Are *they* desirable? Do the expected results of being in the game match what is available to sitting it out, playing something else, or even acting as a predator on the game itself? The distribution need not be equal, indeed "skew" is possible, even relatively large skew, but the expected benefits of joining in must be better than the alternatives

Such a requirement is recognized in mechanism design analysis, which terms it "individual rationality." In the words of Parkes (2001, 34): "Another important property of a mechanism is *individual-rationality*, sometimes known as "voluntary participation" constraints, which allows for an idea that an agent is often not forced to participate in a mechanism but can decide whether or not to participate." Parkes suggests a criterion for such a decision: "a mechanism is individual-rational if an agent can always achieve as much expected utility from participation as without participation, given prior beliefs about the preferences of other agents" (35).

In a dynamic social milieu, the context within which this analysis takes place is itself a matter for shaping. The alternatives of a potential player may be manipulated by others, particularly by those with a big upside in the game in question, who can seek to make a bad payoff choice the best one available. Extreme forms of such manipulation, particularly if they involve sanctions, can be thought of as *coercion*. We can distinguish this from *enforcement*, where punishment is invoked to keep a player honest within a game or institutional structure that the player has freely chosen to be part of. Because they both involve sanctions, enforcement and coercion are often conflated under the concept of punishment, but they are very different in function, effect on output, and normative content. Enforcement is consistent with free choice; coercion antithetical to it. Enforcement keeps us honest; coercion makes us slaves.

This set of distinctions suggests four working definitions for this fairness approach:

Enforcement: making sure through sanctions that those playing an institutionally constrained game with a particular rule set act in accordance with those rules.

Coercion: using sanctions and other manipulations of the alternatives to prevent other actors from choosing to "opt out" of a particular game and its rule set, for example, slavery, at a societal level.

Free choice: undertaken voluntarily by an actor based on applicable factors including enforcement and alternatives, but not coercion.

Fairness: The evaluation of whether the distribution embedded in the payoff structures of the game under consideration is "individually rational" as a matter of free choice.

So far in this fairness discussion we have referred to "games" in a somewhat casual and imprecise fashion; this is deliberate, if not fully rigorous. We can view choice, enforcement, coercion, and fairness working in "games" of increasing duration and complexity, ranging from individual transactions (the most parsimonious sense), through long-term relations such as partnership or employment in a firm, and on to our participation in an economic system or society as a whole. Precise modeling quickly becomes impossibly complex as we move up this scale; nonetheless, the concepts are still useful for understanding the stability challenges involved and the psychological processes humans use to address them.

Fairness in Humans and Other Animals

Humans are good strategic players, and I believe that we can, albeit imperfectly, calculate individual-rationality. We have a sense of when to "hold them" and when to "fold them" with respect to games ranging from sitting at a rigged poker table to selling our labor in a stratified social structure (e.g., see chapter 5). Our human *sense* of fairness provides a rough heuristic for evaluating the distribution of benefits along the lines laid out above as game theoretic individual rationality. Is participation in the strategic structure on offer, whether one-shot or systemic, worth consenting to on an un-coerced basis?

Some scholars suggest a similar calculus in biological evolution as constraining the degree of resource appropriation or reproductive inequality that can obtain in a hierarchical, social species (see chapters 4 and 5). "Reproductive skew," as the mating opportunity disparity is called, has been carefully modeled using a rational actor approach to balance the strategies of dominant and subordinate group members (e.g., Kokko and Johnstone 1999; Hager 2003). Whether modeled from the standpoint of the dominant or a subordinate, it becomes clear that the options of the subordinates, weighed against the gains of group membership, act as constraints on the ability of the dominant to take a larger share of the mating and reproductive opportunities. Experiments have shown that chimpanzees and capuchins are quite sensitive to the sharing and proportionality of rewards, including those that depend on joint labor to obtain, demonstrating an "inequity response" of considerable strength (see chapters 4 and 5).

If such a distributional calculus is predictable for an abstracted agent in mechanism design analysis and for evolved social animals including primates (chapter 5) and wolves and dogs (Bekoff 2001, 2004), we should expect humans to be able to make similar judgments, possessing a dedicated psychological capacity to comprehend distributional systems and to evaluate them on their individual-rationality. Of course, real people lack the information or the calculating capacity to be perfect rational choosers about rule sets. As with many balancing calculations in human psychology, we would expect there to be heuristics, short-cuts and rules of thumb by which we approximate fairness analysis. For instance, Brosnan, in chapter 5 of this volume, suggests that non-human primates may focus on partner behavior as opposed to the actual distribution of rewards. Establishing the outlines of these heuristics in humans is a project for future empirical work.

Giving a functional content to *fairness*, both as a value and as an expectation, is not new (see, e.g., Rawles 1971; Binmore 2005; and, from an evolutionary approach, Jones and Goldsmith 2005). Many of these approaches, however, focus on ex-ante rule making, decided in some kind of abstraction like Rawls's veil of ignorance. The approach suggested here focuses on actual results for the individual involved, recalculated continuously, although involving both benefits in hand and discounted future expectations. Because results are what count, it is not enough that free enterprise be some kind of neutral opportunity set, played out through the transparent and accurate application of rules agreed on in advance. Rather, it must also have rules for *redistribution* that will give to some and take from others to square up accounts at the end of the day in a way that keeps participation individually rational in *result* as well as in *initial conception*.

Viewed this way, fairness is not just a nice add-on in economic cooperation and exchange; rather, it is a critical metric for the long-term stability and productivity of economic relationships, ranging from dyadic games, which can be modeled with reasonable certainty, to multiplayer games, firms, communities, countries, and societies as a whole. Fairness is a value that free enterprise, with its embrace of free choice and its rejection of coercion, must have at its heart.

Free Enterprise: *Fairness* in Action

This kind of functional approach to fairness illustrates how it, too, is an essential value for free enterprise. Understanding fairness in this fashion suggests a set of corollaries about policies and practices ranging from deal structuring in commerce to tax and social benefit policies in societies as a whole. This chapter concludes with a rough sketch, a cartoon of its own, of how the debates over some of these corollaries might proceed.

Preferring Choice to Coercion. Why should we prefer choice to coercion? This question is probably best either left as an obvious matter needing no examination or treated at some length. Nonetheless, a few reasons are worth summarizing. There is a normative content to choice and freedom that includes, but transcends, a "golden rule" notion that these are things we would each individually prefer. There are also utilitarian considerations. People invest heavily in options they have chosen freely but do a bare minimum under coercion. Systems that use penalties to keep players in games with an unfair skew in the distribution of the benefits of cooperation generally have a lower productive potential, and are inherently unattractive (although the benefited elites may argue otherwise).

Coercive rule sets are thus likely to be less productive than fair ones, particularly in any kind of advanced economy. When reduced to subsistence and coercion, there is little reason for the coerced to seek to create surplus; rather, there are increased incentives to "hide" productivity for one's uncoerced use. Such a strategy can involve pretended stupidity among the exploited when in the company of the exploiters—"playing the Paddy," as the Catholic Irish facing their Anglo overlords have historically called it (see Sheehan 2001). This tactic can have the ironic effect of reinforcing the myths of inferiority on which the exploiters rest the case for their privileged position.

High policing costs also depress the overall productivity of highly coercive solutions. The necessity of having consensual participation increases with the degree of intelligence, care, and autonomy involved in the cooperative tasks. Rough, unskilled labor, like agriculture, ditch digging, and rowing in a galley, can be highly coerced. Running complicated machinery cannot. Fair distributional structures can give a socioeconomic system access to higher solutions on the output landscape.

Transactions. As a former transactional lawyer, I have helped structure many deals with relatively high level of formality. At one level, we drafted careful contracts that sought to create a high degree of predictability and reliability for achieving the payoff structure negotiated by the respective sides. At another, we also acted as mechanism design professionals, and understood the necessity of individual rationality for all sides if the deal was to go forward. A senior partner in my firm, and one of the best deal lawyers I ever worked with, explained to me that he always encouraged his clients not to push to obtain the last particle of possible advantage at the expense of the other party. "Always leave something on the table for the other side to take away," was how he phrased it. In the terms of this chapter, we might reformulate it: "A competent deal maker will attend to the individual rationality of the transaction from the standpoint of *all* participants, and not from her personal perspective."

Fairness in a Socioeconomic Structure. Rigorous formal modeling breaks down once we get beyond a game with a few players and relatively well-defined strategic and payoff options. Still, the participation logic underlying fairness can also be applied to multiplayer, socioeconomic systems. In a contemporary economic context, the demands of fairness can sometimes be at odds with compensating players on the basis of their marginal utility in a particular transaction or employment relationship. Thus, fairness at a societal level may mandate some redistribution of income away from the allocation made strictly by individual bargaining for the value of labor to a particular enterprise. The terms of participating in such a fair system depends not on the marginal utility of its members in the system of production but instead on each member's alternatives, including opting out and predation, and the effect that choosing one of those alternatives will have on the system of production and indirectly on themselves.

To the extent that this reallocation leads to a decrease in the theoretical landscape of efficiency, fairness can indeed limit the overall welfare potential of a social system, as Kaplow and Shavell (2002) correctly identify. The limit, however, is really illusory, an inescapable "part of the deal." As a matter of systemic stability, the "unfair" welfare landscape may simply be unavailable in an un-coerced society, and, because of the limits of productivity under coercion, may never be available in a highly coerced social structure either.

Stability of this kind counts in firms, too. Keeping employees is a critical challenge. Fairness heuristics embedded in human psychology may underlie decisions of firms and workers that do not fully square with marginal utility or short-term rationality. For instance, queuing heuristics, recognized in animals, may underlie seniority systems. Being on the short end of inequality will be tolerated in the short run if there is a predicable progression, or queue, in which a participant moves up the ladder to a higher position with relative predictability (Kokko and Johnstone 1999).

Redistribution. The likelihood of discontinuity between these two methods of pricing—marginal utility and participatory fairness—can lead to instability, causing the system to unravel, or it must be redressed, either by coercion of otherwise unwilling participants or by a redistribution rule that "corrects" the allocations made by the production-maximizing rules of property, contract, and labor. Making up the gap between the wage paid by an individual employer and the participation share that would benefit society as a whole is a classic public goods problem, amenable for redress by taxation and governmental intervention.

As noted above, this redistribution can be a drag on efficiency/utility/welfare (pick your favorite designation for output), as Kaplow and Shavell (2002) point out. But since stability must also be solved, not all theoretically possible points on the output landscape can be reached in a manner that is structurally stable. As Parkes puts it: "Essentially, individual rationality places constraints on the *level* of expected utility that an agent receives from participation" (2001, 34; emphasis added).

Redistribution Often Cheaper Than, and Certainly Preferable to, Coercion. We have asserted that the stability costs of a "fair" redistributive system, in terms of lost production of measurable goods and services, appear likely to be less than coercion. Provided that this is the case, then moderately redistributive systems would be expected to enhance output maxima. This is essentially an empirical proposition, susceptible to testing and falsification. Data from actual economic systems appear to bear this out. Zak and Knack (2001) show that economically fair societies typically have higher trust and faster growth than excessively skewed systems. The personal preference for living in a fair system, rooted in free choice and a lack of coercion, is also a "good" to be added to the calculus that would further increase welfare beyond an analysis based on monetary measures.

Is redistribution coercion for those from whom a share is taken? Not as defined here. Provided it is set out as part of the rules, it joins institutions like contract or property that a potential player can evaluate and rely on. It thus becomes a convention that will be enforced, but one to which people of all income levels can freely choose to bind themselves, provided it is within the right bounds, that is, matching a value of fairness.

In the cause of stability, top participants will voluntarily take less than the production-side rules stipulate, up to a point. A fair structure produces benefits for the top of the distribution curve, as well as for the bottom. If part of a freely chosen system, taxation for the purpose of redistribution is no more coercion than property or contract. Excessive redistribution, however, is also unstable; social structures can unravel by defection at the top (capital flight), as well as at the bottom (personal flight). Systems of coercive redistribution, such as overly egalitarian communism, are both bad for production and a challenge to the ethos of free choice.

The Politics of Redistribution: Negotiating within the Acceptable Spread. Note that the exit price at the upper distributional margin is not necessarily identical with the exit price for players at the bottom. The space between the limits of the acceptable ranges at the top and the bottom can create a redistributional margin, a set of solutions, more or less favorable to the various players, but all providing similar levels of un-coerced social stability. The negotiation of the actual location within this range can be seen as a description of economic politics in a modern democracy. Where the balance is struck within that margin is a matter for intra-societal negotiation and for finding a local maximum of the stability function as a contemporaneous solution with the production function. Broader factors of social fairness may come into play in this process. If the margin is not big enough to stabilize both ends, a free society, as envisioned here, is impossible. Special cases, such as very high growth or very constrained circumstances, may provide exceptions.

How the redistribution is handled matters; it, too, is an institution, and should be designed with care. Some forms of redistribution can serve as a disincentive to cooperative effort by its recipients. At a societal level, it is probably better to provide redistribution through certain classes of benefits (e.g., education or health care) than through cash or such bare essentials as food and lodging. In this way, redistribution acts as a supplement to, rather than a substitute for, the employment allocation.

Transmissibility Must Be Solved, Too. Finally, it is worth remembering that productivity and stability are not the complete set of simultaneous solutions that must be provided by a successful socioeconomic system. Whether a firm or a society, it must also find a *transmissibility* solution. As players enter, leave, and change their places in the system, the rules must be transmitted with enough fidelity so that they are durable over time. Perceptions of fairness may play a role in this dimension as well, if they bias the willingness of participants to teach and learn the rules of the game.

Conclusion

The psychology of values, based on the recognition of the character of consenting players by other consenting players, assists humans in taking part in productive, consensual interactions. Classical economic modeling gave insufficient attention to the structural requirements of trade and cooperative interaction. A combination of game theory and institutional economics helps us to redress the balance, and leads us to conclude that values play an important role in many kinds of institutions, and are of *fundamental* importance in interactions that are not subject to complete structuring through such alternatives as law, physical mechanisms, or institutionalized markets.

Values such as honesty and trustworthiness can be very effective in transactional contexts, helping in the restructuring process of mechanism design and changing the dominant solutions in interactions from those with poor cooperative outcomes to those with higher mutual potential. Fairness, by contrast, often plays a different role. It can be viewed, at least in part, as a measuring process in which we decide whether participation in the game as designed is "individually rational," that is, desirable as a matter of uncoerced choice. The payoff structures set in place by a bargaining system based on the marginal utility of the players may not match up with their fairness pricing, a problem that could lead to instability in the system. This instability can be resolved by adding an appropriately calibrated redistribution rule to the overall game, whether in a dyadic pairing, a firm, or in society as a whole.

At both the transactional and systemic levels of analysis, a psychology of values is central to a free economic system. Free enterprise can indeed be viewed as values in action.

Notes

In developing the ideas presented in this chapter, I benefited from many interactions in the course of the initiative "Free Enterprise: Values in Action," out of which this volume has grown. I am particularly grateful to Carl Bergstrom, Sarah Brosnan, Monika Cheney, Morris Hoffman, Lynn Stout, and Paul Zak for reading earlier versions of this chapter and providing many excellent suggestions. Professor Bergstrom also introduced me to the concept of mechanism design and suggested its application in this context.

1. Among our working group, there has been considerable discussion over the usefulness of "free enterprise" as a term to describe the contemporary economic system of America, Europe, Japan, and most other developed countries in the world, This system is not monolithic; it comes in various flavors. As a generalized matter, it calls to mind a high level of private business activity and a low level of government intervention, particularly in the setting of prices. It is also typified by private ownership of the means of production and by active and free capital markets. Other labels in frequent use are "capitalism," which has the drawback of focusing on issues of enterprise ownership, or "free market economy," which gives prominence to markets, again only a piece of the picture. Although "free enterprise" has exactly the rhetorical baggage discussed in this chapter, as well as some antigovernment freight essentially outside the scope of this discussion, the expression has the virtues of common usage and avoiding the limits of more focused terms, and I find it to be a serviceable and intelligible shorthand.

2. A more common formulation of the hunting game is the "stag hunt," which embodies a simultaneous, one-shot interaction; see Skyrms 2004.

3. In real life, the international trade problem has often been solved with very little influence of law by converting it into a series of transactions between well-known, repeat play actors backed by reputation and commitment through the classic "payment against documents" structure involving shippers, insurers, inspectors, and banks. See Folsom, Gordon, and Spanogle 1996.

References

Badcock, Christopher (2000). *Evolutionary Psychology: A Critical Introduction.* Polity, Cambridge.

―――. (1998) Reciprocity and the law. 22 *Vermont Law Review* 295.

Bergstrom, C. T., Antia, R., Számadó, S., and Lachmann, M. (2001). *The Peacock, the Sparrow, and the Evolution of Human Language.* Technical report, Santa Fe Institute. Available at http://www.santafe.edu/research/publications/workingpapers/01-05-027 .pdf.

Binmore, Kenneth (1998). *Just Playing: Game Theory and the Social Contract.* Cambridge, Mass., MIT Press.

―――. (2005). *Natural Justice.* Oxford, Oxford University Press.

Bowles, Samuel (2004). *Microeconomics: Behavior, Institutions, and Evolution*. Princeton, N.J., Princeton University Press.

Bowles, Samuel, and Gintis, Herbert (2000). *The Evolution of Strong Reciprocity*. Available at http://www.umass.edu/economics/publications/econ2000_05.pdf.

Boyd, R., and Loberbaum, J. (1987). No pure strategy is evolutionarily stable in the repeated Prisoners' Dilemma Game. *Nature* 327:58–59.

Boyd, Robert, and Richerson, Peter J. (1985). *Culture and the Evolutionary Process*. Chicago, University of Chicago Press.

Dasgupta, Partha. (1988). Trust as a commodity. In D. Gambetta (ed.), *Trust: Making and Breaking Cooperative Relations*. Electronic edition, University of Oxford, chap. 4, pp. 49–72, at www.sociology.ox.ac.uk/papers/dasgupta49-72.pdf. Available at www.csee.umbc.edu/~msmith27/readings/public/dasgupta-1988a.pdf.

Dawkins, Richard (1976). *The Selfish Gene*. Oxford, Oxford University Press.

———. (1982). *The Extended Phenotype*. Oxford, Oxford University Press.

———. (1999). *Unweaving the Rainbow: Science, Delusion and the Appetite for Wonder*. Boston, Houghton Mifflin.

Dixit, A., and Skeath, S. (2004). *Games of Strategy* (2nd ed.). New York, Norton.

Durrant, Russil, and Ellis, Bruce J. (2002). *Evolutionary Psychology*. Available online at http://media.wiley.com/product_data/excerpt/38/04713840/0471384038.pdf.

Engseld, Peter, and Bergh, Andreas (2005). Choosing opponents in Prisoners' Dilemma: An evolutionary analysis. Working paper. Available at http://econpapers.repec.org/paper/hhslunewp/2005_5F045.htm.

Evensky, Jerry (2005). Adam Smith's *Theory of Moral Sentiments*: On morals and why they matter to a liberal society of free people and free markets. *Journal of Economic Perspectives*, 19(3): 109–130.

Fehr, Ernst, and Gächter, Simon (2002). Altruistic punishment in humans. *Nature* 415:137–140.

Fehr, Ernst, Fischbacher, Urs, and Gächter, Simon (2002). Strong reciprocity, human cooperation and the enforcement of social norms. *Human Nature* 13:1–25.

Fisher, Helen (1992). *Anatomy of Love*. New York, Norton.

Folsom, Ralph H., Gordon, Michael Wallace, and Spanogle, John A. Jr. (1994). *International Business Transactions in a Nutshell* (5th ed.). St. Paul, Minn., West.

Georgakopoulos, Nicholas L. (1999). *Meinhard v. Salmon* and the economics of honor. *Columbia Business Law Review* 1999:137–164.

Gintis, Herbert (2000). *Game Theory Evolving*. Princeton, N.J., Princeton University Press.

Gintis, Herbert, Bowles, Samuel, Boyd, Robert, and Fehr, Ernst (eds.) (2003). *Moral Sentiments and Material Interests: On the Foundations of Cooperation in Economic Life*. Cambridge, Mass., MIT Press.

Goodenough, Erwin R. (1961). *Toward a Mature Faith*. New Haven, Yale University Press; reprinted 1988, Lanham, Md., University Press of America.

Goodenough, Oliver R. (2001). Law and the biology of commitment. In Randolph M. Nesse (ed.), *Evolution and the Capacity for Commitment*. New York, Russell Sage.

———. (1995). Mind viruses: Culture, evolution and the puzzle of altruism. *Social Science Information* 34:287–320.

Goodenough, Ursula (1998). *The Sacred Depths of Nature*, New York, Oxford University Press.

Hager, Reinmar (2003). Reproductive skew models applied to primates. In C. B. Jones (ed.), *Sexual Selection and Reproductive Competition in Primates: New Perspectives and Directions* (65–101). Norman, American Society of Primatologists.

Hobbes, Thomas (1660). *The Leviathan*, available at http://oregonstate.edu/instruct/phl302/texts/hobbes/leviathan-contents.html.

Hoffman, Morris B. (2004). The neuroeconomic path of the law. *Philosophical Transactions of the Royal Society* B, 359:1667–1676.

Hrushka, Daniel, and Henrich, Joseph (2006). Friendship, cliquishness, and the emergence of cooperation. *Journal of Theoretical Biology* 239(1): 1–15.

Jackson, Matthew O., and Watts, Alison (2005). *Social Games: Matching and the Play of Finitely Repeated Games*, Working paper. Available at http://129.3.20.41/eps/game/papers/0503/0503003.pdf.

Jones, Owen D., and Goldsmith, Timothy H. (2005). Law and behavioral biology. *Columbia Law Review* 105:405–501.

Judson, Olivia (2002). *Dr. Taiana's Sex Advice to All Creation*. London, Chatto and Windus.

Kahneman, Daniel, Knetsch, Jack L., and Thaler, Richard H. (1986). The behavioral foundations of economic theory. *Journal of Business* 59(4), pt. 2: S285–S300.

Kaplow, Louis, and Shavell, Steven (2002). *Fairness versus Welfare*. Cambridge, Mass., Harvard University Press.

Kokko, Hanna, and Johnstone, Rufus A. (1999). Social queuing in animal societies: A dynamic model of reproductive skew. *Proceedings of the Royal Society, Biology* 266: 571–578.

Kosfeld, Michael, Heinrichs, Marcus., Zak, Paul. J., Fischbacher, Urs, and Fehr, Ernst (2005). Oxytocin increases trust in humans. *Nature* 435:673–676.

Laland, Kevin N., and Brown, Gillian R. (2002). *Sense and Nonsense: Evolutionary Perspectives on Human Behavior.* Oxford, Oxford University Press.

Maynard Smith, John (1991). Must reliable signals be costly? *Animal Behavior* 47:1115–20.

Maynard Smith, J., and Parker, G. A. (1976). The logic of asymmetric contests. *Animal Behavior*,24:159–175.

Maynard Smith, John, and Zathmáry, Eörs (1995). *The Major Transitions in Evolution.* Oxford, Oxford University Press.

McCabe, Kevin A. (2003). Reciprocity and social order: What do experiments tell us about the failure of economic growth? USAID Forum Series Papers. Available at http://www.mercatus.org/pdf/materials/274.pdf.

McCabe, Kevin, and Smith, Vernon (2000). A two person trust game played by naïve and sophisticated subjects. *Proceedings of the National Academy of Sciences* 97: 3777–3781.

McTeer, Bob (2006). *Free Enterprise Primer.* Available at http://www.tamus.edu/offices/chancellor/essays/primer.html.

Meinhard v. Salmon. 249 N.Y. 458; 164 N.E. 545; 1928 N.Y. LEXIS 830; 62 A.L.R. 1 (NY 1928).

Nesse, Randolph M. (ed.) (2001). *Evolution and the Capacity for Commitment.* New York, Russell Sage.

North, Douglass C. (1981). *Structure and Change in Economic History*. New York, Norton.

Nozick, Robert (1974). *Anarchy, State and Utopia*. Oxford, Basil Blackwell.

Papadimitriou, Christos H. (2001). *Algorithms, Games and the Internet*. Working paper. Available at http://www.cs/Berkeley.edu/~christos/.

Parkes, David C. (2001). Iterative combinatorial auctions: Achieving economic and computational efficiency. Unpublished dissertation. Available at http://www.eecs .harvard.edu/~parkes/diss.html.

————. (2004). On learnable mechanism design. In Tumer, Kagan, and Wopert, David, eds., *Collectives and the Design of Complex Systems*. New York: Springer Verlag.

Rawles, John (1971). *A Theory of Justice*. Cambridge, Mass., Harvard University Press.

Rohwer, Sievert (1982). The evolution of reliable and unreliable badges of fighting ability. *American Zoologist* 22:531–546.

Rohwer, Sievert, and Rohwer, Frank C. (1978). Status signals in Harris' sparrows: Experimental deceptions achieved. *Animal Behaviour* 26:1012–1022.

Rubin, Paul (2000). Hierarchy. *Human Nature* 11:259–279.

Shavell, Steven M. (2004). *Foundations of Economic Analysis of Law*. Cambridge, Mass., Belknap Press of Harvard University Press.

Sheehan, Helena (2001). Irish television drama in the 1980's. In *Irish Television Drama: A Society and Its Stories* (revised) (chap. 6). Available at www.comms.dcu.ie/ sheehanh/80s-itvd.htm.

Skyrms, Brian (2004). *The Stag Hunt and the Evolution of Social Structure*. Cambridge, Cambridge University Press.

————. (2000). Game theory, rationality and evolution of the Social Contract. In Leonard. D. Katz (ed.), *Evolutionary Origins of Morality*, Thorverton, U.K.: Imprint Academic.

Smanniotto, Rita C. (2005). "You scratch my back and I scratch yours" versus "Love thy neighbour": Two proximate mechanisms of reciprocal altruism. *The Agora online* 13: 6–8. Available at http://www.uni-leipzig.de/~agsoz/agora/text/textv131_a.htm.

Smith, Adam (1759). *The Theory of Moral Sentiments*. Variously reprinted, including Amherst, New York, Prometheus Books (2000).

————. (1776). *An Inquiry into the Nature and Causes of the Wealth of Nations*. Variously reprinted, including Amherst, New York, Prometheus Books (1991). Available at http://www.econlib.org/library/Smith/smWN.html, from which the quotations here are drawn.

Smith, Eric A., and Bird, Rebecca Bliege (2005). Costly signaling and cooperative behavior. In Gintis, Herbert, Bowles, Samuel, Boyd, Robert and Fehr, Ernst (eds.), *Moral Sentiments and Material Interests: The Foundations of Cooperation in Economic Life* (115–148). Cambridge, Mass., MIT Press.

Smith, Vernon (2002). Constructivist and ecological rationality in economics. Nobel Prize lecture. Available at http://nobelprise.org/nobel-prizes/economics/laureates/ 2002/smith-lecture.pdf.

————. (1998). The two faces of Adam Smith. *Southern Economic Journal* 65:1–19.

Somanathan, E., and Rubin, Paul H. (2004). The evolution of honesty. *Journal of Economic Behavior and Organization* 54:1–17.

Stake, Jeffrey Evans (2004). The property instinct. *Philosophical Transactions of the Royal Society* B, 359:1763–1774.

Templeton, John Marks, and Ellison, James (1987). *The Templeton Plan*. New York, Harper Collins.

Trivers, Robert L. (1971). The evolution of reciprocal altruism. *Quarterly Review of Biology* 46:35–57.

von Neumann, J., and Morgenstern, O. (1944). *Theory of Games and Economic Behavior*. Princeton, N.J., Princeton University Press.

Young, H. Payton (2001). *Individual Strategy and Social Structure: An Evolutionary Theory of Institutions*. Princeton, N.J., Princeton University Press.

Zahavi, Amotz, and Zahavi Avishag (1997). *The Handicap Principle: A Missing Piece of Darwin's Puzzle*. Oxford, Oxford University Press.

Zajac, Edward E. (1995). *Political Economy of Fairness*. Cambridge, Mass.: MIT Press.

PART V

Values and the Economy

Twelve

Values and Value
Moral Economics

Paul J. Zak

> We need . . . an evolutionary theory of morals . . . and [its] essential feature will be that morals are not a creation of reason, but a second tradition independent from the tradition of reason, which helps us adapt to problems which exceed by far the limits of our capacity of rational perception.
> —Friedrich von Hayek, in a speech at Oesterreichisches College, Wien, 1985

J. Clifford Baxter graduated from New York University and then served in the Air Force, obtaining the rank of Captain in 1985. After leaving the military, he completed an MBA at Columbia University and began a business career. In 1991, he joined a small oil and gas pipeline company in Houston, Texas, called Enron. Cliff Baxter steadily climbed the corporate ladder, becoming the vice chairman of Enron in October 2000. During Cliff Baxter's tenure, Enron began trading energy contracts and grew to become the nation's seventh largest corporation. Toward the end of the 1990s, Baxter began complaining to Enron's CEO Jeff Skilling about their business practices. Former Enron vice president Sherron Watkins, in a letter to Chairman Kenneth Lay, said that "Cliff Baxter complained mightily to Skilling and all who would listen about the inappropriateness of our transactions" (CNN.com, 2002). Baxter resigned from Enron in May 2001 but continued to work for the company as a consultant for the next half-year. Between October 1998 and his resignation, he exercised $22 million in stock options.

By all accounts, Cliff Baxter lived by a code of high moral values. He served with distinction in the military; he had a happy family life with his wife and two children. He was among the few employees at Enron who left a paper trail establishing criticisms of the company's ethical transgressions and legal abuses. Early in the morning of January 26, 2002, Baxter wrote the following note to his wife:

Carol,
    I am so sorry for this. I feel I just can't go on. I have always tried to do the right thing but where there was once great pride now it's gone. I love you and the children

so much. I just can't be any good to you or myself. The pain is overwhelming. Please try to forgive me.

Cliff

This note was found in his car, parked near his house, with a self-inflicted fatal gunshot wound to the head.

Cliff Baxter was driven to succeed and worked incredibly hard to achieve success at Enron. But even resigning from Enron did not sufficiently erase the pain he appeared to have felt for violating a moral code, and he chose to end his life rather than endure the pain of this violation.[1] There must have been many Enron employees like Cliff Baxter who were fine citizens, driving the speed limit, paying their taxes, and who would not think of stealing from their local grocery store. For example, Enron treasurer Jeff McMahon wrote to the CEO that "my integrity forces me to negotiate the way I believe is correct," when discussing questionable shell corporations that were set up apparently to directly benefit senior management (McLean and Elkin, 2003, p. 210). Yet, most employees uncomfortable with Enron's business practices did not resign from Enron, and no others are known to have killed themselves.[2]

There are three leading explanations for the unethical and likely illegal behavior of a large number of Enron employees. The first is that the process of economic exchange values greed and self-serving behaviors, and inadvertently produces a society of rapacious and perhaps evil people. An extension of this line of argument is that all those living in modern economies are dehumanized, a view popularized by socialists like Karl Marx. The amount of anonymous charitable giving each year by individuals living in developed economies belies this claim. As I write this, Hurricane Katrina had recently hit the Gulf Coast of the United States, devastating New Orleans and many other cities. Donations to the Red Cross and other relief organizations topped $1 billion within weeks of this disaster. Similar aid was provided worldwide for victims of the Asian Pacific tsunami of December 26, 2004. If humans in modern economies are so greedy, why all the charity?

A second explanation for the behavior of Enron employees is that there could be a selection bias in which amoral greedy people were hired in key posts, and this behavior filtered down to other employees. An implication of this view is that government regulation is critical to keep these "bad eggs" in check. After the accounting scandals in the first years of the twenty-first century, the United States passed the Sarbanes-Oxley Act to tighten accounting standards and to hold corporate board members personally liable for corporate misdeeds. This additional regulation is costly to both the federal government and to firms, potentially reducing aggregate economic productivity. Is such regulation necessary if the proportion of violators is small? For example, the fraction of those who cheat on their taxes in the United States is small, even though the chance of being prosecuted for tax fraud is nearly zero (Slemrod,

1992). Indeed, why don't all or most employees steal, cheat, and lie to get ahead? Neither the chance of being caught nor the Marxist view of the world adequately explains the mostly decent behavior of employees, managers, and citizens in most societies.

A third explanation is that Enron had a particular institutional environment that encouraged immoral behaviors. For example, Enron employees cheered when forest fires that were destroying people's homes and lives drove up energy prices. This explanation nests the two previous ones—senior managers at Enron devised the company's procedures and compensation that pitted employees against one another for survival and provided incentives to violate accounting standards. One way this was done was to break up decisions into parts so most individuals were only responsible for moving the decision forward and could not claim ultimate responsibility for an action. Given this environment, many employees acquiesced to the incentives that resulted in unethical and possibly illegal behavior. This explanation is more subtle than the first two, as its base assumption is that most people behave ethically most of the time. Nevertheless, in the right circumstances, many people can be induced to violate what seems to be an internal representation of values that holds unethical behavior in check.

So which explanation is right? Perhaps more important, is there a mechanism at work that sustains economic order in a highly decentralized, moderately regulated economy like that in the United States? This chapter surveys neuroscientific research and discusses recent experiments from my lab on the physiologic basis for interpersonal decision making to support the following thesis: most people, most of the time, behave ethically, and a set of shared values is essential to the functioning of modern economies. I call this "moral economics." Next, I draw implications for law, institutional design, and public policy that follow from moral economics. My aim is to demonstrate that values are a critical ingredient in producing the historically high living standards in many nations, and are a prerequisite to economic growth in the developing world.

2. Economics and Values

2.1. What Are values?

Values are guides to action. They are enduring beliefs more basic then heuristics but can be seen as building blocks for heuristics. For example, Aristotle identified eight values: courage, temperance, liberality, magnificence, proper pride, good temper, modesty, and friendliness. Two thousand years later, Benjamin Franklin recognized thirteen values: temperance, silence, order, resolution, frugality, industry, sincerity, justice, moderation, cleanliness, tranquility, chastity, and humility. Values can be thought of as the constituents of a person's character traits. Some values appear to be universal (for more on this, see chapters 2 and 8 in this volume). Schwartz (1994) identified a set of broad values that appear to be held in nearly all societies from a survey of forty-four countries. A value to respect the lives of others, if generally held, offers one protection (in most circumstances) for one's own life. Other values similarly support social probity. Following one's values may generate internal rewards as well. Psychologically, one may consider oneself a "good" person when one conforms to shared values.

Rokeach (1973) distinguished between *instrumental* and *terminal* values. *Instrumental values* are the means to achieve a goal. Examples include being ambitious, courageous, honest, and loving. *Terminal values* are end goals, and these include freedom, security, pleasure, and prosperity. Values—especially instrumental values—can be further broken down into those that are primarily personal and those that are interpersonal or social. The latter are often called *moral values*, where "morality" comes from the broad acceptance or prohibition of a behavior within or even across societies. As I discuss below, engaging in, or observing violations of, moral values typically produces a strongly felt physiologic response. Values are therefore a motivation for behavior.

Values, especially in young people but also in adults, evolve with experience. Nevertheless, values are typically stable within individuals, though variation across individuals is common. A defining characteristic of values is that they constrain our choices; that is, values are exercised at a cost. For example, most people, absent psychopathology, have a deeply held prohibition against killing others. This means that the person talking on the cell phone who is holding up traffic cannot be killed (even though we might fleetingly desire this; David Buss [2005] reports that 91 percent of men and 84 percent of women have had at least one vivid homicidal fantasy). In most societies, the prohibition against killing is lifted for soldiers fighting other soldiers, police officers chasing dangerous criminals, and anyone in imminent mortal danger from another. Values, then, proscribe behavior. Values may be limited to those in one's social group; "enemies," "slaves," and "savages" have, at certain times in history, been considered subhuman, and therefore values can be violated when interacting with them. In any situation, several values may interact or even conflict when deciding how to behave. Further, the weight put on instrumental versus terminal values varies across individuals (Schwartz, 1994) but less so across settings.

This leads one to ask how values might have evolved. One hypothesis for the rapid and extraordinary growth of the human brain is that this occurred to support increasingly complex social behaviors (Reader & Laland, 2002). In particular, cooperation with nonkin is the hallmark of modern civilizations. Cooperation between unrelated individuals enabled the specialization of labor and the generation of surplus in societies, fueling technological advances and increasing living standards (Diamond, 1997). If this supposition is correct, we must ask how humans evolved the ability to behave cooperatively with nonkin. Neuroscientist Cort Pedersen (2004) argued that the same mechanism that in mammals facilitates care and attachment for offspring permitted early humans to "attach" to and cooperate with extended family and eventually nonkin. Pedersen hypothesized that cooperation with nonkin in humans is facilitated by the neuroactive hormone oxytocin, as this molecule promotes attachment to offspring and, in monogamous mammals, attachment to reproductive partners. Most humans are serially monogamous, so extending the attachment role of oxytocin to nonkin in humans is reasonable. Further, humans have an unusually long period of adolescence, requiring a strong and long-term mechanism of attachment (chimpanzees, our closest genetic relatives, are sexually mature at about eight years of age, whereas humans need at least twelve years before they can reproduce and, in modern societies, typically much longer).

As a result of this powerful physiologic attachment in humans, we attach to friends, coworkers, pets, and even ideas (Zak, 2007). An example may clarify this. Most of us, if we were to see a stranger trip and break an ankle, would have an immediate physiologic reaction in which we would internally represent the other's pain. This can be called empathy and, for many people, will motivate an offer to help this stranger. Indeed, it is difficult, though not impossible, to suppress this response. A paramedic or physician observing the broken ankle would quickly suppress the feeling of empathy and get to work assisting the victim medically.

The shared representation of another's situation appears to be the causal mechanism though which we cooperate with strangers. Bowles and Gintis (2004) have called this the evolution of strong reciprocity—the ability to cooperate with others with the expectation that they will reciprocate. This impulse to reciprocate is based on an internal guidance system, most likely utilizing oxytocin. The associated activation of brain regions is the neural representation of values. The neural mechanisms supporting values are detailed in sections 2.4 and 2.5 below.

2.2. Values Are Costly but Also Have Benefits

Values can be construed to impose a cost on individuals. If one of my values is concern for the environment, then I may choose to purchase a hybrid gas-electric vehicle for a premium price, or perhaps to buy other "green" products that may be more expensive than standard products (at least initially). For example, I once served as a consultant to the electric power industry on the market prospects for emerging renewable energy appliances such as solar air conditioning. Even if the cost of electricity went up fivefold, most of these appliances would have to run for thirty to fifty years before they would be more cost-efficient than existing technologies. This is partially because of the high

cost to build each unit (which is high because the demand for units is low)—
a vicious circle. But this could be altered, for example, by government subsidies to adopt these technologies, although the subsidies would be very
expensive for many years. As a society, we may choose to subsidize renewable
energy appliances, even though this is economically inefficient, if that is an
overriding value of the majority (in a democracy). In other words, societies
may be willing to pay to follow their values. A simpler example is that in the
United States we are willing to spend large sums of money to save single individuals who are in immediate danger, for instance, the stranded hiker trapped
on a snowy mountain or the child who falls down an abandoned well. Yet, at
the same time, as a society we have chosen not to offer explicit health insurance to a large segment of the population, undoubtedly resulting in accelerated mortality for these people. This is probably due in part to the American
value of self-reliance, but it has costs.

Because values are costly, there must be associated benefits if values are to
be followed. Individuals may follow their values for various reasons—to obtain
the internal reward associated with following values; to avoid the internal
pain associated with violating values; to acquire external benefits of exhibit-
ing values, such as increased social status within one's peer group; and to be
able to influence others to acquire or follow a certain set of values. Some brain
imaging evidence indicates that choosing to follow one's values may generate
activation in regions of the human brain associated with reward (O'Doherty,
Kringelbach, Rolls, Hornak & Andrews, 2001). As a result of these factors,
individuals appear willing to have what appears to be an internal mechanism
in which the cost of following one's values is compensated by an internal or
external benefit (more on this point in section 2.4 below).

Values may also have social benefits. In general, if most employees follow
generally accepted social values (do not steal, lie, etc.), then monitoring costs
by firms and the government are reduced and substantial deadweight losses are
avoided. As a society, we seek to balance the costs of monitoring or not monitoring individuals or businesses with the expected losses due to violations.
For example, the audit rules for the Internal Revenue Service are implicitly
based on the understanding that, in most cases, direct monitoring is unnecessary. What is needed is some likelihood that random audits and other methods to discover tax cheats are able to capture some proportion of violators.

This insight into human behavior—that most people will follow the law
most of the time—was only formalized by the U.S. federal government in the
last twenty-five years. In the United States, a cost-benefit analysis of new
regulation and legislation prior to enactment was implemented by Executive
Order 12291, signed by President Ronald W. Reagan on February 17, 1981.
This order was strengthened by further executive orders over the next decade
by Presidents Reagan and George H. W. Bush. The requirement of cost-
benefit analysis was extended to a wider set of legislation by President William

J. Clinton in 1993 (Executive Order 12866). One goal of these directives is to use the optimal amount of regulatory oversight, since this is costly and sometimes unnecessary. An analogy to understand the cost and benefits of regulatory monitoring is manufacturing quality control. Typically only a small subset of manufactured objects is monitored for quality, often using sophisticated algorithms to minimize sampling costs while maintaining a high average quality.

Although cost-benefit analyses of monitoring are a move toward moral economics, they typically ignore how laws might counterintuitively actually encourage violations. An example is the rule that one must pick up one's child from day care no later than a certain time of day. If this rule is violated, a teacher must stay late to watch the child, and the child may be stressed by having to wait for a parent. In a recent experiment at two day care centers in Israel, both with a rule in place that parents must pick up their children no later than 4:00 PM, one imposed a fine of $3 for each time the child was picked up late. The other simply depended on the parents' following the rule with no sanction for failing to do so. At the center that imposed a fine, parents' mind-set apparently changed; the fine seemed to remove the implicit social sanction associated with being late, because now one just had to pay a penalty. Over a three-week period, the day care center with the fine saw twice as many parents arriving late, and the proportion of latecomers remained steady thereafter (even after the fines where terminated!) (Gneezy & Rustichini, 2000). The lesson here is that oversight and penalties may crowd out the good behaviors that most people, most of the time, follow.

2.3. Moral Sentiments

Of Adam Smith's two great books, *The Theory of Moral Sentiments* (TMS) is typically considered much less important than *The Wealth of Nations*, though this view is starting to change. In TMS, Smith identified the social nature of human beings and sought to understand the nature of, and motivation for, morality. Smith used the notion of sympathy as a primary psychological feature that guided moral behaviors. He argued that we "feel bad" when we violate a moral value, and this produces a desire to avoid this feeling in the future. The source of sympathy, according to Smith, is the ability to identify with the emotions of others.

> Man, say they, conscious of his own weakness, and of the need which he has for the assistance of others, rejoices whenever he observes that they adopt his own passions, because he is then assured of that assistance; and grieves whenever he observes the contrary, because he is then assured of their opposition. But both the pleasure and the pain are always felt so instantaneously, and often upon such frivolous occasions, that it seems evident that neither of them can be derived from any such self-interested consideration. (Smith, 1759)

Smith suggested that this emotional correspondence was innate and impossible to suppress. The quote from Hayek, the epigraph to this chapter, identifies a similar noncognitive mechanism for moral values.

Smith's view has two primary parts: first, the mechanism that supports virtuous behaviors is primarily affective, not cognitive; and, second, there appear to be universally shared moral behaviors that are the backbone for a well-functioning society. It is worth noting that Smith had these exceptional insights 250 years ago, ideas that were essentially ignored until the last 50 years. Direct tests of Smith's proposed affective mechanism for moral values are less than 10 years old.

2.4. Neural Mechanisms of Moral Sentiments

Several brain mechanisms seem to function together as a moral compass, guiding us on appropriate modes of behavior in our daily interactions with other humans. The first mechanism occurs when we observe others' motor movements. Observing others' movements involuntarily activates in the observer's brain regions associated with the planning and execution of such movements, even when the observer is not moving at all. These neurons have been called "mirror neurons" (Rizzolatti, Fadiga, Gallese & Fogassi, 1996), as they appear to produce an internal simulation in the observer of the occurring action. Mirror neurons have been found in primates, including humans (Rizzolatti, Fogassi & Gallese, 2001; Grezes & Decety, 2001), and in some birds.

A second mechanism helps us to infer the cognitive state of others. It has been called "mentalizing" or having a "theory of mind" (ToM). ToM permits us to forecast the beliefs and intentions of others by putting ourselves into the other's place and asking what we would do in such a situation. This is an extraordinarily useful ability when interacting with others, especially strangers. Children under five years old are unable to mentalize, and impairments in ToM are common in those with autism. ToM has been localized to the medial prefrontal cortex (Frith & Frith, 2003), and has been shown by neuroeconomists to affect economic decision making (McCabe, Houser, Ryan, Smith & Trouard, 2001; Zak, 2004; Camerer & Bhatt, 2005).

A third way the brain interprets what others are doing is through affective representation. This can occur when we observe an action occurring (for example, a nail going into a hand), by observing a facial expression such as fear or disgust, or simply by the knowledge that another is being harmed in some way (Singer, Seymour, O'Doherty, Kaube, Dolan & Frith, 2004a; Wicker, Keysers, Plailly, Royet, Gallese & Rizzolatti, 2003; Canli & Amin, 2002; Decety & Chaminade, 2003). The areas activated vary by study, but all include limbic regions associated with emotional responses. This internal representation of others' emotions can be called *empathy*.

An example of a representations of others' emotional states is an informal experiment I did with the members of my lab. A week after Hurricane Katrina, I asked who had donated money to the relief effort. Several students raised their hands, and then I asked them why they had given money. Each person could offer an explicit reason, often with an accompanying image of human suffering when recounting their motivation, and each clearly displayed emotion on the telling. Even with spatial and temporal distance, others' emotions are felt in ourselves and influence our behavior.

Notably, emotional responses can be provoked absent direct viewing of an individual. For example, when a person makes an intentional monetary sacrifice signifying that he or she trusts a stranger, the brain of the person being trusted produces a surge in the neuroactive hormone oxytocin (Zak, Kurzban & Matzner, 2004, 2005). In numerous studies of rodents, oxytocin has been shown to facilitate attachment to offspring and in some monogamous mammals pro-social behaviors toward unrelated conspecifics. Oxytocin is a physiologic signature of empathy, and appears to induce a temporary attachment to others (Zak, 2007; Carter, Ahnert, Grossmann & Hrdy, 2006). Infusing the human brain with a moderate dose of exogenous oxytocin can induce people to trust strangers with one's money to a much greater degree relative to those receiving a placebo (Kosfeld, Heinrichs, Zak, Fischbacher & Fehr, 2005). High densities of oxytocin receptors are located in regions of the brain associated with emotions (Zak, 2007).

Oxytocin appears to facilitate a representation of what another is feeling. These mechanisms are automatic and largely beyond our conscious control. That is not to say that these mechanisms are not modulated by the external and internal environment. For example, during episodes of extreme stress, other-regarding behavior is typically suppressed as survival of the individual becomes paramount. This is analogous to starving Hurricane Katrina victims breaking into convenience stores to obtain food—there is no moral (or legal) prohibition against this in times of crisis. This may be one explanation for the lack of moral behavior by those at Enron—the enduring stress of "making the numbers" caused them, at some point, to make up the numbers. This moral violation appears to be modulated when others nearby are doing the same thing. This explanation is given for "ordinary" German citizens who tortured and killed Jews under the Nazi regime—their neighbors were doing the same thing so it became acceptable. The same argument holds for the genocides in Cambodia, Rwanda, Kosovo, Armenia, and so on. Social psychologist James Waller (2002) calls this "moral disengagement."

There is great heterogeneity across people in the ability to empathize with others and in the associated neural activation during emotionally engaging activities (Canli, Zhao, Desmond, Kang, Gross & Gabrieli, 2001; Singer, Seymour, O'Doherty, Kaube, Dolan & Frith, 2004a). For example, I studied 212 subjects making trusting decisions and showed that their brains released oxytocin approximately proportional to the intentional monetary signal of trust received from a stranger: the stronger the signal of trust, the more oxytocin is released. Approximately 98 percent of these subjects also had proportional behavioral responses: the higher their oxytocin levels, the more they shared money with the person who initially demonstrated trust in them. But the other 2 percent of subjects, though their brains produced a surge of oxytocin, were untrustworthy, keeping all or nearly all the pot of money they controlled (Zak, 2005). Two percent is roughly the proportion of sociopaths in the population, and these subjects' psychological profiles had elements of sociopathy. A discussion of the neural mechanisms that produced this behavior is beyond the scope of this chapter, but it is worth noting that the relative contributions of nature and nurture are not well understood (see Zak, 2007). There are also interesting gender differences in physiologic responses associated with distrustful behaviors and aggression that, in men, are driven by testosterone (Zak, Borja, Matzner & Kurzban, 2005; Dabbs & Dabbs, 2000; Toussaint & Webb, 2005). The point here is that affective responses to social stimuli vary across individuals and quite likely within an individual as circumstances change.

2.5. Tests of the Neural Mechanisms of Moral Sentiments

Explicit tests of the brain regions active during moral decisions have been carried out recently by a number of researchers. I present a brief summary of this work to emphasize that there is direct evidence for the neural substrates of moral emotions. The relevance of this section to the chapter's thesis is that many values draw upon moral emotions.

The first neuroimaging study of moral decisions was led by Joshua Greene and colleagues (2001). They asked subjects to answer questions about how they would behave in a series of personal and impersonal moral dilemmas drawn from philosophy. Personal moral dilemmas take the following form: "Would you do something directly to harm another in order to save a group of people from certain harm?" Impersonal dilemmas are similar, but the proposed action occurs at a distance: Would you put into motion actions that would harm one person to save others? Contrasting neural activation in the personal versus impersonal dilemmas, these researchers found that prefrontal brain regions associated with theory of mind, as well as evolutionarily older subcortical regions related to emotions, have greater activation, whereas areas of the brain associated with working memory have reduced activations. These authors concluded that making decisions in personal moral dilemmas differentially draws on affective representations of outcomes.

In a related study, Moll, de Oliveira-Souza, Bramati & Grafman (2002) had seven subjects view pictures of emotionally charged scenes and asked them to rate their moral content. Comparing neural activity in moral violations (e.g., a man with a knife to the throat of a woman) versus those that were simply unpleasant (e.g., a mutilated body), Moll and colleagues found the strongest activation in cortical regions associated with emotional processing. Several additional contrasts (e.g., egregious versus weak violations) also show subcortical processing of moral scenes, especially in the amygdala. These neural signals often have a peripheral somatic basis as measured by skin conductance or heart rate (Scheman, 1996).

Sanfey, Rilling, Aronson, Nystrom, and Cohen (2003) asked subjects in an MRI scanner to engage in a simple strategic interaction called the Ultimatum Game (UG). In the UG, there is a proposer and a responder. The proposer is given a sum of money, say $10, and is instructed to offer a split of this to the responder. If the responder accepts the split, the money is paid. If the responder rejects the split, both parties earn nothing. Behaviorally, offers by the proposer less than 30 percent of the total are almost always rejected in industrialized societies (this regularity does not hold for "small-scale" nomadic, agrarian, and pastoral societies; see Henrich, Boyd, Bowles, Camerer, Fehr & Gintis, 2004). The question Sanfey and colleagues asked was why people reject good money to punish another person for being stingy. They demonstrated that neural activity, when a responder received a low offer of the split in contrast to an equal or hyperfair offer, was greater in regions of the brain associated with emotions, especially a region known to activate with visceral disgust (the insular cortex). Subjects who received stingy offers appeared to be disgusted by them and were motivated to punish the transgressor even at a cost to themselves.

Similarly, de Quervain, Fischbacher, Treyer, Schellhammer, Schnyder, Buck and Fehr (2004) asked subjects to play a game that admitted cooperation and defection, with the ability to directly punish those who defected (either at a cost or symbolically without cost). The design of the experiment, however, was that subjects knew they interacted only once, so that costly punishment would not benefit the punisher's earnings from the game (though punishment might be viewed as benefiting others). Costly punishment (as compared to symbolic punishment) produced strong activation in mid-brain limbic regions associated with rewarding behaviors. Subjects in this study punished because it felt good to do so.

Nonhuman social primates also act in ways which suggest that appropriate social behaviors have a physiologic basis. As discussed in chapter 5 of this volume, monkeys appear to have innate values of fairness and equity. They are willing to forgo food to punish another monkey who has violated these expected norms. Since monkeys have quite small prefrontal cortices relative to humans, this behavioral evidence suggests that these values are at least partially felt, not thought out. It is also notable that some monkeys understand symbolic exchange, that is, the use of money. Thus, the evolutionary basis for exchange is ancient, as are, apparently, the value of a fair division from exchange.

In sum, a large number of researchers have demonstrated that the neural representation of moral values is automatic and difficult to suppress, and often utilizes affective representations in the brain. I propose that values in economic transactions utilize similar neural mechanisms. In addition, values can be learned; that is, they may use potentiated pathways in the brain, biasing choices a certain way for a given environment. Long-term potentiation requires experience with this choice, whereas initial choice may have a genetic basis. We can therefore conclude that values have a neural representation but need to be reinforced. Rick Shreve of Dartmouth College, who contributed to the conferences that this volume reports, related an example of a violation of moral behavior and its instantiation. During a medical supply relief effort to Vilnius, Lithuania, in which Rick participated in 1991, Latvian medical school faculty and students were asked to unload the supplies because of the fear of stealing. Shreve observed a Lithuanian professor take some toothbrushes from a partially open box. Later, when he discussed this with colleagues, he was told that he had observed *Homo sovieticus*, who, as a result of years under the depersonalized and capricious rule of the Soviets, lacked the ability to make moral decisions. This suggests that the internal representation of values can be muted under extreme or chronic duress. Conversely, those of us with children typically spend a substantial amount of time inculcating what we see as proper values.

3. The Mechanics of Moral Economics

3.1. Modifying the Standard Economic Model

The argument thus far is that a subset of values is universally held, and there are consistent neural representations of such shared values. Some values, especially personal moral values, draw on affective responses to stimuli. The emotions produce coarse but quick reactions to stimuli. Emotions also influence impersonal decisions but are more important when making decisions directly involving others.

In a standard economic model of constrained optimization, we can consider values to place two types of constraints on utility maximization and also to produce associated utility flows. The first is a constraint on achievable outcomes. This would function like any other constraint (budget, time, etc.) that limits the goods one can consume and may use a terminal value. For example, suppose you are given a box of chocolates. You might decide not to consume the whole box yourself and instead distribute them to friends or family because you value sharing. This limits your own consumption, and increases others' consumption. It may also produce internal and external rewards (utility), for example, when a box of chocolates is given to a romantic partner on Valentine's Day.

A second constraint is placed on economic maximization through instrumental values. Frey and colleagues (Frey, Benz, & Stutzer, 2004) have recently introduced the notion of "procedural utility," namely, that how something is done provides its own utility flow. A straightforward example is the UG that we discussed earlier. Most subjects prefer more money to less, but if offered a 10 percent split of a sum of money, will reject it out of the obvious inequality and stinginess of the proposer. (A subject's typical response to a low offer is "Who the hell does he think he is!"). The utility gain from receiving a small sum of money is less valuable than the process of accepting it. It is worth reemphasizing that following one's values is often costly, even though it "feels right." In my experiments, non-reciprocators appear to have normal neural activity guiding them toward reciprocation but appear to ignore or suppress these signals. These subjects have aspects of antisocial personalities but also make the most money in cooperation experiments (Zak, 2005).

The model I propose is similar to the mathematical model of pro-social emotions by Bowles & Gintis (2003). In this model, agents engage in a public goods–type game; and are both self-interested and receive utility flows associated with pride, guilt, empathy, shame, and regret. In equilibrium, agents display a higher degree of cooperation than in the standard model absent social emotions. Pro-social emotions are the levers through which values are brought to decision making. The Bowles and Gintis conception of emotions producing utility flows is similar to Frey and his colleagues' conception of procedural utility. Overall, the Bowles and Gintis model captures nicely the notion that affect-laden values guide economic decisions.

3.2. Are Values Necessary in Economics? A Discussion of Generalized Trust

Here I demonstrate the predictive importance of shared values by critiquing a well-known paper I coauthored. Figure 12.1 shows survey data on the proportion of people in forty-two countries who answered yes to the question, "Generally speaking, would you say that most people in this country can be trusted?" The data vary from 3 percent answering yes in Brazil to 65 percent in Norway (!). Zak & Knack (2001) build a fairly standard mathematical model of purely self-interested economic principals who seek to make investments over time using agents as intermediaries, where agents have asymmetric information about market returns. Principals can monitor agents at a cost, and the principals' degree of trust is measured by their decision regarding how much not to monitor agents. The formal model predicts that environments where contract enforcement is high, social ties are tight, people are similar (e.g., in income, language, ethnicity, etc.), and incomes are higher will have high levels of trust. Indeed, in empirical tests of the theory, these four factors explain 76 percent of the order-of-magnitude variation in the data. A great model, right?

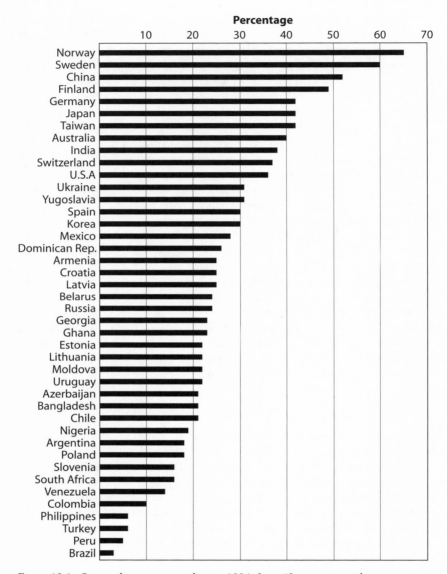

Figure 12.1. Survey data on trust, taken in 1994, from 42 countries with varying institutional environments.

This model, however, has a critical shortcoming: the formal and informal enforcement of investment contracts is specified only *on average*. Why an individual in the model does not cheat on a specific contract is not known. Nor does the model specify how agents that cheat are penalized, either formally or informally, or if they feel something (e.g., guilt, shame, or anger) when they cheat or are cheated. These shortcuts were used primarily because they were analytically convenient. But the omission is not trivial: in this study we also showed that generalized trust is among the strongest predictors that economists have ever found of whether a country would have increasing or decreasing living standards. Low-trust countries have stagnating or declining per capita incomes, whereas high-trust countries enjoy steady income growth. So, we better figure out what trust is and what supports it.

Although the model of Zak & Knack (2001) does provide some important insights into the social and institutional foundations for economic growth, it says nothing about how quotidian transactions raise living standards. The inclusion of shared values can enable us to understand human behaviors that occur in high- and low-trust countries. A parsimonious microeconomic model of trust would include the instantiation of the value of trustworthiness in societies where generalized trust is high. This provides the novel prediction that high-trust countries will have moderate, rather than high, degrees of formal monitoring of contracts. Because of the strong expectation, based on people's experience, that most people in Norway fulfill their contracts, there need not be a policeman (or lawyer) involved in each and every contract. People have a sense of what is expected and appropriate in transactions, and most of time they follow this. This may be why countries whose citizens are more similar have higher trust—the cognitive and affective mechanisms that induce the understanding of another's intentions may simply be easier to read when those around us are similar. Zak & Knack (2001) pointed out that countries with more dissimilarities, like the United States, need stronger formal enforcement of contracts. But the model does not explain at the individual level why, with only sporadic enforcement, most of the time most contracts are fulfilled—even in one-shot transactions. Only an appeal to some other mechanism can do this. I would like to suggest that values are one such device— stochastic punishment is insufficient since that would predict more cheating than is observed in high-trust countries. Further, not only do values do the job, they are, as I argued above, consistent with a large neuroscience literature that shows that values have a clearly identifiable physiologic basis.

4. Applications of Moral Economics

Moral economics is not free of formal institutions to enforce contracts and mediate disputes. It does require that formal institutions reasonably impose impartial and fair outcomes when asked to do so, or the institutions will be inconsistent with shared values and will not be followed. The seminal neuro-jurisprudence thinker Margaret Gruter (1991) emphasized that laws were for-malizations of social behaviors that typically had long histories. The promulgation of laws that run counter to these informal norms—which are based on shared values—are doomed to be ignored. For example, the U.S.-style laws that were instituted in Eastern Europe and Russia in the 1990s were largely ignored in many countries, as the population had already developed informal systems to solve economic problems that worked well enough. Elinor Ostrom and colleagues (Dietz, Ostrom, & Stern, 2003) have documented even more pernicious outcomes when norms of sharing a public resource, such as water in Africa, are changed by a national government or international or-ganization, leading to a lack of cooperation, extreme waste, and conflict.

A likely physiologic mechanism at work when institutions undermine shared values, or when social-political environments are so chaotic that forward-looking decisions are difficult to make at all, is the inhibition of oxy-tocin release during economic transactions and an associated reduction of interpersonal trust. As discussed in the previous section, a sufficient level of trust is necessary for investments that occur over time to be executed. Physi-ologic stress from living in an unstable environment inhibits oxytocin re-lease; people in this situation are in "survival mode" and are typically present- and self-oriented rather than drawing on shared values that support cooperative behaviors and capture gains to trade. There are spectacular ex-ceptions of selflessness in the face of mortal danger, of course, but the argu-ment here is about economic transactions in a bodega in São Paolo, Brazil, or Lagos, Nigeria. Preliminary evidence that environmental factors influence the physiology of trust at the country level is reported in Zak & Fakhar (2006).

This physiologic argument is consistent with individuals responding to environmental conditions in small-scale societies in the study, discussed ear-lier, by Henrich, Boyd, Bowles, Camerer, Fehr & Gintis (2004). In these no-madic, agricultural, and pastoral societies, offers in the Ultimatum Game varied from zero to nearly 100 percent of proposer's endowment. Rejections of offers were equally widespread, including some rejections of offers of 80 per-cent and above. This occurred even though the subjects interacting in the UG were not identifiable to one another. Variations across societies in the UG were associated with typical society-specific social norms. Behavior in these societies had presumably been optimized for the social environment, and only responded imperfectly to the instructions given for the UG during the experiment. Perhaps most intriguing, Henrich and colleagues identified a single factor that explained the incidence of fair offers (those near 50 percent of the total) across societies: the exposure to market exchange. Hold on—isn't it true that market traders are all about getting the best deal, the most for me, the least for you? I hope that, by now, the error in this cartoon of market exchange is evident. According to Henrich and colleagues, those in societies that traded appeared to have understood that trade freely entered into necessitates the acquisition of reasonable gains for *both* parties. Repeated trade is most likely to arise when parties work out a fair distribution of gains. This research suggests that markets reinforce the value of equal sharing of gains.

How does this occur? In the elegant experiment of Kimbrough, Smith, and Wilson (see chapter 13), traders were allowed to invent their own rules. At first, these were quite individually focused, with each trader maximizing his or her gain in a single transaction. But competition among traders quickly led to stable groups that maximized social surplus and worked out a fair division of it. Markets, and the institutions that underpin exchange, appear to support a value of near-equal sharing. Appropriate laws may do the same, moving people toward a change in behavior (and perhaps values) that then make the law itself unnecessary—or the behavior may lead to the law, for example, the law against smoking indoors. When I lived in San Diego, California, as an undergraduate in the mid-1980s, it was common for those in bars where young people congregated to go outside to smoke. This was before the state of California formally banned indoor smoking. Yet, going outdoors to smoke was not common in many other states prior to laws prohibiting this behavior (and in San Diego this was uncommon in bars frequented by "older" people—those thirty and above). Anti-smoking laws are now being adopted throughout the United States and by European countries with higher proportions of smokers than in the United States The underlying value issue is the trade-off between the personal freedom to smoke indoors, with the imposition on others to breathe another's smoke. The former value is individually focused, and the latter is other-focused. Donald Eliot at Yale Law School pithily called this "the didactic value of law" at the third conference convened in the Values and Free Enterprise series upon which this book is based.

The move from personal exchange to modern, mostly impersonal exchange in markets is the key to the division of labor that caused the rapid gains in productivity and wealth since the Industrial Revolution (Smith, 2003). Because the instantiation of values varies somewhat across both individuals and environments, violations of values must have consequences. Enforcement in traditional societies is personal—you cheat me, then I hurt, or ostracize, you. The incentives to cheat, free ride, and steal are rampant during impersonal exchange, necessitating an enforcement body that all accept, namely, a government. As Lynn Stout discusses in chapter 8 in this volume, the mistake legal scholars made, which can be traced to Oliver Wendell Holmes, is the belief that humans are value-free, simply weighing costs and benefits of an action when making choices. This has produced laws that, like in the example of the Israeli day care center, view punishment as a price one pays to engage in a behavior rather than a violation of mutually shared values, and this may increase rather than reduce violations.

The main implication for economics is that laws regulating market exchange, if well designed, should take into account innate predilections among
most of the populace in order to be maximally effective. This suggests, in particular, that economies which are moderately regulated will both create the most
wealth and have the most personal freedom. The former follows because all regulations have costs (direct costs and transactions costs), and increasing costs reduce exchange. When exchange is curtailed, the opportunities to create wealth
are fewer. Having a police officer on every corner would drastically reduce
crime, but at a very high cost that is mostly unnecessary. Competition raises efficiency, and moderate physiologic stress increases cognitive skills. Competitive
stress is good for both the person and the society. But inordinate stress leads to
physical decline and values violations that cause poverty. Further, when every
action is dictated from above, the freedom to create, which is necessary for
technological innovation, is absent. Heavy-handed oversight of social behaviors (in markets and otherwise) also likely crowd out the working of values. This
was shown experimentally by Fehr & Gächter (2002). Douglass North (2005)
made a similar argument when analyzing how institutions evolve (North, 2005).

If my thesis here is correct, it means that values are not specific to the
West or the East, nor are there broadly distinct Western and Eastern economic institutions. Rather, values across all cultures are simply variations on a
theme that is deeply human, strongly represented physiologically, and evolutionarily old. Similarly, the kinds of market institutions that create wealth
and enable happiness and freedom of choice are those that resonate with the
social nature of human beings who have an innate sense of shared values of
right, wrong, and fair. Modern economies cannot operate without these.

Notes

1. At the time of his death, Cliff Baxter was taking a number of psychotropic
medications, including antidepressants.

2. Daniel Watkins, a consultant to Enron and other companies, who worked for
Arthur Andersen, killed himself in December 2002.

References

Bhatt, M., & Camerer, C. 2005. Self-referential thinking and equilibrium as states of
    mind in games: fMRI evidence. *Games and Economic Behavior* 52:424–259.
Bowles, S., & Gintis, H. 2003. Prosocial emotions. Santa Fe Institute Working Paper
    #02-07-028.
Buss, D. 2005. *The Murderer Next Door: Why the Mind is Designed to Kill.* New York: Pen-
    guin Books.
Canli, T., Zhao, Z., Desmond, J. E., Kang, E., Gross, J., & Gabrieli, J. D. E. 2001. An
    fMRI study of personality influences on brain reactivity to emotional stimuli. *Be-
    havioral Neuroscience* 115(1): 33–42.

Canli, T., & Amin, Z. 2002. Neuroimaging of emotion and personality: Scientific evidence and ethical considerations. *Brain and Cognition* 50(3): 414–431.

Carter, C. S., Ahnert, L., Grossmann, K. E., & Hrdy, S. B. 2006. *Attachment and Bonding: A New Synthesis*. Cambridge, Mass.: MIT Press.

CNN.com. 2002. Former Enron exec dies in apparent suicide. Available at http://archives.cnn.com/2002/US/01/25/enron.suicide/.

Dabbs, J. M., & Dabbs, M.G. 2000. *Heroes, Rogues, and Lovers: Testosterone and Behavior*. New York: McGraw-Hill.

Decety, J., & Chaminade, T. 2003. Neural correlates of feeling sympathy. *Neuropsychologia* 41:127–138.

de Quervain, D. J., Fischbacher, U., Treyer, V., Schellhammer, M., Schnyder, U., Buck, A., & Fehr, E. 2004. The neural basis of altruistic punishment. *Science* 305 (5688): 1254–1258.

Diamond, J. 1997. *Guns, Germs, and Steel: The Fates of Human Societies*. New York: Norton.

Dietz, T., Ostrom, E., & Stern, P. C. 2003. The struggle to govern the commons. *Science* 302:1907–1912.

Fehr, E., & Gächter, S. 2002. Altruistic punishment in humans. *Nature* 415:137–140.

Frith, U., & Frith, C. D. 2003. Developments and neurophysiology of mentalizing. *Philosophical Transactions of the Royal Society B* 358(1431): 459–473.

Frey, B. S., Benz, M., & Stutzer, A. 2004. Introducing procedural utility: not only what, but also how matters. *Journal of Institutional and Theoretical Economics* 160 (3): 377–401.

Gneezy, U., & Rustichini, A. 2000. Is a fine a price? *Journal of Legal Studies* 29:1–17.

Greene, J. D., Sommerville, R. B., Nystrom, L. E., Darley, J. M., & Cohen, J. D. 2001. An fMRI investigation of emotional engagement in moral judgment. *Science* 293:2105–2108.

Grezes, J., & Decety, J. 2001. Functional anatomy of execution, mental simulation, observation and verb generation of actions: A meta-analysis. *Human Brain Mapping* 12:1–19.

Gruter, M. 1991. *Law and the Mind: Biological Origins of Human Behavior*. Newbury Park, Calif.: Sage.

Henrich, J., Boyd, R., Bowles, S., Camerer, C., Fehr, E., & Gintis, H., eds. 2004. *Foundations of Human Sociality: Economic Experiments and Ethnographic Evidence from Fifteen Small-Scale Societies*. Oxford: Oxford University Press.

Kosfeld, M., Heinrichs, M., Zak, P. J., Fischbacher, U., & Fehr, E. 2005. Oxytocin increases trust in humans. *Nature* 435(2): 673–676.

McCabe, K., Houser, D., Ryan, L., Smith, V., & Trouard, T. 2001 A functional imaging study of cooperation in two-person reciprocal exchange. *Proceedings of the National Academy of Science USA*, 98(20): 11832–11835.

McLean, B., & Elkin, P. 2003. *Smartest Guys in the Room: The Amazing Rise and Scandalous Fall of Enron*. New York: Portfolio Hardcover.

Moll, J., de Oliveira-Souza, R., Bramati, I. E., & Grafman, J. 2002. Functional networks in emotional moral and nonmoral social judgments. *Neuroimage* 16 (3): 696–703.

North, D. C. 2005. *Understanding the Process of Economic Change*. Princeton, N.J.: Princeton University Press.

O'Doherty, J., Kringelbach M. L., Rolls, E. T., Hornak, J., & Andrews C. 2001. Abstract reward and punishment representations in the human orbital frontal cortex. *Nature Neuroscience* 4:95–102.

Pedersen, C. 2004. How love evolved from sex and gave birth to intelligence and human nature. *Journal of Bioeconomics* 6:39–63.

Reader, S. N., & Laland, K. N. 2002. Social intelligence, innovation, and enhanced brain size in primates. *Proceedings of the National Academy of Science* USA, 99(7): 4436–4441.

Rizzolatti, G., Fogassi, L., & Gallese, V. 2001. Neurophysiological mechanisms underlying the understanding and imitation of action. *Nature Reviews Neuroscience* 2:661–670.

Rizzolatti, G., Fadiga, L., Gallese, V., & Fogassi, L. 1996. Premotor cortex and the recognition of motor actions. *Cognitive Brain Research* 3:131–141.

Rokeach, Milton. 1973. *The Nature of Human Values*. New York: The Free Press.

Sanfey, A. G., Rilling, J. K., Aronson, J. A., Nystrom, L. E., & Cohen, J. D. 2003. The neural basis of economic decision-making in the Ultimatum Game. *Science* 300 (5626): 1755–1758.

Scheman, N. 1996. Feeling our way toward moral objectivity. In Larry May, Marilyn Friedman, & Andy Clark, eds., *Mind and Morals: Essays on Cognitive Science and Ethics*, 221–236. Cambridge, Mass.: MIT Press.

Schwartz, S. H. 1994. Are there universal aspects in the structure and contents of human values? *Journal of Social Issues* 50(4): 19–45.

Singer, T., Seymour, B., O'Doherty, J., Kaube, H., Dolan, J. D., & Frith, C. 2004a. Empathy for pain involves the affective but not sensory components of pain. *Science* 303(5661): 1157–1162.

Singer, T., Kiebel, S. J., Joel, S. W., Dolan, J. D., & Frith, C. 2004b. Brain responses to the acquired moral status of faces. *Neuron* 41(4): 653–662.

Slemrod, J., ed. 1992. *Why People Pay Taxes: Tax Compliance and Enforcement*. Ann Arbor: University of Michigan Press.

Smith, V. 2003. Constructivist and ecological rationality. *American Economic Review* 93(3): 465–508.

Toussaint, L., & Webb, J. R. 2005. Gender differences in the relationship between empathy and forgiveness. *Journal of Social Psychology* 145(6): 673–85.

Waller, J. 2002. *Becoming Evil: How Ordinary People Commit Genocide and Mass Killing*. Oxford: Oxford University Press.

Wicker, B., Keysers, C., Plailly, J., Royet, J.P., Gallese, V., & Rizzolatti, G.. 2003. Both of us disgusted in my insula: The common neural basis for seeing and feeling disgust. *Neuron* 40:655–664.

Zak, P. J., 2004. Neuroeconomics. *Philosophical Transactions of the Royal Society B*, 359:1737–1748.

Zak, P. J. 2005. Trust: A temporary human attachment facilitated by oxytocin. *Behavioral and Brain Sciences* 28(3):368–369.

Zak, P. J. 2007. The neuroeconomics of trust. In *Renaissance in Behavioral Economics*, ed. Roger Frantz. New York: Routledge.

Zak, P. J., Borja, K., Matzner, W. T., & Kurzban, R. 2005. The neuroeconomics of distrust: Physiologic and behavioral differences between men and women. *American Economic Review* 95(2): 360–363.

Zak, P. J., & Fakhar, A. 2006. Neuroactive hormones and interpersonal trust: International evidence. *Economics and Human Biology*, 4:412–429.

Zak, P. J., Kurzban, R., & Matzner, W. T. 2004. The neurobiology of trust. *Annals of the New York Academy of Sciences* 1032:224–227.

———. 2005. Oxytocin is associated with human trustworthiness. *Hormones and Behavior* 48:522–527.

Zak, P. J., Park, J. Ween, J., & Graham, S. 2006. An fMRI study of trust and exogenous oxytocin infusion. *Society for Neuroscience*. Abstract Number 2006-A-130719-SfN.

Thirteen

Building a Market
From Personal to Impersonal Exchange

Erik O. Kimbrough, Vernon L. Smith, and Bart J. Wilson

Adam Smith identified two key components of wealth creation in human societies: exchange and specialization. Voluntary exchange between individuals is a positive-sum activity in and of itself, simply because individuals would never engage in a transaction voluntarily if they did not believe that there was something to be gained from doing so. However, when exchange occurs between individuals who have specialized in those activities in which they have a particular comparative advantage, immense wealth can be created. Despite the long years of acceptance enjoyed by this truism, relatively little is understood about the underlying process by which people build exchange systems and discover comparative advantage. In *The Wealth of Nations* (1776), Adam Smith observed that the extent of the market determines the degree to which people can specialize and thus create wealth (Smith, 3:26). But what are the social mainsprings that give rise to the market?

Personal social exchange among kin and neighbors long precedes the advent of impersonal market exchange. In the form of reciprocal trading of favors, personal exchange is a highly visible kind of cooperation that depends on explicit trust. The cooperation and trust among individuals that initiates and sustains impersonal market exchange, however, is implicit and not readily observable. Both personal and impersonal exchange are founded on a system of largely unspoken values. These "guides to action" promote cooperative behavior by providing a common frame of reference that individuals rely on to support various forms of exchange (see chapter 12). Research on nonhuman primates indicates that value-driven behavior may be an evolutionary adaptation guided more by emotion than by explicit rules and logic (see chapter 5).

In this chapter we report on a pilot experiment where we introduced opportunities for the evolution of long-distance trade into a system of interconnected virtual villages. The experiment allowed us to observe values in our subjects' decisions and conversations to discover how and when the impersonal grows out of the personal.[1] This subject is often thought of as cooperative versus noncooperative behavior, or personal versus impersonal exchange, but with this experiment we sought to observe the trade-off between engaging in personal and impersonal exchange.

Our experiment built on a report by Crockett, Smith, and Wilson (2006; hereafter, CSW) that examined the efficiency of an economic environment capable of supporting specialization so long as the participants discover and develop some self-organized system of exchange. CSW created an experimental model village that produced only two goods and was closed to all external interaction. The participants in the experiment had to discover not only their own comparative advantages but also their ability to gain personally from those advantages by exchanging with one another. The authors found that a small majority of subjects either immediately settled into autarky or started to specialize in the good in which they had a comparative advantage. The other subjects typically followed an erratic development over time, trading between appropriately specialized subjects with complementary comparative advantages. By the end of the experiment, half of the six, two-person economies found and achieved the welfare-maximizing competitive equilibrium. Most extraordinary is that once the subjects discovered exchange, they almost immediately established the competitive equilibrium price. The other three pairs remained content to live in autarky in which they tended to achieve efficient home production but failed to discover the far more efficient exchange equilibrium with another person.

When CSW doubled the size of the economy to four subjects per session, the subjects discovered and achieved complete specialization in one out of six sessions. One session remained at autarky, and the remaining four sessions varied from 25 to 75 percent of the full efficiency at the competitive equilibrium. Even though exchange was discovered quite early in many of the four-person economies, the existence of more potential partners was detrimental to the full development of exchange and specialization with one partner. When CSW again increased the size of the economy to eight people per session, not a single session achieved complete specialization: three sessions remained locked in autarky, two achieved 20 percent of the complete specialization efficiency, and one achieved nearly 60 percent. This contrasts with the conventional folk wisdom that the larger the potential market, the more likely it is to produce the competitive outcome. Instead, the authors found that the individuals in these eight-person sessions were remarkably autonomous and hence, perhaps disturbingly so, inefficient. Finding a trading partner was observed to be far more difficult in that only three of the seven other people in the village were suitable partners with complementary productive advantages. Contrary to the authors' expectations, in the four- and eight-person economies, to the extent that exchange was observed, it was increasingly bilateral over time, not the multilateral sort of exchange generally associated with markets.

CSW concluded that three stages in learning were needed to achieve competitive equilibrium in these economies: (1) discovering the ability to exchange, which may require "mind-reading" (inferring intentions from words and actions) and imitation; (2) finding a suitably endowed trading partner with whom one can benefit from exchange through specialization; and (3) building the relationship by increasing specialization over time. In these model economies, however, no market, as it is commonly thought of, ever emerged. People either did not trade or remained firmly entrenched in bilateral personal exchange. Even more intriguing, in the eight-person treatment, where folk wisdom would most strongly predict the creation of a market, the level of exchange and specialization observed was relatively minimal.

In a subsequent experiment, the authors further explored stages (2) and (3) above by gradually building the eight-person economies from smaller groups. CSW began each forty-period session with four two-person economies, merged these into two four-person economies after twenty periods, and finally merged these into one eight-person economy after thirty periods. This design significantly reduced the transaction/search costs of finding a suitable trading partner. Also, the gradual growth of the size of the economy introduced new potential trading partners who, having had the opportunity to discover their own comparative advantage, and being in the process of increasing their rate of specialization, conceivably could compete with the other participants as potential trading partners. Finally, if individuals had not discovered trade, this allowed them to imitate, or be taught, the innovations of others in exchange and specialization.

Full specialization, and thus full efficiency, often occurred in these economies, but all the exchange remained bilateral and fundamentally personal. Our objective in the new experiment reported here is to introduce new possibilities for the development of more market-like exchange.

Experimental Design

Our study involved twelve subjects, four in each of three virtual villages. Three goods were available for consumption in the world, but each village could produce only two of them. For simplicity, we called the goods *red*, *blue*, and *pink*. There is a *red-blue* village, a *blue-pink* village, and a *pink-red* village.

TABLE 13.1
Multiplier for the Third Good (Pink)

# of Pink	1	2	3	4	5	6	7	8	9	10
Multiplier	1	1.21	1.41	1.57	1.7	1.82	1.92	2.01	2.1	2.17
# of Pink	11	12	13	14	15	16	17	18	19	20
Multiplier	2.24	2.31	2.37	2.43	2.48	2.53	2.58	2.62	2.66	2.7
# of Pink	21	22	23	24	≥25					
Multiplier	2.74	2.78	2.82	2.85	2.88					

Within each village, there are producers and merchants: producers can make twice as much of each good as merchants can, but merchants can also travel to a secondary area that remains unseen to the producers where they can trade with merchants from the other two villages. Producers and merchants are further divided into odds and evens. To prevent this description from becoming too muddled, we discuss this in the context of the *red-blue* village and then extrapolate to the other villages.

Each subject in the *red-blue* village prefers to consume red and blue in strict complementary proportions. But, for each unit of blue, odds must consume exactly 3 units of red to earn 3 cents, and evens must consume 2 units of blue for each unit of red to earn 2 cents. Furthermore, if the merchants from this world are able to acquire the third good, pink, the consumption of each unit of pink acts as a multiplier on that individual's earnings from consuming red and blue. Table 13.1 shows the value of the multiplier for each unit of pink consumed.

This same pattern is followed in the other villages; odds always need a 3 to 1 ratio of the first good in their village's name, and evens always need a 2 to 1 ratio of the second good. Consumption of the third good always acts as a multiplier with the same value across all villages. In addition to their differences in consumption preferences, odds and evens also differ in their production functions. Each person in each village is given a ten-second production period at the beginning of each day. Using a scroll-wheel at the top of their screens, the subjects can allocate a certain percentage of that time to the production of each of the two goods available in their village. Again, to give a concrete example, we will stick to the *red-blue* village and extrapolate from there to the other villages. Table 13.2 summarizes the production tables for each type in the *red-blue* village at 5 percent intervals, that is, 50 percent indicates five seconds dedicated to producing blue and five seconds to producing red. As mentioned above, producers and merchants also have different production functions. Producers can make exactly twice as much of each good compared to merchants. In individual autarky, odd (even) producers who wish to maximize their earnings should optimally spend 56 percent (51 percent) of their time producing red, thus producing and consuming 30 (13) reds and 10 (26) blues, and earning 30 (26) cents each period. Spending the same amount of time producing each good, odd (even) merchants will earn a total of 15 (12) cents.[2]

Without inter-village trade, that is, with internal trade but without the importation of pink, the competitive equilibrium brings significantly higher earnings than is available in home-production autarky. The odd producers specialize completely in the production of the red good, producing 130 units by scrolling their production-time wheel to 100 percent, and the even producers specialize in the blue good, producing 110 units. The competitive price of a blue unit is 4/3 of a red unit, and after exchanging at this price, the odd producer consumes 90 red and 30 blue units, and the even producer consumes

TABLE 13.2
Production Functions for Producers

Odds					Evens			
Blue %	Red %	Blue	Red		Blue %	Red %	Blue	Red
0%	100%	0	130		0%	100%	0	25
5%	95%	1	114		5%	95%	0	24
10%	90%	2	100		10%	90%	1	23
15%	85%	3	87		15%	85%	2	22
20%	80%	5	74		20%	80%	4	20
25%	75%	6	63		25%	75%	7	19
30%	70%	7	53		30%	70%	10	18
35%	65%	8	44		35%	65%	13	16
40%	60%	9	36		40%	60%	18	15
45%	55%	10	29		45%	55%	22	14
50%	50%	11	23		50%	50%	28	13
55%	45%	12	18		55%	45%	33	11
60%	40%	14	13		60%	40%	40	10
65%	35%	15	9		65%	35%	46	9
70%	30%	16	6		70%	30%	54	8
75%	25%	17	4		75%	25%	62	6
80%	20%	18	2		80%	20%	70	5
85%	15%	19	1		85%	15%	79	4
90%	10%	20	0		90%	10%	89	3

40 red and 80 blue goods. Following CSW, we chose these parameters so that the competitive profits (90 cents for odd producers, 80 cents for even producers) are roughly three times greater than autarky profits. Remember that the odd and even merchants, in this situation, trade with one another at the same prices but will earn exactly half the amount producers earn.

Finally, at the competitive equilibrium with inter-village trade, both producers and merchants will remain fully specialized in their production, but they use a portion of that production to purchase the third good from other producers and merchants in the other villages. Since the ratio of each good to

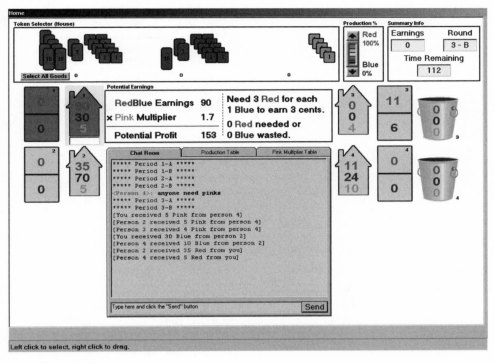

Figure 13.1

Note: In the actual experiment, the tile shapes in the top left, middle, and right areas were colored red, blue, and pink, respectively. The top, middle, and bottom numbers in the house and buckets were also colored red, blue, and pink, respectively. The top and bottom numbers in the fields were red and blue, respectively.

every other good in the world is exactly 1 to 1, the competitive price of each good in terms of each other good in the inter-village arena is 1. Odd (even) producers trade with their counterpart within their own village in order to consume 72 (32) reds and 24 (64) blues; they take the remaining 26 reds (22 blues) and give them to a merchant to exchange for an equal number of pinks from the other villages. They then consume those pinks to achieve maximum earnings of 208 (178) cents. Because of the nature of the multiplier, the merchants' earnings in competitive equilibrium will not be exactly half the producers' earnings; rather, the odd merchants earn 85 cents each and the even merchants 72 cents. The other villages mirror this with their own respective goods. Thus, the equilibrium total earnings in each village are 543 cents per period.

The subjects were presented the interface displayed in figure 13.1. Each subject owns both a house and a field, but half the subjects in each village, the merchants, own a bucket. The house displays what will be consumed at the end of a period and the field (the domino shape) displays what is produced during the production period. The bucket displays the goods that the merchant can carry with him when he goes to the travel screen. The total quantity of each good contained within the house, field, or bucket is displayed on the icon itself at all times. However, to view a moveable icon of each good, the subjects must click on their field, house, or bucket. This highlights that icon and makes all the goods contained within visible in the upper-left section of the screen.

The experimental session takes place over a series of "days," with each day divided into two phases, A and B. This experiment lasted for 34 of these days. Phase A is the 10-second production period during which each subject produces goods in his field using the scroll-wheel described above. This scroll-wheel can be adjusted at any time to allocate production time for the next day. As production occurs, icons representing each of the goods appear in the subject's field.

Phase B is the 120-second exchange and consumption phase during which these icons can be dragged and dropped into any of the houses, fields, or buckets on the screen. A subject's own house and field are green, and those of others are gray. To earn cash, the subjects consume by dragging and dropping the red, blue, and pink icons into their houses. Anything not in a house at the end of a day and any goods not consumed in the proportions specified above are wasted; that is, unconsumed goods do not roll over to the next day. In contrast to the CSW experiment, our subjects are aware that they can move goods to whomever they want, but they are not told that they ought to trade. At any time during the experiment, subjects may communicate with the other people in their village in a central chat room. They can discuss whatever they want so long as they do not use inappropriate language, discuss side payments, or make threats. Every seventh day is a 130-second rest period, during which no production or consumption takes place, but subjects may still use the chat rooms at their leisure. Also, a space is available beneath each house where the owner may leave a one-line message that is always visible to other villagers.

In addition to the abilities mentioned above, merchants can click on a button labeled "travel" on the right side of their screen at any time to move to a secondary area (shown in figure 13.2) with two more chat rooms. When merchants travel, their buckets and the contents of the buckets travel with them. Only on this screen can individuals from the different villages interact, so any inter-village trade that is to occur must take place at this time. Goods acquired on the travel screen must be taken back to the villages and placed in a house to be consumed. Anything left on the travel screen at the end of a day is wasted.

Our subjects were 12 George Mason University undergraduates; 10 were recruited randomly from the 180 subjects in the CSW experiment and 2 were recruited randomly from the student body as a whole. They interacted through visually isolated computer terminals and read self-paced instructions prior to the beginning of the first period. In addition to $5.00 for simply showing up on

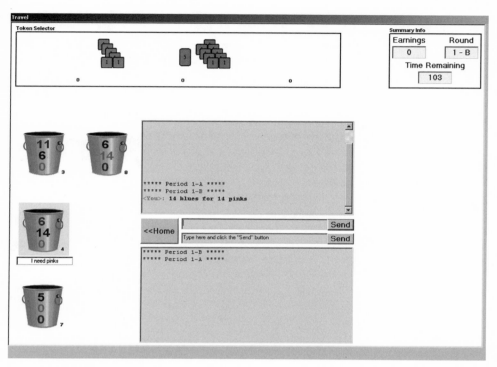

Figure 13.2
Note: See the caption note to figure 13.1 for a guide to the coloring scheme.

time, they received cash payments for their earnings in the experiment. The average total earnings in the session were $18.55, ranging from $5.75 to $46.00.

Results

Chat Room Transcripts

The chat room transcripts of the individual villages and the secondary merchant area were real-time conversations that revealed the qualitative attitudes of the participants toward the people with whom they interacted through their choice of words. We observed in the transcripts that *absolutely no conversation* occurred among the merchants during any of the 4 rest days (days 7, 14, 21, and 28). During these rest days, all the merchants remained at home in their village conversing with their producers. An example from day 14 is illustrated in table 13.3. Producers are identified as Persons 1, 2, 5, 6, 9, and 10 and Persons 3, 4, 7, 8, 11, and 12.

TABLE 13.3
Chat-Room Transcript (A)

Red-Blue Village	*Blue-Pink Village*	*Pink-Red Village*
***** Day 14*****	***** Day 14*****	***** Day 14*****
Person 1: ok	5: i'm still looking for pinks	9: lemme get re
4: 55 blues	5: OK. Is anyone good at producing pink?	9: man
1: i can make 130 reds		11: 10 where you at
4: usually	5: I can produce a lot of blue if we want to trade, that way.	10: yo
1: which is what i've been doing	6: are we allowed to ask what earnings we all have?	11: who you know is the business man
2: correction, 110 blues	8: i am..sometimes	10:10
1: ah, right	7: 194	11: nah i get those great deals
1: wait	5: I'm not sure	11: those meal deals
2: ok, what about 3 and 4?	7: oh I dont know	10: if i had that bucket i wouldve bankrupted every other chat room
3: what r these pinks for?	6: ok well dont say incase we aren't	
1: i can actually make 130 red		11: produce something
4: pinks are a multiplier	7: can anyone else travel	11: doing all that typing
2: ok so $110+55=165$ blues	8: i can	11: produce something
1: the pinks multiply profits	6: no	11: do something
4: yea	5: you know. I was wondering about that	10: naw
2: 3, how many reds can you make?	7: what is it for	10: im chillin
3: 70	6: does anyone have any red yet?	9: yeah i wish i had that bucket
3: how can i increase it	7: nope	11: you wouldnt know what to do with it
2: ok so $70+130=200$ red	5: I haven't seen any red at all	9: i would do better then y'all 2
1: the production scroller on the top	8: i think you can trade with other groups..but im not really sure	11: yall probably just wear it on your head
2: me and 4 need more blues than red	7: I bet we have to ravel to get red	10: i come back with at least 1 million blue
4: right	8: the people in the travel thing have red	11: and get no deals
2: 1 and 3 need more red than blue		

TABLE 13.3 (*continued*)
Chat-Room Transcript (A)

Red-Blue Village	Blue-Pink Village	Pink-Red Village
***** *Day 14*****	***** *Day 14*****	***** *Day 14*****
3: two of us can make more reds and 2 can amke more blues	7: oh k makes sense	11: im helping everyone come up
4: we need extra blues to trade for pinks also	6: i have a 1 next to where it says xred multiplier	11: when i eat everyone eat
1: what should we change for next round?	7: me too	11: you know who said
3: then we can excahnge if required	5: yeah me too	10: oh for real
2: 1 and 3 maximize reds	7: and it stinks	
1: i'm alrady at red producing max		
2: me and 4 will max blues		
1: okay		
2: ok		
1: #3 you got it?		
3: ok		

Notice the personal and casual nature of the conversations between the producers and merchants. In the *red-blue* village, the discussion is about how "*we* need extra blues to trade for pinks" and "what should *we* change for next round?" This indicates a rather obvious sense of community among the producers and merchants within a village. The village members spend their rest day coordinating their production decisions for explicit mutual benefit.

Out of this sense of community they develop a narrative to frame their experience. The tone of the conversation in the *pink-red* village reveals outgroup and in-group attitudes among the village members. Person 10 muses, "if i had that bucket i would've bankrupted every other chat room," and, after some playful exchanges, Person 11 adds the in-group statements that "i'm helping everyone come up" and "when i eat everyone eat." Each villager contextualizes his experience (some frame the experiment as a contest, others as a cooperative venture), and this context with all its implicit values defines the limits of his behavior (see chapter 1).

TABLE 13.4
Chat-Room Transcript (C)

Red-Blue Village	*Pink-Red Village*
***** Day 15*****	***** Day 15 *****

[Person 4 receives 20 Red from person 1.]	9: 10 u got some reds for me?
2: ok so lets to 25 reds and 50 blues for me and 4	[Person 9 received 40 Red from person 10.]
[Person 2 received 20 Red from person 1.]	11: yall need to gimme some pinks
3: i made only 61 this time on 100% reds	9: wow not too many
2: and 25 blues and 75 red for 1 and 3	11: 30 reds
1: oh, ok	
2: does that work out?	9: give me more then 2 blues
3: ok	[Person 11 received 30 Red from person 10.]
1: right, but we now only need to decide, of those red to you and 4, how many i give	9: and u can have mad piks
1: and how many #3 gives	9: pinks*
[Person 1 received 30 Blue from person 2.]	12: some1 gimme pinks and i'll get blues
4: extra blues?	[Person 11 received 20 Pink from person 9.]
2: hmm	10: let me get 20 pink
4: 2 i need the extras	[Person 10 received 15 Pink from person 9]
2: how about 20 blues and 40 reds for me and 4	
[Person 4 received 40 Blue from person 2.]	12: 9 gimme some pinks
1: work your bucket magic	[Person 12 received 15 Pink from person 9.]
1: are you talking to me?	[Person 12 received 2 Blue from person 11.]
1: #2	
2: just throwing out some ideas	[Person 10 received 2 Blue from person 11.]
2: actually 20 reds and 40 blues for me and 4	[Person 9 received 1 Blue from person 11.]
1: oh ok, i was just wondering	
1: right	
2: 20 blues and 60 red for 1 and 3	
[Person 1 received 5 Pink from person 4.]	
[Person 4 received 5 Pink from person 4.]	
[Person 2 received 5 Pink from person 4.]	

The nature of the exchanges within this village is also very similar to the CSW economies that build four two-person economies into one eight-person group. In the six sessions of that experiment, the word "give" is used, on average, forty times per session, or once a period. Similarly, within our three four-person villages, "give" is used fifty-nine times to explicitly invoke the personal nature of giving in order to receive.

A flavor of the personal tone of the conversations is evident in the transcripts for day 15 for the *red-blue* and *pink-red* villages (table 13.4).

Notice all the unspoken trust in the *red-blue* village between the two producers, Persons 1 and 2, and the merchant, Person 4. Person 1, without prodding or discussion, begins the day by immediately moving 20 reds to Person 4, and, upon return from the merchant area, Person 4 moves 5 pinks to Person 1. There is no discussion of how to allocate the pinks that Person 4 brought back from the merchant area. Persons 1, 2, and 4 all contribute units for trade in the merchant world, and, upon Person 4's return, each promptly receives 5 pinks, despite the different amounts of blue and red that each contributed to the inter-village trade. Similarly, in the *pink-red* village, Person 11 solicits pinks, and, upon return, moves blue to Persons 9, 10, and 12. Again, there is no discussion of assurances of delivering blue, and there is no bargaining for the third good. It is simply given to the village members.

The discussion in the merchant area, however, differs quite markedly in tone and substance from the village conversations, as the transcript in day 15 illustrates (table 13.5).

First, note that the use of the word "we" is plainly missing in the transcript. There is no discussion of how *we* coordinate production and or how *we* allocate units among the villagers. The merchants in this chat room appear to be only self-interested: "i need 5 reds"; "i have 12. u want em all?"; "Person 11 can i get some pinks for 10 reds." They are also less personal as they attempt to get each other's attention ("yo 4"). Thus, the merchant area transcripts clearly lack idle chitchat and are filled with price quotes: "20/20" and "7 pinks for 7 blues." The merchants arrive at the area, make their deals, and return to their village community. The cooperation here, as for example in "20/20", is implicit—when *I* give 20 pinks, *you* will give 20 blues, and you and I will each be better off. There is no discussion of how the consumption of the goods generates earnings, nor is it visible how *we* will be better off from the transaction.

Production and Consumption

Figure 13.3 displays the average flow of goods between producers and merchants within each village, and between the villages over the last six days. The volume of trade is clearly largest between the *red-blue* and *pink-red* villages, and, not surprisingly, these are the two richest villages. The price of blue in terms of pink and the price of pink in terms of red are both 1; however,

TABLE 13.5
Chat-Room Transcript (C)

Merchant Area

***** Day 15 *****

3: any pinks for reds

8: yea i got it

3: how many

7: I need 5 redsa

8: i have 12..u want em all

7: ok

7: yes

3: pinks please

7: person 8 what do you want

3: can i get some pinks

8: red

12: 7 pinks for 7 blues?

8: 3..ill give u 12 pink for 12 red

4: got pinks for blues

11: 20 20

3: person 11 can i get some pinks for 10 reds

11: yo 4

7: who has reds

11: 20/20

12: 7 pinks for 7 blues 4?

8: person 3..i can give u 12 pinks

the price of blue in terms of red is about 2.6. We explain this in terms of the relative scarcity of blue. The *blue-pink* village never manages to specialize, but when its merchants begin to trade in the inter-village arena, the other villages are prepared to buy their goods, even at a relatively high price. The higher price creates an incentive for the *blue-pink* village to specialize to a larger extent. Yet, by the end of the experiment, the opportunities are not exploited.

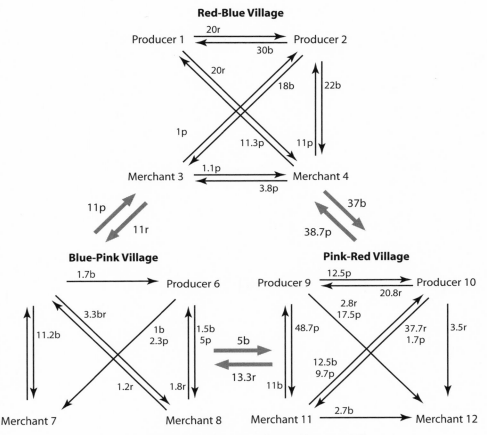

Figure 13.3. Flow of Exchanges within and between Villages

The distribution of earnings within the villages was far from homogeneous. In the *red-blue* village, a single producer managed to fully specialize and
consume red and blue at a level equivalent to the competitive equilibrium
without inter-village trade, *and* he was able to acquire pink on top of that.
Bringing in a total of $46, he was far and away the highest earner in the ex-
periment. Everyone else in his village also specialized, and though they allowed him a disproportionate share of the wealth, each earned at least $20.

Yet, in the *pink-red* village, a single merchant managed to monopolize
the inter-village trade for both his producers, and, as a result, he was the
highest earner in that village, making nearly $34. The other merchant in this
village was left to flounder essentially alone; the wealthy merchant always
gave him a share of the imported third good, but he failed to trade effectively

with the other members of his village. He was the second-lowest earner in the experiment.

The Third World (pun intended), the *blue-pink* village, failed both to specialize internally and to engage effectively in inter-village trade. No one in this village earned more than $14. Notably even though they witnessed trade in the other goods in quantities almost unheard of in their own village, neither merchant from this village ever asked members of the other villages how they created so many goods to trade. Thus, no information about specialization was ever exchanged, and the members of the *blue-pink* village wallowed in (relative) poverty.

In terms of total earnings, the benefits of all the trade in this experiment were distributed nearly evenly between producers and merchants, even though they were so unevenly distributed among the villages. If trade had occurred relative to their ability to produce, as described in the "Experimental Design" section above, the producers would have received two-thirds of the profits and the merchants would have received one-third. The explanation for the general equality is not, however, that the participants consciously smooth consumption. Rather, as they explore new ways to organize consumption and production, occasionally they hit on something that seems to work, and behavioral inertia sets in. Thus, patterns of production and consumption that are not the *best* possible arrangement for all involved, but are satisfactory, appear to become more or less set in stone. That is not to say, however, that all dynamism is wiped out of the economies; instead, change is slow-moving and

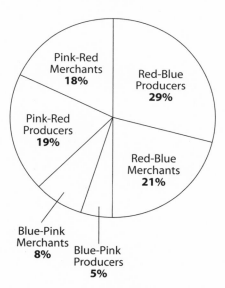

Figure 13.4. Distribution of Earnings

incremental. Figure 13.4 displays the percentage of total earnings captured by both producers and merchants in each village.

When designing this experiment, we focused on an issue raised by Thomas Sowell in his book *Race and Culture* and by Friedrich Hayek in *The Fatal Conceit*. Both argue that, historically, merchants tend to be abused or discriminated against when they receive their profits, because they do not produce visible value but only transfer goods across time and place. Their profits appear unjust to those who do not personally experience, and thus fail to consider, the important role transporters play in matching producers with consumers. As we debated whether merchants would take a margin for themselves, we considered the possibility that this might provoke the ire of their fellow villagers. Nowhere in the transcripts is this trend apparent, however; the personal nature of the exchange among the members of the individual villages appears to mitigate any such behavior. Merchants may take their cut, but the role they play in their villages is closer to that of a fellow villager returning from a successful hunt than that of a broker fulfilling orders for clients at the Chicago Mercantile Exchange.

Efficiency

Figure 13.5 indicates that the overall efficiency in our economy followed the pattern of growth of the economies reported in the CSW study. We measured

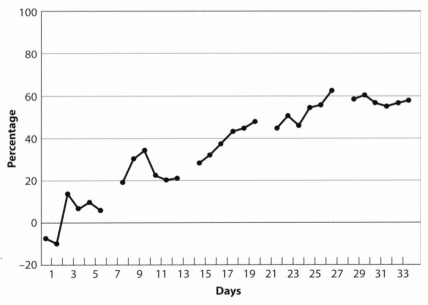

Figure 13.5. Overall Efficiency Relative to Complete Autarky

efficiency as the rate of profits realized by the villagers relative to the maximum profits calculated above, counting the profit with each individual at autarky as the baseline (0 percent). The growth over the first six days was fueled by the participants learning to optimize their autarkic decisions and specialize within the village. Following the first rest day, inter-village trade took off and the *red-blue* and *pink-red* villages became increasingly specialized. This growth continued and then stabilized after day 28, hovering around 58 percent of maximum possible gains from trade.

Figure 13.6 explains the source of the inefficiency, as it plots the efficiency of each village relative to a baseline in which each subject is fully specialized but without any inter-village trade. Hence, the near 100 percent efficiency of the *red-blue* village indicates that this village was fully specialized and extracting nearly all the possible gains from trading for pink from the other two villages. Every village started out with negative (below baseline) efficiency in the first six-period block; two started to break erratically into positive efficiency in

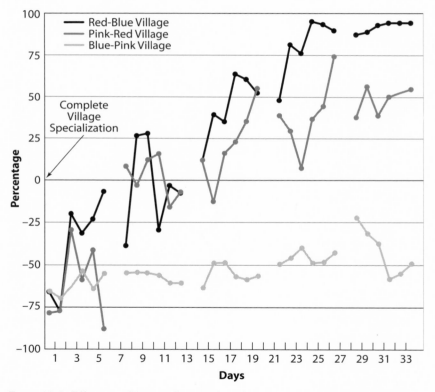

Figure 13.6. Efficiency of Inter-Village Trade Relative to Isolated but Fully Specialized Village

the second six-block period, and thereafter create wealth well above the levels achievable in the village-only economy. The stagnating −50 percent efficiency of the *blue-pink* village reflects the dearth of specialization in that village combined with a minimal level of inter-village trade (see figure 13.3, above). The low blue production in the *blue-pink* village limited the extent to which the *red-pink* village could increase its welfare.

Conclusion

Since the eighteenth century we have known that exchange and knowledge specialization create wealth and that the latter is, in turn, limited by the extent of markets. The mystery is how impersonal market exchange and long-distance trade grew out of highly localized forms of personal exchange that fostered only local specialization in small communities. An earlier experiment examined the development of exchange and specialization in an austere virtual village community; there the participants had to discover their comparative advantage in production, and also discover and create their own forms of exchange, with explicit instruction only about how they could produce and consume. Although exchange and specialization did emerge in that community, it did so to a highly variable degree, and only in the form of "giving" blue for red in strictly bilateral voluntary pairing by individuals. Many individuals failed to break out of home production and consumption, and there was no multilateral exchange that we associate with the emergence of markets.

This chapter reported on our first experimental session in which we introduced merchants into each of three villages, expanded the commodity space from two to three items, and allowed for the possible development of long-distance trade between three independent virtual villages. As in the earlier study, a chat room allowed messages to be exchanged locally within each village, and also locally in the area where merchants travel. Our primary findings can be summarized as follows:

- We observed no conversation among the merchants during any of the four rest days between the six-period production/consumption sequences in a session, although such consolidating discussion remained active in the villages.
- The personal and casual nature of the conversations among both the producers and merchants was evident; for example, in the *red-blue* village, the discussion was about how "*we* need extra blues for pinks." Hence, we saw the development of an obvious sense of community among the producers and merchants *within* a village.
- The tone of the conversation between producers and merchants sometimes revealed out-group and in-group attitudes among the village members.

- As in the initiating study, within our three four-person villages, the word "give" was used repeatedly, explicitly invoking the perceived personal nature of giving in order to receive.
- Trust in exchange within the villages was evident but unspoken.
- The use of the pronoun "we" was plainly missing in the merchant chat room. There was no mention of how "we" coordinate production and or how "we" allocate units among the villagers. The emphasis was on more plainly self-interested expressions: "I need 5 reds" and "i have 12. u want em all?"
- In terms of total earnings, the benefits of all trade in this experiment were distributed nearly evenly between producers and merchants. This was not because the participants were consciously smoothing out their consumption, but rather because each village explored new consumption and production arrangements and hit upon a plan that increased the village's earnings, and then stayed with what worked.
- The wealth amassed by the three villages was highly variable. The red-blue village achieved nearly 100 percent efficiency, indicating that it was fully specialized and extracted nearly all the possible gains from trading for pink from the other two villages. Every village started out with negative efficiency in the first six-period block; two villages began to break erratically into positive efficiency levels in the second six-block period, and thereafter created wealth well above the levels that could be achieved in a village-only economy. The blue-pink village stagnated at 50 percent *below* the baseline, indicating little specialization and a minimal level of inter-village trade.

Notes

1. See North (1981, 2005) for discussions of personal and impersonal exchange and how they relate to the development of long-distance trade and the process of economic change.

2. Based on the production functions described, it would seem that even merchants should earn 13 cents, but because their consumption preferences require that they consume a 2 to 1 ratio, one blue is wasted each turn.

References

Crockett, Sean, Vernon L. Smith, and Bart J. Wilson. 2006. "Exchange and Specialization as a Discovery Process," Working Paper, Interdisciplinary Center for Economic Science, George Mason University. Available at SSRN: http://ssrn.com/abstract=930078.

Hayek, Friedrich A. 1988. *The Fatal Conceit*. Chicago: University of Chicago Press.

North, Douglass C. 1981. *Structure and Change in Economic History*. New York: Norton.

————. 1995. *Understanding the Process of Economic Change*. Princeton, N.J.: Princeton University Press.

Smith, Adam. 1776. *Inquiry into the Nature and Causes of the Wealth of Nations*. New York: Knopf, 1991.

Sowell, Thomas. 1994. *Race and Culture*. New York: Basic Books.

Fourteen

Corporate Honesty and Business Education
A Behavioral Model

Herbert Gintis and Rakesh Khurana

Ever since the mid-1970s neoclassical economic theory has dominated business school thinking and teaching regarding the nature of human motivation. Although the theory has been valuable in understanding competitive product and financial markets, it employs an incorrect, *Homo economicus* model of human behavior that treats managers as selfish maximizers of personal wealth and power. The *Homo economicus* model implies that a firm's board of directors can best further stockholders' interests by selecting managerial personnel who focus almost exclusively on personal financial gain, and by inducing managers to act as agents of the stockholders by devising incentives that minimize differences between financial returns to stockholders and to the firm's leading managers. Moreover, while neoclassical financial theory, in the form of the efficient markets hypothesis, is generally an insightful portrayal of financial markets, neoclassical theory implies that a firm's stock price is the best overall measure of the firm's long-term value. A further implication is that managerial incentives should be tied to stock market performance, as this will best align the interests of managers and stockholders. This implication is invalid, however, when managers can manipulate information flows that influence short-term stock price movements.

Neoclassical economic theory thus fosters a corporate culture that ignores the personal rewards and social responsibilities associated with managing a modern enterprise, and encourages an ethic of greedy materialism where managers are expected to care only about personal financial reward, and where human character virtues such as honesty and decency are deployed only contingently in the interests of personal material reward. Of course, we cannot say that the corporate environment encouraged by business schools is responsible for the high level of managerial misconduct witnessed in recent years. We can say, however, that business education is deeply complicit, because it has failed to provide a consistent and accurate alternative to the *Homo economicus* model.

This is hardly the first time that this charge has been leveled against business education in America. Business schools have widely responded to criticism by adding a course on "business ethics" to the MBA curriculum. Although this is a welcome move, it cannot compensate for the generally incorrect and misleading characterization of human motivation, based on the neoclassical *Homo economicus* perspective, that is promulgated in courses on managerial behavior and corporate culture. This perspective must be directly attacked and replaced by a more accurate model of human motivation.

The *Homo economicus* model is known to be invalid because a wide range of experiments based on behavioral game theory have shown that honesty, integrity, intrinsic job satisfaction, and peer recognition are powerful motivators, and lead to better results for contracting parties than reliance on financial incentives alone. Many individuals place a high value on character virtues such as honesty and integrity for their own sake, and are more than willing to sacrifice material gain to maintain these virtues. We suggest that business schools should develop and teach a professional code of ethics similar to that promoted in law, education, science, and medicine; that the staffing of managerial positions be guided by considerations of moral character and ethical performance; and that a corporate culture based on character virtues, together with stockholder-managerial relationships predicated in part on reciprocity and mutual regard, could improve both the moral character of business and the profitability of corporate enterprise.

Ironically, stockholders themselves are among the victims of a corporate culture centered on greed, run by managers who no more care about real interests of current and future stockholders than they do about any other stakeholder in the operation of the enterprise.

Several scholars have recently voiced their concerns that the model of human behavior taught to business school students, a model whose roots can be directly traced to *Homo economicus*, has contributed to the diffusion of poor management practices. Ghoshal and Moran (1996), for example, argued that the prescriptions that flowed from the *Homo economicus* model have led managers and stockholders to favor purely *extrinsic* motivators, including pay and threats, over *intrinsic* motivators, such as excellence or quality, for mobilizing employee behavior. Such choices, they argue, actually undermine long-term organizational performance. Recently, other researchers have maintained that current business school teachings socialize students into an ethic of self-ishness and limited accountability (Ghoshal 2005; Khurana, Nohria, and Penrice 2005; Ferraro, Pfeffer, and Sutton 2005). Khurana and colleagues (2005), for example, have suggested that contemporary business education has legitimized the idea that managers should solely orient their action and behavior around maximizing shareholder value without regard for the effects of this approach on other employees, the community, or customers. Shiller (2005) wrote that much of the economic and finance theory that students learn in business schools is not only empirically flawed but impairs a student's ethical compass: "The view of the world that one gets in a modern business curriculum can lead to an ethical disconnect. The courses often encourage a view of human nature that does not inspire high-mindedness." Ghoshal's (2005) most recent criticisms were even more pointed; he argued that the contemporary business school curriculum and professors have directly contributed to the institutionalization of corrupt managerial practices, including obtaining excessive executive compensation and routinely taking ethical short-cuts. Current models of economic relationships teach students that managerial and employee contracts cannot be based to any significant degree on trust or trustworthiness. This view, in turn, sets in motion a self-reinforcing cycle in which students come to see opportunistic behavior, such as lying and cheating, as unavoidable and hence morally acceptable. Ghoshal writes:

> Business schools do not need to do a great deal more to help prevent future Enrons; they need only to stop doing a lot they currently do. They do not need to create new courses; they need to simply stop teaching some old ones. . . . [B]usiness school faculty need to own up to our own role in creating Enrons. Our theories and ideas have done much to strengthen the management practices that we are all now so loudly condemning. (2005:75)

Changing the dominant culture of an occupational group is, of course, a difficult task, but it is not impossible; it involves a joint deployment of new incentives and a novel cultural initiative. An example of a successful initiative of this type is the one begun in 1974 by the Hong Kong government, which had endured a culture of corruption going back hundreds of years. In 1974, the government pronounced yet another of its period anti-corruption campaigns, but this time the new Independent Commission Against Corruption (ICAC) was surprisingly successful (Hauk and Saez-Marti 2002). The reason for this success was that new incentives were supplemented by an intense anti-corruption socialization campaign in the schools. The result was that, in 1986, 75.1 percent of the 15–24 age group, the first generation to experience this socialization campaign, believed that corruption was a social problem, compared to only 54 percent of the 45–64 age group. Similarly, in 1977, 32 percent believed that "tipping" government employees to encourage them to perform their prescribed duties was illegitimate, compared to 72 percent nine years later, in 1986. In the same vein, in 1977, 38 percent believed that under-the-table kickbacks were legitimate in business dealings, but in 1986 only 7 percent held this view. Ten years further into the campaign, surveys in 1998 and 1999 revealed that about 85 percent of respondents in the 15–24 age group said they would not tolerate corruption either in government or business.

Of course, "socialization" could take the simple form of altering the expectations of self-regarding agents, but we do not think that this is what occurs when honesty and integrity are promoted in social affairs. Rather, such programs draw upon the innate desire of most citizens to lead a moral life and to have those who behave immorally punished. The evidence presented here supports this alternative, moral perspective.

1. Stockholder Resurgence: Agency Theory as a Tool for Controlling Managers

For the better part of the twentieth century, managers were treated in much the same manner as scientists, physicians, and other professionals. They were all given a long leash and expected to apply their craft in a professional and honorable manner. This changed in the last quarter of the century. Attacks on the professionalism of managers were broad-based, but with respect to business schools two factors stood out. First, the economic decline of the 1970s and the accompanying lack of competitiveness of U.S. firms in the face of foreign competition created a context for a thorough critique of American management, including American business education, and gave rise to competing prescriptions about what could be done to improve America's corporations. Second, takeover firms and activist institutional investors presented themselves forcefully as parties injured by incompetent corporate management, and as defenders of the individual shareholder. These newly empowered actors from outside the firm sought greater control over the corporation.

Using the poor corporate performance of the 1970s as their backdrop, these takeover artists successfully recast the image of corporate managers and executives not as wise corporate statesmen trying to adjudicate the competing concerns of various corporate constituents but rather as a self-enriching, unaccountable elite, primarily interested in taking advantage of weak shareholders to attain leisurely life-styles and exaggerated material gains. The idea that American society could depend on managers disciplining themselves to achieve the desired outcome where corporations contribute to the well-being of society was increasingly replaced by the view that managerial effectiveness could only be attained through the overarching authority of the market. This was an ironic turn of events that could only be truly appreciated by progressive reformers like Edwin Gay, Louis Brandeis, and Walter Lippmann, who saw enlightened managerial professionalism as society's best defense against a soulless and socially indifferent investor class.

The revisionism surrounding managerialism that occurred during the 1980s profoundly influenced business education.[1] It represented an institutional shift away from the basic managerial framework that had defined and informed business school education and animated the managerial professionalization project from the start. This revisionist approach eventually replaced the basic managerial framework with a new conception that is never fully specified but that broadly conceives of management as an agent of shareholders, where the corporation is a nexus of individual contracts with the primary purpose of maximizing shareholder value. The rhetoric and rationale undergirding this new conception of management, with its emphasis on shareholder value, was a variant that originated in the neoclassical model of *Homo economicus*, which was particularly influential in business education over the last two decades, especially in conceptualizing the role of management, and the definition and purpose of the corporation.

During this period, many leading business schools imported neoclassical economic theory practically whole cloth. There is little doubt that the results, on the whole, were extremely effective. Business school students learned modern decision theory, risk assessment, portfolio management, and efficient market theory, and were exposed to an array of analytical tools that are indispensable in modern corporate management.

Among these tools was the neoclassical *principal-agent model* that explained the economic relationship between principal individuals who engage agents to act on their behalf, but no contract fully specifies the services the agents will deliver; such principal-agent relationships are typified by an employer (the principal) and an employee (the agent), an owner (the principal) and a manager (the agent), a lender (the principal) and a borrower (the agent).

The principal-agent model is built on the notion that individuals have a personal *preference ordering* that they attempt to maximize, subject to the informational, institutional, and financial *constraints* they face, and relying on their personal *beliefs* concerning the value of various outcomes and how their actions can produce those outcomes. This is known, in the literature, as the *rational actor model.* However, the term "rational" is subject to flights of interpretive fancy, as discussed below, and so we prefer to call it the more neutral *preferences, beliefs, and constraints* (PBC) model.

The PBC model is the very heart of modern economic theory, and, in its absence, the explanatory success of modern economics would not have been possible. The principal-agent model flows logically from the PBC model, and suggests that the agents will maximize their personal payoffs, subject to whatever constraints are placed on them by the principals. The principal's task, therefore, is to devise a system of *incentives* that will lead a maximizing agent to behave in a manner favorable to the principal's own goals.

The principal-agent model was brought into business schools in the form of *agency theory* by Jensen and Meckling (1976) to directly particularize and elucidate the relationship between stockholders (the principals) and managers (the agents). It is important to note that the validity of the PBC model and of agency theory does not depend on the *Homo economicus* model of the individual. Indeed, Jensen and Meckling argued from the beginning that managers were not simply "money maximizers," and that agency theory applies even where agents included altruistic and social purposes among their personal objectives (Jensen and Meckling 1994). However, neither these nor other agency theorists provided an alternative to *Homo economicus* that was sufficiently concrete and serviceable as an alternative to the ubiquitous *Homo economicus* paradigm. Agency theory in business school practice, by default, imported *Homo economicus* and became a theory of how to motivate selfish money maximizers. We suggest that agency theory be retained but that the underlying model of human agency be thoroughly overhauled.

2. Neoclassical Economic Theory

Following the implementation of the Ford Foundation and Carnegie Foundation reforms during the 1950s and 1960s, in the 1970s neoclassical economics asserted itself as the dominant discipline in business schools. A quick check of major undergraduate or graduate microeconomics textbooks will convince skeptics of the ubiquity of the *Homo economicus* model: *never* does this model make any other assumption than that individual objectives are to maximize some combination of leisure and monetary reward. In this sense, neoclassical theory is a powerful attack on the legitimacy of managerial authority as it was constructed during the founding era of business schools and then revised in the post–World War II period. Proponents of the theory acknowledge the difficulties associated with the broad dispersal of stockholdings and the rise of management, which the *Homo economicus* model addresses by viewing the separation of ownership and control as a condition in which managers have no concern for shareholders' interests, except if compelled to do so by their formal contract of employment.

In a series of papers following Jensen and Meckling (1976), Fama and Jensen (1983), Jensen (1998), and others proposed a theory that the purpose of the corporation is to maximize shareholder value, and, because managers' interests differ from those of stockholders, a major practical challenge is monitoring these managers under conditions of broad stock dispersal. They argued that because their efforts are not easily observable, managers will perforce fail to work toward stockholder goals. The challenge, they concluded, is to create an "alignment of incentives" where managers' personal financial interests will closely correspond with those of owners. Although Jensen and Meckling often stressed the limitations of the *Homo economicus* model, their arguments were effectively drowned out in the mass adoption of traditional economic theory by the business school community.

Much of the discussion in those foundational papers focused on the mechanisms by which owners can effectively align these interests. The research by Jensen, Meckling, Fama, and their colleagues emphasizes three mechanisms: the monitoring of managerial performance, the provision of comprehensive economic incentives, and the promotion of an active market for corporate control. Monitoring managerial behavior involves complex accounting practices and the appointment of a professional board of directors whose members operate in the stockholders' interests by virtue of their need to maintain their personal reputations. The alignment of incentives involves remunerating management in the form of company stock and stock options, so that managers and owners face exactly the same incentives, and hence self-interested managers will maximize shareholder value as a by-product of maximizing their own material gain. The market for corporate control leads to stock prices reflecting firm fundamentals, and it ensures that poorly performing "insiders" will be threatened and ultimately replaced by efficiency- and profit-oriented "outsiders."

Neoclassical theory quickly created a unified approach to organizations and corporate governance in American business schools, catalyzing academic revolutions in corporate finance, organizational behavior, accounting, corporate governance, and the market for corporate control. Neoclassical theory also spearheaded a new paradigm in organizational theory. Unlike much of the earlier scholarship in business schools, many of the core ideas of neoclassical theory were derived not from inductive observation but through what seemed to be logically compelling or received wisdom. In the early 1970s, economists brought a high degree of theoretical rigor and analytically powerful modeling to business school education and research, the absence of which had concerned the Ford and Carnegie Foundations and haunted business education from the start. Drawing on the legitimacy of the economics discipline, business schools had the authority to redefine the character of managerial expertise.

The intimate bonding of disciplinary knowledge to its implications for professional identity is a fundamental postulate in the social sciences. Of course, people actively use knowledge to advance their influence and privilege. But as Michel Foucault has noted, the process is more profound, because the classification of knowledge, grounded in behavioral science, creates distinctions that are seen as *natural*, thus limiting our thinking by providing scripts and preconstituted habits of thought. The classifications we create to run our lives come to rule over us like alien beings; they trivialize culturally induced motivations and drive out of intellectual discourse notions such as honesty, trustworthiness, and fairness as meaningful aspects of managerial motivation.

Business economists often dismiss the arid mathematical formulations of their counterparts in university economic departments and favor a textured description of personal motivation and organizational behavior. However, the abstract formalization defines the conceptual basis for the business school rendition of microeconomic theory, and thus leaves no room for more than minor embellishments around the edges. As a result, even those who do not accept the validity of the neoclassical model, but simply use it as a structural backdrop to a more nuanced description of the business world, are in fact prisoners of it.

3. Neoclassical Theory: A Simple Model

Before offering a critique of the *Homo economicus* model, it is useful to present a simple scenario illustrating the major outlines of the model and its assumptions. Suppose the stockholders hire a manager to run the firm, and the firm's expected profits π are 100 times the manager's effort e, minus the salary s of the manager ($\pi = 100e - s$). We may think of e not simply as how hard the manager works but also how careful he is to choose profit-maximizing strategies versus strategies that aggrandize his position or favor other goals. Suppose the manager's effort varies between 0.1 and 1 ($0.1 \le e \le 1$), so that the stockholders' profit lies between 10 and 100 minus s. Suppose, further, for the sake of concreteness, that the subjective cost of effort to the manager is given by the schedule $c(e)$, where $c(0.1) = 0$, $c(0.2) = 1$, $c(0.3) = 2$, $c(0.4) = 4$, $c(0.5) = 6$, $c(0.6) = 8$, $c(0.7) = 10$, $c(0.8) = 12$, $c(0.9) = 17$, and $c(1) = 21$. No other effort levels can be chosen by the manager.

If the manager is self-regarding, caring only about his net salary, which is $s - c(e)$, then no matter how much the stockholders pay the manager, he will choose $e = 0.1$ to maximize his return, and the firm will make profits $\pi = 100 \times 0.1 - s = 10 - s$. If the manager must be paid at least $s_o = 1$ to take the job, the stockholders will set $s = s_o = 1$, ending up with profit $\pi = 10 - 1 = 9$.

However, agency theory suggests that the manager's incentives must be aligned with the owners, which is accomplished by paying the manager a salary that increases with the owners' profits. For instance, if the manager is paid 10 percent of the firm's profits, the manager earns $11e - c(e)$, which is maximized when $e = 0.3$, giving the manager the payoff $11(0.3) - c(0.3) = 1.3$, but the firm now earns $\pi = 89(0.3) = 26.7$. Paying the manager 46 percent of profits leads the manager to set $e = 1.0$, so his payoff is $46(1) - c(1) = 46 - 21 = 25$, and the firm earns $\pi = 54(1) = 54$. You can check that the owners maximize profits by paying the manager 24 percent of profits, leading the manager to set $e = 0.8$, so his payoff is $24(0.8) - c(0.8) = 7.2$, and the firm earns $\pi = 76(0.8) = 60.8$.

In the real world, of course, the information will never be even remotely so clean. Profits will be more like $\pi = ae + b$, where a may have a mean of 100, but will be a random variable whose value depends on many complex economic conditions, and b may similarly be a random variable whose value is generally much larger than ae. Since stockholders know only π, not its breakdown into managerial and non-managerial components, basing managerial compensation on π may not provide very effective incentives. Moreover, the manager is likely to be extremely risk-averse simply because his wealth is so much less than the firm's profits that variations in earnings may swamp his expected salary. Stockholders can deal with risk-aversion by diversifying their portfolios, but the manager cannot without seriously weakening the alignment of incentives with that of the stockholders. Finally, there is no single, unambiguous measure of the firm's health for which managerial behavior is responsible. By emphasizing one aspect (e.g., stock prices) in managerial compensation, stockholders are inducing self-regarding managers to sacrifice the firm's overall health on the altar of those aspects upon which his compensation is based.

If managers are purely financially motivated, there is no alternative to neoclassical theory's quest for aligning stockholder and managerial incentives, no matter how imperfect the alignment turns out. But why do we assume that managers are purely financially motivated? In other fields, where the subtlety and complexity of agent expertise render explicit contracts manifestly incapable of aligning the interests of agents with those of their principals, very different employment conditions are generally deployed. For instance, we do not consider physicians, research scientists, or college professors as agents who will maximize their income subject to whatever incentives and constraints they face. Rather, we assume that such agents have a *professional ethic* to which they subscribe, the maintenance of which is more valuable than material reward. Of course, appropriate safeguards must be in place to apprehend the occasional agent who violates professional norms, and professional norms themselves may not be perfectly aligned with social needs. Nevertheless, if the health system, scientific research, or higher education were run on the principle that the highest-level decision makers are motivated solely by material reward, and if the training of professionals in these fields stressed that there are no binding ethical rules, and that obeying laws is subject to cost-benefit calculation, there is little doubt that such systems would fail miserably.

One might suggest that the analogy of business leaders with their counterparts in health, science, and education is grossly overdrawn, if only because business is explicitly based on ruthless competition, whereas these other fields are generally service-oriented and have an altruistic element. However, a brief exposure to the competitive pressures in health care delivery or scientific research is generally sufficient to convince open-minded students of their highly competitive nature, and of the massive gains that can accrue to those who violate their professional ethics. One can become massively rich in any of these fields before being banished from the field, and even the latter contingency is relatively unlikely.

Perhaps it is not surprising that business schools would model their problems following economic theory, because both business and economic theory deal with "the economy" in a much more direct way than do the service areas where professionalism is duly recognized. But why are economic theories based on maximizing personal financial gain?

3. Why Does Economic Theory Model the Individual as *Homo economicus*?

Maximizing personal financial gain became an overriding principle in economic theory because, in anonymous market exchanges, individuals generally act precisely in that way. The great strength of traditional economic theory lies precisely in its capacity to explain competitive markets. For this reason, criticism of the *Homo economicus* model should not be seen as a wholesale rejection of economic theory as relevant for business education. For instance, neoclassical economic theory holds that, in the marketplace, the equilibrium price for a product or service is at the intersection of the supply and demand curves for the good. Indeed, it is easy to see that at any other point a self-regarding seller could gain by asking a higher price, or a self-regarding buyer could gain by offering a lower price. This situation was among the first to be approached experimentally, with *the neoclassical prediction virtually always receiving strong support* (Smith 1962; Smith 1982; Holt 1995). A particularly dramatic example is provided by Holt, Langan, and Villamil (1986) (reported by Charles Holt in Kagel and Roth 1995).

Holt et al.'s experiment established four "buyers" and four "sellers," and the good was a chip that the seller could redeem for $5.70 but that the buyer could redeem for $6.80 at the end of the game. In analyzing the game, we assume throughout that buyers and sellers maximize personal financial gain. In each of the first five rounds, each buyer was informed privately that he could redeem up to four chips, and eleven chips were distributed to sellers (three sellers were given three chips each, and the fourth was given two chips). Clearly, buyers are willing to pay up to $6.80 per chip for up to four chips each, and sellers are willing to sell their chip for any amount at or above $5.70. Total demand is thus sixteen for all prices at or below $6.80, and total supply is eleven chips at or above $5.70. Because there is an excess demand for chips at every price between $5.70 and $6.80, the only point of intersection of demand and supply curves is at the price $p=\$6.80$. The subjects in the game, however, have absolutely no knowledge of aggregate demand and supply, as each knew only his own supply of or demand for chips.

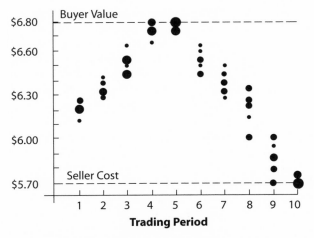

Figure 14.1. Simulating a Market Equilibrium: The Double Auction. The size of the circle is proportional to the number of trades that occurred at the stated price.

Here are the rules of the game: at any time a seller can call out an asking price for a chip, and a buyer can call out an offer price for a chip. This price remains "on the table" until either it is accepted by another player or a lower asking price, or higher offer price, is called out. When a deal is made, the result is recorded and that chip is removed from the game. As shown in figure 14.1, actual prices in the first period of play were about midway between $5.70 and $6.80. Over the succeeding four rounds the average price increased, until, in period 5, prices were very close to the equilibrium price predicted by neoclassical theory.

In period 6 and each of the succeeding four periods, buyers were given the right to redeem a total of eleven chips, and each seller was given four chips. In this new situation, it is clear (to us) that there is an excess supply of chips at each price between $5.70 and $6.80, so the only place supply and demand intersect is at $5.70. Whereas sellers who previously made a profit of about $1.10 per chip in each period must have been delighted with their additional supply, succeeding periods witnessed a steady fall in price. In the tenth period, the price was close to the neoclassical prediction, and now the buyers were earning about $1.10 per chip. A more remarkable vindication of the neoclassical model would be difficult to imagine.

4. Expanding the Rational Actor Model: Beliefs, Preferences, and Constraints

Perhaps because of the stunning success of the twin principles of rationality and self-interest in the analysis of market behavior, economists are wont to conflate the two concepts. The great economist Francis Ysidro Edgeworth (1881) proclaimed at the dawn of the neoclassical era that, "the first principle of economics is that every agent is actuated only by self-interest." Indeed, at least until recently, economists have considered that maximizing material gain is an essential aspect of rationality. In fact, however, the concepts are quite distinct.

The economist's rational actor model can be shown to apply over any domain where the individual has *consistent preferences*, in the sense that if he prefers A to B and B to C, then he prefers A to C. Given this assumption and a few technical conditions, it is easy to show that we can model the individual as *maximizing a preference function* subject to constraints. Moreover, if the individual's choice space is sufficiently rich, a few simple additional assumptions allow us to conclude that the individual has *consistent beliefs* concerning the probability of the states of affairs that determine his payoffs (Kreps 1990). Because "rationality" imposes limited requirements on preferences and beliefs, and because the term *rational* has so many diverse meanings in philosophy, the humanities, and the behavioral sciences, it is best to abandon the traditional term *rational actor model* in favor of the more descriptively accurate term *the beliefs, preferences and constraints (BPC) model*.

The Nobel laureate economist Gary Becker pioneered the extension of the BPC model to such non-market areas as drug addiction, racial discrimination, crime and punishment, the family, education, and fertility (Becker 1976; Becker, Tommasi, and Ierulli 1995). Although Becker and his coworkers virtually identified rationality with self-interest, clearly consistent preferences are compatible with the application of ethical values to individual choice. This point has been made with great clarity by Andreoni and Miller (2002), which we now describe.

In the Dictator Game, the experimenter gives a subject, the dictator, a certain amount of money with the instruction to give any desired portion of it to an anonymous second subject. The dictator keeps whatever he does not choose to give to the second subject. Obviously, a selfish dictator will give nothing to the recipient. But suppose the experimenter gives the dictator m dollars, and tells him that the price of giving some of this to the recipient is p, meaning that each point the recipient gets costs the dictator p points. For instance, if $p=4$, then it costs the dictator 4 points for each point he transfers to the recipient. The dictator's choices must then satisfy the *budget constraint* $\pi_s + p\pi_o = m$, where π_s is the amount the dictator keeps and π_o is the amount the recipient gets. The question, then, is simply, is there a preference function $u(\pi_s, \pi_o)$ that the dictator maximizes subject to the budget constraint $\pi_s + p\pi_o = m$? If so, then it is just as rational, from a behavioral standpoint, to behave charitably toward the recipient as it is to behave selfishly.

The economist Hal Varian (1982) showed that the following Generalized Axiom of Revealed Preference (GARP), which can be applied to analyze the Dictator Game, is sufficient to ensure not only rationality but that individuals have non-satiated, continuous, monotonic, and concave utility functions—the sort found in traditional consumer demand theory. To define GARP, suppose the individual purchases bundle x_i when prices are p_i. Consumption bundle x_s is said to be *directly revealed preferred* to bundle x_t if $p_s x_t \leq p_s x_s$; that is, x_t could have been purchased when x_s was purchased. We say x_s is *indirectly revealed preferred* to x_t if there is some sequence, $x_s = x_1, x_2, \ldots, x_k = x_t$, where x_i is directly revealed preferred to x_{i+1} for $i = 1, \ldots, k-1$. GARP then is the following condition: If x_s is indirectly revealed preferred to x_t, then $p_t x_t \leq p_t x_s$; that is, x_s does not cost less than x_t when x_s is purchased.

The general validity of GARP was shown in an experiment with the Dictator Game by Andreoni and Miller (2002), who worked with 176 economics students and had them play the game eight times each, with price (p) values of $=0.25, 0.33, 0.5, 1, 2, 3$, and 4, and dollar values of tokens (m) equaling $=40, 60, 75, 80$, and 100. The investigators found that only 18 of the 176 subjects violated GARP at least once, and, of these, only 4 were at all significant violations. In contrast, if choices were randomly generated, one would expect that 78–95 percent of subjects would violate GARP.

As to the degree of altruistic giving in this experiment, Andreoni and Miller found that 22.7 percent of subjects were perfectly selfish, 14.2 percent were perfectly egalitarian at all prices, and 6.2 percent always allocated all the money so as to maximize the total amount won, that is, when $p > 1$, they kept all the money, and when $p < 1$, they gave all the money to the recipient.

We conclude from this study that *we can treat altruistic preferences in a manner perfectly parallel to the way we treat money and private goods* in individual preference functions. There is thus nothing "irrational" about a BPC model in which individuals have altruistic preferences..

5. Strong Reciprocity in the Market for Managers

Strong reciprocity is the predisposition to cooperate in a group task, to respond to the cooperative behavior of others by maintaining or increasing one's level of cooperation, and to respond to the noncooperative behavior of others by punishing the offenders, even at personal cost, even when one cannot reasonably expect future personal gains to flow from such punishment. When other forms of punishment are not available, the strong reciprocator responds to defection with defection.

The strong reciprocator is thus neither the selfless altruist of utopian theory nor the selfish *Homo economicus* of neoclassical economics. The cooperative aspect of strong reciprocity is commonly known as *gift exchange* and the punitive side is *altruistic punishment*, the altruistic element coming from the fact that strong reciprocators promote cooperation even in groups with many selfish players, by making the incentive to cooperate compatible for such players (i.e., it costs less to cooperate than to defect and get punished).

Akerlof (1982) suggested that many puzzling facts about labor markets could be better understood by recognizing that employers sometimes pay their employees higher wages than necessary, in the expectation that workers will respond by providing greater effort than necessary. Of course, the relationship between employer and employee is of the same principal-agent type as that between stockholder and manager. Hence, if Akerlof is correct, trust and reciprocity would be expected to play a role in the market for managers. Fehr, Gächter, and Kirchsteiger (1997) performed an experiment that validated this *gift-exchange* model of the market for managers (see also Fehr and Gächter 1998).

The experimenters divided a group of 141 subjects (college students who had agreed to participate in order to earn money) into "owners" and "managers." The rules of the game are as follows: if an owner hires a manager who provides effort e and receives a salary s, his profit is $\pi = 100e - s$, as in section 3 above. The payoff to the manager is then $u = s - c(e)$, where $c(e)$ is the same cost of effort function as in section 3.

The sequence of actions is as follows. The owner first offers a "contract" specifying a salary s and a desired amount of effort e^*. A contract is made with the first manager who agrees to these terms, and an owner can make a contract (s, e^*) with at most one manager. The manager who agrees to these terms receives the salary s and supplies an effort level e, which *need not equal the contracted effort, e^**. In effect, there is no penalty if the manager does not keep his promise, so the manager can choose any effort level, $e \in [0.1, 1]$, with impunity. Although subjects may play this game several times with different partners, each owner-manager interaction is a one-shot (non-repeated) event. Moreover, the identity of the interacting partners is never revealed.

If managers maximize personal financial gain, they will choose the zero-cost effort level, $e = 0.1$, no matter what salary they are offered. Knowing this, owners will never pay more than the minimum necessary to get the manager to accept a contract, which is 1 (assuming only integral salary offers are permitted).[2] The manager will accept this offer and will set $e = 0.1$. Since $c(0.1) = 0$, the manager's payoff is $u = 1$. The owner's payoff is $\pi = 0.1 \times 100 - 1 = 9$.

In fact, however, this self-regarding outcome rarely occurred in this experiment. The average net payoff to managers was $u = 35$, and the more generous the owner's salary offer to the manager, the higher the effort provided. In effect, owners presumed the strong reciprocity predispositions of the managers—making quite generous salary offers and receiving greater effort—as a means to increase both their own and the manager's payoff, as depicted in figure 14.2. Fehr, Kirchsteiger and Riedl (1993, 1998) observed similar results.

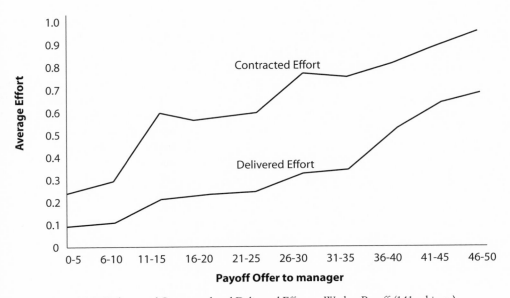

Figure 14.2. Relation of Contracted and Delivered Effort to Worker Payoff (141 subjects)
Source: Gehr, Gächter, and Kirchsteiger (1997)

Figure 14.2 also shows that, although most managers are strong reciprocators, at any salary rate a significant gap still exists between the amount of effort agreed upon and the amount actually delivered. This is not because a few "bad apples" are found among the managers but because only 26 percent of managers delivered the level of effort they promised! Strong reciprocators, therefore, are inclined to compromise their morality to some extent.

To see if owners are also strong reciprocators, the authors extended the game by allowing owners to respond reciprocally to the *actual effort choices* of their workers. At a cost of 1, an owner could *increase* or *decrease* his manager's payoff by 2.5. If owners were self-regarding, they would, of course, do neither, since they would not (knowingly) interact with the same worker a second time. However, 68 percent of the time, owners punished managers that did not fulfill their contracts, and 70 percent of the time, owners rewarded managers who overfulfilled their contracts. Indeed, owners rewarded 41 percent of managers who fulfilled their contracts *exactly*. Moreover, managers *expected* this behavior by their owners, as shown by the *significantly increased* effort levels when their bosses gained the power to punish and reward them. Underfulfilling contracts dropped from 83 percent to 26 percent of the exchanges, and overfulfilled contracts rose from 3 percent to 38 percent of the total. Finally, allowing owners to reward and punish led to a 40 percent increase in the net payoffs to all subjects, even when the payoff reductions resulting from owners punishing managers are taken into account.

A conclusion that can be drawn from this study is that the subjects who assume the role of "manager" conform to internalized standards of reciprocity, even when they are certain that no material repercussions occur from behaving in a self-regarding manner. Subjects who assume the role of owner also expect this behavior, and they are rewarded for acting accordingly. Finally, owners reward good behavior and punish bad behavior when they are allowed, and managers expect this behavior and adjust their own effort levels accordingly. In general, then, subjects follow an internalized norm not only because it is prudent or useful to do so, or because they will suffer some material loss if they do not, but because they desire to do so *for its own sake*.

6. Gift Exchange and Reputation with Incomplete Contracts

"An economic transaction," Abba Lerner (1972) wrote, "is a solved political problem. Economics has gained the title of queen of the social sciences by choosing solved political problems as its domain." Lerner's observation is correct, however, only insofar as economic transactions are indeed *solved* political problems. The assumption in the theory of competitive markets that yields this result is that *all economic transactions involved contractual agreements that are enforced by third parties (e.g., the judiciary) at no cost to the exchanging parties.* However, some of the most important economic transactions are characterized by the *absence of third-party enforcement*. Among these, of course, is the relationship between the stockholders and the managers of a large corporation.

As shown in the previous section, most experimental subjects are strong reciprocators who offer a high level of effort without the need for a complete, exogenously enforced contract, provided they trust their employers to reward their successes. In fact, an experiment conducted by Brown, Falk, and Falk (BFF), in 2004, clearly shows that when contracts are incomplete, owners prefer to establish long-term relationships with managers based on trust and gift exchange rather than on contractually aligning the incentives of managers with their own. The result in this experiment is a market for managers dominated by long-term trust relationships.

BFF used fifteen trading periods with 238 subjects and three treatments. The first treatment was the standard complete contract condition (C) in which managerial effort was contractually specified. The second treatment was an incomplete contract condition (ICF) with exactly the same characteristics, including costs and payoffs to owner and manager, as in section 5. In addition, however, managers received a payment of 5 points in each period that they were out of work. In both conditions, subjects had identification numbers that allowed long-term relationships to develop. The third treatment (ICR) was identical to ICF, except that long-term relationships were ruled out, and subjects received shuffled identification numbers in each experimental period. This treatment is identical to the gift-exchange model in section 5, except for the 5-point "unemployment compensation."

All contracts formally lasted only one period, so even long-term relationships had to be explicitly renewed in each period. If managers are self-regarding, it is easy to see that in the ICR treatment all managers will supply the lowest possible effort $e=1$, and owners will offer salary $s=5$. Each firm then has a profit of $10e-5=5$, and each manager has payoff $s-c(e)=5-c(0)=5$. This outcome will also occur in the last period of the ICF treatment, and hence, by backward induction, this outcome will hold in all periods. In the C treatment with self-regarding agents, it is easy to show that the owner will set $s=23$ and require $e=10$, so managers get $s-c(e)=23-c(10)=5$ and owners get $10e-s=100-23=77$ in each period. Managers are, in effect, indifferent in all cases to being employed or unemployed.

The actual results were, not surprisingly, at variance with the self-regarding preferences assumption. Figure 14.3 shows the path of salaries over the fifteen periods under the three treatments. The ICR condition reproduces the result of Section 5, salaries being consistently well above the self-regarding level of $s=5$. If the C condition were a two-sided double auction, we would expect salaries to converge to $s=23$. The ICF condition gives the highest salaries after the fourth period, validating the claim that under conditions of incomplete contracting, long-term trust and gift-exchange relationships will prevail, and the distribution of gains will be more equal between buyers and sellers.

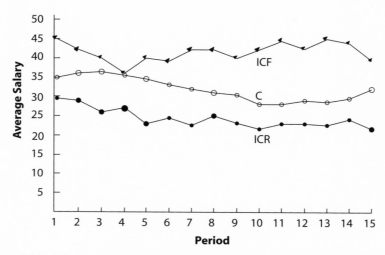

Figure 14.3. Salaries over Fifteen Periods. The C treatment is complete contracting; the ICF treatment is incomplete contracting; and the ICR treatment is incomplete contracting with no long-term relationships permitted.
Source: Brown et al. 2004

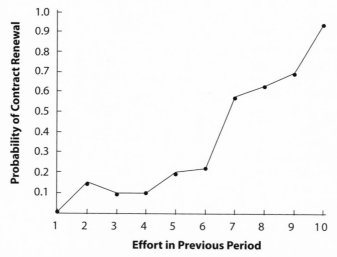

Figure 14.4 Trust and Reputation Provide Powerful Incentives in the ICF Condition

By paying high salaries in the ICF condition, owners were capable of effectively threatening their managers with dismissal (non-renewal of contract) if they were dissatisfied with managerial performance. Figure 14.4 shows that this threat was, in fact, often carried out. Managers with effort close to $e = 10$

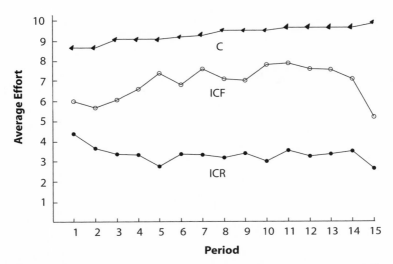

Figure 14.5. Managerial Effort over Fifteen Periods. The C treatment is complete contracting; the ICF treatment is incomplete contracting; and the ICR treatment is incomplete contracting with no long-term relationships permitted.

were not renewed only about 5 percent of the time, whereas managers with effort below $e = 7$ were rarely renewed.

Figure 14.5 shows that different contracting availabilities strongly affect the level of system productivity, as measured by average effort levels. Under complete contracting, effort levels quickly attain near-efficiency ($e = 10$) and remain there. Contingent renewal of long-term relationships achieves between 80–90 percent efficiency, with a significant end-game effect, as the threat of non-renewal is not very effective on the last few rounds. The gift-exchange treatment (ICR), while supporting effort levels considerably above the self-regarding level, is considerably less efficient that either of the others, although, predictably, it suffers a smaller end-game effect that the ICF condition.

An intriguing pattern emerging from this study is the interaction between gift exchange and threat in the owner-manager relationship. One might think that they would be mutually exclusive, as one cannot feel both charitable toward one's employer and threatened by him at the same time. Yet, many of us will recall from personal experience this seemingly ambiguous coexistence of good will and fear. In this study, the importance of gift exchange in the long-term relationship is exhibited by the fact that, even in the last two periods, where the threat of dismissal is weak or absent, effort levels are considerably above those of the pure gift-exchange condition. Thus, gift exchange appears to be stronger when accompanied by the capacity of the owner to harm, as though the fact that the owner has not exercised this capacity increases the willingness to supply effort.

7. The Character Virtues

Character virtues are ethically desirable behavioral regularities that individuals value for their own sake, but they also facilitate cooperation and enhance social efficiency. The character virtues include *honesty, trustworthiness, the keeping of promises,* and *fairness.* Unlike such other-regarding preferences as strong reciprocity and empathy, these virtues operate without concern for the individuals with whom one interacts. An individual is honest in his transactions because this is a desired state, not because he has any particular regard for those with whom he interacts. Of course, the sociopath *Homo economicus* is honest only when it serves his material interests to be so, whereas the rest of us are at times honest even when it is costly to us, and even when no one but us could possibly detect a breach.

Common sense indicates, as do the experiments described below, that honesty, fairness, and promise keeping are not absolutes. If the cost of virtue is sufficiently high, and the probability of detection of a breach of virtue sufficiently small, some, if not most, individuals will behave dishonestly. When we find that others are not virtuous in particular areas of their lives (e.g., marriage, paying taxes, obeying traffic rules, or accepting bribes), we are more likely to allow our own virtue to lapse. Finally, the more easily we can delude ourselves into inaccurately classifying a non-virtuous act as virtuous, the more likely we are to allow ourselves to carry out such an act.

One might be tempted to model honesty and other character virtues as *self-constituted constraints* on one's set of available actions in a game, but a more fruitful approach is to include the state of being virtuous in a certain way as an argument in one's preference function, to be traded off against other valuable objects of desire and personal goals. In this respect, the character virtues are in the same category as ethical and religious preferences, and are often considered subcategories of the latter.

The importance of character virtues, which appear to be confined to our species, is ignored in neoclassical economics, which prefers to ground social cooperation in *enlightened self-interest.* This predilection goes back to Bernard Mandeville's "private vices, public virtues" (1924 [1705]) and Adam Smith's "invisible hand" (1991 [1776]). The Darwinian struggle for existence may explain why the concept of virtue does not add to our understanding of animal behavior in general, but, by all available evidence, it is a central element in human behavior. The reasons for this are the subject of some speculation (Gintis 2003a, 2006b), but they come down to the plausible insight that human social life is so complex, and the rewards for pro-social behavior so distant and indistinct, that adherence to general rules of propriety, including the strict control over such deadly sins as anger, avarice, gluttony, and lust is individually fitness-enhancing (Simon 1990).

Numerous experiments indicate that most subjects are willing to sacrifice material reward to maintain a virtuous character, even when outsiders cannot penetrate the soul to investigate its contents. Sally (1995) undertook a meta-analysis of 137 experimental treatments, finding that face-to-face communication, where subjects can make verbal agreements and promises, was the strongest predictor of cooperation. Of course, face-to-face interaction violates anonymity and has other effects besides the ability to make promises. However, Bochet, Page, and Putterman (2006) as well as Brosig, Ockenfels, and Weimann (2003) report that only the ability to exchange verbal information accounts for the increased cooperation.

Consider, for example, consider the following Trust Game, first studied by Berg, Dickhaut, and McCabe (1995). Here, each subject is given a certain endowment, say, $10. Subjects are then randomly paired, and one subject in each pair, player A, is told he can transfer any number of dollars, from zero to ten, to his (anonymous) partner, player B, and keep the remainder. The amount transferred will be *tripled* by the experimenter and given to player B, who can then give any number of dollars back to player A (this amount is not tripled). When player A transfers a large amount, he is called *trusting*, and when player B returns a large sum to player A, he is called *trustworthy*. In the terminology of this paper, a trustworthy player B is a strong reciprocator, and a trusting player A is an individual who expects his partner to be a strong reciprocator.

Clearly, if all individuals have self-regarding preferences, and if each player A believes his partner has self-regarding preferences, then player A will give nothing to player B. But if player A believes that player B is inequality averse, he will transfer all $10 to player B, who will then have $40. To avoid inequality, player B will give $20 back to player A. A similar result would obtain if player A believes that player B is a strong reciprocator. On the other hand, if player A is altruistic, he may transfer something to player B in the belief that the money is worth more to player B (since it is tripled) than it is to himself, even if player A does not expect anything back. It follows that several distinct motivations can lead to a positive transfer of money from A to B and then back to A.

Berg and colleagues (1995) found that, on average, $5.16 was transferred from A to B, and, on average, $4.66 was transferred back from B to A. Furthermore, when the experimenters revealed this result to the subjects and had them play the game a second time, $5.36 was transferred from A to B, and $6.46 was transferred back from B to A. Both sets of games exhibited a great deal of variability in that some A's transferred everything, some transferred nothing, and some B's more than fully repaid their partner while some gave nothing back.

As noted above, the Trust Game does not really draw on character virtues, as Player B made no explicit promises to Player A. To add character virtue to the mix, Ben-Ner and Putterman (2005) first implemented a standard Trust Game with 194 subjects, all college students. Each player was given $10, and Player A was allowed to pass as much of his $10 to Player B, after the experimenter first tripled the amount. Player B was then permitted to return to Player A whatever fraction of his newly acquired money he pleased. Just as reported above, even without communication, most subjects exhibited some level of cooperative behavior, and, indeed, even a strictly selfish A stood to gain from passing all his money to B. The experimenters then ran the game a second time, now allowing newly paired partners to communicate proposals and counterproposals via their computer screens prior to playing the one-shot, anonymous Trust Game. The result was much more cooperation; the average amount B sent to A increased from $5.50 to $8.10, even though there was no way that B could be punished for breaking a promise. A's were justified in trusting, because 74 percent of B's actually kept their promises, and most of the others returned at least something to A.

A striking finding was that when B is allowed to make a promise, the modal exchange was that in which A sent all $10 to Player B, who then kept $10 of the $30 given to him by the experimenter, and returned $20 to A, so that both parties ended up with $20.

8. The Role of Community: Altruistic Third-Party Reward and Punishment

Pro-social behavior in human society occurs not only because those directly helped and harmed by an individual's actions are likely to reciprocate in kind, but also because general *social norms* foster pro-social behavior and many people are willing to bestow favors on someone who conforms to social norms but punish someone who does not, even if they are not personally helped or hurt by the individual's actions. In everyday life, third parties who are not the beneficiaries of an individual's pro-social act, will help the individual and his family in times of need, will preferentially trade favors with the individual, and otherwise will reward the individual in ways that are not costly but nonetheless greatly benefit the cooperator. Similarly, individuals who have not been personally harmed by the selfish behavior of another individual will refuse to aid the offender, even when it is not costly to do so, and will shun the offender and approve of the offender's ostracism from beneficial group activities.

Community culture means the sharing of cultural norms among community members and the willingness of community members to reward or punish community members who adhere to or violate these norms, even if they are not personally the object of the resulting beneficence or harm. Professional standards are the cultural norms of occupational groups and, once established, lead members to reward or punish one another informally, without the need for costly litigation, legislation, and incarceration. By abjuring professional standards for managers in favor of a culture of greed, business schools that have promoted the neoclassical model of stockholder-manager relations probably have undercut the culture of professional honor among managerial personnel to such a degree that informal third-party punishment and reward is rare, thus contributing to morally deficient behavior by contemporary managerial personnel.

Strong experimental evidence supports this line of reasoning. Although self-regarding actors will never engage in third party reward or punishment if such behavior is at all costly, experimental subjects routinely punish and reward others who have affected the welfare of other group members but not themselves. Fehr and Fischbacher (2004) addressed this issue with an experiment using a series of third-party punishment games, for example, the Prisoner's Dilemma and the Dictator Game.[3] The investigators implemented four experimental treatments, and, in each, subjects were grouped into threes. In each group, in stage 1, subject A played a Prisoner's Dilemma or Dictator Game with subject B as the recipient, and subject C was an outsider whose payoff was not affected by A's decision. In stage 2, subject C was endowed with 50 points and allowed to deduct points from subject A, such that for every 3 points deducted from A's score, C lost 1 point. In the first treatment (TP-DG) the game was the Dictator Game, in which A was endowed with 100 points and could give 0, 10, 20, 30, 40, or 50 points to B, who had no endowment.

The second treatment (TP-PD) was the same, except that the game was the Prisoner's Dilemma. Subjects A and B were each endowed with 10 points, and each could either keep the 10 points or transfer the them to the other subject, in which case the experimenter tripled the number of points. If both cooperated, each earned 30 points; if both defected, each earned 10 points; and if one cooperated and one defected, the cooperator earned 0 points and the defector earned 40. In the second stage, C was given an endowment of 40 points and was allowed to deduct points from A or B or both, just as in the TP-DG treatment.

To compare the relative strengths of second- and third-party punishment in the Dictator Game, the experimenters implemented a third treatment, S&P-DG. Here, subjects were randomly assigned to player A and player B, and A-B pairs were randomly formed. In the first stage, each A was endowed with 100 points and each B with 0 points, and the A's played the Dictator Game as before. In the second stage of each treatment, each player was given an additional 50 points, and the B players were permitted to deduct points from A players on the same terms as in the first two treatments. S&P-DG also had two conditions: in the S condition, a B player could only punish his *own* dictator, but in the T condition, a B player could only punish an A player *from another pair*, to which he was randomly assigned by the experimenters. In the T condition, each B player was informed of the behavior of the player A to which he was assigned.

To compare the relative strengths of second- and third-party punishment in the Prisoner's Dilemma, the experimenters implemented a fourth treatment, S&P-PG. This was similar to the S&P-DG treatment, except that now, in the S condition, an A-B pair could only punish each other, whereas in the T condition, each agent could punish only a randomly assigned subject from another pair.[4]

Because subjects in the first two treatments were randomly assigned to positions A, B, and C, the obvious fairness norm is that all should have equal payoffs (an "equality norm"). For instance, if A gave 50 points to B, and C deducted no points from A, each subject would end up with 50 points. In the Dictator Game treatment (TP-DG), 60 percent of third parties (C's) punished dictators (A's) who gave less than 50 percent of the endowment to recipients (B's). Ordinary least squares regression showed that for every point an A kept for himself above the 50-50 split, he was punished an average of 0.28 points by C's, leading to a total punishment of $3 \times 0.28 = 0.84$ points. Thus, a dictator who kept the whole 100 points would have $0.84 \times 50 = 42$ points deducted by C's, leaving a meager gain of 8 points over equal sharing.

The results for the Prisoner's Dilemma treatment (TP-PD) were similar, but with a twist. If one partner in the A-B pair defected and the other cooperated, the defector would have on average of 10.05 points deducted by C's, but if both defected, the punished player lost only an average of 1.75 points. This shows that third parties (C's) care not only about the intentions of defectors, but also about how much harm they caused or how unfair they turned out to be. Overall, 45.8 percent of third parties punished defectors whose partners cooperated, whereas only 20.8 percent of third parties punished defectors whose partners defected.

Turning to the third treatment (T&SP-DG), second-party sanctions of selfish dictators are found to be considerably stronger than third-party sanctions, although both were highly significant. In the first condition, where recipients could punish their own dictators, they imposed an average deduction of 1.36 points for each point the dictator kept above the 50-50 split, whereas they imposed a deduction of only 0.62 points per point kept by third-party dictators. In the final treatment (T&SP-PD), defectors were severely punished by both second and third parties, but second-party punishment was again found to be much more severe than third; cooperating subjects deducted on average of 8.4 points from a defecting partner, but only 3.09 points from a defecting third party.

This study confirmed the general principle that punishing norm violators is very common, but not universal, and individuals are prone to be more harsh in punishing those who hurt them personally, as opposed to violating a social norm that hurts others.

9. Re-professionalizing Business Education

Although neoclassical economic theory is central to contemporary management education, research and insights from game theory and behavioral economics indicate that many of the theory's core assumptions are flawed. The widespread reliance on the *Homo economicus* model cannot be said to have definitely *caused* the serious problems of observed managerial malfeasance, but it may well have, and, in any case, it surely does not act as a healthy influence on managerial morality. Students have learned this flawed model, and in their capacity as corporate managers they doubtless act daily in conformance with it. This, in turn, may have contributed to the weakening of socially functional values and norms, including honesty, integrity, self-restraint, reciprocity, and fairness, to the detriment of the health of the enterprise. Simultaneously, the *Homo economicus* perspective has legitimized, or at least not de-legitimized, such behaviors as material greed and guileful optimization decisions. This model, as noted, has become highly institutionalized in business education, but, fortunately, a significant potential exists for moving away from it, and thus this chapter can end on a more optimistic note for the future of business education.

Recently business school scholars, such as Jensen (2006), have thought about how to introduce character virtue as a central element in the creation of economic value. Jensen recently proposed a framework for value creation that resonates with one of the key character virtues associated with professionalism. He argues that *integrity* is a necessary condition for maximizing value, because an economic entity has integrity when it is "whole and complete and stable." Jensen then defines the behavioral characteristics of economic entities (agents inside organizations, groups inside organizations, and organizations themselves) that operate with integrity: "nothing is hidden, no deception, no untruths, no violation of contracts or property rights, etc. For those who chose not to play by these rules, integrity requires you to make this clear to all others." This framework is then applied to financial and capital markets and suggests that many of the recent corporate scandals can be traced to the institutions, like investment banks, analysts, and auditors, purposefully setting aside the integrity imperative.

Recent research has demonstrated that business schools have a rich treasure of wisdom, idealism, and vision on which to draw to move away from the dominant neoclassical model (Khurana, Nohria, and Penrice 2005). The obligation to draw away is an inheritance rooted in the mission of the larger university as social institution charged with advancing the public good, and in the original professionalizing mission of the university-based business school charter. However, the notion that those who lead and manage our society's major private economic institutions might provide, or be responsible for providing, a public good is quite foreign to our current way of thinking about management. Yet, as alluded to at the start of this chapter, this idea was often voiced by those who led American business schools in the early decades of their existence. For example, in a speech titled "The Social Significance of Business" that he delivered at Stanford University's School of Business shortly after its founding in 1925 (subsequently published as an article in *Harvard Business Review* of the same title), Wallace B. Donham, the second dean of the Harvard Business School, declared that the "development, strengthening, and multiplication of socially minded businessmen is the central problem of business." As Donham went on to say:

> The socializing of industry from within on a higher ethical plane, not socialism nor communism, not government operation nor the exercise of the police power, but rather the development from within the business group of effective social control of those mechanisms which have been placed in the hands of the race through all the recent extraordinary revolutionizing of material things, is greatly needed. The business group largely controls these mechanisms and is therefore in a strategic position to solve these problems. Our objective, therefore, should be the multiplication of men who will handle their current business problems in socially constructive ways.

Business schools did not begin with neoclassical theory. The founders of these schools never envisioned that the sole purpose of the corporation was to serve only one master, the shareholder. Nor could they have ever imagined that the model for training students to have a worldview would turn out to draw managers as self-interested agents with no consideration of any other values or imperatives but their own wallets. Those of us who study society, coach soccer, help our neighbors, raise children, and look within ourselves, all know that such a view of human behavior is not true; more important, if such a vision were fully realized, it would not offer the foundation for creating a sustainable society. Business schools need to recover what the sociologist Ferdinand Tönnies called *Gemeinschaft*, or *community*. To this end, business school faculties and deans have an institutional responsibility to socialize students to a model of behavior that inspires them to respect other institutions in society, especially the basic units of family and community, and to inspire students to accept the responsibilities and obligations that come with occupying society's most powerful positions.

Notes

1. Business schools were largely insulated from the student protests that deeply affect most of the university, partly because of the conservative attitudes and beliefs of the students and faculty drawn to business.

2. This is because the experimenters created more managers than owners, thus ensuring an excess supply of managers.

3. The Prisoner's Dilemma is a two-player game where both players gain if they cooperate, but each player has an incentive to defect no matter what the other player does. Thus, self-regarding players never cooperate in the Prisoner's Dilemma, although experimental evidence indicates that most experimental subjects actively prefer to cooperate, provided their partners do so as well.

4. It is worth repeating that the experimenters never use value-laden terms such as "punish" but, instead, use neutral terms such as "deduct points."

References

Akerlof, George A. 1982. "Labor Contracts as Partial Gift Exchange." *Quarterly Journal of Economics* 97, 4 (November): 543–569.

Andreoni, James, and John H. Miller. 2002. "Giving according to GARP: An Experimental Test of the Consistency of Preferences for Altruism." *Econometrica* 70, 2:737–753.

Becker, Gary. 1976. *The Economic Approach to Human Behavior.* Chicago: University of Chicago Press.

Becker, Gary, Mariano Tommasi, and Kathryn Ierulli. 1995. *The New Economics of Human Behaviour.* Cambridge: Cambridge University Press.

Ben-Ner, Avner, and Louis Putterman. 2005. "Trust versus Contracting: Who Takes the Plunge, and When?" Minneapolis: University of Minnesota.

Berg, Joyce, John Dickhaut, and Kevin McCabe. 1995. "Trust, Reciprocity, and Social History." *Games and Economic Behavior* 10:122–142.

Bochet, Olivier, Talbot Page, and Louis Putterman. 2006. "Communication and Punishment in Voluntary Contribution Experiments." *Journal of Economic Behavior and Organization* 60, 1 (May): 11–26.

Brosig, J., A. Ockenfels, and J. Weimann. 2003. "The Effect of Communication Media on Cooperation." *German Economic Review* 4:217–242.

Brown, Martin, Armin Falk, and Ernst Fehr. "Relational Contracts and the Nature of Market Interactions." *Econometrica* 72, 3 (May): 747–780.

Edgeworth, Francis Ysidro. 1881. *Mathematical Psychics: An Essay on the Application of Mathematics to the Moral Sciences.* London: Kegan Paul.

Fama, Eugene F., and Michael Jensen. 1983. "Separation of Ownership and Control." *Journal of Law and Economics* 26:301–326.

Fehr, Ernst, and Simon Gächter. 1998. "How Effective are Trust- and Reciprocity-based Incentives?" in Louis Putterman and Avner Ben-Ner, eds., *Economics, Values and Organizations*, 337–363. New York: Cambridge University Press.

Fehr, Ernst, and Urs Fischbacher. 2004. "Third Party Punishment and Social Norms." *Evolution and Human Behavior* 25:63–87.

Fehr, Ernst, Georg Kirchsteiger, and Arno Riedl. 1993. "Does Fairness Prevent Market Clearing?" *Quarterly Journal of Economics* 108, 2:437–459.

———. 1998. "Gift Exchange and Reciprocity in Competitive Experimental Markets." *European Economic Review* 42, 1:1–34.

Fehr, Ernst, Simon Gächter, and Georg Kirchsteiger. 1997. "Reciprocity as a Contract Enforcement Device: Experimental Evidence." *Econometrica* 65, 4 (July): 833–860.

Ferraro, Fabrizio, Jeffrey Pfeffer, and Robert I Sutton. 2005. "Economics Language and Assumptions: How Theories Become Self-fulfilling." *Academy of Management Review* 30, 1:8–24.

Ghoshal, Sumantra. 2005. "Bad Management Theories Are Destroying Good Management Practices." *Academy of Management Learning and Education* 4, 1:75–91.

Ghoshal, Sumantra, and Peter Moran. 1996. "Bad for Practice: A Critique of the Transaction Cost Theory." *Academy of Management Review* 21:13–47.

Gintis, Herbert. 2003. "The Hitchhiker's Guide to Altruism: Genes, Culture, and the Internalization of Norms." *Journal of Theoretical Biology* 220, 4:407–418.

———. 2007. "A Framework for the Unification of the Behavioral Sciences." *Behavioral and Brain Sciences* 30, 1–16.

Hauk, Esther, and Maria Saez-Marti. 2002. "On the Cultural Transmission of Corruption." *Journal of Economic Theory* 107, 2:311–335.

Holt, Charles A. 1995. *Industrial Organization: A Survey of Laboratory Research.* Princeton, N.J.: Princeton University Press.

Holt, Charles A., Loren Langan, and Anne Villamil. 1986. "Market Power in an Oral Double Auction." *Economic Inquiry* 24:107–123.

Jensen, Michael C.. 1998. *Foundations of Organizational Strategy.* Cambridge, Mass.: Harvard University Press.

———. 2006. "Putting Integrity into Finance Theory and Practice: A Positive Approach." Paper presented at the Harvard Business School, Harvard University, Cambridge, Mass., October 18.

Jensen, Michael C., and W. H. Meckling. 1994. "The Nature of Man." *Journal of Applied Corporate Finance* 7, 2 (summer): 4–19.

———. 1976. "Theory of the Firm: Managerial Behavior, Agency Costs and Ownership Structure." *Journal of Financial Economics* 3:305–360.

Kagel, J. H., and A. E. Roth. 1995. *Handbook of Experimental Economics.* Princeton, N.J.: Princeton University Press.

Khurana, Rakesh, Nitin Nohria, and Daniel Penrice. 2005. "Management as a Profession." In Jay W. Lorch, Leslie Berlowitz, and Andy Zelleke, eds., *Restoring Trust in American Business.* Cambridge: MIT Press.

Kreps, David M. 1990. *A Course in Microeconomic Theory.* Princeton, N.J.: Princeton University Press.

Lerner, Abba. 1972. "The Economics and Politics of Consumer Sovereignty." *American Economic Review* 62, 2 (May): 258–266.

Mandeville, Bernard. 1924 [1705]. *The Fable of the Bees: Private Vices, Publick Benefits.* Oxford: Clarendon.

Sally, David. 1995. "Conversation and Cooperation in Social Dilemmas." *Rationality and Society* 7, 1 (January): 58–92.

Shiller, Robert J. 2005. "How Wall Street Learns to Look the Other Way." *New York Times,* February 8.

Simon, Herbert. 1990. "A Mechanism for Social Selection and Successful Altruism." *Science* 250:1665–1668.

Smith, Adam. 1991 [1776]. *Inquiry into the Nature and Causes of the Wealth of Nations.* New York: Knopf.

———. 2000 [1759]. *The Theory of Moral Sentiments.* New York: Prometheus.

Smith, Vernon. 1962. "An Experimental Study of Competitive Market Behavior." *Journal of Political Economy* 70:111–137.

———. 1982. "Microeconomic Systems as an Experimental Science." *American Economic Review* 72 (December): 923–955.

Varian, Hal R. 1982. "The Nonparametric Approach to Demand Analysis." *Econometrica* 50:945–972.

Fifteen

What's a Business For?

Charles Handy

Could capitalists bring down capitalism, wondered the *New York Times* in the wake of the series of corporate scandals in both America and Europe in recent years? They concluded that a few rotten apples would not contaminate the whole orchard, that the markets would eventually sort out the good from the bad and that, in due time, the world would go on much as before.

Not everyone was so complacent. Markets rely on rules and laws, but those rules and laws in their turn depend upon truth and trust. Conceal the truth or erode that trust and the game becomes so unreliable that no one will want to play. The markets will empty then and prices collapse, as ordinary people find other places to put their money, into their houses, maybe, or under their beds. The great virtue of capitalism, that it provides a way for the savings of society to be transformed into the creation of wealth, will have been eroded and we shall all be the poorer. Either that, or we shall have increasingly to rely on government for the creation of our wealth, something that they have always been conspicuously bad at doing.

In the recent scandals truth seemed to be too easily sacrificed to expediency and the need, as the businesses saw it, to reassure the markets that profits were on target. John May, a stock analyst for a US investor service, pointed out that the pro forma earnings announcements by the NASDAQ 100 companies overstated their profits by $100 billion in the first nine months of 2001 as compared with their audited accounts when these finally appeared. Even those audited accounts, it now seems, were able to make things seem better than they really were.

Trust, too, is a fragile thing. Like a pane of glass, once shattered it can never be quite the same again. To many it has recently seemed that corporate executives were no longer running their companies for the benefit of the consumers, or even of their shareholders and employees, but for their personal ambitions and financial gains. A Gallup poll early in 2002 found that only 18 percent of Americans thought that corporations looked after their shareholders a great deal and 90 percent felt that those running the corporations could not be trusted to look after the interests of their employees. 43 percent, in fact, believed that senior executives were only in it for themselves. In Britain, that figure in another poll was 95 percent.

What has gone wrong? It is tempting to blame the people at the top. Keynes once said that Capitalism is the astounding belief that the most wickedest of men will do the most wickedest of things for the greatest good of everyone. Keynes was exaggerating. Personal greed perhaps, a lack of sufficient scrutiny of the company's affairs, an insensitivity or an indifference to public opinion, these charges could be levelled against some corporate leaders, but few, thankfully, are guilty of deliberate fraud or wickedness. At worst they were only playing the game according to the rules as they understood them. It is these rules or, more particularly, the new goal posts that have distorted capitalism.

In the current Anglo-American version of stock market capitalism, the criterion of success now is shareholder value, as expressed by the share price. The directors of a company may insist that the purpose of the business is more than the share price but what is measured is too often what counts. There are many ways of influencing the share price, of which increasing productivity and long-term profitability is only one. Cutting or postponing expenditures that are geared to the future rather than the present will push up the profits immediately even if it imperils the longer term. Buying and selling other businesses is another favoured strategy It is a far quicker way to boost your balance sheet and your numbers than relying on organic growth, and, for those at the top it can be much more interesting. The fact that most mergers and acquisitions do not, in the end, add value has not discouraged others from trying. One result is an inevitable shortening of horizons. Paul Kennedy is not alone in believing that companies are mortgaging their futures in return for a higher stock price in the present, but he may be optimistic in sensing the coming end of the obsession with shareholder value.

The stock option, that new favourite child of stock market capitalism, must share a large part of the blame. Whereas in 1980 only about 2 percent of executive pay was tied to share options it is now thought to be over 60 percent in the United States. These executives, not unnaturally, want to realise their options as soon as they can, rather than relying on their successors to deliver. The stock option has also acquired a new popularity in Europe as more and more companies go public. To many Europeans, however, hugely undervalued stock options are just another way of allowing executives to steal from their companies and their shareholders. Europeans raise their eyebrows, sometimes in jealousy but more often in outrage, at the levels of executive remuneration under stock market capitalism. Reports that CEOs in America earn over four hundred times the earnings of their lowest workers make a mockery of Plato's ideal, in what was then, admittedly, a smaller and simpler world, that no person should be worth more than four times another. Why, some wonder, should business executives be rewarded so much better financially than those who serve their societies in all the other professions? The suspicion, right or wrong, that business takes care of itself before it cares for others only fuels the latent distrust.

Europeans look at America with a mix of envy and trepidation. They admire the dynamism, the entrepreneurial energy and the insistence on everyone's right to decide their own life. They approve of the fact that America has never used its power to remove sovereignty from other nations or to occupy their territories, as the European empire builders did, preferring instead to use economics as the foundation of its international influence. They worry now, however, that the flaws in the American model of capitalism are contagious and will infect their own economies, as they watch their own stock markets follow Wall Street downhill.

This is not just a question of dubious individual ethics or of some rogue firms fudging the odd billion. The concern is that the whole business culture of America may have become distorted. This was the culture that enraptured America for a generation—a culture which argued that the market was king, that the shareholder always had priority that business was the key engine of progress and that, as such, its needs should prevail in any policy decisions. It was a heady doctrine, one that simplified life with its dogma of the bottom line. and in the Thatcher years it infected Britain. It certainly revived the entrepreneurial spirit in that country. But it also contributed to a decline in civic society and to an erosion in the attention and money paid to the non-business sectors of health, education and transport a neglect whose effects have now returned to haunt the successor government.

Continental Europe was always less enthralled by the American model. It left out many of the things that Europeans take for granted—free health care for all, housing for the disadvantaged, quality and free education for all and a guarantee of reasonable living standards in old age, sickness or unemployment. These things that Europeans see as the benefits of citizenship are conspicuously lacking in the land of the free, because they are seen as the price that must be paid for liberty and opportunity. Nevertheless, the accusations from across the Atlantic of a lack of dynamism in Europe, of sclerotic economies bogged down in regulations and of lacklustre management were beginning to hurt, and the culture of stock market capitalism was beginning to take hold. Now, after a series of Europe's own examples of skulduggery at the top and a couple of high profile corporate collapses due to overambitious acquisition policies, many are not so sure.

With hindsight, in the boom years of the nineties America had often been creating value where none existed, valuing companies at up to 64 times earnings. Her consumers' level of indebtedness may well be unsustainable along with the country's debts to foreigners. Add to this the erosion of confidence in the balance sheets and in the corporate governance of some of the country's largest corporations and the whole system of channelling the savings of citizens into fruitful investment begins to look in question. That is the contagion that Europe fears.

Capitalist fundamentalism may have lost its sheen, but the urgent question now is how best to retain the energy produced by the old model without its flaws. Better and tougher regulation would help, as would a clearer separation of auditing from consulting. Corporate governance will now surely be taken more seriously by all concerned, with responsibilities more clearly defined, penalties spelt out and watchdogs appointed. But these will be plasters on an open sore, they will not affect the disease that lies at the core of the business culture which centres around the question: Who and what is a business for? It is a question to which the answer once seemed clear, but which now needs rethinking in a time when ownership has been replaced by investment and when the assets of the business are increasingly to be found in its people not in its buildings and machinery. Are there things that the American business model can learn from Europe, just as there have been valuable lessons that the Europeans have absorbed from the dynamism of the Americans?

Both sides of the Atlantic would agree that there is, firstly, a clear and important need to meet the expectations of the theoretical owners, the shareholders. It would, however, be more accurate to call most of them investors, perhaps even punters. They have none of the pride or the responsibilities of ownership and are, if they are truthful, only there for the money. Nevertheless, failure to meet their financial hopes will only result in a falling share price, expose the company to unwanted predators and make it more difficult to raise new finance. But to turn this need into a purpose is to be guilty of a logical confusion, to mistake a necessary condition for a sufficient one. We need to eat to live, food is a necessary condition of life, but if we lived mainly to eat, making food a sufficient or only purpose of life, we would become gross. The purpose of a business, in other words, is not to make a profit full stop. It is to make a profit in order to enable it to do something more or better. What that 'something' is becomes the real justification for the existence of the business. Owners know this. Investors needn't care They are anxious only for their share of that profit.

To many this will sound like quibbling with words. Not so. It is a moral issue. To mistake the means for the ends is to be turned in on oneself, what Saint Augustine called one of the greatest of sins. Deep down, the suspicions of capitalism are rooted in a feeling that its instruments, the corporations, are immoral in that sense—they have no purpose other than themselves. This may be to do many of them a great injustice, but if so they are let down by their own rhetoric and behaviour. It is a salutary process to ask of any organization—"if it did not exist would we invent it?" "Only if it was doing something more useful, better, or different than anyone else" would have to be the answer, with profits as a means to doing just that.

The idea that those who provide the finance are the rightful owners, rather than just the financiers, dates from the early days of business when the financier was genuinely the owner and usually the chief executive as well. There is, however, a second and related hangover from earlier times, namely the idea that a company is a piece of property, subject to the laws of property and ownership. This was true two centuries ago when company law originated and when a company largely consisted of a set of physical assets. Now that the value of a company largely resides in its intellectual property, in its brands and patents and in the skills and experience of its workforce, it can seem unreal to think it right to treat these things as the property of the financiers, to be disposed of if they so wish. This may be the law but it hardly seems like justice. Surely those who carry this intellectual property within them, who contribute their time and talents rather than their money, should have some rights, some say in the future of what they also think of as their company.

It is worse than that. The employees of a company are treated, by the law and the accounts, as the property of the owners and are recorded as costs not assets in the books. This is demeaning, at the very least. Costs are things to be minimized, assets things to be cherished and grown. The language and the measures of business need to be reversed to recognize this. A good business is a community with a purpose Communities are things you belong to, not things you can own. They have members, members who have certain rights, including the right to vote or express their views on major issues. It is ironic that those countries that boast most stridently of their democratic principles derive their wealth from institutions that are defiantly undemocratic, with all serious power held by outsiders and power inside wielded by a dictatorship or, at best, an oligarchy. Company Law in both America and Britain is out-of-date. It no longer fits the reality of business in the knowledge economy. Perhaps it never did. In 1944 Lord Eustace Percy, in Britain, said this: "Here is the most urgent challenge to political invention ever offered to statesman or jurist. The human association which in fact produces and distributes wealth, the association of workmen, managers, technicians and directors, is not an association recognized by law. The association which the law does recognize—the association of shareholders, creditors and directors—is incapable of production or distribution and is not expected by the law to perform these functions. We have to give law to the real association and to withdraw meaningless privileges from the imaginary one." Sixty years later, the European management writer Arie de Geus was arguing that companies die because their managers focus on the economic activity of producing goods and services and forget that their organization's true nature is that of a community of humans. Nothing, it seems, has changed.

The countries of mainland Europe, however, have always regarded the firm as a community, whose members have legal rights, including, in Germany for instance, the right for their employees to have half, minus one, of the seats on the supervisory board, as well as numerous safeguards against dismissal without due cause and an array of statutory benefits. These rights certainly limit the flexibility of management but they help to cultivate a sense of community which, in the long-term can prove beneficial, generating a loyalty and commitment that can see the company through bad times as well as the sense of security that makes innovation and experiment possible. The shareholders, in this view of the company, are seen as trustees of the wealth inherited from the past. Their duty is to preserve and, if possible, to increase that wealth so that it can be passed on to successor generations.

It is made easier in mainland Europe because of the more closed systems of ownership and the greater reliance on long term finance from the banks. Hostile takeovers are rare and difficult since private shareholders hold only small proportions of the total equity capital while pensions are mostly kept inside the firm, on the liabilities side of the balance sheet, and used as working capital. Pension funds, therefore, are small and most equity holdings are in the hands of other companies, the banks, or family networks. The precise proportions and governance structures differ from country to country but, in general, it can be said that the cult of the equity is not prominent in mainland Europe.

Countries are shaped by their history. There is no way that the Anglo-Saxon countries could adopt any of the different European models, even if they wished to. Both cultures, however, need to restore confidence in the wealth-creating possibilities of capitalism and in its instruments, the corporations. In both cultures some things need to change. More honesty and reality in the reporting of results would help, for a start but when so many of the assets of a business are now literally invisible, and therefore uncountable and when the webs of alliances, joint ventures and sub-contracting partnerships are so complex, it will never be possible to present a simple financial picture of a major business or to find one number that will sum it all up. America's new requirement that Chairmen to attest to the truth of the financial statements may concentrate their minds wonderfully, but they can hardly be expected to double-check the work of their own accountants and auditors.

If, however, the effect of this new requirement is that the accountability for truth-telling is passed down the line, some good may result. If a company took seriously the idea of a wealth-creating community with members rather than employees, then it would only be sensible for the results of their work to be validated by those members before placing them before the financiers, who might, in turn, have greater trust in their accuracy. And if the cult of the stock option waned with the decline of the stock market and companies decided to reward their key people with a share of the profits instead, then those members would be even more likely to take a keen interest in the truth of the financial numbers. It would seem only equitable, in fact, that dividends should be paid to those who contribute their skills as well as to those who have contributed their money, most of whom have not, in fact, paid any to the company itself but only to previous owners of the shares.

It may only be a matter of time Already those whose personal assets are highly valued people such as bankers, brokers, film actors, sports stars make a share of the profits, or a bonus, a condition of their employment. Others, such as authors, get all their remuneration from a share of the income stream. This form of performance-related pay, where the contribution of a single member or a group can be identified, seems bound to grow, along with the bargaining power of key talent. We should not ignore the examples of those organizations, such as sports teams or publishing houses, whose success has always been tied to the talents of individuals and who have had, over the years or even the centuries, to work out how best to share both the risks and the rewards of innovative work. In the growing world of talent, businesses employees will be increasingly unwilling to sell the fruits of their intellectual assets for an annual salary.

As it is, a few smaller European corporations already allocate a fixed proportion of after-tax profit for distribution to the workforce. This then becomes a very tangible expression of the rights of the members. As the practice spreads it will then only make sense to discuss future strategies and plans in broad outline with representatives of the members so that they can share in the responsibility for their future earnings. Democracy of sorts will have crept in through the pay packet, bringing with it, one would hope, more understanding, more commitment and more contribution.

That may help to remedy the democratic deficit in capitalism but it won't repair the image of business in the wider community. It might, in fact, be seen as only spreading the cult of selfishness a little wider. Two more things that need to happen, may actually already be starting to happen.

The ancient Hippocratic oath that doctors used to swear on graduation included the injunction Above all, do no harm. The anti-globalisation protesters claim that global business today does more harm than good. If their charges are to be rebutted, and if business is to restore its reputation as the friend not the enemy of progress around the world, then the chairmen of those companies need to bind themselves with an equivalent oath. Doing no harm goes beyond meeting the legal requirements regarding the environment, conditions of employment, good community relations or ethics. The law always lags behind best practice. Business needs to take the lead in areas such as environmental and social sustainability instead of forever letting itself be pushed onto the defensive.

John Browne, the CEO of BP, the oil giant, is one person who is prepared to do some of the necessary advocacy. He summarized his views in a public lecture on BBC radio. Business, he said was not in opposition to sustainable development; it is in fact essential for delivering it, because only business can produce the technological innovations and deliver the means for genuine progress on this front. Business needs a sustainable planet for its own survival, for few businesses are short-term; they want to do business again and again, over decades. Many would now agree with him and are beginning to suit their actions to their words. Some are finding that there is also money to be made from creating the products and services that sustainability requires.

Unfortunately, the majority still see concepts such as sustainability and a social responsibility as something that only the rich can afford. For most, the business of business is business and should remain so. If society wants to put more constraints on the way business operates, they argue, all it has to do is to pass more laws and enforce more regulations. Such a minimalist and legalistic approach leaves business looking like the potential despoiler who has to be reined in by society and, given the legal time lag, the reins will always seem to many to be too loose.

In the new world of the knowledge economy, sustainability has to be further interpreted at a more human level, as concerns grow over the deteriorating work-life balance for key workers and the stress of the long hours culture. An executive life is, some worry, becoming unsustainable in social terms. One would have to be the modern equivalent of a monk, forsaking all else for the sake of the calling. If the modern business, based on its human assets, is to survive, it will have to find better ways to protect its people from the demands of the jobs it gives them. A neglect of environmental responsibility may lose customers, but a neglect of this type of social responsibility may lose key members of the workforce. More than ever, a modern business has to see itself as a community of individuals, with individual needs as well as very personal skills and talents. They are not anonymous human resources.

The European example of five- to seven-week annual holidays, of legally mandated parental leave for both fathers and mothers together with a growing use of sabbaticals for senior executives and, in France, a working week restricted to 35 hours averaged over the year helps to promote the idea that long work is not necessarily or always good work, and that it can be in the interests of the organization to protect the over-zealous from themselves. Many French firms were surprised to discover that productivity increased when they were forced to implement the new reduced working hours. It is one manifestation of the concept of the organization as a community or, in its original and literal meaning, as a company of companions. The growing practice of individualized contracts and development plans is another.

More corporate democracy and better corporate behaviour will go a long way to alter the current business culture for the better in the eyes of the public, but unless they are accompanied by a new vision of the purpose of the business they will be seen as mere palliatives, a way to keep the world off their backs. It is time to raise our sights above the purely pragmatic. Article 14(2) of the German constitution states "Property imposes duties. Its use should also serve the public weal." There is no such clause in the American Constitution, but the sentiment has its echoes in some company philosophies. Dave Packard once said,

I think many people assume, wrongly, that a company exists simply to make money. While this is an important result of a company's existence, we have to go deeper and find the real reasons for our being. . . . We inevitably come to the conclusion that a group of people get together and exist as an institution that we call a company so that they are able to accomplish something collectively that they could not accomplish separately—they make a contribution to society.

The contribution ethic has always proved to be a strong motivating force in peoples' lives. To survive, even to prosper, is not enough for most. We hanker to leave some footprint in the sands of time, and if we can do that with the help and companionship of others in an organization so much the better. We need a cause to associate with in order to provide real purpose to our lives. The pursuit of a cause does not have to be the prerogative of charities and the not-for-profit sector. Nor does a mission to improve the world make business into some kind of social agency.

Business has always been the active agent of progress, through innovation and new products, by encouraging the spread of technology, or lowering costs through productivity, by improving services and enhancing quality, and thereby making the good things of life available and affordable to ever more people. This process is driven by competition and spurred on by the need to provide adequate returns to those who risk their money and their careers, but it is, in itself, a noble cause. We should make more of it. We should, as charitable organizations do, measure success in terms of outcomes for others as well as for ourselves. Charitable organizations need money just as much as businesses do, but they talk more of their products and services and less of their financial housekeeping.

George C. Merck, the son of the company's founder, always insisted that medicine was for the patients not for profits. In 1987, in line with this core value, his successors decided to give away a drug called Mectizan which cured river-blindness in the developed world but which none of its sufferers could afford to buy. The shareholders were probably not consulted, but had they been, many would have been proud to be associated with such a gesture. Business cannot often afford to be so generous to so many, but doing good not necessarily rules out making a reasonable profit. Maximising profit, on the other hand, often does deny one the chance to do the decent thing.

You can, however, also make money by serving the poor as well as the rich. As C.K. Prahalad has pointed out, there is a huge neglected market in the billions of poor in the developing world. Companies like Unilever and Citicorp are beginning to adapt their technologies to enter this market. Unilever can now deliver ice creams at just two cents each because it has rethought the technology of refrigeration, while Citicorp can now provide financial services to people, also in India, who only have $25 to invest, again through rethinking the technology. In both cases they make money, but the driving force was the need to serve these neglected customers. Profit often comes from Progress.

There are more such stories of enlightened business in both American and European companies, but they still remain the minority. Until and unless they become the norm, capitalism will still be seen as the rich man's game, interested mainly in itself and its agents. High-minded talent may start to shun it and customers desert it. Worse, democratic pressures may force governments to shackle the independence of business, constraining its freedoms and regulating the smallest details of its actions. Capitalism will have become corroded and we shall all be the losers.

Index

The letter *f* or *t* following a page number refers to a figure or table on that page.

International Perspectives on Educational Diversity and Inclusion

The inclusion of minority groups within mainstream education in a way that serves principles of social justice and equity is a familiar one for educators worldwide. *International Perspectives on Educational Diversity and Inclusion* is innovative in its exploration of how globalization impacts on these challenges. With chapters from authors in America, Britain, Europe and India, the book addresses the issue of inclusion within the framework of diversity, and models of comparative education. The editors draw on the extensive experience of the wide-ranging contributors, who examine:

- Accounts from cross-cultural cognitive psychology on the special interests and educational needs of certain ethnic groups
- Research on social class divisions, neighbourhood poverty and school exclusions in Britain
- Educational developments for inclusion of minorities in Europe, Greece and Eastern Europe
- India's educational policies surrounding its struggle to achieve 'education for all' in a nation at the threshold of economic prosperity.

International Perspectives on Educational Diversity and Inclusion is unique in its breadth, in presenting accounts of attempts to include diverse ethnic and social groups, and children with special needs within inclusive educational systems. Different countries, all at different stages of development with contrasted minority populations, face these issues of policy and practice with varying degrees of success. The book should provide stimulating insights into modern concepts of globalization and its impact on educational policy for students of sociology, comparative education and psychology. Readers will learn how the educatic inclusion of diverse ethnic and social groups has received setbacks in America, h been achieved in Britain and some European countries, and is still str achieved in India.

Gajendra K. Verma is Emeritus Professor of Education, U

Christopher R. Bagley is Emeritus Professor of Southampton, UK.

Madan Mohan Jha is Commissioner and Secretary for Edu   .nar, India.

International Perspectives on Educational Diversity and Inclusion

Studies from America, Europe and India

Edited by
Gajendra K. Verma,
Christopher R. Bagley and
Madan Mohan Jha

Routledge
Taylor & Francis Group

LONDON AND NEW YORK

First published 2007
by Routledge
2 Park Square, Milton Park, Abingdon, Oxon OX14 4RN

Simultaneously published in the USA and Canada
by Routledge
270 Madison Ave, New York, NY 10016

Routledge is an imprint of the Taylor & Francis Group, an informa business

© 2007 Selection and editorial matter, Gajendra K. Verma, Christopher R. Bagley and Madan Mohan Jha; individual chapters, the contributors.

Typeset in Times New Roman by
GreenGate Publishing Services, Tonbridge, Kent

Printed and bound in Great Britain by The Cromwell Press, Trowbridge, Wiltshire

British Library Cataloguing in Publication Data
A catalogue record for this book is available from the British Library

Library of Congress Cataloging in Publication Data
Verma, Gajendra K.
  International perspectives on educational diversity and inclusion : studies from America, Europe and India / Gajendra K. Verma, Christopher Bagley and Madan Mohan Jha.
    p. cm.
  ISBN 978-0-415-42777-7 (hardback) -- ISBN 978-0-415-42778-4 (pbk.)
  1. Multicultural education. 2. Minorities--Education. 3. Educational equalization. 4. Globalization. I. Bagley, Christopher. II. Jha, Madan Mohan, 1951- III. Title.
  LC1099.V49 2007
  370.117--dc22

                                              2006036670

ISBN 978–0-415–42777–7 (hbk)
ISBN 978–0-415–42778–4 (pbk)
ISBN 978–0-203–96133–9 (ebk)

Contents

vi *Contents*

Tables

Editors and contributors

Dr Gajendra K. Verma is Emeritus Professor of Education at the University of Manchester, UK, where he held the Sarah Fielden Chair of Education for several years. He has published widely on educational issues and is an expert and government adviser on multicultural issues, including membership of the influential Swann Committee. He has been co-director of a number of EU-funded projects on multicultural teacher training, in collaboration with partners in Finland, France, Netherlands, Germany, Greece and Israel.

Dr Christopher R. Bagley is Emeritus Professor of Social Science, University of Southampton, UK. He has held Chairs in Child Welfare and Social Work in Calgary, Canada, and Hong Kong, and has published widely in the fields of multiculturalism, international child welfare, mental health, and cross-cultural psychology.

Dr Madan Mohan Jha is a member of the Indian Civil Service (IAS), and obtained his doctorate in education from the University of Oxford, UK. He is currently Commissioner and Secretary for Education in the State of Bihar, India, where he is designing and implementing plans for restructuring the educational system with a view to ensuring the fundamental right to education for all children within a framework of inclusion, equity and excellence. He is author of *School Without Walls* (Heinemann, 2002).

Laura Engel is a PhD student in Educational Policy Studies at the University of Illinois at Urbana-Champaign, USA. Her research interests include globalization and educational policy production. Her doctoral research addresses issues of educational policy in Spain, within the context of EU policy.

Dr Carl A. Grant is Hoefs-Bascom Professor of Teacher Education at the University of Wisconsin-Madison, USA. He has written or edited twenty books and monographs on multicultural education, including *The Dictionary of Multiculturalism* (with Gloria Ladson-Billings), Phoenix, Arizona: Oryx Press, 1997 and *Making Choices for Multicultural Education* (with Christine Sleeter), New York: John Wiley, 2003.

Dr Hilary Gray of the British and East European Psychology Group is an experienced psychologist and educator, responsible for several cross-cultural programmes involving Britain and East European countries.

Dr Adamantios Papastamatis is a lecturer in education at the University of Macedonia, Thessaloniki, Greece. His doctorate, on aspects of student self-concept, is from the University of Manchester, UK. He has been involved in a number of EU-funded collaborative research projects.

Dr Fazal Rizvi is Professor in Educational Policy Studies at the University of Illinois at Urbana-Champaign, USA, where he directs the Global Studies in Education program (gse.ed.uius.edu). His next book, co-authored with Bob Lingard, *Globalizing Educational Policy*, will be published by Routledge early in 2008. He has served on a number of government committees, including an international panel on the UK's RAE 2008.

David Rutkowski is a research student in Educational Policy Studies at the University of Illinois at Urbana-Champaign, USA. His doctoral research focuses on a range of theoretical, technical and political issues surrounding the development of global performance indicators through a detailed analysis of the World Educational Indicator (WEI) project.

Dr Rupam Saran obtained her doctorate for a thesis on curriculum policies in New York schools. She has been a teacher in the New York public school system, and is currently Assistant Professor in the Department of Education, Manhattanville College, New York, USA.

Dr Mohammed Akhtar Siddiqui is Professor and Director of the Institute of Advanced Studies in Education in the Faculty of Education, Jamia Millia Islamia University, New Delhi, India. He is also Director of the Academic Staff College at this university, with responsibility for the instruction of university teachers.

Jason Sparks is a doctoral student in Educational Policy Studies at the University of Illinois at Urbana-Champaign, USA. His thesis research focuses on policies concerning the teaching of English and the ways in which governments might seek to regulate its provision by private providers.

Dr Prachi Srivastava is Lecturer in International Education at the Sussex School of Education, University of Sussex, UK. She was ESRC Research Fellow at the Department of Educational Studies, University of Oxford, researching school choice and privatization in economically developing countries. Her co-edited book, *Private Schooling in Less Economically Developed Countries: Asian and African Perspectives*, was published in 2007 by Symposium Books.

Dr Derek Woodrow is Emeritus Professor of Education at Manchester Metropolitan University, UK. Originally specializing in the teaching and learning of mathematics, he is author and co-author of several books and papers on multiculturalism and the impact of society and culture on achievement in minority groups.

Part I

Globalization and diversity in education

1 Equality and the politics of globalization in education

Fazal Rizvi, Laura Engel, David Rutkowski and Jason Sparks

Introduction

Equality has long been a major goal of education around the world. As early as 1948, Article 26 of the United Nations' Universal Declaration of Human Rights stated that, "everyone has the right to education" and, "education shall be free, at least in the elementary and fundamental stage". In line with this declaration, most governments profess a commitment to equality, and have taken various steps to provide at least basic education to all of their citizens regardless of the ways in which governments have interpreted the notion of equality, as well as their limited ability to fund measures working towards the goal of equality in education. The production of social and human capital has often been cited as one of the main reasons for supporting the goal of equality in education. And indeed, there is a great deal of credible evidence to suggest that an investment in education not only provides personal benefits to individuals, in terms of their earning capacity, but also has the potential to benefit whole communities, in both economic and social realms. There has therefore been a major push by intergovernmental organizations (IGOs), like the World Bank, the OECD and UNESCO, and non-governmental organizations, for universal access to primary education, while the demand for secondary and tertiary education has also grown rapidly.

Over the past decade, this call for more education has been made within a broader discourse about the changing nature of the global economy, which is characterized as "knowledge-based", and which is said to require greater levels of education and training than ever before. In the so-called "knowledge economy", educational systems have been asked to produce a workforce adequately prepared to meet the challenges of globalization. It has been suggested that social and economic development is no longer possible without policies that encourage greater participation in education. The goal of access to education is thus reiterated, but is now articulated within a broader discourse about the changing global context within which education takes place. In this way, the rhetoric of access and equality in education and the politics of globalization have become inextricably related (Scholte, 2000).

In this chapter, we explore the nature of this relationship by discussing some of the ways in which globalization is affecting policy priorities in education. We argue that the effects of globalization on educational and social equality for different groups and communities vary greatly, creating considerable disparity around the world, with some communities benefiting enormously from globalization, but others encountering major disruptions to their economic and cultural lives. Moreover, we suggest that globalization has transformed the discursive terrain within which educational policies are developed and enacted, and that this terrain is increasingly informed by a range of neo-liberal precepts that affect the ways we think about educational governance – indeed, about its basic purposes. Along these lines, a particular way of interpreting globalization has become globally hegemonic, which undermines, in various ways, stronger democratic claims to equality in education.

The politics of globalization

The concept of globalization has been widely used in recent years to rethink the imperatives driving educational changes, even if globalization remains poorly understood. While little consensus exists about its meaning, the concept of globalization does appear to encompass some of the profound social and economic changes that are currently taking place around the world. Many of these changes have been driven by recent revolutions in information and communication technologies, which have resulted in a world that is more interconnected and interdependent than ever before. Paradoxically, global processes have themselves created some of the conditions by which the idea of globalization has become seemingly ubiquitous, used widely in both policy and popular discourses to explain the nature of recent changes. It has been used to refer to a set of social processes that imply "inexorable integration of markets, nation-states and technologies to a degree never witnessed before – in a way that is enabling individuals, corporations and nation-states to reach round the world farther, faster, deeper and cheaper than ever before" (Friedman, 2000).

Such integration, however, is far from complete; its nature can be understood in a variety of ways, and it clearly benefits some communities more than others. Globalization is thus a highly contested notion, which articulates historically with a range of colonial practices, on the one hand, and socially with recent technological developments in transport, communication and data processing, on the other. These developments have transformed the nature of economic activity, changing modes of both production and consumption. They have also altered the nature of international relations, and the work of intergovernmental political institutions such as the World Bank and the United Nations. Moreover, these developments have propelled an enormous growth in the movement of people, information and ideologies, leading to an enormous increase in cultural interactions and the hybridization of cultural practices.

David Harvey (1989) provides perhaps one of the best descriptions of economic globalization. He argues that globalization describes "an intense period of time–space compression that has had a disorientating and disruptive impact on political–economic practices, the balance of class power, as well as upon cultural and social life" (p. 8). In this new era, global capitalism has become fragmentary, as time and space are rearranged by the dictates of multinational capital. Improved systems of communication and information flows and rationalization in the techniques of distribution have enabled capital and commodities to be moved through the global market with greater speed. The rigidities of Fordism have been replaced by a new organizational ideology that celebrates flexibility and efficiency as its foundational values, expressed most explicitly in ideas of subcontracting, outsourcing, vertical disintegration, just-in-time delivery systems and the like. In the realm of commodity production, argues Harvey, the primary effect of this transformation has been an increased emphasis on instrumental values and the virtues of speed and instantaneity.

Castells (1996) characterizes the global economy as informational, networked, knowledge-based, post-industrial and service-oriented. He argues that cultural and political meanings are now under siege by global economic and technological restructuring. Castells speaks of an "informational mode of development" through which global financial and informational linkages are accelerated, convert places into spaces and threaten to dominate local processes of cultural meanings. According to Castells, networks constitute "the new social morphology of our societies"; and "the diffusion of networking logic substantially modifies the operation and outcomes in the processes of production, experience, power and culture". The new economy is "organized around global networks of capital, management, and information, whose access to technological know-how is at the roots of productivity and competitiveness" (ibid. 1996). All industries, including education, are trapped within the networking logic of contemporary capitalism, subject to the same economic cycles, market upswings and downturns and segmented global competition.

The global economy has also led to a new conception of governance, requiring a radically revised view of the roles and responsibilities of national governments, minimizing the need for their policy intervention, with greater reliance on the market (Strange, 1996). This interpretation of the declining role of the state in policy development dislodges one of the central tenets of the modern nation-state system – the claim to distinctive symmetry and correspondence between territory and legitimacy. While nation-states fiercely protect their sovereignty, in the age of globalization the exclusive link between territory and political power appears to have been broken. As Held and McGrew (2000) argue, "the state has become a fragmented policy-making arena, permeated by transnational networks (governmental and non-governmental) as well as by domestic agencies and forces". So, while the modern state retains some of its authority, it now needs to negotiate forces beyond its control – not only of international organizations and regimes but also of transnational capital. This applies to educational policy as much as it does to economic policy, as educational priorities become implicated in global power systems.

Within these systems, there is now an ever-increasing level of cultural interaction across national and ethnic boundaries. With the sheer scale, intensity, speed and volume of global cultural communication, the traditional link between territory and social identity appears weakened, as people can more readily choose to detach their identities from particular times, places and traditions (Risvi, 2005). Not only the media but greater transnational mobility has a "pluralizing" impact on identity formation, producing a variety of hyphenated identities which are less "fixed or unified" (Hall, 1992). This has led to the emergence of a "global consciousness", which may represent the cultural basis of an "incipient civil society" (Falk, 1995). This development suggests the need to interpret globalization both descriptively and normatively – as an objective set of social processes, but also as a subjective or phenomenological awareness by people and states of recent changes in global economy and culture.

Despite a recognition of its cultural dimensions, one of the main problems with most accounts of globalization is that they draw attention "disproportionally upon the global economy, presenting it as a pre-given 'thing', existing outside of thought" (Smith, 2000), whose developmental logic has the capacity not only to explain the development of policies but also, it is assumed, to determine the subjectivity of people, without ever interrogating what those people are up to. As Smith (2001) points out, this interpretation of globalization presents contemporary global processes not as an ever-changing product of human practices but as an expression of a deeper economic logic. In so doing, globalization is conceived as historically inevitable, representing a juggernaut with which we simply have to come to terms and negotiate as best as we can.

An increasing number of scholars and activists have, however, begun to challenge this view of globalization. They have interpreted globalization not as an expression of inexorable historical processes, but as an ideology serving a particular set of economic and political interests. Theorists like Bourdieu (2003) have suggested that globalization represents a deliberate, ideological project of economic liberalization that subjects states and individuals to more intense market forces. This project, often referred to as 'neo-liberal', is thus based on a politics of meaning that seeks to accommodate people and nations to a certain taken-for-grantedness about the ways the global economy operates and the manner in which culture, crises, resources and power formations are filtered through its universal logic. It thus "ontologizes" the global market mentality, creating global subjects who in turn view the world and the policy options they have through its conceptual prism. This prism is constituted by an emphasis on market principles and production of profits; a minimalist role for the state; deregulated labor market; and flexible forms of governance. From this perspective, the term "globalization" designates certain power relations, practices and technologies, playing a "hegemonic role in organizing and decoding the meaning of the world" (Scharito and Webb, 2003).

In recent years, educational policies have been deeply affected by this neo-liberal view of globalization, as educational systems have sought to realign their priorities to what they perceive to be its imperatives. While the authority for the development of education policies remains with sovereign governments, they nonetheless feel the need to take global processes into account. However, the relationship between the global processes and policy production at the national level is highly complex, because governments do not simply have the freedom to "pick and choose" from a global menu of policies; rather, their deliberations are framed by the ideological discourses circulating around the world, often through international organizations such as UNESCO and the OECD, as well as media and a global class of policy experts. The political structures beyond the nation-states thus become relevant to national policy deliberations, as does the globalizing cultural field within which education takes place. In the process, a new discourse of educational purposes emerges, sidelining education's traditional concerns with the development of individuals and communities.

Shifting purposes of education

This new discourse highlights the need for education to achieve the objectives of global economic integration, by producing efficient and effective workers to meet the requirements of the global economy. David Labaree (1997) has observed that education has traditionally been thought to have three distinct, but sometimes, competing, purposes: democratic equality, social mobility, and social efficiency. While these purposes of education are not mutually exclusive, one of these has often been highlighted over the others. For example, in the post-World War II Keynesian period, the idea of democratic equality became dominant in many parts of the world, interpreted in Western countries from a liberal-democratic perspective, while in socialist countries it acquired a different meaning. Some countries promoted social mobility and meritocracy, while others stressed a more egalitarian outlook. In many postcolonial countries, the idea of equality became an ideological mantra, even if it was seldom realized in education. In recent years, however, under the conditions of globalization, it is the idea of social efficiency that is more prized by an increasing number of citizens, corporations and intergovernmental organizations, as well as governments.

For Labaree, the concept of democratic equality has long suggested the need for education to facilitate the development of democratic citizens who can participate in their communities in a critically informed manner. It is a view central to John Dewey's philosophy of education (Dewey, 1916). Its focus is on equal access and equal treatment of all citizens, and on regarding education as a public good. This suggests that maximum benefit to society can only be realized if every member of a community is educated equally to realize their full potential. The primary purpose of education is then the creation of productive citizens, and not necessarily efficient workers, able to maximize personal fulfillment. This does not mean that vocational training is unimportant. Nonetheless, it is to insist that such training must be located within the broader role education must play in the development of a socially cohesive democratic community. The purposes of education are thus more social and cultural than economic, focused more on community than on the individual.

In contrast, the social efficiency view of educational purposes focuses more on individuals, but requires education to play a more important, instrumental, role in developing workers able to contribute to the economic productivity of nations and corporations alike. It judges educational systems in terms of their efficiency – their capacity to make an adequate return on investment, assessed in terms of their contribution in producing workers with knowledge, skills and attitudes relevant to increasing productivity within the knowledge economy. In this way, education is viewed both as a public and a private good: public because it contributes to the economic well-being and social development of a community; and private because it serves individual interests within the competitive labor market. However, it is important to stress that the notion of public good that the social efficiency view promotes is markedly different from the social democratic conception, which regards education as intrinsically good, and not linked instrumentally to organizational efficiency, economic outcomes and productivity.

In recent constructions of globalization, the focus on social efficiency has become a key and perhaps the overriding goal of education. Much of what is now regarded as educational reform is based on the ideological belief that social and economic "progress" can only be achieved through systems of education geared more towards fulfilling the needs of the market. It is assumed that educational systems have, for far too long, been inefficient and ineffective in ways that have prevented them from realizing this functional objective. Popular media and corporations have, in particular, propagated this opinion and have called on governments to pursue reforms that are not only more socially and economically efficient but are also cognizant of the new "realities" of the knowledge economy in an increasingly globalized world. This has required the purposes of education to be more instrumentally defined, in terms of education's capacity to produce workers who have grounding in basic literacy and numeracy, are flexible, creative, and multiskilled, have adequate knowledge of new information and communication technologies, and are able to work in culturally diverse environments.

Of course, this account of educational purposes does not imply that social efficiency has entirely displaced concerns for equality and social mobility. However, it is worth noting that both equity and social mobility have been incorporated within the broader discourse of social efficiency. For example, it has been argued by international organizations such as the OECD that a focus on efficiency can in fact lead to greater equality and opportunities for social mobility. It is suggested that without workers who are able to perform effectively in the global labor market, the potential for social mobility is severely reduced; and that since the global economy requires appropriate social conditions for capital accumulation and economic growth, equity concerns cannot be overlooked by policymakers committed to social efficiency. As the OECD (1996) has suggested:

> A new focus for education and training policies is needed now, to develop capacities to realize the potential of the 'global information economy' and to contribute to employment, culture, democracy and, above all, social cohesion. Such policies will need to support the transition to 'learning societies' in which equal opportunities are available to all, access is open, and all individuals are encouraged and motivated to learn, in formal education as well as throughout life.
>
> (OECD, 1996)

What this discourse suggests is that social efficiency must now be regarded as a "meta-value", subsuming within its scope educational aspirations such as the goals of social equality, mobility and even cohesion. In the process, the meaning of equality is weakened, re-articulated to suggest formal access to the institutions of education, rather than stronger claims to equality of treatment and outcome.

Strong and weak concepts of equality

Access to education, of course, is important to all forms of educational outcomes, including economic well-being, health, employment, and productive citizenship. Without access, the chances of achieving social and economic equality are negligible. However, simple formal access to schools has never been sufficient to realizing the potential of education, because unless families have an adequate economic base at home to support students attending schools, the students are unlikely to be able to take advantage of formal access. This, of course, complicates the relationship between access to education and equity outcomes. While a commitment to formal access is entirely consistent with the idea of social efficiency, it is not enough to achieve democratic equality. For this to become a reality, attention needs also to be paid to the social conditions necessary for learning, to instructional quality and to the resources that are necessary to support effective programs. Formal access to schooling does not always translate into effective equity outcomes.

Indeed, simple access can be counter-productive, setting up expectations which, if not realized, have the potential to create considerable social alienation among those who have invested time and effort into education, without the promised rewards. Without good teachers, who have adequate training and professional attitudes, access can undermine equality, even if it meets some of the standards of efficiency. Access can also be counter-productive if the curriculum and instruction are not linked to local cultures and traditions, and are inappropriate to the community in which they are offered. This requires a more complex "stronger" view of access and equality than is suggested by the "weak" social efficiency view. Education has a whole range of purposes; it is not simply for producing efficient workers for the changing global economy. If this is so, then social efficiency has to be reconciled with the broader cultural concerns of education, linked to issues of class, gender and ethnicity.

That simple access is not sufficient for achieving equality in education can be further demonstrated by addressing issues relating to the education of girls. In recent years, IGOs, such as the OECD, the World Bank, and UNESCO have repeatedly emphasized the importance of gender equity in education. And indeed much has been done to provide girls greater access to education; and the number of girls attending school has never been greater. However, the neo-liberal arguments for gender equity reveal a weak conception of equality, cast largely in terms of social efficiency, and the requirements of the global economy. According to the World Bank (2004), for example, "research has also shown that women and girls work harder than men, are more likely to invest their earning in their children, and are major producers as well as consumers". UNESCO (2001) states, "Educating girls yields the highest return in economic terms". Finally, the OECD (2000) urges that "Investing in women (with respect to education, health, family planning, access to land, etc.) not only directly reduces poverty, but also leads to higher productivity and a more efficient use of resources". Each of these views links gender equity to economic consumerism and efficiency. This instrumentalist logic is arguably sexist, as it views women as a means to certain economic ends, rather than as people who participate in education for a huge variety of reasons, some economic, others social and cultural.

A stronger claim to gender equity in education, on the other hand, must address issues not only of their access but also of economic and social outcomes of education, resulting from globalization. Here the picture is decidedly mixed. Recent data show that while girls are participating in education in larger numbers than ever before, the outcomes of their education are not socially and economically proportional to their efforts. For example, in recent years, there have been many more opportunities for women to utilize their education in paid work. However, this work has been predominantly in the service economy of global information, global communication, global retailing and global finance (World Bank, 2004). Each of these areas has been characterized by "flexible" labor conditions and poorer career prospects, perpetuating and sometimes deepening gender hierarchies. Despite the growing level of access of women to higher education, their participation in the fields of the natural sciences and engineering is far from gender parity. With growing importance attached to these fields within the global economy, associated with technological innovation and technical expertise, this inequality is more significant than it might first appear, since it suggests that the growing access of women to tertiary education is in areas that do not enjoy the same high economic rewards, social status and prestige.

What this analysis indicates is that gender equity beyond access requires a radical overhaul of the educational and social processes that perpetuate gender inequalities. This aspiration is clearly informed by a different purpose of education. While the social efficiency view demands better utilization of the human resources that women represent, the democratic equality view seeks a social transformation through which gender relations are totally reconfigured. This latter view not only highlights the importance of access and social inclusion, but also underlines the importance of rethinking the terms of this inclusion. It envisages societies that have potentially been economically, politically and socially transformed in gender terms. This requires changes not only to the ways education is administered but also to the curriculum and pedagogy, especially in the context of globalization, with its potential to reshape patterns of both economic and social relations.

Shifts in the curriculum

Any comprehensive overview of recent shifts in the curriculum is clearly beyond the scope of this chapter. However, it is perhaps important to note that there has been more rhetoric in recent years about the need to rethink issues of curriculum in the light of changes represented by globalization than actual changes in practice. Primary and secondary curriculum remains remarkably unaltered in most countries, even if there have been some changes in the ways in which pedagogy has been approached, especially at the primary level. More child-centered and humanistic ways of thinking about and dealing with young children appear to have become popular, not only in countries with liberal democratic traditions but elsewhere as well. There have also been significant changes in which students are assessed, and teachers are now held accountable to educational systems in much more rigorous, and sometimes even punitive, ways. New technologies of accountability have been established not only at the national level, but also transnationally. Practices of benchmarking and comparing achievements and educational performance on a wide variety of indicators have arguably become common, as a result of an increasing level of regional and international collaboration, much of which is driven by the international organizations such as the OECD and UNESCO. Programs like PISA (2005) and TIMMS (2005) for example, have begun to provide a framework within which educational policy deliberations at the national level are now conducted.

Some of these developments have clearly been facilitated by recent developments in information and communication technologies (ICT), even if it is the technologies that have driven educational changes, rather than the changes driving the technologies. Either way, the need to understand the possibilities of ICT in order to develop more efficient and effective ways of delivering education has become a major feature of contemporary thinking. Also significant have been the efforts to include ICT into the curriculum. Indeed, computer education and the teaching of English language represent perhaps two of the most important new initiatives in the curriculum, responding directly to the pressures of globalization. However, each is problematic in its own way, and raises a whole range of issues about the ways in which it is promoted by governments and IGOs, and relates to issues of equality in education.

Since the early 1990s, policymakers around the world have recognized the curricular significance of ICT. Significantly, in 2000, the Group of Eight (G8) met in Japan to "seriously address the challenges of ICT in education", producing the Okinawa Charter on Global Information Society, a document that describes ICT as "one of the most potent forces in shaping the twenty-first century", and speaks idealistically of an "IT-driven economic and social transformation" impacting "the way people live, learn and work". The Charter calls for a "stronger partnership among developed and developing countries, civil society including private firms and NGOs, foundations and academic institutions, and international organizations" to develop a "solid framework of IT-related policies and action" aimed at insuring that ICT serves a range of goals, such as: creating sustainable economic growth and enhancing public welfare. The Charter states its commitment to the principle of inclusion, mentions democratic values, human development, and respect for diversity, and the potential in ICT for social and economic opportunities worldwide.

While these fine political sentiments are indeed laudable, it is less clear from the Charter how they are to be translated into effective educational reforms that address each of these values, and not simply those sustaining economic growth. Even its discussion on "Bridging the digital divide" calls for policies that lead to the development of human resources capable of responding to the demands (read economic demands) of the information society, a goal to be achieved by supporting effective programs in ICT literacy and skills through education (Plomp *et al.*, 2003). The G8 Charter's characterization of the information society is based on range of neo-liberal assumptions about the global market, and the human resources needed to make it efficient. The broader discourse of the inevitability of rapid economic and social change in the Charter (G8, 2000) is underpinned by what Ulrich Beck (2000) refers to as the "ideology of rule by the world market", reducing the "multidimensionality of globalization to a single, economic dimension". In this way, the Charter regards participation in the global economy as a universal good, the requirements of which need to be understood and enacted by nation-states. It is assumed that all efforts to align education with the needs of the global economy are necessarily beneficial to society; and that not to do so is to exclude students from ICT-driven economic and social transformation. Here, again, the notion of equity is re-articulated in the language of the market.

Much of this call for curriculum reform is thus located within the social efficiency view of educational purposes. ICT is viewed as a vehicle for making education more efficient and effective, leaving intact some of the deeply held assumptions about education and its role in supporting and sustaining the mechanisms of the global economy. Education is conceived as a means to achieve the G8 goal stated earlier: the "development of human resources capable of responding to the demands of the information age" (G8, 2000). It is considered necessary for fully participating in the age of globalization, but the concept of globalization itself is viewed narrowly, as linked to current economic transformations driven by a neo-liberal ideology, which defines social relations in terms of competition between individuals and nations. So long as this view prevails, it is difficult to imagine how the so-called "digital divide" between communities and nations can ever be bridged, except on the edges, for, in a context of global competition, developing countries will continue to struggle to achieve parity within the highly stratified world economic community.

Just as the interest in integrating ICT into the curriculum has been framed within the social efficiency view of education, as contributing to the needs of the changing global economy, so too have been the arguments put forward by policymakers around the world in support of greater emphasis on the teaching of English. The teaching of English is assumed to be crucial in any thoughtful response to the pressures of globalization. For example, a 2002 UNESCO report on curriculum changes in the Asia-Pacific region notes that: "Facing the challenges of globalization trends, curriculum of countries in the region have paid special attention to foreign languages, first and foremost it is English" (UNESCO, 2002). The report goes on to say that the choice of language in education policy is "largely driven by the demands of the international labor market, in particular in the field of ICTs and science".

Similarly, a 2004 APEC document, echoing the UNESCO sentiment above, asserts:

> As English has become the most common medium for communication in a global world, it is the language that provides job opportunities, access to higher education and a broader flow of information, as well as facilitates diplomatic discussions and business negotiations. English has also become the primary medium for communication in science and technology.
>
> (APEC, 2004)

In this discourse, there is an almost unproblematic construction of, and appeal to, the demands of a global economy, which disregards what Tollefson (1991) describes as the very local impact of language policies on "access to economic resources, to policymaking institutions, and to political power". In transforming language use and language education into commodities for a global marketplace, such discourse takes a particular stance with regard to what Pennycook (1999) calls "the cultural, political, social and economic implications of language programs". For example, this stance seems unconcerned with the role English might play in perpetuating global inequalities, as well as globalization's tendencies for homogenizing cultural traditions.

What this discussion clearly reveals, however, is the degree to which the emerging "consensus that professional development of FL [foreign language] teachers is one of the most important and challenging issues that all economies face" (APEC, 2004) is embedded within the social efficiency view of education, and the extent to which curriculum reform is now framed by perceptions concerning educational purposes being inextricably and, perhaps exclusively, linked to the labor market considerations of the global economy. In speaking of economies, and not societies or nations, for example, the quote from the APEC report (2004) above barely hides its neo-liberal assumptions, linked to human capital needs, which are now driving the language policy shifts within the Asia-Pacific region.

These shifts are based on a particular "reading" of global processes as necessarily economic, articulated in ways that subordinate political and cultural concerns that have traditionally been given at least an equal emphasis in policy deliberations. An outcome of this economic reductionism is that it inherently perpetuates the global inequalities, as English-speaking countries and those countries which can afford to develop levels of English proficiency mark themselves out as better able to profit from the global economy. In this way, not only does the increasing use of English worldwide carry the risk of homogenizing local cultures and traditions but it also becomes a marker of social and national differentiation.

Shifts in governance

Differentiation is, however, also reproduced by a range of other developments driven by neo-liberal notions of globalization. Most notably, in recent years, there has been much rhetoric about "good governance", a phrase that masks an underlying shift in educational ideology. Debated under the rubric of "good governance" are issues concerning transparency of decision-making processes, forms of devolution, technologies of measuring educational performance, international benchmarking, mechanisms of quality assurance, appropriate accountability regimes, sources of educational funding, effective uses of public resources, and so on. Even this short list shows how most of these concerns relate to social efficiency, defined mostly in terms of the extent to which educational systems are responsive to the labor market needs of the global economy.

In this way, the idea of good governance has not been left to the local communities to define, even if devolution is assumed to be essential for making educational systems more efficient and effective. The idea of devolution has of course been used in a wide variety of ways in political theory, from radical democratic notions of citizen participation in decision-making to weaker administrative notions of managerial decentralization. It is the latter view of devolution that has gained ascendancy in contemporary global ideologies of governance. It is functional and fiscal decentralization, rather than political devolution, that has been highlighted as a defining characteristic of good governance of education. Under this definition of decentralization, local institutions are permitted to make decisions, but only in ways that are aligned to both national goals and standards, which are increasingly linked to a broader technology of public administration. This also involves the ways in which expenditure on education is allocated, distributed and monitored.

Often such allocation is based on generalized performance criteria that do not always take into account particular needs of communities. This has a negative impact on rural and lower income areas, increasing regional disparities, especially when there are limited financial resources and preparation for local governance (UN, 2004). In addition, an emphasis on fiscal decentralization is linked to political conditions in which privatization is viewed as its logical outcome. Educational managers at local and district levels struggle to manage their own education programs, particularly those that cannot be easily accommodated within the broader national frameworks directing performance-based funding regimes.

The global trend towards privatization of education, not only at tertiary but also at primary and secondary levels, has intensified inequalities in a number of ways. While governments around the world have highlighted the importance of higher levels of education, they have either been unwilling or unable to fund growth in demand for educational participation. The use of the rhetoric of privatization has thus become widespread around the world, along with an emphasis on the notions of quality, efficiency, and productivity. With the scaling back of government funding igniting a rise in privatization, the role of the private sector in education has also grown, blurring the lines between government and private responsibilities over education. The number of private higher education institutions has grown rapidly. These developments have had major implications for educational equity, as private interests have increasingly assumed a greater, often self-interested, role in policy development in education. This has also led to education becoming increasingly viewed more in terms of an individual investment, rather than a social investment.

In very broad terms, the idea of privatization refers to the transfer of services provided by the public sector to a range of private sector interests. As a political construct, the idea of privatization emerged in the late 1970s as an attempt by a number of Western countries, like the United States and Britain, to separate decision-making in the areas of public policy from the provision of services. Three decades later, as we have noted, it has become globally pervasive, increasingly assumed to be the only way to ensure that public services, including education, are delivered efficiently and effectively. It has come to symbolize a new way of looking at public institutions and the role of the state in managing the affairs of its citizens.

This way of looking at governance is based on a set of assumptions that include the view that the power of private property rights, market forces and competition brings out the best in public sector employees; that when the public sector is forced to compete against private contractors then the service delivery is necessarily more efficient; and that when public institutions are thrust into market environments they become much more organizationally agile and innovative, with a greater commitment to reform. Economic arguments in favor of privatization also view it as necessary for growth, for meeting increasing levels of demand for particular services, including education. Such arguments necessarily assume the welfare state to be "withering away", no longer capable of meeting the requirements both of society and individuals who are increasingly interested in managing their own affairs and do not trust the state to look after them.

While, in recent years, many of these arguments have become commonplace, few, if any, can be substantiated with hard data. So, for example, the contention that private contractors are more efficient and cost-effective in delivering services without compromising on quality is one that has repeatedly been shown to be both groundless and perhaps even unverifiable; yet this does not seem to stop advocates of privatization from asserting it repeatedly. The fact is that economic arguments alone cannot justify privatization. To try to do so is to grossly underestimate its political character, and to misunderstand its role as an ideology. In the end, the political context in which privatization is promoted is inherently ideological, based on an assumption that the private sector is intrinsically more efficient and productive than the public sector.

However, the notion of efficiency is highly problematic because it cannot be interpreted in some neutral fashion, without reference to the more fundamental moral and political criteria against which it might be measured. Nothing is efficient in its own right. We need to ask the more basic question, "Efficiency in terms of what?" As the philosopher Alasdair MacIntyre (1981) points out, there are strong grounds for rejecting the claim that efficiency is a morally neutral concept. Rather, it is "inseparable from a mode of existence in which the contrivance of means is in central part the manipulation of human beings into compliant patterns of behavior" (MacIntyre, 1981). In an organizational setting, efficiency drives always involve control over people, achieved either through sanctions or hegemonic compliance.

However, such a focus on efficiency often has a negative impact on the educational opportunities available to marginalized groups and communities who have traditionally benefited from public investment in education. In the context of declining levels of public expenditure on education, families are often forced to pay for the education of their children. This might be fine with those families who can afford it, but privatization has disastrous consequences for marginalized groups, especially for girls in certain cultural traditions. There is considerable evidence to show that when parents, in developing countries in particular, are required to make a difficult choice, they frequently prefer to pay for the education of boys. While privatization might be efficient in some respects, as neo-liberal theorists suggest, it often has negative consequences for those who rely on the public provision of education, as well as on gender equity in education and, by implication, on the broader goals of social equality.

Mobility and trade in education

Just as new modes of governance, driven by global neo-liberal ideologies, have negatively impacted equality in education, so has the increasing levels of mobility, not only of capital, information and ideologies, but also of people. Globalization has affected considerable movement from rural and regional centers of population to cities, especially global cities which occupy, as Sassen (1991) has pointed out, a special place in the global economic division of labor, and which operate as nodes of global circulation of capital, goods and people. This has created conditions for increased mobility of people from regional and rural areas to metropolitan centers where there are greater possibilities of employment. The requirements of city life have always determined educational priorities of nation-states, but with cities of national significance becoming global, a new cultural geography has emerged, affecting all aspects of social and cultural life, including education.

The awareness of the changing nature of the global economy and of the global labor market, however imprecise and speculative, has created a growing demand for international education at the tertiary level, especially in the cities, among those who can afford it. Of course, the idea of international education, itself, is not new. There has always been international mobility of students and researchers in search of new knowledge, and training where this was not available within the nation. In the past, international education helped to create the expertise needed to develop the social, administrative and economic infrastructure of the developing countries. It was concerned with the development of skills, attitudes and knowledge so that, upon their return, graduates could make a robust contribution to national development in the image of their sponsors. The purposes of international education were thus defined in terms of the need to increase intercultural knowledge, and to enhance the level of international cooperation. In this way, equal weight was given to the economic, political and cultural purposes of education.

However, in recent years, a new discourse of internationalization has emerged. This discourse is linked not only to perceptions concerning the emerging labor market stipulations, and the need for people to acquire multicultural and cosmopolitan sensitivities in the era of globalization, but also, and perhaps more importantly, as a matter of global trade in education. The discourse of internationalization of education has thus shifted in recent years, with the introduction of a set of market principles to guide its practices. It is now increasingly viewed as "an export industry", driven by a growing demand for an education abroad within the developing countries, enabling countries like Australia and the United Kingdom to set themselves up as major suppliers. According to the OECD's Center for Educational Research and Innovation (CERI, 2004), the growing demand for international education is simultaneously, "a cause, consequence and symptom of globalization". It responds to the need of industries at the cutting edge of the knowledge economy, such as ICT, financial management, science and engineering, in which the demand for globally mobile labor is growing at a rapid rate.

Not surprisingly, therefore, it is the World Trade Organization (WTO) that has in recent years been a major advocate for what Jane Knight (2002) has called the "trade creep" in higher education. The WTO's General Agreement on Trade in Services (GATS, 2004) has sought to specify a range of conditions under which trade in education is to be pursued. These conditions include such matters as: transparency of rules; liberalization of markets; elimination of practices acting as barriers to trade and student mobility; and the development of rules for resolving disputes. Now, while in one sense these rules appear perfectly sensible, from another perspective they serve a more ideological function, of institutionalizing a particular way of looking at international education, defining it in terms of the efficiency of the global markets in education, rather than in its more political, social and cultural purposes.

In broader terms, the heavily commercial character of international education serves only to reproduce global inequalities. Under earlier regimes of international education, universities in the developed countries provided access to a large number of students from poorer, less developed countries. Under a market regime, the number of financially sponsored students has dwindled markedly, further widening the skills gap that now exists between the newly industrializing countries and poorer Pacific countries, whose economic prospects have steadily declined. Moreover, international education reflects the globally uneven and asymmetrical nature of student flows within the global market of international education. For example, Marginson (2003) has noted the magnetic attraction of American higher education, and has argued that the UK, Australia, Canada and New Zealand sit "in the American slipstream, operating on a more entrepreneurial basis than American institutions. These countries gain the referred power as lesser English-language providers and sites for migration, often in a transitional stage in passage to the USA".

These developments represent a form of economic and social haemorrhaging of poorer countries caused by the new global geometry of power. This haemorrhaging is further perpetuated by the "brain drain" of the highly talented international students who can make a significant contribution to the national development of their own countries but are seduced by the opportunities presented by the richer countries. In so far as government policies in developed countries view international students as potential immigrants in areas of skill shortage, they accelerate this pattern of "brain drain". It is estimated that more than 60 per cent of international students from developing countries qualify for immigration to a developed country and are granted permanent residence, even if they do not abandon their citizenship and plan instead to work in a transnational space (Rizvi, 2005). This situation is further complicated by the fact that many students who do return to their own country either seek or are recruited into well-paying jobs in transnational corporations, depriving national institutions of their expertise. In these ways, international education has increasingly become a hand-maiden to corporate globalization, providing the new global economy the human resources it needs to expand into new markets rather than to contribute to broader social and cultural goals.

Conclusion

In this chapter we have discussed how the politics of globalization, or more accurately a particular construction of globalization, has, in recent years, reconfigured the discursive terrain within which educational policy is developed and enacted; and how this reconfiguration has undermined, in various ways, the goal of equality and social inclusion in education. We have suggested that the hegemonic dominance of the neo-liberal conception of globalization has greatly benefited some communities, while it has had disastrous consequences for others. In educational policy, the politics of globalization has had the consequence of making the social efficiency goals of education become dominant over its more traditional social and cultural concerns with the development of the individual and needs of the community. In line with these goals, a global conception of educational governance has emerged, associated with functional and fiscal decentralization and privatization, which has encouraged global trade in education. None of these developments bring us closer to the goal of equality. Instead, they have perpetuated social hierarchies within and across national boundaries.

These developments have also left many educators and educational systems feeling disenfranchised, especially when they are expected to conform to unrealistic accountability regimes, and deliver outcomes for which they have not been adequately funded or resourced. At the same time, the policy shift towards privatization has compromised the goals of access and equality and has widened inequalities across gender, class and nations. The excessive emphasis on efficiency has resulted in greater focus on the operational requirements of the systems rather than upon the lives of people and their communities. This has happened as a result of the balance between competing purposes of education becoming tilted towards social efficiency, undermining the potential of education to build democratic communities.

There is clearly no turning back from globalization. However, globalization need not necessarily be interpreted in neo-liberal terms. It must be possible to recognize that the world is more interconnected and interdependent than ever before, without accepting entirely the logic of the market, and the technocratic solutions to the problems of education. The new global times require us to think and act imaginatively, both locally and globally, if we are to tame the excesses of the market, and work with globalization in ways that are more creative, while remaining committed to the potential that education has for building democratic communities, committed to the ethical idea of equality in its stronger sense.

References

APEC (2004) *Strategic Plan for English/Foreign Language Learning, Sub Theme Paper 1*, Santiago, Chile. At: http://biblioteca.mineduc.cl/documento/English_ APEC _strategic _plan_final.

Beck, U. (2000) 'What is globalization?' In: Held, D. and McGrew, A. (eds), *The Global Transformations Reader: An Introduction to the Globalization Debate*. Cambridge: Polity Press.

Bourdieu, P. (2003) *Firing Back: Against the Tyranny of the Market*. London: Verso.

Castells, M. (1996) *The Rise of the Network Society*. Oxford: Blackwell.

CERI (2004) 'Cross-border post-secondary education in the Asia-Pacific region'. In: *Internalization and Trade in Higher Education*. Paris: OECD.

Dewey, J. (1916) *Democracy and Education: An Introduction to the Philosophy of Education*. New York: Macmillan.

Falk, R. (1995) *On Humane Governance: Towards a New Global Politics*. Cambridge: Polity Press.

Friedman, T. (2000) *Lexus and the Olive Tree: Understanding Globalization*. New York: Farrer, Straus & Giroux.

G8 (2000) *Okinawa Charter on Global Information Society*, G8 Kyushu-Okinawa Summit Meeting's Declaration. At: http://lacnet.unicttaskforce.org /Docs /Dot% 20Force/ Okinawa%20Charter.

GATS (2004) *General Agreement on Trade in Services*. New York: World Trade Organization. At: www.wto.org/english/tratop_e/gats_factfiction1_e.htm.

Hall, S. (1992) 'The question of identity'. In: Hall, S., Held, D. and McGrew A. (eds) *Modernity and its Futures*. Cambridge: Polity Press.

Harvey, D. (1989) *The Conditions of Postmodernity*. Oxford: Blackwell.

Held, D. and McGrew, A. (eds) (2000) *The Global Transformation Reader: An Introduction to the Globalization Debate,* second edition. Cambridge: Polity Press.

Knight, J. (2002) 'Trade creep: implications of HATS for higher education policy'. *International Higher Education,* 28, 4.

Labaree, D. (1997) *How to Succeed in School Without Really Learning*. New Haven: Yale University Press.

MacIntyre, A. (1981) *After Virtue*. London: Duckworth.

Marginson, S. (2003) 'The phenomenal rise in degrees down under'. *Change,* May–June: 1–7.

OECD (1996) *Lifelong Learning for All*. Paris: OECD.

OECD (2000) *Gender and Economic Development: The Work of Diane Elson*. Paris: OECD. At: www.oecd.org/LongAbstracts/0,2546.

Pennycook, A. (1999) 'Development, culture and language: ethical concerns in a postcolonial world'. Presented to Fourth International Conference on Language and Development, October 19 1999. At: www.languages.ait.ac.th/hanoi_ proceedings/ pennycook.htm.

PISA (2005) *Program of International Student Assessment*. At: www.pisa.oecd.org/ pages/0,2987,en.

Plomp, T., Anderson, R., Law, N. and Quale, A. (2003) *Technology Education: Cross-National Information and Communication Technology Policy and Practices in Education*. Greenwich, Connecticut: Information Age Publishing.

Rizvi, F. (2005) 'International education and the production of cosmopolitan identities'. *Research in Higher Education, International Series 9,* Hiroshima, Japan: 77–92.

Sassen, S. (1991) *The Global City: New York, London and Tokyo*. Princeton, NJ: Princeton University Press.

Scharito, T. and Webb, J. (2003) *Understanding Globalization*. London: Sage.

Scholte, J. (2000) *Globalization: A Critical View*. London: St Martin's Press.

Smith, M. (2000) *Transnational Urbanization: Locating Globalization*. Oxford: Blackwell.

Strange, S. (1996) *The Retreat of the State*. Cambridge: Cambridge University Press.

TIMMS (2005) *Trends in International Mathematics and Science Study*. At: http://nces. ed.gov/timss.

Tollefson, J. (1991) *Planning Language, Planning Equality: Language Policy in the Community*. New York: Longman.

UN Declaration of Human Rights 1948–1998. At: www.un.org/overview/rights.html.

UN (2004) *Human Development Report 2004*. New York: United Nations. At: http://hdr. undp.org/.

UNESCO (2001) *All for Girls' Education! Why it is Important*. Paris: UNESCO. At: http:// portal.unesco.org/education/en/ev.php.

UNESCO (2002) *Building the Capacities of Curriculum Specialists for Educational Reform,* Final Report of the Regional Seminar, Vientiane, Lao PDR, September 9–13, 2002. Paris: UNESCO.

World Bank (2004) *Gender and Development*. Washington, DC: The World Bank.

2 Diversity and multicultural education

Cross-cutting issues and concepts

Gajendra K. Verma

> The fundamental change that is necessary is the recognition that the problem fac-
> ing the education system is not how to educate children of ethnic minorities, but
> how to educate all children – (ours) in a multiracial and multicultural society and
> all pupils must be enabled to understand what this means.
>
> (*The Swann Report,* DES, 1985, p. 363)

The above quotation is from the Swann Report (DES, 1985), an influential report
on the education of children from ethnic minority families in the UK. Published
under the title of *Education for All,* the report emphasised the need to ensure that
education not only addressed the particular learning needs of young people
regardless of their ethnicity, but it also emphasised the need to teach all young
people how to respect ethnic and cultural differences and to accept them as part of
the cultural richness of life in our modern world, and not as a threat.

Twenty years on, the messages from Swann remain vital. We live in an increasingly interdependent and globalized world, in which we work and interact with
groups of different ethnicities and cultures – some beyond national borders – as
well as with people of varying cultural, linguistic, and religious groups *within* the
nation-state. In such a world of increasing interdependence – economically,
socially and politically – multicultural education can play an important role in
challenging stereotypes, prejudices and ethnocentric perspectives of both individuals and groups in national and international exchanges.

Consider a recent definition of globalization:

> Globalization results from the abolition of borders for all kinds of economic,
> financial and cultural activities. It affects not only the economic and finan-
> cial sphere but also national cultures and services, including education. In
> education it leads to an increased concern for quality.
>
> (Caillods, 2003, p. 1)

Globalization confronts societies and individuals with new learning challenges
that educational planners often do not know how to tackle. Paradoxically, with

globalization comes increased localization of educational enterprise. Decentralizing educational administration is the second major issue which according to Caillods (2003) has profoundly transformed planning practices. Caillods' phrase 'abolition of borders', apart from its literal meaning, also serves as a symbolic reminder of the ethnic diversity to be found within modern societies, often stemming from massive immigration since 1945.

Caillods' definition of globalization indicates two important parts of an equation, namely that it impacts not just on societies, but also on individuals. Furthermore, the reference to 'increased decentralization' occurring hand-in-hand with globalization also provides an indication of the intricacies of serving needs at both societal and individual levels, through localized provision. Finally, the 'learning challenges' posed by globalization represent ones facing policy makers and planners, teacher educators and, crucially, young people.

Ethnic diversity

Ethnic diversity is reflected in differences arising from linguistic styles, cultural and religious values and traditions which influence the behaviour, cognitive styles, attitudes and values of groups of people in a given society (Verma, 1989). The benevolent acknowledgement of ethnic diversity is not new but is accentuated by a greater political consciousness among groups wishing to retain or emphasise their identity within the country in which they now live (Gollnick and Chinn, 2002; Verma, 1989). In the UK, for example, there has been a long history of immigration over the last few centuries, yet ethnic diversity has only become 'an issue' of consequence in the wake of the large scale immigration and settlement that occurred in the 1950s to 1970s (Verma, 1986).

Differences in socio-economic status and tensions between ethnic groups have been the product of minority groups finding themselves subject to discrimination in access to employment and housing, and even sometimes to violence, and finding their life chances impaired by gross inequalities in the system (Verma, 1999).

Carl Grant (1995) argues that the concept of diversity demands the awareness, acceptance and affirmation of cultural and ethnic differences. In addition, Grant suggests that diversity promotes both the appreciation of human differences and the belief that in order for students to think critically – especially about life circumstances and opportunities that directly or indirectly impact their lives and the lives of their family members, community and country – they must affirm both *social* diversity (cultural pluralism) and *human* diversity.

It is interesting that Grant makes the distinction between the social and the human dimensions of diversity. He does so, it would appear, because of a legitimate concern about the individual, especially in an ethnically diverse society. He suggests that the term 'at risk', used in the report *A Nation at Risk* (National Commission on Excellence in Education, 1983), had subsequently been used, not to refer to the 'nation', but to those students, 'who are most often of color and poor and whose first language isn't English' (Grant, 1995, p. 4). One of the effects of this is negative stereotyping. He later asserts that certain characteristics of student diversity (such as colour, language, ethnicity, and socio-economic class) can label an individual as a potential educational winner or loser.

> The term often sticks to a student and clouds teachers' perceptions of that person as he or she moves through the grades. Finally, whether or not the student needs a great deal of assistance, the 'at risk' label ... can bring forth a self-fulfilling prophecy.
>
> (Grant, 1995, p. 11)

I would argue that this is not a phenomenon that is unique to American society. It is one that has considerable universality, certainly as far as ethnically diverse societies are concerned. Rivlin and Fraser (1995) remind us of the importance of approaching people in such ethnically diverse societies on an individual basis. Individual differences are not simply a matter of one's ethnicity. Regardless of ethnic origins, every individual reflects in his or her lifestyle, the socio-economic class of which the person is a part. 'This is not to say that all lower-class, upper-class or middle-class persons are carbon copies of one another – but it is to recognize the influence of social class on the individual' (Rivlin and Fraser, 1995, p. 375). The social class element is a factor of increasing relevance in multicultural societies in which ethnic groups are well-established, with an ever-increasing proportion being of working age, born and educated in the country of settlement, rather than born overseas.

Multicultural education

Multicultural education should not be considered simply as something that ought to permeate the education of young people, not just as another requirement that we place on schools and on further and higher education institutions, but as reflecting a set of values which inform all social institutions. Key to the implementation of multicultural education are the recognition and acceptance of the right of different cultural groups to exist and share equally in the differential rewards of social institutions (Gollnick and Chinn, 2002).

Two broad strategies are required in the planning and provision of multicultural education in a plural society. By 'plural society', I mean a society that respects and accommodates ethnic differences and in which there is equality of opportunity, regardless of an individual's origins. The first strategy relates to the philosophy that should underpin the work in all schools and institutions. The second relates to particular educational provision made to meet particular educational needs of children and young people from different ethnic minority groups. Neither strategy can be effective without the other (Verma, 1993).

This second strategy is one employed in schools where a significant proportion of the student population is of ethnic minority origin. In Britain, such schools are all too often located in the most socially-economically disadvantaged urban areas, especially the inner city ones. Very often too, teachers face the greatest challenges, not just because of ethnic diversity among their students, but also because of the hardships and prejudice experienced by their families, and because of poorly resourced schools and large classes. Furthermore, apart from poor physical resources, such schools may face real staff recruitment problems, with a high staff turnover, and a high proportion of newly-qualified and relatively inexperienced teachers. While there may be the need for special provision to provide extra support for children from ethnic minority families, especially those who have recently arrived in the country, such provision cannot be considered to represent all that a true multicultural education requires. It might provide some alleviation, but it alone cannot provide the basis for a long-term solution.

The 'Education for All' recommended by the Swann Report (DES, 1985) was the product of a widely recognised need for social justice and for equality of opportunity. In the UK context, this meant finding a way to prepare *all* children and adolescents for life in a multicultural society. Although this provided the broad philosophical framework for addressing equality of opportunity and social justice, the Committee proposed a number of more specific measures designed to alleviate the disadvantages experienced in school by students from minority groups. These included addressing scholastic underachievement (especially of students from Bangladeshi, Afro-Caribbean, and Pakistani backgrounds). There were also elements in the Education for All philosophy propounded by the Swann Report that were closely related to inclusiveness, and one which was also part of the philosophy of the Warnock Report (1978) into provision for children with special needs.[1] Elements of such a philosophy ought to underpin formal education provision in any civilised country.

The first strategy required is one aimed at preparing people to cope with diversity, so that they do not feel challenged by or feel 'under threat' from ethnic/cultural diversity: thus the dominant groups should come both to understand and respect value systems that differ from their own.

The second strategy offers measures that are responsive to the needs of groups/individuals who are experiencing disadvantages in the education system, as a result of being brought up in a culture that is some way distant from the mainstream culture. Within the school system, the objective is to provide children from different cultural backgrounds with access to the same personal opportunities as those from main cultural groups. This may necessitate some bilingual support in key transition phases. The goal of multicultural education should not simply be to recognise and appreciate cultural diversity as practised in most western democratic societies, for this can amount to mere tokenism. People must understand the significance of a culture's history and tradition as part of the dynamic and multifaceted culture of any contemporary society. The education system therefore ought to develop curricula and pedagogies that integrate and understand cultural process and cultural continuity, and changes within a framework of the complex national identity of a plural society.

The delivery of effective multicultural education is heavily dependent on the quality and training of the teaching force recruited to deliver it (Verma, 1993). It is sometimes argued that multicultural education strategies may reduce the present inequality which exists within the education system. Yet when analyzed at the macro level, many of the factors contributing to inequality transcend the boundaries of the education system itself, and reflect socio-economic inequalities of society. Most Western European states claim, and probably believe that they espouse, equality – thinking of it as a central pillar of their law and administration. Unfortunately, however, arguments can readily be adduced to show that this is not so. It can be argued that states, by their laws and administrative processes, are concerned to ensure the perpetuation of inequality so that those who have, retain their privileges and those who have not, continue to be deprived of them. Such an arrangement serves the social stability of the state well. In Britain social class divisions and the unequal division of wealth, with many ethnic minorities being 'the poorest of the poor', means that schools serving ethnic minorities struggle with lack of resources, larger classes, and impermanence of teaching staff (see Bagley's critique of the British educational system in the final chapter of this volume).

Teacher education

The Swann Report (DES, 1985) was the work of a public committee of inquiry set up by the UK government to investigate the education of children from ethnic minority families. Among the report's findings were a number of failings on the part of the teaching profession in addressing the needs of children from ethnic minority families. Research evidence, various reports and the findings of the Swann Report clearly show that the factors contributing to underachievement of ethnic minority pupils are:

- stereotyped attitudes in teachers;
- low expectations among teachers;
- a eurocentric/anglocentric curriculum;
- biased assessment and testing procedures;
- poor communications between school and home;
- racism in the educational system;
- racial prejudice and discrimination in society at large.

(Pumfrey and Verma, 1993)

Teacher education needs not only to focus on the classroom, that is, on the 'mechanics' of teaching and learning, but also on the impact of these on classroom interaction. Teaching–learning processes are not culturally neutral, but are heavily value-laden (Verma, 1993). This has an important bearing not only on what is taught and on how effectively it is taught, but it also has an important bearing on how students perceive themselves, their fellow students and others around them. It is important that teachers understand more about how the cultural messages, implicit in their teaching processes, affect students from diverse backgrounds.

There is a moral obligation implicit in the task of teacher education to consider how best to prepare teachers to work in schools so that they will be:

- conscious of the ethnically and culturally diverse nature of the societies they live in;
- capable of recognizing their own prejudices;
- able to identify discrimination in others and in the institutions they work in;
- prepared to act as agents of change in the education of a diverse and pluralistic, but harmonious, society.

There is some evidence to suggest that many, if not most, trainee teachers have quite well-formed ambitions of the kinds of schools they wish to teach in, the priorities they have in seeking to develop young minds, and the kinds of youngsters they intend to work with. However, as maturing young adults undergoing a course of professional preparation, it is reasonable to expect these intentions to be subject to change and refinement.

A teacher education course should take account of the needs of a teacher to:

- be equipped to prepare young people for a life in a multicultural and harmonious society;
- have an awareness and understanding of racism, both historically and in contemporary society, and to be conscious of the various forms in which racism manifests itself;
- have an awareness of intercultural relations and of their social and economic contexts;
- be able to teach with sensitivity and skill, recognising the particular needs of ethnic minority students;
- interact effectively with colleagues in the institutional framework in relation to these issues.

Also, it is important that these issues permeate the *whole* training curriculum, and that they are not simply 'added on' to the training package, nor merely offered as an optional study module.

From a multicultural education perspective, teacher training programmes need to achieve the following:

- to raise the awareness of students in their critical approach to cultural bias, prejudice, racism and stereotyping in teaching schemes, school texts and other teaching materials, and the ways in which they are used;
- to adopt an approach to all subjects in the school curriculum which avoids an ethnocentric view of the world;
- to recognise the values of teaching which identify and acknowledge effectively the aspirations of all students, and seek to enhance their chances of maximizing their potential;
- to prepare all elements of the course with multicultural and anti-racist considerations, in both theoretical and practical components;
- to identify and use effective strategies for working with students whose mother tongue is not the language of instruction.

Objectives such as these derive their legitimacy from the ideals of a true Education for All. All students should arrive at an awareness of the cultural identity and belief systems of different ethnic groups, with at least a critical respect for their values. This should be regarded as being of personal benefit to them as individuals and as essential for a humane and just society. Without this awareness, and without an understanding of how racism operates to their disadvantage, young people intending to be teachers are not adequately equipped to guide and help form the attitudes of children and adolescents as they prepare them for life in the society in which they are growing up.

Moreover, in this age of globalization, with its implications for migration and increasingly complex ethnic diversity and competition in many societies, it is important to recognise that multicultural education embraces not only a local perspective, but also has worldwide implications (see, for example, the comparative study of teacher training for multicultural awareness in the UK, Finland, Greece, Germany, France and Israel described below).

Such issues place very heavy demands on teachers. Among the complex tasks teachers face, they must be able to recognise racism and ethnocentrism, counter it in their teaching, and design new curricula that deal creatively with the controversies in the competition between shared values, and plural ways of seeing the world.

> Furthermore, educators cannot operate effectively without multiple partnerships. These enable them to draw on the skills of parents and the community to assist in diversifying the curriculum, affirming diversity rather than ignoring or devaluing it, and improving social relations between students.
>
> (Hickling-Hudson, 2003, p. 5)

Densmore (1995), writing of America, draws attention to the purpose of multicultural education, and reminds us that there is still much to do before it permeates the whole education system, a necessary pre-condition for creating a system that offers equality of opportunity for all students: 'Even though conservative critics have recently been trying to create the impression that ethnic-centered curricula have been imposed in schools nationwide, in fact, changes in the ethnic diversification of curriculum content are not yet widespread' (Densmore, 1995, p. 490). Multicultural education is intended, in Densmore's model, to affirm the worth and dignity of those students who have been historically marginalized.

Teacher education and technology

Research supported by the European Union has investigated the effectiveness of Web-based learning and tuition in helping train both student and qualified teachers in order to enhance their intercultural understanding, and their teaching strategies for working in multicultural classrooms. Quite apart from being able to access materials from the Web, those following such programmes have also been encouraged to engage in dialogue with fellow trainees in other European countries.

The *Immigration as a Challenge for Settlement Policies and Education: Evaluation Studies for Cross-Cultural Teacher Training* (ECT) project involved a partnership between teacher training institutions in Finland, Israel, France, Germany, Greece, and the UK (Pitkänen *et al.*, 2002, 2006). Another partnership between Finland, Germany and the UK completed a further experimental programme, building on the lessons learnt from the evaluation for cross-cultural teacher training (ECT) project. These EU-funded projects made extensive use of the Web to facilitate cross-cultural communication between student teachers in the participating countries.

The overriding lesson to be learnt from such projects is the fact that in an increasingly globalized world it is now possible to offer new opportunities in teacher education for multicultural societies. Technology makes it possible to bring together trainee and practising teachers and to encourage them to engage in dialogue with colleagues in other parts of the world, from their own homes, while offering support materials and tutorial help and support through the same electronic medium. Use of the Web creates new learning opportunities with a multifaceted interface: access to tutors, as well as other trainees on the same programme, but working in a different cultural context, seeking to enhance their intercultural understanding. This work is generating new dialogues, and facilitating understanding across cultures.

The European Union is not alone in making such developments possible. A 2002 UNESCO seminar on Open and Distance Learning (ODL) emphasised to its delegates that they would '"have an important role in achieving the great vision which motivates UNESCO's work, the vision of a world in which everyone can get an education", the vision of Education for All' (D'Antoni, 2003).

Ethnic diversity, and multicultural education are not challenges that face particular areas of the world only, but need to be recognised as *global* ones. While we may accept that challenges may vary in intensity from area to area (both nationally and regionally) because of local politico-cultural circumstances, their impact is global.

Conclusions

One of the effects of globalization is a form of 'cultural reductionism'. This appears to be a product of an increasingly global media (especially advertising and the pop culture) that increasingly penetrates our lives. This presents a challenge to the existing cultural frames within which modern societies operate, and to the values we hold. The younger generation seems most susceptible to the pressures of a superficial international popular culture. Over twenty-five years ago, in a book calling on French academics to fight to preserve the French language and culture in the face of the pressures of the English-speaking world, Gobard (1976, p. 122) referred scathingly to the risk of, 'Peoples in their infinite diversity becoming transformed into a horde of customers in the same international supermarket' (translated from the French).

Globalization poses considerable challenges, not least as far as education and influence upon the young in ethnically diverse societies, is concerned. In this, technological progress is a two-edged sword. The one edge offers new opportunities for peoples to meet and interact on a scale that was previously impossible.

This is so, whether we sit at our computers and explore the Internet, a minor cultural revolution in itself, or whether through the opportunities for travel now available to more people than ever before.

However, ever-increasing globalization of capitalist enterprises might well contribute to inequalities in the treatment and life chances of different sectors of the population in any societies stratified by ethnicity and social class. Increasingly, there will be additional pressures on national education systems to provide minimum-level, transferable skill training for sectors of young people who are useful for international capitalism. In ethnically diverse societies, tensions may become acute when these educational policies are not applied even-handedly, favouring the more privileged ethnic groups. Hence it is all the more important that the education system plays its full part in tackling inequalities. Hernes and Martin (2003) argue that education has both the potential of either easing or exacerbating ethnic conflict through the way it is organized and delivered to different ethnic groups:

> School is where life's chances are distributed – often unequally – and thus may either favour or hamper the social mobility of different ethnic groups. School is also the place where 'socially constructed' attitudes towards other ethnic groups may be formed or reassessed and its functioning thus determines the rules of ethnic interaction.
>
> (Hernes and Martin, 2003, p. 1)

Otherwise, there is a risk of internal unrest, which may further jeopardise the capacity of national governments to attract the inward investment needed to remain competitive and to offer good standards of living for all sectors of the local population. Thus there is an economic imperative that stands alongside the moral one. Educators must ensure that they offer the very best education possible for *all* young people regardless of their ethnic background, and wherever they live. Education can play an important role in leading the battle against inequality (Verma, 1993):

> Education in the twenty-first century can become an essential contributor to integration, to a culture of peace, and to international understanding. Through this we can assure respect for diversity, whether diversity of behaviour, or diversity of philosophical or religious belief.
>
> (Verma, 1997, p. 337)

Notes

1 Ironically, Warnock (2005) has argued that the principles of inclusion that she advocated have not been fulfilled, to the extent that Britain now stands in violation of the UNESCO Salamanca Statement principles on equal and fair treatment of pupils with 'special educational needs'.

References

Caillods, F. (2003) 'The changing role of the state: new competencies for planners'. *International Institute for Educational Planning Newsletter,* 21, 2.

D'Antoni, S. (2003) 'Open and distance learning: technology is the answer but what was the question?' *International Institute for Educational Planning Newsletter,* 21, 14.

Densmore, K. (1995) 'An interpretation of multicultural education and its implications for school–community relationships'. In Grant, C. (ed.) *Educating for Diversity: An Anthology of Multicultural Voices.* Boston: Allyn & Bacon.

DES (1985) *Education for All: Report of the Committee of Inquiry into the Education of Children from Ethnic Minority Group (The Swann Report).* London: HMSO Department of Education and Science, Cmnd 9453.

Gobard, H. (1976) *L'Aliénisation Linguistique: Analyse Tetralogique.* France: Flammarion.

Gollnick, D. M. and Chinn, P. (2002) *Multicultural Education in a Pluralistic Society.* New York: Prentice Hall.

Grant, C. A. (1995) (ed.) *Educating for Diversity: An Anthology of Multicultural Voices.* Boston: Allyn & Bacon (sponsored by the Association of Teacher Educators).

Hernes, G. and Martin, M. (2003) 'Planning education in multi-ethnic and multicultural societies'. *International Institute for Educational Planning Newsletter,* 21, 1.

Hickling-Hudson, A. (2003) 'Teacher education for cultural diversity and social justice'. *International Institute for Educational Planning Newsletter,* 21, 5.

National Commission on Excellence in Education (1983) *A Nation at Risk: The Imperative of Educational Reform.* Washington DC: US Government Printing Office.

Pitkänen, P., Verma, G. K. and Kalekin-Fishman, D. (2002) (eds) *Education and Immigration: Settlement Policies and Current Challenges.* London: Routledge-Falmer.

Pitkänen, P., Verma, G. K., and Kalekin-Fishman, D. (2006) (eds) *Increasing the Multicultural Understanding of Student Teachers in Europe.* Oxford: Trafford Press (for The Manchester Educational Research Network).

Pumfrey, P. and Verma, G. K.(1993) (eds) *Cultural Diversity and the National Curriculum: Vol. 1 – The Foundation Subjects and RE in Secondary Schools.* London: The Falmer Press.

Rivlin, H. N., and Fraser, D. M. (1995) 'Ethnic labeling and mislabeling'. In Grant, C. (ed.) *Educating for Diversity: An Anthology of Multicultural Voices.* Boston: Allyn & Bacon.

Verma, G. K. (1986) *Ethnicity and Educational Attainment.* London: Macmillan.

Verma, G. K. (1989) (ed.) *Education for All: A Landmark in Pluralism.* London: The Falmer Press.

Verma, G. K. (1993) (ed.) *Inequality and Teacher Education: An International Perspective.* London: The Falmer Press.

Verma, G. K. (1997) 'Inequality and intercultural education'. In Woodrow, D., Verma, G., Rocha-Trindada, M., Campani, G. and Bagley, C. (eds) *Intercultural Education: Theories, Policies, and Practice.* Aldershot: Ashgate, England.

Verma, G. K. (1999) 'Inequality and education: implications for the psychologist'. *Education and Child Psychology,* 16, 6–16.

Warnock, M. (1978) *Children and Young People with Special Educational Needs.* London: HMSO.

Warnock, M. (2005) *Special Educational Needs: A New Look.* London: The Philosophy of Education Society.

Part II

Inclusive education: conceptual issues

3 Barriers to student access and success

Is inclusive education an answer?[1]

Madan Mohan Jha

Introduction

The principle of basic education as a human right has been accepted internationally. However, the experience in many developing countries shows that a large number of children are not able to complete the minimum number of school years. They face a variety of barriers before coming to school and even within the school. Does 'inclusive education' offer a solution? This chapter attempts a response to this question by analyzing the origin, concept and practices of inclusive education, and also the nature of barriers which children, particularly those at risk and from the disadvantaged sections, have to confront when they want to access school education.

Origin of the concept 'inclusive education'

The 'Salamanca Statement' adopted at the 1994 World Conference on Special Needs Education: Access and Quality urged all governments to: 'Adopt as a matter of law and policy the principles of *inclusive education,* enrolling all children in regular schools, unless there are compelling reasons for doing otherwise' (UNESCO, 1994, Salamanca Statement, p. ix).

There are two distinct perspectives on inclusive education. First are those emerging largely from the developed countries; and second are those referring to the felt need and circumstances prevailing in the developing world. In richer developed countries, education is largely inclusive of girls, the disadvantaged, and all ethnic minorities. Children challenged by disabilities, sensory, cognitive and physical, were previously educated in separate 'special' schools, but are now being recommended for admission to regular schools with an inclusive orientation. Discourse on inclusive education in developed countries mostly centres on the extension of special education, or at most a reform of special educational practice. The underlying approach in this perspective has been the assumption that children's disabilities are due to medical factors that need to be addressed in order to adapt them for the conventionally organized school, its curriculum and pedagogy.

However, a plethora of critical literature has emerged recently, re-examining the concept of inclusive education from an educational reform perspective.

Schools in this critical perspective should respond and adapt to the needs of *all* children, regardless of gender, physical, cognitive and sensory needs, ethnicity, and religious and cultural background, and fit themselves to children's learning styles and needs, and not the other way round. Ferguson (1996), Udvari-Solner (1996), Thomas *et al.* (1998), Ainscow (1999) and Mittler (2000) have extensively discussed the school reform perspective in order to develop the concept and practice of inclusive education.

Sebba and Ainscow (1996) have offered the following definition of inclusion:

> Inclusion describes the process by which a school attempts to respond to all pupils as individuals by reconsidering its curricular organization and provision. Through this process, the school builds its capacity to accept all pupils from the local community who wish to attend and, in so doing, reduces the need to exclude pupils.
>
> (p. 9)

The presumption in this definition is that most students from the local community would 'wish to attend' the neighbourhood regular schools. Those who do not may be going either to special schools or to private (including boarding) schools. In the UK, some seven per cent of pupils attend private schools of one kind or another.

Inclusion in developing countries

Developing countries demand a different approach to the concept of inclusive education. In such countries a high proportion of children may rarely attend school, or leave primary education prematurely, for a range of reasons including social and economic disadvantage. The 1994 UNESCO World Conference also understands this situation when it argues that a school should accommodate all children regardless of their physical, intellectual, social, linguistic or other conditions. This should include disabled and gifted, street and working children, children from remote or nomadic populations, children from linguistic, ethnic, or cultural minorities and children from other disadvantaged or marginalized areas and groups (UNESCO, 1994, pp. 11–12).

These inclusive schools: must recognize and respond to the diverse needs of their students, accommodating both different styles of learning and ensuring quality education to all through appropriate curricula, organizational arrangements, teaching strategies, resource use and partnerships with their communities (ibid.).

Special educational needs

From the UNESCO 'Salamanca Statement' and the 'Framework for Action on Special Needs Education' (1994) there does not appear to be any ambiguity in regard to approach and perspectives on inclusive education. Some confusion presumably has arisen, however, from the terminology 'special needs education' used for the title of the world conference, leading to the Framework for Action 'on principles, policy and practice in special needs education'. A similar term, and the concept 'special educational needs' or SEN was introduced in Britain by the Report of the Warnock Committee (1978), later enshrined in The Education Act of 1981 as follows:

> A child has 'special educational needs' if he/she has a learning difficulty, which calls for special educational provision to be made for him/her. A child has learning difficulty if she/he: (a) has significantly greater difficulty in learning than the majority of children of the same age; (b) has a disability which either prevents or hinders the child from making use of educational facilities of a kind generally provided for children of the same age in schools within the area of the local authority.
>
> (See Jha, 2002, p. 64)

The SEN concept represented some progress on educating children with disabilities in the UK, which earlier was mostly in separate schools of poor quality, as a matter of policy. Warnock abolished the eleven categories of special educational need which existed at the time, but increased the proportion of children needing special educational treatment from two per cent to twenty per cent. She considered a variety of factors that might contribute to learning difficulties, but was 'forbidden to count social deprivation as in any way contributing to educational needs' (Clough and Corbett, 2000, p. 4). In developing countries such as India, the aspects of social and economic deprivation cannot be ignored, and indeed in many developing countries a majority of children may be said to have special educational needs. The concept of SEN and identification of children with special educational needs under a statutory code of practice has been criticized by a number of commentators and educationists in Britain on a variety of grounds (Tomlinson, 1982; Galloway *et al.*, 1994; Vlachou, 1997; Booth *et al.*, 1998; Mittler, 2000).

Mittler (2000) sees the identification of children with 'special educational needs' as labelling and discriminatory. Ainscow (1999) sees the very concept of 'special educational needs' as a barrier to inclusion. Mittler argues:

> I think the concept of special educational needs, particularly as it is seen in this country [UK] becomes another barrier. I don't think it has a productive contribution to make to the inclusive education agenda. If anything, it is one of the barriers to moving forward.

The Salamanca Framework for Action did refer to a move from the term 'special educational needs' to 'inclusive education' when it concluded:

> In the context of this Framework, the term 'special educational needs' refers to all those children and youth whose needs arise from disabilities or learning difficulties ... There is an emerging consensus that children and youth

with special educational needs should be included in the educational arrangements made for the majority of children. This has led to the concept of the inclusive school.

(UNESCO, 1994, p. 6)

However, without significant changes in the policies and curricular arrangements in the schools including those in the West, the ultimate objectives of inclusive education cannot be achieved. Tomlinson comments on this difficulty:

> There is considerable anxiety that despite rhetoric of inclusive education, education policies in developed countries continue to ensure that vulnerable and disadvantaged groups are often excluded from the forms of education regarded as most valuable, and from gaining qualifications that can be exchanged for good employment, income and security. There is, in particular, a growing awareness that creating competitive markets in education, with schools competing for the most desirable pupils and resources, is incompatible with inclusive education.

(Tomlinson in Foreword to Jha, 2002)

The World Conference on Special Needs Education (1994) noted the need for reforms in school education, in both developing and developed countries:

> Special Needs Education – an issue of equal concern for countries of the North and South ... has to perform part of an overall educational strategy and, indeed, of new social and economic policies. It calls for major reform of the ordinary school.

(UNESCO, 1994, pp. iii–iv)

Barriers in schools

There are walls between schools and children before they get enrolled; they face walls with curriculum inside classrooms; and finally 'they face more walls when they have to take examinations which determine how successful they will be in life' (Jha, 2002). On walls and barriers confronting school systems today, it is further observed:

> Removing barriers and bringing *all* children together in school irrespective of their physical and mental abilities, or social and economic status, and securing their participation in learning activities leads to the initiation of the process of inclusive education. Once walls within schools are broken, schools move out of their boundaries, end isolation and reach out to the communities. The distance between formal schools, non-formal schools, special schools and open schools will be eliminated.

(Jha, 2002, pp. 15–16)

At most, school systems are confronting two types of barriers, external and internal. Children face external barriers before coming to and getting enrolled in schools. The nature of such barriers could be the physical location of schools, social stigmatization or the economic conditions of children. Sometimes non-availability of a school or its location in areas that cannot be accessed can become the major barrier for children seeking education. Children with physical, cognitive or sensory challenges face barriers if the building has not been constructed with their mobility needs in mind.

Schools offer a variety of reasons, particularly in countries which do not have strong neighbourhood school policies adapting to local needs, for rejecting admission of students whom they perceive to be difficult or unpleasant to teach, for reasons ranging from their alleged behavioural or disruptive problems to their unpopular ethnicity. In Europe for example, Gypsy and Roma children are frequently rejected by schools on the grounds that they are difficult to teach, slow to learn, and will soon move on. Exclusion of undesirable children by fees, examinations and administrative fiat is likely in educational systems where there is a strong private element, even when such schools are partially funded by the state. In India for example, forty-six per cent of secondary schools are private but are also state-aided, and their inclusion policies are often erratic and idiosyncratic. It takes a degree of political maturity for a country to organize education on a neighbourhood basis so that comprehensive schools admit all permanent and temporary (e.g. Roma) residents of the neighbourhood, and retain pupils without the need for expulsion or exclusion.

Children face barriers within schools and classrooms because of curriculum factors and teaching methodologies. Those who are visibly different can be isolated within schools, and even relegated to different classrooms in order to receive discriminatory treatment. In England, under the existing policy more than twenty per cent of children are identified through 'statements', a dossier of alleged problem behaviours, achievement failures, and the special demands such a child may place upon overworked teachers who have to service large classes. Statemented children are particularly likely to be excluded from schools in Britain. In developing countries too, the curriculum is not child-friendly, but relies on strict discipline and didactic style by the teacher, who is given no training or incentive to be flexible to the needs of individual children.

The realization is coming in many countries, such as the economically developing countries of east and south Asia that the present system of school organization and its associated curriculum may not be able to cope with the demands of globalization for a highly educated and flexible work force. A *Time Magazine* survey of east Asian schools (Beech, 2002) found that:

> Japan is completing its radical (educational) restructuring, abolishing Saturday classes, encouraging volunteerism and allowing schools to experiment with different curricula; Taiwan is scrapping its university entrance exam system in favor of a more holistic approach that considers grades, essays

and extracurricular activities, and South Korea is picking up a third of incoming college students not based on their test scores but for their unique talents.

These innovative educational models have yet to be adopted by India. Examination scores judge, in a rather arbitrary manner, success in the present model of schooling in India. This model of selection is a barrier for many promising students:

> Examinations also drive out many children, particularly the rural, the disadvantaged and the disabled, out of school. It is a great filtering mechanism. It suits the system, since only a select few students, largely from the urban middle class, get high scores, thanks to the system of tuitions and coaching, in order to get admission into higher academic institutions, which have limited seats.
>
> (Jha, 2002)

Inclusion: a solution to barriers which prevent success?

Inclusive schools are designed with a vision and principle that believe in the culture of rights, social justice and equity. The concept is that all children are not the same, and it accepts diversity as a strength, and not a problem. It believes in a basic pedagogy which asserts that children learn in different ways, and relates success more to the learning of life and social skills, than scoring high marks in narrowly defined examination curricula. The admission policy of such schools would accept children from a diverse community rather than rejecting them on the grounds of test scores, or other physical, social and economic factors.

Inclusive schools offer flexible curricula that would respond to the diverse needs of children. Child-centred pedagogy, and the application of Gardner's (1993) Multiple Intelligence (MI) principles, are other major departures from traditionalism that inclusive schools will follow. The UNESCO Framework has again highlighted the need of child-centred pedagogy:

> The challenge confronting the inclusive school is that of developing a child-centred pedagogy capable of successfully educating all children, including those who have serious disadvantages and disabilities. The merit of such schools is not that they are capable of providing quality education to all children; their establishment is a crucial step in helping to change discriminatory attitudes, in creating welcoming communities and in developing an inclusive society.
>
> (UNESCO, 1994, p. 6)

Traditional schools offer scope for the use of only two types of intelligence – linguistic and logical–mathematical learning styles. This singular approach can create learning barriers for many children, particularly those belonging to the first generation to be schooled, and members of various minority groups. Gardner

(1993) has identified seven types of intelligence: linguistic or verbal; logical–mathematical; spatial or visual; musical; kinaesthetic; interpersonal; and intra-personal. Schools encouraging the identification of these different intellectual styles, and fostering a child's talents over multiple intellectual styles, should be able to remove unseen and internal barriers that children face in traditional schools.

Inclusive schools use a variety of innovative practices to get children involved and participating in learning processes. Some of the inclusion strategies I propose (Jha, 2002, p. 140) are as follows:

- whole class inclusive teaching
- group/co-operative learning
- peer tutoring/child-to-child learning
- activity-based learning
- team approach/problem solving
- equity in assessment and examinations.

Inclusive education and its evolution in school systems as a process for removing barriers to access is a growing phenomenon. The strategies suggested above have been developed in many schools across different countries and also have conceptual and pedagogical backing. However, it is yet to be shaped into a reform movement or as a replacement of the traditional system within a country or state.

Quality with equity

There is one more dimension to the inclusion concept. It addresses the issues of quality in consonance with equity. In traditional styles of schooling quality and excellence are usually divorced from principles of equity of admission, teaching styles and curriculum development. The institution of the school as a public system for mass education has its origin in the industrial era, and was developed to create workers who could serve the first phase of globalizaton, that of preparation for colonial supremacy, and the domination of capitalist modes of production. This model of education serves the interests of social control over the masses, and still exists in many parts of Europe and Asia. Schools borrowed ideologies and vocabularies from industry, with ideas of 'performance', 'standard', 'quality control through testing' and so forth. In the information age schools adapt their curricula to serve the needs of global capitalism for a stable and co-operative work force. Skrtic (1991), Lipsky and Gartner (1999) and Lloyd (2000) have questioned the 'adequacy, relevance and appropriateness' of the public education system that was shaped and influenced by the needs of the industrial era, and its subsequent developments in a world of globalizing trade.

The post-industrial, modern work place based on rapid technological change and communications development requires a more collaborative, problem-solving and team-work basis. 'Collaboration means learning collaboratively with and from persons with varying interests, abilities, skills and cultural perspectives' (Skrtic, 1991). Equity, therefore, becomes a pre-condition for post-industrial era schools. Skrtic observes further: 'The successful schools in the postindustrial era will be ones that achieve excellence and equity simultaneously – indeed ones that recognize equity as the way to excellence' (p. 223).

Open education

Open education is characterized by the removal of 'restrictions, exclusions and privileges' (Richardson, 2000). It provides an alternative curricular route to students who are not able to cope with the rigid curriculum and fixed timeframe of the traditional school system. For many students and parents, however, the alternative educational approach is a second choice, considered only when they have not been able to access the traditional school system. However, the growth of information and communication technology in recent years and its application to education is reducing the distance between open, interactive learning systems and the traditional rigidities of the more formal type of schooling. Children in regular schools are accessing information with the help of modern educational technology and the Internet. They are becoming active partners in knowledge production, as they would do in the open system. Teachers are changing their role and are becoming facilitators rather than stern dominies. Schools are becoming learning places for dialogues and exchanges.

Inclusive education in its philosophy, and also in its practice, is closer to the open education system. In India's 'national open school' for instance, students have demolished the myth that 'open school' must correspond to the mode of 'distance education', whereby students should not assemble daily at a place and teachers should not be on hand to help them. Many children with disabilities in special schools are opting for open school curricula. Such open schools are removing barriers to access for a cross section of students and are assuring success that might have been denied by the traditional school system.

Case examples of open schooling

Two schools in India have been studied closely as examples. They have addressed the issues of equity and quality simultaneously and are close to the concept of inclusive schooling, though they remain under the administration of school boards.

Loreto Day School (Sealdah – Kolkata) is affiliated with the West Bengal State School Board, but is unlike many other private or partially aided schools in the country. In 1979 it contained ninety poor, non-fee paying students out of a total of 790 on its roll. In 1998 the school had a roll of 1,400 students, 700 of whom were non-fee paying. These students were subsidized by the fee-paying students, sponsors and donors and by the West Bengal government. This increase in non-fee paying students stems from a vision and value system that the school has created for itself. Its other programmes include the 'Rainbow School' – a school-within-school for street children. This is not a tag-on afternoon programme to address equity issues in a token fashion, but is a structured and integral programme of curriculum development, and child-to-child teaching and learning.

The street children are individually tutored by 'regular' pupils from classes Five to Ten as part of their work experience. Many 'rainbow children' succeed in becoming enrolled in regular schools, and others have found secure jobs. The school runs many other programmes and activities to reach out to the community.

> Loreto challenges a fixed view of school and its structures by seeking to live out a set of values which continually challenges parents, teachers and pupils of the school to build an outward looking community, to be flexible, and to live in simplicity ... flexibility places utmost value on people ... simplicity places the resources at Loreto's disposal in the broader context ... it therefore stands against acquisitiveness, consumerism and the trappings of modern life in favor of valuing people and relationships.
>
> (Jessop, 1998)

The school has also maintained the conventional academic performance of its students, with fifty per cent having first class marks in the Year 12 public examinations conducted by the school board. Loreto has succeeded in breaking the conventional mindset that creates barriers to access by poor students, as well as the structural features (e.g. fees) which bar entry. There is an expectation that all students can succeed, regardless of their status or social origins. 'There are lessons for all schools, worldwide, rich and poor, in the boundary breaking strategies which Loreto has adopted to maximize its resources' (ibid.).

There are many schools in Kolkata and other projects in India which bring better-off children face to face with children of economically poor parents, though not to the extent and manner that Loreto does. The point made is that breaking the barriers to access need not be an isolated strategy but could become a systemic strategy to bring inclusion, equity and redefined quality as a wider vision of education.

St. Mary's School in New Delhi developed an inclusion policy with the admission of Komal Ghosh, a student with severe cerebral palsy, who was earlier in a special school. 'Komal's presence helped school become more humane,' said the Principal. Since then the school has opened its gates to other types of children with disabilities, along with orphans, and economically poor students. Priority in admission is given to neighbourhood students and all children learn together in the same classroom. Teachers have evolved a variety of teaching methodologies to involve children in a variety of learning activities, both group and individual. The school has focused on social integration and the skills development of all students, regardless of background and handicap, and, overall, students do not have a strong profile in public examinations. Teachers meet frequently to share their experiences, seeking to develop problem-solving strategies for individuals and groups, addressing the learning needs of all children. In addition, the school has outreach programmes whereby it helps children and adults from underprivileged groups to achieve literacy and other skills. These two examples suggest that adoption of the inclusion process by schools can develop in natural ways that can vary according to ecological setting and populations served.

As a matter of policy, Indian law requires that children with disabilities be educated in regular schools as far as possible. Many schools, including some private schools, are following this policy by giving admission to these children. But, in the absence of a vision and orientation, children get isolated in schools and many times are segregated in special units. Even when they are in mainstream classes they can be ignored, especially when the class is very large, the teacher has no aides and no training or structural support for programmes of inclusive education. The concept of inclusion, initiated as an educational policy for all children, must go beyond the idea of special education, particularly when one addresses policy development in developing countries. Inclusive education takes into its fold the vulnerable, and children at risk for whom access is not just a question of physical availability of space in schools and services of teachers; the success of such programmes is not measured merely by success in public examinations.[2]

Conclusion

Barriers to access can be viewed in physical as well as structural terms. But more than that, it is the curriculum, the pedagogy, the examination and the school's approach, which create barriers. Unless these unseen barriers are taken care of, access for all children and an assurance of success for all will remain an unachieved goal. The inclusive education movement, combined with technological developments and new approaches to open schooling, has come at a crucial time. Countries, and school systems choosing a holistic approach to access and success are more likely to succeed in reaching Education for All.

Notes

1 Updated version of a paper presented to the Commonwealth of Learning Conference, Durban, South Africa (June, 2002). At the time of writing the author was a joint secretary to the Government of India in the Ministry of Human Resource Development. Views expressed do not necessarily represent those of the Government of India.
2 For a detailed account of inclusive education practices and policies in three New Delhi schools, see Jha (2006).

References

Ainscow, M. (1999) *Understanding the Development of Inclusive Schools.* London: Falmer.
Beech, H. (2002) 'School daze'. *Time Magazine,* 159, No. 14, April 15th, 2002.
Booth, T., Ainscow, M. and Dyson, A. (1998) 'England: inclusion and exclusion in a competitive system'. In Booth, T. and Ainscow, M. (eds) *From Them to Us: An International Study of Inclusion in Education.* London: Routledge.
Clough, P. and Corbett, J. (2000) *Theories of Inclusive Education.* London: Chapman.
Ferguson, D. (1996) 'Is it inclusion yet? Bursting the bubbles'. In Beres, M. (ed.) *Creating Tomorrow's Schools Today: Stories from Inclusion, Change and Renewal.* New York: Teachers College Press.
Galloway, D., Armstrong D. and Tomlinson, S. (1994) *The Assessment of Special Educational Needs: Whose Problem?* London: Longman.

Gardner, H. (1993) *Frames of Mind: Theories of Multiple Intelligence.* London: Fontana.

Jessop. T. (1998) *A Model for Best Practice at Loreto Day School, Sealdah, Calcutta.* New Delhi: Education Sector Group, DfID.

Jha, M. (2002) *School Without Walls: Inclusive Education for All.* Oxford: Heinemann.

Jha, M. (2006) 'The inclusion of children with "special"educational needs in India: a case study of three schools in New Delhi'. Unpublished D. Phil. thesis, University of Oxford.

Lipsky, D. and Gartner, A. (1999) 'Inclusive education; a requirement of a democratic society'. In Daniels, H. and Garner, P. (eds) *World Yearbook of Education 1999: Inclusive Education.* London: Kogan Page.

Lloyd, C. (2000) 'Excellence for all – false promises! The failure of current policy for inclusive education and implications for schooling in the 21st century.' *International Journal of Inclusive Education,* 4, 133–151.

Mittler, P. (2000) *Working Towards Inclusive Education: Social Context.* London: Fulton.

Richardson, J. (2000) *Researching Student Learning: Approaches to Studying in Campus Based and Distance Learning.* Buckingham: Open University Press.

Sebba, J. and Ainscow, M. (1996) 'International development in inclusive schooling: mapping the issues'. *Cambridge Journal of Education,* 26, 5–18.

Skrtic, T. (1991) *Behind Special Education: A Critical Analysis of Professional Culture and School Organization.* Denver: Love Publishing.

Thomas, G., Walker, D. and Webb, J. (1998) *The Making of the Inclusive School.* London: Routledge.

Tomlinson, S. (1982) *A Sociology of Special Education.* London: Routledge.

Udvari-Solner, A. (1996) 'Theoretical influences on the establishment of inclusive practices'. *Cambridge Journal of Education,* 26, 19–30.

UNESCO (1994) *The Salamanca Statement on Special Needs Education.* Paris: UNESCO.

Vlachou, A. (1997) *Struggle for Inclusive Education.* Buckingham: Open University Press.

Part III

Diversity, equality and education in the United States

4 Diversity and inclusion in the United States

The dual structures that prevent equality

Carl A. Grant

Introduction

Macedo (2000) declares that "Diversity is the great issue of our time". He observes that nationalism, religious sectarianism, a heightened consciousness of gender, race and ethnicity, a greater assertiveness with respect to sexual orientations, and a reassertion of religious voices in the public square, are but a few of the forms of particularity that stubbornly refuse to yield to individualism and cosmopolitanism.

And, let me add, racism as another "form of particularity" that refuses to yield. The purpose of this chapter is to discuss diversity and inclusion in the United States, particularly in education. I begin with definitions, some history, and contextual framing of diversity and inclusion in the United States. Next, I address the conceptual perspective that I use to frame the discussion. Third, I discuss the status of diversity and inclusion in Kindergarten to Grade 12 education by pointing out how dual structures are maintained in two areas of education: school desegregation/integration and curriculum.

Definitions, historical context, and theoretical perspective

Diversity, as used in this chapter, refers to the differences among people. Although there are many differences among individuals, I am referring to group differences. Traditionally, in the United States, the national (e.g. political, media and educational) focus on group difference has been on race and ethnicity and, since the 1970s, it has included gender equity (Bem, 1993; Watkins *et al.*,1993) and disability (Education for All Handicapped Children Act of 1975). Currently, the discourse on group diversity is broadening, but not without opposition, to include religions other than Christian (Sacks, 2002) and sexuality (Kumashiro, 2002). Inclusion, as used in this chapter, is the bringing into the intra-structures of societal institutions race and ethnic groups, and other cultural groups who are located on the margins of public discourse, practice, and action. Here I am not addressing inclusion as a form of racial and cultural assimilation, but as "social participation". Social participation does not assume that those who are becoming included must assimilate into the dominant norms of society's institutions (Tinto, 1993).

Before colonial time, cultural and racial diversity was a social reality on the continent of North America where the United States is now a nation. When the Pilgrims (English Separatists) arrived in 1620 at Plymouth in New England, American Indians were living in the area. It was because of the benevolence of the American Indians that the Pilgrims were able to survive their first harsh New England winter. Diversity and inclusion was off to a good start in the "New World" and in order to show appreciation for the great deeds of the American Indians, and to celebrate the Pilgrims' accomplishment, a day of thanksgiving was proclaimed. Historians report that a major feast was held and the Pilgrims and Indians had a joyous celebration. A celebration of thanks – Thanksgiving Day, a national holiday – takes place each November in the United States to remind the nation of the occasion and the generosity of the American Indians. Thanksgiving Day, not withstanding the positive perspective on diversity and inclusion that the Pilgrims and American Indians initiated, did not last. Instead, early White Americans put in place a dual structure: one way to treat White people and another way to treat people of color.

Ringer (1983) informs us that since colonial times White America's response to and treatment of perceived "racial" minorities has had a dual character, which stressed the separateness of the races and the inferiority of Non-White peoples. This dual structure also aids and abets patterns of knowledge production within society, including the rules that frame and guide how people think and act in socially stereotyped ways. Popkewitz (1998) calls this framing and guiding "systems of reason". In the early nineteenth century, Alexis de Tocqueville observed and reported on the dual structure that existed between Whites and Non-Whites and the systems of reasoning that kept it in place. Tocqueville stated: "The prejudice rejecting the Negroes seems to increase in proportion to their emancipation and inequality cuts deep into mores as it is effaced from the laws" (Tocqueville, 1848, p. 316).

In 1903, W. E. B. DuBois made a similar but stronger observation about diversity and inclusion. DuBois (1903/1968) claimed: "The problem of the twentieth century is the problem of the color line – the relation of the darker to the lighter races of men in Asia and Africa, in America and the islands of the sea". More recently, Loury (2002) reminds us what Gunnar Myrdal had brought to our attention in 1944. Myrdal (1944) pointed out that the power in what he described as "vicious circles" of cumulative causation served as a self-sustaining process in which the failure of Blacks to make progress justified for Whites the very prejudicial attitudes that, when reflected in social and political action, served to ensure that Blacks would not advance (Myrdal, 1944).

This duality, Ringer claims, is mainly derived from the two fold process of colonization and colonialization that was generated by White Europeans' conquest and settlement of the New World. The duality, which is deeply rooted in America's past, holds firm through legal mandates and the attitude and behavior of White Americans. Governmental institutions enact and enable this duality. These institutions include the following: courts (e.g. U S Supreme Court: *Plessy* v. *Ferguson* 1896 "separate but equal" doctrine); legislative bodies (U S Congress: Exclusion

Act of 1882 and Immigration Act of 1924, denied Chinese and Japanese immigrants entrance into the United States); presidency (e.g. President Jackson's Indian removal policy that forced the Cherokees to go west of the Mississippi river and resulted in the infamous "trail of tears").

Besides governmental policy serving as an expression and instrument of duality, and reinforcing the systems of reason that keep inclusion and diversity on the margin in the United States, society's mores, customs, individual and collective attitudes, and behaviors keep in place the dual structures. For example, in 1958, not forty years ago, in the middle of the night, in the bedroom of their Virginia home, newlyweds Richard and Mildred Loving, White American man and Black American woman, woke up to blinding flashlights and police. The couple was arrested and charged with violating the ban on marriage for interracial couples. It was not until 1967, in *Loving* v. *Virginia,* that the United States Supreme Court struck down the remaining interracial marriage laws across the country and declared that the "freedom to marry" belongs to all Americans.[1]

The United States was, in legal terms, a dual society, until the following legislation was passed: *Brown* v. *Board of Education* in 1954, which ended legal racial segregation in schools; the Civil Rights Act of 1964, which declared that discrimination based on race, color, religion, or national origin in public establishments was prohibited; and the Voting Rights act of 1965, which declared that no citizen should be denied the right to vote based on race. Although the legal structure was struck down, the systems of reason that support the dual structure remain in place and keep progress toward diversity and inclusion moving at a slow pace.

Dual structures today

Once the legal dual structures were struck down in the 1950s and1960s, the Civil Rights Movements of African Americans, American Indians, Asian Americans, Latino/a Americans and American Women pushed for structural changes to bring about equity and equality of opportunity throughout society for all people. In addition, members of the different civil rights groups demanded the inclusion of their history and culture in societal institutions, such as schools, museums, and concert halls.

Has the dual structure given way to one inclusive structure? Have diversity and inclusion been achieved by culturally and racially diverse groups and women to their full potential? Three recent articles, one in the *Washington Post* (December 5, 2004); the second in *Time Magazine* (December 13, 2004); and the third in the *New York Times* (January 14, 2005) are good barometers on the state of racial diversity and inclusion in the United States. The title of the *Washington Post* article provides the context: "In college football, a glaring disparity: only 2 Blacks among 117 head coaches". The motivation for the article was the firing of an African-American coach, Tyrone Willingham, after only three seasons at the highly respected University of Notre Dame. In Notre Dame's history, no football coach had been fired without completing the full five years of his contract. College football in the United States is a "cash cow", and a source of fun and relaxation that also provides bragging rights for fans of teams that are victorious. Here at the University of Wisconsin, and at many college stadiums across the country on any given Saturday during football season, 80,000 or more fans paying approximately $40 per ticket (student tickets do not cost as much) are in the stands. College football coaches may make well over $1,000,000, whereas a senior professor may make $90,000 or a bit more or less. The author of the article, Liz Clarke, a *Washington Post* staff writer, makes three points in the article that are pertinent for understanding diversity and inclusion in the United States:

1   Research shows that when A1 schools (major universities) hire Black coaches, it has typically been to rebuild troubled programs. While ousted White coaches often get re-hired by rival schools, no Black coach who has been fired from an A1 school has ever gotten hired by another;
2   A reason for the back-pedalling in diversifying college coaching is the fear that donations from well-heeled boosters (university alumni) will drop off;
3   Most Blacks are hired on coaching staff as recruiting coordinators – jobs that often lead nowhere other than the rough neighborhoods that produce so many top high school prospects. This is so because they are seen to have the gift-of-the-gab and legitimism to sell the athletic program to African-American kids.

Another illustration of the status of diversity and inclusion in the United States comes from *Time Magazine*. *Time*'s Christopher John Farley interviewed Tavis Smiley, an African-American talk television/radio host, about leaving National Public Radio (NPR). NPR programming is designed to attract an intelligent, progressive, and well-read audience.

> *Farley:* "Why did you decide to leave NPR?"
> *Smiley* replied they had agreed on the destination they were to arrive at, but somewhere along the line NPR wavered in the journey. "Our show is the most multiracial in NPR's entire history; it has the youngest demographic of any show in NPR's history, so progress was being made. My concern was the pace the network was moving at – it wasn't fast enough."
> *Farley:* "Is it true you got angry letters from listeners when you started at NPR?"
> *Smiley:* "I can't begin to tell you the hate mail that I received when I started three years ago ... They didn't like the way I talked, the way I sounded. Because my whole style was so antithetical to what the traditional NPR listener had been accustomed to."
>
> (*Time Magazine,* December 13th, 2004, p. 8)

Tavis Smiley's departure from a public radio show illustrates that the show's producers and some of the listeners place narrow boundaries around programming which is considered to be progressive and diverse. In addition, inclusion is mainly for those who are viewed as intelligent and progressive according to the norms of White middle class culture and ways of thinking and behaving.

The final article, *"Macy's settles complaint of racial profiling of $600,000"*, comes from the *New York Times*. Andrea Elliott reports that Macy's, an old established and highly respected department store, agreed to settle a $600,000 complaint over racial profiling. Racial profiling, or to be identified and victimized for a possible wrong doing because of race or ethnicity, is a common experience for many Blacks and Latinos; it indicates to them that full inclusion into the mainstream of United States society and the acceptance of multiracial diversity has yet to take place. The complaint, reported in the *New York Times*, contends that Macy's New York department stores engaged in racial profiling and the unlawful handcuffing of Black and Latino customers detained on suspicion of shoplifting. The article reports that most people detained at a sampling of Macy's twenty-nine stores around New York state were Black and Latino, a disproportionately high number when compared with the percentage of Blacks and Latinos who shopped at the stores.

Women have made considerable strides toward gender equality over the past fifty-plus years; however, they still do not have full membership in United States society (e.g. Bem, 1993; Watkins *et al.*, 1993). In the field of higher education, for example, much diversity and inclusion work remains to be done in the areas of gender discrimination, collegial inclusion, and understanding of innate ability.

The Massachusetts Institute of Technology in a report of 1999 acknowledged that female faculty were victims of pervasive discrimination, and that although female faculty members have grown to 34 percent, up from 28 percent in 1975, the gap between salaries for females and males faculty actually widened during that period (Goldberg, 1999). Also, a study at Princeton University in 2003 reported that women faculty members in science and engineering feel less satisfaction and sense of inclusion than men; and although women faculty achieve tenure equal to men faculty they earn less and are not promoted as readily (Arenson, 2003). Finally, Lawrence H. Summers, the president of Harvard University, set off a fire storm within the academic community and United States society in general in January 2005. Summers suggested that the number of women in mathematics and science is less than the number of men because women may be innately less able to succeed in math and science careers (Dillon and Rimer, 2005).

Discussions about diversity and inclusion can be readily located in many discourses taking place in society, but these discussions most often show that the dual structure and the systems of reason that serve as a barrier to inclusion and diversity are firmly in place. Also some discussions about inclusion and diversity have reached the point where the concepts are trivialized. Cheryl Lieberman, a professional diversity consultant for the business community, highlights this point in addressing the attention given to diversity and inclusion within the business community (Lieberman, 2004). She observes that over the past decade, diversity and inclusion have become "the business buzzwords". The management shelf of any bookstore is filled with titles devoted to "understanding", "managing", "increasing", "building", "focusing", "working towards", "achieving", "leveraging", "exploiting", "creating", and "mastering" diversity and inclusion (Leiberman , 2004, p. 1).

Such trivialization of diversity and inclusion is another means by which the dual structure in society is maintained, thereby keeping non-White racial groups and women on the margin. When diversity and inclusion become buzzwords, or "habits" that effective managers use to monitor and arrange their employees or products, then true commitment to the hard work and coalition building still needed dissipates and then disappears. Speaking of diversity and inclusion as if they are only consultant buzzwords negates the realities that many people of color and women in the United States endure. These realities, such as many people living in poorer conditions with fewer opportunities, are negated because there is less and less space in public discourse to note this continuous lack of participation. Diversity and inclusion are just trends, not necessities for society. In addition, trivialization of diversity issues, such as that noted in the phrase "I am colorblind" seeks to offer a false norm, and overlooks a person's culture, race, and ethnicity; and seeks to maintain the status quo of a White male-dominated society (Sue, 2003).

Diversity and inclusion in education: school desegregation and curriculum

By examining two areas in education – school desegregation/integration and curriculum – one can get a reading on the status of diversity and inclusion in the United States' educational system. To begin, a brief description of the historical context of diversity and inclusion in United States education is provided. Schools in the United States have long been a site of struggle over our priorities and values as a nation and the vision of ourselves that we want to pass on to the next generation. The question of whose history, language, literature, and concepts of science and mathematics should be taught in our public schools rests principally on our perspective of diversity and our willingness to be inclusive. Also, arguments over who can and should be educated have been at the core of the nation's beliefs about diversity and inclusion. For decades after the nation was established, public schooling in the United States was not accepting of racial diversity and inclusion, and women received a second class education – one that mainly prepared them to be good home-makers.

The first 250 years or more saw every European colony in the (now) United States, and many other states which were later admitted to the Union, prohibit or stridently restrict teaching free and enslaved African Americans the fundamentals of reading and writing (Span, 2003). In 1740, South Carolina passed the first compulsory illiteracy law, making it a crime to teach enslaved African Americans. Anti-Black literacy laws were passed in many other Southern states including Georgia, North and South Carolina, and Virginia. In addition, several of these states imposed fines, public whippings, and/or imprisonment on anyone caught teaching enslaved or free African Americans how to read and write (Span, 2003). In the North, African Americans were treated somewhat better, but diversity and inclusion as practices were not supported by Whites. In Northern states, such as New York, Massachusetts, and Connecticut, separate schools for Black and White were set up. A White supremacist ideology, instead of an ideology of diversity and inclusion, remained legally in place in the South until it was struck down by the US Supreme Court in *Brown* v. *Board of Education* in 1954.

Inclusion in the schools:
the status of school racial desegregation

When Chief Justice Earl Warren stated for the Supreme Court of the United States in the *Brown* decision (1954) "we conclude – unanimously – that in the field of public education the doctrine of 'separate but equal' has no place. Separate educational facilities are inherently unequal", Black people and other people of color believed that racial inclusion was here, or just around the corner. Such has not been the case. May 17th, 2004, marked the fiftieth anniversary of *Brown* v. *Board of Education.* The fiftieth anniversary was a time of major speeches and recognization for the following: Thurgood Marshall and other members of the National Association for the Advancement of Colored People (NAACP) legal defense team who argued the *Brown* case; Chief Justice Warren who wrote and delivered the unanimous Supreme Court decision; and many young African American students, such as Ruby Bridges who rode the buses and walked through crowds of screaming segregationists to make the judgment in the case a reality (Bridges, 1999).

However, another prevailing theme among the speeches and the published articles for the fiftieth anniversary was that *Brown* had produced "fifty years of broken promises". Geneva Gay (2004) captures the theme well in her article "The paradoxical aftermath of *Brown*":

> Our children (and, for some, their parents and grandparents as well) have waited far too long for educational equality to become a functional reality 'at all deliberate speed'. Another 50 – or even 5 – years should not elapse with us being in the same place where we are now in equalizing educational opportunities and outcome for ethnically, racially, culturally, socially and linguistically diverse students.
>
> (p. 17)

In addition, for some African Americans, the belief that Brown will make education better for Black students and other students is rapidly fading or has faded. These people believe the dual structure, both legal and attitudinal, which historically served as a barrier to diversity and inclusion, are still very much in place. Legal decisions in such court cases as *San Antonio Independent School District* v. *Rodriguez* (1973) and in *Milliken* v. *Bradley* (1974) have not served well for diversity and inclusion within the education system. In Rodriguez, the court rejected demanding equality in educational spending. In other words, schools where students of color and of working class attend do not have to receive an equal amount of revenue as the schools where White middle and upper class students attend. Another court case, *Milliken* v. *Bradley* (1974) thwarted Northern metropolitan school desegregation plans, to integrate urban students with suburban students.

Here the court argued that since suburban jurisdictions were not normally legally responsible for segregation in cities, court-ordered desegregation remedies would be confined to city limits.

The dual structure has been kept in place by the attitudes and perspectives of several United States presidents toward racial diversity and inclusion in schools. President Nixon developed a Southern Strategy, which was an appeal to Southern Whites who opposed court-ordered busing to end school desegregation, in order to win the presidential election in 1968. In addition, the Office of Civil Rights during the Nixon administration did very little to enforce the *Brown* decision (Ashmore, 1994; Tushnet, 1994). President Reagan rescinded the Emergency School Act of 1972. This Act provided the only significant source of public money earmarked for the educational and human relations dimensions of desegregation plans (Orfield and Eaton, 1996).

In the 1980s and early 1990s, the dual structure and the systems of reason that resist diversity and inclusion were refortified as the Supreme Court released schools from desegregation orders. In the *Board of Education of Oklahoma* v. *Dowell* (1991), the United States Supreme Court ruled that formerly segregated school districts could be released from court-ordered busing once they had taken all "practicable" steps to eliminate the legacy of segregation, even if students of color remained isolated or segregated. Also, in *Freeman* v. *Pitts* (1992), the United States Supreme Court ruled that Federal district courts had the discretion to order incremental withdrawal of court supervision over school districts that were ordered to comply with desegregation orders. Futher, in the latter half of the 1990s, the Court began to use strict scrutiny of racial classification to strike down even voluntary efforts by local communities to address de facto (non legal) racial segregation. In *Eisenberg* v. *Montgomery County Public Schools* (1999), the U S Supreme Court let stand a 1999 decision striking down the use of race in a student transfer policy aimed at reducing White flight (Kahlenberg, 2004).

Even more revealing about the status of diversity and inclusion in the United States is how White and Non-White students are segregated along racial lines. Gary Orfield and Susan Eaton (1996) report that school segregation has increased steadily over the past two decades, especially in non-Southern states. Also, Orfield (2001) claims that although the South is still much more integrated than it was before the Civil Rights Movement, it is moving backwards at an accelerating rate. The proportion of Black students who attended White majority schools was 2.3 percent in 1964, 44.5 percent in 1988, but had declined to 34.7 percent in 1996, and the percentage is still decreasing. In addition, Orfield and Eaton (1996) claim that 80 percent of Latino students attend predominantly minority schools, compared with 42 percent in 1968. Furthermore, they claim that in New Jersey, New York, Texas, Illinois and California, 38 percent of the Latino students attend school where the student population is less than one tenth White. Finally, the North West Regional Educational Laboratory (1997) argues that "segregation in schools is more than racial separation; it also separates students by class, family, and community educational background" (p. 1).

Curriculum

Schools and classrooms are the arenas in which battles are fought to describe who we are as a people, our history, and our future. Curriculum and textbooks – the major conveyors of curriculum – are the major sites of struggle over how a nation tells its historical and cultural narrative. Because curriculum is the conveyor of the country's narrative, it receives the major attention in regard to diversity and inclusion in education. Schools' curriculum, as I look over my grandson Gavin's third grade curriculum, and other grade level school curriculum, conveys to Gavin and other students which racial and ethnic groups are part of the American narrative, their role in the narrative, which groups (ethnic and gender) are not included in the narrative and how women are constructed in the narrative. It tells students which groups contribute to U S society, as well as how much they contribute. In addition, the curriculum indicates how racial and gender groups as well as other groups (e.g. physically challenged) are perceived by others and themselves.

Gavin's curriculum and the curriculum used throughout US schools do a far better job of addressing diversity and inclusion than the curriculum his mother, Alicia, used in the 1970s, and his grandfather used in the 1950s. Why does Gavin's curriculum pay more attention to diversity and inclusion? Is the dual structure that keeps diversity and inclusion only a dream for people of color diminishing, as shown by this increased acknowledgment of a broader range of groups of people?

The dual structure and the systems of reason that act as a barrier to diversity and inclusion in societal spaces received a major below during the Civil Rights era of the 1960s and 1970s. Groups of people of color and women crusaded for their rights. During this Civil Rights era, curriculum and textbooks were evaluated for two sins: the sins of omission, which included the failure to recognize the contribution of individuals from particular groups; and the sins of commission, which depicted various groups in negative and stereotypical ways. Geneva Gay (1990) observes that early efforts to desegregate school curricula (i.e. implement multicultural education) were designed to address two simultaneous challenges:

> One was to correct the sins of omission – when the members, heritages, contributions, and experiences of minority groups were excluded from instructional materials. The other was to undo the sins of commission – the repeated presentation of stereotypical images and biased views of racial minorities.
>
> (Gay, 1990, p. 58)

Evaluations of kindergarten through twelfth grade curriculum textbooks from the 1960s through the 1990s point out that Latinos/as and Asian Americans were rarely portrayed in the curriculum. At best they are figures on the American landscape with virtually no history or contemporary ethnic experiences. In addition, there is no sense of ethnic diversity within each group. Asian Americans and Latinos/as are discussed as just one monolithic group (Grant and Grant, 1981; Sleeter and Grant, 1991). American Indians are portrayed as "fierce savages" or as "noble savages" and are mostly set within a historical context (Jetty, 1999; Pearce, 1988). African Americans, while given more attention in curriculum than other racial and ethnic groups except Whites, are shown in a much more limited role than European Americans. In addition, only a sketchy account of Black history is included and very little sense is provided of African Americans in contemporary life (Sleeter and Grant, 1991). In addition, women of color are rarely shown as active agents in political, social, and economic struggles for equality (Sleeter & Grant, 1991). Gavin, his mother Alicia, and her female friends of color learned from their curriculum that they should remain in traditional, passive, female roles. As students of color, they learned that Whites are the dominant group that maintains both power and privilege. In addition, Gavin, Alicia, and friends received very little information about their American-Indian, Latino/a, and Asian-American friends (Sleeter and Grant, 1991).

In the late 1980s and 1990s, Cornbleth (1995) argues the "inclusion diversity" debate in curriculum was characterized as the "E pluribus 'v.' unum" critique. *E pluribus unum* is a Latin motto meaning "one out of many". The motto can be seen on any US nickel. Her critique argued that the curriculum overemphasizes the *pluribus* at the expense of the *unum*. Critics of diversity and inclusion argued against curriculum inclusion on the grounds that this would challenge or change existing accounts of history in the curriculum, or would replace European (Western) authors with authors of color, and more women authors. These critics contended that such diversity and inclusion would weaken United States democracy and lead to national disunity and chaos (Schlesinger 1991) Also, they claimed that diversity and inclusion in the curriculum would devalue America's European heritage (Ravitch, 1990; Schlesinger 1991) and that people of color and White women are not only being included in the curriculum, but were also diverting attention from White males (*New York Times,* November 18, 1987). In addition, authors such as Ravitch (1990) and Schlesinger (1991) claimed that there was already enough diversity and inclusion in the curriculum, and that ethnic bias by and large had been eliminated through the curriculum and textbook reform efforts of the 1970s and 1980s. Therefore, there was no longer any need to continue changing the curriculum.

Some proponents of a Western curriculum have argued that American schools are failing because they have too much diversity and inclusion. E. B. Hirsch, in his national best seller *Cultural Literacy,* asks:

> Why have our schools failed to fulfill their fundamental acculturative responsibility? In view of the immense importance of cultural literacy for speaking, listening, reading, and writing, why has the need for a definite, shared body of information been so rarely mentioned in discussions of education?
>
> (p. 19)

Allan Bloom was one of the leading spokespersons for the anti-diversity and inclusion argument. In his bestselling book *The Closing of the American Mind: how higher education has failed democracy and impoverished the souls of today's students,* Bloom (1987) argued for a college curriculum based upon "Great

Books" of the Western tradition. He also argued for a curriculum that is guided by the fundamental work of Western philosophy, especially ancient Greek philosophy. Bloom states:

> Men may live more truly and fully in reading Plato and Shakespeare than at any other time ... The books in their objective beauty are still there, and we must help protect and cultivate the delicate tendrils reaching out toward them through the unfriendly soil of students' souls.
>
> (Bloom, 1987, p. 381)

Bloom's curriculum suggests a narrow view of the culture of the United States, and the world, since, for example, he excludes the contributions of Chinese and Indian thought to rational inquiry. Diversity and inclusion are denied because of this parochialism, which may well lead both teachers and students to forget that the essential activity of education takes place as students inquire, debate and discover.

Frederick Crews (1992) addresses this point when he claims that Allan Bloom and other proponents of only or mainly a Western curriculum are "Cultural nostalgics" who implicitly subscribe to a "transfusion" model of education, whereby the stored-up wisdom of the classics is considered a kind of plasma that will drip beneficially into our veins if we only stay sufficiently passive in its presence. Crews argues that:

> My own notion of learning is entirely different. I want keen debate, not reverence for great books; historical consciousness and self-reflection, not supposedly timeless values; and continual expansion of our national canon to match a necessarily unsettled sense of who 'we' are and what we ultimately care about.
>
> (Crews, 1992, p. xv)

There are other important areas in education where diversity and inclusion are kept on the margin. Special education is one such area, where African-American students, especially males, are often assigned to special education classrooms and/or are isolated or separated in regular classrooms (McCray *et al.*, 2003; Valles, 1998). Assignment to honors or higher track classes is another area where diversity and inclusion are kept on the margin. Honors classes are sometimes not offered in schools where students of color mostly attend, and in schools where students of color do attend, African-American and Latino students often are not students in the high honor classes (Gandara *et al.*, 1998).

Conclusion

Apologists might look at President Bush's multiracial and gender-conscious Cabinet and conclude that inclusion and diversity are well spoken for in the United States. Such is not the case in spite of the positive recognition that can be attributed to President Bush for selecting a multicultural, gender-fair cabinet.

Diversity and inclusion in the United States have come a good distance since the *Brown* decision in 1954, with passage of the Civil Rights Act of 1964, Voting Rights Act of 1965 and the Women's Movement of the 1970s. However, the "color line" as Du Bois noted in 1903, and gender inequity as Bem (1993) argues, are still very much with us today as evidenced in the Tyrone Willingham, Tavis Smiley and the racial profile stories; along with the stories of gender inequity in higher education. Whereas the Willingham and Smiley cases demonstrate the subtle but powerful effect of race (and racism), the Macy's racial profiling complaint, Orfield's report on the resegregation of schools, and the salary inequities between male and female professors, demonstrate a blatant disregard for diversity and inclusion. And all are reminders that the systems of reason and dual structures that keep diversity and inclusion marginalized are still very much in place in the United States.

Note

1 Marriageequalityca.org/history_marriage 2004

References

Arenson, K. W. (2003) 'Uneven progress is found for women on Princeton science and engineering faculties'. *The New York Times,* September 30th, Section B, p. 5.

Ashmore, H. S. (1994) *Civil Rights and Wrongs: A Memoir of Race and Politics 1944–1994.* New York: Pantheon Books.

Bem, S. (1993) *The Lenses of Gender: Transforming the Debate on Sexual Inequality.* New Haven, CT: Yale University Press.

Bloom, A. (1987) *Closing of the American Mind,* New York: Simon and Schuster.

Bridges, R. (1999) *Through My Eyes.* New York: Scholastic Press.

Clarke, L. (2004) "In college football, a glaring disparity: only 2 Blacks among 117 head coaches". *The Washington Post,* December 5th, p. 1, 11.

Cornbleth, C. (1995) "Controlling curriculum knowledge: multicultural politics and policy making". *Journal of Curriculum Studies,* 27, 165–185.

Crews, F. (1992) *The Critics Bear it Away: American Fiction and the Academy.* New York: Random House.

Dillon, S. and Rimer, S. (2005) "President of Harvard tells women's panel he's sorry". *The New York Times,* January 25th, Section A, p. 19.

DuBois, W. E. B. (1968) *Souls of Black Folks.* New York: Blue Heron Press.

Elliott, A. (2005) "Macy's settles complaint of racial profiling for $600,000". *The New York Times,* January 14th, Section B, p. 1. At: Nytimes.com/2005/01/14/nyregion/14macys.html.

Gandara, P., Larson, K, Rumberger, R. and Mehan, H. (1998) "Capturing Latino students in the academic pipeline". *California Policy Seminar Brief Series,* May 1998. At: www.ucop.edu/pipeline.

Gay, G. (1990) 'Achieving educational equality through curriculum desegregation'. *Phi Delta Kappa,* 71, 56–62.

Gay, G. (2004) "The paradoxical aftermath of *Brown*". *Multicultural Perspectives,* 6, 12–17.

Goldberg, C. (1999) "M.I.T. admits discrimination against female professors". *The New York Times,* Section A, p. 1, March 23rd, 1999.

Grant, C. A. and Grant, G. W. (1981) "The multicultural evaluation of some second and third grade textbook readers – a survey analysis". *Journal of Negro Education,* 50, 1, 63–74.

Hirsch, E. B. (1987) *Cultural Literacy: What Every American Needs to Know,* Boston: Houghton Mifflin.

Jetty, R. (1999) "They're supposed to be sovereign nations: hegemonic constructions of contemporary American Indian issues in current United States textbooks". Unpublished dissertation. University of Wisconsin-Madison, Wisconsin.

Kahlenberg, R. (2004) *Beyond Brown: The New Wave of Desegregation Litigation,* Guilford, CT: McGraw-Hill.

Kumashiro, K. K. (2002) "Against repetition: addressing resistance to anti-oppressive change in the practices of learning, teaching, supervising, and researching". *Harvard Educational Review, 72,* 67–92.

Lieberman, C. (2004) "Diversity and inclusion: the concepts behinds the words". *Connections.* Cambridge, MS: Cambridge Chamber of Commerce's newsletter. *At:* www.cornerstone consults.com/diversity.html.

Loury G. C. (2002) *The Anatomy of Racial Inequality.* Cambridge, MA: Harvard University Press.

Macedo, S. (2000) *Diversity and Distrust: Civic Education in a Multicultural Democracy.* Cambridge, MS: Harvard University Press.

McCray, A. D., Webb-Johnson, G.C. and Neal, L. I. (2003) "The disproportionality of African Americans in special education". In Carol Camp Yeakey and Ronald D. Henderson (eds), *Surmounting All Odds: Education, Opportunity, and Society in the New Millennium,* Greenwich, CT: Information Age.

Myrdal, G. (1944) *An American Dilemma: The Negro Problem and Modern Democracy.* New York: Pantheon.

North West Regional Eduational Laboratory (1997) "Blacks and Latinos face segregation in school". www.nwerl.org/cnorse/infoline/may97/article2.html.

Orfield, G. (2001) "School more separate: consequences of a decade of resegregation". *Rethinking Schools Online,* 16. At: www.rethinkingschools.org.

Orfield, G. and Eaton, S. E. (1996) *Dismantling Desegregation: The Quiet Reversal of Brown* v. *Board of Education.* New York: The New Press.

Pearce, R. H. (1988) *Savagism and Civilization: Study of the Indian and the American Mind.* Berkeley, CA: University of California Press.

Popkewitz, T.S. (1988) "Educational reform: rhetoric, ritual and social interest". *Educational Theory,* 38, 77–93.

Ravitch, D. (1990) "Multiculturalism E pluribus plures". *American Scholar* (Summer) 37–354.

Ringer, B. B. (1983) *"We the People" and Others: Duality and America's Treatment of its Racial Minorities.* New York: Tavistock.

Sacks, J. R. (2002) *The Dignity of Difference: How to Avoid the Clash of Civilizations.* London: Continuum.

Schlesinger, A. (1991) *Disuniting of America: Reflections on a Multicultural Society.* New York: Norton.

Sleeter, C. E. and Grant, C. A. (1991) "Race, class, gender, and disability in current textbooks". In Apple, M. and Christian-Smith, L. (eds) *The Politics of the Textbook.* New York: Routledge.

Span, C. (2003) "Knowledge is light, knowledge is power: African American education in antebellum America." In C. Yeakey and D. Henderson (eds), *Surmounting All Odds: Education, Opportunity and Society in the New Millenium.* Greenwich, CT: Information Age.

Sue, D. W. (2003) "Dismantling the myth of a color-blind society – last word". *Black Issues in Higher Education* November 6th. At: www.looksmart.com.

Time Magazine (2004) "10 Questions of Tavis Smiley", December 13th, p. 8.

Tinto, V. (1993) *Leaving College: Rethinking the Causes and Cures of Student Attrition* (2nd ed.). Chicago: University of Chicago Press.

Tocqueville, Alexis de (1848) *Democracy in America.* New York: Harper & Row.

Tushnet, M. V. (1994) *Making Civil Rights Law: Thurgood Marshall and the Supreme Court, 1956–1961.* New York: Oxford University Press.

Valles, E. C. (1998) "The disproportionate representation of minority students in special education: responding to the problem", *Journal of Special Education,* 32, 52–54.

Watkins, S. A., Rueda, M. and Rodriguez, M. (1993) *Introducing Feminism.* London: Icon Books/Faber.

5 Reconceptualizing multiculturalism in American urban education

Rupam Saran

Introduction

Contemporary postmodern American urban education is facing the challenge of incorporating multiculturalism in public schools socially and academically. American society is increasingly becoming multicultural and the time demands that educators and policymakers simultaneously honor multiculturalism, and create a meaningful environment for learners. Concurrently, the most important task of the modern urban education system is to instill a sense of Americanism along with the maintenance of cultural identity and diversity in heterogeneous society. The prevalent belief among education leaders and education reformists is that by embracing the concept of multiculturalism, educational establishments can help students maintain American identity, and nurture unity within diversity in schools. At the present time, the thrust towards appreciation for multiculturalism is an important struggle for the public schools. Students are being exposed to their cultural heritage, and the histories of many groups are being incorporated with varying degrees of success into urban school curricula.

The United States' history is culturally diverse, and American identity is constantly struggling with the idea of who and what is American. To gain the fullest understanding of end-of-the century multicultural conditions, one must understand the role of diversity in American society.

In multiracial and multicultural American society the thread of diversity works as both a unifying and separating force. Understanding American culture and American identity involves grasping the nuances of the cultural, economic, social, and political diversity of Anglo-American society and understanding migrant-dominant relationship patterns. The impact of continuous immigration upon American culture, economy, politics, and education has been an issue of debate and tension throughout American history.

Contradictions of American urban education

In the context of challenges, the contradictions of contemporary American urban education, and the role of multiculturalism, one must understand the complex nature of underlying forces of elitism, racism, internal colonialism, and capitalism that have shaped the American urban education system. The emergence of mass free elementary education in the United States was due to the need to Americanize the common population. The idea of common schooling for the mass population saw the infancy of urban education. By the mid-nineteenth century the United States had a greater number of students attending free elementary and high schools than many European countries. Before the twentieth century, the main function of education was to breed an elite class in society, or to promote religious belief and literature. Owing to industrialization, the role of education gradually changed.

In the United States changes in education were associated with the economy, immigration, overcrowding of cities, unattended children of newly-arrived immigrants, and the need to breed good citizens. In order to fulfill the demands of many new social conditions, compulsory schools were established in cities. The urban education system developed as an inclusive, free, and mandatory system for everyone regardless of their political, cultural or socioeconomic status. Equality lay behind the ideology of urban education; urban schools were not intended to differentiate among citizens and immigrants.

The democratic principles of the United States shaped its mass education system. The development of mass education was the reflection of an ideology that an open education system was an essential element of a modern political system. Consequently, in the United States, the basic foundation of urban education emphasized the equality of race, ethnicities, genders, socioeconomic status, and opportunities. In reality, however, urban education did not practice this doctrine, as its main goal had become to promote social control by transforming the young immigrant population into a skilled, compliant workforce for industries. Unfortunately, the contemporary urban education has turned into the largest institution to promote the process of assimilation and adaptation among immigrant children, with the implicit purpose of maintaining social control. In the United States, urban education has largely been the result of the search for a single universal system suitable for all types of populations – a system where people followed one discipline, were taught obedience and were made to act in a uniform manner (Harris and Duane, 1874; Draper, 1899; Tyack, 1974; Nasaw, 1979).

One of the main functions of urban schools has been to transmit the dominant culture of the host country to future generations. Urban schools became channels through which the traditions of a dominant culture based on Protestant ethics and values were transmitted. The curricula of these schools did not acknowledge the cultural differences within the immigrant population, and consequently, students learned under the norms and the hegemony of the mainstream culture. Thus urban education at the time became a subtle way of destroying a variety of treasured cultural heritages, even in segregated schools teaching only African Americans.

Americanization of the immigrant and established ethnic minority populations was therefore the main goal, both explicit and implicit, of urban education in the United States (Tyack, 1974; Richman, 1905). The argument in favor of Americanization was the need for greater uniformity within American society. In urban schools immigrant children of diverse cultures and ethnic backgrounds were molded to assimilate into American culture and were transformed into committed "Americans", accepting of and subservient to dominant Anglo-American culture and language.

The years between 1820 and 1930 were very important in the history of immigration to the United States since approximately 35 million people migrated to the country. Mass immigration from Europe, Asia, Canada and Mexico became a great challenge for the American education system and "constituted an educational problem unparalleled in human history" (Callahan, 1962). The immigrant population threatened the "host" Anglo-Americans by bringing with them a diversity of language, culture, ethnicities, and values. Consequently, the dominant American society demanded that urban schools Americanize the immigrant population by teaching them the English language, preparing them for citizenship, and the American way of life (Mohl, 1997, p. 113). In this model, public schools became the "melting pot" that fostered Americanization and assimilation. Regarding the goals of Americanization in 1909, one of the leading American educators, Ellwood P. Cubberly, urged that it was the task of education "to break up these groups or settlements, to assimilate these people as our American race, and to implant ... the Anglo-Saxon concept of righteousness, law and order ... and to awaken them to a reverence of our democratic institutions" (Cubberly, 2004, pp. 15–16). The politics of urban education for most of the twentieth century focused on the complete assimilation and integration of immigrants into mainstream of American life. Simultaneously, this notion promoted the idea of a common economic and political life for all Americans and virtually all educational policies revolved around themes of a "melting pot".

American society is predominantly under the misconception that "Anglo-Saxon culture" is superior.[1] Milton Gordon's melting pot idea in which minorities abandon their identities and "melt" in with the dominant groups was endorsed by most Americans due to the belief in "Anglo-Conformity" (Gordon, 1964). However, currently, the practice of assimilation exists only residually, and has proven to be myth rather than reality because subjugated populations were not allowed to assimilate into white dominant culture because people of color were believed to be inferior. On the other side of the spectrum, minorities rejected the idea of assimilation because it was hard for them to abandon their heritage and dissolve into the host society.

Contemporary American society widely endorses the concept of "cultural diversity" and "cultural pluralism" that does not demand total rejection of cultural identities but rather visualizes American society as a salad bowl where all elements stay together retaining their flavor (Glazer and Moynihan, 1970). The ideals of "cultural diversity" and "cultural pluralism" contributed to the transformation of the mission of urban schools that believed in the superiority of Anglo-Saxon culture. The philosophy of urban schools changed to such an extent that public schools were expected to be sensitive to the cultural identities and heritage of all immigrant children. The idea of "cultural diversity" and "cultural pluralism" became the seeds of multiculturalism.

Multiculturalism in American urban education

In neo-conservative American society, the term multiculturalism implies two interpretations. First, multiculturalism stresses the idea that different cultural groups maintain their cultural identity in a dominant culture. Second, multiculturalism refers to the idealistic belief that all cultural and ethnic values should be respected and understood by the dominant culture. Unfortunately, the true meaning of multiculturalism has been reduced to a slogan. The vision of multiculturalism is viewed ambiguously in postmodern American society. While postmodern education is excited about educational opportunities offered by multiculturalism, the term is defined very superficially and narrowly. The moral vision of multiculturalism varies according to the political beliefs of the group that is defining the concept. Consequently, there are many definitions of multiculturalism that in turn can serve the interests of different groups who may be in contest for limited social and political ends (McLaren, 1997).

The theoretical perspective of multicultural education promises acknowledgment and opportunities to all cultural groups and ethnicities. Multicultural curricula require schools to preserve diversities, negate persisting stereotypes and prevent formation of new ones (Gonzalez, 1995). In addition, it is schools' responsibility to enable students of diverse cultures to maintain their identity in the dominant culture and provide tools to survive socially and economically in wider society (Arnowitz and Giroux, 1991). In summation, multiculturalism asserts that no group is inferior to the other, although in practice hidden conflicts within schools often mean that one ethno-cultural group is dominant, at the expense of minorities.

In a society where dominant groups make every effort to maintain their hold over minority groups, the success of pluralism and multiculturalism is often in doubt. While dominant groups demand total subjugation of minorities and acceptance of dominant-group ideology that labels them as inferior, marginalized cultures can often react in a spirit of pluralism and multiculturalism. Although American political institutions now seem to embrace the concept of multiculturalism by taking a stand to help maintain marginalized cultures, languages, ethnicities, and customs, through bilingual and bicultural programs, the effort is minimal compared to the subtle but profound efforts addressed to the Americanization of the marginalized immigrant population. Officially, educational reform policies are not concerned with the linguistic needs of this marginalized population and consequently bilingual education is not encouraged.[2] The concept of monoculturalism advocates the superiority of Western patriarchal culture, and acts against the spirit of true multiculturalism (Kincheloe and Steinberg, 1997).

Postmodern American education demands the implementation of multiculturalism that would honor all cultures and enable schools to discourage all forms of prejudice and discrimination. Until the present time, however, urban educators have developed a multicultural curriculum that is a hollow promise of academic excellence and equity for all students. Under the dominance of a traditional

Eurocentric monocultural curriculum, the infusion of multiculturalism in curricula has been reduced to a shallow knowledge of minority groups without respecting their cultural heritage and acknowledging their contributions to American culture. Though times have changed, Edward Ross's idea of "social control through education" is still alive and his idea of "order and regulation" in education still dominates the contemporary education system. Contemporary education leaders and education reformists often still try to maintain Fredric W. Taylor's idea of scientific management that advocated management of schools based on the efficiency model of the corporate world (Kliebard, 1995).

In the context of urban education, given the extent of competing political interests, facile multiculturalism has become a ritualized, decoration-piece of the urban school curricula and has been interpreted as a requirement to portray minority groups as "subjugated but equal" and their knowledge as "indigenous knowledge". The cultural content of the urban school curriculum serves to promote or diminish some groups at the cost of others by presenting distorted realities. Discourses of labeling and stereotyping of cultural groups in curricula often create conditions for "inter-ethnic rivalry" that leads to a "culture of victimization", and negative competitive trends within people of color. For instance, the portrayal of Asian immigrants as model minorities has created a culture of self-victimization within ethnic groups (McLaren, 1997). The label of "model minority" has led to a backlash against some groups of Asian students in schools and colleges. Education policymakers and politicians use the accomplishments of Asian students to justify the notion of meritocracy and the slogan of "equal opportunities for all". Simultaneously, examples of model minorities are used as an excuse to discredit and ignore low achieving or disadvantaged ethnic groups. Thus, in the guise of multiculturalism, conservative bureaucrats promote negative social relations, and a sense of powerlessness and failure within some ethnic groups, such as Hispanics and African Americans.

Why reconceptualize existing ideas of multiculturalism?

There is a need for a socio-cultural educational environment that can recontextualize multiculturalism in ways that would challenge traditional modes of schooling and knowledge production. There is a need to pursue democratic education and create critical conceptions of "true multiculturalism" with global and local perspective in a monocultural, Eurocentric urban education arena (McLaren, 1997). Cameron McCarthy views multiculturalism as representing a "cultural truce" between white liberals and black liberals, and demands restructuring of school knowledge and "rearticulation" of minority failure, cultural characteristics, and language proficiency. Multiculturalism should promote cultural understanding, cultural competence, and cultural emancipation (McCarthy, 1993). True multiculturalism should be examined from multiple perspectives. The goal of such multiculturalism should be to confront cultural conflicts and contradictions, expose power relations within Anglophone society, empower students to find their voices and ethnic identities, bring an understanding of racism, legitimize subjugated knowledge, and improve academic and economic success of minority students. True critical multiculturalism should not be limited only to the issues of cultural diversity: rather it should attempt to reach diverse constructs that work for social justice and it should take into account "constituencies" such as poor working-class whites, that have not supported social justice in the past but currently need attention due to joblessness and poverty (Kincheloe and Steinberg, 1997). Though the term multiculturalism is primarily used when speaking of race, other issues such as socio-economic class, language, gender, culture, sexual preferences, and disabilities are addressed within the expanded horizon of critical multiculturalism.

What is critical multiculturalism?

Critical multiculturalism is the response to cultural, economic, and political contradictions of American society and their relationship to pedagogy. Simultaneously, it challenges the monocultural education system that values the Cartesian notion of superiority of Western knowledge and considers all cultures subjugated except a few Western European cultures. With the commitment to emancipatory pedagogy that demands social justice, critical multiculturalism exposes persisting social inequalities that are the consequences of power and oppression. Critical multiculturalism emphasizes that ways of knowing are inherently and culturally cultivated and persuaded. In this context, the central concern of critical multiculturalism built upon the critical consciousness of class, racial, and economic diversities, is to examine the power of patriarchy, white supremacy, and class elitism. In addition it exposes hidden forces of racism, sexism, and class biases of dominant culture that creates prejudice and discrimination. Simultaneously, it identifies and legitimizes contributions of oppressed groups to the dominant culture. The monocultural hegemonic curricula promote dominant perspectives of seeing and evaluating minority groups. Such an approach of pedagogy is confronted by critical multicultural curricula by providing positive conditions for identity formation in school settings.

Current political and social conditions of American society demand unity within diversity. In this context the most challenging question is: how can unity be maintained in a society where diversity is increasing constantly? The question of American identity, i.e. who is an American, has created tension for decades. Given the context of American identity, Roger Collins asserts the case for an American identity that honors the integrity of its marginalized cultural elements (Collins, 1993). Collins takes a stance for an identity that involves mediation between native and acquired cultures. Concurrently, he suggests that assimilation can occur without sacrificing one's cultural integrity and that the dominant culture can be more tolerant of different cultural identities. In this regard, the concern is why the dominant group should have the right to use their knowledge and culture as norms for a much larger and diverse population of the United States (Apple, 1982). Under such tension, the concept of critical multiculturalism provides answers to questions of identity, political possibilities of emancipatory actions and the historical situatedness of subjugated and indigenous knowledge. In the context of dominant and subordinate knowledge and culture, there is an urgent need for a critical pedagogy that would encourage critical understanding of social realities and contribute to "building emancipatory curriculum" (Giroux, 1988).

In the spirit of emancipatory curricula, critical pedagogy involves cultural formation in which students are actively involved in cultural literacy. In response to Cartesian dualism, critical multiculturalism provides cultural studies as an alternative method of gaining insight into the interrelationship of race, class, and gender within the dominant society. In this regard, cultural studies are an alternative to mainstream Western values and add "intellectual diversity" to Eurocentric monocultural curricula. For Arnowitz and Giroux, the role of critical pedagogy is to expose forms of subordination that create inequalities and reject undemocratic ideas in society. The task of critical pedagogy in a multicultural environment is to transform teachers and students to be critical viewers of "political, social, and cultural enterprise", and understand their limitations and capacities (Arnowitz and Giroux, 1991, 1997).

Notions of "whiteness" and critical multicultural education

The understanding of critical multiculturalism and postmodern conditions leads to insights into white positionality and issues of power and powerlessness between white and non-white population. Monoculturalism demands that people of color surrender to the supremacy of "whiteness" and assimilate, rejecting their cultural capital. Giroux refers to cultural capital that is the legitimation and distribution of certain forms of knowledge. The notion of superiority of the whiteness creates internal colonialism in society. People of color do not get the respect and recognition for their academic achievements that they deserve. In a general sense, whites of European heritage consider themselves superior. In the context of education, critical transformative pedagogy has the power to expose the vanity of "whiteness", and build an open and intellectual "community" in a multicultural, multiethnic, and multiracial classroom where all voices would be heard and all knowledge would be legitimized and respected (Hooks, 1994). Hooks asserts that for the emergence of such a community, it is crucial that the multiple effects of "whiteness" should be explored, analyzed, and discussed and an unbiased perspective of multiculturalism should be practiced in classrooms. The demography of the United States predicts that in the near future "whiteness" might not be the mainstream ethnicity in many American classrooms. Although in many American inner city schools students of color have majority over white students, such schools still conform to Eurocentric monocultural curricula. The invisible power of whiteness has shaped all aspects of schooling and has justified the privileges of the white population. Critical multicultural education can study the effects of whiteness and enable minority ethnic groups to be liberated from its grip.

Racism and critical multiculturalism

The dominant Eurocentric establishment tries to deny the significance of racial and ethnic diversity that is a vital force in American society. The term "people of color" categorizes all non-whites in one category and ignores their individual racial identities. The Eurocentric curriculum denies African-American, Latino, and Asian-American students their identities, and treats them as minority students. Ironically, all these racial minorities are dominant forces in shaping the American economy. Racism has been institutionalized in the American curricula. As a result, racial inequalities persist in the American education system. The non-critical curriculum does not recognize "racism" and racism is not discussed. In the mainstream curriculum, the histories of African Americans, Latinos, and Asians are misrepresented and they are labeled as inferior races.

The notion of criticality attempts to cultivate awareness of racial oppression and racial justice in education, and simultaneously to suggest that curricula should include understanding of different racial perspectives. Although the marginalized population has always been viewed through a Eurocentric lens, it is time for the dominant population to be viewed through the lens of the marginalized. Through this method, white and non-white students would gain "self-knowledge" and be aware of their limitations and strengths.

Mainstream curricula ignore the persisting racial prejudice in American society and deny the existence of racism that has influenced all education policies, and discriminated against African Americans (McCarthy, 1993; Kincheloe, 1993). Broadly speaking, critical multiculturalism can fight racism, support Afrocentrism, and enable oppressed African Americans to take pride in their heritage by integrating Afrocentrism in mainstream curricula. Afrocentrism epistemology has the capacity to understand the intensity of emotional responses, identification, and involvement of black life (King, 1994). King cautions against a "false universalism" in curricula that ignores the experiences of the marginalized and shuns sociocultural and socioeconomic differences; it appeals for multiple perspectives and critical approaches of pedagogy that would enable marginalized people to obtain insight into racism and other social realities. The emancipatory pedagogy and African-American epistemology can influence people of color not to blindly accept the dominant culture's prejudice and racial discrimination. The notion of emancipatory pedagogy is a struggle against racism (McLaren, 1997; Giroux and McLaren, 1993).

Victor Villanueva's (1993) understanding of racial inequalities in American society is that they are caste-like, based on an ascribed birth status. Though there are minorities who manage to assimilate into the mainstream, Latinos, African Americans, Asians, and American Indians often remain outcast because of their color and suffer an imposed racial and cultural inferiority. It is their "subservience" to the dominant group that promotes a caste-like racial system. Although there is no visible colonization within American society, many minorities are treated like the colonized. Villanueva (1993) defines colonization as "The economic, political, and cultural domination of one cultural-ethnic group by another" (p. 31). The hegemonic nature of internal colonialism in American society encourages minorities to stay within their predetermined and predefined social parameters.

Issues of class and critical multiculturalism

In the context of economic domination, critical multiculturalism understands the vital role of class system and power inequalities in the American education system. Mainstream curriculum operates to favour certain classes and "order of knowledge over others" (Apple, 1982). Without an understanding of the true nature of hegemony, and the dominance of class system in the culture, "critical multiculturalism" in educational thinking and planning is incomplete. Critical pedagogy believes that the study of the nature of hegemony, and finding strategies to fight hegemony, is a way of empowerment of low-income subjugated minorities. Although traditional curricula promise equality in education, the hegemonic characteristics of a stratified society produce inequalities. In postmodern American society there is no upward mobility for marginalized minorities, and equal opportunity implies conformity to a class system. The persisting class system in traditional education conforms to the elitist idea that intelligence is the right of elites. The notion of criticality points out that class conflict and prejudice in education are a cultural phenomenon that is rooted in class differences (Kincheloe, 2001).

Unfortunately, throughout the United States schools are labeled on the basis of socioeconomic status of neighborhoods, and the quality of education fluctuates with the economic condition of the communities. African Americans, Latinos, Mexicans, and other students of the underclass are underrepresented in high achievement and magnet schools. There is a wide gap in graduation rates, college admission, and school funding of urban magnet schools and neighborhood schools. Urban magnet schools and schools of high socioeconomic urban areas provide more challenging curricula, qualified teachers, and better resources than neighborhood schools with poverty issues. Consequently, schools of poor districts have a low achievement and graduation rate. The low achievement rate of students of low-income families is a reflection of poor home environments where parents are poorly educated, and face the pressure of supporting a family with low-income jobs. Although American urban schools have the task of being an equalizer in a stratified condition by providing uniform education, schools are often unable to provide conditions that can repair deficits produced by poor economic and social home environments. Children in low-income neighborhoods have many risk factors that add to the deficiencies of school districts. Young kindergarten students entering schools may have parents (or a mother) with less than a high school education; they may be welfare recipients (of food stamps or cash payments); they may come from a single parent family or have parents with two jobs and/or limited English language proficiency (Zill and West, 2000). These risk factors influence students' academic attainments. According to Kozol, poor academic achievement is not a reflection of poor schools: it speaks for academic deficiencies caused by inequalities of family resources before children start kindergarten (Kozol, 1991).

Critical multiculturalism could enable teachers to see influences of poverty on students' academic achievements. The class system dominates schools and supports the interests of the elites by encouraging social reproduction, and maintaining meritocratic social order in American society. Critical multiculturalism rejects the prevailing notion of hegemony in the American education system. The hegemonic practices do not impose overt force on the oppressed, but rather dominate the marginalized by winning their consent or articulating them to believe that they are inferior to the mainstream population. The hegemonic classist assumptions support the deficit ideology that the under-privileged working-class people are "incapable of learning" and that only the elite class has an inherent right to learn. The students of inner city poor neighborhood schools have been victims of this deficit ideology. By internalizing the belief that they are incapable of learning they resign to the domination of the elites. Through this social positioning of inner city marginalized youth as deficient, the schools focus on suppressing youth resistance and "fixing" them with educational policies that encourage social control.

Issues of gender in critical multiculturalism

The concept of critical multiculturalism examines how gender roles and gender relationships affect the political, economic, and social conceptual framework of a patriarchal society. In order to understand how gender politics have shaped the American education system, one needs to examine the history of American education. During the first half of the nineteenth century the discourse of coeducation and the hiring of female teachers influenced all aspects of education. The inclusion of girls in all-boy schools demanded the hiring of more female teachers, who were economically cheaper than male teachers (Tyack, 1974). Thus, the number of female teachers in schools increased enormously. However, the number of male administrators did not change. Anglo-protestant, white middle class men dominated the school administration. Unfortunately, in contemporary education arenas this pattern is still alive. Although this was really an issue of sexual power and economics, the rationale for hiring more female teachers was that they were better in nurturing and civilizing students, and coeducation was better for students' social growth. Today the gender differences are highlighted in schools by a tracking system that encourages boys to enrol in challenging programs such as science and technology, while girls tend to opt for careers such as nursing, education, and other less challenging programs.

Sandra Harding (1998) raises concerns of how gender roles and gender relationships affect the political, economic, and social conceptual framework of a patriarchal multicultural society. Postcolonial multiculturalism analyzes "culturally distinctive histories and practices shaping women's conditions, interests, and desires in different local, national, and transnational culture" (Harding, p. 76). Critical multiculturalism asserts that there are other ways of knowing science and technology than the traditional masculine ways of knowing. The patriarchal notion of knowledge suppresses a feminine domain of knowledge. Patti Lather asserts that feminism and research practices for the empowerment of women should be employed in postmodern multicultural education. Feminist research can help men see the world with a woman's perspective that can eliminate gender-based inequalities (Lather, 1991). Given the extent of gender-based exploitation of women, a critical multicultural curriculum articulates social dynamics related to the gender issue: empowerment of women, exposing forms of subjugation caused by male dominance in society, and gender injustices (Kincheloe and Steinberg, 1997).

Issues of linguistic needs in critical multicultural curriculum

Critical multicultural education advocates democratic inclusive education that is sensitive to the cultural, academic, and linguistic needs of poor and non-white students. Though there is bilingual education, the monolingual traditional education does not provide articulated support for the language diversity of urban students. Immigrant students with poor academic achievement in English have major problems in schools because they do not get language development programs essential for academic success. For students with English language deficiencies, learning English often means rejecting their cultural identities and native language, and assimilating norms, values and folkways of American society that are often destructive to their confidence and identity (Gumperz, 1982, 1994). Linguistic issues are cultural biases that ignore metalinguistic awareness and endorse "English only" curricula (Villanueva, 1993). Linguistic issues play a very important role in standardized tests because they target minorities with language barriers. Students who are not proficient in language perform poorly in standardized tests. Recently, the college entrance test, the Scholastic Aptitude Test (SAT), added mandated essay-writing as one of its components. Students with low English language proficiency will likely fail these tests. Urban schools have failed to prepare students who have linguistic problems to pass the SAT. It is the responsibility of multicultural education to construct curricula to understand complexity of language diversity and its implications in education.

Indigenous knowledge and critical multicultural curriculum

Critical multicultural education provides opportunities for inclusion of socially constructed, indigenous knowledge in traditional curricula. This critical epistemological consciousness proposal asserts that all knowledge is socially constructed and no knowledge is indigenous or subjugated because all cultures produce knowledge that has ontological legitimacy and validity. Knowledge is constructed in cultural contexts and these contexts facilitate our understanding of the world (Bagnall, 1999). Inclusion of indigenous knowledge in traditional Eurocentric curricula would generate an awareness of different perspectives and enable dominant mainstream groups to see that there are multiple perspectives on all issues. Indigenous knowledge is the product of life-experiences of subjugated culture that helps students understand their own and other cultures. It helps them rationalize knowledges they are given in school and simultaneously ignites "double consciousness" that enables them to critically rethink and recontextualize the "knowledge" with which they are presented (Kincheloe and Semali, 1999).

Critical multicultural education focuses on unmasking hegemonic processes of power inequalities that shape one's consciousness and create social and political conflicts in society. The Eurocentric curricula evade knowledge produced by the colonized and implicitly discredit all non-Western knowledge. The knowledge productions of subjugated cultures always have some validity in Western-Eurocentric curricula (Kincheloe and Steinberg, 1997). For example, Western mathematics education discredits Asian and African knowledge of mathematics, considering it subjugated and arguing that mathematics is purely "European and androcentric" (Kincheloe and Steinberg, 1997). Contemporary urban education restricts higher mathematics education for minority students and women owing to the belief that they are generally incapable of learning higher mathematics. The traditional curricula restrict advanced mathematics education for minorities. Concurrently, maths education is used to weed out marginalized populations (Moses and Cobb, 2001). Urban school mathematics instruction has low expectations of students of color, and from blue collar backgrounds. Critical education challenges these hegemonic processes in academic discipline and can encourage "democratic education for all".

A multicultural science education would expose the false claims of the Eurocentric science curriculum that claims that all scientific knowledge was discovered by Western culture. By including indigenous knowledge in curricula, critical science education would reveal the exploitation of subjugated cultures by Western colonizers who stole their knowledge and disguised it as original Western Eurocentric knowledge. It is the responsibility of multicultural science education to point out a Europology of modern science that has demonstrated a "pattern of scientific ignorance" and clarify that the "European miracle" was not a miracle at all because European scientific knowledge and gains were "borrowed" from colonized, subjugated cultures (Harding, 1998). In a critical paradigm, a multicultural science curriculum can bring understanding that all subjugated knowledge has validity, and has made significant contributions to the dominant culture.

Conclusion

Throughout US history, schools and curricula have been shaped by the sociopolitical and economic conditions, and ideologies of dominant groups in American society. Political theorists argue that schools and curricula are instruments to promote dominance and sociopolitical oppression in society through encouraging racism, poverty, social stratification, and capitalism. Various dominant ideologies, customs, cultural practices, rituals, beliefs, and values that schools nurture are accepted by a marginalized population as natural because they have no control over the curricula they are supposed to follow. Marginalized students and their parents are voiceless in curricula construction. Elite education policymakers ignore their interests. In this condition, the concept of critical multicultural education demands a fundamental rethinking of urban education and reconceptualization of schooling. Critical multiculturalism is a pro-democratic movement that acknowledges the anti-democratic forces of American urban education. It asks for emancipatory transformative democratic education and simultaneously tries to form counter-hegemonic practices in education milieu. The critical multicultural curricula is grounded on the notion of social transformative education that aims to expose persisting inequalities based on race, class, gender, ethnicity, and religion and recognizes the cultural capital of the marginalized.

The most important aspect of multicultural education is that it validates subjugated knowledge and examines dominant group through the lenses of the subjugated, analysing dominant ideologies and structures in order to understand how these are perceived by marginalized populations. In the realm of cultural disequilibrium and cultural manipulation, critical multiculturalism identifies with the German tradition of *Buildung* which stresses emancipation from oppression, an individual's self determination, and the need to build a democratic social education (Sunker, 1994). In the context of multicultural education, the democratic intent of *Buildung* is relevant because it is committed to a person's development as a mature individual with understanding of existing manipulative socioeconomic and political forces (McLure, 1997). The concept of critical multiculturalism focuses on empowerment of educators, teachers, students, and parents for the purpose of creating a democratic education that might reduce cultural manipulation and cultural disequilibrium of mainstream curricula. The catalytic multicultural pedagogy can analyze the dynamics of hegemony that is evident in every aspect of urban education, and simultaneously pursue the contemporary pedagogy to take a diagnostic approach to the paradoxical reality of equality.

Notes

1 The majority of British immigrants to America were not in fact "Anglo-Saxon", but "Celtic" of Scots, Irish and Welsh origins. In any case, only a minority of Britons are purely of "Anglo-Saxon" descent, the majority having mixed Viking, Norman, Anglo-Saxon and Celtic origins.
2 Canada, with its official policy of bilingualism in French and English, and its many bilingual programs in Canadian public schools, provides an interesting contrast to America's official monolingual policy, in the presence of a large and increasing Spanish-speaking minority.

References

Apple, M. W. (1982) *Cultural and Economic Reproduction in Education.* New York: Routledge.
Apple, M. W. (2001) *Educating The "Right Way": Markets, Standards, God, and Inequality.* New York: Routledge-Falmer.

Arnowitz, S. and Giroux, H. (1991) *Postmodern Education: Politics, and Culture, and Social Criticism.* Minneapolis: University of Minnesota Press.

Bagnall, R. (1999) *Discovering Radical Contingency: Building a Postmodern Agenda in Adult Education.* New York: Peter Lang.

Callahan, R. (1962) *Education and the Culture of Efficiency.* Chicago: University of Chicago Press.

Collins, R. (2003) 'The long way home'. *Obsidian III: Literature in the African Diaspora,* 4, 39–60.

Cubberly, E. P. (2004) *The History of Education.* Kila, MT: Kessinger Books.

Draper, A. (1899) "Common schools in the larger cities". *The Forum,* 27, 385–87.

Giroux, H. (1988) *Schooling and the Struggle for Public Life.* Minneapolis: University of Minnesota Press.

Giroux, H. and McLaren, P. (1993) *Between Borders: Pedagogy and The Politics of Cultural Studies.* New York: Routledge.

Glazer, N. and Moynihan, D. (1970) *Beyond the Melting Pot,* 2nd ed. Cambridge, MA: MIT Press.

Gonzalez, N. (1995) "Processual approaches to multiculturalism". *Journal of Applied Behavioral Science,* 3, 233–44.

Gordon, M. (1964) *Assimilation in American Life.* New York: Oxford University Press.

Gumperz, J. L. (1982) *Language and Social Identity.* Cambridge: Cambridge University Press.

Gumperz, J. L. (1994) *Discourse Strategies.* Cambridge: Cambridge University Press.

Harris, W. and Duane D. (1874) *A Statement of the Theory of Education in the United States Approved by Many Leading Educators.* Washington DC: GPO.

Harding, S. (1998) I*s Science Multicultural? Postcolonialisms, Feminisms, and Epistemologies.* Bloomington: Indiana University Press.

Hooks, B. (1994) *Teaching to Transgress: Education as the Practice of Freedom.* New York: Routledge.

Kincheloe, J. (2001) *Getting Beyond the Facts: Teaching Social Studies/Social sciences in the Twentieth Century.* New York: Peter Lang Publications.

Kincheloe, J. (1993) *Toward a Critical Politics of Teacher Thinking: Mapping the Postmodern.* Westport, CT: Bergin and Garvey.

Kincheloe, J. and Semali, L. (1999) *What is Indigenous Knowledge? Voices from the Academy.* New York: Routledge Falmer Press.

Kincheloe, J, and Steinberg, S. (1997) *Changing Multiculturalism.* Buckingham: Open University Press.

King, J. E. (1994) 'The purpose of schooling for African American children: including cultural knowledge.' In Hollins, E., King, J. and Hayman, W. (eds) *Teaching Diverse Populations: Formulating a knowledge Base.* New York: State University of New York Press.

Kliebard, H. (1995) *The Struggle for the American Curriculum: 1893–1958.* New York: Routledge.

Kozol, J. (1991) *Savage Inequalities: Children in America's Schools.* New York: Crown.

Lather, P. (1991) *Getting Smart: Feminist Research and Pedagogy with/in the Postmodern.* New York: Routledge.

McCarthy, C. (1993) "After the canon: knowledge and ideological representation in the multicultural discourse on curriculum reform". In McCarthy, C. and Crichlow. W. (eds) *Race, Identity, and Representation in Education.* New York: Routledge.

McLaren, P. (1997) *Critical Pedagogy and Predatory Culture: Oppositional Politics in a Postmodern Culture.* New York: Routledge.

McLure, M. (1997) *Expertise Versus Responsiveness In Children's Worlds: Politics in School, Home and Community.* New York: Routledge.

Mohl, R. (1997) "Cultural pluralism in immigrant education: the YWCA's International Institutes 1910–1940". In Majagij, K. and Sprat, M. (eds) *Women Adrift: The YMCA and YWCA in the City.* New York: New York University Press.

Moses, P. and Cobb, E. (2001) *Radical Equation: Civil Rights from Mississippi.* Boston: Beacon Press.

Nasaw, D. (1979) *Schooled To Order: A Social History of Public Schooling in the United States.* New York: Oxford University Press.

Richman, I. B. (1905) *Rhode island: A Study in Separatism.* Boston: Houghton Mifflin.

Sunker, H. (1994) "Pedagogy and politics: Heydorn's survival through education and its challenges to contemporary theories of education – *Buildung*". In Media, S., Baste, G. and Wardekke, W. (eds) *The Politics of Human Science.* Brussels: VUB Press.

Tyack, D.B. (1974) *The One Best System: A History of American Urban Education.* Cambridge, MA: Harvard University Press.

Villanueva, V. (1993) *Bootstraps from an American Academic of Color.* Urbana, IL: National Council of Teachers in English.

Zill, N. and West, J. (2000) "Entering kindergarten: a portrait of American children when they begin school,". In *The Condition of Education 2000,* NCES 2000–062, Washington, DC: US Department of Education.

Part IV

Diversity and educational equality in Britain and Europe

6 Multicultural education

A European perspective

Gajendra K. Verma and Adamantios Papastamatis

Introduction

Cultural diversity has been a fact of life in countries of east and west, north and south for centuries, and has led to various kinds of co-operation or coercion between the different cultural, religious and ethnic blocs contained within national boundaries (see Bagley, 1973). Since the process of migration and the history of different nations vary considerably, the cultural profile of different regions and ethnic groups within any particular country also varies. However, contemporary recognition of the value of cultural, linguistic and religious diversity and their implications for social justice have marked the post-war, and the post-Soviet, eras.

A society (or country) may be regarded as a system of interrelationships which connects people from diverse backgrounds in various forms of co-operation, domination, or attempted exclusion. Rarely can cultures exist without a stable and ordered nation-state, although there are notable exceptions. Before the foundation of Israel, Jews of the Diaspora existed as separated cultural or religious groups, as do Roma people whose Diaspora and wandering is part of their cultural heritage. Each cultural group has an identity and value which must, if principles of social justice are to be served, be respected and preserved not only by their nation-state, but by the international community as well (Kyuchukov *et al.*, 1999).

By the nineteenth century, cultural diversity in most European countries was already marked in terms of differences in the religious, linguistic and ethnic profiles of their inhabitants. Political and social forces frequently attempted to assimilate divergent groups into the dominant culture and language and, when this proved difficult, to eliminate groups such as Jews, Gypsies and Roma through expulsions and pogroms. Some European nation-states have allowed the new settlers to retain their distinctive cultures which often differed from the dominant group e.g. the plural society of The Netherlands (Bagley, 1973). Many countries however, have failed to recognize and support the right to individual difference and identity amongst different cultural groups.

The end of World War II was a watershed when minority ethnic/cultural communities throughout the world started asserting their rights in Europe; and in Africa and Asia the process of decolonization began. Both national groups and minorities became conscious of the fact that their identities had been eroded or suppressed because of deliberate social and educational policies. This awareness caused them to challenge the gap between the declared values and beliefs of democratic societies, and the realities of such policies in practice. Aboriginal peoples in North America and Australasia are cases in point.

Over the last four decades the classical concept of culturally homogeneous society has been openly challenged. For example, there is the process of European integration – political, social and economic – of laws and procedures which are likely to affect the daily lives of citizens. At the same time, cultural, linguistic and religious differences between communities have come under scrutiny and there is now political tension between nation-states as the enlarged EU seeks to frame a 'super-constitution' imposing binding laws and a framework of rights and obligations on nations who may have little in common. Ironically within this enlarged, quasi nation-state of the EU, countries themselves become cultural groups or blocs within the supra-national society which the EU has become. Continued immigration of people with different cultures, religions and languages across the EU adds to this complexity. Most EU governments have outlined policies highlighting the importance of equality and programmes opposing racial and religious discrimination. Despite debates, discussions and actions during the past few years no European nation-state has succeeded in executing effective policies for the successful integration of all immigrant cultural minorities and their dependents. There are complex reasons for this lack of success, but the problems are not, ultimately, insoluble.

One step taken towards a collective European Policy was the Amsterdam Treaty that took effect on 1 May 1999. The Amsterdam Treaty emphasizes the basic rights of immigrants and the principle of anti-discrimination. Signatories of the Treaty have a commitment

> to maintain and develop the Union as an area of freedom, security and justice, in which the free movement of persons is assured in conjunction with appropriate measures with respect to external border controls, asylum, immigration and the prevention, and combating of crime.

Within five years of the ratification of the Treaty, there were to be no controls for citizens of the EU or for nationals of third countries when crossing internal borders of the EU (Section 73, EU, 1999).

This will undoubtedly have far-ranging effects on approaches related to the integration of immigrant cultural minorities. The main question is whether a specific settlement policy implies a monolithic view of society or a pluralistic one. When the aim is unilateral acceptance of newcomers of a given social structure and its sets of norms, the settlement policy can be characterized as assimilation, forced or otherwise. By contrast, when the goals are pluralistic, the aim is to develop a society in which members of cultural minority groups are given opportunities and support to maintain and develop the fundamental characteristics of their own cultures (Verma, 1997, 2002). At the same time they should have the scope to participate as equals in the political, economic, social and civic institutions of the majority society (Verma, 1990; Pitkänen *et al.*, 2002).

Dimensions of culture

The argument of this chapter is that society is equivalent to the nation-state, and increasingly amongst the population of each nation are cultural groups who co-operate or compete in varying degrees of amity or enmity. Sometimes these competing factions are based on longstanding religious differences (e.g. Catholic and Protestant in the United Kingdom and The Netherlands) or on language, as in Belgium. In the present century new cultural groups emerge in societies as a result of patterns of migration or the rise of individual consciousness in some ethnic groups. Groups subjected to profound prejudice and discrimination by the host society are particularly likely to develop a militant group consciousness as a result of their alienation (Bagley and Verma, 1979).

Cultural values and ways of acting are learned, communicated and shaped through individual attempts to master, and participate in, the life of the nation-state. Thus culture is constructed and reconstructed through the process of social interaction of self-defined members of the cultural group with the major institutions of society. There are many aspects of an individual's identity which develop through this interactive socialization process. For the migrants who are an ethnic minority in a new country, adaptation can be complex and stressful, especially when racist social structures have to be negotiated (Furnham and Bockner, 1987). This process is influenced by family structures, by schooling and experience in the wider society. These forces contribute to the development of an individual's identity which consists of specific behaviours, values, lifestyles, attitudes and world views. Components of identity sometimes come into conflict with generally accepted societal norms. There is evidence that many of the cultural disparities that students experience are caused by conflicting values, beliefs and behaviours that they encounter in their home or cultural community, contrasted with those of the school (Pumfrey and Verma, 1990).

Some writers adopt a different view of culture. For example, Greene (1994) argues that culture is not a package of knowledge, attitudes and customs which can be parcelled up, handed to the child and then passed on intact to the next generation: 'The culture of any nation or community is constantly evolving and exists in the thought and actions and creations of its participants'. The concept of 'culture' is both complex and elusive. Some writers tend to imply that individuals or groups belong to and/or live in an easily identifiable and stable cultural milieu. This is a simplistic view of an individual's complex characteristics. While individuals can be born into a more or less clearly defined cultural group, through a process of change and development the nature of their cultural socialization changes, and is essentially dynamic. Individuals can acquire more than one cultural identity (e.g. being both Moslem and British) and will balance dual loyalties through a continuous process of dialogue and adaptation. 'Culture' in its broadest sense embraces every aspect of human development (Verma, 1989).

In most writings, culture is treated as an institutionalized group phenomenon on the assumption that there is homogeneity within a given social group (Verma *et al.*,1994). But such homogeneity rarely exists for very long. Regional characteristics, religion, socio-economic background, occupation, language and personal experience all contribute to the formations of changing subcultures within a social group. It is possible that, in spite of those differences between the subcultural groups, they may seem to share elements of a major, 'symbolic culture' such as those of the nation-state itself (Verma, 1999).

Cultural diversity and education

Educationalists and social reformers develop theories and models of cultural pluralism which imply that each cultural group has the right to develop distinctive characteristics within the framework of the wider society. Built into such a model are assumptions about equality, since the model assumes the existence of a continuous dialogue in democratic forums between various cultural groups, and the governors of the nation-state. Such a dialogue can have validity only if it is between equals. For example, in schools in various countries containing ethnic minority students, theories of so called 'multicultural education' have been developed and put into practice based on the concept of cultural integrity of groups within the schools, and their rights to participate in a curriculum which addresses both their needs, and the demands for them to develop into 'good citizens'.

Educational change in the context of the plural society rarely takes place in a tidy and uniform fashion. Issues such as ethnicity, culture, identity and religion have generated debates about the nature of society and its future social cohesion in many Western countries. There seems to be a gradual acceptance by policymakers in many countries that cultural diversity in society is a contextual imperative for all citizens. The educational system must aim to find ways to help children and young people become comfortable with diversity, and accept it as a normal part of existence and not as exotic and novel (cf. Booth and Ainscow, 1998).

Studies and reports published in various Western countries in the last forty years or so have implications for the ways in which we perceive European society in educational terms (DES, 1985; Verma, 1989). These have contributed to the debates about the way society and its various institutions are organized, the common values that should be upheld, and the ways in which the past, present and future of the nation-state should be appraised. In a culturally diverse society such a debate inevitably involves contentious issues surrounding racial discrimination, ethnicity, language rights and religious freedom. These issues have long been the source of considerable controversy, particularly in the field of education (Zmas, 2004). Whenever there are covert or overt disagreements over different values and beliefs surrounding multiculturalism, there are tensions within educational institutions ranging from teacher education colleges to schools at all levels.

Diversity is reflected in differences arising from cultural and religious values and traditions which influence the behaviour, attitudes and values of people in particular cultural groups, and in recent decades there has developed greater political consciousness among cultural groups wishing to reclaim or emphasize their identity within their society or nation-state (Verma, 1989). In most Western societies for example, there has been a continual history of immigration over the last few centuries and, since 1945, from their former colonies. Yet ethnic diversity has only become 'an issue' of consequence in democratic societies in the wake of large scale immigration and settlement that occurred after the 1930s.[1]

Differences in socio-economic status and tensions between ethnic groups in any society have often been the product of minority groups finding themselves subject to discrimination and prejudice, and even violence, and finding their life changes impaired by gross inequalities in the various social systems of society (Bagley and Verma, 1979; Gay, 2000; Verma, 1999).

Multiculturalism: the Greek case

To illustrate issues of complexity and social change in framing multicultural policies, we offer the case of Greece. A popular European stereotype is that Greece is an ethnically and linguistically homogeneous country whose main influence on the multicultural stage has been the continued emigration of its citizens, paralleled by the frequent but irregular arrival of tourists who stay but briefly. But on closer examination, Greece is a dynamic and complex society in terms of culture, language and immigration. There has been, in Makri's (2003) analysis, 'immigration of great masses of people to Greece'. These have been both economic and political refugees in post-war years. It is salutary that there remain about one million Greek-speaking individuals in the countries of the former Soviet Union who are not allowed to legally migrate to Greece, whatever their desire to do so.

In a series of papers, Sofia Dascalopoulos (2003) outlines the ethnic and linguistic complexities of Greek society. A particularly interesting group in Greece are the 150,000 Romani people, some 20 per cent of whom are Moslems. Other minority groups making up Greece's complex plural society are Turks, Valachs, and Arvanites as well as various groups speaking Slavic languages. Dascalopoulos observes that how the concept of multicultural education and plural cultural inputs to Greek life will apply to those speaking Slavic languages is unclear: an impressive number of Greeks from the former Soviet Union have high school and even college education in Russian, while the Polish immigrant community in Greece is positively accepted.

Makri's (2003) valuable analysis of multicultural education in Greece contrasts several distinct approaches, which have their analogue in countries such as Britain. The first approach which pertained until the 1960s was an *assimilationist* approach, requiring newly arrived groups to learn to assimilate to the dominant culture's values and language. The later *integrationist* approach separated and treated newly-arrived groups with some consciousness of their special cultural needs and rights. In the 1970s the *multiculturalist* approach emerged; this recognized aspects of racism in the social system which impaired the occupational and cultural progress of long-settled immigrant groups and their children. Finally, the *intercultural* model reconizes fully the 'democratic equality of rights' of all ethnic groups in the nation-state, both minorities and majorities.

Cultural diversity and teacher education

While the curriculum is an important area in terms of responding to cultural diversity and the opportunities for inclusiveness it can offer to children and young people, there is an even more important aspect of the educational process, namely the teachers. Many of us would, we suspect, when reflecting back on our education tend to recall first and foremost, those who taught us, rather than the curriculum we studied. This reminds us of the centrality of teaching and teachers in the educational process. If an inclusive education is to be offered to young people then the training of the teaching force is of paramount importance: a curriculum is only as good as those delivering it. Thus the delivery of effective education in a culturally diverse society is heavily dependent on the quality and training of the teaching force (Gagliardi and Mosconi, 1995; Pitkänen *et al.*, 2006).

In the context of the response to cultural and ethnic diversity, it is important that teachers are trained in a way that develops in them an awareness of cultural nuance and its effects on the behaviour, cognitive styles and bearing of children in the classroom by equipping them with relevant knowledge and operational skills (Appelbaum, 2002; Kesidou, 2004). Education should strengthen the cultural identity of all individuals in a particular society, and offer individual students the ability to see the world in unity within its plurality. The development of a cultural identity and a sensitivity to others are not built up automatically; they can only be outcomes of a curriculum designed to achieve them and delivered by dedicated teachers with the appropriate competencies.

Another crucial aspect of the teaching process is the 'hidden curriculum'. Formal changes are likely to come from the prescribed curriculum process which may teach knowledge, facts and competencies, but the hidden curriculum has a complementary and powerful part to play in bringing out changes in values, attitudes, and beliefs to complete the process of personal adaptation. For example, it may be damaging to the self-esteem of ethnic/cultural minorities if school expectations are insensitive to, and dismissive of, cultural and religious backgrounds (e.g. symbols of dress, and religious diets in various communities) (Pitkänen *et al.*, 2002).

In this respect it is important to develop teachers' capacity for multicultural education, that is, to provide them with information about pupils' communities and train them in how to use this information in teaching. Multicultural teacher training also needs to develop the teachers' capacity for communicating with pupils from different cultures. Teachers should also be trained in strategies for analyzing pupils' learning obstacles. They should know how to teach about sustainable development and how to train pupils to improve the quality of life (Gagliardi and Mosconi, 1995). The answer then, seems to be professional development. Teachers must grow through their work over time to develop the awareness, skills, and knowledge necessary to implement a democratic, pluralist educational programme (Appelbaum, 2002).

Note

1 Germany and Austria under the Nazis were fundamentally undemocratic, and developed a set of racist policies which, far from absorbing different cultural groups, suppressed them to the point of genocide. The USSR (see Gray, Chapter 8 in this volume), despite its monocultural theory of the State, did allow cultural and language retention in many of its diverse ethnic groups.

References

Appelbaum, P. (2002) *Multicultural and Diversity Education: A Reference Handbook.* Santa Barbara, California: ABC-CLIO.

Bagley, C. (1973) *The Dutch Plural Society: Comparative Studies in Race Relations.* London: Oxford University Press.

Bagley, C. and Verma, G. (1979) *Racial Prejudice, The Individual and Society.* Aldershot: Saxon House.

Booth, T. and Ainscow, M. (1998) *From Them to Us: An International Study of Inclusion in Education.* London: Routledge.

Dascalopoulos, S. (2003) *Multicultural and 'Plural': The Case of Greece.* School of Social Sciences, University of Aegean. At: www.aegean.gr/culturaltec/dasc/

DES (1985) *Education for All: Report of the Committee of Inquiry into the Education of Children from Ethnic Minority Groups (The Swann Report).* London: HMSO. For the Department of Education and Science.

EU (1999) The Text of the Treaty of Amsterdam, www.eurotreaties.com/amsterdamtext.html.

Furnham, A. and Bockner, S. (1987) *Culture Shock: Psychological Reactions to Unfamiliar Environments.* London and NY: Methuen.

Gagliardi, R. and Mosconi, P. B. (1995) 'Teacher training for multicultural education in favour of democracy and sustainable development: the territorial approach'. In Gagliardi, R. (ed.) *Teacher Training and Multiculturalism.* Paris: UNESCO.

Gay, G. (2000) *Culturally Responsive Teaching.* New York: Teachers College Press, Columbia University.

Greene, S. (1994) 'Growing up in Ireland: development in an Irish Context'. *The Irish Journal of Psychology* 15, 354–371.

Kesidou, A. (2004) 'Intercultural education – main aims and practices'. In Terzis, N. (ed.) *Intercultural Education in the Balkan Countries – Balkan Society for Pedagogy and Education, Education and Pedagogy in Balkan Countries 4.* Thessaloniki: Kyriakidis Brothers.

Kyuchukov, H., Van Driel, B. and Batelaan, P. (1999) 'The Education of Roma Children', Special Issue of *European Journal of Intercultural Studies,* vol. 10, no. 2.

Makri, V. (2003) 'Intercultural and multicultural educational policy in Greece'. Presented to LSE Symposium on *Current Social Science Research in Modern Greece.*

Pitkänen, P., Kalekin-Fishman, D. and Verma, G. (2002) *Immigration Settlement Policies and Current Challenges to Education.* London: Routledge-Falmer.

Pitkänen, P., Verma, G. and Kalekin-Fishman, D. (2006) *Increasing the Multicultural Understanding of Student Teachers in Europe.* Oxford: Trafford Publishing.

Pumfrey, P. and Verma, G. K. (1990) *Race Relations and Urban Education: Promising Practices.* London: The Falmer Press.

Verma, G. K. (1989) (ed.) *Education for All, A Landmark in Pluralism.* London: The Falmer Press.

Verma, G. K. (1990) 'Pluralism: some theoretical and practical considerations'. In CRE Publications *Britain: A Plural Society.* London: Commission for Racial Equality.

Verma, G. K. (1997) 'Inequality and intercultural education'. In Woodrow, D., Verma, G., Rocha-Trindada, M., Campani, G. and Bagley, C. (eds) *Intercultural Education: Theories, Policies and Practice.* Aldershot: Ashgate.

Verma, G. K. (1999) 'Inequality and education: implications for the psychologist'. *Education and Child Psychology,* 16, 6–16.

Verma, G. K. (2002) 'Migrants and social exclusion: a European perspective'. In *Integrated Approaches to Lifelong Learning.* Kuala Lumpur: Asia–Europe Institute, University of Malaysia.

Verma, G. K., Zec, P. and Skinner, G. (1994) *The Ethnic Crucible: Harmony and Discord in Secondary Schools.* London: The Falmer Press.

Zmas, A. (2004) 'Educational dilemma, in the context of intercultural approaches'. In Terzis, N. (ed.) *Intercultural Education in the Balkan Countries – Balkan Society for Pedagogy and Education, Education and Pedagogy in Balkan Countries 4.* Thessaloniki: Kyriakidis Brothers.

7 The impact of culture in creating differential learning styles

Derek Woodrow

Introduction

Underwriting any educational system are two alternative principles. The first principle is that of providing the right cadre of leadership and knowledgeable governance for management of the nation's structures. This leads to the provision of a necessary elite group able to move the economy and services forward. This principle is often that needed in the early days of developing an organised society when resources are limited and insecure. This does, however, maintain the dominance of those who have access to the education system and does not encourage the development of a democratic and equal opportunities society. Modern societies need more than a small educated elite if they are to flourish and thus the commitment to universal education has arisen. It was in this context that the second principle of 'education for all' arose and with it the notion of equality of access and opportunity. It is from this second principle that the commitment to educating all citizens to the best of their abilities emerged, and of attempting to combat discrimination in terms of opportunities brought about by class, caste, wealth, religion or hereditary power. It is this commitment which promoted and encouraged the development of comprehensive schooling in the UK and common schooling in India. It also encouraged for a time the introduction of 'mixed ability' classes in English secondary schools, though the wide variations in achievement proved too difficult to manage in some subjects, notably mathematics and foreign languages. Within primary education, however, most classes remain 'mixed ability' and a sense of common community is engendered which can be retained in neighbourhood secondary schools.

Inevitably, however, differential abilities and differential opportunities do arise. This is in part a consequence of Bourdieu's notion of 'cultural capital' in that those who hold advantage and power generated by certain forms of 'cultural knowledge' tend to increase in power and to sustain that form of knowledge (Apple, 2001; Woodrow, 2001). This cultural capital is gained in many ways, partly by heritage, partly by wealth, partly by ascription of power by the culture in which a person lives. In this analysis the most privileged students do not only owe the habits, behaviour and attitudes which help them directly in pedagogic tasks to their social origins, 'they also inherit from their knowledge and savoir-faire, tastes and a 'good taste" (Bourdieu and Passeron, 1964, p. 30).

In Western Europe the Marxist educationalists of the 1960s and '70s clearly established the role of education as a vehicle for socialisation, for confirming and continuing the social order and for conditioning the population to their varied roles. In more recent times education has been similarly viewed, with fewer overt political overtones, as a vehicle for enculturation by both majority and minority cultures. Of course, this assertion about 'fewer political overtones' is in itself a Western, even perhaps 'Anglo-American', view of education that it should ideally be non-political. For many cultures and societies the interweaving of education with politics and religion represents the ideal, a holistic and comprehensive view of the world and people. Certainly Islam would reject such separation as not reflecting the importance of dedicating to God the whole of one's life-actions. Many Muslim countries are very clear about the significance of education in providing shared values and shared beliefs. The early Catholic invaders of South and Central America found education a powerful tool for conversion. However, education can be a tool of emancipation as well as one of coercion: Gerdes (1985) on Mozambique, D'Ambrosio (1990) on Brazil, and Vithal and Skovsmose (1997) on South Africa, write and talk movingly about the role of education (in these cases mathematical education) in promoting just and fair societies.

Varying cultural and national identities have had a clear impact on formal education and different societies have different informal child-rearing and adult-initiation practices. Formal education is itself socially created and generally only becomes universal with the rise of the urban industrial dwelling. Different societies have different perceptions of authority and respect for elders, different perceptions of freedom (especially for children), and different assumptions about gender roles and gender relationships, all of which have a significant influence on educational practices. Even within a single society these assumptions change over time and within subgroups and lead to changes in educational practices. Clearly the impact of religion is significant. Some religions prioritise social groups and social dependency, in particular, Islam and Christianity both rely upon regular gatherings of the believers whereas other religions such as Buddhism and Hinduism stress more the individual path to enlightenment. These contextual features affect the person's view of the nature of knowledge and learning. Different societies develop differing attitudes to authority and autonomy, and developing a common system of schooling which facilitates all these preferred modes of living is not an easy task. In complex multi-group societies one consequence of state education and common schooling that needs to be addressed is the development of a style of curriculum and assumptions about learning which favour and advantage particular subgroups. These are the issues which this chapter seeks to explore.

Cultural mores and beliefs relate not just to social behaviours and interests but affect assumptions about ways of learning; even the meaning of 'learning' may be different within different social constructs. Bourdieu (1977) with his concept of *habitus* and Kelly's construct theory (1973), both emphasise the impact of cultural context on thinking and learning, with different communities providing different cultural capital to their offspring. On a macro level the dialectic between culture and learning presents problems in that different societies (often unwittingly) misunderstand each other. On the micro level it can create a mismatch between local subculture and that of the wider society within which that subculture exists, leading individual learners to a sense of dissonance and classroom unease.

There has been a considerable amount of work on individual learning styles, particularly in the USA, the UK and in Australia and Hong Kong. One of the outcomes of this research has been to establish the impact of different cultures on these learning styles, and to explore the tendency for some subgroups of a society to have different styles and means of learning. For a country like India such variations in learning styles are of considerable significance, with different dominant religions, different cultures sustained through its caste system, different ethnic groups, large urban and rural differences and environmental variations which support the development of differential lifestyles, all having their effects. That such variations in individual learning styles exist is beyond dispute, but how the school curriculum should respond to such supposedly innate characteristics is one of the critical issues that needs to be faced. Do you compensate for apparent 'weakness', such as spatial awareness, or do you find ways of revaluing or discarding inappropriate criteria? If the curriculum is biased towards certain intellectual strengths, do you change the curriculum to enable learning to develop through alternative strands, or do you work to strengthen, if possible, the necessary abilities in the individual. At the root of the learning styles' interaction with education and curricula lies this fundamental question, and its response often seems to be rooted in the notions of autonomy and authority, of individual freedom of social responsibility which a society holds.

More recently Western psychologists and academic-curriculum creators have provided new 'process-driven' categories of cognitive style, which mirror the move of the Western school curriculum away from content and facts towards strategy and process. This has been marked in mathematics and science by the concern for problem-solving and other generic skills and in literacy by more respect for oral communication and fluency of expression. Much of the theoretical framework related to these mental processes is reminiscent of attempts during the 1960s and early 1970s to identify and value 'creativity' as a specific domain (see, for example, Guilford, 1967; Gardner, 1993). This involved tests of 'divergent' and 'convergent' thinkers, and other bi-polar measures such as whether a person adopted a 'scanning' approach to a problem or a 'focused' approach, whether people are reflective or impulsive, holistic or serialists, and the development of the learning styles literature (see Riding and Cheema, 1991, for an authoritative overview; Adey *et al.*, 1999, for a good descriptive treatment; and Coffield *et al.*, 2004, for a critical review). The best established general descriptor is that provided by Witkin (1967) of a single 'field dependent' versus 'field independent' cognitive processing style. Once again it is important to recognise the social and historical determinants of much of psychological theorising and the classification of behaviours (Popkewitz, 1998). Thus notions of 'wholistic/analyser, verbaliser/imager' are only meaningful within the constructs derived from recent process-driven curriculum frameworks. The extent to which these categorisations are meaningful with non-Anglo-American students is debatable and these terms need to be exemplified and an appropriate language developed to characterise these students.

A note of caution needs to be made at this point. Since the thrust of the argument presupposes the impact of culture on providing, even understanding, differences in learning and thinking, it follows that research in the USA or the UK cannot be transposed to India. It is indeed very difficult to escape from the assumptions which are built into one's own imagery and experience. We assume, too readily, that since learning is a universal human characteristic it must take the same form everywhere. Recent comparisons of Japanese and Chinese education with that of the USA and the UK illustrate how fundamental are the underlying principles and beliefs which drive the act of learning. The research produced by the West needs to be re-interpreted and re-analysed by Indian educationalists if it is to have significance in an alternative culture. There are at best likely to be parallel outcomes or related concepts which need to be addressed by Indian researchers. It is in this context that the following analysis is offered.

Achievement and strengths of different groups of students

Within most school systems there are groups of pupils who appear to succeed and others who appear, more often, to fail. Within the UK there is currently considerable concern that groups of Pakistani and Bangladeshi pupils are not apparently achieving as highly as their peers, and Black male students appear equally disadvantaged by the school system. Conversely, pupils of Indian and Chinese origin appear to perform better than their peers. Some of this is undoubtedly due to economic factors; poverty universally has a depressing effect on educational achievement and there is some evidence for the proposition that Indian immigrants to the UK derive from higher economic and social groups than do the Pakistani immigrants. Research reported below suggests that different learning styles, imposed by different cultures, might also be a significant feature of the problem. It is not just that students have different learning styles, but that the curriculum has embedded within it assumptions about learning, and teachers in their planning often implicitly and unwittingly embed assumptions about learning. If these resonate with an individual pupil's learning styles then the curriculum is a success but where there is a mismatch then problems arise. Black researchers, in particular, have drawn attention to the problems of mismatch between the curriculum and pupils with a different philosophy. Conversely, of course, the success of Indian and Chinese students in the UK might be related to their complementary learning style; the curriculum emphasises group work and constructive psychology whereas their own style is more rooted in practice and memorising, and the alternative styles in this case provide a strength. All educational systems need to take such variations in achievement seriously and particularly any system of schooling which aims for equality of opportunity.

Equally important is the recognition that certain groups of students are attracted differentially to study particular subjects. UK statistics on university entrance (e.g. UCAS (2000) Table 4) shows substantial variations in recruitment to English universities amongst various ethnic groups (these groups are self-declared from a list offered) which have persisted over a number of years. For the percentages of students accepted to study *mathematics and informatics* in UK universities in 2000, see Table 7.1.

This supports a common perception that Asian students, in particular, are drawn to the subject, but so also to some degree are the Black students. In fact, the recent large increases in recruitment to this subject area are almost entirely comprised of applications for computing (informatics) and the rapid growth of computing skills amongst young people in India (currently being attracted by comparatively large salaries into the USA and UK where there are shortages of such skills) would seem to suggest a cultural inclination for the subject as well as a financial incentive. A recent study from the Royal Society of Chemistry and The Institute of Physics confirms the strong ethnic bias in subject choice in students entering university. Students of Chinese and Indian origin are most likely to read science, mathematics and computer science, while those of Afro-Caribbean origins favour arts and humanities. Students from Bangladeshi and Pakistani origins are more likely than others to opt for business studies or law. Asian-origin students, in general, are more likely to opt for medicine and degrees allied to medicine (Garner, 2006).

There are a number of other interesting variations between ethnic groups and genders. Applications by Asian and Black students represented only four per cent of applications for Education (including teacher training) compared to the maths and informatics proportions of about 32 per cent. Differences between the genders in subject choice are similarly clear and equally persistent. Three times as many women study languages as men; six times as many men study engineering. In the UK teaching is a predominantly female profession, whereas in Cameroon (for example) teaching is mainly done by males. It is also noteworthy that Black students are reluctant to take up teaching, whereas they are very keen on social work.

It is tempting to begin to draw speculative conclusions to explain these differences, though the evidence for any particular explanation of these differential recruitment figures is not strong. Some of these were explored in a previous article (Woodrow, 1996) including the suggestion that learning styles might have a significant influence, and recent research by Jarvis (Jarvis, 2002; Jarvis and Woodrow,

Table 7.1 Accepted students for maths and informatics (2000)

	Males %	*Females %*
Of all the Asian students	33	12
Similarly for Black students	22	8
Whereas for White students	13	3
Overall for all students	16	4

2002) supports this idea. Jarvis has shown that university students in different subjects do indeed exhibit different learning styles, and in particular that the learning styles shown by mathematics students is significantly different from that of students studying many other subjects. It is also clear from research with Chinese students (see below) that the learning style characteristics of mathematics students mirror many of those held by Chinese students, and it is conjectured that Indian students will hold slightly different but comparable learning preferences.

In the Jarvis study, in order to discover if there were distinct disciplinary differences in students' beliefs about knowledge and their learning preferences, a questionnaire was devised, based primarily on Biggs' (1987) Study Process Questionnaire which relates to the notions of 'deep' and 'surface' learning. The questionnaire also incorporated ideas based on work by Schommer (1990) and Vermunt (1996) relating to beliefs about knowledge and the regulation of study strategies. The questionnaire was a fifty-item Likert-style questionnaire with five categories ranging from 'strongly agree' to 'strongly disagree'. It was multidimensional and factor analysis confirmed the existence of five distinct dimensions concerning the learning preferences of students. Details of the five dimensions derived are given in Table 7.2 with examples of the foci of the related questions.

Altogether, 384 undergraduates from five disciplines were questioned. In a parallel project 483 graduate teacher trainees in twelve different subject areas were also surveyed, and their outcomes followed closely that of the undergraduates. For the purposes of clarity in Table 7.3, the scales have been normalised to provide for each of the five factors a spread of ten points. This shows more clearly

Table 7.2 Five dimensions of student learning preferences

Scale title	Meaning of low score
Interaction and Participation (14 items)	Students are keen to interact and exchange ideas. Examples are that students show enthusiasm for group work and making presentations, and are confident in asking and answering questions in class.
Approach to learning (12 items)	Students show intrinsic motivation and use a 'deep' approach to learning. Examples are that they prefer to learn ideas rather than memorise facts. They do not want to be given exact instructions by their tutors.
Instructional preferences (6 items)	Students favour theoretical work and text over diagrams. They seek knowledge other than that provided by the tutor and are not motivated by a future career.
Beliefs about knowledge (7 items)	Students have a relativist view of knowledge. They do not expect their tutors to be able to transfer a body of knowledge intact to them. Examples are that students like to work on their own ideas and connect them to real-life situations.
Regulation of (8 items)	Students self-regulate their own studies. Examples are that students take control of their own learning by preparing for lessons and reading around the subject.

Reliability was confirmed by a Cronbach's alpha score (value 0.81).

Table 7.3 Ratings on the five scales

Under-graduate subjects	Number of students	Mean score overall	Partic-ipation	Approach to learning	Instruct-ional preference	Beliefs/ knowledge	Type of study
					Measure of		
English	75	0.00	0.00	1.00	3.98	0.00	2.85
Business Studies	75	3.16	0.63	0.00	9.68	4.24	0.00
Art	89	3.69	0.30	9.48	3.36	5.22	8.49
Science	47	4.24	10.00	1.64	0.00	10.00	6.21
Maths	98	10.00	6.44	10.00	10.00	9.84	10.00

the extent to which English and mathematics students are so often at the extremities. As is common with large scale Likert-style questionnaires, the trainees' interpretations of the questions were not probed; exploring meanings requires a different qualitative methodology. However, the way in which students interpreted the questions makes it clear that English and mathematics students respond quite differently to the survey.

Clearly mathematics students, as well as science students to a marked degree, tend to be driven by extrinsic factors and adopt a surface form of learning dependent on memorising and skill development. They like to be given exact instructions by their tutors. They focus narrowly on what is required and are motivated by concerns about their future career. They expect their tutors to transfer a body of knowledge to them. These students like to be disciplined by their tutors in their work. This could well be a factor in the differential recruitment of ethnic groups, and even an issue in gender discrimination.

Culturally imposed learning styles

Little research appears to be available regarding Indian students' learning preferences. The origins of some of the work in Hong Kong and its population's identification as a clearly differentiated ethnic group has meant there is a quantity of research on Chinese students, and this is presented as an example of how distinctive learning styles do reside in subgroups of pupils. In a recent study of Chinese pupils in English secondary schools (Sham and Woodrow, 1998; Verma *et al.*, 1999; Woodrow and Sham, 2001) significant differences were found in the assumptions and learning styles of Chinese pupils, even though many had been in England for a considerable period and many had been born there. The overwhelming conclusion from this research was the extent to which the British-Chinese pupils were conditioned by traditional Chinese behavioural rules. The family context was overwhelming and totally dominant, so that even those who were born in England were immersed (submerged) in their family context. The two fundamental rules of 'respect for superiors' and 'loyalty and filial piety' provide a framework within which they create expectations and attitudes with regard to their education.

One possible reason for this strong familial influence lies in the lack of a centralised religion in Chinese culture. Not only has the communist influence weakened religious aspiration in China, but the tradition is not for religious meetings or for religious 'gurus'. The lack of religious leaders makes it more difficult for the transitions to a local culture to be negotiated and recognised. In addition, Confucianism, the most popular surviving religious tradition in Chinese cultures, emphasizes private rituals honouring ancestors, rather than any kind of temple worship.

Not unexpectedly, British-European pupils have the 'right' cultural capital and have learning styles which seem compatible with the teaching styles they experience. The individual autonomy that is emphasised resonates with the social assumptions of parents and the stress in British society on individual rights and freedoms. There are many opportunities for the pupils to think and work independently, problem-solve and make up their own minds. Generally, the Chinese pupils would much prefer to work on their own rather than in a group and would prefer a quiet classroom. For Chinese pupils the purpose of group work seems little understood, and being questioned in class is embarrassing and makes them nervous. Discussions with their peers, a common feature in English schooling, are to them irrelevant when it is the teacher who holds the knowledge. Solving problems or making up their own minds is the most difficult learning strategy, yet this is at the root of most English classrooms:

> This attention to generalised process skills as the central feature of English education is not true of learners in some other countries, where knowledge is still rooted in facts, and where the investment possible in education makes very large classes inevitable, and teacher knowledge precarious. Where factual knowledge and algorithmic skills are precarious they maintain their central importance. Learning in this situation is inevitably 'book bound' and rote-learned skills are not just valued but found useful and are indeed valuable. Students and teachers believe that success comes from being told what to learn and this can then be memorised for success.
>
> (Woodrow, 1997a, p. 39)

One interesting outcome of a similar study in Hong Kong was that one of the most distinctive differences between Hong Kong classrooms and English classrooms was in how often pupils are praised by their teachers. There are few English teachers who will consider the proposition that praising pupils is irrelevant to learning – even when it is pointed out that most praise given is for qualities over which pupils have no control and which can cause disenchantment and disbelief. Yet when comparing pupils in Hong Kong and Manchester we found that whilst over 80 per cent of Manchester pupils had been praised, less than 20 per cent of Hong Kong pupils had ever been praised – yet their expressed enjoyment of school was higher.

Learning to read with meaning is difficult if the language being learnt is not the mother tongue. This tends to lead to reading being learnt entirely phonically without a concern for understanding. This early introduction to a style of learning is important, as in a similar way is the learning of Chinese calligraphy, a painstaking and practised art which teaches patience, neatness, visual acuity and motor control as learning virtues. The high value placed on calligraphy promotes by association the use of memory and repetition as a means of learning, and as suggested by Tang and Williams (2000) leads to the development of a more sophisticated (even perhaps different) form of learning method and cognitive style (Bagley, 1996).

Dunn (1990) considered a group of twenty-one elements in a 'Learning Styles Inventory' (LSI) which revealed some interesting differences between Mexican-American, Chinese-American, African-American and Greek-American fourth-, fifth- and sixth-grade pupils. The LSI included personal construct items (such as responsibility/conformity, authority, self-motivation, parent and teacher motivation) together with methodological issues (such as learning alone, preferring a variety of approaches, tactile and kinaesthetic approaches) and contextual issues (such as morning/afternoon working, noise and temperature preferences). The African-American and Chinese-American profiles proved to be consistently opposite, almost perfect mirror images in their preferred learning styles. Aloneness was a strikingly strong positive for the Chinese and an equally strong negative for the African; indeed fifteen of the twenty-one items were statistically significantly different. Chinese-Americans seem to require a variety of instructional approaches, whereas African-Americans prefer established patterns and routines to their learning.

A Black colleague has emphasized to me the importance he feels of 'vibes', intuitive responses that most of White (Greek-derived) academia rejects with distrust. This issue is also discussed by Asanti (1987) and Collins (1990), amongst other Black writers, who stress the 'spirituality' of African thinking and the holistic view of reality this provides. In contrast to Western, either/or dichotomous thought, the traditional African world-view is holistic and seeks harmony (Collins, 1990, p. 212). The myth of objectivity and the use of a methodology of objectification is one aspect of universalism as an expression of the European driving force and as a tool of Western cultural imperialism: 'Objectification becomes a means of claiming universality where there is none. European cultural imperialism is therefore an inherent part of European objectification (scientism)' (Ani, 1994, p. 411).

The factors that create these differences would appear to be culturally or socially based and will lead to the prioritisation by different groups of different descriptions of the learning act as better or worse descriptions of how their learning takes place. Problems only arise when the systems (or the teacher) unnecessarily and discriminatorily prioritise some factors above others, and hence some pupils above others (see Lewis, 2002).

The impact of social contexts

The impact of religion on attitudes and aspirations relating to education was clearly shown by Singh-Raud (1997) in researching young 'British-Asians' and in particular comparing women from the Hindu, Sikh and Muslim religious communities. It was interesting to note, for example, that the Muslim women pushed hard for separate and single-sex schools whereas the Hindu and Sikh women were against the notion of separate denominational schools and preferred co-educational schools. These different Asian religious groups had differing attitudes towards education, employment, marriage and settlement. Singh-Raud noted that whilst in most cases responses were influenced by religious upbringing, in other cases religion was probably acting as a marker for clearly differentiated cultures. It was, indeed, extremely difficult in Asian communities to distinguish between religion and culture, since they were so closely interwoven. Singh-Raud makes the point that variance does exist due to religious faiths in UK education and if these differences are not addressed then there is the danger of being discriminatory on a group level, i.e. *'creedist'*. It is not positive 'discrimination' that is called for here but rather positive 'action'.

Whilst cognitive techniques such as memory are universal, the way they are used to develop concepts and solve problems differs from person to person and group to group. Different societies value and utilise these skills in different ways, with rational, emotional, empathetic and interpersonal skills being differently prioritised. When social groups are in effect locally- (family-) based these discrepancies can be accommodated and adjustments made to ensure individual self worth and esteem. Sham (1996) describes how Chinese families manage and cope with children with learning disabilities in a different way from the typical English response, being able to support and contain the needs within the wider family structure; for the Chinese the family will provide a protective overcoat to members with disabilities so that the disability becomes less relevant. The causes, meanings and responses to such disabilities are radically different in different cultures and are almost impossible to place in correspondence. As societies have become larger, families smaller and more mobile and structured in more complex ways, they are no longer adaptable to individuals with unusual needs in the same way. The days in which the village 'simpleton' was accepted and socially nurtured by a small supportive community are long gone in England.

As people move from supportive 'villages' to large urban towns so conditions change, and the urbanisation of India, for example, will have consequences. Those with distinctive needs must be assimilated and they can only try to conform as best they can. This need for citizens to conform and live routine lives makes eccentrics and other individuals less acceptable, much as all world markets begin to look the same – the same stores, the same merchandise, and little attempt to localise the product. Thanks to television and impressive marketing, young people all over the world seem to wear the same jeans and tank tops. It probably has quality but does it have character? In Ritzer's (1993) evocative phrase, it represents the McDonaldisation of Society, a complex process of pseudo-individualisation of a unitised universal product.

One consequence of this trend towards uniformity, represented in the UK by the National Curriculum and its consequences, is that different cognitive styles and strengths result in more discrimination and inequity. Schools have always had to cope with the complexity of varying learning styles, and individual pupils have needed careful support. Clearly, however, where there is a dominant assumption about how the students are learning then any pupil dissonance from that assumption will lead to disadvantage and lack of development consistent with those expectations. As educational valuations in England have become more overt and incontestable so more pupils have been excluded.

Constructivism – a culture-bound theory of learning

Many of the concerns within education lie in conflicting notions of authority and correctness, whether it be in instilling 'morals', establishing the nature of 'proof', following grammatical rules or in decision-making. Cultures which have strong respect for ancestors and elders will tend to have a view of knowledge which is heavily based on the notion of a 'body of knowledge' rather than knowledge as a creative and individual voyage of discovery. The source of authority is critical. Individual identity as contrasted with belonging to a societal group (be it family, ethnic or cultural) will have a fundamental bearing on such issues. The growth of 'constructivist' theories in both mathematics and science education relates to the rejection of 'bodies of knowledge' and extrinsically created truth and authority which challenge the supremacy of the individual and self-determinism. It would appear to be an interesting paradox that this development of 'constructivist theory', with its stress on individual conceptions of knowledge, should have taken place in mathematics and science. These two curriculum subjects are traditionally perceived as being the least related to individual pupil contribution and creative activity, compared with concerns for external truth, facts, rules and objectivity. Yet it was probably the very neglect of individual autonomy that led researchers to focus on this omission from the academic portfolio of these subjects.

In radical constructivist theory it is held that there is no knowledge other than that which is owned by the individual. The role of the teacher is therefore to create situations or experiences that present the learner with new ideas to rationalise. It is a 'teaching for meaning' psychology in which metaphor and language explorations are the vehicles for development. This places enormous emphasis on the images and constructs that the pupil owns, and many of these will be focused within, and derived from, the pupil's own culture rather than that of the teacher or society at large:

> It crucially removes from the teacher the position of arbiter of knowledge, the only person in the classroom with authority. In some societies this removal of authority is unacceptable, or unimaginable, and makes this particular psychology of learning irrelevant and inapplicable.
>
> (Woodrow, 1996, p. 32)

It is related too to the roles and responsibilities accorded to the teacher and pupils. It assumes not only the possibility of a negotiated position between teacher and pupils but also one in which pupils have autonomy and rights. Children appear to have adult rights, and adult responsibilities for their learning. Any such 'negotiations', of course, take place within cultural assumptions which may leave little room for variance or re-definition; the pupils may simply not allow the teacher to abdicate the role of knowing authority.

It is also perhaps no accident that the 'constructivist' theory should have arisen largely within US and English education as a response to the commitment of the culture and society to capitalistic and self-reliance philosophies. One of the basic outcomes of the right wing 'Thatcher/Reaganite' policies was that the 'state' ('there is no such thing as society' said Mrs Thatcher, talking to *Women's Own* magazine, October 31 1987) was no longer responsible for individuals. The current position of individuals as employed or unemployed, rich or poor, and by implication literate or illiterate is their responsibility and all they need do is to exert their entrepreneurial talents. Guilt is passed onto the individual rather than being the responsibility of collective society. They are not enterprising enough or just do not have the right internal language. It is interesting also to note that in countries where there is a clearer concern for individual rights in learning (i.e. Western countries), blame for failure to learn by an individual pupil is often attributed outside the control of the individual to 'the teacher' or to 'not being clever enough', whereas where more social methods of teaching are the norm (e.g. in Japan and Taiwan) the blame for non-achievement is more often accepted by the pupil who will assert that they did not work hard enough (Stevenson and Stigler, 1992).

Constructivism becomes more problematic as a theory when family rather than self is the identity unit and social responsibility rather than self-aggrandisement is the motivating force. Social constructivism attempts to address this issue by considering the individual pupils within a social context, and looking for social interaction as a support for individual self-concept development. In social constructivism too, however, the premise remains that there is no knowledge except that known by the pupil and it is individual self-exploration which is central. This also becomes a difficult theoretical position when the teacher's role is founded in a culture which values authority and leadership.

The conflicts can be seen in the paradox contained within a recent doctoral thesis in which a Kenyan author expressed a firm commitment to constructivist (and hence individually focused) theory but felt constrained to interview pupils in groups, since it was so abnormal for a Kenyan teacher to talk with a pupil individually (Wanjada, 1996). Where authority, rather than autonomy, is valued then it is likely that traditional approaches to mathematics and science will cause fewer stylistic conflicts and constructivist theories will not find favour. With large classes of pupils and few resources, discourse is problematic. Johnson (1997) tells of teaching in Lithuania where partly because of lack of books, the subjects are taught by lectures where students take detailed notes which they learn to reproduce for formal examinations. He describes 'recitation' as the main teaching mode in Lithuanian schools. In the context of India the variations in the pre-eminence of authority or autonomy will clearly be affected by the local social contexts, by the context of large school classes and limited resources, and above all by the differing religious affiliations.

The imperialism of individualism

The influence of the contrast between individual rights and social responsibility on the fundamental concept of 'democratic' education is discussed further in Woodrow (1997b), but the position is far from clear and paradox is ever present. The notion of democracy would seem to be inimical to isolation and individualism; it is impossible without the interaction of people and without reference to 'society'. The value of radical constructivism is in its contribution to maintaining the debate, and highlighting the conflict, between individuals and society. An interesting early debate about the paradox which must be sustained between 'rights of individuals' and their empowerment contrasted with the 'identification of needs', which more paternal societies try to fulfil, can be found in Rappaport (1981), who maps the move in the USA away from social paternalism in community support to individual responsibility. At the core of current Western dogma lies the notion of individual autonomy, and the promotion of self assertion and self decision, yet democracy depends upon the denial of this (whether voluntary or majority enforced) when social cohesion and the social good are implied. On such contradictions reality is created. This same paradox needs to be sustained (accepted but not resolved!) in learning theories where contradictory notions such as individual construction and bodies of knowledge just have to co-exist. Statements that assert that individuals 'exist' and that individuals 'construct their own thoughts' are rather banal unless they are juxtaposed with other notions such as 'bodies of knowledge' and 'social mores' or 'citizenship'. There is a delicate balance to be held between the autocracy of tradition and the anarchy of existentialism, and it is easy for democracy and justice to vanish or become misrepresented through imbalance towards either position.

Cultural capital and imperialism

The discussion in this chapter provides a rich annotation of the notions of Pierre Bourdieu (1977), whose use of 'cultural capital' to denote the outputs of culture on social power and dominance matches much of the underlying concern. Both home and school provide students with 'capital', forming a richly developing *habitus* in which they operate. Some of their home 'capital' is also valuable within the school economy; the acceptance of authority without overt questioning makes them 'good' pupils. By contrast the attributes of some pupils appears to generate a much less valued cultural capital. The ability/commitment/ application of some to memorising knowledge is positively powerful in the school market and the absence of over-desire for leisure activity and a habit of working are seen as useful school currencies. The late 1990s saw a shift of valuation in the dominant English cultural field towards these currencies and there is now less commitment to the currencies of the 1980s, namely, problem solving, peer interaction and democratic debate within classrooms. It is still evident, however, that the *habitus* of the English classroom is focused around individual rights, individual responsibilities and individual choices as the significant currencies.

Sociability, being liked, 'belonging to the club', are still dominant, to repeat a quote from earlier:

> The most privileged students do not only owe the habits, behaviour and attitudes which help them directly in pedagogic tasks to their social origins, they also inherit from their knowledge and savoir-faire, tastes and a 'good taste'.
>
> (Bourdieu and Passeron 1964 p. 30)

In England it is assumed that as the generations of immigrants pass through, the *habitus* within which individuals exist will be more and more affected by the ambient social milieu and this will provide more usable symbolic currency for the young assimilated people. Indeed, the 'British-Asian' culture is already becoming more assertive in creating its own cultural nexus with a unique set of values concerning self and others. It is assumed that such a strong alternative culture will in its turn affect British traditional culture.

There are, however, two dangers: one, that British Asians remain (or are constrained to remain) within their own field rather than emerging and succeeding within the dominantly White culture (degree courses in mathematics and information technology becoming a cultural home for Asian students); the other is that success comes from real assimilation and the distinctive, and internally valuable, attributes of these minorities are dissipated. Indian society faces its own divisions and subgroups, some of whom are privileged and others under-privileged. It is the aim of common schooling to tackle these issues and to try to provide a more equal opportunity for these different groups. It too must struggle with maintaining or destroying variety and difference.

An idealistic objective would be that what is required is a curriculum that can respond to variety and variation, since there is clearly as much of these within all social and ethnic groups as there is between them. Whilst ethnic origins and family life may affect the *habitus* of an individual, so too do their own characteristics, their extroversion or introversion, their excitability or pacifity, and many other variables which make individuals individual. Assumptions about how students learn almost inevitably discriminate for or against particular learning preferences. Teachers often excuse themselves in terms of 'if only I had known I wouldn't have done that', when in practice you can never know enough and must teach in a way which doesn't depend upon knowing and that allows for individual learning traits.

This assumes, of course, that education really can be an altruistic, empowering agent for all individuals rather than a vehicle for pre-determined enculturation and social control. For example, it is evident that much of the altruistic Western empowerment agenda of social development policies during the 1980s served to empower the powerful more effectively than it did the underclass it was promoted to advantage. The 'headstart' curriculum, introduced in New York in the 1970s to improve the achievement of pupils from ethnic minorities, was also used by the strong middle classes to promote their own offspring, leading not to the catching up but to the falling further behind of the children it was intended to help. According to Bourdieu (see Grenfell and James, 1999, pp. 20–21) as the subcultures become symbolically richer and have more capital, the governing society will intuitively change the exchange rates and work to devalue the currencies in which the subcultures have saved. Ways of teaching and the messages passed on by curriculum assumptions are an essential part of that maintenance of cultural dominance. Power changes are a slow process without a total collapse in the market. Without appropriate political intervention the rich get richer and the poor get poorer, and without appropriate educational policies the gulf between the 'intelligentsia' and the 'illiterate' also widens.

On the wider stage, assumptions of constructivist principles of learning reinforce Western valuations of individual knowledge, individual rights and individual autonomy compared to 'book knowledge', traditional bodies of knowledge and authority which depend upon social valuations. There is a concomitant commitment to social interaction and debate as the form of academic self-validation and justification, rather than reference to traditional texts, authority and expert opinion. Valuations such as these are determined by the dominant participants in the field of operation. They legitimise the symbolic exchange rates in which the educational economy trades, defining the power and influence which the social capital represents. It is vital, therefore, that India develops its own curriculum, its own theories of learning that tune into, and are resonant with, its own society and its own values.

References

Adey P., Fairbrother R. and William D. (1999) *Learning Styles and Strategies: A Review of Research*. London: King's College.

Ani, M. (1994) *Yurugu, an Afrikan Centred Critique of European Cultural Thought and Behaviour*. New Jersey: Africa World Press.

Apple, M. W. (2001) 'Comparing neo-liberal projects and inequality in education', *Comparative Education*, 37, 4, 409–423.

Asanti, M. (1987) *The Afrocentric Idea*. Philadelphia: Temple University Press.

Bagley, C. (1996) 'Field independence in group-oriented cultures: comparisons from China, Japan and North America', *Journal of Social Psychology*, 135, 523–6.

Biggs, J. (1987) *Student Approaches to Learning and Studying*. Hawthorn: Victoria, Australian Council for Educational Research.

Bourdieu, P. (1977) *Outline of a Theory of Practice*. Trans. Nice, R., Cambridge: Cambridge University Press.

Bourdieu, P. and Passeron, J. (1964) *Les Heritiers*. Paris: Les Editions de Minuit.

Coffield, F., Moseley, D., Hall, E. and Ecclestone, K. (2004) 'Should we be using learning styles? What research has to say to practice', *Learning and Skills Research Centre (online)*. At: http://www.lsda.org.uk/files/PDF/1540.pdf (Accessed 08 August 2005).

Collins, P. (1990) *Black Feminist Thought*. London: Routledge.

D'Ambrosio, U. (1990) 'The role of mathematics education in building a democratic and just society', *For the Learning of Mathematics*, 10, 20–23.

Dunn, R. (1990) 'Cross-cultural differences in learning styles', *Journal of Counseling and Development*, 18, 68–93.

Gardner, H. (1993) *Frames of Mind: The Theory of Multiple Intelligences.* London: Fontana.

Garner, R. (2006) 'Chinese and Indian pupils most likely to pick science', *The Independent Online,* May 9th, 2006.

Gerdes, P. (1985) 'Conditions and strategies for emancipatory mathematics education in undeveloped countries', *For the Learning of Mathematics,* 6, 10–17.

Grenfell M. and James D. (1999) *Bourdieu and Education: Acts of Practical Theory.* London: Falmer.

Guilford, J. P. (1967) *The Nature of Intelligence.* New York: McGraw Hill.

Jarvis, J. (2002) 'Learning preferences of undergraduate and PGCE students in relation to subjects of study', unpublished PhD Thesis. Manchester Metropolitan University.

Jarvis, J. and Woodrow, D. (2002) 'Learning preferences and academic culture', *Proceedings of the 7th ELSIN Conference: Reliability and Validity.* University of Ghent, Belgium, June 2002.

Johnson, M. (1997) 'Classroom relationships in three cultures, comparative classroom ambience in the UK, France and Lithuania', in Woodrow, D. *et al.* (eds) *Intercultural Education: Theories, Policies and Practices.* Aldershot: Ashgate.

Kelly, G. (1973) *A Theory of Personality.* New York: Norton.

Lewis, L. K. (2002) 'Cultural development group work: an evaluation', unpublished PhD Thesis. Manchester Metropolitan University.

Popkewitz, T. S. (1998) 'Dewey, Vygotsky, and the social administration of the individual: constructivist pedagogy as systems of ideas in historical spaces', *American Educational Research Journal,* 35, 535–570.

Rappaport, J. (1981) 'In praise of paradox: a social policy of empowerment over prevention', *American Journal of Community Psychology,* 9, 1–25.

Riding, R. J. and Cheema, I. (1991) 'Cognitive styles – an overview and integration', *Educational Psychology,* 11, 193–215.

Ritzer, G. (1993) *The McDonaldisation of Society.* London: Pine Forge Press.

Schommer, M. (1990) 'Effects of beliefs about the nature of knowledge on comprehension', *Journal of Educational Psychology,* 82, 498–504.

Sham, S. (Yuen Mei) (1996) 'Reaching Chinese children with learning disabilities in Greater Manchester', *British Journal of Learning Disabilities,* 24, 104–109.

Sham, S. (Yuen Mei) and Woodrow, D. (1998) 'Chinese children and their families in England', *Research Papers in Education,* 13, 203–226.

Singh-Raud, H. (1997) 'Educational attitude and aspirations of Asian girls', Unpublished PhD Thesis. Manchester Metropolitan University.

Stevenson, H. and Stigler, J. (1992) *The Learning Gap – Why Our Schools are Failing and What We Can Learn from Japanese and Chinese Education.* New York: Summit Books.

Tang T. and Williams J., (2000) 'Who Have Better Learning Styles – East Asian or Western Students?' Proceedings of the 5th ELSIN Conference. University of Hertfordshire, UK, June 2000.

Verma, G., Chan, Y.-M., Bagley, C., Sham, S., Darby, S., Woodrow, D. and Skinner, G. (1999) *Chinese Adolescents in Britain and Hong Kong: Identity and Aspirations.* Aldershot: Ashgate.

Vermunt, J. (1996) 'Metacognitive, cognitive and affective aspects of learning styles and strategies: a phenomenographic analysis', *Higher Education,* 31, 25–40.

Vithal, R. and Skovsmose, O. (1997) 'The end of innocence: a critique of ethnomathematics', *Educational Studies in Mathematics,* 345, 131–157.

Wanjada, E. K. (1996) 'Secondary school pupils' errors in algebra,' Unpublished PhD Thesis. University of Leeds.

Witkin, A. (1967). 'A cognitive-style approach to cross-cultural research', *International Journal of Psychology,* 2, 237–238.

Woodrow, D. (1996) 'Cultural inclinations towards studying mathematics and sciences', *New Community,* 22, 23–38.

Woodrow, D. (1997a) 'Social construction of theoretical beliefs', in Woodrow, D., Verma, G., Rocha-Trinidade, M., Campani, G. and Bagley, C. (eds) *Intercultural Education: Theories, Policies and Practices.* Aldershot: Ashgate.

Woodrow, D. (1997b) 'Democratic education; does it exist and can it exist for mathematics education?' *For the Learning of Mathematics,* 17, 11–16.

Woodrow, D. (2001) 'Cultural determination of curricula, theories and practices', *Pedagogy, Culture and Society,* 9, 5–27.

Woodrow, D. and Sham, S. (2001) 'Chinese pupils and their learning preferences', *Race Ethnicity and Education,* 14, 377–394.

8 Diversity, inclusion and education

The educational needs of children from severely disadvantaged socio-cultural groups in Europe

Hilary Gray

Introduction

The Salamanca UNESCO Forum (1994) called for inclusive principles to operate in education, and recognised that no matter how dedicated the teachers, there are serious hazards in segregated schooling even for pupils with significant learning difficulties (UNESCO, 1994). In 2000, the Dacca Forum called for national education systems to take account of the poor and most disadvantaged 'including working children, remote rural dwellers and nomads, and ethnic and linguistic minorities, children, young people and adults affected by conflict, HIV/AIDS, hunger and poor health, and those with special learning needs' (UNESCO, 2000). The issue of satisfactory schooling for severely socio-culturally disadvantaged pupils unites various of these agendas. I will discuss two approaches that have contributed to some progress regarding the complex and often stubborn problems of meeting their educational needs.

All modern education systems must respond to the fact that the generation and decay of information are today simply too fast for knowledge mastery to be a sufficient aim of education systems. Effective education must facilitate in younger pupils the skills for simple exploration of information and use of evidence, whether verbal, graphic, personal or artefact. Older pupils need to progress to more complex research skills. This implies a movement away from memorising putative 'facts' and it underlies the satisfactory education of all pupils, but is of particular and direct relevance to the education of pupils from severely disadvantaged and/or excluded groups. Further, it has been recognised for some time in the UK and North America that the educational needs of children from socio-culturally disadvantaged families are often different from those of children with upper or middle class parents, and it was recognised decades ago that children from ethnic minority groups are often caught up in these intergenerational and vicious cycles (Rutter and Madge, 1976). The most severe examples link both race and class, and it is in the ethnic majority–minority forum that many of the principles for improvement have been developed. For some time, the UK government has maintained detailed and publicly available information about the school achievements of pupils. Thus there is objective information that, on average, a high proportion of ethnic minority pupils and many from the white working class underachieve (Gillborn and Mirza, 2000). The underachievement is particularly serious among Gypsy-Traveller children, children from Bangladeshi families, and refugee children. That the problem is intergenerational and affects certain class and/or ethnic groups more than others does not imply genetic causes, but rather reflects its intergenerational nature, where majority group prejudices are built up often over centuries and where members of the disadvantaged groups come to feel that they and their children will never be among society's successful.

Apart from the human right of all children to develop to their potential (UNICEF, 1989) governments and international bodies have pragmatic reasons for anxiety when severe disadvantage is blatantly linked to one socio-cultural group. This is partly because of risk of social disorder between the more and less privileged, especially when the disadvantaged group is easily visibly distinguished as when there is an ethnic element. Also there is economic loss when a large, unskilled section of the population cannot contribute to the GDP, and because welfare payments such as unemployment benefits absorb large slices of the national income. A current example of such governmental anxiety is the European Union's strong concern regarding the plight of Roma children and families who, all over former communist central Europe, suffer grave educational, health and employment difficulties. Thus the EU has established its only support fund that is dedicated to a single ethnic group (EU, 1999; Sarkar and Jha, 2000).

Traditionally, it was thought helpful for such pupils to join segregated special schools. However, it is increasingly recognised that early categorisation of children is dangerous. Intelligence is 'plastic': children change, often dramatically. For example, Hindley and Owen (1980) found the relative positions of one-quarter of a sample of ordinary London children moved up or down a minimum of one-sixth of the entire range of normal intelligence between the ages of five and eight.[1] A more positive approach was initiated by the 1960s US government. Black American children were massively handicapped vis-à-vis schooling, and in consequence the government made pre-school provision available through the 'Head Start' programme. Monitoring the effectiveness of this provision throughout the schooling of a sample of children showed that good preschool education led to significantly better school performance with a low 'wash-out' factor. In fact, gains tended to last well into their school years, and even into their adult lives, with better employment rates and health, more stable relationships, and fewer crime convictions still discernible in the follow-up statistics at age twenty-seven (Schweinhart & Weikart, 1993). Thus the national investment in their early education was amply recouped via their income taxes and also via welfare savings. Schweinhart *et al.*'s (1986) monitoring of Head Start programmes also identified the characteristics of delivery of a planned, age-appropriate preschool curriculum that predicted best gains. These were experiential, participant learning, and involvement of the parents in their children's education. These two principles will inform the rest of this chapter.[2]

Engaged, participant learning

Universally, children's concentration and their language skills are most critical for their access to the curriculum and their educational progress. How best can we facilitate these characteristics? It might seem obvious that a child who concentrates longer on the content of a lesson will learn that lesson better, but this ignores the quality of the concentration. Effective learning involves learning to use and assess evidence, and also involves flexible mastery over the components of the learned skill so that the learning generalises to contexts other than those in which it was first mastered. Such learning comes via teachers who deliver the curriculum in a way that actively engages the pupils' understanding, rather than requiring that they passively rehearse material whose meaning may be unclear to them.

A year after the 1917 communist revolution in Russia, the Supreme Soviet set out the Basic Principles of Uniform Schooling for Workers, namely that education be free, compulsory, gender- and ethnicity-equal,[3] secular, and 'uniform' (Council of Deputies, 1918). In practice, this concept of 'uniform' education implied teaching the curriculum to whole classes without allowance for individual differences, that is 'undifferentiated', which in any case was then the teaching style in most of Europe and North America.

Teaching through projects or specific topics has been one approach to ensuring that pupils are actively engaged rather than passively exposed to curriculum material. This method also readily lends itself to differentiating the curriculum so that pupils with different learning needs can remain in the same class but still benefit from the curriculum that is delivered. The essence of the approach is that some topic or project is defined which the child or children explore, with the teacher providing relevant source materials in the form of books, artefacts, discussions with relevant persons inside or outside school, and so on. The children actively participate as they accrue and assess evidence about the topic. At the same time they are practising their basic skills of language, literacy, mathematics on the topic material. After the armed conflict that fully established the 1917 communist revolution and a massive, compulsory campaign to redress the extreme adult illiteracy inherited from the Tzars (Tomiak, 1972, p. 13), Soviet pedagogues began to experiment with teaching by project. However, influenced by Marx's typically materialist concept of 'technological education' (e.g. Marx, 1867, Book 1) which later led to the prioritisation of science over the arts and humanities, pupils in these early communist schools literally engaged with technological production of goods and services and their projects were literally within factories. One can, for example, see how projects monitoring operational effectiveness or quality control would help rehearse the mathematics learned in the classroom. But unsurprisingly, managers in industry soon objected on safety grounds, and after a few years Stalin condemned teaching via projects as a 'Pedological Perversion', (Central Committee of the Communist Party, 1936). Soviet education withdrew to its 'uniform', undifferentiated style. A standard lesson plan was developed in which the teacher engaged the whole class with the lesson content, firstly by setting written work, then by following it up with appropriate questions aimed at clarifying the task and drawing out deeper aspects (Tomiak, 1972, p. 63). The technique involved discussion with individual children, the other class members listening in (Alexander, 2000). The question is the extent to which this uniform delivery lived up to the true communist ideals of providing for each according to his needs. PISA international comparisons (OECD 2000) suggest that the methods were not satisfactory.

This style of curriculum delivery largely remains in Russia and instruction is very similar in the east central European states which became communist in the 1940s. So far as this concerns the severely disadvantaged Roma pupils, they often enter school with some weaknesses in their own Roma language and without skills in the language of instruction which was always the majority language except in the small Yugoslavian Republic of Macedonia (Poulton, 1998). As well as literacy and mathematics, those countries' national curricula tended to prioritise grammatical correctness, a sure way for teachers who are inclined to criticise to find ways of doing so. The children failed formal tests in or even before school and the slower pace of a very similar curriculum in the special schools failed to stimulate them. Thus the 'mental handicap' of this entire ethnic group appeared to be confirmed (e.g. European Roma Rights Centre, 1998).

Teaching by project and group learning in the UK and US

Teaching by topic or project was widely adopted in the US and the UK in the 1970s and 1980s. Rationales included pupil motivation and the facility for differentiating the curriculum which suited 'comprehensive' classes with wide ability ranges and pupils with different patterns of strengths and weaknesses. Also, the approach represents an example of our collective commitment to empiricism because rote learning is reduced and the children have experience of management of evidence. At one extreme each individual child might work for a month or more on his or her chosen topic that has age-appropriate relevance to history, geography and science, with the reading and writing and maths involved serving as rehearsal for these basic academic skills. At the other extreme, a whole school might adopt the same topic, pupils and classes engaging with it in different ways and at different academic levels. The success of this style of teaching depends utterly on the teacher's input, which includes design of conceptual maps or 'webs' to draw out the learning opportunities of the topic and thus guide the provision of materials, which may be very simple indeed (e.g. UNICEF, 2000). For example, the topic 'water' could be used for volume calculations in mathematics, for the rain cycle and endless explorations of the essentials of life in geography and science, for natural disasters in history, for writing poetry and stories, and depends on collections of graphics, natural or artefactual, to support the children's learning.

The opponents of this teaching method argue that it involves a waste of learning time, e.g. in physically moving around the classroom to consult reference material; unrealistic reliance on the children to manage their own needs, e.g. in group discussions; and neglect of rote learning where automatic mastery is required, e.g. in some literacy sub-skills. Alexander (2000) argues thus in his large study of the effectiveness of teaching styles in five countries including India, Russia and England. He also argues that the soviet collective social ethos ensured that the whole class were actively (but silently) engaged when the individual pupil was targeted as the teacher's discussant in the standard whole-class lesson sequence (see above) equally as much as English children working on their topics. Certainly the UK's weak scores in the mid-1990s' international comparisons of literacy and mathematics skills (OECD, 2000; IAEEA,1995) suggest caution regarding the topic/project method as a total approach to education, and UK schools sharply swung away from the method following these comparisons. As for the former Soviet Union, there was no educational outcome data (Coolahan 1998; Bakker, 1999) but recent international comparisons of literacy skills (OECD, 2003) suggest that Russian teaching is still far from satisfactory. In any case, it must be repeated that exploration of materials and assessment of evidence will remain vital in modern education. Thus project and topic work continue to play a part in most UK schools, with some primary schools still successfully delivering the entire National Curriculum in this way, though with secure methods in place to ensure that the children efficiently also master the sub-skills of literacy and maths.

Positive role models and working with families and communities

The intergenerational nature of socio-cultural disadvantage implies that pupils from severely affected groups often especially lack the educational support from their families that pupils from more advantaged backgrounds can take for granted. Teachers' understandable frustration at this situation demands highly professional management.

Ethnic minorities and/or children from a rigid class system who are stereotyped because they are easily 'categorised', need to actually witness educational and employment success of adults from their own or a closely related 'category'. Similarly, the old excluding processes will only fade as more privileged groups witness the success of members of the excluded groups. Teachers who are themselves members of the disadvantaged groups can be marvellous role models, but reversing educational disadvantage may often take decades before there are enough graduate members of these groups to have much impact. However, teaching assistants from disadvantaged groups can also act as powerful evidence that educational and employment success is not just a dream. All education authorities that seriously want to reverse the effects of long-term negative categorisation should consider such appointments.[4] The assistants can help the teacher with materials that are culturally more appropriate and support struggling pupils. They can also liaise with families. Very careful selection and training of the assistants regarding teaching methods and curriculum content are needed, and in particular with regard to their position as a delicate bridge between previously hostile cultures.

Relationships with families

The problem of disadvantaged families' support to their children's education has a historic and a contemporary source. The parents themselves were not encouraged by their own families; but weak commitment to their children's education also arises from feelings of inferiority, which can be reinforced, often unintentionally, by the teacher's manner. We should never underestimate how 'ordinary' respect from a professional can lift the spirits of a disadvantaged parent who is without the skills that are currently thought acceptable by society. Nor should we underestimate how negative reports about their children can damage parents' confidence if they are insensitively delivered: if negative information must be given, positive reports should always be included.

Some support schemes for disadvantaged parents aim to empower them through health-care education including contraception and AIDs awareness, and/or general parenting classes, and/or by redressing their deficits in basic literacy and numeracy.[5] Grandparents or other adult members of the disadvantaged community can provide culturally relevant material, artefacts, or stories on an enormous variety of themes ranging from past national disasters to culturally sensitive issues such as the circumstances of exclusion in their own childhoods. Since such oral histories are relevant and understandable to the children, they can powerfully motivate writing or reading practice and they are a rich source of material that can help bridge the inter-cultural and/or inter-class divide so that privileged children understand more about their classmates, and the disadvantaged child feels that his or her heritage is valued. We should, however, remember that children differ, and that some of them may dislike school's attention to their family's social and/or ethnic status; but many children flourish when their families contribute in this or similar ways.

Schools serving disadvantaged communities need to establish good relationships with formal or informal community leaders. In the UK, the expertise of these representatives is often valued by the lay management panel of schools, as they are in the best position to advise about cultural practices that affect the children's education. They may also be well placed to work with particular parents, and they can contribute to the curriculum, for example, by taking classes about cultural issues. Work that involves families and communities as well as children is known as 'Multilevel' work, and frequently the local education authority is the leading partner for such developments. The European Union project – Developing Intercultural Education through Cooperation between European Cities (DIECEC) – explored a wide variety of schemes for multilevel work in its nineteen member cities with their large populations of disadvantaged ethnic minority communities. The DIECEC handbook of community support schemes (Green, 2000) includes many ideas for working with families and communities. Importantly, it also shows the local education authority's leading role in initiating, monitoring and supporting these schemes.

Conclusions

Where do the above approaches leave the severely socio-culturally disadvantaged pupils such as the Roma in central Europe, the Travellers in many parts of Europe, and Bangladeshis in the UK, and many other groups worldwide? When the UK Department of Education 'paid for' the restrictions it imposed by the Commonwealth Immigration Act (UK, 1974) by providing help in school for ethnic minority pupils, there was a great deal for schools to learn, and the very unassertive Bangladeshi community, and constantly moving Gypsy-Travellers, were very low priority. In the 1980s, a few local education authorities began to develop various of the above ways of engaging these children and their families in education, but government recognition that some problems are more difficult to solve waited until the 1990s (e.g. UK Ofsted, 1996). Now some progress is being made, and in their detailed comparison by ethnicity and class of average public examination results, Gillborn and Mirza (2000) report that when education achievements of the UK's main ethnic groups are ranked for each local authority, every group (including the Bangladeshis) have the best average achievements in at least one authority.

The totalitarian, centralised bureaucracies of east central Europe have actively disempowered the entire Roma ethnic group, which officially simply did not exist. At the same time, teaching policy prioritised the learning of rules and facts rather than participant, experiential teaching to which Roma children could have more easily related. As documented above, undifferentiated delivery of an inappropriate curriculum to children, many of whom had some initial weaknesses, led to their exclusion to unstimulating special schools from which there was no return. Since the 1989 revolutions in east Europe, most former communist states are reforming their education systems (Illner, 2001). In the interests of both privileged and disadvantaged children, national curricula are being reformed to reduce memory of facts and rules and to increase participant, engaged learning and assessment of evidence (as above). Increased inclusivity is also an aim, often with the Roma predicament particularly in mind. Teacher training establishments are slowly recognising these priorities; Roma classroom assistants and home–school liaison officers now feature regularly in areas with high Roma populations; post-statutory education is being opened up to adults needing literacy, mathematics and vocational skills.

In these countries, as the DIECEC shows (Green, 2000), the local education authorities are also developing their coordinator techniques so that developments can happen at the most appropriate levels, and/or as opportunity presents. This is crucial because, while many teachers would say that experiential, participant learning and involvement of the parents in their children's education are part of responsible education for most children, and not only the severely disadvantaged, the difference is that to break the mould of centuries often demands more than one of the above approaches, and thus it is necessary for the issues to be treated as among the education system's leading priorities.

Notes

1 Hindley and Owen found a minimum 0.67 SD change in quartile of their ordinary London sample whose scores changed most. Many made larger positive or negative changes.
2 For recent information about the US Head Start programme which was restarted with updated rules see the large collection of policy and evaluation material at: http://www.acf.hhs.gov/programs/core or: http://www.acf.hhs.gov/programs/core /ongoing _research/ehs/ehs_intro.html.
3 The Russian Empire and the USSR included well over 100 ethnic minority groups (referred to as 'national minorities' because they had some rights in respect of their ethnicity, e.g. to education in mother tongue and the use of mother tongue in courts. In the late 1920s and early 1930s, the USSR Ministry of Education (Narcompros) ordered the translation and printing of some of the centrally published school textbooks into more than 120 languages (Tomiak,1972). The initial intention was that mother tongue be the language of instruction throughout statutory schooling, at that time eight years. However, Russian as the language of the soviet brotherhood and other pressures on schools, e.g. to teach science and technology, meant that mother tongue was used less and less, and often only in primary school (the first three statutory years). By 1990 only eighteen 'national' groups received education in their 'national' language, these mainly being the large groups in the now independent western and southern republics of Ukraine, Kazakhstan, etc.
4 Mentoring schemes are also fruitful. Here, successful young adult members of a disadvantaged group befriend one or more pupils from the same group, helping with homework, providing interesting experiences, etc. 'Buddy' schemes are similar, where children within the school offer support to weaker children from their ethnic or class group.
5 See for example, Save the Children's Women's Literacy Strategy, with classes also in numeracy and health care, at: http://www.savethechildren.org/education/literacy.asp; or Save the Children's Uganda programme which also includes AIDS awareness for parents and adolescents, at: http://www.savethechildren.org/countries/africa/ uganda.asp.

References

Alexander, R. (2000) *Culture and Pedagogy.* Oxford: Blackwell.
Bakker, S. (1999) 'Educational assessment in the Russian Federation', *Assessment in Education,* 6, 291–303.
Central Committee of the Communist Party (1936) *Pedological Perversions* (cited in Tomiak, 1972, p. 18).
Coolahan, J. (ed.) (1998) *Reviews of National Policies for Education.* Paris: OECD.
Council of Deputies, (1918) *All-Russian Central Executive Committee, 16 10: Basic Princi ples of Uniform Schooling for Workers* (cited in Tomiak, 1972, p. 12).
European Roma Rights Centre (1998) *A Special Remedy: Country Series 8.* Budapest: ERRC.
EU (1999) 'European Commission: Support for Roma communities in central and eastern Europe'. At: http://europa.eu.int.
Gillborn, D. and Mirza, H. S. (2000) *Educational Inequality: Mapping Race, Class and Gender.* London: Office for Standards in Education. At:http://www.ofsted.gov.uk/ public/docs00 /inequality.pdf.
Green P. (2000) *Raise the Standard.* Stoke on Trent: Trentham Books.

Hindley, C. B. and Owen, C. F. (1980) 'The extent of individual change in IQ for ages between six months and 17 years, in a British longitudinal sample', *Journal of Child Psychology and Psychiatry,* 19, 329–350.

IAEEA (1995) *International Association for Evaluation of Educational Achievement: Third International Maths and Science Study.* At: http://timss.bc.edu/timss1995i /TIMSS Publications.html.

Illner, M. (2001) *Public Consultation on Educational Policy in the Czech Republic.* Paris: OECD.

Marx, K. (1867) *Das Kapital: Book 1.* New York: Schmidt.

OECD (2000) *Literacy in the Information Age: Final Report of the International Adult Literacy Survey.* At: http://www.oecd.org/publications.

OECD (2003) *Further Results from PISA (Programme for International Student Assessment) 2000.* At: http://www.oecd.org/dataoecd/59/30/2960583.pdf.

Poulton, H. (1998) 'Linguistic minorities in the Balkans', in Paulston, C. B. and Peckham, D. (eds), *Linguistic Minorities in Central & Eastern Europe.* Clevedon: Multilingual Matters.

Rutter, M. and Madge, N. (1976) *Cycles of Disadvantage.* London: Heinemann.

Sarkar, B. and Jha, S. (2002) 'Hungarian minorities: issues and concerns', *International Studies,* 39, 2, 139–164.

Schweinhart, L., Weikart, D. and Larner, M. (1986) 'Consequences of three pre-school curriculum models through age fifteen', *Early Childhood Research Quarterly,* 1, 15–45.

Schweinhart, L. J. and Weikart, D. P. (1993) *A Summary of Significant Benefits: the High Scope Perry Pre-School Study Through Age Twenty-Seven.* Ypsilanti MI: High Scope.

Tomiak, J. (1972) *The Soviet Union.* Newton Abbott: David & Charles.

UK (1974) *Commonwealth Immigration Act.* London: Her Majesty's Stationery Office.

UK Ofsted (1996) *Education of Travelling Children.* London: Office for Standards in Education, HMR 12/96/NS. At: http://www.ofsted.gov.uk/public.

UNESCO (1994) *The Salamanca Statement and Framework for Action on Special Needs Education.* Paris: UNESCO Special Education Program. At: http://portal.unesco.org/ education.

UNESCO (2000) *World Education Forum Final Report.* Paris: UNESCO Special Education Program. At: http://www.unesco.org/education/educnews/20_10_16/ report.htm.

UNICEF (1989) *Convention on the Rights of the Child. At:* http://www.unicef.org/crc/crc.

UNICEF (2000) *Ideas for Active Learning. At:* http://www.unicef.org/teachers/resources/ mod 5.htm #sugarcane.

9 The Greek common school system

Bridges and barriers to inclusion

Adamantios Papastamatis

Introduction

This chapter begins by outlining the general structure and function of the Greek education system and providing information on its administration and management. After that, it provides an evaluation of the system and indicates some proposals and means for its improvement so that it may become more democratic, inclusive and diverse.

Greece occupies the southern end of the Balkan Peninsula and some 2,000 islands (of which about ninety are populated) in the south-east of Europe. The country shares common boundaries with Albania, FYROM (the former Yugoslav Republic of Macedonia), Bulgaria and Turkey. The land is so mountainous and stony that out of the 50,147 square miles that constitute the area of Greece, only 25 per cent is considered arable. Greece joined the European Union as a full member in 1981 but it is characterized by unique socio-economic, cultural and educational patterns. Social inequality is pronounced in Greece, and the educational system is over-centralized. The official language is modern Greek and about three per cent of the total population consists of linguistic and cultural minorities, mainly Muslim. A considerable number of economic immigrants, mainly from Albania and Eastern European countries, have recently been added to the indigenous population (Terzis and Moutsios, 2000). According to the population census of 2001 there were 762,191 non-Greek citizens of all ages present in a total population of just under eleven million.

General structure of education

The declared aim of the Greek educational system is to contribute to global, harmonious and balanced development of the intellectual psycho-physical abilities of pupils so that, independently of their sex and origin, they have the opportunity to become integrated personalities and live creatively (Greek Ministry of Education, 1985). More specifically, official policy states that education should be aimed at:

- developing independent, responsible and democratic individuals;
- creating individuals who are able to protect their national independence and democracy;

- helping pupils to understand social values and the equivalence of intellectual and manual work;
- the development of creativity and cooperation with other nations.

Education in Greece is divided into three cycles corresponding to primary, secondary and tertiary education. After one or two years of nursery education, the child spends six years in primary school, followed by three years in the Gymnasium (equivalent to Junior High School in other systems). After this the student will attend either a Comprehensive Lyceum, or Technical or Vocational schooling. From the Lyceum a student may proceed to university, or to Technology Institute if they have attended Technological secondary education.

Table 9.1 presents data on the 2000–01 and 2001–02 academic years as regards the number of pupils, teaching staff and the types of educational institutions throughout the country.

Nursery education is optional, while primary education is compulsory. Infants entitled to attend nursery schools are between four and six years of age.

Table 9.1 Educational institutions in Greece

Types of educational institutions	Schools		Pupils/students		Teaching staff	
	Public Sector	*Private Sector*	*Public Sector*	*Private Sector*	*Public Sector*	*Private Sector*
Nursery (2001–02)[a]	5,647	111	138,544	5,024	9,973	322
Primary (2001–02)[a]	5,739	373	594,639	45,775	47,998	3,185
Gymnasium (2001–02)[a]	1,768	112	321,674	19,054	35,221	2,301
Comprehensive Lyceum (2001–02)	1,182	98	219,269	16,814	21,454	1,879
Technical and Vocational Education (2001–02)	413	77	122,581	6,502	15,973	1,399
Institutes of Initial Vocational Training (2000–01)	138	76	36,857	13,232	16,536	1,981
Technological Educational Institutions (2000)	14		129,683		7,686	
Universities (2000)	20		276,902		10,459	

Source: Greek Ministry of Education (2002) Operational Research and Statistics Branch, Education Statistics.

Notes
a Compulsory education

The curriculum is based on cross-curriculum themes. According to the cross-thematic approach to learning the educational knowledge has:

- to be provided in a unified form in order to offer holistic views of reality;
- to be linked with children's experiences in order to be perceived and related to children's daily life; and
- to be constructed gradually through children's relevant research activities.

(Avgitidou and Botsoglou, 2003)

Responsible for the development and introduction of the curriculum is the Pedagogical Institute (a policymaking body).

Primary schooling is of six years' duration and is for all children who are admitted to the first class at six years of age. This means that all children of the same age are exposed to the same materials irrespective of their individual differences. Textbooks are prescribed by the National Pedagogical Institute and they are provided free in the state schools and universities. However, pupils have to pay for other materials, such as notebooks, pens, and pencils. Traditionally, both nursery and primary schools operated for only half a day. However, it is now accepted that the number of women in full time employment has increased dramatically and this has resulted in the increase of child care problems since both parents are often out of the home. Consequently, all-day nursery and primary schools are beginning to operate full-time in order to facilitate parents who both work and who can not afford to pay for child care. Among the various facilities provided is special help for children with learning difficulties. In 2001 the Ministry operated 2,482 all-day nursery and primary schools (Greek Ministry of Education, 2002).

Secondary education covers the age range of twelve to eighteen years. It is divided into two cycles: the lower (Gymnasium) and the upper (Lyceum), each lasting for three years. The former is compulsory, whereas the latter is optional. Textbooks, as in primary schooling, are prescribed by the Pedagogical Institute and a more recently established agency, the Center for Educational Research. In addition to the above, special education and multicultural education are provided. The former is for physically and mentally handicapped children unable to benefit from ordinary schools. It is given either in special schools or in special classes at the normal schools. The function of the special education is to provide an environment which helps children to overcome their learning difficulties and to grow into self-reliant and active members of society in so far as their handicaps allow. In the school year 2001–02 the total number of pupils with special educational needs (SEN) attending separate Schools of Special Education or Inclusion Classes was 18,585 (Greek Ministry of Education, 2002).

Multicultural education is for children from a different cultural origin and/or ethnic minority as well as children of Greek immigrants abroad. The function of multicultural education is to help pupils to recognize and value cultural diversity found in the society. This education is normally provided in separated institutions, called multicultural schools. For these schools or classes the Pedagogical Institute prepares additional rules and learning materials other than the prescribed ones. In the school year 2002–03 the total number of pupils from a different cultural origin and/or ethnic minority was 98,241 and the total number of pupils of Greek immigrants was 31,873 (Gotovos and Markou, 2003).

There also exists a fourth post-secondary level, consisting of a network of public and private Institutes of Initial Vocational Training (IEK) attended by students who do not follow studies at tertiary education. Studies at IEK have a duration of three years. The aim of IEK is to equip students properly for working life.

The tertiary level includes the universities and the Higher Technological Educational Institutes (ATEI). The latter are oriented toward the application of knowledge as opposed to the former that have academic orientation. Studies in universities last four to six years depending on the subject areas and in ATEI four years. All tertiary education is free as well and students are provided with free textbooks. Tertiary education is the responsibility of the state and thus private tertiary institutions are not formally recognized. Degrees received from private universities and colleges abroad are therefore not formally recognized by the Greek state (Patrinos, 1991). It must be said here that there are no restrictions for the operation of private primary and secondary schools.

For the provision of adult education and lifelong learning, the responsible body is the General Secretariat of Adult Education based in the Ministry of Education. This is concerned with adult literacy, the organization of some vocational courses and other special programmes. However, there is not yet a developed system of adult education integrated into the formal educational system as post-tertiary education. It is also true that adult education has not produced a national pre-service teacher-training scheme. Thus, many who enter adult education have no qualification in the education of adults at all. The Government, however, is planning to establish special institutions for adult education and lifelong learning in universities.

In the academic year 2004–2005 there were ten centres of Adult Education with 10,507 students. There were also thirty-two Second Chance schools. Individuals from low socio-economic classes who fail to finish the compulsory nine years' education because of personal and/or antisocial problems, are offered Second Chance schools which cover the needs of adults who wish to finish their basic education.

Administration and management

The administration and management of the Greek educational system is highly centralized. The curriculum, teaching methods, appointment of teachers, salaries, entrance to university and numbers of students in universities, are all decided by the Ministry of Education. Greece is geographically and administratively divided into thirteen regional divisions, headed by the divisional directors and these divisions are subdivided into fifty-four prefectures headed by the 'director of education'. In each of the fifty-four prefectures there are also a number of educational advisers. The main decision body at the school level is the Teachers' Association which consists of all teachers working in the school, and is chaired by the head.

However, since educational policy is determined centrally and money for educational development is allocated entirely by the central government, Teachers' Associations rarely do little more than implement central government official policy. Universities, however, are exempt from this type of control and have the freedom to decide on key issues such as what will be researched, although they are under tight financial control. A nominally self-governing Senate, elected by the Faculties, is bound by law a of 1932, and its decisions and membership has to be approved by the Ministry of Education.

The organization of the Greek educational system is keyed to academic disciplines, and it is difficult for teachers to develop a sense of personal responsibility for pupils' total development. There is not much room for school-based curriculum development or for its adaptation to the particularity of the local community. Teachers and pupils are expected to meet the demands of a predetermined syllabus within a time limit that assumes the importance of the product rather than the process. As a result it has been the norm to stick with the common curriculum and traditional teaching methods, regardless of the developmental levels, motivation and unique learning styles of particular groups of students. Hence the child is adjusted to the curriculum and not the curriculum to the child.

Programmes should be such that they can be adapted to the needs of the individuals. It would not be surprising therefore if some children, particularly those at the extremes of the ability range or with a different social and cultural background, felt alienated from school. These adverse effects are more significant, because the system is highly centralized and there is therefore little room for teachers to exercise their autonomy in making decisions about how to teach. Yet teaching is in some degree an art and requires an able teacher with autonomy to adapt general instructional principles to his or her classroom situation. Despite this, teachers are excluded from decision-making in curriculum development. However, the content of decisions is perhaps the most important factor in the effective implementation of any innovative changes.

Public spending on education is rather low. As a percentage of the GDP, Greece not only spends less than the average European country, but even less than the average developing country. In the year 2003, public educational expenditure in Greece was estimated at 3.5 per cent of GDP for all educational levels; this was the lowest proportion of the EU countries. This lack of investment fosters demand for selective private schools, which charge high tuition that constitutes their main source of finance since they do not receive state subsidies. This situation benefits the rich to a greater extent as they are more likely to enrol their children in such schools. The size of the private sector, however, is not important, probably because education is free, and there is also a vast number of private institutions ('frontistiria') which aim to prepare students for passing university entrance examinations. In addition to this, a number of teachers who work in the state or private schools give private tuition to prospective university candidates. Only about five per cent of children attend private schools (see Table 9.1).

Proposals for improvement

If education is to be equal for all individuals it must be tailored for them and their needs. This invokes the whole area of relevance of discerning basic needs and re-designing curricula which challenge traditional disciplines and practice. To achieve education for all, it must involve planning *by all* as well. Gone are the days when educational planners had the luxury of selecting all their statistics and inputs, creating data with their computers, with education ministry heads churning out a national plan for education. Today that process must become much more interactive and therefore inevitably less centrally organized, less uniform and less predictable.

There are, of course, many reasons that make this so:

- the change in the role of the state in the face of increasing decentralization to local governments;
- the budgetary restrictions which call for innovative programmes which do more with less, which call for greater community participation and commitment in the delivery of education;
- the more active role of the private sector, non-government organizations; and
- the ubiquity of microcomputers and the dissemination of relevant databases (see Ordonez, 1991).

In Greece, education has been looked upon as the main instrument for individual and economic development and as the major social force for equalization of opportunities. From this point of view the education system has been partially successful in expanding the possibility of making the same opportunities available for more students. However, the rapid growth in enrolments over the past twenty years and increased expenditure have failed to keep pace with the phenomenal social demand for higher education. This growth of resources and the educational innovations have failed to live up to some of the hopes held for them regarding 'equal opportunities'.

Thus, whereas the Greek educational system allows most pupils to flourish, it has also created victims who are systematically excluded from its benefits. For this reason, it is important that the system redistribute resources in a way that will directly challenge intrinsic inequality. That means that we could accept the logic of the need to resocialize individuals who from their early life are marked out for failure and massively enhance the resources for schools operating in demoralized communities. A democracy is served neither by a narrow and legislated national curriculum nor, in the long run, by teachers and pupils who feel unable to take risks, to innovate, to see beyond what counts as convention. In this context, researchers such as Katsikas and Kavvadias (1994) and OECD (1996) argue that the failure of the Greek educational system is due to its uniformity and the Procrustean method which follows. The system should acknowledge cultural diversity in all schools and not just in multicultural ones, and accommodate this diversity in instruction.

Despite this, the curriculum remains firmly within Greek–Christian ideology. Thus, according to the Organization for Economic Cooperation and Development

(OECD, 1997) the curriculum is excessively 'Grecocentric' and, despite the teaching of the English language, it gives insufficient place to the European dimension and also does not convey much in terms of knowledge and understanding of other people and other cultures.

Unfortunately, despite the efforts of successive governments to revise the curricula and adapt them to the needs of the modern society, the content has not overcome its excessively classical humanistic and literary nature. Of course, Greek–Christian ideology is part of the Greek culture and should be transmitted to new generations. It could be argued, however, that its over-emphasis could lead to undesirable results. Greek children will have to grow into adults able to interact with various other groups, particularly those in the European Community. In order to develop harmonious relationships with other societies it is necessary to include in the curriculum elements which will lead to an appreciation of other societies and the values of cultural pluralism. Yet, over-emphasis of Greek–Christian ideology seems to be an obstacle to the country's attempts to achieve the technology of advanced society.

In order to achieve technological advance there is also a need to educate the whole population to a higher standard than ever before, since there are no longer many unskilled jobs, and schools are expected to play their part in preparing people for employment in a world where there is intense international competition for trade. New technology has transformed most industries and is in the process of transforming schools. It is also important to educate people for living in a rapidly changing world where what happens in one country affects many others, and where societies are increasingly pluralistic (see Dean, 1999, p. 7). In a world of accelerating change, learning must be a continuing process from birth to death and society therefore must provide educational resources and services throughout people's lives.

Despite this, adult education and lifelong learning has often been perceived and created to have a remedial role, signified by the language 'Second Chance schools'. While this aim may be worthy, it is contained within, and perpetuates a discourse of, individual failure. This concept ignores the fact that learning is a continuum and can create a sense of inferiority in an adult who decides to follow further studies.

In addition, the centralized hierarchical system of Greece, contrary to its democratic aims, promotes dogmatism, conformity and subordination, to the extent that teachers do not have the autonomy to organize lessons according to their own particular teaching–learning situations. Thus, the system makes teachers simply agents. However, if the government is really concerned with preparing future citizens for life in a democratic society then it is time that the educational philosophy inherent in the highly centralized system be re-examined in order to put forward the necessary changes for teachers. More specifically, it requires a degree of decentralization (Papastamatis, 1988).

Finally, it is also true that there does not exist any national system for the evaluation of teachers, lectures and administrators that would guarantee standards and good practice in educational institutions, although the development and the application of such a system is proposed by the operational programme 'Education and Initial Vocational Training 2000–2006'. Evaluation should be inherent in the teaching profession. It is not possible to meet the demands of teaching without planning, organizing, monitoring and evaluating the activities that are carried out. Evaluation is one of the most important tools available to teachers in the development of their teaching and their ability to facilitate their students' learning. Insights gained from evaluation provide teachers with a guide and indication for developing and improve teaching methods since it is possible through evaluation to develop skills and qualities for effective teaching.

Conclusions

It is increasingly being argued that Greece is faced with a crisis that demands radical rethinking of how education is to develop. In particular, multicultural education, lifelong learning, decentralization and evaluation are keys for improvements and so forms of education are required that are capable of fostering such schooling. These forms involve not only administrative and organizational elements of education, but also instructional content and materials, teaching and learning strategies, and evaluation. The Greek education system will need to make adjustments to their pedagogic methods in order to deliver continuing and multicultural education in a flexible and effective way.

The above reality generated much debate in the late 1980s and early 1990s regarding the quality of education provided in schools. In response, a number of new, non-compulsory, educational programmes with a cross-curricular character were introduced. These provided an opportunity to study a variety of modern issues through a teaching and learning process based primarily on field work and involving active experiences in the real world, beyond the traditional classroom boundaries (see Giannakaki, 2004). To be effective, the Greek educational system has to take account of the very varied life experiences, assumptions and interests of different pupils and different groups.

The tensions and problems created by the major changes in the education system have made it easy for the public to feel that standards of education in Greece have declined. It is difficult to provide a definite answer to this in the absence of systematic research, apart from theoretical rhetoric characterized by banality and over-simplification. This is despite the fact that research is very important for the Greek educational system in which schools have to follow a predetermined curriculum imposed by the Pedagogical Institute whose members may have been away from active teaching for many years. More important is the fact that this committee introduced ideas often without testing them in some pilot schools, although such schools have been established.

In concluding this chapter, it may be said that the Greek educational system is in crisis. This may be true in varying degrees of all systems around the world. The Greek system, however, is under stress; its teaching methods are outdated and cannot cope with a technologically-based and multicultural society where high standards of numeracy and technical skills, tolerance and empathy are needed to do even the most modest jobs. The over-centralized character of the system discourages educational administrators and teachers from taking initiatives and to promote diversity. Neither is there any systematic evaluation to give the relevant feedback. As a result, the Greek education system cannot respond appropriately to the demands imposed on it by society.

References

Avgitidou, S. and Botsoglou, K. (2003) 'Play care and education. Early childhood curriculum in Greece: trends and policies at the outset of a new century', *Researching Early Childhood*, 5, 7–19.

Dean, J. (1999) *Improving the Primary school*. London: Routledge.

Giannakaki, M. S. (2004) 'An examination of the link between school management and curriculum innovations: a comparative study of public and private primary schools in Greece'. Unpublished PhD Thesis, University of Strathclyde.

Gotovos, A. E. and Markou, G. P. (2003) *Greek Immigrants and Foreign Pupils in Greek Education: General Outline* (in Greek). Athens: Greek Ministry of Education.

Greek Ministry of Education (1985) *Education Act 1566/85* (in Greek). Athens: OEDB.

Greek Ministry of Education (2002) Operational Research and Statistics Branch. Athens: Greek Ministry of Education.

Katsikas, C. and Kavvadias, (1994) G. *Inequalities in Greek Education* (in Greek). Athens: Gutenberg.

OECD (1996) *Review of the Greek Educational System*. Athens: Greek Ministry of Education.

OECD (1997) *Reviews of National Policy for Education: Greece*. Paris: OECD.

Ordonez, V. (1991) *The Reform of Educational Systems to Meet Local and National Needs*. Manchester: University of Manchester.

Papastamatis, A. (1988) 'Teaching styles of Greek primary school teachers'. Unpublished PhD Thesis, University of Manchester.

Patrinos, H. (1991) 'Financing higher education in Greece and the privatization option', in Turner, D. (ed.) *The Reform of Educational Systems to Meet Local and National Needs*. Manchester: University of Manchester.

Terzis, N. P. and Moutsios, S. (2000) 'The Greek educational system', in Terzis, N. P. (ed.) *Educational Systems of Balkan Countries: Issues and Trends – Education and Pedagogy in Balkan Countries 1*. Thessaloniki: Kyriakidis Brothers.

Part V

Diversity and educational equality in India

10 The right to education

Developing the common school system in India

Madan Mohan Jha

Introduction

This chapter addresses the issue of rights with regard to basic or elementary education in India. Any issue, for example education, can be addressed from a variety of perspectives. We approach basic education from a *rights* perspective in contrast to charity, humanitarian or need-based perspectives. This approach is therefore rights-based (RBA) and not needs-based (NBA). We go onto further examine the policy and structural framework of the Indian school system. Tracing development of the framework, it is argued that a common school, or the common school system in the Indian context, is the only hope for realization of the right to education. A common school is understood to be a school which does not select or sort children on any criteria, and offers equal opportunity to all in terms of admission from the neighbourhood.

What is rights-based education?

Rights in the modern Western conception may be traced to the English Magna Carta, to the US Declaration of Independence, and to the French Declaration of the Rights of Man and the Citizen. In the wake of the wartime Holocaust, on 10th December 1948, the UN General Assembly proclaimed a Universal Declaration of Human Rights, which included the right to 'life, liberty and security of person', 'freedom of movement', 'nationality', 'freedom of thought, conscience and religion', 'freedom of peaceful association and assembly', and 'freedom to take part in the government'. The Indian concept of rights developed during the freedom movement, with a demand for self-governance and total independence from the colonial rule, and culminated in Part III of the Indian Constitution as 'Fundamental Rights'. The Indian Fundamental Rights are close to the UN's declaration on Human Rights. The most important fundamental right impacting the quality of life of common people in India happens to be Article 21, which guarantees 'right to life and personal liberty'. It declares: 'No person shall be deprived of his life or personal liberty except according to procedure established by law'.

Historical development of the right to education

The first attempt to regard elementary education as a matter of right, though obliquely, was made way back in 1909 when G. K. Gokhale introduced a Bill under the Indian Council Act of 1909 to make primary education compulsory, with state funding. However, the Bill was defeated by a large majority. While addressing the legislatures Gokhale made the emotional observation that the issue would keep coming back again and again until all children realized their right to free and compulsory education.

In 1950 India gained its own Constitution, which provided Fundamental Rights to equality, to freedom, against exploitation, to freedom of religion, to constitutional remedies and cultural and educational rights. The right to free and compulsory education was retained in Part IV of the Constitution that incorporates Directive Principles of State Policy. Article 45 of the Constitution declares: 'The state shall endeavour to provide, within a period of ten years from the commencement of this Constitution, for free and compulsory education for all children until they complete the age of fourteen years'.

The distinction between Fundamental Rights and Directive Principles of State Policy is well settled under the Indian Constitution. While the former is absolute and legally enforceable, the latter is a policy directive of the State.

However, the 1980s and 1990s saw a very liberal interpretation of Article 21 of the Indian Constitution by the Indian judiciary. The most relevant of these judgments from an educational point of view was the Supreme Court's Unnikrishnan Judgment (1993). The court ruled that Article 45 of the Directive Principle of State Policy must be read in harmonious conjunction with Article 21 since right to life and personal liberty loses its meaning if a child is deprived of elementary education (*Unnikrishnan* v. *State of Andra Pradesh,* Article 1993 Supreme Court of India 217). Another liberal interpretation of Article 21 relates to environment protection and public health, the right to food and shelter and the right to rehabilitation in the case of bonded labourers. According to the court verdicts these freedoms are vital to life and liberty of a person. In addition to making the right to free and compulsory education as good as a fundamental right, the Unnikrishnan Judgment ruled against state commercialization of education. The Supreme Court also held that economic and financial constraints could be a ground for restricting the state from making provisions of post-basic and higher education, but not in the case of elementary education.

The Unnikrishnan Judgment stimulated several civil society groups to demand incorporation of the right to education as a Fundamental Right in Part III of the Constitution. The government finally agreed to bring a new FR (Federal Right) marked 21A in December 2002, which reads: 'The State shall provide free and compulsory education to all children of the age 6 to 14 years in such manner as the state may, by law, determine.'[1]

Notably, the amendment was introduced after Article 21, keeping in view the spirit of interpretation of this article by the Supreme Court of India.

Many activists have, for two reasons, criticized this amendment. First, it is argued that 21A gives power to the state to decide the 'manner' for providing 'free and compulsory education'. Second, it restricts the 'right' to the age group six to fourteen, unlike the original Article 45 of the Directive Principle of State Policy, which referred to 'all children until they complete age 14' (see Sadgopal, 2003, 2004).

I want to revisit both of these criticisms. Many legal luminaries and educationists have emphasized the wide ramifications of the right to education. For example, Justice J. S. Verma, former Chief Justice of India and also former Chairman of the National Human Rights Commission, observes that providing free elementary education is an 'essential sovereign function' of the welfare state.[2] Justice (retd.) V. R. Krishna Iyer (2005) has observed that education is a cardinal component of human dignity, and access to it is enshrined in the Indian Constitution. The right to education is absolutely fundamental and 'judicial construction cannot jettison this right, based on the subconscious impact of the dubious mantra of privatization' (*The Hindu*).

It seems to me that the expert group set up by the Government of India (GOI) after the Unnikrishnan Judgment, with economist Tapes Manmade as chair, chose the rights-based approach or the 'RBA' to elementary education – as the report said. For, being an incremental developmental goal in the process of education for all, universalisation of elementary education has in consequence of the Unnikrishnan Judgment, now become the legal right of every Indian child:

> entitlements sanctioned by the Constitution cannot be deferred by the State at its convenience. The State has to make the necessary reallocation of resources, by superseding other important claims, if necessary, in a manner that the justiciable entitlement can become a reality.
> (Unnikrishnan v. State of Andra Pradesh, Article 1993 Supreme Court of India 217)

As regards the perception that the introduction of the phrase 'in such manner as the state may, by law, determine', would give unfettered power to the government to control or dilute the scope of elementary education for all children, it is argued that Article 13 (2) bars the state from taking away or abridging any right contained in Part III of the Constitution. Furthermore, the Supreme Court in the same judgment ruled that, after the age of fourteen years, the fundamental right to education continues to exist but is 'subject to limits of economic capacity and development of the State' as per Article 41. In other words, financial reasons cannot be offered as a fundamental or final excuse for not providing free and elementary education to all children. The essential sovereign duty of each Indian state is to secure 'equality of status and opportunity', 'the dignity of the individual'; social justice laid down in the preamble of the Constitution, is likewise not limited by the financial capacity of the state.[3]

By implication, legislative operationalization of 21A does not give a free hand to the state, and it is fair to expect that the law made in this regard would only extend the right further, rather than restrict it. I hold the view that the introduction of a new Article 21A in the Constitution provides a renewed opportunity to reduce the increasing inequality in education at the elementary level and achieve the goals of justice – social, economic and political – as pledged in the preamble. However, the import of this new fundamental right has yet to be properly understood by policymakers and academics, and has yet to appear on the agenda of genuine social and political activists. The fundamental right to free education of children aged six to fourteen as granted by Article 21A since December 2002 has yet to acquire the stature of other Fundamental Rights.

It is understandable therefore that those who drafted the recent report of the CABE (Central Advisory Board of Education) committee on the free and compulsory education bill have argued that the right to education which Article 21A seeks to confer, is different from other Fundamental Rights – while the earlier Fundamental Rights had no or insignificant financial implications for the state:

> the Right to Education has major financial implications... such artificial classification and hierarchy in Fundamental Rights is the product of the gaps in the class characteristics of those who control education and those who are being deprived of the equal opportunity.
>
> (CABE, 2005)

But it is difficult to deny that the federal state is spending huge amounts on police and higher judiciary to protect citizens' right to life and liberty, and equality before the law. Therefore, the argument of 'financial implications' for avoiding the obligation by a state to ensure the right to universal elementary education as lower in status than other rights, is flawed. Thus, I would argue that just as police are an important guarantee of Fundamental Rights with regard to the protection of life, and the judiciary is meant to secure justice and equality before the law, so schools and teachers need to be regarded as a guarantee of the right to elementary education. This guarantee seems possible only within the framework of common schooling, in which quality education is offered, without charge, to children of all citizens. Extension of this argument would mean less and less scope for fee-charging private schools, since all children, including those who opt for private schools, should have the right to free education. It is argued that many parents go to private schools because of the absence of, or deficient functioning of, government schools in the neighbourhood.

Institutional framework for realization of the 'Right to Education'

School provides part of the institutional framework for realization of the universal right to quality education. The school as an institution is the product of an industrial age. While the need for mass education to respond to the industrial age remained confined to basic literacy and numeracy for the masses, schooling for enhancing life chances remained confined to the select few even in the West. For example, in Britain, a full system of vocational and academic post-primary education was not introduced until the early 1950s. Up to that time many British children's only education was in all-age primary schools, which they left for the world of work at age 14.[4]

India had 'indigenous schools' spread over thousands of villages in the nineteenth century that produced 'professionals' required during that period (DiBona, 1983). However, locally relevant education was meant for the masses in Pathshalas. Only a select few belonging mainly to the Brahmins experienced scholarly Sanskrit education in the Gurukuls. A great debate on Indian education began at that time between the anglicists supporting a Western-style education and the orientalists favouring an education system based on Indian values and cultures. The debate ended with the famous minute of Macaulay on 2nd February 1835, intending to create 'a class of persons Indian in blood and colour, but English in tastes, in opinions, in morals and in intellect' (cited in Fagg, 2002: 19). Over the years, the system became 'hierarchical and elitist, top heavy with higher education at the expense of primary education' (Steele and Taylor, 1994,).

Gandhi's 'Basic Education' and after

Gandhi's 'Basic Education' was the first official policy in India 'to change the established structure of opportunities for education' (Kumar, 1994: 508). It was 'contemporary not modern, ideal not practical, and it might have achieved limited success but ultimately failed' (Jha, 2002). Fagg (2002) has contested these perceptions in his study based on the primary sources on the Basic Education movement, and he argues that the education system unveiled by Gandhi in 1937 influenced government policy for the next thirty years, until the Education Commission (1966) replaced it by 'work education' as a subject for study in Indian schools.

The common school system

The Education Commission (1966), popularly known as the Kothari Commission, coined the term 'common school system' for the first time. While in England, the 'comprehensive struggles' had a larger objective of ending 'two nations in education' divided by grammar and secondary modern schooling (Tomlinson, 2001: 14), in India they were part of a report to improve school administration and remove the 'caste' system in school management; also to reduce bureaucratic control. The recommendations have remained largely ignored.

For equal opportunity the CSS (Common Schooling System) introduced the concept of neighbourhood schools. The Kothari Report said: 'Each school should be attended by *all* children in the neighbourhood irrespective of caste, creed, community, religion, economic condition or social status, so that there would be no segregation in schools' (quoted in Sharma, 2002). If the report were to be written today one would have expected terms like 'disability' and 'special needs' to be included in the "all". Arguing for the neighbourhood school the Commission advanced two arguments. First, a neighbourhood school would provide 'good' education to children because sharing life with common people would be an essential ingredient of good education. Second, the establishment of such schools would compel rich, privileged and powerful classes to take an interest in the system of public education and thereby bring about its early improvement.

The Commission seems to have given an educational theory behind the neighbourhood school system for 'good' education. However, as Archer (1979: 4) argues, 'There is no such thing as an educational theory ... there are only sociological theories of educational development'. The developments in school education post-Kothari demonstrate that the 'rich, privileged and powerful classes' did not 'take an interest in the system of public education', as Kothari had hoped. The growth of private schools for the privileged, at the cost of public education (or government schools), in recent years confirms another theory offered by Archer (1979: 2) that: 'Education has the characteristics it does because of the goals pursued by those who control it'.

The 1968 national policy on education accepted the Commission's recommendation on the Common School System (CSS) aiming at the implementation of the neighbourhood school concept within twenty years. After some twenty years, however, in 1986 the new policy maintained the rhetoric of the CSS but in reality had abandoned it.

CSS in the National Policy for Education (NPE) 1986/92

The NPE 1992 is a modified version of the original policy announced in 1986. The 1986 policy shifted the CSS from the 'education for equality' chapter to a new chapter called the 'national system of education'. It said the 'concept of a National System of Education implies that up to a given level, all students, irrespective of caste, creed, location or sex, have access to education of comparable quality' (MHRD, 1998:5). The Education Commission (1966) chaired by Kothari, however, had recommended the neighbourhood school concept for *all* children irrespective of caste, creed, community, religion, economic conditions and social status. The 1986 policy dropped the phrase 'economic conditions and social status'.

Further, the 1986 policy also promised to take 'effective measures' to implement the CSS. The Programme of Action 1992 meant to implement the policy made no mention of the common school system (MHRD, 1996). Thus dilution of the commitment on 'education for equality' coincided with the direction in the Indian economy towards privatization, and at the international level the rhetoric of 'education for all' from the Jomtien Conference of 1990.

In 1990, the government set up the Ramamurti committee (summarized by Sharma, 2002)[5] to review the 1986 policy. The committee outlined the reasons for the CSS not gaining ground: low investment in government schools because the elites and privileged class were not sending their children to government schools; lack of political will; the 'craze' for English-medium (private) schools; and the growth of institutions like central schools for specified categories of children. The committee expanded and extended the scope of the CSS as 'a first step in securing equity and social justice'. It recommended that the CSS be extended to private schools and selection of children by these schools even at the primary stage be dispensed with. These recommendations had the potential to change the face of the school education system of India, and could have removed increasing disparities in access to schools. But none of these recommendations was incorporated in the modified policy of 1992.

The CABE committee on policy (MHRD, 1992), while reviewing the Ramamurti committee report, expected the 'privileged schools' to accept 'social responsibility by sharing their facilities and resources with other institutions, and facilitating access to children of the disadvantaged groups' (MHRD, 1992: 16). There is an interesting consequence of this policy, namely, that private schools began running 'centres for the underprivileged' in the afternoon, thereby 'doing excellence in the forenoon and equity in the afternoon'[6] (Jha, 2004a). Skrtic (1991: 233) had argued that 'The successful schools in the postindustrial era will be ones that achieve excellence and equity simultaneously – indeed ones that recognize equity as the way to excellence'.

Non-formal education and equity

The growth and glamorization of non-formal education (NFE) was another design whose net effect was to undermine the implementation of the policy of a common school system providing a quality education for all. The 'non-formal programme'[7] was designed for education of the 'dropouts', children from habitations without schools, working children and girls who could not attend schools for the whole day (MHRD, 1998: 14). These arrangements were expected to be transitory in nature, to be phased out when the formal system could admit all children. However, as many commentators observe, 'some education' was offered through the parallel non-formal system to the majority of the disadvantaged, while the formal system catered to a small minority to prepare them for higher education (Ahmed, 1975; Beare and Slaughter, 1993; Watkins, 2000). The Ramamurti Committee pointed out that even at the time of policy formulation, the population of out-of-school children was half that of the school-going age (MHRD, 1990: 123).

The committee recommended that the formal system itself should be 'non-formalized' to include all children within its fold. However, a committee of the CABE on policy constituted to look into these recommendations commented that it was not 'desirable to overload the [formal] school system with yet another formidable challenge of meeting the educational needs of children with severe para-educational constraints' (MHRD, 1992: 31).

The Indian Planning Commission in its evaluation of the Non-Formal Education (NFE) system in 1998 concluded that: 'The NFE system has not made any significant contribution to the realization of the goal of the UEE' and 'elementary education needs to be delivered primarily through the formal education system' (MHRD, 1998). However, almost at the same time the central government accepted it as a part of its national programme of the *Sarva Siksha Abhian* (universal elementary 'Education for All'), to be offered to groups of children not necessarily belonging to the categories earlier defined under the NFE. Many attractive names, including the 'education guarantee scheme',[8] have been given across the states, and what were non-formal arrangements earlier are formal arrangements now, parallel and inferior tracks within the public education system for the poor and the disadvantaged.

The rhetoric of the Education for All (EFA) lobby following the Jomtien Conference (1990), and the entry of the international agencies and NGOs into primary education, has distracted government commitment to education for equality in educational policy, and the policy has turned into something like: 'literacy for your children and education for mine' (Jha, 2004b). Shotton (1998: 21) notes that 'literacy' (and not education) is regarded as the need of the 'new era' of the Indian global economy, as multinational companies require literacy in order for people to read the 'labels' of their products.

To sum up, while the policy began with the aim of introducing a common school system of public education in 1968 that could address education for equality, it has since deflected policy into creating a parallel track with unequal categories of a common education system on the one hand, and a contribution to the growth of private sector education on the other.

Private schools

Private schools in India have played a major role in the development of school education in terms of tradition and numbers. The characteristics of private schools from the management point of view and also from the perceptions of the clients they service, are heterogeneous. From the management perspective, they fall into three categories: recognized and aided by the government; recognized but unaided and also called independent or 'public' schools; and unrecognized schools. As noted by many researchers, over the years aided schools have become an integral part of the government school system, because of the conditions laid down by the government on aspects of management, including teacher recruitment and service conditions (see Kingdon, 1996; De *et al.*, 2000).

There is a perception that fee-charging private schooling is an urban phenomenon confined to the privileged class, but evidence from field-based studies in rural and semi-urban areas does not support this (see Kingdon, 1996; PROBE, 1999; Jha and Jhingran, 2002). Casual labourers, members of scheduled castes and slum dwellers have been reported in these studies as sending at least one of their children, preferably the male child, to low-fee-charging private schools. This, however, does not mean that there is no divide between socio-economic backgrounds of children going to private schools and state run-schools. The parents who cannot afford even low fees send their children to government schools, but even then there is no legal requirement for a parent to send a child to school.

A major characteristic of private schools is their independence in matters of student admission, which they invariably manage by tests and selection, fixing the quantum of fees according to market forces, and by hiring the best teachers. The schools, particularly in urban areas, apply selection criteria including interview of parents even at the nursery stage of admission. Contrary to this, in her study in the rural area schools of Uttar Pradesh, Srivastava (2005, and the next chapter) found parents bargaining for lower fees, with the school often acceding to this as the supply side has apparently outgrown the demand. Recent studies on low-fee-paying private schools in rural and urban areas suggest that their growth is due to an increased demand for education, and the non-expansion as well as inefficient functioning of the government schools (Kingdon, 1996; De *et al.*, 2000; Srivastava, 2005).

Many do not accept the argument that unaided private schools are totally independent of government subsidies. For example, a recent study submitted to the Central Ministry of Education takes note of the benefits accrued to private schools from the government in terms of concessions for income tax, wealth tax and property tax; direct subsidies towards the cost of land allotted; concessions in electricity charges, and other items (Bhatnagar and Omer, 2004). Other hidden subsidies available to private schools, it is argued, are the employment of state-trained teachers, and services from the curriculum and affiliating bodies, most of them being supported by the state.

It should be noted that if private educational institutions in India are to be registered as not-for-profit organizations under the Societies Registration Act, or as a Trust, then profit-making and commercialization in educational provision is not permissible under existing Indian policy. The 1986/92 policy states: 'Non-government and voluntary effort ... will be encouraged ... at the same time, steps will be taken to prevent the establishment of institutions set up to commercialize education' (MHRD, 1998: 35). This policy was further enforced by the judicial verdict,[9] which ruled against commercialization of educational institutions.

The tradition of opening private schools in India was once considered to be philanthropic, even religious, for the larger benefit of the society. However, in recent decades, particularly in the 1990s since the Indian economy 'opened up', the market argument has prevailed over other arguments. Kumar (2003: 5165) notes a new trend of opening elite private schools and advertising 'facilities which are identical to those offered by five-star hotels', and which serve not only children of Indian elites but overseas children as well.[10] This type of school is adding a new layer in the already existing hierarchy of schools.

Some advocate the desirability and growth of private schools on the ground of choice. But ethically the choice has to be available to all, regardless of income level. It should not be restricted only to an elite.[11] Indian society is both heterogeneous and unequal, as reflected in a variety of aspects of educational provision, including schools serving different social, economic, gender and 'special needs' groups. Private schools are contributing to the social and economic divide by 'perpetuating inequalities' in education (Panchmukhi, 1983; Tilak and Sudarshan 2001).

It is argued that any school following the state curriculum and entering pupils for public examinations should not charge fees for the six to fourteen age group of children. This restriction should apply to the private schools also, as they function, in this regard, as an instrumentality of the state.[12] The Supreme Court of India in earlier verdicts has observed that any agency discharging state function as its 'instrumentality' is bound by the constitutional provision. By a similar logic the Law Commission earlier in 1998 had recommended that private schools should admit 50 per cent of children without charging fees. It is thus seen that while there is a strong constitutional foundation for rights and equity in education, at least at the elementary stage, there is very little appreciation in policymaking to address this question.

In most studies and reports, the growth of private schools is attributed on the one hand to dysfunctional government schools with poor infrastructure and lack of teacher motivation and accountability, and on the other to perceived 'quality' education given by the private schools (Jha and Jhingaran, 2002; Ramchandaran, 2002; MHRD, 2003). However, 'quality' in private schools is not uniform. This is confirmed by PROBE (1999: 104), which did not find any overall difference between the government and private schools, but the 'selling point' of the latter was the fact that English was a medium of instruction.

Private school students often out-perform students in government schools, although this may reflect their middle class backgrounds rather than the quality of the fee-paying schools. This is an acknowledged finding of most of the quantitative studies, in many countries, including India (see Kingdon, 1996; Tilak and Sudarshan, 2001). It is argued, however, that there is no level playing field between the two: private schools have far more autonomy and management flexibility than government schools, and they select students to show better performance at the board examinations (Qamar and Zahid, 2001).

Hierarchy in schools

PROBE (1999) has reported forms of social discrimination operating in the Indian school system. A system of 'multiple tracks' has been identified, which provide different types of schooling opportunities to different sections of the population. The poor and the disadvantaged go to government schools, and the well-off students go to private schools; some children from economically poorer backgrounds go to formal schools, but those for whom the formal system is not 'suitable' are sent to the 'informal' or non-formal educational centres. There is thus a hierarchy of schools catering to the allegedly different groups. Some such school groups are:

- growing numbers of elite schools offering international certifications
- private fee-charging schools for upper middle and rich classes
- schools for the children of staff in central government, public undertakings and defence (heavily subsidized)
- schools for 'talented' rural children
- low-fee private schools in rural areas
- government and municipal schools for lower middle classes
- NFE, EGS (Education Guarantee Scheme), SSA (Sarva Shiksa Abhiyan), alternative schools for the poor and disadvantaged
- schools for child labourers (non-formal type)

- government schools for the scheduled tribes (residential but sub-standard)
- special schools for children with disabilities outside the mainstream education system
- schools offered by Christian missionaries for Dalit and Tribal children, sometimes on a 'low fee' basis.

Conclusions

The existing system of education seems to have been impacting the quality of governance, and not the other way around. The seeds of superiority, hierarchy and discrimination against certain groups, the poor in particular, are sown at a very early age in the existing school system in India, and this is reflected very strongly in the schools' governance, at each level. The system reinforces compulsion, comparison and competition that restrict options, individuality and cooperation. There is, however, evidence to suggest that mixing children of different abilities and socio-economic backgrounds can enhance school standards for all (Kahelnberg, 2001).

In India we are in the information age of the twenty-first century, and in a democracy. The nature of workplaces in particular, and the social system in general, is changing very fast. Hence, the school system needs to change. The three Cs: compulsion, comparison and competition of industrial-age schooling need to be replaced by another three Cs relevant for twenty-first century schooling: choice, consideration and collaboration. The latter group respects the rights of all children, rather than creating parallel systems on perceived 'needs' of the poor and the disadvantaged, decided by those in power. Implementing a common school system that provides quality education for children of all citizens is a major but exciting challenge for India.

Notes

1 The Constitution (86th Amendment) Act 2002.
2 Observation made at the national convention on the Right to Education Bill 2005 organized by the People's Campaign for the Common School System in New Delhi on December 9th, 2005.
3 Observed by Justice Verma in the above convention.
4 Personal communication from Professor Christopher Bagley, who attended such a school in rural Oxfordshire.
5 The Congress government in 1986 had announced the NPE. In 1990, the non-Congress government set up the Ramamurti committee to review the policy, but by the time the committee submitted its report, the Congress had come back to power (in 1992).
6 Many urban private schools in Delhi run learning centres in the afternoon for the disadvantaged as a charity.
7 The centre was to be run in a shed or place provided by the community, where a group of children could be taught for a couple of hours by a local untrained youth engaged at a small salary on contract.
8 The name is given as it claims to guarantee education to a community if they felt their children were not receiving it otherwise. Instead of opening regular schools they are given centres in the NFE pattern.

9   Supreme Court of India: Unikrishnan v. Andhra Pradesh, 1993.
10 A BBC documentary in April, 2006, gave examples of British parents sending their children to elite boarding schools in India because of the high curriculum standards and levels of discipline. Indian school fees plus airfares meant that costs were often less than sending a child to an elite-level boarding school in the UK.
11 The problem of whether elite schools should offer scholarships only to very able children, regardless of parental income, has yet to be addressed.
12 A submission was made on behalf of the Public Study Group – a Delhi-based group of academics and activists before the CABE meeting on 14 July 2005 to reconsider a draft bill on the 'right to education' on this ground.

References

Ahmed, M. (1975) *The Economics of Non-formal Education.* New York: Praeger.
Archer, M. S. (1979) *Social Origins of Educational Systems.* London: Sage.
Beare, H. and Slaughter, R. (1993) *Education for the Twenty-First Century.* London: Routledge.
Bhatnagar, D. and Omer, K. (2004) *Public Utility of Private Schools.* New Delhi: Government of India, Ministry of Human Resources Development.
CABE (2005) *Report of the Central Advisory Board of Education Committee on Free and Compulsory Education and Other Issues Related to Elementary Education,* New Delhi: Central Advisory Board of Education.
De, A., Majmudar, M., Samson M., and Noronha, C. (2000) *Role of Private Schools in Basic Education.* New Delhi: NIEPA and Ministry of Human Resources Development.
DiBona, J. (ed.) (1983) *One Teacher, One School: The Adams Report on Indigenous Education in Nineteenth Century India.* New Delhi: Biblia Index.
Education Commission (1966) *Report of the Education Commission, 1964–66: Education and National Development.* New Delhi: Government of India.
Fagg, H. (2002) *Back to Sources: A Study of Gandhi's Basic Education.* New Delhi: National Book Trust.
Iyer, V. R. Krishna (2005) *The Hindu,* 26th November p. 11.
Jha, M. M. (2002) *School Without Walls: Inclusive Education for All.* Oxford: Heinemann.
Jha, M. M. (2004a) 'Inclusive education and the common school in India', in Mohapatra, C. S. (ed.) *Disability Management in India: Challenges and Commitment.* Secunderabad: National Institute for the Mentally Handicapped.
Jha, M. M. (2004b) 'Too many nations in education', *The Indian Express.* New Delhi, November 4th.
Jha, J. and Jhingran, D. (2002) *Elementary Education for the Poorest and Other Deprived Groups: The Real Challenges of Universalisation.* New Delhi: Centre for Policy Research.
Kahelnberg, R. D. (2001) *All Together Now: Creating Middle Class Schools through Public School Choice.* Washington DC: Brookings Institution Press.
Kingdon, G. G. (1996) 'Private schooling in India: size, nature and equity effects', *Economic and Political Weekly,* December 21st, 3306–3314.
Kumar, K. (1994) 'Mohandas Karamchand Gandhi', in Morsy, Z. (ed.) *Thinkers on Education,* Vol. 2. Paris: UNESCO.
Kumar, K. (2003) 'Judicial ambivalence and new politics of education', *Economic and Political Weekly,* December 6th, 5163–5166.
Ministry of Human Resource Development (MHRD) (1990) *Towards an Enlightened and Humane Society: NPE 1986 – A Review.* New Delhi: Government of India.

Ministry of Human Resource Development (MHRD) (1992) *Report of the CABE Committee on Policy.* New Delhi: Government of India.

Ministry of Human Resource Development (MHRD) (1996) *National Policy of Education 1986: Programme of Action 1992.* New Delhi: Government of India.

Ministry of Human Resource Development (MHRD) (1998) *National Policy on Education (as modified in 1992) with National Policy on Education, 1968.* New Delhi: Government of India.

Ministry of Human Resource Development (MHRD) (2003) *Education for All: National Plan of Action.* New Delhi: Government of India.

Panchmukhi, P. R. (1983) 'India: a century of effort', in Simmons, J. (ed.) *Better Schools: International Lessons from Reform.* New York: Praeger.

PROBE (Public Report on Basic Education) (1999) *Public Report on Basic Education in India.* New Delhi: Oxford University Press.

Qamar, F. and Zahid, M. (2001) 'Cost, equity, quality and resource use efficiency in senior secondary schools: some policy imperatives', in *National Conference: Focus Secondary Education.* New Delhi: NIEPA (National Institute of Educational Planning and Administration).

Ramachandaran, V. (2002) 'Hierarchies of access', in Ramachandaran, V. (ed.) *Gender and Social Equity in Primary Education.* New Delhi: The European Commission.

Sadgopal, A. (2003) 'Exclusion and inequality in education: the state policy and globalisation', *Contemporary India: Journal of the Nehru Memorial Museum and Library,* 2, 3.

Sadgopal, A. (2004) *Globalization and Education: Defining the Indian Crisis.* New Delhi: Zakir Husain College.

Sharma, R. N. (2002) *Indian Education at the Cross Roads.* New Delhi: Shubhi.

Shotton, J. (1998) *Learning and Freedom.* New Delhi: Sage.

Skrtic, T.M. (1991) *Behind Special Education: A Critical Analysis of Professional Culture and School Organisation.* Denver: Love Publishing.

Srivastava, P. (2005) 'The business of schooling: the school choice processes, markets, and institutions governing low-fee private schooling for disadvantaged groups in India', unpublished D. Phil. thesis, University of Oxford.

Steele, T. and Taylor, R. (1994) 'Against modernity: Gandhi and adult education', *International Journal of Lifelong Education,* 13, 33–42.

Tilak, B. G. and Sudarshan, R. M. (2001) *Private Schooling in Rural India.* New Delhi: National Council of Applied Economic Research.

Tomlinson, S. (2001) *Education in a Post-Welfare Society.* Buckingham: Open University Press and McGraw-Hill.

Watkins, K. (2000) *The Oxfam Education Report.* Oxford: Oxfam International.

11 Low-fee private schooling
Challenging an era of education for all and quality provision?

Prachi Srivastava

Introduction

The widespread emergence of what is termed here *low-fee private* (LFP) school-ing in India heralds the need to look beyond international and national rhetoric framed by various *Education for All* (EFA) targets and campaigns, in order to closely examine emerging private sectors of schooling in economically develop-ing countries facing the problem of increasing educational demand, constrained public budgets, and the deteriorating actual or perceived quality of state educa-tion. The significance of the LFP sector is critical, not only because it is uniquely characterised as a private sector of formal provision targeted to a clientele with persistent schooling gaps and low levels of participation, but also because it necessitates an examination of the changing nature of provision for the schooling of disadvantaged groups. Paradoxically, the increased marketisation and privati-sation of the schooling arena for disadvantaged groups point to an alteration in the way that schooling is delivered to and accessed by these groups, in an era of increased outward commitment to the EFA goals of access, equity, and quality in schooling provided by the State.

This chapter is envisioned as a starting point in the analysis of the context in which private provision of schooling for disadvantaged groups is emerging and operating in India, with reference to Uttar Pradesh. Building on a recently com-pleted study of LFP schooling in Uttar Pradesh (Srivastava, 2005), it provides an analysis of the EFA dialogue in India by focusing on two debates that are most closely linked with the changing nature of schooling provision for disadvantaged groups: quality schooling and increased private provision.

Following a presentation of the research strategy, the chapter examines the EFA debate and strategy in India in general, and in Uttar Pradesh in particular. Third, it considers how 'quality' has been addressed in Indian schooling provi-sion and delivery. By applying the under-analysed District Information System for Education (DISE) data to Uttar Pradesh, the chapter then highlights the need for an expanded set of indicators to assess quality for disadvantaged groups. The fifth section is devoted to disentangling the main sectors of formal public and private provision while locating the LFP sector within the broader schooling arena. Sixth is a reassessment of public and private delivery in the new context of the LFP sector. The chapter ends with a consideration of some implications on the provision and delivery of schooling for disadvantaged groups in this new context.

Defining the low-fee private sector and outlining the research strategy

The emergence of LFP schooling in Uttar Pradesh is not atypical. Recent studies on formal schooling in India have documented the growth of this sector in the country (De *et al.*, 2002; Mehrotra *et al.*, 2005; Tooley and Dixon, 2005). Its potential impact has been noted regarding EFA targets in the context of constrained public resources (De *et al.*, 2002; Panchamukhi and Mehrotra, 2005). However, despite its reported emergence, analysis on the LFP sector is scarce. Furthermore, the sector has neither been officially defined by the State nor operationally defined by researchers.

For the purposes of the study, the LFP sector was defined as occupying a part (often unrecognised) of the heterogeneous private unaided sector. The private unaided sector in India is privately funded and run. LFP schools were defined as those that saw themselves targeting disadvantaged groups; were entirely self-financing through tuition fees; and charged a monthly tuition fee not exceeding about one day's earnings of a daily wage labourer at the primary (grades one to five) and junior levels (grades six to eight), and two days' earnings at the high school (grades nine and ten) and higher secondary levels (grades eleven and twelve).

The discussion in this chapter is based on a household, school, and state-level study on LFP schooling in Lucknow District, Uttar Pradesh (see Srivastava, 2006; Srivastava, forthcoming). Uttar Pradesh along with Andhra Pradesh, Bihar, Madhya Pradesh, Rajasthan, and West Bengal, is classed as one of the most 'educationally backward' states in India. These six states plus Assam account for three-quarters of the country's out-of-school children (Mehrotra and Srivastava, R., 2005). Furthermore, with a literacy rate of 57.4 per cent, Uttar Pradesh was ranked thirty-first of the thirty-five states and territories in the latest census (Government of India, 2001). At the same time, it had the second highest distribution of private school enrolments in elementary education in the country at 57.6 per cent[1] (Panchamukhi and Mehrotra, 2005, p. 236).

Data were collected between July 2002 and April 2003. The study examined the school choice processes and schooling patterns of disadvantaged households accessing LFP schooling; the internal organisational structures of LFP schools and the nature of local school markets in which they operated; and the formal and informal institutions, or regulatory frameworks, which governed their interaction within the LFP sector and the State. Data were collected through 100 formal interviews with sixty households (thirty urban and thirty rural), ten case study schools (five urban and five rural), and government officials; numerous informal interviews with school owners/principals and officials; official and 'grey' school and government documents; and school and state-level observations.

The historical struggle to provide Education for All

To better understand the role of the LFP sector in the new schooling market, it is important to sketch its place within the national policy context for education provision and EFA. The concern with providing free and compulsory elementary education is not new to India. It predates the most recent international initiatives of the 1990 Jomtien Conference and the 2000 World Education Forum setting the EFA agenda, by almost 100 years.

Balagopalan (2004) traces the history of the movement for free and compulsory education in India back to 1893 with an educational experiment that began in Baroda, Gujarat; and then to 1910 when a resolution for free and compulsory education in areas with a male school-aged population of at least 33 per cent was made but rejected. Similarly, in 1909, a bill introducing free and compulsory education was introduced by Gokhale in the Legislature following the Indian Council Act but was rejected in the Parliament (Drèze and Sen, 1995).

The Uttar Pradesh Primary Education Act of 1919 was instituted to introduce compulsory primary education through municipal boards for children aged six to eleven in the state (UPPEA, 2001, p. 623). Two key features of the Act were the introduction of compulsion for parents with a fine if school-aged children were not sent, and the establishment of formal basic education as a distinct system covering a specific age group. Furthermore, Section 4 of the Act stated that boards must satisfy the State Government in making 'adequate provision in recognised primary schools for such *compulsory primary education free of charge*' (emphasis added) (UPPEA, 2001, p. 624). Subsequently, the United Provinces District Boards Primary Education Act, 1926, was passed to ensure that: 'universal, free and compulsory primary education for boys and girls should be reached by a definite programme of progressive expansion' (UPDBPEA, 2001, p. 628).

In the post-independence period, while adopting the Indian Constitution in 1950, Article 45 of the Directive of Principles of State Policy in Part IV further gave a policy direction to all states with the duty to provide free and compulsory education to all children until the age of fourteen within a period of ten years (Mehta, 1998; Rao, 2002). Two landmark education platforms, the Kothari Commission (of 1964–66) and the Acharya Ramamurthi Committee (in 1990) were launched to identify the best strategies to advance the goals of free education provision. While the Government accepted the Kothari Commission's recommendations and announced the National Policy of Education (1968), the Ramamurthi Committee report was subjected to further scrutiny through the Central Advisory Board of Education committee in 1992. Disconcertingly, most of its major recommendations regarding the Common School System, quality, and equity in schooling were rejected when announcing the latest National Policy of Education (1992).

The concern for free compulsory education has recently been enshrined as Article 21A in the Eighty-sixth Amendment Act 2002 of the Indian Constitution which states: '21A. Right to Education – The State shall provide free and compulsory education to all children of the age of six to fourteen years in such manner as the State may, by *law,* determine' (Government of India, 2004). Some government officials in this study claimed that the insistence of including the term 'by law' was to mark India's outward compliance in the international politics of education agenda-setting, as the Supreme Court had already declared basic education up to age fourteen a fundamental right in 1993.

Following from this, the Free and Compulsory Education Bill of 2004 (the latest at the time of writing), has not been without debate. In its insistence on expanding education provision to meet targets increasingly influenced by international rhetoric, the 2004 Bill enshrined what critics claimed to be a two-tier or parallel system through 'approved schools' and 'transitional schools', both with different standards. For example, while trained teachers would provide instruction in approved schools, instructors who only completed high school and received thirty days' training, were to impart instruction at transitional schools (Balagopalan, 2004, p. 3). The only compulsion in the Bill was that parents whose children did not attend any school had to send them to a transitional school or be penalised. Hence, according to Balagopalan (2004),

> The 'compulsory' provisions in the draft Bill thus serve to institutionalise a parallel system that poor parents have no recourse to reject. The reason that this idea of 'compulsion' does not provoke more outrage is because the middle class strongly believes ... that the primary reason that children are not in school is because of parental encouragement of child labour ... 'compulsion' takes precedence over quality of schooling issues.
>
> (p. 4)

The Bill has since been withdrawn and a Central Advisory Board of Education committee has been delegated the task of drafting a new bill.

India's struggle with achieving free and compulsory education is highlighted above. However, to further the debate, three issues should be considered when assessing persistent schooling gaps in the Indian context. First, there is a lack of focus on the quality of schooling to be delivered in the mass education system. There was no mention in Article 21A or in the draft Bill about minimum quality standards that should be provided at approved or transitional schools, although the implications of having teachers less qualified in the latter do not seem to advance the notion of quality schooling.

Second, the thrust on 'mobilisation' in central campaigns such as *Sarva Shiksha Abhiyan* and state campaigns such as Uttar Pradesh's *School Chalo Abhiyaan!*, presents disadvantaged groups with homogeneous views on schooling, i.e. they are unable to see its relevance and are in need of 'mobilisation'. This view obscures the fact that for many disadvantaged parents the motivation to access schooling is conditional on a positive assessment of their options (Srivastava, 2006). According to the results of this study, disadvantaged households often did not see the benefit of sending their children to school if their only option was a perceived malfunctioning state school.

Finally, the Indian EFA discourse has focused on state provision and, failing that, on non-formal education or alternative school models (i.e. transitional schools and

Education Guarantee Scheme centres). This obscures a focus on models of formal education delivery outside the state sector, such as the LFP sector. While there is some recognition that disadvantaged groups have begun accessing a segment of the private sector (De *et al.*, 2002; Duraisamy *et al.*, 1997; Majumdar and Vaidyanathan, 1995; PROBE, 1999; Tilak and Sudarshan, 2001), an examination of the LFP sector is largely ignored when assessing EFA strategies.

This is not to say that LFP schools will necessarily be better providers than the state sector or will address the schooling needs of disadvantaged groups in the long run; there is currently insufficient evidence to support such a view despite claims to the contrary (e.g. Tooley and Dixon, 2005). But without concerted examination of the LFP sector, it is not possible to accurately understand the schooling behaviours of disadvantaged groups, and the new schooling arena within which schooling is increasingly being delivered to and accessed by them.

Examining 'quality' in the provision and delivery of schooling

Quality provision and delivery of schooling assumes importance as one of the key issues framing the larger educational debate in India. The proliferation of LFP schools has been attributed to a very low level of quality in the state sector (De *et al.*, 2002; Tooley and Dixon, forthcoming). Govinda (2002) stresses that: 'government schools were never marked as of especially poor quality in comparison with their private counterparts as is done today with little exception' (p. 11). This is despite the State's focus, in principle, on increasing quality provision in state schooling. However, a closer examination of what 'quality provision' constitutes, reveals an inadequate definition of quality fraught with competing goals and target levels within the same policy.

Part 3.2 of the National Policy for Education 1992 states: 'The concept of a National System of Education implies that, up to a given level, all students, irrespective of caste, creed, location, or sex, have access to education *of a comparable quality*' (emphasis added) (Government of India, 1998, p. 5). While there is an insistence on some level of comparable quality throughout the policy, 'quality' itself is not defined. Instead, several measures (summarised in Table 11.1) are proposed in the National Policy and its Programme of Action to minimise the level of variance in such a large education system.

Nonetheless, while Part 3.13 of the National Policy states that the Central Government should 'promote excellence at all levels of the educational pyramid throughout the country' (Government of India, 1998, p. 7), elsewhere in the same policy, the level of quality drops to 'satisfactory' (Part 5.12), or rests at 'substantial improvement' (Part 5.5).

Perhaps the most important measures in the National Policy (highlighted in Table 11.1) can be seen as attempting to address quality by setting certain standards to be met in priority areas. These areas can be extrapolated from the National Policy for Education 1992 and its Programme of Action as: universal elementary education (grades one to eight), matching skills in the labour market, and equality of opportunity for various disadvantaged groups.

Table 11.1 Compilation of key proposed measures in the National Policy for Education 1992 and Programme of Action 1992

Proposed measures	*Summary*
Common educational structure	• 5 years of primary; 3 years' junior; 2 years' high school • Recommendation to incorporate the + 2 (intermediate) level as part of 'school education'
Common curricular core in addition to other flexible components	• History of India's freedom movement; constitutional obligations; components nurturing national identity • Promotion of common values: common cultural heritage; egalitarianism, democracy and secularism; equality of the sexes, protection of the environment; removal of social barriers; observance of the small family norm; and inculcation of a scientific temper
Provision of equal opportunity	• Equal access and conditions for success • Awareness of equality of all through the core curriculum • Although not stated in the NPE 1992, the national norms of 1 km radius for primary and within 3 km radius for junior as well as a maximum teacher–student ratio of 1:40 may be indicators here
Minimum levels of learning	• To be laid out for each stage of education
Essential school facilities	• To provide 3 all-weather, reasonably sized classrooms under Operation Blackboard • Blackboards, maps, charts, toys, and other necessary aids • Minimum of 3 teachers increasing to one per class as soon as possible • At least 50% of teachers recruited should be women
Education for equality	• Various provisions for women and girls and schedule caste, scheduled tribe, other backward caste, and minority groups • Incentives for families to send children to school; scholarships for specific groups; recruitment of teachers from scheduled caste groups; establishing residential schools; focusing on indigenous languages

Sources: Government of India, 1996; Government of India, 1998.

However, the notion of educational quality is confounded with standardisation in the area of achievement. A continuing debate in more economically advantaged countries such as the USA and the UK, this concern is emerging in the Indian context because of a number of increasing factors, partly due to the new Indian economy and an increasingly competitive labour market. Labour market forces have intensified in addition to existing competitive selection procedures in many professional fields through public exams. Furthermore, parents of all socio-economic groups are increasingly sending their children to private tuition centres because they feel that teachers are not imparting the required instruction in state schools (Majumdar and Vaidyanathan, 1995). These concerns have emerged as public confidence in education (particularly the state sector) is decreasing (De *et al.*, 2002; PROBE, 1999).

The introduction of 'minimum levels of learning' in the curriculum cloaked as 'a strategy for improving the quality of elementary education [in] an attempt to combine quality with equity' (Government of India, 1996, p. 41), can be interpreted as an attempt to address that lack of confidence by establishing basic 'competencies' that all students should acquire regardless of the sector they access. According to the Government of India, a minimum level of learning: 'Lays down learning outcomes in the form of competencies or levels of learning for each stage of elementary education' (Government of India, 1996, p. 41). However, focusing on basic competencies without addressing other areas of quality or equity in the curriculum, treatment of children at school, or broader issues of access particularly regarding the state sector, has been criticised as being rigid and promoting teachers to 'teach to the test': 'The slogan of "competency-based learning" has made little difference to curricula and textbooks, which have religiously followed the unrealistic list of "contents", only flimsily disguised as "competencies"' (PROBE, 1999, p. 79).

If India is to address quality concerns in its three priority areas in the state sector, raising public confidence in a sector which is increasingly being characterised as malfunctioning should be of paramount importance. As found in this study and others (Aggarwal, undated; Balagopalan and Subrahmanian, 2003; Bashir, 1994; PROBE, 1999), parental perceptions of inferior state school quality; an iniquitous system rife with issues such as teacher absenteeism; little public accountability; teachers over-burdened with other state duties resulting in frequent school closures and minimal teaching activity; and teachers ridiculing lower caste children, all undermine notions of educational quality.

Ironically, the perceived growth of these 'inequalities and dysfunctions' (Datt, 2002; Govinda, 2002) come at a time when the focus on EFA and access to and quality of schooling have officially been the utmost guiding concern. As stated in the National Policy for Education 1992:

> The new thrust in elementary education will emphasise three aspects: (i) universal access and enrolment, (ii) universal retention of children up to 14 years of age and (iii) a *substantial improvement in the quality of education* to enable all children to achieve essential levels of learning (emphasis added).
>
> (Government of India, 1998, p. 13)

In this regard, the Government of India launched *Sarva Shiksha Abhiyan* in 2000, a national EFA campaign with 'its central objective of mobilising all resources, human, financial and institutional, necessary for achieving the goal of UEE [universal elementary education]' (Government of India, 2002a, p. 55). Its main goals are to ensure completion of elementary school by children aged six to fourteen and to bridge gender and social gaps in elementary education by 2010, with a 'focus on elementary education of satisfactory quality' (ibid.). Significant financial outlay has been released to all states according to their District Elementary Education Plans for items such as the construction of new schools and the establishment of Education Guarantee Scheme centres. Nonetheless, while *Sarva Shiksha Abhiyan* officially affirmed the State's commitment to universal elementary education, it simultaneously threatened the quality of schooling by reducing the minimum required number of teachers in a school from three in the 1992 Programme of Action, to two.

Assessing quality using Uttar Pradesh as a case

The recent establishment of the District Information System for Education (DISE), an educational management information system, is one attempt at advancing quality assessment. The DISE database is to be updated on a yearly basis to collect time-series data on three groups of indicators: school, enrolment, and teacher-related. It is the result of a government effort to improve on existing surveys used for educational analyses (e.g. All India Education Survey, National Council of Educational Research and Training [NCERT]) which are not updated at regular intervals; and on household survey data (e.g. National Sample Survey Organisation data) which reportedly present relatively more accurate accounts of household school enrolment than the NCERT surveys, but do not report much on the characteristics or numbers of schools in a given area.

At the time of writing, DISE data were collected across eighteen Indian states, resulting in the 2003 DISE report (Mehta, 2004). Disappointingly, however, while data from unrecognised schools were gathered at the village level (Mehta, 2004, p. 3), only data from recognised schools were reported in statewide analyses of the DISE report.[2] This has serious implications when considering results about the private unaided sector, since it is estimated that a good proportion of private unaided (and LFP) schools are unrecognised (e.g. Tooley and Dixon, 2005). Nonetheless, the report yielded some interesting results.

The total number of schools was 853,601 in the eighteen states. While Uttar Pradesh is officially characterised as one of the most educationally backward states in India, due to its size it had the highest number of schools at 119,443 (Mehta, 2004, p. 32). Construction efforts under *Sarva Shiksha Abhiyan,* in combination with the long-running District Primary Education Programme (from 1994 to 2003), were attributed to the opening of 161,279 new schools across the country since 1994 (Mehta, 2004, p. 53). Uttar Pradesh saw the second highest number of new schools at 33,452. The percentage share of all schools in rural areas was 87 per cent nationally, and 91.5 per cent in Uttar Pradesh. Table 11.2 presents some DISE data extracted from the report to assess quality across the three categories of indicators.

School-related indicators on facilities reveal that the picture for Uttar Pradesh may not be as bleak in certain areas as expected, relative to other states. The indicators show that it ranks quite highly (within the top three) for the percentage of primary schools with access to drinking water, separate girls' toilets, book banks, and playgrounds. However, in absolute numbers, particularly regarding girls' toilet facilities, the percentage of schools was low at only 40.9 per cent. Nonetheless, it seems that compared with other states, Uttar Pradesh invested some money into its schools as only 9.9 per cent of primary schools had classrooms in need of major repair, and 66.3 per cent and 65.5 per cent of primary schools received

Table 11.2 Quality comparison of schools in Uttar Pradesh and nationally [a]

	National (All districts in the 18 states)	Uttar Pradesh	Rank[b]
School-related indicators			
Total number of schools	853,601	119,443	1
New schools built since 1994	161,279	33,452	2
Ratio of primary to upper primary	3.18 Prescribed norm: 2[c]	5.24	17
% Private unaided schools	6.74	10.12	3
% Share of schools in rural areas	87.0	91.51	4
% Share of private unaided schools in rural areas	4.6	7.93	3
% Schools with 3 or more classrooms	36.89	51.48	5
Level of highest enrolment as a per cent in primary education	26.91% schools with 21–60 children enrolled	32.26% schools with 141–220 children enrolled	n/a
% Schools > 3 teachers	35.85	42.38	5
Average number of instructional days in elementary schools/sections	215	209 State norm: 220 days[c]	Tied at 14
% Schools with a pupil–teacher ratio >100	8.94	24.2	3
% Schools with pupil–teacher ratio >60	25.5	49.1	2
% Single-classroom schools	15.7	2.7	17
% Single-teacher schools	19.1	15.9	10
% Schools with drinking water facility	71.9	91	1
% Schools with separate girls' toilet	15.64	40.88	2
% Schools with book bank	40.76	63.98	3
% Schools with playground	42.22	59.22	3
% Schools with classrooms needing major repair	27.3	9.9	10
% Schools received School Development Grant	48.81	66.3	5
% Schools received Teaching–Learning Material Grant	39.69	65.47	2
Enrolment-based indicators			
Gender parity in enrolment (primary)	0.89	0.90	12
Gender parity in enrolment (junior)	0.79	0.71	14
% Enrolment in single-teacher schools	12.2	13.2	5
% Enrolment school classroom ratio > 60	25.7	49.1	2
% Under-age children in primary	10.16	6.56	14
% Over-age children in primary	5.52	2.01	18
% Transition rate from primary to junior	Male: 65.96 Female: 62.73	Male: 40.22 Female: 36.30	16 17

	National (All districts in the 18 states)	Uttar Pradesh	Rank[b]
Teacher-related indicators			
Average number of teachers in private schools	4.88	4.41	10
Average number of teachers in government schools	2.47	2.51	10
% Female teachers	34.4	27.5	Tied at 13
% Trained teachers	44.4	57.3	5
% Of para-teachers to total teachers	11.03	9.71	6
Pupil–teacher ratio	46	67	2

Source: DISE data for 2002–2003 reported in Mehta (2004), *Elementary Education in India, Where do we stand?*

Notes

a These were the latest DISE data available at the time of writing. Data combine all school types, unless otherwise stated. All data are for primary schools unless otherwise stated. Primary schools comprise grades one to five. Junior schools comprise grades six to eight. Elementary education in India refers to the combined primary and junior cycles.
b Ranks added by the researcher to enable general comparison with the eighteen states in the DISE report.
c Researcher's comments.

School Development and Teaching–Learning Grants respectively. However, while the percentage of single-teacher and single-classroom schools was quite low, improvement in physical access to schools was required. The ratio of primary to junior schools/sections was 5.24:1, even though the prescribed norm was 2:1, placing it seventeenth of the eighteen states.

Examination of school-related indicators on classroom activities and enrolment and teacher-related indicators for primary schooling also reveals a less rosy picture. While the gender parity index for Uttar Pradesh was higher than expected at 0.90, it only ranked twelfth out of eighteen states on this measure. Furthermore, Uttar Pradesh was fourteenth in the country for the average number of instructional days, which at 209 days was lower than the 220 prescribed by the State. Also, while the percentage of single-teacher primary schools was comparably lower than in other states, the percentage of enrolment in those schools was 13.2 per cent, the fifth highest. Transition rates in elementary education for boys and girls from primary to junior were also among the lowest in India, at 40.22 per cent and 36.30 per cent. Interestingly, however, the failure rate for Uttar Pradesh decreased across elementary education as grade level increased.

Finally, teacher indicators revealed that female primary teachers accounted for only 27.5 per cent of the total primary teaching pool in Uttar Pradesh, placing it thirteenth. Furthermore, while Uttar Pradesh ranked fifth highest on the percentage of trained primary teachers, this only corresponded to 57.3 per cent of the teaching force. This is surprising, given the clear insistence in state norms against hiring untrained teachers in state and recognised private schools, but it indicates a possible lack in the availability of trained teachers in Uttar Pradesh and nationally. The relatively high rank in the percentage of primary para-teachers according to DISE data is in line with data collected from state officials and principals in this study who claimed that there was an insistence in Uttar Pradesh to hire *shiksha mitra*[3] to cover teacher shortfall, particularly in rural areas.

Most disturbingly perhaps, Uttar Pradesh ranked second in the country for the largest average pupil–teacher ratio at 67:1 for the primary level. When combined with the fact that 49.1 per cent (second highest) of the total enrolment was accounted for in schools with a classroom ratio greater than 60, this highlights issues of real concern over adequate delivery at the classroom level. This is further stressed as the percentage of schools with a pupil–teacher ratio greater than 100 was 24.2 per cent, the third highest among the eighteen states.

The variability in the different quality measures (i.e. Uttar Pradesh's relatively high ranks in facilities versus its low ranks on classroom–teacher ratios or enrolment), highlights the importance of employing an array of indicators. The DISE school-based, enrolment, and teacher indicators are a first step. While a valuable contribution to providing an overview of the state of schools in India, the DISE report examined only how recognised schools measured against these derived indicators and not how they fared against prescribed norms. The depth of analysis could be greater if comparisons were made with state or central-level norms. Furthermore, including data on unrecognised schools is of utmost importance for a more complete examination of the sources and types of variation across schools.

Finally, there is a need to expand the set of indicators currently employed. Aggarwal (2002) suggests examining the internal efficiency of India's education system through retention, transition, completion, and drop-out rates. While this is a necessary area of analysis, indicators relating to children's *lived experiences* at school, or Stephens' (1991) notion of 'something more' beyond efficiency and relevance, cannot be ignored as they may be primary factors encouraging families of first generation learners to enrol their children at all.

Lloyd *et al.* (2000) assert that: 'Few studies of school quality have examined those aspects of schooling that are most conducive to *encouraging initial enrolment and retention*' (emphasis added) (p. 113). For example, indicators focusing on household–school relationships or students' experiences at school beyond achievement should also be considered. If lessons are to be learned from studies documenting parents' level of dissatisfaction and the preferential treatment of students according to gender or caste particularly in state schools (e.g. Balagopalan and Subrahmanian, 2003; Duraisamy *et al.*, 1997; PROBE, 1999), then a more encompassing notion of quality is required. Furthermore, with the expansion of the private sector and the emergence of the LFP sector in particular, examining schooling provision for disadvantaged groups must extend beyond new methods of quality assessment to include analyses of the recognised and unrecognised private sectors.

Disentangling public and private sectors of education in India

Complicating private sector analyses in India is a continued conflation in the identification of the specific school types within the broad sectors of education provision. The boundary within and between public (traditionally Central Government and state) and private (traditionally private-aided and private unaided) provision is blurred. Data from this study suggest that this is partly because of the funding mechanisms associated with public and private provision, and also because of the practical modes of classification by government officials in their administrative work. In an effort to broadly map out a typology of formal schooling provision (with reference to Uttar Pradesh), Table 11.3 is provided as a starting point.

The table presents a typology of the three overarching school types by their primary financing, management, and accountability structures. Admittedly, the construction of such a broad typology is problematic (and will likely be contested) as the specific school types subsumed under the different 'public' and 'private' sector classifications are highly heterogeneous.

For example, the common usage of the term 'government schools' in Indian education discourse, obscures their heterogeneity. Typically, the term is used to refer to schools run by state governments through their Departments of Education or local bodies.[4] However, some Central Government departments also operate a small number of schools such as the Department of Tribal Welfare (for tribal groups), the Ministry of Labour (targeted for child labour), the Ministry of Defence (*Sainik* schools), and the Ministry of Social Justice (for children with disabilities).

The Central Government has also established three types of schools located in most states: *Kendriya Vidyalayas* or Central Schools mainly for employees of the Central Government, *Navodaya Vidyalayas* for talented rural students regardless of socio-economic status, and Tibetan Schools for Tibetan refugees. These are all centrally-funded and administered. In practice, however, state officials in this study explained that most statistics collected by them either did not differentiate between government school types, or that data collected at the district level on government schools only included schools run by the Department of Education and local bodies. Thus, when 'state schools' are referred to in this discussion, they are conceptualised as those run by the state's Departments of Education and local bodies.

When assessing the private sector, the issue is further complicated owing to the system of private-aided and private unaided schools. Private aided schools are classified here as public–private hybrids. While they are privately managed, the majority of their funding comes from the state government. Up to 95 per cent of a school's budget could be through state government grant-in-aid (Kingdon, 1996a; Tilak and Sudarshan, 2001). Most state funding covers teachers' salaries equivalent to those in state schools, as well as recurrent spending on non-teacher inputs (Panchamukhi and Mehrotra, 2005), while management must ensure that teachers meet state qualifications. Private-aided schools must raise their own funds for initial

Table 11.3 Typology of school sectors by primary financing, management and accountability structures

	School Type			
	Government and State	*Private-Aided*	*Private Unaided*	
			Recognised	*Unrecognised*
General Classification	• Public	• Public–Private Hybrid	• Private	• Private
Financing	• Central and state governments (directly and through centrally or state-sponsored schemes) • Very slight parental contributions	• State government (up to 95%) • Private (typically parents through parent–teacher associations)	• Private sources: e.g. parents, individuals, charitable trusts, NGOs and other agencies • Some state scholarships for children from scheduled caste, scheduled tribe, and other backward caste groups (in principle)	• Private sources: e.g. parents, individuals, charitable trusts, NGOs and other agencies
Management	• Central/State government structures • Relevant boards	• District and state-level committees • School committee • Relevant boards	• Owners (in practice) • School Committee of Management and managing society (in principle) • Network if part of a chain	• Owners (in practice) • Managing society if registered (in principle) • Network if part of a chain
Accountability Structure	• District State and Central governments (as applicable) • Relevant board	• District • State government • Boards • Parents (secondary) (in principle)	• Parents (market) • Network if part of a chain • District office and State through the concerned Board (in principle)	• Parents (market) • Network if part of a chain

Note: This typology was constructed using data from the study and from existing literature on Uttar Pradesh. Specific regulations and structures will vary according to state.

and ongoing costs, typically through parents' contributions to schools' parent–teacher associations. Because of the nature and amount of state intervention in the management and funding of private-aided schools, Kingdon (1996a) and

Tilak and Sudarshan (2001) assert that they could be called 'semi-government' or 'government-aided' schools.

Owing to the public–private hybrid, there is some confusion in the literature on some of the finer administration issues of private-aided schools. For example, concerning recruitment procedures and staff management, Kingdon (1996a) states that private-aided schools 'cannot recruit or dismiss their own staff' and that in Uttar Pradesh, the 'U[ttar] P[radesh] Government Education Service Commission selects and appoints their staff' (p. 3306). However, according to Panchamukhi and Mehrotra (2005), 'the decision to hire teachers lies with the management [of private-aided schools], who can also finance additional teacher posts and other recurrent expenditure from their own funds' (p. 230). Like Kingdon, the researchers also comment on the existence of a recruitment board or committee, but seem to accord a different balance of power to private-aided and government representatives. Kingdon does not mention the role of private-aided members, implying that they have little say in the recruitment process of staff. Panchamukhi and Mehrotra (2005) on the other hand, describe the private-aided recruitment committee as having only one government representative (ibid.).

Private unaided schools are autonomous, privately managed, and free of state financing, though in principle, recognised schools are more accountable to the state and their respective boards than unrecognised schools. If the private sector is conceptualised as comprising schools that are both financially independent of the state and privately managed, then the true private sector is composed only of recognised and unrecognised private unaided schools. To reiterate, LFP schools, as defined here, are part of this sector. Private unaided schools span a range of varying fee structures that are run by voluntary organisations, missionaries, philanthropic bodies, or individual owners as business enterprises (Tilak and Sudarshan, 2001). This is despite a 1993 Supreme Court ruling (*Unnikrishnan PJ and Others* v. *State of Andhra Pradesh and Others*) that schools should not be run for profit. LFP school owners in this study unofficially claimed profits from their schools.

Typically, the literature has accepted the claim that private schools start off as part of a cycle from unrecognised private unaided to recognised private unaided schools, en route to achieving private-aided status due to the appeal of state funding (Kingdon, 1996a; Panchamukhi and Mehrotra, 2005). However, this study revealed that owing to the complexity of individual state regulations governing grants-in-aid and private unaided schools, a more nuanced understanding is required when examining private school sectors. While the cycle described above may have earlier been the case, it no longer holds in principle, at least in Uttar Pradesh. As of 1996, officially, the State Government stopped disbursing grants-in-aid for an indefinite period because of insufficient state funds. Furthermore, LFP school owners in this study asserted that they would not avail themselves of this provision because of the severe restriction of autonomy through stringent state control by becoming a private-aided school.

Under the Indian Constitution, private unaided schools may exist regardless of whether or not they are recognised (Balagopalan, 2004; De *et al.*, 2002;

Majumdar and Vaidyanathan, 1995). However, state legislation applies in governing private schools on this point. In principle, for a private unaided school to be recognised it must conform to regulations of the board with which it seeks affiliation. Recognition criteria for state boards vary on the particulars but cover such areas as norms for infrastructure, teacher qualifications, language of instruction, and fees. The main incentive for LFP schools (like other private unaided schools) to seek recognition is that only recognised schools can issue official documentation such as 'transfer certificates', or officially send their students as 'regular candidates' for exams. This increases a school's credibility and reputation in local school markets. The results of this study showed that, in practice, LFP schools were able to obtain recognition without meeting norms due to corrupt practices (Tooley and Dixon, 2005, had similar findings). Furthermore, like other unrecognised LFP schools, case study schools followed complex informal norms and procedures to circumvent official recognition norms, ensuring that their students gained benefits similar to those at recognised schools (Srivastava, 2005).

Assessing the size and nature of private provision in India and Uttar Pradesh is also compounded by the difficulty that much of the literature refers to both private-aided and unaided schools when speaking of the 'private sector'. There are further inaccuracies in statistical data due to a lack of regularly updated time-series and the exclusion of unrecognised private schools in most data sets. However, when examining the LFP sector, further complications arise. In addition to the exclusion of unrecognised schools in educational databases, no household survey or educational database disaggregates the private unaided sector by level of fees. Since the sector is highly heterogeneous, this presents a subtle yet crucial methodological point.

Anyone familiar with the educational context in India will agree that there has been an increase in the number of LFP and private unaided schools. However, since traditional statistical data show increases in the total number of private unaided schools, this increase cannot be as easily attributable to increased LFP provision alone, as some may suggest . This is because available databases *do not present data on the private unaided sector by level of fee charged*. Therefore, until new datasets include unrecognised private unaided schools and disaggregate private-unaided sector data by fee-level, it is impossible to accurately assess the amount of variation across the sector by high-, medium-, or low-fee sub-sectors. This makes accurate assessments of each sub-sector's growth over time and the proportion of enrolment claimed by each, difficult at best.

Reassessing public and private education delivery

Colclough (1993) argues that the main case for the public provision of education, particularly in economically developing countries, is due to the following concerns with market provision:

(a) private provision would result in under-provision of schooling because of externalities which are social as well as individual;

(b) 'merit goods' such as education may be under-supplied if left to the market;

(c) investment in education has a long gestation period which the market may not be able to adapt to, leading to inefficiency;

(d) a concern with economies of scale that mass provision can meet;

(e) *increased equity costs* since the private purchase of schooling is beyond the means of the disadvantaged;

(f) further *aggravating household cost-benefit analyses* which may compel even lower participation by disadvantaged communities; and

(g) *low private demand particularly for disadvantaged groups* facing social and cultural barriers to enrolment, calling for increased subsidies, and not increased costs.

(emphases added) (pp. 1–2)

While not specifically focusing on private sectors serving disadvantaged groups, earlier studies have examined the prevalence of private provision at all education levels in economically developing countries. James (1993) noted that:

1 there were systematically higher proportions of private secondary school enrolments in economically developing countries compared with more economically advantaged countries, and

2 there was a seemingly random distribution of private and public enrolment in economically developing countries at a given educational level and state of development.

(p. 574)

James's data showed that the percentage of enrolment in private primary schools in economically developing countries ranged from 100 per cent in Lesotho to one per cent in Algeria and Kenya, and 25 per cent for India.[5] The spectrum for more economically advantaged countries ranged from one per cent in Japan and Sweden, with England and Wales at 22 per cent, and the USA at 10 per cent. Thus, according to James's data, India shared a larger percentage of its enrolment at the primary level in private schools than England and Wales and the USA.

Such findings have led some researchers, mainly economists, to focus on the relative efficiency of public and private schools in economically developing countries (Cox and Jimenez, 1990; James *et al.*, 1996; Jimenez *et al.*, 1989; Jimenez *et al.*, 1991; Jimenez *et al.*, 1988; Salmi, 2000). Proponents of the expansion of private schooling counter arguments for further public education by insisting that private schools are more cost-effective, leading to greater efficiency of the education sector as a whole. Another argument is that private expansion will allow countries with constrained public resources to meet increasing educational demand. Some studies on India have looked at the possibility of the private sector meeting increased educational demand in view of universal elementary education goals (De *et al.*, 2002; Mehrotra *et al.*, 2005; Tilak and Sudarshan, 2001).

More recently, studies have reported on private models of schooling specifically for disadvantaged groups in a number of economically developing countries. While such private schooling models may be different from the LFP sector in India, reports on Ghana, Kenya, and Nigeria (Tooley and Dixon, forthcoming), Haiti (Salmi, 2000), Indonesia (Bangay, forthcoming), and Pakistan (Alderman *et al.*, 2001; Kim *et al.*, 1999) examine expanding private sectors for disadvantaged groups. This suggests the merits of analysing such forms of private provision to ascertain its different models, the school choice behaviours of disadvantaged households, and the implications for EFA targets in wider international and national policy contexts.

While some studies on private schooling in India acknowledge the heterogeneity of the private sector (De *et al.*, 2002; Kingdon, 1996a; 1999b; Panchamukhi and Mehrotra, 2005; Tilak and Sudarshan, 2001), assumptions about the nature of private schooling in most of the published literature on schooling in economically developing countries are less clear. In fact, Colclough (1993) noted that most studies suffer from two inadequacies: the first is the paucity of data available, not allowing a full analysis of value-added elements; and the second is that they do not account for the heterogeneity in the private and public sectors. Thus, claims of relative efficiency or effectiveness (Cox and Jimenez, 1990; James *et al.*, 1996; Jimenez *et al.*, 1988; Jimenez *et al.*, 1989; Jimenez *et al.*, 1991) should be treated with caution.

The present analysis would add two further caveats taking the LFP sector into consideration. First, as previously noted, value-added measures from the school effectiveness framework without quality assessments based on an expanded set of indicators encompassing 'something more' (e.g. similar to those used by Lloyd *et al.*, 2000, and Lloyd *et al.*, 2003), will add to incomplete analyses of the private sector (both at different levels of schooling and how it is accessed by different socio-economic groups). Second, merely noting the heterogeneity of the private sector is an insufficient condition for a more complete analysis. A more focused and detailed emphasis on models of private schooling accessed by disadvantaged groups, such as the LFP sector, is necessary for two reasons.

First, it is necessary to distance the private schooling debate from traditional assumptions that it is accessed only by the elite, and to focus on the schooling preferences and choices of disadvantaged groups as they actually are. Second, it is necessary to refocus the debate on how the existence of private schooling for disadvantaged groups can compel and challenge existing public systems to better meet the needs of the most disadvantaged.

Private schooling models targeting disadvantaged groups, and specifically the LFP sector as examined in this study, seem to challenge Colclough's assertions (e), (f), and (g) (see p. 153 above) critiquing private provision. Increased equity costs due to the inaccessibility of private schooling for disadvantaged groups, a low private demand for it among this group, and aggravated household cost-benefit analyses, seem to apply only to a limited extent in relation to the LFP sector. This is because these critiques are made on assumptions that apply to high-fee models of private schooling and their associated concerns for disadvantaged groups in contexts where low-fee options do not exist. Furthermore, they do not take into account the fact that low-fee models exist in the context of a (perceived or real) malfunctioning state sector, which itself raises issues of equity, low demand, and aggravated household costs for disadvantaged groups.

For example, the share of enrolment in government schools[6] in Uttar Pradesh shows that it ranked among the lowest in the country at every school level (see Table 11.4). While not focused on the LFP sector, results from the 2003 DISE report (Mehta, 2004) show that the share of recognised private unaided schools in Uttar Pradesh ranked the third highest of the eighteen states at 10.1 per cent of its total schools. Furthermore, contrary to traditional assumptions, the prevalence of private unaided schools was not just an urban phenomenon. The report showed that 7.9 per cent of all rural schools in Uttar Pradesh were private unaided, again ranking the third highest.

This points to the necessity of reassessing traditional analyses of private schooling applied to economically developing countries. While certain elements of older analyses may be applicable, studies working from a new set of assumptions must be undertaken to adequately assess the possible contribution of private schooling and its interface with public provision (see papers in Srivastava and Walford, forthcoming, for some new studies on this topic).

Table 11.4 School and enrolment-based indicators on private unaided and government schools

	National (All Districts)	*Uttar Pradesh*	*Rank[a]*
School-Related Indicators			
Total number of schools	853,601	119,443	1
% Private unaided schools	6.74	10.12	3
% Share of schools in rural areas	87.0	91.51	4
% Share of private unaided schools in rural areas	4.6	7.93	3
Enrolment-Based Indicators			
(% Enrolment in government schools)			
Primary	89.9	87.5	13
Primary with junior	77.6	17.4	18
Primary with junior and secondary/high school	34.9	23.4	17
Junior only	80	77.7	11
Junior with secondary/higher secondary	60.1	14.7	15

Source: Extracted from Mehta (2004)

Note
a Ranks were added by the researcher to enable general comparison with the eighteen states in the DISE report.

Implications for the schooling of disadvantaged groups

Increased marketisation and privatisation targeted to groups with historically low participation rates in schooling necessitate examination in the context of EFA and quality provision, because they question the State's fundamental responsibility of upholding children's universal right to education. In the case of India, this is mandated by its own constitution. The issue, scaffolded by the EFA framework of increased access, equity, and quality in schooling, is one of an outward affirmation to this commitment by the State on the one hand, and increased private provision on the other. Since LFP schools as defined in this study target groups earning between one to two dollars a day, the question is: what does the increased segmentation of the schooling market mean for them and for those earning less than one dollar a day?

First, while traditional arguments favouring privatisation hinge on raising the public sector's efficiency and effectiveness through increased competition, they do not take into consideration systems where the public sector has no incentive to compete. While EFA Goals 2, 5, and 6 focus on increasing the quality of schooling offered to vulnerable groups (UNESCO, 2000), most international and national funding mechanisms are not contingent on quality performance. Thus, if the state sector continues to be funded under the international EFA banner without conditionalities of quality improvement, it has no incentive to compete to increase its performance. Ultimately, in this scenario, LFP schools have to be only marginally better than 'malfunctioning' state schools (e.g. Balagopalan and Subrahmanian, 2003; Datt, 2002; PROBE, 1999) to be considered 'better', leaving disadvantaged parents with few real options for quality schooling.

Second, according to Hirschman's (1970) classic identification of 'inert' and 'alert' clients, those accessing the LFP sector would be classed as quality-conscious alert clients in relation to the state sector. This has fundamental implications for the future of children whose parents either cannot or do not access the LFP sector. If the state sector is as malfunctioning as it is perceived and documented to be, and there is cream-skimming of clients from among disadvantaged groups, then the future of schooling for the most disadvantaged does not seem promising. From this perspective, while in the short run, the LFP sector may be desirable for some disadvantaged children faced with only malfunctioning local state schools, in the long run it is likely to be highly iniquitous for the most disadvantaged if it has no recuperation effect for the state sector.

Further analyses should examine the extent to which the LFP sector is considered by the State when developing policies or initiatives for disadvantaged groups. If these groups are increasingly attracted to the LFP sector, it is critical to understand its implications for EFA. Currently, EFA initiatives such as *Sarva Shiksha Abhiyan* are either centrally-sponsored or have a strong external funding component to them. Dyer (2000) notes that such centrally-funded schemes are: 'A very powerful way of setting the direction of the development of education' (p. 19). It seems that the government's response to EFA goals has been a push towards meeting targets of quantity rather than a combined approach addressing quality issues as well. This would best be done by adequately assessing disadvantaged households' changing schooling choices and behaviours. For example, even though Uttar Pradesh's *School Chalo Abhiyan!* was geared to attract children who had never been enrolled, it did not incorporate disadvantaged parents' conceptions of quality or assess their reasons for not enrolling their children in the first place (see Srivastava, 2006). Thus, government officials in this study estimated that half of those enrolled through the initiative in June and July withdrew by September.

It is clear that systematic quality comparisons between state and LFP schools are needed. If these are the two sectors most accessible to disadvantaged groups then it is crucial to assess what, if anything extra, parents get for their money. Such inquiries should assess 'quality' by employing indicators beyond expenditure and results to include, for example, students' gendered school experiences, facilitation of home-school support, and teachers' attitudes. Furthermore, future statistical analyses should include data on unrecognised private unaided schools and disaggregate private unaided sector data by level of fees, so that more detailed characterisations can be made of the LFP sector. As the analysis here showed, available statistical data are insufficient as, at best, they can approximate growth, enrolment, and expenditure trends in the recognised private unaided sector as a whole without specifying variation across this heterogeneous sector, or pinpointing the LFP sector's position. Given the changing nature of the schooling arena, research agendas excluding such analyses will ignore the reality of schooling conditions for the most disadvantaged children.

Notes

1 This refers to elementary education enrolments (primary and junior) in recognised and unrecognised private unaided schools through a survey carried out in eight states by Panchamukhi & Mehrotra for UNICEF. They compared UNICEF survey figures with data for a further eight states from the 1998 National Sample Survey Organisation household survey.

2 Since the time of writing two more DISE reports have been published (see Mehta 2005a; Mehta, 2006) covering more states. Also, since the time of writing, the first report on unrecognised schools has been released using 2005 DISE data for Punjab (see Mehta, 2005b).

3 These are para-teachers on ten-month temporary contracts qualified at higher secondary level and provided with one month's training. In principle, *shiksha mitra* are hired to teach children in the lower primary grades.

4 Local bodies are institutions developed for local governance at district, sub-district, and village levels created under the Indian Constitution.

5 James's data do not specify whether the reported private sector corresponded to private-aided plus private unaided schools, and, further, whether it includes recognised and unrecognised schools. The corresponding figures for secondary enrolment in the study are 52% for India, 16% for England and Wales, and 9% for the USA. Note the big jump in private enrolment from primary to secondary in India and the larger gain over England and Wales and the USA. Tooley's (1999) data show 42% enrolment at the private secondary level. He claims this to be from middle schools including private-aided, but does not state whether or not unrecognised schools are included. 'Middle schools' are also not defined.

6 The term 'government schools' is used when speaking of secondary data reporting on public sectors since it is unclear if the data combine figures from schools run by state governments and local bodies and Central Government.

References

Aggarwal, Y. (2002) *Regaining Lost Opportunity: The Malaise of School Inefficiency.* New Delhi: Department of Elementary Education and Literacy and NIEPA. Available online at: http://www.dpepmis.org/downloads/lost.pdf (retrieved August 20, 2004).

Aggarwal, Y. (undated) *Quality Concerns in Primary Education in India: Where is the Problem?* Available online at: www.dpepmis.org/downloads/quality1.pdf (retrieved August 23, 2004).

Alderman, H., Orazem, P. F. and Paterno, E. M. (2001) 'School quality, school cost, and the public/private school choices of low-income households in Pakistan', *Journal of Human Resources,* 36, 304–326.

Balagopalan, S. (2004) 'Free and Compulsory Education Bill 2004', *Economic and Political Weekly,* August 7, 2004.

Balagopalan, S. and Subrahmanian, R. (2003) 'Dalit and Adivasi children in schools: some preliminary research themes and findings', *IDS Bulletin,* 34, 43–54.

Bangay, C. (forthcoming) 'Cinderella or ugly sister?: What role for non-state education provision in developing countries?', in Srivastava, P. and Walford, G. (eds), *Private Schooling in Less Economically Developed Countries: Asian and African Perspectives.* Didcot, Oxon: Symposium Books.

Bashir, S. (1994) 'Public versus Private in Primary Education: comparisons of school effectiveness and costs in Tamil Nadu', unpublished PhD thesis, London School of Economics and Political Science, University of London, UK.

Colclough, C. (1993) *Education and the Market: Which Parts of the Neo-Liberal Solution are Correct?* Florence: International Development Centre, Innocenti Occasional Papers, Economic Policy Series, No. 37.

Cox, D., and Jimenez, E. (1990) 'The relative effectiveness of private and public schools: evidence from two developing countries', *Journal of Development Economics,* 34, 99–122.

Datt, R. (2002) 'Educational policy and equity', in Rao, K. (ed.) *Educational Policies in India: Analysis and Review of Promise and Performance.* New Delhi: National Institute of Educational Planning and Administration (NIEPA).

De, A., Majumdar, M., Noronha, C. and Samson, M. (2002) 'Private schools and universal elementary education', in Govinda, R. (ed.) *India Education Report: A Profile of Basic Education.* New Delhi: Oxford University Press.

Drèze, J. and Sen, A.K. (1995) *India: Economic Development and Social Opportunity.* New Delhi: Oxford University Press.

Duraisamy, P., James, E., Lane, J. and Tan, J. P. (1997) *Is there a Quantity-Quality Trade-Off as Enrolments Increase? Evidence from Tamil Nadu, India.* Washington, DC: World Bank, Working Paper No. 1768. Available online at: http.//econ.worldbank.org/docs/466.pdf. (retrieved August 17, 2004).

Dyer, C. (2000) *Operation Blackboard: Policy Implementation in Indian Elementary Education.* Didcot, Oxon: Symposium Books.

Government of India (1996) *National Policy on Education 1986: Programme of Action 1992.* New Delhi: Department of Education, Ministry of Human Resource Development.

Government of India (1998) *National Policy on Education 1986: As Modified in 1992 with National Policy on Education 1968.* New Delhi: Department of Education, Ministry of Human Resource Development.

Government of India (2001) 'State of literacy', *Provisional Population Totals:* Census of India. India, Series 1, Paper 1 of 2001. Web edition. Available online at: www.censusindia.net/results.html (retrieved September 2003).

Government of India (2002a) *Annual Report 2001–2002* New Delhi: Department of Elementary Education and Literacy, and Department of Secondary and Higher Education, Ministry of Human Resource Development.

Government of India (2004) *Free and Compulsory Education Bill, 2004, Revised.* Available online at:http://www.education.nic.in/htmlweb/ssa/free_compulsory _edu_ bill_2004_intro.htm. (retrieved July 20, 2004).

Govinda, R. (2002) 'Providing education for all in India', in Govinda, R. (ed.) *India Education Report: A Profile of Basic Education,* New Delhi: Oxford University Press.

Hirschman, A. O. (1970) *Exit, Voice, and Loyalty: Responses to Decline in Firms, Organizations, and States.* Cambridge, MA: Harvard University Press.

James, E. (1993) 'Why do different countries choose a different public–private mix of educational services?' *The Journal of Human Resources,* 28, pp. 571–592.

James, E., King, E. M. and Suryadi, A. (1996) 'Finance, management, and costs of public and private schools in Indonesia', *Economics of Education Review,* 15, 387–398.

Jimenez, E., Lockheed, M., Luna, E. and Paqueo, V. (1989) *School Effects and Costs for Private and Public Schools in the Dominican Republic.* Washington, DC: World Bank, Policy Planning and Research Working Papers 288.

Jimenez, E., Lockheed, M. and Paqueo, V. (1991) 'The relative efficiency of private and public schools in developing countries', *World Bank Research Observer,* 6, 205–218.

Jimenez, E., Lockheed, M., and Wattanawaha, N. (1988) 'The relative efficiency of private and public schools: the case of Thailand', *The World Bank Economic Review,* 2, 139–164.

Kingdon, G. G. (1996a) 'Private schooling in India: size, nature, and equity effects', *Economic and Political Weekly,* 31, 3306–3314.

Kingdon, G. (1996b) 'The quality and efficiency of private and public education: a case study of urban India', *Oxford Bulletin of Economics and Statistics,* 58, 57–82.

Kim, J., Alderman, H. and Orazem, P. (1999) 'Can private schools subsidies increase schooling for the poor? The Quetta Urban Fellowship Program', *World Bank Economic Review,* 13, 443–65.

Lloyd, C. B., El Tawila, S., Clark, W. H. and Mensch, B. (2003) 'The impact of educational quality on school exit in Egypt', *Comparative Education Review,* 47, 444–467.

Lloyd, C. B., Mensch, B. and Clark, W. H. (2000) 'The effects of primary school quality on school dropout among Kenyan girls and boys', *Comparative Education Review,* 44, 113–147.

Majumdar, M. and Vaidyanathan, A. (1995) *The Role of Private Sector Education in India: Current Trends and New Priorities.* Thiruvanthapuram, Kerala: Centre for Development Studies.

Mehrotra, S. and Srivastava, R. (2005) 'Elementary schools in India: producing human capital to unleash human capabilities and economic growth?', in Mehrotra, S., Panchmukhi, P., Srivastava, R. and Srivastava, R. *Universalizing Elementary Education in India: Uncaging the 'Tiger' Economy.* New Delhi: Oxford University Press.

Mehrotra, S., Panchmukhi, P., Srivastava., R. and Srivastava, R. (2005) *Universalizing Elementary Education in India: Uncaging the 'Tiger' Economy.* New Delhi: Oxford University Press.

Mehta, A. C. (1998) *Education for All in India: Enrolment Projections.* New Delhi: NIEPA.

Mehta, A. C. (2004) *Elementary Education in India, Where Do We Stand? Analytical Report 2003.* New Delhi: Government of India and NIEPA. Available online at: www.eduinfoindia.net/Report/ar2003/ar2003.pdf. (retrieved 23 August, 2004).

Mehta, A. C. (2005a) *Elementary Education in India: Analytical Report 2004.* New Delhi: NIEPA. Available online at: http://www.dpepmis.org/webpages/reports &studies.htm.

Mehta, A. C. (2005b) *Elementary Education in Unrecognised Schools in India: A Study of Punjab on 2005 DISE Data.* New Delhi: NIEPA. Available online at: http://educationforallinindia.com/study%20on%20unrecognised%20schools%20in%20india%20a%20case%20study%20of%20punjab.pdf.

Mehta, A. C. (2006). *Elementary Education in India: progress towards UEE. Analytical Report 2004–2005.* New Delhi: NIEPA. Available online at: http://www.dpep mis.org/Downloads/AnalyticalReport2005/AnalyticalReport2005.pdf.

Panchamukhi, P. R. and Mehrotra, S. (2005) 'Assessing public and private provision of elementary education in India', in Mehrotra, S., Panchmukhi, P., Srivastava, R. and Srivastava, R., *Universalizing Elementary Education in India: Uncaging the 'Tiger' Economy.* New Delhi: Oxford University Press.

PROBE Team (1999) *Public Report on Basic Education in India.* New Delhi: Oxford University Press.

Rao, K. S. (2002) 'National education policy: analysis and review', in Rao, K. (ed.) *Educational Policies in India: Analysis and Review of Promise and Performance.* New Delhi: NIEPA.

Salmi, J. (2000) 'Equity and quality in private education: the Haitian Paradox', *Compare,* 30, 163–178.

Srivastava, P. (2005) *The Business of Schooling: the School Choice Processes, Markets, and Institutions Governing Low-Fee Private Schooling for Disadvantaged Groups in India,* unpublished doctoral thesis, University of Oxford, UK.

Srivastava, P. (2006) 'Private schooling and mental models about girls' schooling in India', *Compare,* 36 (4), 497–514.

Srivastava, P. (forthcoming) 'For philanthropy or profit? The management and operation of low-fee private schools', in Srivastava, P. and Walford, G. (eds) *Private Schooling in Less Economically Developed Countries: Asian and African Perspectives.* Didcot, Oxon: Symposium Books.

Srivastava, P. and Walford, G. (eds) (forthcoming) *Private Schooling in Less Economically Developed Countries: Asian and African Perspectives.* Didcot, Oxon: Symposium Books.

Stephens, D. (1991) 'The quality of primary education in developing countries: who defines and who decides?', *Comparative Education,* 27, 223–233.

Tilak, J. B. G. and Sudarshan, R. (2001) *Private Schooling in Rural India.* New Delhi: National Centre for Educational Research and Training, Working Paper Series, No. 76.

Tooley, J. and Dixon, P. (2005) 'An inspector calls: the regulation of 'budget' private schools in Hyderabad, Andhra Pradesh, India', *International Journal of Educational Development,* 25 (3), 269–285.

Tooley, J. and Dixon, P. (forthcoming) 'Private education for low-income families: results from a global research project', in Srivastava, P. and Walford, G. (eds) *Private Schooling in Less Economically Developed Countries: Asian and African Perspectives.* Didcot, Oxon: Symposium Books.

UNESCO (2000) *The Dakar Framework for Action. Education for All: Meeting Our Collective Needs.* Paris: UNESCO. Available online at: http://unesdoc.unesco.org/images /0012/001211/121147e.pdf.

United Provinces District Boards Primary Education Act, 1926 (UPDBPEA) (2001) in R. K. Jain (ed.), *H. S. Nigam's Uttar Pradesh Education Manual: An Encyclopaedia of Education Laws in U P from Primary Education to Higher Education* (5th ed.), pp. 628–633. Allahabad, India: Alia Law Agency.

Uttar Pradesh Primary Education Act, 1919 (UPPEA) (2001) in R. K. Jain (ed.) *H. S. Nigam's Uttar Pradesh Education Manual: An Encyclopaedia of Education Laws in U.P. from Primary Education to Higher Education* (5th ed.) pp. 623–627. Allahabad, India: Alia Law Agency.

12 Inclusive education for working children and street children in India

Mohammed Akhtar Siddiqui

Introduction

Progress in society affects individuals differently and unequally. For many, the globalizing economy can herald substantial changes in countries such as India, with improvement in the lives of citizens through increased access to socio-economic opportunities. But for substantial numbers in the developing world increased economic prosperity remains an irrelevant process and sometimes for minority groups it even enhances difficulties and hardships, particularly for their children. The very poor remain very poor, in a country where the majority may become moderately prosperous. The children of the very poor are the most vulnerable, and continue to face the most difficult circumstances in the form of hunger, poverty, insecurity, high infant and child mortality, illiteracy, and exploitation and abuse of varied kinds. India is no exception, and the most vulnerable include working children, street children, those living in slums and resettlement colonies, children of sex workers, children of prisoners, children living in institutions, and children of construction workers and other migrant labourers.

Fast-paced and unplanned urbanization in the recent past, often without commensurate increase in services, has only multiplied the numbers of these children and further aggravated their sufferings, particularly in terms of their general educational deprivation. It is they who deserve the most immediate attention of the planners, administrators and educationists. In this chapter I confine an overview to educational rights and deprivations of the two largest and most important sections of deprived Indian children, namely, working children and street children. Despite some policy declarations in the past for out-of-school children, the Government has not been able to ensure their constitutionally guaranteed right to education. In democratic India the State has a constitutional duty to ensure that they are able to develop into healthy, efficient, responsible and productive adults, and thus may effectively contribute to the development and healthy survival of society in a highly competitive globalized world. Education is the single most powerful medium that can help people achieve this cherished goal.

However, it is not the need for growth and development of society alone which calls for proper attention to the education and training of its children and future citizens; rather, in a civilized world, it is the human entity of the child who in his or her own right deserves full and equitable attention by society's institutions. Today's human rights-conscious society is obliged to recognize this aspiration and the fundamental right of each child in terms of education and social welfare. Neglect of children not only arrests their growth and development but that of the nation as a whole (Bhagwati *et al.*, 1987).

Indian society, like many others, has failed to fully realize these twin values associated with the optimizing of children's development, and this is evident from the fact that in India there are perhaps the highest number of out-of-school children in the world, most of whom are working or street children, at the mercy of the exploitative adult world that surrounds them. In commenting on this situation, Justice Krishna Iyer appropriately quoted the Nobel Laureate Gabriel Mistal of Chile, who says that:

> Many of the things we need can wait. But the child cannot. Right now is the time his bones are being formed, his blood is being made and his sense being developed. To him, we cannot answer, 'Tomorrow'. His name is 'Today'.
>
> (Iyer, 1979)

Elementary Education for All

As a democratic society, having subscribed to the principles of equality and social justice, the Indian State over the past five decades has been consciously trying to provide education to the masses and particularly elementary education to all children aged six to fourteen. The first notable step was taken by the Constitution framers in 1950 by incorporating Article 45 in the Directive Principles to the State Policy of the Constitution which declared that 'the State shall endeavour to provide within a period of ten years from this Constitution for free and compulsory education to all children until the age of 14 years'. To realize this constitutional goal and to universalize elementary education, massive efforts were launched. One finds that in the period from 1950 to 2001 there has been a tremendous growth in elementary education: in 1950 22.3 million children aged six to fourteen years were enrolled in schools; by 2001–02 that figure had risen to 158.7 million, a quantum leap in as far as enrolments are concerned.

But this 'quantum leap' is illusory: when one compares enrolment figures with completion rates of elementary education, the real picture emerges. According to the Government of India enrolment for grades one to eight in September 2003 was initially 84.9 per cent of the eligible population, but this was seriously marred by a high dropout rate of the order of 52.2 per cent in 2003. That a major part of this dropout occurred in grades one to three only makes the situation more serious. Thus by the age of fourteen more than half of all children have dropped out of school – many well before the nominal school-leaving age. Either they have never joined any school or they have dropped out permanently.

Out-of-school-children

It is difficult to say with certainty how many children below fourteen years have dropped out of school. Lieten (2006) points out that the estimates of Government and NGOs vary. According to government figures the number has decreased over the past two decades from approximately 21 million in 1980 to 9 million in the year 2000. Lieten estimates, however, that in India around 80 million children who have not been counted in government child labour statistics do not go to school. They have either never attended school or have dropped out permanently. These children cannot be found amongst the statistics of working children nor amongst the statistics of school-going children.

According to the Ministry of Human Resources Development (MHRD) statistics for 2000–01, at least 24 million children in the age group six to fourteen years are out of school, of whom 60 per cent are girls. The National Plan of Action for Education for All (EFA), formulated in 2003, estimated the number of out-of-school children in this age group at 35 million (Kanth, 2005).

Zutshi (2000) has computed the out-of-school children population in the age group five to fourteen years, using population projections for 2000 prepared by the Expert Committee of Census of India for this purpose. According to the census of 1991, India recorded a child population (five to fourteen years) of 209.98 million. Of these, around a half (105.72 million) were out of school. The estimated child population in the year 2000 was 242.11 million. It was then projected that the proportion of out-of-school children in 2000 would come down from 50 per cent to about 30 per cent, and so the estimated population of out-of-school children would be some 72.63 millions (Zutshi, 2000).

Out-of-school children include those who stay at home to care for cattle, look after younger children, collect firewood, work in fields, cottage industries, restaurants, roadside tea stalls, motor mechanics' workshops, or as domestic servants in middle class homes. They may also become prostitutes or live as street children, begging or picking rags and bottles from trash for resale. Many are bonded labourers and work for local land owners (Weiner, 1991).

Those children in urban areas endure the most difficult circumstances. The urban population has grown from 159 million in 1981 to about 315 million in 2005, one-third of whom live below the poverty line, and lack access to basic facilities and services. It was also estimated that in 2005, children aged six to fourteen in urban areas would number 65 to 70 millions, of whom almost 20 millions were children experiencing extreme poverty (Kaushik, 2005). However, the National Plan of Action for Education for All (EFA) (2003) puts the figure of out-of-school children in India in the age group six to fourteen years at 35 million, with 10 million of them being in urban areas. Even if this most conservative estimate of out-of-school children is accepted, the magnitude of the problem is profound.

The problem of large populations of street and working children is accentuated by other problems of continued rural to urban migration: the mushrooming of slums and unauthorised habitations in subhuman conditions; gross socio-economic inequalities; exploitation of various kinds; slow and unbalanced development of the disadvantaged classes and their children; lack of helping resources and their unequal distribution; and an inefficient and haphazard management of educational programmes. Extreme poverty, the main reason behind expanding slums and urban populations, remains largely responsible for the increase in the population of street and working children.

A substantial section of out-of-school children consists of those who live in the most difficult circumstances, and they need more urgent and special attention. They include working children, street children, children in slums and resettlement habitations, children of sex workers and prisoners, children of construction workers who live within the shell of the buildings they construct, and migrant labourers.

Working children

Defining 'working children' is problematic. Experts have tried to draw a distinction between household work and economic labour, since some activities of the child may not fall in the category of labour but may still be called work within a household where economic activity takes place. Some experts have classified working children into six categories, based on distinct features of activities children are made to undertake.

These include:

1   domestic work
2   non-domestic and non-monetary work
3   bonded child labour
4   external wage labour
5   commercial and sexual exploitation
6   child combatants – a new type of child exploitation and a difficult challenge of the current century.

(Kaushik, 2005)

However, Burra (1995) has simplified this classification into four categories:

1   those children who work in factories, workshops and mines
2   children under bondage in agriculture or industry
3   street children mostly found in service sectors
4   children who work as a part of family in agriculture, industry or home-based work.

However, many children may be classified in more than one category.

There are several determinants of that status of 'working children' which may include demand- as well as supply-side factors. Abject poverty of the family is the main reason. Poor people tend to send their children out for work to supplement a meagre family income. As many as 25 per cent of people living below the poverty line force or require their children to engage in economic activities. Poverty also leads to pledging of children as security for a loan. Socio-economic factors like female literacy, family size, adult wage rates, diversification of the rural economy, and female work participation are also important determinants of child labour. Lack of educational facilities in an area may also increase the supply of child labour. Working children or child labour may increase in number due to demand-side factors also. Employers' preference for children due to their favourable physical features, low wages, ease of discipline, etc. are demand-side factors. Economic development and better access to schooling is a supply-side factor which may reduce the supply of working children.

Child labour deprives children of the opportunity for education, play and recreation which in turn arrests their physical as well as emotional growth, and thwarts their preparation for adult responsibilities. It also causes physical hazards to children. Absolute abolition of child labour in India will, however, take many years, and requires a multi-pronged strategy. There are two schools of thought on how to address the problem. The first school argues that it is the abject poverty of parents which is the main cause of children being withdrawn from school, and entered into the labour market. Proponents of this school advocate regulation of labour, progressively eliminating the possibility of child workers. Thus by regulation of employment in selected industries, improving working conditions, reducing working hours, ensuring minimum wages, and providing adjunct health and education facilities, the plight of child labourers can be eased. These advocates feel that sudden elimination of child labour would further bring down the standard of living of already impoverished families.

In contrast, the second school argues that it is due to lack of both access to quality education and rewards for education that children are working; therefore child labour should be completely prohibited, with steps taken to provide compulsory primary education of adequate quality. This group does not distinguish between child labour and child work and also does not believe in non-formal education, which it sees as an illusion. In its policy statement, a multi-focused strategy is the best answer to the problem of child labour which includes legislative measures, educational interventions and social mobilization for children's rights.

Street children

The term 'street children' identifies those who spend considerable time on the street in connection with a job, or without it. They live on the street with or without a family. UNICEF (2004) has defined street children as those for whom the street (in the widest sense of the term, i.e. unoccupied dwellings, wastelands, etc.) has become their real home, a situation where there is no protection, supervision or direction from responsible adults.

UNICEF has classified street children into three groups:

1   Children on the street: these are children who have family connections of a more or less regular nature. Their focus in life is still the home. Most of them

return home at the end of each working day and have a sense of belonging to the local community where their home is situated.

2   Children of the street: this group is smaller but more complex. Children in this group see the street as their home, and it is there that they seek shelter, food and a sense of family among companions. Family ties exist but are remote and their former home is visited infrequently.

3   Abandoned children: this group may appear to be a part of the second group and in daily activities is fairly indistinguishable from it. However, by virtue of having severed all ties with a biological family these children are entirely on their own not just for material but also for psychological survival.

These three categories are found particularly in all developing countries but more so in South Asian countries including India. Children of the street have been divided further into two groups, firstly, 'Roofless' who live and work on the street (i.e. in abandoned buildings, under bridges, railway stations, bus stands, in door-ways, in public parks) yet maintain occasional contacts with the families who may live in the same city or in other cities or rural areas. They see the street as their home. The second type, 'Rootless' children, live and work on the street (in the widest sense of the term) and have no family contacts.

Street children are susceptible to drug/alcoholic addiction including the use of inhalants, such as cobbler's glue, correction fluid, gold/silver spray paint, nail pol-ish, rubber cement and gasoline – which offer them an escape from reality and take away hunger. In return they risk a host of physical and psychological prob-lems including hallucinations, pulmonary oedema, kidney failure and irreversible brain damage. They sniff glue because it also gives them the 'courage' to steal and engage in survival sex. These children are routinely detained illegally, beaten and tortured by police and by employers, to extract maximum labour out of them (Human Rights Watch Asia, 1998). This is a consequence of several factors including the inadequacy and non-implementation of legal safeguards, and the level of discretion that police enjoy in administering welfare legislation. These children invite such extreme reaction only because they are viewed as vagrants and criminals. No doubt, street children are sometimes involved in petty thefts, drug trafficking, prostitution and other criminal activities, yet very few attempts are made to examine the root cause for such activities, or to provide care and reha-bilitation. They are, on the contrary, easy targets of police atrocities. These children are young, small, poor, alone and ignorant of their rights and often have no family members who will come to their rescue or defence. It does not take much time or effort to detain and beat a child to deter any formal complaints about these atrocities.

The issues and needs of the children living on the streets of the cities have attracted worldwide attention. The most effective response to this, of course, is prevention through general support to families in poverty, creating broad-based awareness among the parents and society, addressing the factors underlying fam-ily disintegration. Other preventive measures may include employment for adults, support in times of crisis, strong childcare programmes, relevant schooling, and efforts to address the roots of domestic violence to keep families intact so that they are able to fulfil responsibilities towards their children.

Children on the street need psychological support, relationships and a role in society, along with other basic issues related to their survival, security, and protection of their civil rights like food, money, shelter, clothes, health care, and education. They should have the right to live in dignity, to health and education, to protection from abuse, exploitation and violence and to voice their own feelings and sufferings (UN, 1998).

Street children (as opposed to working children) are an exclusively urban phenomenon. There are no exact data on their numbers. According to one estimate 18 million children live and/or labour in the streets of Indian urban centres (*International Herald Tribune,* 26th June 2005). An updated estimate by the UN (2003) indicates that there is a population of street children worldwide numbering 150 million in the age range of three to eighteen years, with the number still rising. About 40 per cent of them are homeless. The other 60 per cent work on the street to support their families. They are unable to attend school. India has the dubious distinction of having the largest population of street children. In urban areas alone there are at least 11 million children 'on the streets' (Kaur, 2003).

This survey in ten cities of India found that a sizable number of street children includes abandoned children who are the direct consequence of the rapid advance of industrial growth and persistent rural and urban poverty. The cycle of events leading to abandonment begins from the migration of a family to the city, abject urban poverty, then disintegration of the family, which often begins with the father's abandonment of it. Pressures on deserted mothers in maintaining their families, dependency of abandoned mothers on new partners, and rejection of these mothers' children by stepfathers can lead to the final abandonment of children.

India's street children, both 'on the street' and 'of the street' face a myriad problems which include lack of religious socialization and moral development, malnutrition and exposure to infection, drug addiction, loss of personal development, forced involvement in theft, pickpocketing, child sex, being trapped in crime rings, early entry into labour markets with physical, mental and psychological hazards, high risk of catching HIV/AIDS, police torture and illegal confinements (Zutshi, 2000). During the course of illegal detention and torture, police even murder some of these children without being subject to any action or censure. In fact the most common complaint mentioned by the street children is that they live in a state of continual fear of the police who often round them up, lock them up and torture them on the pretext of suspicion which is always unfounded. Police do this to fill the 'quota' they are expected to achieve (Kaur, 2003). According to estimates, 50 to 80 per cent of these street children are school dropouts and that too before completing their first grade. Police detention also results in dropout of street children who are, against the odds, enrolled in primary schools (Lieten, 2006).

Like working children, street children also need both preventive and supportive initiatives. The most important need is to prevent them from falling into the trap of the street by alleviating the poverty of their families, addressing the factors underlying family disintegration that leads to life on the street with preventive measures leading to some check on child abandonment. These preventive measures include: increased adult employment, support in times of crises, containing domestic violence, strong child care programmes, and effective schooling. Supportive steps for those who are already street children include security and health care of children on the street, protection of their civil rights, respecting their sense of freedom as members of street peer groups and providing them with a secure and supportive environment. Their rehabilitation is a greater need than restoration to their family which may often not be willing to take back their responsibility, and the children may also have developed close peer bonds on the street. Their education, health care, shelter and regular counselling, in fact, need greater and more immediate attention.

Educational issues

The educational issues of street and working children in the Indian context may be analyzed in various ways:

1   Children's access to good quality and relevant education remains a lead issue.
2   There are several barriers to access, enrolment, continuation and transition from one class to another. For example, for immigrating families a persistent demand for identity proof, birth certificate, transfer certificate, etc. continues to block their children's easy access to education.
3   Quality of education is a very serious issue that is directly responsible for creating the large number of working and street children. This issue demands immediate attention. Quality of education does not merely mean providing infrastructure, water and toilet facilities, buildings, etc. The measure of quality needs to be clearly understood. Its scope extends beyond learning outcomes. Thus, quality has also to be seen in terms of students' happiness, relevance of their education, capacity building, confidence development and concrete skill development.
4   Teachers' inappropriate, indifferent and rather hostile attitude towards students is another serious issue which causes distance between them and their students and makes them less effective in guiding students in learning and personality development. There also exists a social distance between teachers and their students. Teachers generally come from middle class backgrounds, and municipal school students are generally drawn from very poor families, from lower and 'untouchable' castes.
5   There is a conspicuous absence of any teaching manual that could guide innovative ways of educating this disadvantaged group, and which is contextualized and friendly to these children.
6   These children often suffer from very low attention levels, especially those who have experienced sensory damage through addictions. It is often difficult to gain the concentration of the child for more than fifteen minutes. They

need a different pedagogic approach through which their attention can be engaged for longer teaching–learning encounters.

7   There is a high dropout rate due to high demand of child workers in the employment market, poor quality of education and educational environments, insecurity, police round-ups and irregular attendance. Work alongside studies can undermine the child's concentration and effort in school.

8   Simply rounding up 'out of school (working) children' and forcing them into formal schools results in illegal confinement, compulsion and trauma for the children involved, and their families (Reddy, 2004).

9   In mixed population schools in cities the question of the medium of instruction is raised which requires deployment of multilingual teachers. Teacher deployment and preparation is a critical issue.

10   Getting dropout students back to school after a Bridge course is a crucial but difficult task. However, getting them back to school is not enough. These mainstreamed children often tend to drop out quickly for various reasons.

11   Urban schools have diverse groups coming from different socio-economic and cultural backgrounds. A large number of street and working children in many urban slums are Dalits and Muslims. They have diverse educational needs as special groups.

12   Often the quality of learning in urban municipal schools is low due to overcrowding in classes and poor quality of teaching.

13   Street and working children need to acquire employable skills early in life owing to their peculiar socio-economic background. Yet we find that there is little or no arrangement for vocational education at elementary school level.

14   Non-formal education (NFE) and alternative schools are poorer and weaker options for children.

15   The dichotomy between private fee-paying schools and local government schools is another factor for keeping deprived urban children out of school. The former are seen as a hub of quality and better than the latter, though this may not always be true (Kaushik, 2005).

16   It is often difficult to make the child realize the importance of education for his/her development. This feeling is mainly due to the attitude of the parents towards education. Hostile attitudes of parents who would prefer their child to work the whole day rather than study for even one hour contributes to children's lack of interest in education.

In search of solutions

Several measures have been undertaken to deal with the problem of street children and working children, which include legal approaches and state policies for the more than 90 million working and street children in India (Lieten, 2006). Distinctions will have to be drawn between demand-side and supply-side factors of child labour. While on the one hand, attention will be paid to educational factors, it is equally important to pay attention to poverty alleviation, female literacy, fertility rate, adult wage rates, diversification of rural economy, female work participation rate, better opportunities for adult labour, etc. to check the supply of child labour (Dev, 2004).

Some policy advocates feel that education is a well established alternative strategy to finally deal with the problem of working and street children. But this strategy will work only if we are able to ensure full enrolment and retention in the formal education system. As long as poor enrolment and high dropout persist, there is the chance that a high proportion of child labour will persist. This is due to several reasons including non-availability of proper schools, poor and irrelevant course curriculums, lack of teachers and other relevant infrastructure like buildings and furniture, non-availability of text books, teaching material, lack of employment opportunity and of further education after completing elementary education (Aggarawal, 2004). Poor or little training of teachers, their indifferent and rather hostile attitude towards children from stigmatized groups such as Dalits, is also responsible for poor quality of education and high dropout of children. Scholars like Burra (1995), Weiner (1991) and Satgopal (2003) who believe that the answer to the problem of working and street children lies in compulsory elementary education, assert that non-formal education which implies education with work (earning while learning) is a myth as it is neither feasible nor desirable. Satgopal rightly observes that there has been a dilution of the policy commitment to the principle of 'education of equitable quality' in the last fifteen years, by instituting parallel layers of educational facilities. He feels that recourse to the multi-layered education system will only ensure maintenance of social hierarchies of class, caste, culture and gender (Satgopal, 2003). What needs to be done is to put in place the Common School system with neighbourhood schools as envisaged in the Kothari Education Commission's Report (1964–66) and committed to by Parliament in approving the national education plans in the years 1968, 1986 and 1992.

It is the pressure of increasing privatization and commercialization of education that is continuously pushing the implementation of the concept of the Common School system to the margins and delaying implementation of the principles of quality education for all (Annan, 2001). While emphasizing proper formal education, in 1994 the UNESCO Salamanca Statement and Framework for Action on Special Needs Education had resolved that ordinary schools should accommodate all children regardless of physical, intellectual, emotional, social, linguistic or other conditions (Annan, 2001).

Steps taken to address the issues

Several measures have been initiated during the last few decades to address the problems faced by children in difficult circumstances. Some have begun to see fruition. Soon after India's independence, the country invoked State Constitutional support for the cause of children's educational rights. Article 45 in the Constitution was the first manifestation in independent India of the State's will to provide elementary education to all children irrespective of their conditions and background. The recent attempt of the Ninety-third Constitutional Amendment contained in

Article 21A has strengthened the Constitutional commitment of the State, since it is now recognized that education is a fundamental right of the child. The amendment is in line with the school of thought which asserts that proper provision of universal elementary education of good quality will attract children to education and this might solve the problem of working and street children. However, there are certain other provisions of law which due to their restrictive approach are withholding the benefits of this amendment to reach all children. The Child Labour Act (1986) is one such example in point; it prohibits employment of children only in hazardous industries. The Act does not prohibit child labour completely and by implication permits child labour in 'non-hazardous industries'. Two provisions, Article 21A of Constitutional Law (1993) and the Child Labour Act of 1986, do not seem to be fully compatible; they cannot both go together so far as education of equal quality for all children is concerned.

Looking at the international minimum age standards for employment we find that they are also directly linked to schooling. The ILO Minimum Age Convention 1973 (No. 138) which was built on the ten instruments adopted before the Second World War, expresses this tradition by stating that the minimum age for entry into employment should not be less than the age of completion of compulsory schooling (UN, 2000). This link was clearly aimed at ensuring that the child's human capital is developed at least to its minimum level of potential. The ILO Minimum Age Convention 1973 (No. 138) was declared (if not enacted by many countries) on 19 June 1976. Paragraph two of its Article 2 further clarified that the minimum age of beginning full-time work shall not be less than the age of completion of compulsory schooling, and in any case not less that of fifteen years. Only in exceptional cases may the governments of the member states be allowed to specify a minimum age of fourteen years.

The UN Convention on the Rights of the Child was adopted by the General Assembly of the United Nations on 20 November 1989, and the Government of India acceded to it on 11 December 1992. Article 32 of the UN Convention recognises the right of the child to be protected from economic exploitation and from performing any work that is likely to be 'hazardous' or to interfere with the child's education, or to be harmful to the child's health or physical, mental, spiritual, moral or social development.

In the 1990s it was increasingly debated whether there were certain forms of labour that were so inhumane that they could no longer be tolerated, and so in Convention 182 of ILO in 1999 it was finally proposed that many types of child labour should be done away with immediately. Recommendation 1990 of UNICEF (UNICEF, 2005) also recommended priority attention to preventing children from undertaking hazardous work (Lieten, 2006). This suggests that even the UN does not find it immediately feasible to completely abolish child labour in all forms.

Educational policies and programmes

Educational initiatives are recognized as among the best strategy to protect working and street children from all kinds of social, moral, emotional, and economic exploitation and deprivation (Weiner, 1991). The initiatives for education of street and working children were incorporated in the National Policy on Education 1986 and its revisions carried out in 1992. In the Programme of Action (1986) it was envisaged that in order to provide special support for the education of children in urban slums, working children and children in under-served areas, hill areas and tribal areas, Non Formal Education (NFE) centres should be opened with state support from both government and NGOs. These NFE centres were to have features of organizational flexibility, relevance of curriculum, diversity in learning activities to relate them to learners' needs, and decentralization of management (MHRD, 1986). While expressing its satisfaction over the opening of 272,000 NFE centres with an enrolment of 6.8 million children in 1992, the Revised Programme of Action (RPOA) declared that NFE schemes must be strengthened further and will serve those children who cannot attend formal schools (MHRD, 1992).

The Eighth Five Year Plan also placed greater emphasis on opening non-formal education centres for out-of-school children and children with special needs. As a result, by the turn of the century the country had as many as 297,000 NFE centres covering 7.42 million children in twenty-four States and Urban Territories. Of these, 238,000 NFE centres are being run by the State Governments and 58,788 centres are being run by 816 NGOs (Kanth, 2005). In the year 2000 the Government launched a massive scheme of 'Education for All' called *Sarva Shiksha Abhiyan* (SSA) in mission mode, which also incorporated many important strategies for out-of-school children with special circumstances. It included the idea of Alternative and Innovative Education (AIE) centres, Education Guarantee Scheme centres (*Balika Shivirs*), and back-to-school camps for bridge courses for children who had dropped out of school (MHRD, 2001). A ten-year programme of University Elementary Education (UEE) was to be completed in 2010 with the hope that all children, including working and street children, would be enrolled in schools or alternative education institutions. Many alternative education centres, have been opened in rural and urban areas, but the infrastructure and teaching resources in them are not comparable with formal schools by any reckoning.

Special action for street children

In 1993 the Ministry of Social Justice and Empowerment of the Government of India launched an Integrated Programme for Street Children 'for full and wholesome development of street children without homes and family ties and for prevention of destitution and withdrawal of children from a life on the street and their placement into national mainstream'. The essential components of the programme include provision of shelter, nutrition, health care, sanitation and hygiene, safe drinking water, education, recreational facilities and protection against abuse and exploitation to destitute and neglected street children (MSJ, 1993).

The scheme includes a wide range of programmes and initiatives:

- contact programmes offering counselling, guidance and referral services to destitute and neglected children;
- establishment of 24-hour drop-in shelters for street children with facilities for night stay, safe drinking water, bathing, latrines, first aid and recreation;
- non-formal education programmes imparting literacy, numeracy and life education;
- programmes for reintegration of children with their families;
- programmes for enrolment of these children in schools including full support for subsistence, education, nutrition, recreation, etc.;
- programmes providing facilities for training in meaningful vocations;
- programmes for occupational placement;
- programmes aimed at mobilizing preventive health services and providing access to treatment facilities;
- programmes aimed at reducing the incidence of drug and substance abuse, HIV/AIDS and STIs and other chronic health disorders;
- programmes aimed at providing recreational facilities;
- programmes for capacity building of NGOs, local bodies and state government to undertake related responsibilities;
- programmes for advocacy and awareness-building on child rights.

The scheme is already operational in thirty-seven cities across the country. In 1999–2000 there were 32,451 beneficiaries of the scheme through 103 NGOs and voluntary organizations which have been provided with a 90 per cent grant. The Ministry has provided grants of 69.50 million Rupees to the NGOs to provide non-institutional and institutional support to the 32,451 street children (MSJ, 2000). Excellent though these initiatives are, they are thinly spread and inadequately funded, and reach only a small minority of street children, although further funding with US aid is promised (ILO, 2006).

Special project for working children

The Government of India passed the Prohibition of Child Labour Act in 1986 which was followed by the National Child Labour Project, begun in 1998. This initiated several action-oriented programmes in order to withdraw children from hazardous work, prevent them from entering again into the labour market and to rehabilitate them successfully. The following responsibilities under the project have to be taken up:

- effective enforcement of child labour laws;
- identification of areas for starting NFE centres through opening new schools/centres;
- creating public awareness through adult education, and income generation;
- creating employment opportunities for the target families.

Under the project, ten hazardous industries[1] were identified with a high incidence of child labour. By June 2000 the scheme was being implemented in ninety-three districts of the country. Special schools under this scheme are run in two types, one with fifty students and the other with 100 students. The Ministry of Social Justice and Empowerment has opened such centres in 2,571 schools with an enrolment of 155,250 children under the NCLP scheme (MoL, 2000; ILO, 2006). Welcome as these initiatives are, they clearly meet the needs of only a small number of the many millions of street and working children in India.

Initiatives by NGOs

The following are a few examples of the involvement of NGOs and the sharing of responsibility by them in this area (Butterflies, 2001, 2003):

- *Child in Need Institute* (CINI) has the mission to improve the life of urban disadvantaged children through education and mobilization. This group establishes Child Centres in co-operation with the community to educate out-of-school children, mainstreaming them into age-appropriate classes in formal schools. Their experience suggests that a sartorial approach does not bear much fruit unless it is coupled with the activities of other sectors such as night shelters, protection from abuse, food and health care. CINI operates through community-based preparatory centres, residential camps for working children and also coaching centres for children studying in formal schools.
- *Pratham* is another NGO established in 1994 in Bombay and has since branched out to twenty-six other cities including Delhi, Patna, Ahmadabad, Pune, Bangalore and Vadodra, and five rural districts in nine states. Pratham's interventions are with pre-school children, out-of-school children, in-school children and working children. Pratham has also started an out-of-school children's programme called *Akhar Setu.* This was set up in response to the difficulties faced in mainstreaming particular groups of children: those who are working or supporting their parents economically and are therefore unable to attend school; older children who cannot be admitted to their age-specific classes; and those who have no schools nearby.
- *Balajyothi,* a part of the National Child Labour Project, was one of the first in Andra Pradesh to address the issue of child labour and education in an urban context covering 9000 children in 150 slum clusters.
- *Naandi Foundation* has been supporting elementary education in Hyderabad city through its schemes of Support our Schools (SOS) and the Midday Meal scheme involving 60,000 government school children. The efforts have boosted attendance rates by 25 per cent which has significantly reduced health-related absenteeism (Kanth, 2005).
- *Prayas* is another large NGO which believes that for working children Alternative Education is the first step towards mainstream schooling; in the process it weans children away from any form of child labour. Annually,

they educate and mainstream about 3,000 children. Prayas also has an out-reach programme that provides educational opportunities to working children, street children, pavement dwellers, and children in conflict with the law. Teaching and learning happens in places of work on streets, pavements, railway platforms – wherever children want to learn. To supplement efforts for mainstreaming it also provides supplementary nutrition and health care facilities to all children enrolled in its AE centres. Prayas has been identified by the Government of India as one of only twenty NGOs in the country to implement vocational education and training of twelve to sixteen-year-old street children and working children, including girls and sometimes women also. Girls and women are also encouraged to form self-help groups. To help and facilitate its vocational students as well as surrounding communities to get placements it liaises with corporate houses, and small and medium sized business units. Prayas also arranges funds for those starting their own ventures. Thus it also helps in vocational preparation of street and working children and their placement in society.

Excellent though these various initiatives are, they still reach only a minority of the estimated 90 million street and working children.

Suggestions for improving education of deprived children

Although several steps both by the State and civil society have been taken to address the educational issues concerning street and working children, they nevertheless continue to face complex social, economic, developmental, security, and other problems (Weiner, 1991). Further attention needs to be paid to the following aspects in order to improve the educational status of these deprived children:

- Teachers' roles are extremely important in both retaining the children in schools and non-formal education (NFE) learning centres and in helping them acquire a desire for learning. So teachers need to be fully sensitized towards the special features, problems and needs of these children during their pre-service training as well as during in-service training programmes, and be encouraged to empathize with these children in order to understand their problems and needs.
- Teachers and NFE instructors should also be trained for preparing need-based teaching – learning material, keeping the socio-emotional context of deprived children in mind and organizing learner-centric learning experiences in and outside the classroom.
- A school's whole appearance and teaching–learning equipment and facilities require additional allocation of funds for improving infrastructural and working conditions.
- Preventive and creative approaches of supervision in schools and learning centres have to be adopted and expert guidance to teachers in a friendly

environment have to be given, so as to bring qualitative improvement in learning and achievement of these children.

- The government should arrange for education of all children in a locality in the neighborhood schools through the common school system. This is a long-pending recommendation and resolve of the national education policy.
- Professional development of schoolteachers and NFE/AIE (Non Formal Education/Alternative and Innovative Education) instructors should be accepted as an ongoing activity. This will keep them informed and motivated and will have favourable effects on their performance.
- Ambience and infrastructure of NFE centres should offer basic facilities and attractive learning environments to learners. This will make the much-criticized NFE centres more acceptable and useful for children attending them.
- NFE instructors need to be better oriented in making teaching–learning more interesting. Their training, salaries, use of teaching aids and methodology of teaching also need a serious review in order to ensure their better and more motivated participation in their responsibility.
- Students attending NFE/AIE centres merely acquire literacy and numeracy whereas they really need to learn life skills also. It is important that an alternative curriculum for NFE/AIE centres is developed keeping the socio-economic and emotional needs of street and working children in mind, and that instructors are properly oriented to this curriculum.
- Locally relevant teaching–learning material has to be developed to make teaching–learning more interesting and meaningful and compatible with the socio-cultural situation of the learners.
- There is an urgent need for integration of programmes of different ministries for working children and street children into a single well-resourced programme so as to increase the collective impact of these initiatives and programmes. This should be done through some joint committee with representatives from all concerned ministries and some experts.
- Special emphasis on girls' education has to be made for their self-sufficiency, independence and self protection. It may be done by organizing corner meetings, campaigns, media presentations and discussions, etc.
- A partnership between employers, government departments and NGOs needs to be promoted, as has been required in a 2001 order by the Supreme Court, so that their concerted efforts lead to better results (Sharma, 2003).
- Improving and strengthening the monitoring of State-sponsored programmes and schemes meant for these children should be emphasized for their better and timely implementation. Monitoring mechanisms, norms and procedures at different levels should therefore be evolved and enforced.
- For better returns of non-governmental participation in the education of deprived children it seems necessary for networking NGOs to establish resource centres for teachers at strategically located points in the cities. These could provide technical, academic and professional support to teachers of NFE centres and regular schools.

- Funding for NFE/EGS (Education Guarantee Schemes) centres also needs to be enhanced to improve the quality of the teaching–learning environment in these centres.
- Gradually all formal and non-formal centres should be developed into fully fledged formal schools so that the deprived children joining them are also able to acquire education of a comparable quality.
- By involving NGOs in organizing mass contact programmes for parents of working children they should be persuaded to help their children join education centres and spare them for studies.

Conclusions

The goal of universal elementary education in India which was to be realized in 1960 has yet to be achieved. Although many state-level and non-governmental initiatives have been taken and constitutional and legal provisions have been made, many millions of children are still out of school and many millions educationally deprived. Despite some policy announcements and the launch of heavily funded, centrally sponsored schemes for 'education for all' these children have generally remained out of the ambit of education. Either they don't have access to education as yet or they have dropped out of it. Education for them is often irrelevant, unattractive or actively rejecting. These out-of-school children are dominated by two major categories, namely, street children and working children who are forced by family, social, economic and educational circumstances to survive on the streets or remain heavily engaged in different kinds of work. Despite a constitutional amendment which grants the Right to Education as a Fundamental Right of the child, not much seems to be moving in the direction of the kind of inclusive education which encompasses all categories of deprived children, and provides education that is relevant and of acceptable quality. In fact, the Child Labour Act 1987 still permits child labour in 'non-hazardous' industries. The deprived street and working children are predominantly found in urban areas where migration of the rural poor and their destitute families is taking place continuously. An additional problem is that of child labour in rural areas in which children are hidden away in locked workshops working from dawn to dusk – making, for example, matches, which involves skinned and blistered fingers and breathing sulphur, in children as young as five (Dhariwal, 2006). One danger of increased legislation against child labour is to drive it either underground or out of the cities.

Handling the educational problems of these children in a vast country like India is a major challenge. There is little possibility of controlling or curtailing the pace of urbanization. It is only a thoughtful and imaginative multi-focused handling of the problem of out-of-school children, implementation of the long-pending state-supported common school system and concerted efforts to deal with the educational problems of working and street children in their own socio-economic contexts, that India may achieve the goal of basic education for all children. An empathic and humane approach to the education of deprived children is the need of the hour. There is an urgent need that all concerned with organization and delivery of education have to be thoroughly reoriented and sensitized towards the special circumstances and peculiar needs of these children so that they approach them with a true sense of concern and commitment to improve their educational lot and thereby empower them socially and economically.

At the same time the pending Compulsory Primary Education Act has to be passed without any further delay. To protect children from economic exploitation, suitable amendments to the Child Labour Act 1987 need to be expedited. Similarly, the draft Bill of Offences Against Children Act 2006 has to be passed immediately, with more comprehensive provisions than those currently proposed. All these and other similar initiatives must ensure and translate the *de jure* equality of educational opportunity into *de facto* equality, and help realize the long cherished dream of universal elementary education. A lengthy and important ILO report (2006) was entitled *The End of Child Labor Within Our Reach.* This is possible in many countries, but as the ILO data show, the demand for child labour will only end when average incomes in a country exceed $500 per annum for the majority of workers. While India has in statistical terms passed this critical income level in terms of GDP per head, national wealth is distributed very unequally, and the large majority of workers earn less than $500 per annum. Until the majority of workers achieve this crucial income level, the use of children as an income supplement for their parents will continue. While we expect the situation to improve year by year, up to 2003 the situation was one of 'small change' with millions of children still employed in 'hazardous' industries (HRW, 2003).

Note

1 These 'hazardous industries' include the manufacture of fireworks, footwear, cigarettes, matches, bricks, silk and glass as well as work in building and mining (ILO, 2006).

References

Aggarawal, S. K. (2004) 'Child labour and household characteristics in selected states', *Economic and Political Weekly,* 38, 173–185.

Annan, K. (2001) *We The Children.* New York: UNICEF.

Bhagwati, P. N. Justice, N. and Pathak, R. S. (1987) 'The Children's Aid Society', *Supreme Court Journal,* 1, Part II, 12.

Burra, N. (1995) *Born to Work: Child Labour in India.* New Delhi: Oxford University Press.

Butterflies (2001) *In Search of Fair Play.* New Delhi: Mosaic Books.

Butterflies (2003) *My Name is Today,* Vol. II. New Delhi: New Delhi Butterflies.

Dhariwal, D. (2006) 'Child labour – India's cheap commodity', *BBC News Online,* June 30th, 2006.

Dev, S. M. (2004) 'Female work participation and child labour', *Economic and Political Weekly,* 39, 736–744.

Human Rights Watch Asia (1998) *Police Abuse and Killings of Street Children in India.* New York and New Delhi: Human Rights Watch.

Human Rights Watch (2003) *Small Change: Bonded Labor in India's Silk Industry.* New York: Human Rights Watch.

ILO (2006) *The End of Child Labor Within Our Reach.* Geneva: International Labor Organisation, Report I (B). At: www.ilo.org/declaration.

Iyer, K. V. R. Justice (1979) *Law and Life.* New Delhi: Deep and Deep Publications.

Kanth, A. (2005) 'Education of urban deprived children', in Banerji, R. and S. Surianarain, S. (eds), *City Children, City Schools.* New Delhi: UNESCO.

Kaur, P. (2003) 'The statistical story of Indian street children', *My Name is Today,* Vol II. New Delhi: Delhi Butterflies.

Kaushik, A. (2005) 'Elementary education in urban areas,' in Banerji, R. and Surianarain. S. (eds), *City Children, City Schools.* New Delhi: UNESCO.

Lieten, G. K. (2006) 'Child labour: what happened to the worst forms?' *Economic and Political Weekly,* 41, 103–106.

MHRD (1986) *Programme of Action – 1986.* New Delhi: Ministry of Human Resources Development, Government of India.

MHRD (1992) *Revised Programme of Action – 1992.* New Delhi: Ministry of Human Resources Development, Government of India.

MHRD (2001) *Sarva Shiksha Abhiyan – A Programme for Universal Elementary Education.* New Delhi: Ministry of Human Resources Development, Government of India.

MoL (2000) *Ministry Document 6.6.2000.* New Delhi: Ministry of Labour, Government of India.

MSJ (1993) *An Integrated Programme for Street Children.* New Delhi: Ministry of Social Justice, Government of India.

MSJ (2000) *Annual Report 1999–2000.* New Delhi: Ministry of Social Justice, Government of India.

Reddy, N. (2004) 'Working with children', seminar proceedings, January 2004, 104–109. New Delhi: UNESCO.

Satgopal, A. (2003) 'Political economy of education in the age of globalization', in *My Name is Today,* Vol. I, 30–45. New Delhi: Delhi Butterflies.

Sharma (2003) 'Widespread concern over India's missing girls,' *The Lancet,* 362, 1553.

UN (1998) *Universal Declaration of Human Rights 1948–1998.* New York: The United Nations. At: www.un.org/Overview/rights.html.

UN (2000) *Minimum Age Convention (1973) No 138,* New York: Office of the United Nations High Commissioner for Human Rights.

UNICEF (2004) *Wheel of Change.* Kathmandu: UNICEF Regional Office for South Asia.

UNICEF (2005) *One in Twelve of the World's Children are Forced into Child Labour.* London: UNICEF, UK.

Weiner, M. (1991) *The Child and The State in India: Child Labor and Educational Policy.* Princeton, NJ: Princeton University Press.

Zutshi, B. (2000) *Situational Analysis of Education for Street and Working Children in India.* New Delhi: UNESCO.

13 Dalit children in India

Challenges for education and inclusiveness

Christopher R. Bagley

Introduction

Like many Western observers, I regard the cultural richness of India with affection and awe. In Britain, Indians form a major pillar of a plural society and are important supporters, through their professional commitment, of Britain's medical services. They are also major contributors to Britain's economic prosperity. For overseas Indians caste seems to have little relevance.

Caste is a relatively recent institution in the long history of Indian civilization. But at one time caste was said to form the basis of an ideal, harmonious society in which functional tasks were performed in a co-operative manner by individuals born into various vocational roles. Nevertheless, in its most venerable form, Hinduism did not incorporate a caste system, as Gandhi emphasized in advocating a modern, Hindu state in which caste as the basis for a proscribed occupational career was abolished, and all religions were given both freedom and equal status within a modern, democratic society (Dumont, 1970). Caste may have been introduced as a concept which consolidated the power of an elite group, and was then cloaked in a metaphysical idea concerning alleged sins in a previous existence, so-called fate or karma (Ramaiah, 1994).

Hinduism's most venerable philosopher, Shankara (c. 800CE), was a preacher and mystic who was influenced by the Buddha's teachings, and denounced caste divisions as both foolish and meaningless. Shankara declared that all human beings, regardless of caste, could acquire *Brahman,* the singular basis for understanding the divine, through a combination of religious contemplation, asceticism and good works. Individuals, he insisted, were not born into a *Brahman* state, but could achieve this within their lifetime through the spiritual exercises that he prescribed (Isaeva, 1993).

Although caste divisions have been formally abolished in the post-independence Indian constitution, the organization of life by caste remains powerful. In particular, the large group of Dalits (formerly 'Untouchables') below the fourfold caste system (Priests and Scholars; Soldiers and Leaders; Craftsmen; General Workers) endure lives of profound poverty and deprivation. Some modern scholars have compared the situation of Dalits in India to the slavery and segregation endured by African Americans in previous centuries, notably in the classic American text by Oliver Cox (1948) on *Caste, Class and Race.* The comparison with 'racial segregation' of African Americans with Dalits is not exact, however. Segregated they are, but colour lines are complex. Usually Dalits have darker skin than upper-caste Hindus, but this varies by region within India's vast continent.[1] Particularly in southern India higher caste groups may have very dark skin. More recently Rajeshekar (1997), in a study published in America, entitled his book *Dalit: The Black Untouchables of India,* and in sociological terms the comparison with African Americans, who still experience huge economic disadvantages often based on living in involuntary segregation, is apt.

The struggle of Dalits can be compared to that facing African Americans in post-war years, and the struggles for voter registration and integrated schooling. The State of Rajeshekar offers a typical scenario of murder which can be compared to America. In 1955, in Mississippi, a fourteen-year-old African American boy was beaten to death for whistling at a white woman. In 2000, a Bihari Dalit teenager was beaten to death for picking flowers on land owned by an upper-caste member. Press reports of the murders of Dalits appear occasionally in the West, but these events are mostly ignored. In 2002 the London *Daily Telegraph* reported that the Ranvir Sena, a private army raised by landowners in Bihar, had killed at least 500 Dalits in the previous year, with numbers in excess of 6,000 in the previous decade in Bihar (Bedi, 2002). In one event in 1997 sixty-one Dalits, including children and pregnant women, were hacked to death by agents of landowners. In 2006 in Bihar six members of a family, including five children, were burned to death because they would not withdraw their claim that a landowner had stolen their buffalo (Foster, 2006). Mungekar (2001) in a pessimistic analysis, compared the status of Dalits with that of slaves in post-bellum America – liberated in name but, in fact, cut off from virtually all mainstream institutions, and with little legal protection.[2]

Dalits, within the occupational hierarchy of traditional Hinduism, are allocated the roles of digging graves and rubbish pits, disposing of dead animals, sweeping and cleaning streets and rubbish dumps, and disposing of faeces. In rural areas especially, Dalits live in segregated zones, although even in Mumbai there are Dalit colonies. Dead animals and faeces are often dumped into Dalits' living space (Sainath, 1996). Most Dalits in rural areas are landless peasants, working as day labourers, earning less than one US dollar a day. There are frequent reports of Dalits being required to dig graves without payment (HRW, 1999). Anyone who refuses is likely to be beaten. Each year more than 1,000 Dalits are murdered by members of higher castes, although such violence is frequently unreported because of police apathy, connivance or corruption (HRW, 1999).

Dalit is a term meaning 'crushed' or 'stepped on', and the concept owes much to the civil rights pioneer Dr B. R. Ambedkar, himself an 'Untouchable'. Dr. Ambedkar was a unique individual who obtained his degree from Columbia University, and many advocates of Dalit emancipation draw on his writings. In 1955 Dr Ambedkar, along with millions of Dalit followers, converted to Buddhism (Omvedt, 1994). Since the Buddha's Enlightenment in what is now the State of Bihar, many Buddhists have left for countries elsewhere in Asia. In recent times many Dalits have renounced their affiliation within Hinduism, converting mostly to Buddhism, but also in large numbers to Islam and most recently to Christianity. This means that a large but unknown proportion of religious minorities in India are still regarded as untouchable, despite their conversions (Sainath, 1996).

Estimates of the numbers of Dalits in Indian society vary from 150 to 250 million out of India's one billion inhabitants. The varying numbers in estimates given apparently reflect the fact that sometimes numbers for Scheduled Tribes (descendants of aboriginal people in remote rural areas – see Joshi and Kumar, 2002) are included in their numbers, as well as some members of scheduled caste groups such as 'untouchable' labourers employed in 'unclean' tasks such as dying and curing leather. In addition, some estimates may include Dalits who have converted to other religions, but who are nevertheless still recognized by caste labels (Shahabuddin, 2002). However, in the most recent census 'untouchables' and 'Christians' were counted as mutually exclusive categories (Trapnell, 2004)[3]. 'Untouchability' as a practice was outlawed in 1989, but like many enactments this has never been enforced: the segregation and oppression of Dalits continues apparently unabated (Borooah and Iyer, 2002).[4]

The political situation of Indian Dalits

Before the problems of education facing Dalits can be analyzed, the political problems of Dalits and the numerous acts of everyday violence and discrimination that they face must be understood. For this understanding we rely strongly on the profoundly important analysis carried out by the international organisation Human Rights Watch (HRW, 1999). I rely heavily on the HRW accounts, being acutely aware that a Westerner in India, besides having restricted access, can actually endanger the safety and even the lives of individuals if he is observed making obvious 'human rights' inquiries within a larger community which practises a totalitarian system of social control over stigmatized groups. Human Rights Watch, a New York based organization, has used local workers to produce hundreds of case histories detailing human rights abuses against Dalits. One I found particularly compelling was this: although Dalits in cities are usually less persecuted than those in the rural areas, in 1997 a statue of the great Dalit civil rights leader Dr Ambedkar, standing in the Dalit 'colony' of Mumbai, was desecrated. Dalits marched in protest, straying outside their colony. Police shot dead ten unarmed protesters, and injured twenty-six more.

Dalits in rural areas often must walk miles to obtain drinking water, being forbidden to use local water taps which they would allegedly pollute. They usually work as day labourers in addition to the imposed tasks of handling the filth that the more prosperous Hindus produce. It is difficult for them to register to vote in the face of massive intimidation. Women and children are frequently mishandled and raped, and HRW present numerous case histories of women who are sexually mutilated following rape. Untouchable they may be, but this does not prevent the frequent rape of Dalit women.

At least 50,000 Dalit girls are removed each year (or are surrendered by their impoverished parents) to serve in brothels, and a large but unknown number of Dalit children (probably in excess of 15 millions) endure conditions of slavery as bonded workers, given up in exchange for family debt. Complaints to police are unlikely to be successful, and are often counterproductive. Complainants about sexual assault or other violence can be imprisoned for years as material witnesses, or are beaten by police for complaining (HRW, 1999).

India's most impoverished state, Bihar, is apparently the scene of some of the worst outrages of violence experienced by Dalits. Past governments of Bihar have failed to implement Acts regarding land reform and minimum wages for day labourers which could have aided impoverished Dalit communities. Mainly in this State a group known as Naxalites has developed (with some Dalit membership). This is a quasi-Marxist group practising violent insurrection. It has achieved no political success for Dalits. Instead private militias (e.g. *Ranvir Sena*) predominantly recruited and armed by upper castes, raid Dalit colonies in villages, killing and mutilating without mercy or fear of legal sanction, the excuse being, apparently, that 'a dead Dalit child can never become a Naxalite' (HRW, 1999).

Corrie (1995) has developed 'a human development index for the Dalit child in India' which measures progress on health and welfare indicators. There is wide variation across Indian states, and progress is 'slow and uneven'. While Bihar has poor scores on this index, there are now grounds for some optimism, given the change of government in Bihar in 2005–6, and its declared aim of improving the education, welfare and economic progress of its citizens. Nevertheless, we must also record the fact that India's new prosperity based on global trade has actually, according to economic analysts, diminished the occupational advancement of Dalits in relative terms since the 1990s (Mungekar, 2001). Benefits of international trade enabled 'the educated unemployed' (Verma *et al.*, 1980) from groups other than Dalits to obtain newly-created jobs. But educated Dalits remain unemployed (Jeffrey and Jeffery, 2004), and despite obtaining high school diplomas or degrees the only forms of employment usually available are the traditional menial tasks, and day-wage labouring. According to Rodrigues (1999) the growing consciousness of their oppression and political attempts to counter it, have resulted in an increased number of physical attacks (including murders) against them, both by the authorities and by private groups of caste Hindus and their hirelings.

The dramatic rise of the Hindu nationalist party (BJP) since the 1980s is connected to the continual oppression of Dalits. Although the BJP's formally declared philosophy offers an inclusive vision of Hinduism in which all have dignity within a religious framework, it is also true that the BJP has attracted many violent and right-wing elements who are particularly opposed to the rights and advancement of Muslim and Dalit minority groups. Some elements within the BJP have argued that it should be illegal for Hindus to convert from the religion into which they were born – meaning, of course, Dalits. The BJP through cross-alliances has held power in the Indian parliament, the Lok Sabha. Today the more moderate Congress party holds power. It is possible that some Muslims (some 12.5 per cent of all Indians) and those Dalits registered to vote, placed their votes strategically in favour of the Congress party in order to undermine the BJP's position.

The influence of Hindu nationalism can be judged by the passing of laws at the State level which require any Dalit wishing to convert from Hinduism to another religion, to appear before a magistrate to make sure he or she has not been 'coerced' in making this decision. Missionaries can be imprisoned if they are judged to be offering inducements or pressure in seeking conversion. Madhya Pradesh is the latest State to introduce such a law (Nelson, 2006). Dalits appearing before magistrates when they want to convert are marked individuals, and run the risk of intimidation and violence from fundamentalist Hindus. It is not coincidental that these laws have paralleled the large increase in Dalit conversions in the past decade, a reflection of the increasing movement for Dalit self-identity and freedom from the oppressions of the caste system.

The educational position of Dalits

The patterns of violence and segregation experienced by Dalits lead to a better understanding of the educational deprivation of this group. When enrolled in school they are likely to be segregated, must eat separately, often have no access to toilets and drinking water which other pupils can access, and are subjected to bullying and violence in school yards. Fifty per cent of Dalit children who enter state primary schools drop out before the age of fourteen (Ramachandran and Saihjee, 2002). Teachers often negatively label Dalit and Tribal children as backward, dull and poorly motivated. Given the profound discrimination which these groups suffer, this is likely to be a self-confirming prophecy. Dalits experience what Shiva (2003) calls 'apartheid in education', reflecting the failure of India's 'Education for All' policies as set out in the National Plans for Education of 1968, 1986 and 1992. Shiva observes that the elites who should have implemented these plans are likely to have sent their children to private schools, ignoring the realities of low-quality public schools. Most Dalits attend one-roomed rural schools with poorly trained teachers. Where there are higher ability groups in the school, Dalits are often purposefully excluded, whatever their ability level. Ambedkar (1992) referred to the lot of Dalits in education and in other areas as 'dungeons of exclusion'; while official apartheid has been superseded in South Africa, it remains strong in India. Mendelsohn and Vicziany (1998) argue that the Indian government is indifferent or hostile to Dalit aspirations for equality, and tacitly tolerates their persecution.

Given the political analysis of the position of Dalits, it comes as no surprise to find that their education status is depressed, compared with all other groups in India (with the exception of Scheduled Tribes). In rural Bihar which has one of the highest proportion of Dalits amongst Indian States, teachers were until recent times often absent from school for ten months in a year, a reflection of their very low pay, and the lack of administrative systems ensuring the quality of education (OWSA, 2006). In Bihar the overall pupil–teacher ratio in publicly-funded primary schools (for six to fourteen-year-olds) was until recently 122:1, three times the national average of 40:1. There is a shortage of classrooms in the poorest States, and even those available are of very poor structural quality with drinking water and toilets atypically available. There has been some improvement in primary education in the 1990s with the implementation of the national government's *Sarva Shiksa Abhiyan* (Education for All) programme (see Chapters 10 and 11 in this volume by Jha and Srivastava), but this seems to have helped Dalit children little (Ramachandran and Saihjee, 2002). The problem remains one of lack of funding for rural primary schools.

Borooah and Iyer (2002) analyze data from a survey of 33,000 rural households in 1,765 Indian villages, predicting from statistical modelling the likelihood of a female child being enrolled in primary education. The proportion of Dalit families, the proportion of Muslim families, and the proportion of Scheduled Castes (i.e. lower castes) were statistically significant predictors, the prediction of low enrolment for females being strongest in the smallest communities. This finding underscores the work of Gobinda and Diwan (2003) which argues that movements for Education for All must be locally based and adapted to local conditions. The campaign in Bihar, for instance, must be different from that in Kerala.

Jeffrey and Jeffery (2004) describe how Dalit pupils are shunted into the lowest quality state schools, and are discouraged from taking any kind of advanced or examination course. Dalits who do succeed in overcoming multiple educational barriers, and make it through secondary school and even to college, face discrimination which becomes evermore rigorous at each stage. The Dalit who graduates from college is unlikely to find commensurate employment, or indeed any employment at all (Jeffrey and Jeffery, 2004; Jeffrey *et al.*, 2004). The Jefferys in their anthropological research based on a semi-rural community 150 km north east of New Delhi describe a situation of 'Degrees without Freedom' in which Dalit graduates without jobs, lacking the social connections which higher castes use to obtain professional employment, translate their grief and frustration at their status into non-violent political rhetoric which nevertheless rejects the Fabian-type approach of the BSP, the main Dalit political party. It is salutary that this is a study of young men: the Jefferys could not find enough Dalit women graduates to study.

The position of Musahars, the Tribal peoples of the Gangetic Plains, is significantly more depressed than that of most Dalits within whom they are sometimes counted (Joshi and Kumar, 2002). Tripathi (2002) argues that Tribal groups such as Musahars require a culturally appropriate educational medium. However, it is likely that the large majority of Dalits prefer an English-medium education which could give them the potential for entering the mainstream of Indian life.

English-language education for Dalit children: the network of voluntary religious schools

For many years Christian missionaries have worked with Dalit communities practising both conversion and the setting up of schools, primarily for Dalit children but open to children of all faiths and backgrounds on a low-fee payment basis. The Christian movement has a dual role: it offers Dalits an ideological escape from their oppressed position within Hinduism; and it offers a solution to the Dalit Freedom Network's demand for English-language education, as opposed to education in local languages (Dalit Freedom Network, 2006). Education in the English medium is likely to offer better economic advancement than literacy in a local language.

The Christian-endowed schools each serve between fifty and 250 children taught by qualified teachers with class sizes of less than forty. These schools are well built and equipped and are heavily subsidized by voluntary donations, mainly from America (DaySpring International, 2005). The larger of these schools can also serve as teacher training institutions for Dalit students aged eighteen to twenty-five (OM India, 2004). Excellent though these schools are, only a fraction of the millions of Dalit children in India have the opportunity of attending them. They could, however, provide an excellent model of 'low-fee private' schools for *all* Indian children regardless of denomination, provided that fees are kept to the level of about one day's pay of a day labourer, per month of schooling. It is to India's credit that she does allow foreign missionaries to work in India, often converting those for whom they provide education. Bhattacharya (2003) comments positively on the work of Christian missionaries with Dalit and Tribal groups, providing educational resources not otherwise available to them. But the number of such schools in India is probably less than 5,000, while the population of Dalit children of primary school age is at least 30 millions. Thus if they attend primary school at all, at least 90 per cent of Dalit children must attend poor quality government schools. Unfortunately, some mission schools attended by Dalits have experienced the same kind of violence experienced by Dalits themselves, with schools being wrecked and burned. Christian missionaries themselves have been attacked and murdered (Nelson, 2006).

The educational position of girls and women in India

In addition to the disadvantaged educational position of minority groups such as Dalits and Moslems, girls in general are educationally disadvantaged in India. Compared with other Asian countries (including Indonesia, Malaysia and China) the proportion of females who acquire the skills of literacy is low, except in Kerala (Borooah and Iyer, 2002). Between 1993 and 1999 the overall proportions attaining literacy in India rose from 68 per cent to 79 per cent (UNESCO, 2003). But the large gender gap remains, and overall only 34 per cent of females had completed education to age fourteen in Bihar (compared with 88 per cent in Kerala). In India as a whole, literacy rates are lowest in Dalit girls and women (Sreedhar, 1999). Poor families are likely to enrol sons rather than daughters in primary school, despite the goals of India's formal Education for All policy (One Country, 2004). Only in 2001 did India's national government formally implement its policy for universal, free primary education (to age fourteen), and it comes as no surprise that implementation is uneven, with boys more likely to be enrolled in primary school (Borooah and Iyer, 2002).

The hidden oppressed: Nepalese Dalit girls in Indian brothels

Each year thousands of young girls aged twelve to seventeen from Nepal are tricked or sold into the slavery of prostitution in India. They are robbed not only of education and career aspirations, but of their dignity and health as well, and when they become infected with HIV they are shipped back to Nepal, where they live out stigmatized lives (Simkhada and Bagley, 2006). According to our research, about half of these girls are of Dalit origin, the other half being mainly of 'Mongoloid' appearance, for which some Indian men seem to have a sexual fetish. Christian ministers who have attempted to 'rescue' these sex workers (some as young as ten) have been physically attacked by Hindu militants (Reynolds, 2006).

Policy solutions?

The disadvantages and persecution experienced by Dalits are both profound and deeply rooted in Indian social structure. Despite the abolition of untouchable status (in Acts of 1955 and 1989), these laws like so many others in India are unevenly administered, and often ignored at the local level. If the status of Dalits in India is to be improved, then profound changes in Indian society must take place. The prospects for such change are not good and Hindu nationalism, which sees Dalits as profoundly inferior and absolutely untouchable, is on the rise rather than on the decline. But if the educational status of Dalit children is to be improved there must be fundamental political changes in India. The 'Education for All' proposals of The Kothari Commission of 1964 were enthusiastically relaunched in 2004, after many false starts. Now is certainly the time for the implementation of quality schooling for Dalit children. Kothari's proposal that India should spend six per cent of its GDP on education has never been implemented, and this may account for the fact that many teacher posts in rural primary schools remain unfilled, because of lack of money for their salaries.

The Christian schools aimed at Dalits provide excellent models of quality education. However, for sensitive political reasons (their funding from America and Europe), and their evangelistic nature, they will remain an option for only a minority of Dalit communities. But the quality of these primary schools can offer a model for education for all of India's children. India currently spends about four per cent of its GDP on education; increasing this to six per cent in a climate of increasing prosperity through international trade could provide funding for the increase in quality of Indian primary schools. Teachers and teacher training are crucial in providing a non-discriminatory climate for delivering education in schools with children from varied social and religious backgrounds.

Should Dalits continue their tradition of converting to religions (Buddhism, Sikhism, Islam, Christianity) which, formally at least, do not discriminate on grounds of 'untouchability'? This traditional movement away from the oppressive forces of Hindu nationalism would be spiritually and ideologically satisfying. But Hindus themselves should defer to the teachings of Shankara and Gandhi for a renewal of Hinduism's core spiritual values which disavow the determinism of being born into an occupational caste. In a truly democratic society, merit should be achieved and earned, and should not be endowed by birth. India claims to be 'the world's largest democracy'. But India will not be a true democracy until the privileges bestowed by caste, and the persecution of certain Hindu groups on the basis of their birth, are abandoned.

The Human Rights Watch Report (HRW, 1999), after their detailed analysis of several hundred cases of human rights violations (including many cases of murder and rape) against Dalits, makes some crucial recommendations:

1   A concerted effort to direct policy measures to alleviate these injustices must be made by the national government of India, making sure that human rights legislation is enforced at the state level.
2   The 'Prevention of Atrocities Act' should not simply lie on the statute book, but should be fully implemented, with police and other officials training at the local level, to ensure enforcement.
3   There should be reform of police procedures and personnel so that violence against Dalits is properly investigated, and laws enforced. More women police officers should be recruited and trained, to investigate the many cases of rape against Dalit girls and women. Dalits themselves should be recruited as police personnel.
4   The practice of bonded labour should be ended through the enforcement of Acts already passed.
5   The right of Dalits to register to vote should be strictly protected.
6   Dalits should not be arbitrarily detained, on minimal excuses. India has extensive legal powers to detain suspected terrorists for lengthy periods. These powers should not be used simply because police suspect a Dalit might harbour negative but non-violent views of existing political systems. It is incumbent on Dalits, of course, not to engage in violence or terrorism of any kind. Private militias recruited by Caste Hindus such as the *Ranvir Sena* should be vigorously prosecuted.
7   Discrimination against women in education and employment should be diminished; this would aid not only Dalit women but women from all economically disadvantaged groups in India.
8   India should ratify and enforce the 1984 UN Convention Against Torture and Other Forms of Cruel, Inhumane or Degrading Treatment or Punishment.
9   World Bank loans and various other international loans and forms of aid to India should be conditional upon the ending of discrimination against groups such as Dalits, Tribal Groups and Muslims. India has a poor record of allowing UN Special Rapporteurs access to minority groups, and the informal research methods used by HRW have had to be employed.

Parallels have been drawn between attempts in India to guarantee government positions and college admission places to educated persons from minority groups, with efforts in the US. As in the United States, this has been a controversial policy, with backlashes and legal challenges (Weisskopf, 2004). But if goals of social justice are to be served, such policies should guarantee Dalits the levels of educational and economic success that they merit.

As HRW (1999) concludes, without major determination by the Indian government the caste system 'in practice relegates millions of people to a lifetime of violence, servitude, segregation, and discrimination'. Mendelsohn and Vicziany (1998) see no end to Dalit poverty, and no end to the widespread continuance of violence expressed against them. It is crucial that scholars and politicians outside of India continually monitor the situation of Dalits. A new Human Rights Watch report should be issued every five years. The Human Rights and Law Unit in New Delhi does continue to issue valuable, updated accounts of 'atrocities' experienced by Dalits (HRLU, 2006).

Conclusions

India is a complex plural society with many cross-cutting lines of ethnicity, power, status and religion, with frequent points of social tension at the intersection of the blocs of this plural society (Bagley, 1979). Given India's social complexity, its huge population size, its great inequalities of wealth, and the crowding of its cities, it is actually surprising that episodes of murder and violence are relatively rare, rather than relatively frequent (Bagley, 1989). Sociologically, it appears that certain minority groups such as Dalits and Muslims are singled out as 'whipping boys' or scapegoats to enable the tensions of communal violence to be temporarily purged, with occasional bouts of mass murder and daily acts of lethal and sub-lethal violence costing only a few thousand lives each year, in a population of one billion.

However, as India enters a new phase of development based on globalization and international trade, it is essential that the national government ensures that social justice reaches all of its citizens in a newly social democratic state. A new government-corporate partnership plans to provide scholarships for 50,000 Dalit and Tribal university and college students by 2009 – approximately 0.2 per cent of the estimated 30 million Dalit and Tribal children eligible to enter primary school. Prasad (2006a) is pessimistic, however, of industry employing the Dalits and Tribals who graduate from these college-level programmes.

Commenting on tardy progress in fulfilling official quotas for recruiting Dalits to government positions, Prasad (2006b) cites Dr Ambedkar, the Dalit equivalent of Martin Luther King, who argued that rights are protected not by law but by the social and moral conscience of society. 'If social conscience is such that it is prepared to recognize the rights which law chooses to enact, rights will be safe ... if the fundamental rights are opposed by the community, no law, no parliament, no judiciary, can guarantee them ...'

Ambedkar was writing in the 1950s. Unfortunately, his observations are still relevant today. The 'social and moral conscience' of India has yet to seriously consider the issue of Dalit equality, and inclusion in mainstream social institutions. Until that time, Dalits remain one of the most brutally oppressed ethnic groups anywhere in the world.

Notes

1 There is some evidence that higher caste Hindus in Northern India are genetically related to European invaders from Central Asia (Bamshad, 2001).
2 Violent attacks, rape and murder of Dalits continue on a daily basis. The most recent case in my file is the fatal beating of a fourteen-year-old Dalit boy in Bihar, accused of theft in September, 2006: www.DailyIndia.com/show/56932.php/Dalit_boy_beaten_to_ death/
3 Effectively, this meant that Dalits who had converted to Christianity were, unwillingly, classified by the Census as Hindus.
4 For a comprehensive bibliography on sociological studies of Dalits see Charsley (2004).

References

Ambedkar, B. (1992) *Writings and speeches: Volume 2.* Mumbai: Government of Maharashtra.

Bagley, C. (1979) 'Social policy and development: the case of child welfare, health and nutritional services in India', *Plural Societies,* 10, 3–26.

Bagley, C. (1989) 'Urban crowding and the murder rate in Bombay, India', *Perceptual and Motor Skills,* 69, 1241–1242.

Bamshad, M. (2001) 'Genetic evidence on the origins of Indian caste populations', *Genome Research,* 11, 994–1004.

Bedi, R. (2002) 'Caste war will go on', *The Daily Telegraph Online,* 31st August, 2002.

Bhattacharya, S. (2003) *Education and The Disprivileged.* New Delhi: Orient Longman.

Borooah, V. and Iyer, S. (2002) *Vidya, Veda and Varna: The Influence of Religion and Caste on Education in Rural India.* Coleraine: School of Economics and Politics, University of Ulster.

Charsley, S. (2004) *A Dalit Bibliography.* Glasgow: Dept of Sociology, Anthropology and Applied Social Sciences, University of Glasgow.

Corrie, B. (1995) 'A human development index for the Dalit child in India', *Social Indicators Research,* 34, 395–409.

Cox, O. (1948) *Caste, Class and Race.* New York: Doubleday.

Dalit Freedom Network (2006) *Abolish Caste, Now and Forever.* Accessed 14th June, 2006, at: www.dalitnetwork.org.

DaySpring International (2005) *The Challenge of India.* Accessed 16th June, 2006, at: www.dayspringinternational.org.

Dumont, L. (1970) *Homo Hierarcus: The Caste System and its Implications.* New Delhi: Oxford University Press, India.

Foster, P. (2006) 'Family of six burned alive by caste killers', *The Daily Telegraph Online,* 4th January 2006.

Gobinda, R. and Diwan, R. (2003) *Community Participation and Empowerment in Primary Education.* London: Sage.

HRLU (2006) *Human Rights of Dalits.* New Delhi: Human Rights and Law Unit, Indian Social Institute.

HRW (1999) *Broken People: Caste Violence Against India's 'Untouchables'.* New York: Human Rights Watch. At: www.hrw.org/reports/1999/India.

Isaeva, N. (1993) *Shankara and Indian Philosophy.* New York: State University of New York Press.

Jeffrey, C. and Jeffery, P. (2004) 'When schooling fails: young men, education and low-caste politics', *Contributions to Indian Sociology,* 39, 1–38.

Jeffrey, C., Jeffery, R. and Jeffery, P. (2004) 'Degrees without freedom: the impact of formal education on Dalit young men in Northern India', *Development and Change,* 35, 963–986.

Joshi, H. and Kumar, S. (2002) *Asserting Voices: Changing Culture, Identity and Livelihood of the Musahars in the Gangetic Plains.* New Delhi: Deshkal Publications.

Mendelsohn, O. and Vicziany, M. (1998) *The Untouchables: Subordination, Poverty and State in Modern India.* Cambridge: Cambridge University Press.

Mungekar, B. (2001) *Dalit Human Rights,* lead address to Human Rights Conference, University of Mumbai, 4th November 2001.

Nelson, D. (2006) 'Missionaries face jail in India', *Sunday Times Online,* 30th July 2006.

OM India (2004) *Dalit Education Centers.* Accessed 4th May 2006 at: www.usa.om.org/omIndia.

Omvedt, G. (1994) *Dalits and the Democratic Revolution: Dr Ambedkar and the Dalit Movement in Colonial India.* London: Sage.

One Country (2004) Proceedings of UNICEF Conference on Education for All. Accessed 4th May 2006 at: www.onecountry.org (Bahai Online).

OWSA (2006) 'Schools in Bihar', *One World South Asia,* 2nd May 2006.

Prasad, C. (2006a) 'Corporate schools', *New Delhi Pioneer Online,* 3rd September 2006.

Prasad, C. (2006b) 'Hail CII proposal', *New Delhi Pioneer Online,* 20th August 2006.

Rajeshekar, V. (1997) *Dalit: The Black Untouchables of India.* New York: SCB Distributors.

Ramachandran, V. and Saihjee, A. (2002) 'The new segregation: reflections on gender and equity in primary education', *Economic and Political Weekly,* 27th April, 1600–1613.

Ramaiah, A. (1994) 'The Dalit issue: a Hindu perspective', in J. Massey (ed.) *Indigenous Peoples.* New Delhi: ISPCK.

Reynolds, J. (2006) 'Ministers attacked who help keep teens out of prostitution'. Lake Forest, CA: Assist. News Service. At: www.assistnews.net.

Rodrigues, E. (1999) 'Dalit assertion and casteist retaliation', *Economic and Political Weekly,* 9th October, 411–412.

Sainath, P. (1996) *Everybody Loves a Good Drought: Stories from India's Poorest Districts.* New Delhi: Penguin Books.

Shahabuddin, S. (2002) 'Comments on Yoginder Sikand's article on Dalit Muslims', *Journal of Muslim Minority Affairs,* 22, 479–481.

Shiva, A. (2003) 'A kind of apartheid in education', *Times of India Online,* 11th November 2003.

Simkhada, P. and Bagley, C. (2006) *The Sexual Trafficking of Nepalese Girls and Women: Health and Policy Issues.* Manchester: The Manchester Educational Research Network.

Sreedhar, M. (1999) 'Reaching the unreachable: enabling Dalit girls to get schooling', *Manushi: Women and Society in India,* 111, 10–20.

Trapnell, J. (2004) *The Indian Census, Hindu-Christian Dialogue, and Perceiving the Religious Other.* Accessed 4th June 2006, at: www.infinityfoundation.com/mandala/s _es_trapn_census.htm.

Tripathi, G. (2002) 'The desire for education', in H. Joshi and S. Kumar (eds) *Asserting Voices: Changing Identity and Livelihood of the Musahars in the Gangetic Plains.* New Delhi: Deshkal Publications.

UNESCO (2003) *UNESCO Education for All Global Monitoring Report.* New York: UNESCO. Accessed 7th May 2004 at: www.efareport.unesco.org.

Verma, G., Bagley, C. and Mallick, K. (1980) *Illusion and Reality in Indian Education.* London and New Delhi: Gower Press.

Weisskopf, T. (2004) *Affirmative Action in the United States and India: A Comparative Perspective.* London: Routledge.

Part VI
Conclusions

14 Crisis, rhetoric and progress in education for the inclusion of diverse ethnic and social groups

Christopher R. Bagley

The assumption of this chapter is that racism in Western societies is a sub-system of societies whose basis is the exploitation of human beings through capitalism. A society based on the accumulation of capital through profitable entrepreneurship needs a stable and largely acquiescent work force, as well as a reserve army of labour which can be laid off in slack times without problems of social unrest. This latter role is usefully filled by assimilated migrant workers, women, and degraded ethnic minorities.

(Bagley, 1985, p. 49)

The values which underpin the practice of inclusive education for diverse individuals and groups are changing – a reflection of diverse social forces – and the naïve Marxist model implied in the quotation above needs to be amended, or developed. The global economy calls for a more educated workforce, but these demands require an education system that is subordinated to the needs of international capitalism. This is strangely reminiscent of the Acts of Parliament in nineteenth-century Britain which provided universal primary education on the ironic premise that 'we must educate our masters'; the newly emancipated voters had to be educated to a minimal standard (Brock, 1978). Today, it is pressures of globalization which have persuaded the Indian economy to finally introduce a policy of Education for All, see Chapters 3 and 10 (Jha) and 11 (Srivastava) in this volume. Even then Dalits, children of the 'untouchable' caste, seem unlikely to gain educationally in ways which could lead to occupational advancement (see Bagley, Chapter 13, in this volume). Marx had characterized the underclass as 'a reserve army of labour' to be called on in times of economic expansion. Unfortunately, this Marxian model does not fit modern-day India, in which Dalits' caste status relegates them to that comparable with African Americans 150 years ago. Ironically, an educated elite of Dalits is emerging, facing both permanent unemployment and greater degrees of relative deprivation than their landless labourer parents (Jeffrey *et al.*, 2004).

Rizvi and colleagues (Chapter 1, this volume) suggest that although the forces of globalization are creating a new discourse and rhetoric in education, the effects of globalization on educational equality and opportunity vary greatly. Information and communication technologies may be transforming the economies of countries such as India, but their benefits mainly affect the educated elite, given that country's long history of 'educated unemployment' in which graduates can wait months or years before ever obtaining employment commensurate with their educational investments (Verma *et al.*, 1980). Globalization of the world's economy and communication systems is not a democratic process, but one which feeds into existing stratification systems, with America as the elite, ruling class and the peasants of Asia and Africa as the ignored masses, noticed only when their assertiveness takes terrorist proportions. As Rizvi *et al.* observed in this volume: 'All industries, including education, are trapped within the networking logic of contemporary capitalism, subject to the same economic cycles, market upswings and downturns and segmented global competition'.

The dominant view of education reflects a neo-liberal ideology, and in many countries educational policymakers have tried to provide systems of training which will create a co-operative and skilled workforce who can add to a country's GDP. But as Rizvi *et al.* (above) observe, 'while the authority for the development of education policies remains with sovereign governments, they nonetheless feel the need to take global processes into account'. These social pressures include the international agencies providing financial aid for education, with the subtext being the need to serve 'the globalizing cultural field within which education takes place'. In this respect we were interested to experience the Christian mission elementary schools for Dalit children, which provide quality education without any clear economic benefit in prospect – education as a goal in itself, albeit with metaphysical overtones (see Bagley, Chapter 13, this volume). Jha (2006) provides detailed case studies of Christian and other schools in India with excellent programmes for children with various physical and sensory disabilities. These latter children do not fit into educational programmes in which the economics of globalization is a driving force, and in many Third World countries a deaf or partially deaf, blind or partially blind child is lucky if they are admitted to a primary school class containing 100 children. They are likely to have highly individual learning needs, which their teacher is unable to meet.

The crisis in British education: coping with diversity through exclusion

Britain, unfortunately, stands low in the European league on the successful inclusive education of children with special needs. A Select Committee Report to the British Parliament (Asthana and Hinsliff, 2006) on education for children with 'special needs' of various kinds identified a system in confusion, in urgent need of reform. The committee found that schooling for the inclusive education of special needs pupils was 'not fit for purpose'. Policies differed dramatically between different local areas, often with no direction or financial support for specialist teachers or equipment. Special schools are being closed and their pupils transferred to mainstream schools where teachers, coping with class sizes larger than thirty are often unable to meet their needs (Bagley, 2006). Warnock (2005) has argued that the treatment of children with special educational needs in mainstream schools in Britain is so negligent that Britain violates the UNESCO (1994) Salamanca Statement principles on inclusive education. But as Frederickson and Cline (2002) show, children with 'special educational needs' could be absorbed with equity into mainstream schools, fulfilling the principles of Warnock's (1978) idealistic report on special needs education.

British teachers, having to cope with the demands of frequent assessments in relation to a National Curriculum, seem keen to remove pupils who have poor academic performance (who might depress average test scores), and they have powers to do this through making permanent and temporary exclusions; according to a Parliamentary Committee 87 per cent of excluded pupils from primary schools, and 60 per cent of those excluded at the secondary stage are pupils with 'special needs'. These needs include autism, dyslexia, emotional and behavioural challenges, as well as the traditional forms of sensory and cognitive challenge. In addition (but sometimes coincidentally), a pool of more than 100,000 pupils leave school more or less permanently before that statutory leaving age (sixteen) forming a street army of delinquent youth who plague systems of social control (Bagley, 2006).

For those struggling to provide universal and free quality education at primary and secondary levels, the British case is paradoxical. Pupils are allowed to stay in school without cost, may take public examinations up to age eighteen, and are encouraged to apply for college and university entrance (Dutton *et al.*, 2005). For the large majority of students from the disadvantaged social classes (comprising about a fifth of the nation), the possibilities of entering tertiary education is, however, often a vain hope (HEFCE, 2005). In addition, an important minority of pupils at secondary level are permanently excluded from school (or permanently exclude themselves) because of their alienation from school and learning; or because of problems of behaviour and under-achievement which make them unacceptable to their schools, which in consequence suspend or permanently expel them.

Cumulatively, at least 1 per cent of students under age seventeen had been permanently or temporarily excluded from school by the end of 2003 (DfES, 2004). The rate of these permanent exclusions is significantly higher in areas of Britain blighted by poor housing, in which a high proportion of parents live in poverty, and experience a high incidence of various indicators of social deprivation (Reed, 2004; Bagley, 2006). The proportion of rejected, dispirited or discouraged pupils who never take any public examinations at age sixteen had, by late 2005, reached a record level in Britain, with 12.6 per cent of adolescents leaving school without any public examination success. These young people either become permanently unemployed or are frequently unemployed and welfare-dependent (Bekhradnia, 2006). According to OECD figures (2004) Britain ranked twenty-seventh out of twenty-nine industrialized nations in terms of young people staying on at school after age sixteen.

Britain is the eighth richest country amongst the 30 OECD States, in terms of GDP per head (OECD, 2006) – but Britain's annual GDP of $2.3 trillions is distributed very unequally compared with the majority of the OECD nations (Hobson, 2001).[1] However, the distribution of this wealth is more unequal than in many other European countries (Bradshaw and Chen, 2002). Britain ranked fifty-first (i.e. has a high score on the Gini Coefficient, which measures unequal distribution of wealth) in the 124 countries in the latest international comparisons (UN, 2005). Nordic and some former Communist countries as well as Japan have the most equal distributions of national wealth. There is a strong correlation in twenty-one major industrialized countries between higher Gini scores (indicating a greater income gap between rich and poor) and diminished life expectancy in the poorest groups within a country (De Vogli *et al.*, 2005). This correlation (0.87) was unaffected by the availability of a universal, free health service in any particular country, including Britain which ranked strongly in terms of inequality of income distribution. In part, this is due to the failure of social security and welfare payments in Britain to have very much influence on chronic patterns of income inequality (Bradshaw and Chen, 2002). It appears too that primary and secondary education in Britain is (like welfare services) significantly under-funded in comparison with other wealthy countries, and the number of students per teacher in publicly-funded schools in Britain is actually rising (Haile, 2005). The proportion of GDP spent by the British government on education fell from 5 per cent in 1995 to 4.7 per cent in 2003 (Smith, 2005); this proportion is slightly below the European average (OECD, 2004). Those countries spending the highest proportion of national wealth on publicly-funded education are the Nordic countries. Significant increases in spending on education in Britain have, however, been promised (Brown, 2006).

A report from the British Centre for Economic Performance in 2004 (Chevalier and Dolton, 2004) has found that the drift of teachers from the profession outstripped new recruits. Main reasons given by departing teachers were low pay, stress, and problems of teaching large classes which contained too many disruptive students. There is also a strong social class bias in Britain in secondary school students who continue on to university studies, ranging from eight per cent of the age group following this path in the poorest region, to 62 per cent in the most prosperous (HEFCE, 2005). In the country as a whole, young people living in the more advantaged areas are more than five times as likely to go on to university than are young people in areas where average family incomes are in the lowest quintile. By the end of 2005 the proportion of children with parents in the highest wealth quintile had increased their chances of university entry to six times the numbers whose parents were in the lowest income quintile (Cassidy, 2005). Research cited by Cassidy indicates that an important mediating factor is the poorer quality of the secondary schools attended by many of the students from economically poor homes. Even when they do enter university, children of the poorest parents tend to have poorer degree outcomes, largely because they have to work part-time because their parents (unlike well-off parents) are unable to provide for their living allowance (Van Dyke and Little, 2005).

In a survey of educational achievements of school students conducted by the Organisation for Economic Co-operation and Development (OECD, 2004)

Britain ranked fifteenth in achievements in reading, science and mathematics amongst the forty-one countries who submitted data, despite Britain being, at that time, eighth in world rankings for wealth per head of population. This statistic is another indicator of the 'poverty of education' in Britain, a pattern of under-funding of an important resource in a wealthy country. In fact, there are grounds for supposing that these figures overestimate the achievements of British pupils (Smithers, 2004). This is because schools deliberately exclude or expel students who are underachieving, apparently in many cases to improve their school's achievement profile. If the achievements of this army of excluded pupils were included, Britain's ranking would, apparently, be closer to twenty-fifth than fifteenth out of the forty-four countries surveyed (Wragg, 2004).

Schools in Britain are subject to a heavy-handed bureaucratic control from central government, with formal examinations for pupils at ages seven, eleven, fourteen and sixteen. Schools are frequently inspected and their performance in the periodic tests is publicly ranked. Far from ensuring higher levels of achievement and learning, the opposite has resulted, with high levels of teacher malaise in under-funded schools, with large classes of dispirited pupils and frustrated teachers. Subjects such as music, games and physical education are increasingly left out of the curriculum to make way for yet more classes of formal instruction in 'basic skills'. Partly because of the publicity surrounding published league tables and the 'shaming' of underachieving schools, there is strong pressure to exclude learning-disabled, disruptive, and emotionally maladjusted pupils. It is perhaps no coincidence that in the summer term in which public examinations are held, the number of excluded pupils reaches a peak; more than 17,000 pupils were permanently excluded from school in 2003, the majority of them in the summer term (Blair, 2004). The main reason given for these expulsions was pupil misbehaviour, but this is intimately linked to failure to take advantage of the instruction offered, and failure on formal tests of achievement. Some of these 17,000 excluded pupils will be admitted to other schools, but many join the small army of the permanently excluded.

The top country in the international league tables on comparable tests of reading, science and mathematics was Finland (sharing honours with several Scandinavian countries). This prompted Curtis (2004) to examine Finnish school policies that might explain this. The idea that schools should be run from the centre, or even have their test results published, is unthinkable in Finland. The only public examinations are those taken by students at age eighteen. Secondary schools are entirely comprehensive, taking all ability bands and those with 'special' educational needs, attempting to teach to their pupils' highest potential, not the lowest as seems the case with Britain's 'comprehensives'. Private schools are unknown in this small country. Teachers themselves have high status and salaries (on a par with lawyers and doctors), and all are qualified at the master's level or beyond. Schools themselves have priority in government funding, and class sizes are much smaller than in Britain.

Finland also has an excellent record in its educational policies for the reception and absorption of children of immigrants and refugees, in comparison with several other European countries, including Britain (Pitkänen *et al.*, 2002). Britain's comparative failure in the integration of children of immigrants is demonstrated by the exclusion statistics (DfES, 2004). These show that the highest rates of exclusion are of Gypsy and Roma children, followed by students with cultural origins in the Caribbean – this latter group being more than three times as likely as any other group to be excluded from school.

Particularly striking with regard to mainstream British students is a report from the Confederation of British Industry (CBI, 2004) of newly-recruited school leavers. Cumulatively since 1997, two million school leavers had insufficient skills in literacy and mathematics to enable them to advance occupationally: they were judged to be fit only for the lowest level of occupation, since they had failed to achieve adequate basic skills in their schooling. Overall, 47 per cent of British firms were dissatisfied with the educational quality of the school-leavers they recruited. These figures do not include the small army of permanently excluded pupils, who rarely enter the job market in any capacity. These figures are consonant with a 1999 report from the Basic Skills Agency (Moser, 1999) which found that one-fifth of British adults had 'severe problems' with basic literacy and numeracy, with skills in these areas lower than in any other European country except Poland and Ireland. A 'National Curriculum' has aimed to improve the reading and mathematical skills of eleven-year-old British children; but by the age of sixteen virtually all of these gains had been lost in children from deprived areas. Official British figures (Smithers, 2006) indicate that some 12 million British workers cannot read beyond the level expected of eleven-year-olds in the national literacy tests. In contrast, in other developed European countries (including Germany) the problem of functionally illiterate school-leavers is virtually absent (Machin, 2005).

What happens to permanently excluded students in Britain? The answer to this important question is not entirely clear, since the government Department for Education and Skills (DfES, 2004) acknowledges that each year educational systems 'lose track' of some 10,000 students before they are aged sixteen. This has been attributed to school policies which aim to enhance achievement profiles by expelling under-performing pupils (Brighouse, 2004). Such pupils are usually referred either to special 'referral units', or to any other school that is willing to take them on. The referral units offer remedial courses in basic skills (reading, writing and arithmetic) and vocational training. However, the atmosphere in these centres is often less than professional, and students are frequently absent. Permanent drop-out is often not followed up. Many of the permanently excluded form a cadre of street youth, alienated and depressed, making money from petty crime from early adolescence onwards, and increasingly becoming prey to drug pushers and those who wish to sexually exploit the young (Bagley and Pritchard, 1998a; Bagley and King, 2003).

School class sizes in Britain

For some time the myth has prevailed in British educational policy that 'class size doesn't matter', and it is the qualifications, experience and the dedication of the teacher that is most important. Of course, well-qualified and highly motivated teachers are important, but unfortunately the morale of British teachers has been undermined in recent years because of poor pay, difficult working conditions and the popular perception that teaching is an unrewarding profession, not just in financial terms.

In Britain some primary school teachers used to have instructional aides, but the employment of these partially qualified part-timers has been curtailed because of funding cuts in education (Smith, 2003). While a legal regulation in 1998 specified that early school classes in Britain should be no larger than thirty, in practice class sizes in the primary school are often much larger than this. In Bangladesh, a prominent NGO the Bangladesh Rural Advancement Committee (BRAC), offers primary education in many areas based on a maximum of twenty-three children a class in rural primary schools (Verma, 2004).

The connection between class size and school exclusions is, I would argue, linked to the fact that teachers in Britain are often unable to address the learning problems of bored, alienated and potentially rebellious students, as well as those with 'special needs'. It is no coincidence that the highest proportion of exclusions from school occur in local authority areas which have the poorest teacher–pupil ratios. Again, there is a vicious circle here: in those areas which have the lowest achieving and most poorly behaved students, teacher turnover is highest and in consequence classes frequently become very large because of chronic teacher shortage. The latest figures on exclusions from British schools (for 2003–4) showed a six per cent increase in the numbers excluded compared with the previous year. The number of exclusions has doubled in a decade. Many of these 344,510 students were approaching the final year examination stage, examinations which they would never take (Halpin, 2005; Smithers, 2005).

In Britain Iacovou (2001) has argued that previous British research on class size and achievement has been flawed, since it failed to control for reasons why pupils have been assigned to small classes; often it has been pupils with educational difficulties, underachievement due to underlying cognitive problems, and/or behavioural maladjustment who have been assigned to very small classes. Including the achievements of these pupils with those who are retained in larger classes gives a skewed result, showing that larger classes contain more highly achieving pupils – but this finding is an artefact of referral procedures. It has been acknowledged by researchers that children in private schools in Britain, where class sizes are on average less than half of those in publicly-funded schools have much higher achievements than pupils in the state system, but this effect has usually been attributed to the social class bias in the intake of private schools.

Iacovou's (2001) British research followed up some 12,100 children in the National Child Development Study (NCDS), a cohort of children born in one week in 1958, and studied systematically at birth and at ages seven, eleven and beyond. First of all, she found as expected that pupils assigned to lower streams in primary schooling had poorer initial reading ability, and these lower streams had smaller numbers of children, and included many children with 'special needs'. Iacovou found that smaller class size – the normal variation in numbers in regular streams, not that resulting from any specific experiment, or allocation for special education purposes – was associated, when streaming policy and a variety of other social factors were controlled for, with *higher* achievement.

The smaller class-size effect accounted for an enhancement of about one-third of a standard deviation in reading test scores, a highly significant result. This important finding suggests that even quite small levels of class size reduction can have positive effects. Furthermore, a reduction in class size of eight pupils below the national average was associated with a highly significant 40 per cent increase (of one standard deviation) in reading scores, slightly larger than the achievement advantage of coming from an advantaged social class, and ten times the advantage bestowed by having a mother with an additional year of completed education. The advantage in reading ability through being in a smaller class at age seven was retained at age eleven, particularly in children from larger families.

While the variation in class sizes in this British study reflected a naturally occurring variation in the policies and resources of different schools, and was not the result of a carefully contrived experiment as in the Tennessee STARS project, the effect size in enhancement of achievements was quite similar to those observed by Achilles (1996) in Tennessee. The American research has shown that halving school class sizes in Grades Kindergarten through Three has long term benefits in achievement and behaviour lasting into college age, and the expenditures involved in halving these class sizes is highly cost effective (Finn and Achilles, 1999; Krueger, 1999).

The British NCDS study reflected an era of very large classes (average primary school class sizes were 35.9), and since that time average primary school class sizes have fallen to a little over thirty. There are strong grounds for supposing, however, that since the Tennessee STARS experiment and the NCDS statistical study produced similar results in school achievement, the social advantages produced by the Tennessee experiment (better student morale, higher self-concept, better behaviour, higher motivation, lower school drop-out) would also occur in pupils in smaller classes in Britain. This suggestion is important for the discussion in a later section on school exclusions and their sequels in our 'two-schools experiment'.

The two schools experiment in educational and social work intervention to prevent school exclusions and the 'cycle of poverty'

I turn now to our own work which offers somewhat optimistic conclusions. There is substantial evidence that schools which serve neighbourhoods with a high proportion of indicators of deprivation and social problems (poverty and unemployment; overcrowded and impermanent housing; child welfare interventions; high delinquency and crime rates; and high rates of mental illness) have, on average, significantly poorer achievement in their students, and much higher rates of school exclusions than in schools in stable or prosperous neighbourhoods.

Farrington and Welsh (2007), in important British research on 'delinquent careers' concluded:

> The whole process is self-perpetuating, in that poverty ... and early school failure lead to truancy and lack of educational qualifications, which in turn lead to low status jobs and periods of unemployment ... all of which make it harder to achieve goals legitimately.

The experiment described below was funded through the British Home Office 'Safer Cities' programme, and aimed through focusing on school-based social work, to reduce pupils' disruptive behaviour and expulsions, and to increase their motivation to achieve legitimate goals. In this we also attempted to replicate the experimental English work of Rose and Marshall (1975) which showed that social work interventions at the school level could have a strong role in reducing delinquency.

Our experimental study (Bagley and Pritchard 1998a, 1998b; Pritchard, 2001) selected two schools (linked primary and secondary, serving some 1,300 children) in a city in southern England and matched them with two similar schools in another area of the city. In both experimental and control school settings there were similar levels of deprivation, with poverty rates of 60 per cent (judged by the proportion of pupils receiving free school lunches). The neighbourhoods serving these two school areas had well above average proportions of social service interventions, unwanted pregnancies, and criminal convictions.

Inputs over three years in the experimental schools were an additional teacher in the primary school, a half-time additional teacher in the secondary school, and a project social worker who operated with families and children attending both primary and secondary schools. The additional teachers worked in the areas of instruction and counselling, and with the project social worker in co-ordinated strategies. The additional primary teacher worked intensively with children in the infant reception classes and with their families, trying to ensure that incipient problems of learning and behaviour could be addressed. In the secondary school the additional teacher focused on both bullying and behavioural problems, seeking a variety of solutions to avoid the need for exclusion of disruptive students.

The social worker ensured that all families received maximum benefit from income and social services, with the focus on preventing family disruption. Families of pupils whose under-performance in scholastic areas reflected their frequent absenteeism were engaged. Again, the focus was on helping the parents to emphasize the need for achieving educational goals by full attendance. Health education in the secondary school focused on risky sexual behaviour and drug use, with a focus on long-term achievements versus short-term gratifications.

Evaluation consisted of self-report questionnaires and tests completed by pupils at the beginning and end of the three-year project. These measures were completed by pupils in the experimental and control primary and secondary schools (Bagley and Pritchard, 1998a). There was a highly significant fall in self-reported delinquency, fighting, experience of bullying, truanting and drug use in the project schools, but the incidence of these events actually increased in the control schools. Positive attitudes to school increased significantly in the project schools, but there was no parallel increase in the control schools. In the project schools, for children's families there was a significant decline in problem behaviours, including movement of children into care, criminality in adult family members, and unwanted pregnancies. Significantly fewer children from the project schools were excluded for any reason.

A follow-up of children from the secondary schools to age nineteen indicated that the positive effects of the school social work experiments were retained, with significantly fewer young people becoming pregnant, delinquent, leaving school early, or being unemployed. Careful estimates of the costs to the public purse of processing delinquents, supporting unmarried mothers, keeping children in care, and processing and maintaining delinquent children in youth detention, indicated that although initially expensive, the intensive social work and educational inputs had, over a five-year period saved the public purse a net sum of £156,310, using the most conservative estimates. Generalizing these figures to the country as a whole we estimated that 'at least a billion dollars' of public expenditure could be saved in the long-term, through early interventions and the reordering of chaotic and wasted lives which were the lot of many of the pupils who graduated from the control secondary school in our experiment (Bagley and Pritchard, 1998b).

Educational failure, poverty and school exclusions in Britain

There is a chronic crisis in British education from the highest to the lowest levels. Universities face a crisis of under-funding, and infant and primary school classes are too large for fully effective teaching. Teacher morale is low, and classes are getting larger. In such contexts alienated pupils and those with special needs are easily ignored or expelled. These policies operate, by default, in an extremely wealthy country, but in one in which incomes and resources are unequally distributed, with degrees of inequality much greater than in many countries with similar or lesser sources of national wealth.

The ecological dimension of unequal schooling means that schools, both primary and secondary, serve deprived areas marked by very high levels of poverty, infant mortality and morbidity, poor housing, unemployment, criminality, and mental health problems (Bagley, 2006). Schools in these areas struggle not only with a high proportion of disaffected and underachieving pupils, but also experience a poverty of resources and a high turnover of teachers who find working in such schools particularly difficult. This in turn leads to chronically larger classes than those – thirty pupils per class – required by current policy.

British research indicates that even relatively small reductions in school class sizes can be reflected in a significant enhancement in reading abilities. American research clearly shows that halving class sizes in primary schools in the early years (to between fifteen and seventeen pupils per class) results in significant and enduring scholastic gains, better behaviour and motivation, better self-concept, less school drop-out, and greater college attendance (Slavin, 1990; Achilles,

1997). The reasons for these improvements seem to be that teachers of small classes in the early years are able to focus more readily on the individual learning, behavioural and social needs of their pupils. Although halving class sizes in the age group five to eight years is expensive, these expenditures are highly cost-effective in the medium-term.

It is not surprising that pupils in Britain's overcrowded classrooms perform, on average, rather poorly on internationally standardized tests of ability, and clearly below the level expected of a nation with Britain's national wealth. Inequalities of income make these problems worse, and children from the poorest families attending the poorest schools are also likely to experience significantly higher rates of illness and premature death (from infections, accidents, and incidents of abuse), as well as neglect, delinquency, underachievement, and school exclusions. Economic and social disadvantage in Britain is often transmitted between generations, and upward mobility rates are low compared with several other countries. In other words, being born into a disadvantaged social class tends to be a deterministic status (Bagley, 2006).

A review of our experimental programmes aiming to prevent school exclusions and improve the welfare of families and children from poverty neighbourhoods shows that despite their initial expense, these programmes can be highly cost-effective in preventing children moving into a cycle of family poverty in which their own children are neglected, demotivated, marked down for careers of petty and sometimes major crime, unemployment, and drug-taking. Vigorous interventions which are school-based and family-oriented can be successful in breaking the deterministic patterns of being born into a disadvantaged family in an underprivileged neighbourhood.

This is the dilemma of social policy of Britain today. A rich nation could afford to vastly improve the quality of education and the welfare of families and children. Far from being expensive this would actually be cost-effective in the medium- to long-term, saving the public purse many millions of pounds. But governments seem reluctant to make major social investments whose return might not be measurable within the normal life of a parliamentary five-year term.

American education in crisis? Ways forward

America has its own crises and dilemmas in inclusive and multicultural education, as Grant and Saran show in Chapters 4 and 5 of this book. The movement to desegregate American schools is under legal and administrative challenge, and in 2006 was under contest before the US Supreme Court in a suit brought on behalf of a group of Seattle parents who objected to African-American children being bused into (high quality?) mainly Euro-American schools, while some white children are bused into (low quality?) mainly Afro-American schools (Dillon, 2006). The Supreme Court decision will have far-reaching effects for the more than 1,000 school districts which operate some form of busing in order to achieve equity in enrolments. Arguments to the US Supreme Court were scheduled for November 2006 (after this book went to press): a federal judge had previously ruled that the educational system did not require quotas, arguing that other factors such as new geographical boundaries and increased quality of individual schools could address the issue of 'segregated education equals poor quality education for some ethnic minorities'.

If the Supreme Court Justices rule against busing policies for integration, then the 'creative multiculturalism' advocated by Saran in Chapter 5 of this volume will be doubly important. To the present time, multiculturalism in American urban education has served the hegemony of the European-American ruling class. Lip-service has been paid to the concept of multiculturalism, but the fundamental value has been one of 'America first' with what is effectively a mono-cultural, Eurocentric curriculum. It must be acknowledged, however, that the 'critical consciousness' which such a multicultural policy evokes faces many roadblocks, stereotypes and backlashes (Pitner and Sakomoto, 2005). Seeking equity of educational treatment in dynamic, changing societies is a constant challenge, as Verma and Papastamatis argue in Chapter 6.

Booth's (2006) 'Index of Inclusion' can be a valuable guide for teachers fostering social inclusion in the face of potential backlash against policies for inclusion and school-based multiculturalism. Booth's 'Index of Inclusion' has been used in twenty-five countries, including India, Brazil, South Africa, England and countries in the Middle East and North Africa, and offers ways in which teachers can include various minorities in mainstream activities. Ainscow *et al.* (1998) also offer a valuable account of the 'hidden voices' of schoolchildren with disabling conditions, and the meanings which these accounts hold for teachers.

Adapting cognitively and culturally, following migration

Should schools offer a common curriculum underpinned by a set of values which enable all of its students to participate successfully in society, co-operating despite their individual backgrounds and aspirations in a 'civic society' marked by tolerance rather than conflict? The alternative model is one in which career tracks reflect a cultural (as opposed to a multicultural) curriculum with simple common elements socializing all students for life in American, British or European society. In this latter model the individual cognitive style of students may be addressed by teachers who try and instruct them according to their alleged individual culturally determined needs. As Woodrow observes in Chapter 7 of this volume, the dominant culture transmits through its educational systems the 'cultural capital' of learning styles which serve the hegemony of privileged classes. Social class and immigrant groups which cannot access this cultural capital may become alienated and doubly disadvantaged by what transpires in learning environments.

Woodrow draws attention in his chapter to the work on cognitive style by Herman Witkin (Witkin and Goodenough, 1981). Witkin contrasted the 'field independent' cognitive style, in which learning and perception is not dependent on cues in the individual's social milieu or wider environment, with the 'field dependent' cognitive style in which group settings and external cues for learning and motivation are crucially important. The first kind of cognitive style is common in individualistic (mainly industrialized) cultures. In contrast, group-oriented (mainly pastoral) cultures foster a field dependent cognitive style, in which individuals have poorly developed perceptual skills as measured by the Embedded Figures Test. However, Asian countries (such as Japan and China) that, despite their industrial status, discourage individualism, challenge Witkin's model.

Before undertaking research with a Japanese colleague comparing British and Japanese children's scores on the Children's Embedded Figures Test (CEFT), we asked Witkin for a hypothesis about likely results. Witkin (personal communication) argued that given Japan's strong cultural emphasis on subordination of individual aspirations to those of the group, Japanese children would be highly field dependent. The *opposite* proved to be the case: Japanese children had excellent skills on the task of perceptual disembedding, and a third of these ten-year-olds achieved the maximum score, equivalent to that achieved by the average American fourteen-year-old on which the test had been normed (Bagley *et al.*, 1983).

We speculated that Japanese excellence in field independent perceptual disembedding skills reflected the fact that becoming literate in Japanese involves 'unpacking' complex symbols for individual words, a task quite similar to that investigated by the Children's Embedded Figures Test. Although Japanese script forms have been simplified, this has not been the case in China. In some parts of rural China some children (mainly girls) receive only a minimal education in learning to recognize the pictographs, each of which represents a single word. But we found that in rural China exposure to schooling was *not* correlated with measured abilities on the Children's Embedded Figures Test (CEFT), on which even partially-schooled Chinese girls had high scores (Bagley, 1996). The search for the causes of Chinese children's excellence in these and other cognitive tasks must continue – but we are reluctant to ascribe specific genetic potential to an ethnic group, and instead point to the numerous pressures to achieve cognitive excellence, even in very poor Chinese families: but these pressures do take their toll on Chinese children's mental health adjustments (Tse and Bagley, 2002).

Comparative research between Jamaica and Canada showed that before migration rural Jamaican children were (as Witkin's model predicted) highly field dependent, with low scores on the CEFT. But following migration to Canadian cities, these rural Jamaican children had within two years acquired levels of perceptual disembedding which equalled those of children born in urban Canada (Bagley and Young, 1983; Bagley, 1988). Living in a complex urban setting means that children rapidly master a variety of perceptual and cognitive skills which give them options within urban education.

Children of Chinese and Asian parents in Britain and Canada are likely to be high achievers, compared to children of European-origin parents (Verma *et al.*, 1999; Bagley *et al.*, 2001). Saran (in Chapter 5 of this volume) warns against a kind of symbolic prejudice which says in effect that since Asian-origin students are doing so well, then there is little structural racism in society, and little bias in the curriculum. In effect, Asian students accept the stereotyped roles in technology, medicine, information science and entrepreneurship which society has prepared them for – but the 'glass ceiling' remains for both Asian-origin males and female graduates from all ethnic groups. We thus address the findings in Woodrow's chapter by agreeing with one of his conclusions: 'as the subcultures become symbolically richer and have more capital, the governing society will intuitively change the exchange rates and work to devalue the currencies in which the subcultures have saved'.

Woodrow identifies further the 'Assumptions [by teachers] about how students learn, which almost inevitably discriminate for or against particular learning preferences'. Our own work supports this opinion, and we have criticized educationists who advise that teachers should match their teaching style to the alleged cognitive style of their students (Bagley and Mallick, 1998). Urban cultures in Europe and America produce children with complex cognitive styles, and teachers should possess multiple skills for working in the multi-ethnic classroom, just as the teacher should have awareness and skills in teaching for 'critical multiculturalism' (see Saran's chapter in this volume). Teachers too should actively engage in delivering an education formed by multicultural values which address issues of inequality, racism and diversity (see Chapter 2 in this volume).

Demise of the Marxian models of education?

In some parts of the European Union, countries such as Greece still deliver educational systems whose values and practice owe more to tradition (see Chapter 9 by Papastamatis in this volume) than to the values of inclusion which have emerged, for example, from the landmark UNESCO Symposium, which urged that inclusive schools:

> must recognize and respond to the diverse needs of their students, accommodating both different styles of learning and ensuring quality education to all through appropriate curricula, organizational arrangements, teaching strategies, resource use and partnerships with their communities.
>
> (UNESCO, 1994)

It was traditional in the Marxist countries of the USSR to offer the ideology to the Western observer that education's major purpose was to overcome class divisions, and to educate each child according to his or her individual talent; this ideology provided, perhaps paradoxically, schools for sporting and scholastic elites, but selected regardless of social backgrounds and with a strong subtext of political socialization within each school for conformism to Marxist-Leninist values. In the socialist economy, there would be no 'reserve army of labour': the talents of all pupils would be developed so that each one could find an occupational role serving a socialist society, in which unemployment does not exist.

Gray, in Chapter 8 of this volume, offers an interesting historical perspective on Soviet education which, despite a centralized, Marxist-Leninist curriculum delivered in Russian, nevertheless allowed education at the primary level to continue in local languages in dozens of officially recognized ethnic groups (Tomiak, 1983).

Gray draws attention to the plight of Gypsy and Roma children who, in all countries of Europe (including Britain), experience greater marginalization and persecution than any other ethnic group, Islamaphobia notwithstanding. In previous decades two European countries (Germany and Austria)[2] sought to achieve a 'final solution' to the 'Gypsy problem' through extermination – the final logic of programmes of failed assimilation. Roma and Gypsy people wish to retain rights of cultural retention, and travel across regional and national boundaries, ideally accessing centres along the way which provide educational, medical, social and cultural support. This lifestyle does not fit well into the requirements of capitalism's new global ethos, which infects all regions of the world. Like Dalits in India, Roma people are not 'a reserve army of labour', but a hindrance and a cost to capitalism, to be removed and relegated by whatever means the world will tolerate.

In Britain a tenth of the population in this class-ridden society forms an under-class ignored or rejected by educational systems, and this subgroup is destined for lifestyles which are hugely expensive for the capitalist system which both creates and controls them. The irony is that the cost of controlling this unruly underclass over their lifetime costs billions of pounds (Bagley and Pritchard, 1998b), and this is much greater than the cost of educating them properly, offering them inclusive rather than exclusive education, and training, counselling and supporting them for productive and rewarding employment. Linked to these policy failures, policy and practice for children, adolescents and young people with a variety of 'special needs' in Britain remain in chaos (Halpin, 2006). The capitalist model of efficiency and reducing costs has failed, both tolerating and fostering the perpetuation of a despairing and highly expensive underclass.

Ways forward for Indian inclusive education

India stands at a threshold of economic change and cultural development. Its 'Education for All' policies are only now, some fifty years after independence, being implemented properly. And only in some voluntary schools are children with 'special needs' being included (Jha, 2002, 2006). Jha offers an idealistic vision in his picture of a 'school without walls' which:

> Removing barriers and bringing all children together in school irrespective of their physical and mental abilities, or social and economic status, and securing their participation in learning activities leads to the initiation of the process of inclusive education. Once walls within schools are broken, schools move out of their boundaries, end isolation and reach out to the communities. The distance between formal schools, non-formal schools, special schools and open schools will be eliminated.
>
> (Jha, 2002, pp. 15–16)

Siddiqui's chapter (12) in this volume, on inclusive education for street and homeless children, shows how schools without walls and part-time schools can to some extent serve the needs of this intensively deprived population. His chapter also shows that state schools should offer free (or highly subsidized) quality education for all social classes, and this requires vigorous interventions by the government of India. Srivastava, in Chapter 11 in this volume, gives a detailed case study from Uttar Pradesh and comparable states of policy options and dilemmas in providing Education for All through the 'low-fee private' system (the 'low-fee' being the average daily wage of a labourer, for one month of schooling). For Dalit and Tribal children, schools provided by Christian missions are valuable, even when the missionaries experience the same violence and persecution as the Dalits themselves.

India struggles to provide 'Education for All' in a rapidly modernising country, but one in which barriers of caste and poverty mean that full social inclusion and 'schools without walls' (Jha, 2002; and Chapter 2 in this volume) are confined to a small number of outstanding examples. America struggles to maintain the movement towards equity and integration of ethnic minorities in the face of the reassertion of power by privileged majority groups. Britain copes with problems of diversity and difference through policies of exclusion, despite case examples that inclusive policies, supported by social service interventions, can be highly cost-effective. Countries of the 'new' European Union struggle against a legacy of past policies which still influence a rigid pedagogy and which often excludes groups such as those with special needs; and still persecutes marginalized peoples, such as Roma and Gypsy ethnic groups (REI, 2005).

Notes

1 How is this high level of national wealth compatible with the poor quality of education for many school-leavers, and their marginal employability? The answer is the employment of a large number of temporary workers with varying degrees of skill, from European countries (Smallwood, 2006).
2 From 1939 to 1945 the Nazi government of Austria and Germany set up extermination camps for Roma and other ethnic groups in all of the European countries they occupied. Between 25 and 50 per cent of Roma people in Europe (in excess of 250,000) were murdered in these camps, and in mobile execution centres (Laqueur and Baumenl, 2001).

References

Achilles, C. (1996) *Summary of Recent Class-Size Research With an Emphasis on Tennessee STAR and Derivative Research Studies.* Nashville: Tennessee State University, Center of Excellence for Research Policy on Basic Skills.
Achilles, C. (1997) 'Small classes, big possibilities', *The School Administrator,* 54, 6–15.
Asthana, A. and Hinsliff, G. (2006) 'Special needs education condemned: parents and children let down by schooling, says damning Commons Committee report', *The Observer Online,* 2nd July 2006.
Ainscow, M., Booth, T. and Dyson, A. (1998) 'Inclusion and exclusion in schools: listening to some hidden voices', in Booth, T. and Ainscow, M. (eds) *From Them to Us: An International Study of Inclusion in Education.* London: Routledge.
Bagley, C. (1985) 'Multiculturalism, class and ideology: a European-Canadian comparison', in S. Modgil, G. Verma and C. Modgil (eds) *Multicultural Education: The Interminable Debate.* Lewes: The Falmer Press.

Bagley, C. (1988) 'Cognitive style and cultural adaptation in Blackfoot, Japanese, Jamaican, Italian and Anglo-Celtic children in Canada', in G. Verma and C. Bagley (eds) *Cross Cultural Studies of Personality.* London: Macmillan.

Bagley, C. (1996) 'Field dependence in children in group-oriented cultures: comparisons from China, Japan and North America', *Journal of Social Psychology,* 135, 523–526.

Bagley, C. (2006) *Achievement and Failure of Disadvantaged Pupils in Britain: Policy Challenges and Initiatives in Education, Health, Income Support and Community Development.* Manchester: The Manchester Educational Research Network.

Bagley, C., Bolitho, F. and Bertrand, L. (2001) 'Ethnicities and social adjustment in Canadian adolescents', *Journal of International Migration and Integration,* 2, 99–120.

Bagley, C., Iwawaki, S. and Young, L. (1983) 'Japanese children: group oriented but not field dependent?' in C. Bagley and G. Verma (eds) *Multicultural Childhood: Education, Ethnicity and Cognitive Styles.* Aldershot: Gower Press.

Bagley, C. and King, K. (2003) *Child Sexual Abuse: The Search for Healing.* London: Routledge-Tavistock.

Bagley, C. and Mallick, K. (1998) 'Educational implications of field dependence-independence: a commentary', *British Journal of Educational Psychology,* 68, 589–594.

Bagley, C. and Pritchard, C. (1998a) 'The reduction of problem behaviours and school exclusion in at-risk youth: an experimental study of school social work with cost-benefit analyses', *Child and Family Social Work,* 3, 219–26.

Bagley, C. and Pritchard, C. (1998b) 'The billion dollar cost of troubled youth: prospects for cost-effective prevention and treatment', *International Journal of Adolescent and Youth,* 7, 211–225.

Bagley, C. and Young, L. (1983) 'Class, socialization and cultural change: antecedents of cognitive style in children in Jamaica and England', in C. Bagley and G. Verma (eds) *Multicultural Childhood: Education, Ethnicity and Cognitive Styles.* Aldershot: Gower Press.

Bekhradnia, B. (2006) *Demand for Higher Education to 2020.* Oxford: The Higher Education Policy Unit.

Blair, A. (2004) 'Violence leads to exclusion of 17,000 pupils in one term', *The Times Online,* 21st July 2004.

Blair, T. (2006) 'New educational policies,' *Daily Telegraph Online,* 25th February 2006.

Booth, T. (2006) 'The "Index for Inclusion" in use: learning from international experience', *EENT (Enabling Education Network) Newsletter,* no. 9.

Bradshaw, J. and Chen, J.-R. (2002) *Poverty in the UK: A Comparison with Nineteen Other Countries.* York: Social Policy Research Unit, University of York.

Brewer, M., Goodman, A., Shaw, J. and Shepherd, A. (2005) *Poverty and Inequality in Britain 2005.* London: The Institute of Fiscal Studies.

Brighouse, T. (2004) *Policy Statement.* London: Office of the Commissioner for London Schools.

Brock, M. (1978) 'We must educate our masters', *Oxford Review of Education,* 4, 221–232.

Brown, G. (2006) *The Chancellor of The Exchequer's Budget Speech to the House of Commons in Full, Guardian Online,* 22nd March 2006.

Cassidy, S. (2005) 'University entrance getting more difficult for poor pupils', *Independent Online,* 30th November 2005.

CBI (2004) *Educational Standards of British School-Leavers.* London: Confederation of British Industry.

Chevalier, A. and Dolton, P. (2004) *Teacher Shortage: Another Impending Crisis?* London: London School of Economics, Centre for Economic Performance.

Curtis, P. (2004) 'Best schools are in Finland and the far east', *Guardian Online,* 7th December 2004.

Dillon, S. (2006) 'Schools' efforts on race await Justices' ruling', *The New York Times Online,* 24th June 2006.

De Vogli, R., Mistry, R., Gnesetto, R. and Cornia, G. (2005) 'Has the relation between income inequality and life expectancy disappeared? Evidence for Italy and top industrialised countries', *Journal of Epidemiology and Community Health,* 59, 158–162.

DfES (2004) *Statistical Profile of School Exclusions.* London: Department for Education and Skills.

Dutton, E., Warhurst, C. and Fairley, J. (2005) *New Britain – Old Politics: Devolved Post-16 Education and Training.* Glasgow: University of Strathclyde, Department of Human Resource Management, Research Paper 57.

Education Guardian (2004) 'Rise in school class sizes', *Guardian Online,* 2nd October 2004.

Farrington, D.P. and Welsh, B.C. (2007) *Saving Children from a Life of Crime.* Oxford: Oxford University Press.

Finn, J. and Achilles, C. (1999) 'Tennessee's class size study: findings, implications, misconceptions', *Education Evaluation and Policy Analysis,* Summer, 97–109.

Frederickson, N. and Kline, T. (2002) *Special Educational Needs, Inclusion and Diversity: A Textbook.* Buckingham: The Open University Press.

Haile, D. (2005) 'Supersize classes blow for school-kids', *Manchester Evening News Online,* 19th May 2005.

Halpin, T. (2005) 'Almost 10,000 pupils expelled as violence against teachers escalates', *The Times Online,* 24th June 2005.

Halpin, T. (2006) 'The House of Commons Education and Skills Select Committee Report on Special Education', *The Times Online,* 6th July 2006.

HEFCE (2005) *Young Participation in Higher Education.* London: Higher Education Funding Council for England.

Hobson, D. (2001) *The National Wealth: Who Gets What in Britain.* London: Harper Collins Business.

Iacovou, M. (2001) *Class Size in the Early Years: Is Smaller Really Better?* Wivenhoe: University of Essex, Institute for Social and Economic Research.

Jeffrey, C., Jeffery, R. and Jeffery, P. (2004) 'Degrees without freedom: the impact of formal education on Dalit young men in Northern India', *Development and Change,* 35, 963–986.

Jha, M. (2002) *School Without Walls.* London: Heinemann.

Jha, M. (2006) 'The inclusive education of children with "special" educational needs in India: a case study of three schools in New Delhi', unpublished D. Phil. thesis, University of Oxford.

Krueger, A. (1999) 'Experimental estimates of educational production functions', *Quarterly Journal of Economics,* 114, 497–532.

Laqueur, W. and Baumenl, J. (2001) 'Genocide of European Roma (Gypsies)', *The Holocaust Encyclopedia.* London: Yale University Press.

Machin, S. (2005) 'Demise of grammar schools leaves poor facing uphill struggle', *The Times Online,* April 25th 2005.

Moser, C. (1999) *Illiteracy and Innumeracy in Britain's Adults.* London: Institute of Education, the Basic Skills Agency Resource Centre.

OECD (2004) *International Survey of Achievement in 15-Year-Olds.* London and Paris: Organisation for Economic Co-Operation and Development.

OECD (2006) *Education at a Glance.* Paris: Organisation for Economic Co-operation and Development.

Picus, L. (2000) *Class-Size Reduction: Effects and Relative Costs.* Eugene ORE: ERIC Clearing House, Policy Report No. 1.

Pitkänen, P., Kalekin-Fishman, D. and Verma, G. (2002) *Education and Immigration: Settlement Policies and Current Challenges.* London: Routledge-Falmer.

Pitner, R. and Sakomoto, I. (2005) 'The role of critical consciousness in multicultural practice: examining how its strength becomes its limitations', American Journal of Orthopsychiatry, 75, 684–694.

Pritchard, C. (2001) *A Family-Teacher-Social Work Alliance to Reduce Truancy and Delinquency – the Dorset Healthy Alliance Project.* London: The Home Office, RDS Occasional Paper No. 78.

Reed, J. (2004) *Toward Zero-Exclusion: Beginning to Think Bravely.* London: Institute for Public Policy Research.

REI (2005) *Separate and Unequal: Combating Discrimination against Roma in Education.* New York: Columbia University, Budapest Law Center, Roma Education Initiative.

Rose, G. and Marshall, T. (1975) *Counselling and School Social Work.* Chichester: John Wiley.

Scottish Office (2005) *For Scotland's Children.* Edinburgh: The Scottish Office.

Slavin, R. (1990) 'School and classroom organization in beginning reading: class size, aides, and instructional grouping', *Contemporary Education.* Reprinted at: www.successforall.com/Resource/research/schorgrd.htm.

Smallwood, C. (2006) '"Reserve army" in defusing demographic time bomb', *Sunday Times Online,* 27th August 2006.

Smith, L. (2003–2005) 'England's schools in funding crisis', *World Socialist Web Site.* At: www.wsws.org/articles/.

Smithers, R. (2005) 'Permanent school exclusions rise', *Guardian Online,* 24th June 2005.

Smithers, A. (2004) *Policy Statement.* Buckingham: University of Buckingham, Centre for Education and Employment Research.

Smithers, R. (2006) 'Twelve million workers have reading age of children', *Guardian Online,* 24th January 2006.

Tomiak, J. (1983) *Soviet Education in the 1980s.* London: Croom Helm.

Tse, J. and Bagley, C. (2002) *Suicidal Behaviour, Bereavement and Death Education in Chinese Adolescents: Hong Kong Studies.* Aldershot: Ashgate.

UN (2005) *United Nations Development Program Report for 2005.* New York: The United Nations.

UNESCO (1994) *The Salamanca Statement and Framework on Special Needs Education.* Paris: UNESCO.

Van Dyke, R. and Little, B. (2005) *Debt, Part-Time Work and University Achievement.* London: University of the South Bank, for Higher Education Funding Council.

Verma, G. (2004) School of Education, University of Manchester, personal communication.

Verma, G., Bagley, C. and Mallick, K. (1980) *Illusion and Reality in Indian Education.* Aldershot: Gower Press.

Verma, G., Woodrow, D., Darby D., Shum, S., Chan, D., Bagley, C. and Skinner, G. (1999) *Chinese Adolescents in Hong Kong and Britain: Identity and Aspirations.* Aldershot: Ashgate.

Warnock, M. 1978) *Children and Young People with Special Educational Needs.* London: HMSO.

Warnock, M. (2005) *Special Educational Needs: A New Look.* London: Philosophy of Education Society.

Witkin, H. and Goodenough, D.R. (1981) *Cognitive Styles, Essence and Origins: Field Dependence and Field Independence.* New York: International Universities Press.

Wragg, T. (2004) *Commentary on the OECD Achievement Figures.* Exeter: University of Exeter, Institute of Education.

Index

THE BASICS

How do I read a poem? Do I really understand poetry?

This comprehensive guide demystifies the world of poetry, exploring poetic forms and traditions which can at first seem bewildering. Showing how any reader can gain more pleasure from poetry, it looks at the ways in which poetry interacts with the language we use in our everyday lives and explores how poems use language and form to create meaning.

Drawing on examples ranging from Chaucer to children's rhymes, Cole Porter to Carol Ann Duffy, and from around the English-speaking world it tackles subjects including:

- how technical aspects of rhyme and measures work;
- how different verse forms poem;
- how poetry relates to everyday language;
- how forms of poetry work, from sonnets to free verse;
- how all the elements of a poem contribute to its meaning.

Poetry: The Basics is an invaluable and easy-to-read guide for anyone wanting to get to grips with reading and writing poetry.

Jeffrey Wainwright is a poet and Professor of English at Manchester Metropolitan University, UK.

You may also be interested in the following Routledge Student Reference titles:

LITERARY THEORY: THE BASICS
HANS BERTENS

POETRY: THE BASICS
JEFFREY WAINWRIGHT

SHAKESPEARE: THE BASICS
SEAN MCEVOY

THE ROUTLEDGE COMPANION TO POSTMODERNISM
EDITED BY STUART SIM

THE ROUTLEDGE COMPANION TO RUSSIAN LITERATURE
EDITED BY NEIL CORNWELL

POST-COLONIAL STUDIES: THE KEY CONCEPTS
BILL ASHCROFT, GARETH GRIFFTHS AND HELEN TIFFIN

FIFTY KEY CONTEMPORARY THINKERS
JOHN LECHTE

WHO'S WHO IN DICKENS
DONALD HAWES

WHO'S WHO IN SHAKESPEARE
PETER QUENNELL AND HAMISH JOHNSON

WHO'S WHO IN CONTEMPORARY WOMEN'S WRITING
EDITED BY JANE ELDRIDGE MILLER

WHO'S WHO IN LESBIAN AND GAY WRITING
GABRIELE GRIFFIN

WHO'S WHO IN TWENTIETH-CENTURY WORLD POETRY
EDITED BY MARK WILLHARDT AND ALAN MICHAEL PARKER

WHO'S WHO OF TWENTIETH-CENTURY NOVELISTS
TIM WOODS

LANGUAGE: THE BASICS (SECOND EDITION)
R. L. TRASK

SEMIOTICS: THE BASICS
DANIEL CHANDLER

POETRY
THE BASICS

jeffrey wainwright

Routledge
Taylor & Francis Group

LONDON AND NEW YORK

First published 2004
by Routledge
11 New Fetter Lane, London EC4P 4EE

Simultaneously published in the USA and Canada
by Routledge
29 West 35th Street, New York, NY 10001

Routledge is an imprint of the Taylor & Francis Group

Typeset in Aldus Roman and Scala Sans by Taylor & Francis Books Ltd
Printed and bound in Great Britain by MPG Books Ltd, Bodmin

British Library Cataloguing in Publication Data
A catalogue record for this book is available from the British Library

Library of Congress Cataloging in Publication Data
Wainwright, Jeffrey.
    Poetry : the basics / Jeffrey Wainwright.
        p. cm.
Includes bibliographical references (p.).
    1. English poetry–History and criticism. 2. American poetry–History and
    criticism. 3. English language–Versification. 4. Poetics. 5. Poetry. I. Title.
        PR502.W27 2004
        808.1–dc22

ISBN 0–415–28763–4 (hbk)
ISBN 0–415–28764–2 (pbk)

TO KEN LOWE

CONTENTS

PREFACE

This book is – and I hope it will seem to be – a work of enthusiasm. My overriding aim is to enhance the pleasure that readers gain from poetry. No special expertise is required to read and enjoy a poem, but, as with most pleasures, it can be greatly enriched by knowledge. This book tries to provide some knowledge and some ideas to all who want to read, study or write poetry.

There is now substantial evidence that poetry has more writers than readers. Formal studies of literature, in universities and else-where, now include composition as well as analysis. The democratization of culture which has encouraged people to be producers of music and visual art has also influenced the language arts, and especially the poem. It is the most practically available literary space. It probably requires paper and certainly some time, but perhaps not as much of either as writing a novel. It needs readers, but not the human and physical resources necessary to realize a play or a film script. Moreover, the development of 'free verse' in the twentieth century has – for good and ill – had the effect of loosening convention, and this, together with the wider availability of knowledge about models in other periods and cultures, has expanded the long-standing practice of verse-making and given the developing poet a great range of possibilities from which to proceed. This book aims to encourage writers, who *must* therefore be readers, and readers who might also practise writing.

It aims to do so by providing knowledge of two kinds. The first is suggested by the topics of the chapters. At the heart of the book are chapters on the most distinctively formal aspects of poetry: the different 'voices' of poetry, the poetic line both measured and 'free', rhyme and stanza. I hope these will help with those technical aspects readers often find daunting. (Besides their explanation in the text, special terms – marked in **bold and italic** – are defined in the Glossary.) Around these are chapters which attempt to associate poetry with wider language-use whilst establishing the special character of what I shall call the 'deliberate space' that a poem occupies. The last chapter explores wider notions about the nature of poetic utterance, 'inspiration', and what it might be to be a poet.

I hope that the second kind of knowledge gained will be a greater familiarity with a wide reach of poets. The range is drawn from the Middle Ages to the present day, and from poetry across the English-speaking world. Each chapter includes sustained discussion of individual poems as well as briefer examples. I hope that these will develop the way readers might read individual poems closely, and draw them on to explore the work of poets, whether new or familiar, who attract them. Whilst I do not believe we all read a 'different' poem, none of us reads a poem exactly like our neighbour. Reading is a process, and the aim of my readings here is to contribute to the reader's own interior and exterior dialogues about the ideas and the whole experience of individual poems.

One particular idea about the nature of the art recurs in this approach to the 'basics' of poetry. This sees writers and readers working in the midst of a perpetual paradox. At one extreme is the desire to use words to *say* something that is meaningful and memorable: for instance, the kind of substantial statement required by grief or love. At the other is the desire to use words to say *nothing*, that is to free language from meaning and revel in the qualities and associations of words, even inventing new words: ''Twas brillig, and the slithy toves / Did gyre and gimble in the wabe.' The creative tension between these poles of interest will be apparent through much of the discussion and I hope that thinking about this will prove part of the pleasure I hope to foster.

ACKNOWLEDGEMENTS

The authors and publishers would like to thank the following for granting permission to reproduce material in this work:

Carcanet Press Limited for a quotation taken from Tom Raworth's poem 'now the pink stripes' found in *Collected Poems* © Carcanet Press Limited; Iona Opie for a quotation from 'Ladles and Jellyspoons' found in *The Lore and Language of Schoolchildren* by Iona and Peter Opie, Oxford University Press 1959, © Iona Opie; University of Nebraska Press for a quotation from 'That's What You Did And ...' translated by Dennis Tedlock found in *Finding the Center: Narrative Poetry of the Zuni Storyteller* © 1972, 1999 by Dennis Tedlock; Carcanet Press Limited for 'Siesta of a Hungarian Snake' by Edwin Morgan found in *Collected Poems* © Carcanet Press Limited; Pollinger Limited for a quotation from 'How Beastly the Bourgeois Is' by D. H. Lawrence found in *The Complete Poems of D. H. Lawrence* © The Estate of Frieda Lawrence Ravagli; Carcanet Press Limited for a quotation taken from 'An Early Martyr' by William Carlos Williams found in *Collected Poems* © Carcanet Press Limited; Carcanet Press Limited for a quotation taken from 'Homage' by William Carlos Williams found in *Collected Poems* © Carcanet Press Limited; New Directions Publishing Corp. for a quotation from 'Homage' by William Carlos Williams found in *Collected Poems: 1909–1939, Volume I* © 1938 New Directions Publishing Corp.; Carcanet Press Limited for a quotation taken from 'Pastoral' by William Carlos Williams found in *Collected Poems* © Carcanet Press

Limited; New Directions Publishing Corp. for a quotation from 'Pythagorean Silence' by Susan Howe found in *Europe of Trusts* © 1990 by Susan Howe; Faber and Faber Limited for a quotation from 'Considering the Snail' by Thom Gunn found in *My Sad Captains*; Faber and Faber Limited for a quotation from 'Portrait of a Lady' by T. S. Eliot taken from *Collected Poems 1909–1962*; Carcanet Press Limited for a quotation taken from 'Death the Barber' by William Carlos Williams found in *Collected Poems* © Carcanet Press Limited; New Directions Publishing Corp. for a quotation from 'Death the Barber' by William Carlos Williams found in *Collected Poems: 1909–1939, Volume I* © 1938 New Directions Publishing Corp.; New Directions Publishing Corp. for a quotation from 'Mid-August at Soughdough Mountain Lookout' by Gary Snyder found in *Earth House Hold* © 1969 by Gary Snyder; J. M. Dent for a quotation from 'Welsh' by R. S. Thomas found in *Collected Poems*; Carcanet Press Limited for a quotation taken from 'Les Luths' by Frank O'Hara found in *Selected Poems* © Carcanet Press Limited; Alfred A. Knopf, a division of Random House, Inc., for a quotation from 'Les Luths' by Frank O'Hara found in *Collected Poems* © 1971 by Maureen Granville-Smith, Administratrix of the Estate of Frank O'Hara; Carcanet Press Limited for a quotation taken from 'Death is a Loving Matter' by Charles Olson found in *Selected Poems* © Carcanet Press Limited; International Music Publications Ltd for a quotation from 'I'm Always True to You in my Fashion' by Cole Porter from *Kiss Me Kate* © 1968 Buxton Hill Music Corp., USA. Warner/Chappell North America, London W6 8BS. All rights reserved; Carcanet Press Limited for a quotation taken from 'Brevity' by Judith Wright found in *Collected Poems* © Carcanet Press Limited; John Harwood for a quotation from 'Long After Heine' by Gwen Harwood found in *Collected Poems* © The Estate of Gwen Harwood, 2004; The Ivor Gurney Trust for a quotation from 'Never More Delight Comes Of …' by Ivor Gurney found in *Ivor Gurney: Selected Poems*, ed. P. J. Kavanagh, Oxford University Press, 1990, © The Ivor Gurney Trust; Dent Publishers and David Higham Associates for a quotation from 'Do Not Go Gentle into That Good Night' by Dylan Thomas found in *Collected Poems*; New Directions Publishing Corp. for a quotation from 'Do Not Go Gentle into That Good Night' by Dylan Thomas found in *The Poems of Dylan Thomas* © 1952 by Dylan Thomas; Faber and Faber Limited for a quotation from 'Sonnet 23' by John

Berryman found in *Collected Poems*; Harvard University Press and the Trustees of Amherst College for a quotation from 'Safe in their Alabaster Chambers' by Emily Dickinson found in *The Poems of Emily Dickinson*, Thomas H. Johnson, ed., Cambridge, Mass.: The Belknap Press of Harvard University Press, © 1951, 1955, 1979 by the President and Fellows of Harvard College; HarperCollins Publishers, Inc. for a quotation from 'Self-Portrait as Hurry and Delay' by Jorie Graham found in *The End of Beauty* © 1987 by Jorie Graham; Carcanet Press Limited for a quotation from 'She Makes Her Music ...' by Matthew Welton found in *New Poetries II An Anthology* © Carcanet Press Limited; Carcanet Press Limited for a quotation from 'Dust' by Judith Wright found in *Collected Poems* © Carcanet Press Limited; Carcanet Press Limited for a quotation from 'Asphodel, That Greeny Flower' by William Carlos Williams found in *Collected Poems* © Carcanet Press Limited; Tony Harrison for 'The Bedbug' found in *Selected Poems* by Tony Harrison, Penguin © 1987 Tony Harrison; Faber and Faber Limited for a quotation from 'How to Kill' by Keith Douglas found in *Complete Poems*; Random House Inc. for a quotation from 'Shield of Achilles' by W. H. Auden found in *Collected Poems* © 1956 by W. H. Auden; Elizabeth Barnett, literary executor, for 'I, Being a Woman and Distressed' by Edna St Vincent Millay found in *Collected Poems*, HarperCollins, © 1923, 1951 by Edna St Vincent Millay and Norma Millay Ellis. All rights reserved; Harcourt, Inc. for a quotation from 'Mr Edwards and the Spider' by Robert Lowell found in *Lord Weary's Castle* by Robert Lowell © 1946, 1974 by Robert Lowell; The Marvell Press, England and Australia, for a quotation from 'Church Going' by Philip Larkin found in *The Less Deceived*; Carcanet Press Limited for 'The End of Love' by Sophie Hannah found in *The Hero and The Girl Next Door* © Carcanet Press Limited; Faber and Faber Limited for a quotation from 'What Are Years?' by Marianne Moore found in *Collected Poems*; Scribner, an imprint of Simon & Schuster Adult Publishing Group, for a quotation from of Random House, Inc. for a quotation from 'What Are Years?' by Marianne Moore found in *The Collected Poems of Marianne Moore*, © 1941, 1969 by Marianne Moore; The Estate of Marianne Moore for ebook reproduction of 'What Are Years?' by Marianne Moore, © 1981 Viking/Macmillan. All rights reserved; Carcanet Press Limited for a quotation taken from 'And Ut Pictur Poesis is her Name' by John Ashbery found in *Selected*

Poems © Carcanet Press Limited; Carcanet Press Limited for a quotation taken from 'From "Contingencies"' by Eavan Boland found in *Outside History* © Carcanet Press Limited; Pan Macmillan Limited for a quotation taken from 'Mrs Icarus' by Carol Ann Duffy found in *The World's Wife*; The author and W. W. Norton & Co., Inc. for a quotation from 'Aunt Jennifer's Tigers' by Adrienne Rich © 2002, 1951 taken from *The Fact of a Doorframe: Selected Poems 1950–2001* © 1971 by W. W. Norton & Co., Inc.; The author and W. W. Norton & Co., Inc. for a quotation from 'Planetarium' by Adrienne Rich © 2002 taken from *The Fact of a Doorframe: Selected Poems 1950–2001* © 1971 by W. W. Norton & Co., Inc.; Harvard University Press and the Trustees of Amherst College for a quotation from 'Why – do they shut Me out of Heaven?' by Emily Dickinson found in *The Poems of Emily Dickinson*, Thomas H. Johnson, ed., Cambridge, Mass.: The Belknap Press of Harvard University Press, © 1951, 1955, 1979 by the President and Fellows of Harvard College; A. P. Watts for a quotation from 'Among Schoolchildren' by W. B. Yeats found in *Collected Poems*; Scribner, an imprint of Simon & Schuster Adult Publishing Group, Inc. for a quotation from 'Among Schoolchildren' by W. B. Yeats found in *Collected Poems*.

I should like to thank the following students at Manchester Metropolitan University who have taken time to read portions of my draft: Claire Milner, Katie Fennell, Catharine Huggett and Katie Watkinson; my colleagues Margaret Beetham, Michael Bradshaw and Michael Schmidt; Jon Glover and Judith Wainwright. At Routledge I wish to thank Liz Thompson for commissioning this book, Liz Thompson and Milon Nagi for their scrupulous and hugely helpful editing, and Susannah Trefgarne. Thanks are due too to the several anonymous readers whose comments have been invaluable. My special thanks and appreciation go to my school English teacher, Ken Lowe, to whom this is dedicated. All final responsibility is of course my own.

BECAUSE THERE IS
LANGUAGE THERE IS
POETRY

> A clock in the eye ticks in the eye a clock
>   ticks in the eye.
> A number with that and large as a hat
>   which makes rims think quicker than I.
> A clock in the eye ticks in the eye a clock
>   ticks ticks in the eye.

Through evolution, the human vocal tract has become able to give voice to a variety of particular sounds and complex combinations of sounds. With these we have created languages which can communicate information of very different kinds and to a very high degree of subtlety. As we acquire them as children we respond to the sounds themselves as we hear, imitate and relish them. Just as we learn how effectively word sounds denote objects in our world and carry information to others, so too we enjoy the reiteration of the sounds themselves in the repetition of favourite new words, and variations upon them. As **Kenneth Koch** (1925–2002) writes, 'Each word has a little music of its own.'

The *sounds* of language are further enjoyed when they are combined in such sequences as a run of the same consonants (***alliteration***), or the repetition of certain words or rhythmic patterns. The lines above from **Gertrude Stein's** (1874–1946) 'Before the

Flowers of Friendship Faded Friendship Faded' show this kind of fascination – as does the title of the poem itself. The intrigue can extend to the surprising juxtaposition of word meanings. Stein called her book *Tender Buttons*. Ponder for a moment what associations come with putting the words 'tender' and 'button' next to each other.

These resources of language, especially **recurrence** – the anticipated pleasure of a sound or shape being repeated – have been used in the pre-literate, oral tradition of all societies for dances, riddles, spells, prayers, games, stories and histories. The work of the American poet/researcher **Jerome Rothenberg** (1931–) provides a wealth of examples of this from every continent and many cultures. We should not assume though that work of this kind from pre-literate cultures is simple. Often, as Ruth Finnegan shows in her anthology *Oral Poetry*, work such as the Malay form of the *pantun*, which we will meet in its English adaptation in Chapter 7, 'Stanza', can be very elaborate.

More familiarly, the early enthusiasm for nursery rhymes, chants, schoolyard games, songs, advertising slogans and jingles all feature the same kind of *gestural* characteristics. *Gesture* is an important concept here. What I mean by gesture in language are those qualities we employ to signal our meaning strongly by emphasizing particular word sounds, rhythmic sequences or patterns. Thus the words will catch our attention not only through a grasp of their dictionary meanings but through their sensuous impression, not unlike, indeed, the way we accompany speech by hand gestures and variations of *tone*. The incantatory, 'musical' qualities of beat, drum and dance are close to this and are part of the close relation between poetry and song. Indeed the term which is still key to both – **lyric** – points to this connexion. Lyric refers to that kind of **verse** most readily associated with the chanted or sung origins of poetry, traditionally to the harp-like stringed instrument known as the lyre. We still refer to the words of songs of all kinds as lyrics, and poetry closest in style and span to songs, as opposed to poems that tell substantial stories or are the medium for drama, is defined as lyric. I shall have more to say about this **genre** of poetry in Chapter 3, 'Tones of voice'.

In poetry without music these qualities have become formalized into what are its most prominent distinguishing features: its *rhythms*, that is the way a sequence of words moves in the ear, and its *metres*, that is the regular patterning of such movement into the poetic line. The character of the many different kinds of poetic line will be explored in separate chapters.

So, while the evolution and use of language has obviously been functional, exchanging information with the necessary clarity, its sounds and shapings, both spoken and written, are also inevitably gestural. Of course, those instrumental uses of language will be as simple and direct as possible, like the bald instructions for using a computer: 'Press Enter'; 'Select the file to be moved'; 'Double-click the mouse icon'. But even the specialized language associated with computers is not literal but *metaphorical*: the mouse, windows, desktop and bin. My computer manual promises 'Right Answers, Right Now', and the simple emphasis of this punchy phrase – repeating 'Right' – is the kind of language I am calling gestural. This snatch of a conversation is invented, but I think it is recognizable:

> So I had to go back to the bank. No sign of it there. Back to the butcher's. No sign of it there. Back to the chemist. No sign of it there. Back to the sweet-shop. No sign. Back to the café. No sign. Where was it? Slap-bang in the middle of the kitchen table.

The speaker wants to express tedium and exasperation at mislaying a purse and these repetitive, truncated phrasings with their slight variations impress this upon the listener. These are the gestural features of language.

So, the argument of this chapter is that poetry is not really a peculiar, demarcated zone out of the mainstream of language-use, but that language is inevitably and intrinsically 'poetic' in the qualities that I'm calling gestural. However, historically, these qualities have been highlighted and formalized for particular uses and occasions. Poetry is a form for special attention and one that calls unusual attention to the way it is formed.

The ancient ceremonial aspect of gestural language persists in our desire for special forms of language for particular occasions. We all know for instance how difficult it is to 'find words' of condolence. In greetings cards, at weddings, funerals, in sorrow and commemoration and in love, wherever we feel the need for heightened, deliberate speech, wherever there is a need for 'something to be said', we turn to the unusual shapes and sounds of poetry. This is also why we might be drawn to write poetry in order to form an utterance that is out of the ordinary and commensurate to the weight or the joy of the occasion. Always at such times we will encounter the familiar difficulty of finding what we know to be the 'right words'.

The deployment of impressive sounds and shapes, the deliberate speech required by that 'something to be said', has been known in the western tradition as **rhetoric**. In this emphasis, from *Paradise Lost* to a local newspaper's *In Memoriam* verses, poetry can be seen to be a part of rhetoric.

However, every experience with language teaches us that communication is frequently less transparent than we would wish. Disappointment at the failure of language to be clear, and at its capacity to mislead and sway us into deception, has marked our thinking about language for centuries. Ambiguity, double meanings, 'equivocation' intended and not intended, all manner of '-speaks', result from, or exploit, the potential anarchy of language. Often, it seems, 'words run away with themselves' and take us with them. This may lead into a cheerful gallimaufry of free association and **word-play**, or into saying things we did not mean. Those 'right words' can be very elusive.

As a form of utterance that is especially sensitive to all the various resources of language in both its **semantic** (i.e. meaningful) and sensuous dimensions, poetry has taken upon itself the freedom and opportunity for word-play, and also its responsibilities. Because language is as it is we might say anything. Because life is how it is we need to watch what we say.

Through language we can convey common information, but also achieve a vast capacity to generalize particulars and to abstract from experience. We can also invent and fantasize and relay any of this to others. Its immeasurable creative flexibility means that 'language enables the promotion of endless associations between any one object/person/event and another', writes the neuroscientist Susan Greenfield. So, since the nature of language itself does not necessarily oblige us to be purpose-like, it also enables associations which may seem purposeless. It is often attractive – even just for the hell of it – to remove its use as far as possible from any externally driven direction.

The philosopher Daniel Dennett observes the essential biological function of language in evolution, but continues, 'once it has arrived on the evolutionary scene, the endowment for language makes room for all manner of biologically trivial or irrelevant or baroque (non-functional) endeavours: gossip, riddles, poetry, philosophy'. On this view, *all* poetry, including the most rhetorically purposeful, might be seen as 'baroque', and we shall look at the philosophical problems surrounding rhetoric and **figurative** language in Chapter 8. But when the owl and the pussycat go to sea in a beautiful pea-green boat, then we can see that language can indeed promote Greenfield's 'endless associations' in ways that seem especially 'baroque'. And delightful.

So language can be deployed 'uselessly', and an alternative emphasis to the rhetorical one would see poetry as the space where the glory and freedom of the possibility to say anything is specialized:

> an orange the size of a melon rolling slowly across the field
> where i sit at the centre in an upright coffin of five panes of glass

Such a relish as this from **Tom Raworth**'s (1934–) poem 'now the pink stripes', may be inspired by a desire to explore the light-spirited freedom afforded by language in a space not subject to instrumental demands. There can be a sheer game-playing element in such poetry, a love of messing with words because we can. Alternatively – or in addition – it may be a response to a disillusionment with the use of language in the 'real' world of affairs, transactions and 'information'. In recent years the American L=A=N=G=U=A=G=E poets have taken such a view, questioning the dominance in modern industrialized, business-oriented society of what **Charles Bernstein** (1950–) calls 'authoritative plain style'. Its increasing standardization claims a monopoly over coherence and excludes tones and styles of speaking and writing that do not conform to 'mannered and refined speaking'. It is a doubt that language simply carries common sense that has even greater depths. In **William Shakespeare**'s (1564–1616) *Twelfth Night* the clown Feste complains that language has been so much discredited by being used to lie and deceive that 'words are grown so false I am loath to prove reason with them'.

All of these pleasures and problems are with us because of the character of human language. What we have come to call poetry gives us a constructed, deliberate space in which to enjoy and to tussle with the experience of language. Reading or writing a poem offers a practising awareness of the problems of language and meaning – specifically of what we must say and how we can best say it. *Because there is language there is poetry*: in the rest of this book I shall try to set out some of the principal ways that poetry in the modern English language has been made across the whole spectrum of rhetoric and nonsense.

Summary
In this introductory chapter we have looked at:

- the character of human language, especially with regard to its sounds;
- the pleasure of rhythm and rhyme from children's verses onwards;
- how language operates functionally, uses *gesture*, but can also work 'uselessly' – and enjoyably;
- the problem of finding 'the right words';
- the basic idea of rhetoric and the challenge of nonsense;
- the continuity between the general use of language and poetry *and* its distinctiveness.

FURTHER READING

Finnegan, Ruth (ed.) (1982) *The Penguin Book of Oral Poetry*, Harmondsworth: Penguin Books.

Hughes, Ted and Heaney, Seamus (eds) (1982) *The Rattle Bag*, London: Faber.

Opie, Iona and Peter (1959) *The Lore and Language of Schoolchildren*, London: Oxford University Press.

Rothenberg, Jerome (ed.) (1968) *Technicians of the Sacred: A Range of Poetics from Africa, America, Asia and Oceania*, New York: Doubleday.

Trask, R. L. (1988) *Language: The Basics*, second edition, London: Routledge; see Chapter 3, 'Language and Meaning'.

DELIBERATE SPACE

Jimmy Fryer Esquire,
Walking along a telegraph wire
With his shirt on fire!

I have used the word *gesture* in Chapter 1 to describe those aspects of the nature of language other than those covered by dictionary definition and word-order. These gestures are similar to the physical gestures and tones of voice in conversation. Obviously it is common and important that we use clear, 'cold' instrumental language in everyday life so that we know to meet inside or outside a cinema, or take two per day after, not before, meals. But, as we have seen, little or no communication is confined to transmitting information in this way.

All the resources of verse emphasize the gesturing elements of words and their combinations in ways that draw attention and impress. These include sound effects like **alliteration** and **assonance**; **reiteration**; **rhyme**; **rhythm** and **metre**; figures of speech or **tropes**; **limericks**, **sonnets**, **haiku** and all the other verse forms that we shall meet later. (All of these highlighted terms are defined briefly in the Glossary.) Even poems that do not manifest such features so obviously – or indeed aim to avoid them altogether – are still *shaping* words in some deliberate form.

This chapter will suggest that poetry can be seen as a particular space, created or adapted by the poet out of the flux of language-use with great deliberation. A poem is a part of the functioning and the gesturing of the words we use every day, but it is also set aside. Just as a prayer mat is made of fabric found everywhere but, once laid out, marks off a space from the surrounding daily world, so does the shape of the poem organize language into a space for pause and for different attention. It is a space in time marked by the **rhythms** of pace and pause, and the sensations and ideas in every word. This space is shaped in the mind of poet and listener, and when the poem is on the page its impression is also in the formation made by the letters.

In later chapters we shall look at the various features which mark out this space and fill it, but first I want to explore what this space might be, and the kind of effects it creates. I shall begin where poetry began, in **oral** recitation and performance. I shall look at children's rhymes and oral poetry from Britain and elsewhere, and move from that to look at how strongly visual spaces are created in the familiar space of the printed page.

ORAL TRADITION AND CHILDREN'S RHYMES

As we can see from

> Jimmy Fryer Esquire,
> Walking along a telegraph wire
> With his shirt on fire!

a prime motive for manipulating words in gestural ways is pleasure: the simple sensuous exhilaration that comes of making up such a combination of sense and sound, or of adapting it, or certainly of uttering it. The delight in this verse comes from the **semantic** absurdity – the picture the words show us of someone walking along a telegraph wire perhaps as yet unaware of his predicament is just wonderfully silly. Besides the relish of others' discomfiture which is so much a part of children's cat-calling, the image's impossibility is also part of its pleasure. Words can be put together in ways that offer a little holiday from reality, a momentary fantasy.

But there is more to it than this. Certain 'manipulations' are more effective than others, and sound, timing and rhythm, for example, are all crucial. Poetry works on the ear. It is a form developed for and in performance, within a long *oral tradition*.

To understand the effects of these oral origins, we can return to the playground and Jimmy Fryer. As we have said, the rhyme offers a certain pleasure through the absurdity of what it describes, but the coincidence of the word sounds is no less striking. The name *'Fryer'* offers the chance of the **rhymes** of *'Esquire'*, *'fire'*, and *'wire'*, although it is just as likely that in earlier versions of the joke the point was to mock the use of the pompous title *Esquire*. It could easily be adapted, for example, to read:

> Jeffrey Wainwright Esquire,
> Walking along a telegraph wire
> With his shirt on fire!

The gender of 'Esquire' would present a momentary problem, but it's easy to imagine such further adaptations as:

> Kylie Fryer Esquire,
> Walking along a telegraph wire,
> With her skirt on fire.

The schoolyard poet would certainly substitute *skirt*, not *dress*. Any polysyllabic word like *overcoat* or *handbag* would certainly make the line a clumsy mouthful, but not all **monosyllables** are equal. The lighter vowel sound and the soft slipping-away –ss of *dress* would dissolve the mocking bite of harder-sounding words like *skirt* and *shirt*. Of such details are successful verses made, and the ear for it is vitally related to the fact that children's verses have an oral existence. Their precision has been honed by repetition and the fact that the playground can be a very critical arena. With one verbal slip the mocker could instantly become the mocked. Imagine, for instance, the embarrassment of a child eager to pass on this rhyme,

> Good King Wenceslas
> Knocked a bobby senseless
> Right in the middle of Marks & Spencer's

who remembers the content, but not the exact words, and puts in *policeman* or *constable* for *bobby*. The rhythm of the second line would stumble disastrously. It is the predicament of the bad joke-teller. Jokes often turn upon features of language, most obviously puns, which makes them cousin to the **word-play** of poetry. But we all know that it's the way you tell 'em, and so it is. The ability to structure and above all *time* a joke is vital to its success. It is the **cadence**, the arrangement of acceleration and pause, the manipulation of time, that matters. The space of these children's rhymes is shaped very precisely to meet these requirements of timing. They choose words with an exact ear for the cadence, the way each sound *falls* in relation to the ones that surround it.

I have emphasized so far the continuity between the gestural features of speech and of poetry, and suggested that in familiar children's rhymes we can see the attractions of verbal gesture in 'poetic' features such as **rhyme** and **rhythm**. Yet just as these verses delight in the peculiarity with which they are constructed, so does their content in the space they occupy in the child's mind. The oddity of King Wenceslas abandoning wherever and whenever it is he comes from and embarrassing the law in a respectable store like Marks & Spencer's is a child's cheerful transgression against adult authority as represented by carol-singing and policemen. The happy coincidence that *Wenceslas* rhymes with *senseless* and *Spencer's* parallels this little disorder because it is in the character of rhyme to be anarchic. By 'anarchic' I mean that because rhyming words have no necessary connection in meaning, following the quest for a rhyme can lead the solemn progress of meaning in quite a different, coincidental direction. (We shall look at this in more detail in Chapter 6, 'Rhyme').

We can see the way the forms of children's verses can coincide with their ideas to tweak the nose of the adult world in what Iona and Peter Opie in their classic study *The Lore and Language of Schoolchildren* call 'tangletalk'. This verse travesties the kind of speech-making 'not unknown in their school halls':

> Ladles and Jellyspoons,
> I stand upon this speech to make a platform,
> The train I arrived in has not yet come,
> So I took a bus and walked.

> I come before you
> To stand behind you
> And tell you something
> I know nothing about.

The space of the poem is deliberately separated from the current of ordinary speech by the way it is shaped. Here the prose norm of conventional speech is broken into the segments as the child heard them to form separate lines.

Often too the poem is marked off from the sphere of daily expectation by the alternative *imaginative* space that it occupies. As the dignitary rises to speak, the children sitting in their obedient rows hear something different in their heads. It probably makes little sense to them anyway so they take this to an extreme by mixing or 'tangling' it up, substituting words that sound similar to produce a nonsensical parody. Sometimes it is a more fantastic space, a realm where pigs take snuff to make them tough and the elephant, 'a pretty bird',

> builds its nest in a rhubarb tree
> And whistles like a cow.

All this kind of fancy, and such bold language strokes, might be seen to have something in common with another delight of children – cartoon animation. Everything about the words is larger, more obvious, and the action improbably dramatic.

Drawing again upon children's rhymes, a simple example of a clapping song can teach a great deal about the fundamental and powerful poetic techniques used in the oral tradition to impress stories upon listeners.

> When Suzi was a baby, a baby Suzi was
> And she went, 'wah, wah, wah, wah.'
> When Suzi was an infant, an infant Suzi was
> And she went scribble, scribble, scribble, scribble.
> When Suzi was a junior, a junior Suzi was
> And she went – 'Miss, Miss, I can't do this
> I've got my knickers in an awful twist.'
> When Suzi was a teenager, a teenager she was

> And she went – 'Oh ah I've lost my bra
> I've left my knickers in my boyfriend's car.'

The rhyme goes on through Suzi as a mother, granny, skeleton and ghost. The remarkable, simple economy of this life-story is achieved by a number of techniques. Fundamentally we have the signposting structures of repetition, 'When Suzi was ... When Suzi was ...', to introduce each phase of her life. Allied to that is the strong beat on these syllables in the 'When' lines,

> When **Suzi was** a **ba-by**, a **ba**by Suzi **was**

which coincide with the claps. Interestingly the inversion of the first phrase of this line in its second phrase is in fact a familiar device in **classical rhetoric** where it is known as **anadiplosis**. But reiteration can also be tedious and variation is necessary if the attention of an audience, and of a performer, is to be held. In 'Suzi' the 'wah wah' and 'scribble scribble' pattern soon looks likely to be boring, hence the shift of phrase and of rhythm with

> And she went – 'Miss, Miss, I can't do this
> I've got my knickers in an awful twist.'

These lines alter the rhythmic pattern and introduce rhyme at the same time as they take the girl out of infancy.

All of these reiterative techniques also work as a stalling device which helps the clappers to be sure they remember the next line. Indeed this whole clapping song is a small-scale instance of the memorizing or **mnemonic** qualities of oral recitation. Poetry is far more ancient than print or even written cultures, and it is devices such as **rhyme**, **beat** and **recurrent** structure that enable singers, clappers and storytellers to organize and remember their material for the audience.

Orality is the major reason for the primacy of the *verse line*. (We shall look at this in detail in Chapters 4 and 5.) A recurrent pattern, defined by repeated sequences of **beats**, together with related aural effects such as rhyme and other formulaic constructions, enabled the pre-literate makers of poetry not only to 'remember their lines', but to hold their audience. We still use *mnemonics* as an aid to memory as in this old history lesson revision aid:

> In fourteen hundred and ninety-two
> Columbus sailed the ocean blue.

and we know that verse in set forms is easier to memorize than prose. So the clappers of 'Suzi', with some twenty-plus lines to remember, have beat, rhyme, reiteration of both phrase and structure to help them through the narrative, which is in any case structured by its progress through the span of a life.

ORAL TRADITION IN EPIC AND NARRATIVE

If we turn to the great stories of ancient Greece and other communities told for generations before they were ever written down, we can understand more about how the space of poetry was organized for large and lengthy purposes. We shall see that 'remembering' is hardly an accurate term for the way in which these oral poems of the past existed within their tradition.

Perhaps the poems best known to us of the ancient oral tradition are the Greek heroic epics attributed to Homer: *The Iliad* which tells the story of the Greeks' long war against Troy, and *The Odyssey* which recounts the protracted voyage home from Troy of its hero Odysseus. These are poems which in present-day conventional English editions will occupy some 450 pages each, but they existed, long before they were written down in the sixth century BC, only in the mind and voice of their poets and narrators. In 1934 the researcher Milman Parry heard a Serbian bard, whose tradition is thought to be related to the ancient Greek manner of Homer, recite a poem as long as *The Odyssey*, and take two weeks to do so in twice-daily sessions each lasting two hours. This would not be done by remembering the poem line by line but by a process of continuous re-composition. The bard was working with a knowledge of the narrative outline and an ingrained sense of the movement of the verse line. This given structure acts as a channel through which he could convey the story's larger and more detailed incidents. These in turn would include an array of formulae which allow set repetitions of such things as the arrival of a messenger, how he is received, and how he delivers his missive. Description of natural objects and persons would also be made in appropriate given forms.

So this huge poem is being re-made as its recitation goes along, but in accordance with strongly established conventions. In his *The World of Odysseus* the historian M. I. Finley reckons that about one third of the *Iliad* and the *Odyssey* is composed of lines or blocks of lines that appear elsewhere in the poem. All of this shows us the importance of creativity in the recitation of oral poetry, and that performance is vital to it. Much nearer to our own time and experience, the process is somewhat similar to the way we might improvise the verses of a pop song whilst returning to the chorus, or the way football crowds adapt songs to hymn their club or particular players. Incidentally the *collective* character of the composition of such songs and chants shows us another continuing feature of the **oral tradition**.

The experience of the anthropologist John Tedlock is interesting here as it highlights a further aspect of the relationship between oral performance, the full meaning of a work and the nature of poetry's deliberate spaces as against conventional prose. In the 1960s Tedlock set himself to record the narrative poetry of the Native American Zuni people in the south-western United States. Transcribing his many hours of tapes Tedlock found himself dissatisfied by pages of written prose which seemed to capture so little of the experience he had heard. Most specifically he writes, 'there is no silence in it', and the varying pauses of the Zuni narrators, together with their varieties of level, are part of the body of the story. Consequently, influenced also by poetry readings in his own culture, Tedlock devised a written version of the Zuni stories set out in '**free verse**' lines to point the pauses and with capitals and different letter sizes indicating voice levels:

> 'THAT'S WHAT YOU DID AND YOU ARE MY REAL
>
>   MOTHER,' That's what he told his mother. At that
> moment
>   his mother
> embraced him
>   embraced him
> His uncle got angry

his uncle got angry.

He beat

his kinswoman

he beat his kinswoman.

That's how it happened.

The boy's deer elders were on the floor.

His grandfather then

spread some covers

on the floor, laid them there, and put strands of turquoise

beads on them.

Tedlock's translations make use of the gestural features of poetry, including the deployment of white space on the page, to render what might otherwise have been thought of – especially had they been transcribed by pen rather than by tape-recorder – as prose narratives. He found that prose doesn't capture performance, where varying emphases and plays of sound and silence are part of the story and its meaning. Verse lines and the deployment of deliberate poetic spaces do. His experience, he writes, 'convinced me that prose has no real existence outside the written page'. Introducing Tedlock's collection of Zuni work, Jerome Rothenberg asserts:

> We have forgotten too that *all* speech is a succession of sounds and silences, and the narrator's art (like that of any poet) is locked into the ways the sound and silence play against each other.

It is exactly this interactive play of varyingly stressed sound and silence that constitutes the deliberate space of poetry.

OUT OF THE ORAL TRADITION – TOWARDS THE PAGE

As these instances suggest, the relationship between oral and written forms of poetries that certainly have their origins in the oral tradition is complex. As the editor and translator Michael Alexander writes of the Old English epic *Beowulf*, 'it is likely that the poem had more than one oral stage and more than one written stage'. This has implications too for our conception of the *authorship* of poetry which I shall discuss in Chapter 8. For the moment however I want to continue to explore the influence of the oral origins of poetry upon its shapes.

In English the '*ballad* tradition' carries all these complexities of origin and transmission. As a consciously defined 'tradition' this usually refers now to the gathering of poems by F. J. Child as *The English and Scottish Popular Ballads* published between 1882 and 1898. However, the form had attracted a lot of literary interest and imitation much earlier, especially in the eighteenth century and most prominently by **William Wordsworth** (1770–1850) and **Samuel Taylor Coleridge** (1772–1834) whose joint volume *Lyrical Ballads* appeared in 1798. Formally the *ballad* is usually a poem that tells a story and is written in short stanzas. The stories tend to have simple plots and straightforward characterization, and are frequently dramatic and violent. They often feature encounters with the supernatural, love tragedies, sons lost at sea, and adultery. Here are some lines from the medieval ballad *Little Musgrave and Lady Barnard*, a tale in which the lady, having fallen in love with a young man at church, and he with her, takes him off to her 'bower at Buckelsfordbery' where 'Thou's lig in mine armes all night'. Told of this by 'a little tinny page', the Lord Barnard discovers them:

> With that my Lord Barnard came to the door,
>   And lit a stone upon;
> He plucked out three silver keys,
>   And he open'd the doors each one.
>
> He lifted up the coverlet.
>   He lifted up the sheet:
> 'How now, how now, thou Little Musgrave,
>   Does thou find my lady sweet?'
>
> 'I find her sweet,' quoth Little Musgrave,
>   'The more 'tis to my pain;
> I would gladly give three hundred pounds
>   That I were on yonder plain.'
>
> 'Arise, arise, thou Little Musgrave,
>   And put thy clothes on;

It shall nere be said in my country
  I have killed a naked man.

'I have two swords in one scabbard,
  Full dear they cost my purse;
And thou shalt have the best of them.
  And I will have the worse.'

Many of these ballads are very lengthy and so it is necessary to
quote at some length to give an idea of how any incident unfolds. In
these lines we can see how the **narrative** is built piece by piece in
each self-enclosed stanza: Lord Barnard arrives at the door; he lifts
the sheet – and here the repetition acts both as part of the
reciter/singer's **mnemonic** and as suspense; Musgrave's reaction;
Barnard's chivalric challenge to a duel in two parts – the command
to Musgrave to dress, and the donation of the better sword.

Each of these stanzas has the same two features to its structure.
First, each rhymes the second and fourth lines: *upon/one;
sheet/sweet; pain/plain*. Rhyme schemes are conventionally
notated, in this example: *abcb*. (See Chapter 5 for a fuller discussion
of **rhyme**.) Second, each has the same **metrical** pattern, that is the
beat or **stress** – like the claps in the clapping song – falls upon syllables in each line in a way that corresponds to related lines. Again we
shall study **metre** more fully later (Chapter 4), but broadly this
means that the first and third lines contain *four beats* and the
second and fourth lines *three beats*:

He **lift**ed **up** the **cover**let,
  He **lift**ed **up** the **sheet**;
'How **now**, how **now**, thou **Litt**le Mus**grave**,
  Does thou **find** my **la**dy **sweet**?'

This then is the marked-out working space of the ballad. To the
singer/reciter this pattern would be grooved as a channel in which
to carry the narrative, a basic framework in which to fit details and
devise variations. Especially since it often includes 'pause' reiterations, it helps the performer recall the content, or if necessary
improvise. In this stanza we can see how it serves the purpose of the

story. The key action is of course the lifting of the bedclothes, so **lift** and **up** are bound to carry stress and their reiteration lends suspense. The last line of the ballad stanza is often the punchline of that part of the narrative and in this stanza the key content is contained in the three words **find**, **la**dy and **sweet**. The rhyme **sheet/sweet** – less expected than ballad rhymes often are – is brilliantly impressive as it reveals the Lord Barnard's sensual, sardonic psychology, the way his mock interest in Musgrave's experience of his wife is sinisterly controlled. Ballads are often represented as painting action and character with a very broad brush, but this detail, resting as it does in that one word 'sweet', is wonderfully resonant.

The five hundred years of the printed word since the flourishing of these ballads have all but eclipsed oral verse. But print has been a centralizing and standardizing force – hence the history of the struggles that have always centred round access to the press. In this context it is not surprising that oral cultures have remained most important to groups with least access to standard publication. The African-American Blues tradition is one prominent instance, and, more strictly in verse than in song, so is the oral poetry of the Afro-Caribbean, especially Jamaica, both on its home ground and in Afro-Caribbean communities in Britain and elsewhere.

This poetry is composed in dialect and is usually *narrative*, often using the manner of conversation to cover material ranging from outraged denunciation to neighbourly gossip. One example is the work of **Louise Bennett** (1919–) who emerged in Jamaica in the 1940s. The poetry persists both in authored instances such as Bennett's and in traditional transmission, often of memorized verses hundreds of lines long. The verse scheme is usually similar to the ballad in that it uses a four-line stanza rhyming *abab*, and either a four-*beat* line or an alternation of four and three beats. But, in context, most important is the dialect, for in the shaped sound-space of the poem the ordinary language of the people proclaims its seriousness and demands respect. It returns pleasure and recognition to its own speakers and reminds 'standard' speakers of the capacities of varieties of English other than their own. The space of the poem is thus doubly deliberate: first in its formal shaping and then, through this, as an act of cultural assertion.

Valerie Bloom (1956–), a successor to Louise Bennett, performs her work in character and often in costume, and mobilizes the apparently lightsome qualities of her tradition to serious purposes. Here she adopts the manner of a street gossip telling a relative the story of a boy shot dead by police in a dispute at a picture house. Skilfully she uses the repeated phrase 'a soh dem sey' ('or so they say') to make an ironic comment on different versions of the incident. These are the last three verses of 'Trench Town Shock (A Soh Dem Sey)' in which the teller describes how the official version is that the boy pulled a knife and was thus shot through the head in self-defence, something that happens often:

> Still, nutten woulda come from i',
> But wha yuh tink, Miss May?
> Di bwoy no pull out lang knife mah!
> At leas' a soh dem sey.

> Dem try fi aim afta im foot
> But im head get een di way,
> Di bullit go 'traight through im brain,
> At leas' a soh dem sey.

> Dry yuh yeye, mah, mi know i hat,
> But i happen ebery day,
> Knife-man always attack armed police
> At leas' a soh dem sey.

SPACE ON THE PAGE

The 'deliberate space' of the poems we have looked at so far is created primarily by patterns of sound and we follow the poems by ear. Sound is nearly always important to poetry, but with the wider dispersal of literate culture, especially following the advent of printing by movable type from the fifteenth century onwards, the poem's effect is complemented by the shape of the space it occupies on the page. Now I want to look at a series of poems in which the visual becomes of increasing importance.

As I wrote in Chapter 1 we often turn to poetry in the midst of strong feeling, especially grief. Several of the poems we shall look at in the next few pages are linked by featuring tears. Although we are primarily concerned with the shapes of these poems, in theme and *imagery* the globe of the tear might be said to occupy its own small space within the poetic tradition.

> Luveli ter of luveli eyghe,   [lovely ... eye]
>   Why dostu me so wo?   [give me such woe]
> Sorful ter of sorful eyghe,   [sorrowful]
>   Thu brekst myn herte a-to.   [you break my heart in two]

This beautiful little refrain from the mid-1300s, possibly a love lyric but certainly part of a devotional poem to Christ, conveys its tender feeling with the most minimal deployment of four simple lines. It alternates four- and three-**beat** lines, though in the first and third lines ensuring that the beats *start* the lines (these stressed **beats** are shown in bold):

> **Luv**eli **ter** of **luv**eli **eyghe** ...
> **Sorful ter** of **sorful eyghe**

This ensures a heavier emphasis, an effect that pushes forward the sense of the expostulation of grief and mimes a tolling measure befitting the mournful subject. **Measure** is a term often applied to poetry and these lines, with their simple solid balance of these key stresses on key words, convey the powerful sense of being measured, in the sense of calibration *and* of restraint, whilst also holding the sense of bursting forth that is weeping. These common Old English words *luveli* and *sorful* (sorrowful) are just allowed to impose their accumulated weight both in the strong sound of consonant and vowel, and of their roots and associations of meaning in *love* and *soreness*. The second and third lines are differently but no less strongly stressed,

> Why **dostu me** so **wo** ...
> Thu **brekst** myn **herte** a-**to**

but that slight difference opens a break in the beat through that unstressed *why* which is almost like a tiny catch in the voice. Short, with obvious beats and reiterated vocabulary, these four lines appear to make little inventive use of their space, and yet these minimal means achieve remarkable emotional power.

Not weeping, but still lamenting, **Geoffrey Chaucer** (?1340–1400) in this poem with the interestingly paradoxical title 'Merciles Beaute' has a refrain of similar affecting economy:

Your yen two wol slee me sodenly;   [eyes will slay me suddenly]
I may the beautee of hem not sustene.   [survive]

In these longer, ten-syllable lines – counting *yen* as two – Chaucer gets the pain into his rhythm by obliging a tiny pause after *yen* and then accelerating the line through the sibilance of *slee* and *sodenly*. The second line has three beats much stronger than the others, on **beautee**, **not** and *sustene*, which intensifies the emotion after that apparently equable **may**. *Sostene* – which is made to rhyme in the whole with *kene*, *grene* and *queen* – seems to me a particularly effective choice because it picks up the *s* sounds of *slee* and *sodenly*. The word in this sense of 'withstand', 'endure', is unusual, and thus gives the sense of the poet casting urgently about to find the proper word to convey the pressure. Again, a small space carrying a potent effect.

But the pleasures of the poetic space can be less doleful:

There once was a poet called Donne,
Who said 'Piss off!' to the sunne:
The sunne said 'Jack,
Get out of the sack,
The girl that you're with is a nun.'

Not all **limericks** are so indelicate, although this one easily could be more so, and, as it has transmuted into a popular joke form, many are. The origins of the form are obscure and it has always, it seems, migrated back and forth between written and oral traditions. The pattern however is broadly the same: five lines, with lines 3 and 4 shorter, usually two beats, and a rhyme scheme of *aa bb a*. Especially because the form, like all joke formulae, has become so well worn, the first two lines can often use their length to produce a knowingly laborious, even pedantic quality. This must however then be recovered by the acceleration of the short lines, and, vitally, by surprise in the final rhyme, even if – as in 'Hickory Dickory Dock' – it repeats the first rhyme word and gains its effect by second-guessing what the listener expects. In this case the deliberate anti-climax might only serve to highlight the real inventiveness that is in the third and fourth lines. This, by **Edward Lear** (1812–88), is an example:

> There was an old man of Thermopylae,
> Who couldn't do anything properly;
> But they said 'If you choose
> To boil eggs in your shoes,
> You shall never remain in Thermopylae.'

But in all cases it is shape that satisfies, a fixed form within which wit can devise surprises. As always the *cadence*, or fall, is a matter of timing.

Another short form that shows the appeal of distinct shape is the *haiku*. Originally a Japanese poetic form developed in the thirteenth and fourteenth centuries, it attracted western imitation around 1900. Although there are variations both in the Japanese tradition and in western practice, the commonest definition of the haiku is that it has seventeen *syllables* distributed over three lines in the pattern of 5–7–5. Its interest for English language poets, especially for **Ezra Pound** (1885–1972) and the *Imagist* poets of the early twentieth century, is that its extreme compression purges verse of dilation and decoration in order to concentrate on a single noticed thing. Pound himself did not in fact write any true haiku, though the influence of the form can be seen in such poems as 'Alba', 'In a Station of the Metro' and the satiric squib 'The New Cake of Soap'. Later, the associations of the haiku with Zen Buddhism attracted the attention of poets exploring eastern religion and philosophy, and the influential translations of R. H. Blythe of the Japanese poets **Basho**, **Buson** and others represent the form in both shape and tone, but cannot always translate it exactly in terms of that syllable structure:

> The coolness:
> The voice of the bell
> As it leaves the bell!
> (Buson)

The precision and pregnancy of the form makes it continually appealing, not least to **parodists**. The Internet might be seen as a contemporary site for the oral tradition and familiar frustrations with its technology are a frequent feature of its exchanges. The haiku seems an appropriate form to express such infuriation, as it suggests the brevity of computer commands and the technology's nearly mystical inscrutability. Here, composed in English and thus able to follow the template exactly, are two of many in circulation:

> Windows NT crashed.
> I am the Blue Screen of Death.
> None will hear your screams.
>
> * *
>
> With searching comes loss
> And the presence of absence:
> 'My Novel' not found.

Press Exit in tears.

We shall look in more detail at the set spaces poets use in Chapter 7 on the **stanza**. However, returning to the theme of tears, the opening of **John Donne**'s (1572–1631) farewell to his lover as he departs overseas, 'A Valediction: of Weeping', is a striking contrast to the simplicity of the haiku. Here we see a poet figuring a more elaborate space in which to work:

> Let me pour forth
> My tears before thy face, whilst I stay here,
> For thy face coins them, and thy stamp they bear,
> And by this mintage they are something worth,
> For thus they be
> Pregnant of thee;
> Fruits of much grief they are, emblems of more,
> When a tear falls, that thou falls which it bore,
> So thou and I are nothing then, when on a divers shore.

The *idea* of an enclosed space, the globe of his tear which contains the reflection of his lover, governs this **stanza**. It is then expanded in the poem's two succeeding stanzas as the tear becomes an image first of the world and then of the seas in which he might drown.

The anguish of parting lies in the sense of fragility: the imminence of their last moment together, and the possibility that the break might never be repaired whether because of disaster or change of heart. The falling tear is an exact image for this fragility, and its enclosure is paralleled by that of the stanza that holds it. It opens with the outbreak of weeping carried by the bursting forth of that short first line, and ends with the dissolution of the tear as it hits the ground – 'thou and I are nothing then'. The space between contains the speaker's urgent thoughts as they tumble forward, stopping and starting as he grasps at connecting words, *whilst, for, when, that, so*, to sort out his logic. Like the first line, the other two short lines are moments when the emotion is most forceful. Their heavy and irregular stresses, and crude *be/thee* rhyme, punctuate the effort in the longer lines to hold back the fall of the tear, and so the parting, by inventing so much to see in it. All the anxiety about the relationship – brevity, fragility – is represented in the instant of the falling tear, held up, as though in slow motion, by the space between line 1 and line 9.

It is interesting to ponder the degree to which such distinct shapes as the limerick and haiku are held in the poet, reader or listener's mind as sound or as a visual shape. My own sense is that since the printed page is so much part of our mental landscape, even the shapes of poem that arise out of the oral tradition like the ballad stanza and the limerick occupy a visualized space in our imagination. Some early sixteenth- and seventeenth-century printings of *lyrics* and sonnets drew a series of arcs down the right-hand margin linking together the rhyme words in order to illustrate that there is a visual, normally an elegantly symmetrical pattern to the poem which complements its harmonies of sounds. Drawing these arcs and then turning the page through ninety degrees, the rhyming of some elaborate stanza forms, such as **John Dryden**'s (1631–1700) 'Song for Saint Cecilia's Day', can be seen to have a nearly architectural structure.

Visual shape, then, is the other dimension of the poem's deliberate space. Such shapes as the **quatrain** and the **sonnet** have a presence on the page and in the ear. But prior to their realization they can occupy a distinct working area within the poet's mind during composition. For a variety of reasons the poet's first motivation may be to write a **sonnet**, or a **villanelle** or some other set form. Less specifically he or she may see the poem in the mind's eye at an early stage of its gestation as composed of *quatrains*, or six-line stanzas or some other configuration. In the same way that the twentieth-century Russian poet Osip Mandelstam said that a poem first of all existed for him as 'a tune in the head' even before the words came to him, so poets might glimpse the beginning of a poem as a shape. All the poetic forms have this visual, even sculptural, dimension.

EMBLEMS AND 'CONCRETE' POEMS

Some poems, like the *emblem* poems of the seventeenth century, foreground the visual dimension by patterning the words on the page so as to present a visual image of the poem's subject. This is **George Herbert**'s (1593–1633) 'Easter Wings':

> Lord, who createdst man in wealth and store,
> Though foolishly he lost the same,
> Decaying more and more,
> Till he became
> Most poore:
> With thee
> O let me rise
> As larks, harmoniously,
> And sing this day thy victories:
> Then shall the fall further the flight in me.

This is one of a pair of poems that can be read this way, and turned through ninety degrees to present the image of two angels standing with wings outspread. *Emblem* poems were often, although not exclusively, religious, and aim to convey their point briefly and vividly. In 'Easter Wings' the outside, longer lines treat of the expansiveness of God while the inside lines waste to the near vanishing point of human frailty. The poem's meaning therefore is carried in both word and image.

In the twentieth century several poets have been drawn to this tradition, among them **Dylan Thomas** (1914–53), with a number of poems both imitating Herbert's wings and inverting them to produce a diamond shape on the page. **Geoffrey Hill**'s (1932–)

'Prayer to the Sun' is composed of three short poems, each in the shape of a cross, stepped diagonally down the page, and **John Hollander**'s (1929–) 'Swan and Shadow' is figured as a swan on the water together with its exactly inverted reflection.

These poems all make use of visual effect to endorse the meaning. *Concrete poetry* of the twentieth century may also do so, but many of its practitioners are most interested in the nature and variety of *text*. This is **Edwin Morgan**'s (1920–) 'Siesta of a Hungarian Snake':

s sz sz SZ sz SZ sz ZS zs ZS zs zs z

The poet and editor Richard Kostelanetz defines text broadly as 'anything reproducible in a book', and since concrete poems have often appeared as posters or other forms of art-work, even his definition may be too narrow. In the twentieth century 'concrete' poets have used the expanding resources of typography and modern printing techniques to emphasize the material nature of the poem and of words themselves. The page consists of (usually) white space and letter and/or number forms. While there is only one kind of empty space there are myriad forms, including – now that photographic means of mechanical reproduction are the norm – handwriting. The emphasis of concrete poetry, or text in Kostelanetz's sense, is to weaken or remove the *semantic* properties of words, letters and figures, that is to reduce their attachment to meaning.

Morgan's poem above might not qualify as a 'pure' concrete poem since its effect depends on at least three things: a rudimentary knowledge of the character of the Hungarian language with its clusters of consonants and apparent plethora of Zs; our understanding of the words *siesta* and *snake*; and upon the *onomatopoeic* convention by which cartoons and comic-strips denote snoozing. All these go together to create a verbal and visual joke.

But **Jose Garcia Villa**'s (1908–) 'Sonnet in Polka Dots' is more inscrutable. The first quatrain reads

```
0 0 0 0 0 0 0 0
 0 0 0 0 0 0 0 0 0 0
 0 0 0 0 0 0 0 0 0 0
 0 0 0 0 0 0 0 0
```

and continues through the whole fourteen lines in the classic 8 and 6 division of the **sonnet** form. Villa's text seems to be playing a game with the abstraction of poetic form: it is in the shape of a sonnet but its content is zero.

Morgan's 'Hungarian Snake' might also be considered a **sound poem**, that is a work that takes the physiological qualities of voice-sounds, whether they are words or parts of words, or onomatopoeic sounds like 'sz', or sounds with no semantic features at all, as its key feature. Tongue-twisters such as

> Betty bought a bit of butter
> But she found the butter bitter
> So she bought a bit of better butter
> To make the bitter butter better

are close to sound poems since the consonants or vowels are so prominent. But they still use words. More radically, and towards the nonsense end of this spectrum, is Morgan's own hilarious perform-ance poem in which Nessie the Loch Ness Monster slowly surfaces and returns to the depths. The poem consists simply of a series of bubblings, exhalations and snorts from Morgan's virtuoso sinuses. Most recently, PC multimedia offers resources whereby a poet might compose a text by making simultaneous use not only of the semantic properties of the words, but their typography, their utterance, together with layerings of other sound and music as well as images.

CONCLUSION

Always the poet is working between the poles of the minimal and the full, between economy and plenitude. The one side draws towards compression – how briefly can I put this? – the other towards expansion, dilation – how richly, how extravagantly can I put this? Both extremes will eventually disappear into nothingness: silence, the blank page, or an endless, shapeless spume of words.

Between these poles poetry uses conventions to demarcate and organize a space for the words to dwell. These bounds draw and intensify attention. Just as the structuring of a joke will have us listening for and expecting the punchline, so the techniques of eloquence will have us straining for the key points of the argument or the point of inspiration. All are verbal spaces, marked out deliberately, *with* deliberation, and *for* deliberation. The poet feels for a space that seems at once demanding and accommodating, whether given by tradition or 'made anew'. It is a space marked for special attention. Indeed, poets accept outlines in order to achieve the powers of concentration and effect that come from limitation. **John Crowe Ransom** (1888–1974) argued, for instance, that formal measures obliged him to think harder, made him reject his first words because they would not fit, and therefore obliged him to discover other more interesting things that he would never have written had he not made himself meet the restrictions he had imposed upon himself. **Thom Gunn** (1929–) puts it like this: 'As you get more desperate, you actually start to think more deeply about the subject in hand, so that rhyme turns out to be a method of thematic exploration.' **Walt Whitman** (1819–92), on the other hand, sought to write 'as though there were never such thing as a poem'.

In practice the very decision to write poems presupposes *some* conception of what a poem is, and this will relate to the *space* and *sound* that it is. This does not mean however – as Whitman demonstrated – that the expectations that compose any genre cannot be expanded and renewed in the work of new poets. Whether these shapes, short or long, are accepted from the existing tradition or invented anew, the poet will discover workable outlines to contain the work: the deliberate space of the poem.

Summary
In this chapter we have considered:

- more of what is meant by *gestural* language;
- the nature of the oral tradition in poetry and the importance of its devices for recitation and memorability;
- children's rhymes, including the relation of clapping and poetic rhythm;
- the oral tradition in epic and narrative poetry both ancient and modern, including the ballad;
- the development of sound and shape as poetry moves from the oral to the page;
- poems as visual artefacts.

FURTHER READING

Bloom, Valerie (2002) *Hot like Fire: Poems*, London: Bloomsbury.

Donne, John (1998) *Selected Poetry*, ed. John Carey, London: Oxford University Press.

Frost, Robert, 'The Figure a Poem Makes' and Maxwell, Glyn, 'Strictures' in W. N. Herbert and M. Hollis (eds) (2000) *Strong Words, Modern Poets on Modern Poetry*, Tarset: Bloodaxe.

Furniss, Tom and Bath, Michael (1996) *Reading Poetry: An Introduction*, Hemel Hempstead: Prentice Hall/Harvester Wheatsheaf; see Part I, 1, 'What is Poetry? How Do We Read It?'

Grigson, G. (ed.) (1975) *The Penguin Book of Ballads*, Harmondsworth: Penguin.

Holub, M. (1990) *The Dimensions of the Present Moment*, tr. D. Young, London: Faber and Faber.

Koch, K. (1998) *Making Your Own Days: The Pleasures of Reading and Writing Poetry*, New York: Simon & Schuster.

Ong, W. J. (1982) *Orality and Literacy: The Technologizing of the Word*, London: Methuen.

Opie, Iona and Peter (1959) *The Lore and Language of Schoolchildren*, Oxford: Oxford University Press.

Pinsky, R. (1988) *Poetry and the World*, New York: The Ecco Press; see 'Poetry and Pleasure'.

3

TONES OF VOICE

My view is
Far too complicated
To explain in a
Poem.

(E. J. Thribb, 'Lines on the Return to Britain
of Billy Graham')

Master Thribb, the satirical magazine *Private Eye*'s spoof schoolboy poet, clearly thinks that poetry is not the appropriate vehicle to do justice to the intricacies of his views on religion. Whether or not some subject-matter is beyond the capacities of verse, we can readily see that the tone struck by a poem needs to be appropriate to its content. Our merriment in reading E. J. Thribb comes from enjoying the disparity either between the dignity of his chosen topics and the banal inertia of his verse, or between the ambition to write high-flown verse and the trivial character of his concerns. It is exactly the disjunction that all **parody** uses to gain its effect, and by recognizing the incongruity we are also reminded of the importance of an appropriate fit between subject and tone of voice.

I use the term 'tone of voice' in this chapter for two reasons. First, in reading poetry, we often have the impression that the poet is 'speaking' to us. This is partly because of the long association of poetry with the spoken word, but more specifically because the predominant poetic mode has become one in which one person is telling – or *singing* to us – often about emotional experience. This **lyric** mode fosters a sense of intimacy, and is the model we also see in words set to music, from Schubert to the popular songs of our own day. As I shall show later, this notion of how a poem 'speaks' needs serious qualification. The second reason however is that there are respects in which the system of tones adopted by poems can be equated with those we employ in daily speech, and it is this variety that I want to explore in this chapter.

WAYS OF SPEAKING

In conversation we adopt different ways of speaking according to circumstance and to whom we are speaking. To people we know well we will speak more allusively, drawing upon shared knowledge and assumptions, whereas with strangers, or in more formal situations, we might strive to be more explicit and precise in vocabulary and sentence structure. We always work with a loose system of formalities, adjusting how we say things according to the situation. For instance, we may only use that kind of intonation of our voice which aims to tell the listener that we mean the opposite of what we are saying – for example, remarking 'Lovely day' when it is cold and raining – if we are confident that they will understand what we really mean. In many languages the same part of speech can mean entirely different things depending on how the spoken sound is pitched. English does not work like that, although contouring the sound ironically, as in '*Lovely* day', is part of the way we pitch the **tone** of our speech. Indeed, intonation can generate meaning all by itself: think how many ways there are of saying 'Yes' and 'No'. These have their counterparts in many kinds of writing. As we shall see later, the **stress** patterns of the verse line are a special resource for deploying **pitch** of this kind.

Linguists often use the term **register** when picking out the formalities appropriate to speaking to various audiences, and in this chapter I want to argue that we can see the different sub-**genres** of poetry as registers of this kind. The lament for the dead in an **elegy**, the inwardness and song-like qualities of the **lyric,** the wryness or vituperation of the **satire**, the considered mixture of public and private concern in the *ode*, all these are examples of how the traditional modes of poetry have evolved to fit a tone of voice composed of word-choice, rhythm and shape to the appropriate occasion. Employing the resources of such modes is akin to finding the appropriate way to speak in public, or to a friend, whether in celebration, condolence, gossip or any other situation.

'NATURAL' AND 'UNNATURAL'

Now, it can be objected that emphasizing the place of traditional modes denies the spontaneity or naturalness of poetry. This takes us back to that valued association between the verse and the poet as speaker. The American poet **Kenneth Rexroth** (1905–82) stated that 'I have spent my life trying to write the way I talk.' Rexroth evidently thought that his verse should approximate in tone to the way he talked. Presumably he valued the idea of consonance between these different speech acts, indeed different parts of his personality. Since we tend to think of our ordinary speech as 'natural', it seems obvious that he wanted to cleanse his verse of 'artificial' expression. Interestingly, however, since Rexroth says that he has spent his life on this quest, it seems the pursuit of the 'natural' is not easy. Settling to work upon a poem, he encounters an expectation for a certain, specialized language and manner, acting strongly upon him. Although he speaks every day of his life, it is a lifetime's work to write in just the way he speaks.

If this suggests that 'natural' is not in fact so natural, it does not eclipse the long-standing demand for poetry to achieve a tone close to 'everyday speech', what **William Wordsworth** (1770–1850) called 'the speech of ordinary men'. Wordsworth was reacting against a poetic manner that he thought had become stultifying, and the changes of poetic styles have often been driven simply by an intuitive sense that poetry can't work that way any more; how close or how distant poetry and common speech should be is a perennial issue. But we should recognize that in speech too audience and occasion will usually influence, even determine, the *tenor* of how we talk, and that within these conventions we will choose our words. There is then no simple distinction here between the 'natural' and the 'artificial'. Artifice – by which I mean considered making – is not only unavoidable, it is 'natural'. None of this is to say that the time will not come when impatience with set forms, transgression, innovation and irreverence will be as appropriate and necessary in poems as it is in life.

But we can ask whether an 'ordinary' tone is what we want in poetry. Might we not want extraordinary speech, choice of words, a **diction**, that we do not encounter every day? At present, after half a century in which the informal, the conversational, the 'unpoetic' manner has dominated poetry in English, we might think so. The following discussion of the different registers that have been used over the preceding centuries might help us decide.

AUTHORITY AND AUTHENTICITY

In considering the different classifications of poetry as 'tones of voice' I shall be making a broad division between 'public' and 'personal' registers. But in both spheres – and of course they are not wholly distinct – one aspect of the 'voice' is bound to concern us. As we encounter a poem we will pick up a **tone** and want to ask, as we do mentally or explicitly with an unexpected phone-call, 'Who is speaking here?' We ask this not in the literal biographical sense, but in search of assurance as to the *authority* of the voice, that is whether we want to give it our trust. This might mean trusting the poem to tell us a story with a secure grasp of its plot and with compelling embellishment of description and character. It might mean being confident that the ideas the poem contains are intelligible and substantial rather than arrantly prejudiced. It might mean being convinced that the emotional world of the poem is plausibly and affectingly conveyed. None of these things necessarily depends upon our knowing the poet's credentials in the sense that she or he was an eye-witness to events, or a profound philosopher, or actually experienced the pain or joy described. Authenticity in a poem is a matter of rhetoric: of how the poet draws us into trust, makes us inhabit the events, ideas or emotions of the poem as we read. Finally this is done only by the words themselves on the page and in the ear. Of course, as with all trusting, we might come to feel we were misled.

PUBLIC VOICES

Although we now often see poetry as the most intimate of the literary genres, through most of its history it has worked as a public medium. Yet again this has to do with its oral origins, the transmission of the poem by voice to an audience.

Narrative, as we have seen with the ballad, is the most obvious of poetry's traditional functions, and storytelling has been the principal purpose of poetry, or a significant component of it, for centuries. In his massive collection *The Canterbury Tales* **Geoffrey Chaucer** (?1340–1400) makes the pleasure and purpose of storytelling the very method and substance of his work. The Host of Southwark's Tabard Inn, from which Chaucer's varied company of characters will leave on their pilgrimage to the shrine of St Thomas à Becket, exhorts each of them to tell a tale or two along the way as the ride will be cheerless if they ride 'the weye doumb as stoon' [dumb as stone]. He himself will give a free dinner as the prize for the best story,

> That is to seyn, that telleth in this caas   [say; case]
> Tales of best sentence and moost solaas   [solace]
>   ('Prologue' l. 797–8)

'*Sentence*' here encompasses theme and significance, and '*solaas*', solace, amusement and pleasure. But there is another dimension to this double purpose that Chaucer takes pains to set out, and that is his aim to represent the tale of each of his companions exactly as spoken. Of course this is a fiction, but it is a significant one, and his bold apology for it is significant:

> For this ye knowen al so wel as I,
> Whoso shal telle a tale after a man,
> He moot reherce as ny as evere he kan   [rehearse; near]
> Everich a word, if it be in his charge,   [every]
> Al speke he never so rudéliche and large,   [rudely]
> Or ellis he moot telle his tale untrewe,
> Or feyne thyng, or fynde wordes newe.   [pretend]
> He may nat spare, although he were his brother;   [flinch]
> He moot as wel seye o word as another.

Crist spak hymself ful brode in hooly writ,
And wel ye woot no vileynye is it.   [you know]
                    ('Prologue' l. 730–40)

[*You know as well as I that anyone who repeats another's tale must keep as close as he can to each and every word, no matter how crude they might be, or else he is being untruthful, pretending and putting in new words. He must not flinch from that even to save his brother's blushes but say one word as straight as another. After all, in the Bible Christ spoke broadly, and as you know there's nothing unfit in that.*]

The poet claims that being true to what is actually said, and *how* it is said, even if the words are 'rudeliche and large', is his first duty even if this 'vileynye' offends against decorum. His licence to do this he takes from the Gospels in which Christ himself speaks 'ful brode'.

The **register** of *The Canterbury Tales* then is presented as natural speech. Implicitly the work is setting itself apart from literary decorum. In the 1300s too it was important that the work is in the English tongue rather than the more prestigious Latin or French. The poem is declaring itself for a 'middle' style of verse, one that will include voices of various social degrees in the context of convivial, improvised storytelling.

Yet Chaucer is assigning himself a greater task because this naturalistic effect needs to vary according to the different manners of his narrators. It is striking how often in the brilliant character sketches of the Prologue the poet remarks upon the vocal character and sometimes precise sound of voice of the pilgrims. Thus the Knight's Squire is 'Syngynge ... or floytynge, al the day' [singing ... fluting]; the Prioress sings the divine service 'Entuned in hir nose ful semely' [properly refined]; the sweet-talking Friar's speech is unparalleled for 'daliaunce and fair langage' [stylish show]; of the fashionable Wife of Bath he writes 'In felaweshipe wel koulde she laughe and carpe', part of the social skills that have brought that 'worthy womann' five husbands; in his cups the Summoner falls into Latin, though only in the form of a few clichéd tag-terms; the venal, arrogant Pardoner, come straight 'fro the court of Rome', is reduced by the observation 'A voys he hadde as small as hath a goot' [goat]. Two contrasting examples might show how Chaucer fits his verse to create the impressions of different registers.

Here the Miller, who can barely stay on his horse, brushes aside the Reeve to insist on being next after the Knight's tale, a challenge to precedent and status which effects a striking cultural shift in the sequence of the *Tales* in itself, and for which Chaucer affects to apologize.

> This dronke Millere spak ful soone ageyn
> And seyde, 'Leve brother Osewold,
> Who hath no wyf, he is no cokewold.   [cuckold]
> But I sey nat therefore that thou art oon;
> Ther been ful goode wyves many oon,   [many a one]
> And evere a thousand goode ayeyns oon bade.
> Thou knowestow wel thyself, but if thou madde.
>
> ('The Miller's Prologue' l. 3150–6)

[*This drunken Miller was soon off again: 'By your leave, brother Oswald, if a man isn't married he can't be a deceived husband. Not that I'm saying you're one – there are a thousand good wives for every bad one. You know that yourself, unless you're mad.'*]

The ribald quip that if you don't have a wife then you can't be cuckolded is followed by some clumsy verse reeking of inebriation. Trying to speak it ourselves we find the metre stumble towards the same rhyme word 'oon' [one], a word that also comes up in the next line. Putting 'Thou knowestow wel thyself, but if thou madde' into a separate sentence makes it one of those 'And another thing ...' of the button-holing drunk, a lurching afterthought banged home with the big stress on 'madde'.

The corrupt Pardoner, whose handsome living depends upon his seductive eloquence selling fake pardons and relics to the faithful, begins the preface to his own tale like this:

> 'Lordynges,' quod he, 'in chirches whan I preche,
> I peyne me to han an hauteyn speche,   [take pains; highflown]
> And rynge it out as round as gooth a belle,
> For I kan by rote that I telle.   [know by heart]
> My theme is alwey oon, and evere was –

Radix malorum est cupiditas.'   [love of money is the
                                                   root of all evil]
                                          ('The Pardoner's Prologue' l. 329–34)

[*'My lords,' he said, 'when I preach in church I take pains to use a
highflown kind of speech and to ring it out clear as a bell for I've it all
by heart. My theme is always and ever has been the same: the love of
money is the root of all evil.*]

The haughtiness of his manner is immediately evident, as is his
own high opinion of his delivery, which contrasts so clearly with
what we have already been told about his goat-like tone. After that
first sentence he rounds off his paragraph with a sophisticated turn
by contriving a couplet mixing English and Latin. His choice of
Latin text shows an effrontery which anticipates irony. As with so
many such characters, the 'solaas' in the Pardoner's self-portrayal
comes of his outrageousness, and the 'sentence' in the poet's depiction of ecclesiastical greed.

So in the different ways Chaucer has these characters speak we
can see their character. The Miller's 'rudelich', knockabout verse
reveals his heedless heartiness, whilst in the Pardoner's affected
smoothness we see a condescending superiority, convinced as he is
that his audience is too stupid to see through his hypocrisy. In these
instances we see a deployment of tone that is akin to characterization in dramatic speech.

Another simple feature of *The Canterbury Tales* is the way
virtually every tale begins with a geographical placement. The scene
may be Oxford, Syria, Holderness, Tartary or Trumpington, but the
audience is always immediately told. Alternatively the provenance
of the tale, for instance Arthurian legend or the Roman authors, is
announced at the outset, but both procedures serve to situate the
audience at once. Thus we know where and when we are, and so are
able to take our bearings, to *'naturalize'* the situation as the
sequence of the tale unfolds. These are two aspects – characterization through speech, and physical placement – of the 'voice' of the
Tales which compose the overall register in which Chaucer endeavours to create a narrative mode suited to actual storytelling before a
varied audience.

EPIC AND THE MUSES

Traditionally the grandest type of narrative poem is the **epic**. The model of epic is a large poem whose story is of great historical significance for the group or nation from which it comes. In the western tradition Homer's tale of the ten-year war against Troy in *The Iliad* (see Chapter 2, 'Deliberate space') and its sequel *The Odyssey*, telling of the return home of the hero Odysseus, also known as Ulysses, form the defining myths of the Greek culture from which they sprang. Similarly Virgil's Latin epic *The Aeneid*, which begins with the escape of Aeneas from the ruins of Troy and tells of its hero's wanderings and the eventual fulfilment of his destiny to found the city of Rome, was deliberately composed to recount and sustain the founding myth of the city and its burgeoning empire.

For poets setting themselves a narrative task of this magnitude which calls so much more for 'sentence' than for 'solaas', the issue of authority weighs to an extraordinary degree. How can a single mind embrace the historical span and spiritual significance of a people's identity and also have the technical skill to engage and excite an audience? Traditionally the answer is **inspiration** (see also Chapter 8, 'Image – imagination – inspiration'). Our common notion of inspiration is of an idea or sense of possibility that comes to us individually and unbidden without conscious mental process, as though from 'elsewhere'. The classical poets looked to the gods and to the **Muses** for aid, and in the Middle Ages the Italian poet **Dante** (1265–1321), lost in the dark wood of doubt and depression, recounts how he is sent divine assistance to enable him to start his imaginary journey and thus his poem. Occasionally the Muses' assistance is sought in drama, as when the Prologue in **William Shakespeare**'s (1564–1616) *King Henry V* yearns for reality to replace the stage's shadow-play:

> O, for a muse of fire, that would ascend
> The brightest heaven of invention;
> A kingdom for a stage, princes to act
> And monarchs to behold the swelling scene!

> (Act I, Scene I, l.1–4)

This is 'a muse of fire' because fire, the lightest of the elements, is associated with poets whose work aspires upwards towards the spiritual realm.

For us the important thing about poetry that presents itself in this manner is that the poets are acknowledging that the process of composition is not a simply individual matter but an effort that requires support, and is in some sense collective. Whether or not the particular poet literally believes in the touch of a Muse, Aphrodite or the Virgin Mary, he or she is seeking a confidence and authority that transcends their own individual voice. The poet is saying that behind these verses is a weight of precedence and knowledge greater than my own, that I am a channel through which this gathered force of knowledge flows. The epic tone therefore aspires to be impersonal. The unnamed figure clad in a long black cloak who traditionally spoke Shakespeare's choruses is therefore an appropriate figure for this register.

What lies behind this kind of writing is well shown in the work of perhaps the last writer in English to attempt it with full seriousness, **John Milton** (1608–74). In *Paradise Lost* Milton sets out to write an epic in the classical manner, but, even more adventurously, in English rather than either of the anciently prestigious languages of Latin or Greek. The poem will retell the foundation story of the Judeo-Christian religion: the loss of innocence of the human race through Adam and Eve in the Garden of Eden. For Europeans of Milton's time there could be no greater story, and since it is told in the sacred text of the book of Genesis, Milton's attempt could well be seen as presumptuous in the extreme. Milton is aware of this, referring to his poem as

> ... my adventurous song,
> That with no middle flight intends to soar
> Above the Aonian mount, while it pursues
> Things unattempted yet in prose or rhyme.
>
> (*Paradise Lost* I, l. 13–16)

In these early lines Milton is doing two things. He is invoking the assistance of the Muses and of the Christian Holy Spirit as metaphysical aid. He is also girding himself with his palpable awareness and knowledge of the poetic and religious tradition – 'the Aonian mount' and other references. But it is also important to Milton as a Christian that he surpasses this classical and pagan literary tradition. The 'Aonian mount' is Mount Helicon, in Greek mythology the sacred home of the Muses, and his own poetic flight is going to take him above that. By these devices he asserts his vision of the transcendence of his religion. He also makes it seem that it is not merely John Milton, born in Bread Street, Cheapside, who sings, but the 'heavenly muse', who is, by implication, no less than God himself. Midway through the poem he calls for renewed strength to continue:

> Descend from heaven Urania, by that name
> If rightly thou art called, whose voice divine
> Following, above the Olympian hill I soar,
> Above the flight of Pegasean wing.
> The meaning, not the name I call: for thou
> Nor of the Muses nine, nor on the top
> Of old Olympus dwell'st, but heavenly born,
> Before the hills appeared, or fountain flowed,
> Thou with eternal Wisdom didst converse,
> Wisdom thy sister, and with her didst play
> In presence of the almighty Father, pleased
> With thy celestial song. Up led by thee
> Into the heaven of heavens I have presumed
> An earthly guest, and drawn empyreal air,
> Thy tempering;

> (*Paradise Lost* VII, l.1–15)

These lines are dense with classical and Christian allusion. Once more he claims to be flying above 'the Olympian hill', or the height reached by Pegasus, the winged horse of legend. Urania is the Greek Muse of Astronomy, thus the most heavenly. She was the Muse Christian writers 'adopted' for inspiration and the one Milton calls upon and associates with his own God. All of this is meant to display his knowledge, his humility and his determination to reach and survive in the most rarefied altitudes of poetry and spirituality. Milton's epic poetic voice is not a singular thing but a composite of what he saw as proper influences. To adopt Coleridge's metaphor, he

College Hill Library (WPL)
07/26/17 04:44PM
Patron Name: BORCHERDING, DANIEL
R

I must be living twice : new and selecte
33020010144789   Date Due: 08/16/17

The road not taken : finding America in
33020010175122   Date Due: 08/16/17

Poetry /
33020009706523   Date Due: 08/16/17

Poetry : the basics /
33020007496549   Date Due: 08/16/17

Item total: 4

Overdue fees are $.20 a day per item.

Renew materials by calling
303-658-2658

For account information, call
303-658-2601
(Please have your library card and PIN)

****

has giants' shoulders to mount on and will thus see further. This is how this author seeks to establish his authority.

For all the apparent command of its manner, *Paradise Lost* is a poem of embattlement and struggle. It is after all a story of loss, of how Adam and Eve's disobedience of God's ban on eating the forbidden fruit 'Brought Death into the World, and all our woe'. The theme of such momentous loss is common in the grand narratives of the world. It figures for instance in a modern poem that can properly be called epic, *The Arrivants* by **Edward Kamau Brathwaite** (1930–). Subtitled *A New World Trilogy*, its matter is the history and the culture of the Caribbean, a subject which from the self-styled metropolises of western culture might appear marginal, but which embodies the massive and enduring themes of a spoiled 'new world': migration, transportation and the effort to find a homecoming.

But unlike Milton, who has those few verses of the book of Genesis to expand upon, Brathwaite has no set preceding narrative. Instead he makes his own creation myth for the islands of the Caribbean in the *image* of a stone thrown across water that 'skidded arc'd and bloomed into islands ... curved stone hissed into reef ... flashed into spray / Bathsheba Montego Bay'. The scattering implicit in this image can also describe his sources which are fragments of a lost history, indeed a lost language: the stories and words of the native American people and of the slaves brought there from Africa. Thus,

> Memories are smoke
> lips we can't kiss
> hands we can't hold
> will never be
> enough for us;

('Prelude', *The Arrivants*, 1973: 28)

Brathwaite sees about him a shattered history, a contemporary culture split into folk fragments and marked by the experience of further emigration. Alongside highly formal western education for a few is mass illiteracy. All this means that a single voice is surpassingly difficult to establish. Within the different sections and sub-sections of his trilogy, therefore, Brathwaite cuts between a great variety of voices, often, as in the lines above, employing a pared, staccato style of stops and starts, as far as possible from the grand progress of Milton's heroic style. Indeed the first lines of the whole sequence seem to invoke a source in unvoiced sound and image,

> Drum skin whip
> lash, master sun's
> cutting edge of
> heat,

and it is out of this that the poet says 'I sing / I shout / I groan / I dream.' To see how different the registers can be in sections barely a page apart, we can compare these passages from the third book of the trilogy, *Islands*, in a section called 'Ancestors'. The first is a simple, apparently autobiographical, recollection:

> Every Friday morning my grandfather
> left his farm of canefields, chickens, cows,
> and rattled in his trap down to the harbour town
> to sell his meat. He was a butcher.

This matter-of-fact voice shifts into a dreamier remembrance of his grandmother 'telling us stories / round her fat white lamp ...':

> And in the night, I listened to her singing
> in a Vicks and Vapour Rub-like voice what you would call the
>   blues

> 3
> Come-a look
> come-a look
> see wha' happen

> come-a look
> see wha' happen

> Sookey dead
> Sookey dead
> Sookey dead-o

(l. 239–40)

The subject of *The Arrivants* is too large and various to be covered in one register alone. Milton strove for an encompassing English voice, believing that he could render the true interpretation of Scripture to his people. For Brathwaite there is no 'scripture' for his subject. Instead there are fragmentary histories, songs, murmurings, intuitions, memories like smoke. Thus the poem comprises many styles and tones, some jagged and furious, others ruminative, some descriptive, others in song. Often, as in passages quoted, the poem moves quite suddenly between the different kinds of English included in what Brathwaite himself has called the 'prism of languages'.

PUBLIC ANGER AND SATIRE

In view of its subject-matter it's not surprising that an angry note is often struck in *The Arrivants*. But this is no modern novelty: historically poetry has frequently been scorched by anger, and whilst we might readily assume that this might be an eruptive, unbridled phenomenon, it can be related once more to the association poetry has with those deliberate styles of public speaking known as **rhetoric**. In **classical** rhetoric there were *ways* of getting mad. In Latin, *vituperatio* – vituperation – was a special, calculated mode of speech to express fury. In this poem **D. H. Lawrence** (1885–1930) may simply seem to be stamping his foot, but in fact he does so in time. It takes its title from the first line:

> How beastly the bourgeois is
> especially the male of the species –
>
> Presentable, eminently presentable –
> shall I make you a present of him?
>
> Oh, but wait!
> Let him meet a new emotion, let him be faced with another
>   man's need,
> Let him come home with a bit of moral difficulty, let life face
>   him with a new demand on his understanding
> and then watch him go soggy, like a wet meringue.
> Watch him turn into a mess, either a fool or a bully.

> Just watch the display of him, confronted with a new demand
>   on his intelligence,
> a new life-demand.

Lawrence is clearly angry but the rage is not inchoate. Structurally the poet marshals it into a series of **anaphoric** clauses, 'Let him … Let him … Watch him … Watch him', together with those **alliterated** 'b' sounds and the refrain 'How beastly the bourgeois is / especially the male of the species'. These are stock rhetorical devices often heard in speeches by which the orator seizes the ear. Of course the vituperative mode might often speak direct to the object of disgust rather than, as here, to a supposed third person. Perhaps, however, Lawrence is being disingenuous. Perhaps he suspects that a good part of his contemporary readers will start to wonder, 'Might he mean me?' Such wiliness is not the most full-on manner of vituperation, though it is outspoken, enraged and brooks no qualification.

A subtler and more extensive version of the critical register is *satire*. The traditional targets of satire are pride and presumption, and the satirist's stance is that of the undeceived observer who relentlessly spies the gap between pretension and reality. The serious satirist has a philosophical view of the limitations of humankind and is enraged by the spectacle of fellow-creatures who persist in vanity and self-deception of every kind. In this poem the anonymous **Miss W—** is provoked by the calumny on women she has seen in **Jonathan Swift**'s (1667–1745) scatological poem 'The Lady's Dressing-Room'. Swift had chronicled, in stomach-churning detail, the realization of a particularly soppy lover that his lady, far from being an ethereal creature, has bodily functions. Miss W— retaliates with an equally scabrous denunciation of the habits of the male sex, 'The Gentleman's Study'. She concludes:

> Ladies, you'll think 'tis admirable
> That this to all men's applicable;
> And though they dress in silk and gold,
> Could you their insides but behold,
> There you fraud, lies, deceit would see,
> And pride and base impiety.
> So let them dress the best they can,
> They still are fulsome, wretched Man.

Although here Miss W– means the male sex by 'Man', her insistence is upon the chasm between appearance and reality. Dress hides nothing from the satirist's X-ray eye.

It is because of this disparity that the ironic tone features so prominently in satire of all kinds. *Irony* is the mode in which what is said is the opposite of what is meant. For example, when the songwriter **Randy Newman** croons, 'Short people got no reason to live', we are taken aback until we recognize that he is making a proposition he knows to be ludicrous in order to mock prejudice. Ironic effect usually involves such a time-lag between our first hearing the idea and realizing its true meaning – rather as in any 'wind-up' joke. In verse the *couplet* (see also Chapter 7, 'Stanza') is a device particularly suited to such delayed revelation of the underlying truth. A couplet is a pair of successive rhyming lines, and our wait to see what the second rhyming word will be is an interval that can be exploited. **Mary Barber**'s (1690–1757) 'An Unanswerable Apology for the Rich' takes this ironic tone in sympathizing with the wealthy 'Castalio':

> No man alive would do more good,
> Or give more freely, if he could.
> He grieves, when'er the wretched sue,
> But what can poor Castalio *do*?
>
> Would Heaven but send ten thousand more,
> He'd give – just as he did before.

The 'apology' is of course a caustic condemnation of Castalio's lack of charity. We are led to wonder if he would be more generous if he had yet *more* money, but when told he would as he did *before*, we realize that means he would do nothing. In these next lines, in which **Alexander Pope** (1688–1744) sardonically skewers the casual habits of judges and jurors, we see the couplet deployed to full effect:

> Meanwhile, declining from the noon of day,
> The sun obliquely shoots his burning ray;
> The hungry judges soon the sentence sign,
> And wretches hang that jurymen may dine.

Although the comfortable progress of the first two lines is becoming disrupted by the signing of the sentence and especially the image of hanging, it is not until the last syllable that the full import of his complaint is clear: the wretches are off to the scaffold because the court wants its lunch. The passage comes to a point, and what is distinctive about the satiric register is its acuteness. The wit to produce surprise in which things become instantly clear is the successful verse satirist's weapon. But the stance is not without its dangers, for seeming to be so perspicacious, so certain and so ready to mock and correct opens such writers to charges of presumption and simple peevishness. Deciding what in the poet's tone is properly critical, and what is merely spiteful, is one of the difficulties and fascinations of reading satire.

PASTORAL AND SIMPLE SPEAKING

Concerned as it is with foibles and manners, often as symptoms of a greater malaise, satire is the most social of poetic modes. *Pastoral* by contrast is a form which criticizes worldly sophistication not by pillorying it but by staying apart from it. The notion of pastoral goes back to classical Greece and the poet **Theocritus**, reckoned to have been at work around 270 BC. It claims its source and inspiration from the simple songs of shepherds, which is why the reed pipe, Milton's 'oaten flute' in his 'pastoral elegy' 'Lycidas', is its familiar emblem. But this attribution should not be taken literally, for pastoral is largely a feigned form, a style employed by poets as a means to criticize their own sophisticated society by contrasting it with the unaffected virtues of the 'humble' shepherd's life.

It is also a mode through which poets can seek another kind of idealized return: to a simpler, less elaborate manner of writing. For Milton the pastoral was the style for the poetic beginner, but for **Wordsworth**, certainly at the outset of his career, it is an ideal in itself. For him it respects the unnoticed lives of the rural poor and provides a purified diction for poetry based in ordinary language. Here in 'We Are Seven' the voice can sound determinedly straightforward:

> I met a little cottage Girl:
> She was eight years old she said;

> Her hair was thick with many a curl
> That clustered round her head.

The short lines, alternating between four and three beats, the predominant use of *monosyllables*, of familiar words and the unremarkable nature of what is described, all go to constitute a register that is defiantly plain. It is indeed a kind of anti-poetry. Elsewhere, and increasingly as his career goes on, Wordsworth does develop a more elaborated style using the longer, unrhymed, *blank-verse* line of Milton, whom he revered. But in his pastoral manner he had first worked to strip and simplify his style to create a voice that sounds closer to that of the subjects and characters in his poems. Thus he does not seem to be speaking from the mountain of poetic tradition, and his 'authority' might appear less forbidding. Indeed, in 'We Are Seven' we hear the encounter of two voices, the poet and the 'cottage Girl'. As she insists, against adult rationality, that there are seven in her family, including the ones living away and the two who 'in the church-yard lie', her view of the world confounds her senior. It is a poem in which Wordsworth seeks to dismantle his own – and others' – educated authority.

A similar turn towards the simple voice can be seen in twentieth-century *modernist* poetry. **William Carlos Williams** (1883–1963) was an American poet writing in urban New Jersey, about as far from the life of humble shepherds as you can get. But he sought to write about the daily life in the streets about him in a deliberately simple way, and he titled several of his poems 'Pastoral'. This poem, 'An Early Martyr', published in 1935, is in this style:

> Rather than permit him
> to testify in court
> Giving reasons
> why he stole from
> Exclusive stores
> then sent post-cards
> To the police
> to come and arrest him
> – if they could –

> They railroaded him
> to an asylum for
> The criminally insane
> without trial

For Williams the development of this manner was involved with how he felt he related – or did not relate – to the accepted poetic tradition. 'From the beginning I felt I was *not* English', he wrote. 'If poetry had to be written I had to do it in my own way' (*I Wanted to Write a Poem*, 1958). In 1913 he had begun a poem, 'Homage', like this:

> Elvira, by love's grace
> There goeth before you
> A clear radiance
> Which maketh all vain souls
> Candles when noon is.

Here, clearly, he is using a voice that he thinks is the one appropriate to verse but one foreign to his sensibility. His shedding of that manner is a search for authenticity, a simple 'natural' style not governed by precedent. But, as with Wordsworth, the style serves immediate, unsung subjects. These lines are from one of those poems he called 'Pastoral' in which he is walking through a poor district:

> the fences and outhouses
> built of barrel-staves
> and parts of boxes, all,
> if I am fortunate,
> smeared a bluish green
> that properly weathered
> pleases me best
> of all colors.

In *I Wanted to Write a Poem*, Williams later described what he was doing like this:

> The rhythmic unit was not measured by capitals at the beginning of a
> line or periods within the lines ... The rhythmic unit usually came to

me in a lyrical outburst. I wanted it to look that way on the page. I
didn't go in for long lines because of my nervous nature. I couldn't.
The rhythmic pace was the pace of speech, an excited pace because I
was excited when I wrote. I was discovering pressed by some violent
mood. The lines were short, *not* studied.

PERSONAL VOICES

It is clear in this discussion of the pastoral register how we have
come back to the notion of the distinctively individual voice and
Rexroth's ideal of writing in the way one speaks. Williams uses the
phrase 'lyrical outburst'. As we saw in Chapter 1, lyrical poetry has
its origins in song – an important sense which survives in the term
'song lyrics'. But as this kind of poetry has evolved, it has also come
to refer to poems which carry an immediately felt emotion compul-
sively expressed. The *lyric* has become predominantly the medium
for the personal voice.

This is an anonymous lyric known, with its music, from the
early sixteenth century, though it may be older:

> Westron wind, when wilt thou blow,   [western]
>   The small rain down can rain?
> Christ, if my love were in my arms
>   And I in my bed again!

We know nothing of the circumstances surrounding this poem but
the melancholy atmosphere created by the images of wind and rain
takes us into the author's sadness. Then the rhythmic surge of the
exclamation 'Christ', followed by the **stresses** on the two key words
'love' and 'arms', carries us into the sense of loss and longing. It is
an exact simple example of that 'immediately felt emotion compul-
sively expressed' and to convey that immediacy is its whole burden.
If we as readers or listeners feel this, it comes entirely from these
particular words, not from anything we know outside of them.

But we could not claim that these lines are 'ordinary speech'.
There is obvious artifice in the use of the *image* of rain and wind, in
the metrical **beat** and in the use of **rhymes**. These are what
constitute the lyric's 'musical' qualities. The poem may be felt, and
we may feel its emotion, but it is *composed*. This, I believe, is the

essential quality of the lyric of emotion: that it can convey its sincerity through artifice.

We might see this working in an idiom much closer to our own in this contemporary lyric by the songwriter **Nick Cave** (1957–). It is a love-song called 'Into My Arms', and the phrase 'Into my arms, O Lord' is its refrain. It is hard to think of a few words that can more directly convey a lover's simple longing. But the song also has a more elaborate scheme. It begins:

> I don't believe in an interventionist God,
> But I know darlin' that you do.
> But if I did I would kneel down and ask Him,
> Not to intervene when it comes to you:
> Not to touch a hair of your head,
> To leave you as you are,
> If He felt He had to direct you
> Then direct you into my arms,
> Into my arms, O Lord,
> Into my arms, O Lord
> Into my arms, O Lord
> Into my arms, O Lord.

The lyrical longing is powerfully there, but built around it is this pondering about belief in a God who steps into our lives. It becomes a prayer, though one perhaps uttered with tongue in cheek since it is phrased conditionally: 'If I did [believe] … If He felt He had to direct you …' The song is intriguing because of its mixture of love-longing and a weighty religious idea, and the possibility that the singer is playing with both these elements ironically.

When we recognize this complexity two questions might occur. First, is he sincere? We might answer that by trying to research the song's background in Nick Cave's life or some other aspect of its sources. But if – as it is – the song is performed by someone else, how can we know if the singer is sincere? Of course we don't expect performers to have felt and experienced everything they sing, but to rely on tracking back from the singer to what we might discover about the composer's life or sources is surely a convoluted approach to evaluating our response to hearing the lyric. First and foremost we must enter the little world the song creates for us and be convinced and entertained by that alone. If we are to connect it to a real life, let it be our own.

Second, is its felt emotion undermined by the song's construction around the religious idea – is it too blatantly artificial? If we think 'sincerity' depends upon conveying a direct, unambiguous feeling as simply as possible, then perhaps so. If, however, we think it might include shades of feeling, and its complication by ideas about that feeling, as I believe Cave's song does, then no. All verbal expression involves artifice. The 'personal voice' of Cave's lyric is the more interesting and enjoyable because of its complexity of structure and feeling.

PERSON AND PERSONA

It was largely because of the too-ready association of the 'I' in the poem and the 'I' who is the author that some poets have cast their poems in the voice of distinct, named characters. This strategy of the *dramatic monologue* is used most notably by **Robert Browning** (1812–89) in such poems as 'My Last Duchess', 'Andrea del Sarto' and 'The Bishop Orders His Tomb'. All these have Italian Renaissance settings, and so while the poems feature an 'I' speaking to us, the reader recognizes the distance from Browning when his Fra Lippo Lippi, painter and monk, says, 'You understand me: I'm a beast, I know.'

Modernist poets of the early twentieth century, especially keen to break the identification of the individual poet with what is spoken in the poem, adapted Browning's example. **Ezra Pound** (1885–1972) used free translation, playing variations upon the literally translated meaning, as another way of achieving this in such poems as 'The River-Merchant's Wife: A Letter' which is taken from the Chinese, and 'Homage to Sextus Propertius' which adapts the Latin poet Propertius. The 'voice' becomes a *persona*, or mask, which enables the poet to explore a personality which might include some indistinguishable part of him or herself but can range more freely, much as a dramatist can in creating a character. Such figures are **T. S. Eliot**'s (1888–1965) anxious, fastidious 'J. Alfred Prufrock', Pound's struggling poet 'Hugh Selwyn Mauberley' and **W. B. Yeats**'s (1865–1939) 'Michael Robartes' and 'Crazy Jane'.

Later in the century some poets – or perhaps more accurately some critics – came to see the casting-off of any mask as a virtue in itself. For enthusiasts of what came to be called the 'confessional' school of poetry the manner of speaking should be open, easy, even slangy, and the openness should reveal personal intensity and pain. This is the ending of 'The Abortion' by the American poet **Anne Sexton** (1928–74):

> *Somebody who should have been born*
> *is gone.*
>
> Yes, woman, such logic will lead
> to loss without death. Or say what you meant,
> you coward … this baby that I bleed.

Major poems have been written by poets often lumped simplisti-cally together as the 'confessional' school of the 1950s and 1960s, usually taken to include John Berryman, Robert Lowell and Sylvia Plath as well as Sexton. However, the critical fashion has tended to insist that the success of the poetry depends upon the guarantee that the experiences in the poem are biographically true. The extremes of painful experience, abortion, madness, suicide, become seen to be the stuff of poetry. This is a perilous model and, in view of the distorted prominence given to the suffering and the suicides of Plath and Sexton, perhaps an especially dangerous one for women poets. The distinction between the self who writes and the self who 'appears' in the poem is put well, I think, by **John Berryman** (1914–72):

> poetry is composed by actual human beings, and tracts of it are very closely about them. When Shakespeare wrote [in Sonnet 144] 'Two loves I have of comfort and despair', reader, he was *not kidding* … but of course the speaker can never be the actual writer who is a person with an address, a Social Security number, debts, tastes, memories, expectations.

'WHAT IS THE LANGUAGE USING US FOR?'

The broad assumption so far in this chapter has been that the poet chooses his or her 'tone of voice' for a poem just as, hopefully, we choose when we speak and write in everyday life. But is this simply true? We have all had the experience of letting our tongue run away with us so that once started on a way of speaking – sarcasm, for example – we say more than we mean because the force of that tone becomes irresistible. Similarly we sometimes strike the wrong note, perhaps make a joke at the wrong time. In these instances the tone we begin with seems to have a power all its own.

This is surely part of the character of language. As we learn to speak and then to write, we learn language's component parts and how to put them together individually and creatively. But at the same time we are absorbing many ways in which it has already been put together in set phrases, sayings and associations. We also speak and write as others have done before us and around us. As we have seen above, *epic* poets turned this to their advantage, but for poets like Wordsworth in the '*Romantic*' period and his successors, who set so much more store by originality, the idea that our 'expression' is not all ours has been more discomfiting.

The twentieth-century philosopher R. G. Collingwood in his *The Principles of Art* wrote this about emotion and expression:

> 'Expressing' emotions is certainly not the same thing as arousing them. There *is* emotion there before we express it. But as we express it, we confer upon it a different kind of emotional colouring; in one way, therefore, expression *creates* what it expresses, for exactly this emotion, colouring and all, only exists *so far as it is expressed*.

So, in a poem, the materials of language like vocabulary and syntax, the 'background noise' we hear of preceding and contemporary language-use, and especially how it has and is used in poetic convention, might drive the poem as much as the emotion or idea which is 'there before we express it'.

It is this sense that language speaks us rather than the other way round that provoked **W. S. Graham** (1918–86) to write

> What is the language using us for?
> Said Malcolm Mooney moving away
> Slowly over the white language.
> Where am I going said Malcolm Mooney.

> Certain experiences seem to not
> Want to go in to language maybe
> Because of shame or the reader's shame.
> Let us observe said Malcolm Mooney.

Notice the little transgressions against the norms in these lines: the *image* of 'moving away' over language and its being called white; the absence of a question mark at the end of the first stanza; the awkwardness of the split infinitive over lines 5 and 6, 'seem to not / Want to go in to ...', and the splitting of 'in to' to mime the reluctance. These, following the slightly paranoid opening line, testify to Graham's sense of tussling with this force called language. Another of his poems, 'Language Ah Now You Have Me', suggests the same unease: 'Here I am hiding in / The jungle of mistakes of communication.'

This jungle is treacherous. Language, especially 'everyday language', is subject to the wear of custom, and words and phrases which once seemed pithy and exact lose their currency and become cliché. The attention to language that goes into the poetic space will always want to reject cliché in favour of new ways to speak. Moreover, that attention will also be aware of the *echoing* of history, social usage and *connotation* in language. All these things affect the '*tone*' we try to adopt. The poet both uses and is used by the conventions of the craft. The voice of the poem is both the poet's own and the voice of other poets. 'Beauty', writes Pound, 'is a brief gasp between one cliché and another.' It makes writing poetry difficult, but is not a reason to despair. The American poet **Susan Howe** (1937–), from 'Pythagorean Silence':

> age of earth and us all chattering
>
> a sentence   or character
> suddenly
>
> steps out to seek for truth   fails
> falls
>
> into a stream of ink   Sequence
> trails off

must go on

waving fables and faces

Summary
In this chapter we have looked at:

- what is meant by 'tone of voice' in poetry and how different *registers* in speech might correspond to different poetic styles;
- the relationship of speech to poetry and what we might mean by 'natural' and 'unnatural' style;
- the concept of the author's authority, and authenticity in the poem;
- public styles for poetry: narrative poetry and characterization;
- the epic and the idea of the Muses in composition;
- poems of anger and satire;
- the simple voice in poetry and the pastoral style;
- poetry as personal expression and the idea of the persona.

FURTHER READING

Brathwaite, E. K. (1973) *The Arrivants: A New World Trilogy*, Oxford: Oxford University Press.

—— (1995) *History of the Voice: The Development of Nation Language in Anglophone Caribbean Literature*, New York: New Beacon Books.

Leech, G. N. (1969) *A Linguistic Guide to English Poetry*, Harlow: Longman; see Introduction and Chapter 1.

Mayes, Frances (1987) *The Discovery of Poetry*, Orlando: Harcourt, Brace, Jovanovich; see Chapter 4, 'The Speaker: The Eye of the Poem'.

Preminger, A. and Brogan, T. V. F. (eds) (1993) *The New Princeton Encyclopedia of Poetry and Poetics*, Princeton: Princeton University Press; see entry on 'Tone'.

Tomlinson, C. (1972) *William Carlos Williams: Penguin Critical Anthology*, Harmondsworth: Penguin Books; see articles in Part I.

4

THE VERSE LINE

Measures

Poets die adolescents, their beat embalms them,
(Robert Lowell, 'Fishnet',
The Dolphin, 1973)

Transcribing her husband's most famous poem, Mary Hutchinson Wordsworth picked up her pen one day and wrote *'I wandered like a lonely ...'* At this point she stopped and realized her mistake. In this small difference between

> I wandered like a lonely cloud

and

> I wandered lonely as a cloud

we *hear* the essential importance of **rhythm** to poetry. In these two versions the sentiment expressed is the same, the image used to convey it is the same, the number of **syllables** and even the placing of the **beats** is the same. Nonetheless, and not only because of familiarity, *'I wandered like a lonely ...'* sounds wrong. Analytically, the reason must be that *like*, though a vital part of speech, is too weak a word to bear a stress at this point in the impetus of the line. Putting it there delays the important idea of loneliness, especially as associated with the *I*, whereas the stresses placed in '*I **wandered lonely** ...*' enable the line to gather its meaning into the long and important syllable **lone**- so that the line pivots upon it in both rhythm and meaning. But '*I wandered like a lonely cloud*' simply sags in the mouth.

THE POETIC LINE

What is often called 'poetic' language is usually marked by a high incidence of **imagery**, **metaphor** and the 'rich' sounding of words. But these features might just as often be encountered in prose (which, for all its apparent spaciousness, would in itself have to be described as a 'form'). What most marks off poetry is the *line*. In Chapter 2 we saw how the claps in the schoolyard rhyme 'When Suzi was ...' defined the verse. Each clap is a **beat**, and the beats are put together in lines. The **rhythm** created in the line is a sound in the head and the ear, and, later, a defined space on a page. These lines of rhythm have been fundamental to the practice and concept of poetry, both as a **mnemonic** and as a device working on the senses. Once more we must recall poetry's **oral** roots. During its speaking, the way that the poem manipulates the time by deployment of pace, length of **syllable**, and emphasis, or **beat**, is decisive. These qualities constitute the **cadence** of the words (see Chapter 2, 'Deliberate space'). 'Cadence' comes from Latin and Italian words meaning to fall, and this description, as when the wistful Orsino in Shakespeare's *Twelfth Night* says of a song 'It had a dying fall' (Act I, Scene i, line 4), is as often used of verse as of music. Under 'cadence', the *Oxford English Dictionary* quotes George Puttenham writing in 1589 of

> the fal of a verse in euery last word with a certaine tunable sound which being matched with another like sound, do make a concord.

Poetry highlights the element of time and timing in how the particular sounds of words fall against each other and so compose 'a concord', or pleasing harmony of sound. It is in this respect of course that poetry is closest to music, and both share such terms as 'rhythm' and 'beat'. It is here too that the question of whether poetry is 'sound' or 'meaning' is most acute. **Ezra Pound**

(1885–1972), a poet who was also a composer, saw the different roles of words as musical 'concord', and words as items which carry meaning in a poem, like this: 'The perception of the intellect is given in the word, that of the emotions in the cadence.' This division may be too simplistic, but it does address the experience of poetry in which we apprehend an idea and feel the sensuous surge we derive from music at one and the same time.

The principal formal means that poetry employs to create its particular cadences is the *measure* of the poetic line. The line gains its effect by *recurrence*, the reappearance of notable features in the language time and again. It is the same principle that makes the chorus of a song its most important and memorable feature – the point where we can all join in.

One way of creating recurrence in early English verse was to use *alliteration*, that is to repeat the same consonant throughout a line, and as the line *recurs* to the left-hand side of the page, or announces itself in the voice, it repeats the trick with another consonant:

> Swart swarthy smiths besmattered with smoke
> Drive me to death with din of their dints.

The alliteration defines the line.

Until twentieth-century explorations of '*free verse*', lines usually recurred in the sense that their lengths and patterns had the same *measure*, or that the same measures recurred within reach of one another. 'Free verse' will be considered in a later chapter (Chapter 5), but here I am concerned with the working of the poetic line as it uses set measures, or *metres* – what is often described as 'formal verse'.

RHYTHM AND METRE

At this point I should distinguish the terms *rhythm* and *metre* since they are often confused and used interchangeably. *Rhythm* refers to the way the sound of a poem moves in a general sense either in part or through its whole length. *Metre* is more specific and refers to a set pattern which recurs line by line:

> Hickory dickory dock,
> The mouse ran up the clock.

A 'free verse' poem will not have a fixed measure like this since measure is one of the things it seeks to be free of. But, unless it is to be quite inert, it *will* have rhythm, as in these other varied lines from Lowell's 'Fishnet':

> The line must terminate.
> Yet my heart rises, I know I've gladdened a lifetime
> knotting, undoing a fishnet of tarred rope;
> the net will hang on the wall when the fish are eaten,
> nailed like illegible bronze on the futureless future.

If we read these two examples aloud one after the other we will hear the regularity in the first and the irregular flow of the second. But the difference is not between a fixed tick-tock and a more liberated 'flow'. A poem written in set measures will have a rhythmical movement of its own which includes the effects of the measure unless it is to sound tediously mechanical. The aim of this chapter is to describe the characteristics of set measures and then point to the ways in which their regularity varies to produce particular rhythmic effects.

DIFFERENT METRES

Traditionally these measures are made by one of four different systems depending on what they count. They might count:

1 **syllables**, that is the segments of sounds that make up individual words (**syllabics**);
2 **quantity**, that is the length of varying syllables (**quantitative**);
3 **beats**, that is where the **stress** or **accent** falls on different syllables in normal speaking patterns (**accentual** or **strong-stress**);
4 the number and pattern of **stressed** and **unstressed** syllables (**accentual-syllabic** or **stress-syllabic**).

When a poem measures its lines by one or other of these systems it is said to have **metre**, and the procedure for identifying and describing their working is **scansion**. In poetry in English by far the most used of these metres is (4) – **accentual-** or **stress-syllabic**.

However, because it is important to understand **syllables** and **stress** separately, I have placed **accentual-syllabic** last in this series so that we can work towards it. Here is a fuller description of each of the four systems.

SYLLABICS

The recognition of **syllables** is crucial to the composition and the study of formal measures. If we are to be absolutely precise the syllable is difficult to define and is in part a concept as well as a concrete item. That we speak of putting something 'in words of one syllable' suggests that that concept has to do with simplicity, of breaking things down to basics. But, in the sounds of a language, the **phoneme**, not the syllable, is the most basic item. Phonemes are the sounds actually used by a particular language that will make a difference to the meaning of a word. Thus if the phoneme that is the *k* sound in *cat* is replaced by the *m* phoneme we have the entirely different meaning of *mat*. The number of different phonemes will vary according to the language, as will the permitted sounds. English has some forty-plus phonemes (there is a margin of variation in practice), some of which will not feature at all in other languages: for instance the *th* sound as in *thin* is not a French phoneme, just as the Welsh *ll* as in *Llanelli* is not an English one. Our respective difficulties in pronouncing such sounds occur because they fall outside the phonological range we learn as children.

A **syllable**, however, might be made up of a number of phonemes: *k/a/t* go to make up the single expressed voicing of *cat*. We do not articulate all its component sounds separately ('*k/a/t* spells *cat*') but when we say *catarrh* we must voice two separate sounds, *cat-arrh*; for *catapult* three: *cat-a-pult*; for *catamaran* four: *cat-a-mar-an*. The difficulty in precise definition comes from occasional doubt as to how many syllables a word has. For instance an especially resonant actor might declaim 'O for a muse of fire' in such a way as to stretch *fire* into two syllables. Similarly many English north-country speakers might shorten *poetry* in a way that makes it sound like *po'try*. Within **polysyllables** too we might not always be sure where the divisions fall: is it *pol-y* or *po-ly*? These are factors which make linguists pause before providing a suitably scientific definition of the syllable.

Nevertheless the sound feature we call the syllable is generally recognizable and the number and the disposition of them is essential to all kinds of verse line. Since this is so, it might seem that the most obvious measure of a line is to count syllables. In fact, however, this method, **syllabics**, has not been much attempted in English poetry because, as we shall see below, stress is so prominent a feature of spoken English that it is bound to become a dominant feature in any sequence.

Where **syllabics** have been attempted, notably in the twentieth century, they have often been employed specifically to disrupt the accentual expectations of the traditional line by letting the stresses fall where they may. But then how do we distinguish a syllable-counted line as poetry from say / the last ten syllables of this sentence? We have seen already (Chapter 1) that the **haiku**, adapted from the Japanese form, is constructed in lines counting syllables in the pattern of 5–7–5. This shaping is one way in which the distinction from consecutive prose is made.

But a prose-like manner is likely to be an aim of syllabic verse and may often emphasize this by eschewing the use of capital letters at the opening of new lines in the way that poems conceived of as verses normally do. This is the first stanza of 'Considering the Snail' by **Thom Gunn** (1929–):

> The snail pushes through a green
> night, for the grass is heavy
> with water and meets over
> the bright path he makes, where rain
> has darkened the earth's dark. He
> moves in a wood of desire.

Careful counting will reveal that each of this poem's lines has seven syllables. It seems as though Gunn is simply being contrary in imposing his chosen count in ways that break up the syntax, for example separating 'green / night', and isolating the first word of a new sentence 'He / moves'. In some other poems he accompanies syllabics with **rhyme** – in 'Considering a Snail' there is the lightest of **half-rhymes**: *green/rain, heavy/he, over/desire* – which increases its difference from prose. But, as he has said in an interview subsequently, he was drawn to use syllabics as a method of changing the way he was writing:

> because after you've been writing metrically for some years, you have that tune going in your head and you can't get rid of it or when you try you write chopped up prose. My way of teaching myself to write free verse was to work with syllabics. They aren't very interesting in themselves. They're really there for the sake of the writer rather than the reader.

Exactly what Gunn was trying to get rid of we shall look at when we study stress-syllable metres, and in Chapter 7, examining stanza-form in a poem by Marianne Moore, we shall see just how much can be made of syllabics. Apparently the most crudely mechanistic of the ways to measure a line, it can be tooled to produce the most singular effects.

QUANTITATIVE – THE CLASSICAL TRADITION

Quantitative measures were the principal way of making verses in Greek and Latin poetry. The measure depends upon the sound lengths, or **quantity**, of different syllables, although the degree to which this followed in actual pronunciation is debated, and it may simply have become a convention. A line is defined by the number of syllables being divided into distinctive arrangements of long and short syllables: *heart* would be a long syllable and *hit* a short one. Educated in Latin and Greek, many European **Renaissance** poets of the fifteenth, sixteenth and seventeenth centuries composed poems in those 'classical' languages and looked to those forms as ideal for both reasons of prestige and an aesthetic liking for their complexity. Some tried to reproduce the **quantitative** method in verses in their own native languages. However, these have been only rarely successful, for certainly in English measuring quantities takes poets too far from the rhythms of the spoken language.

The main reason now to know anything of quantitative metre is because this long-standing devotion to the **classical** tradition has influenced the vocabulary used to describe verse measures. As we shall see later, scansion of lines that are in no way quantitative employs technical terms drawn from this classical tradition. It is useful to know the fundamentals of that terminology so that we can describe metrical features when we encounter them.

There are five principal types of metrical *foot* and their names derive from Greek. Originally a foot is a distinctive arrangement of short and long syllables:

> **Types of quantitative poetic feet**
>
> 1 **iamb**: one short and one long syllable, notated ˘ ¯
> 2 **spondee**: two long syllables, notated ¯ ¯
> 3 **trochee**: one long and one short syllable, notated ¯ ˘
> 4 **dactyl**: one long and two short syllables, notated ¯ ˘ ˘
> 5 **anapest**: two short and one long syllable, notated ˘ ˘ ¯

A classical quantitative line is therefore made up of a set number of feet, the type depending upon the requirements of the poem. The conventional terms for these lengths of line are drawn from Latin:

> **Lengths of quantitative line**
>
> | | |
> |---|---|
> | **Dimeter** | two feet |
> | **Trimeter** | three feet |
> | **Tetrameter** | four feet |
> | **Pentameter** | five feet |
> | **Hexameter** | six feet |

These terms, both for groupings of syllables and lengths of line, have been taken over in order to describe lines based upon *stress* – the main metrical feature of verse in English that we shall approach later. Instead of long and short syllables English scansion has come to recognize *stressed* and *unstressed* syllables, with stress often marked \ and unstressed ˘. In subsequent quotations here I have marked stressed syllables in **bold type**.

STRESS OR ACCENT

This refers simply to the prominence some syllables have over others in speech. In some languages, French or Italian for instance, where the stress falls is sometimes indicated by marked **accents**: *possibilité* (French), *possibilità* (Italian). English rarely uses such marks except for words borrowed from other languages, but any English word of two or more syllables will be accented. Thus we say, **love**-ly, com-**pare**, **summ**-er, **temp**-er-**ate**. The modulations here are not always exact. For instance, do we say 'syllable' by stressing only **syll** and leaving -ab and -le unstressed, or do we put a lesser accent on the last to give **syll**-ab-**le**? I think I would go for the former but there are variations. Do we say 'dis-**trib**-ute' or '**dis**-trib-ute', 'con-trib-ute' or 'con-**trib**-ute'? Some of these are much more marked with different language groups providing quite different accents. English football fans will speak of United's de-**fence**, whereas American sports crowds chant '**de**-fense, **de**-fense'. In the English Midlands the city is **Birm**-ing-ham, but in Alabama it can be Birm-ing-**ham**.

Notwithstanding these differences, **accent** or **stress** is a prominent feature of English speech and therefore of verse. In a sequence of **monosyllabic** words some will be stressed more prominently than others, as in *'we **will** be **glad** to **send** you **cash**'*. Usually the stress will be on the items most important for carrying the meaning with the grammatical items such as a, the, from, at, some, and suffixes like -ed and -ing being unaccented. When listening for stress in verse lines it is helpful to think of that clapping song in Chapter 1: 'When **Su**zi **was** a **ba-by**, a **ba**by **Su**zi **was**.' We clap on the stressed syllable.

In some measured verse stress is the main recurring feature:

> **Why why why** De-li-lah?

These are known as *pure-stress* or **strong-stress** metres. Besides their alliteration those blacksmiths lean heavily on stress to create their recurrent effect:

> **Swart swarthy smiths** be**smatt**ered with **smoke**
> **Drive** me to **death** with **din** of their **dints**.

Here we have a preponderance of stressed syllables, in this case clearly aiming to mime the blows on the anvil. In other instances the line will be defined by having a set number of stresses irrespective of how many syllables the line contains. Here are some two-stress lines (*dimeters*) but with lengths varying between six, five and four syllables:

> There **was** an old **wo**man
> And **what** do you **think**?
> She **lived** upon **noth**ing
> But **vict**uals and **drink**
> **Vict**uals and **drink**.

Similarly these lines each have four stresses, making them *tetrameters*, but again with lengths varying between five and eight syllables:

> **Half** a **pound** of **tup**penny **rice**   (8 syllables)
> **Half** a **pound** of **trea-cle**   (6)
> **Mix** it **up** and **make** it **nice**   (7)
> **Pop goes** the **wea**-sel   (5)

In reciting this, 'treacle' and 'weasel' are made to rhyme and so stretched to give two stresses as against their conventional values of '**trea**cle' and '**weas**el'.

Strong-stress metres can be used for urgent purposes, as in **Percy Bysshe Shelley**'s (1792–1822) indignant sonnet 'England in 1819' which begins:

> An **old**, **mad**, **blind**, des**pised**, and **dy**ing **king** –
> **Prin**ces, the **dregs** of their **dull race**, who **flow**
> Through **pub**lic **scorn** – **mud** from a **mud**dy **spring**;

These are ten-syllable lines but what dominates them is the heavy hitting of certain clustered syllables as in 'An **old**, **mad**, **blind** ...' However, the most used formal line in English poetry is one that employs a set pattern of stressed *and* unstressed syllables.

ACCENTUAL-SYLLABIC OR STRESS-SYLLABLE

Accentual-syllabic metres may vary in the length of lines but have *a set number of syllables and a set number of stresses*. Most often they will alternate like this opening of Shakespeare's Sonnet 12:

> When I do **count** the **clock** that **tells** the **time**.

Or, from Sonnet 9:

> Is **it** for **fear** to **wet** a **wid**ow's **eye**
> That **thou** con**sum'st** thy**self** in **sing**le **life**?

Each of these lines has ten syllables, and each stresses them alternately in the pattern of

> ti-**tum**, ti-**tum**, ti-**tum**, ti-**tum**, ti-**tum**.

If we devise a slightly different version of that first line, still using just ten syllables, *'When I count the clock's strokes to tell the time'*, we don't have the same pattern. Scanned, it might sound like this,

> When I **count** the **clock's strokes** to **tell** the **time**

which sounds simply clumsy and certainly has none of the regular measure of the clock's ticking with its implications of relentlessness.

These stress-syllable lines given above are called ***iambic pentameter***, a term composed from the classical vocabulary referred to earlier. This means that they consist of five *'feet'* (hence **pentameter**), each of which is a pairing of unstressed and stressed syllables (*ti-tum*) which are known as *'iambs'*. This is the line pattern for all of Shakespeare's sonnets, and the staple too of stage-speech in his and his contemporaries' plays in Elizabethan theatre. Indeed, it is the most common formal measure in English verse.

There are, however, a number of other measures. Iambic lines might be longer and stress six of their twelve syllables making a **hexameter**, or, to take the name from the most standard French verse line, an ***alexandrine***, like these translated from Pierre Corneille's *Le Cid*:

> No **day** of **joy** or **tri**umph **comes** un**touched** by **care**,
> No **pure** con**tent** with**out** some **shad**ow **in** the **soul**.

Occasionally it can be even longer, as in **Robert Southwell**'s (c.1561–95) 'The Burning Babe':

> As I in hoary winter's night stood shivering in the snow,
> Surprised I was with sudden heat that made my heart to glow;

Such a long line, the number of syllables make it a *'fourteener'*, has generally been found hard to handle. A verse line needs a tension much as a washing-line does, and longer stretches can sag into incoherence. But the 'fourteener' has another life in which it is split into two segments in the pattern of four stress/three stress. This is also known as **ballad metre** or **common metre** that we met in Chapter 2 and is strongly associated with the **oral tradition**:

> The King sits in Dunferling toune
>   Drinking the blude-reid wine:
> O whar will I get a guid sailor
>   To sail this schip of mine.

Used independently, the four-stress line and the three-stress line are known as **tetrameters** and **trimeters** respectively. Staying with iambic versions of these lines, here is **Jane Cave** (1754–1813) speaking in the voice of her ladyship to 'Good Mistress Dishclout', an understandably sulky kitchen-maid:

> And learn to know your fittest place
> Is with the dishes and the grease;

and **Thomas Campion** (1567–1620) in 'Now Winter Nights Enlarge':

> Much speech hath some defence,
>   Though beauty no remorse.

Yet shorter lines, such as the two-beat **dimeter** or even a line with but one stress, are uncommon, but here is the seventeenth-century poet **Robert Herrick** (1591–1674) showing off his versatility to the senior poet **Ben Jonson** (1572–1637) by devising a stanza that includes lines of one stress up to pentameter:

> Ah, **Ben**!
> Say **how** or **when**
> Shall **we**, thy **guests**,
> **Meet** at those **lyr**ic **feasts**
> **Made** at the **Sun**,
> The **Dog**, the **Triple Tun**,
> Where **we** such **clusters had**
> As **made** us **nobly wild**, not **mad**;
> And **yet** each **verse** of **thine**
> Out**did** the **meat**, out**did** the **frolic wine**.

We can see that not all of these lines are iambic. For example, '**Meet** at those **lyric feasts** / **Made** at the **Sun**' are lines that both begin with a stressed syllable. In this pub-crawling poem, these inject some cheerful impetus by striking into the line without delay.

The metrical foot composed of a *stressed syllable* followed by an *unstressed syllable* – ***tum**-ti* – is called a **trochee**, and thus the measure made of these feet is trochaic. The tale of Simple Simon's encounter is mostly trochaic:

> **Simple Simon met** a **pie**man
>   **Going to** the **fair**
> Said **Simple Simon to** the **pie**man
>   **Let** me **taste** your **ware**.

Trochaic metres, because of their incipient stress, are often used for poems with an urgent tone, sometimes to exhort someone, especially at the beginning. This is often true of hymns such as '**Praise** we the **Lord**', '**Brightly did** the **light** di**vine**',

> **Christ** will **gather in** His **own**
> **To** the **place** where **He** has **gone**
> **Where** our **heart** and **treasure lie**,
> **Where** our **life** is **hid** on **high**.

This is a four-beat measure, making the lines *trochaic tetrameters*, but, as Herrick's lines have shown, trochaic feet can be used in other lengths of line.

Trochaic measures tend to draw attention to the verse's metricality, marking it off very clearly from ordinary speech. **Edgar Allan Poe** (1809–49) wanted poetry to have a mesmeric quality which carries the reader in a nearly musical reverie, blurring the meaning of the words. Thus he makes bold use of alliteration, rhyme, repetition and emphatic metres. In 'The Raven' he employs an eight-stress trochaic line:

> While I nodded, nearly napping, suddenly there came a
>   tapping,
> As of some one gently rapping, rapping at my chamber door –

However, we can see that these very long lines actually have a decided break in the middle:

> As of some one gently rapping, / rapping at my chamber door.

Such a break is called a *caesura* (from the Latin, meaning 'cut') and is a feature we shall meet elsewhere. We might think the lines are really combined tetrameters, but elsewhere in the poem we can see that Poe wants the full continuous effect of the longer line:

> Ah, distinctly I remember it was in the bleak December;
> And each separate dying ember wrought its ghost upon the
>   floor.

Besides the iambic and the trochaic feet there are two other, less common metres that employ not two but three syllables. These are the *dactyl*, which goes \ ˇ ˇ , *tum ti ti*, and the *anapest*, which goes ˇ ˇ \, *ti ti tum*. This American spiritual's refrain uses the dactyl:

> Steal away, / steal away, / steal away / to Jesus

as does **Alfred Lord Tennyson** (1809–92) in 'The Charge of the Light Brigade':

> Flash'd all their sabres bare,
> Flash'd as they turn'd in air.

The anapest by contrast delays the stress, as in this anonymous tale of an 'Australian Courtship':

> But I **got** into **troub**le that **ver**y same **night**!
> Being **drunk** in the **street** I got **in**to a **fight**;
> A **cons**table **seized** me – I **gave** him a **box** –
> And was **put** in the **watch**-house and **then** in the **stocks**.

In lines like these the effect is to strengthen the stress and so give the line a bouncing effect suitable to its knockabout subject-matter. In longer lines the anapest can evoke languor, as here in 'Hymn to Proserpine' by **Algernon Charles Swinburne** (1837–1909) where the delayed stress falls on a series of long vowels:

> Thou art **more** than the **day** or the **mor**row, the **seas**ons that
>   **laugh** or that **weep**;

The beat can also be delayed for stirring purposes as in **Julia Ward Howe**'s (1819–1910) 'Battle Hymn of the Republic' which begins many of its lines with anapests: 'He is **tramp**ling …', 'He hath **loosed** …', 'I have **read** …', 'In the **beau**ty …' The famous first line is, however, iambic,

> Mine **eyes** have **seen** the **glor**y **of** the **com**ing **of** the **Lord**.

though it is possible that scanning it this way is influenced by the marching beat of the tune which is now inseparable from these lines. Might Howe have meant to record a visionary moment, to exclaim in her eagerness 'Mine eyes **have** seen …' which could give us a more nearly anapestic line where the stresses are all on the most semantically important syllables?

> Mine eyes **have** seen the **glor**y of the **com**ing of the **Lord**.

This would make the line closer in measure to those that follow, which after their opening anapests mostly put three unstressed syllables between each decided beat:

He is **trampl**ing out the **vint**age where the **grapes** of wrath are
    stored;
He hath **loosed** the fateful **light**ning of his **terr**ible swift **sword**;
        His **truth** is **march**ing **on**.

These possible variations point to an unavoidable experience when
we seek to scan lines: *regular measure is not absolute and fixed,
and indeed is not wholly regular even in ostensibly regular poems.*
Crucially, for instance, whilst we distinguish between stressed and
unstressed syllables, that simple divide does not take account of the
different weights of stress that we might hear. Returning to Howe's
opening line we might easily hear a stronger beat on some of the
stressed syllables:

Mine **EYES** have **seen** the **GLOR**y **of** the **COM**ing **of** the **LORD**.

If this is how we hear the line, instead of the regular alternation of the
first way that we scanned it, we could say we are giving greater stress
to the beats in capitals than to those in bold. Alternatively, we might
say that the line really only has those *four* beats, and that it bounds
from one stress to the next across *three* unstressed syllables. This
would mean that the line is less like the regular alternation of iambics
and more similar to the strong-stress line we looked at earlier.

All of this points to the obvious fact that the possible variation
within a metrical scheme shows us how crude the simple **stress/**
unstressed analysis of a line is. There are arguably many gradations
of possible stress, several in even a single line. Some readers have
tried to introduce a calibrated system which would give a value to
each syllable ranging, say, from 1 to 4 according to the strength of
the beat. This might give us:

2   3   1   3   1   4   2 1   3   1 2 1   4
Mine **eyes** have **seen** the **glor**y **of** the **com**ing **of** the **Lord**.

But *scansion* does not approach an exact science and very few of
us, I think, would want to read poetry in this way. It does
however draw attention to the *degrees* of stress that any reading
aloud is going to make, and to the amount of variation in formal
measures.

VARIATION IN PRACTICE

This brings us to a vital insight. The purpose of acquiring the technical expertise in scansion is *not* so that we dutifully scan every line we read, but to have the means to describe the general rhythmic pattern of a poem and – most importantly – the significant moments where that pattern *varies*. Almost always such shifts in the pattern are the most important moments.

Shakespeare's Sonnet 94 is about nice, controlled people, but nonetheless is pregnant throughout with indignation. It begins with praise but praise that is guarded, and the guardedness is implied in the opening line's irregularity:

> **They** that have **power** to **hurt** and **will** do **none.**

The stress on the first syllable, **They** – an inversion from the normal ˘ \ of the iamb to the trochee, \ ˘, suggests a finger being pointed, and the further effect of this is to leave *that have* unstressed and so give the pivotal word **power** a very strong beat, one that is equalled by the next stress, **hurt.** *Power to hurt* is the poem's subject and the two words are the fulcrum on which the first line is balanced. However, if we were to hear a beat on *do* at the end of the line, thus giving us three consecutive stresses, then the line does come weighted at that end. So far such an emphasis might be thought of either as a confirmation of the virtue of *they*, or as a gathering irony, that *they* might not be all they seem. This impression is strengthened at line 5 with another opening stress on *They* doubled by the immediate stress on the second syllable:

> **They right**ly are the **lords** and **own**ers of their **face**s.

The line has the regulation five stresses of the iambic pentameter, but is irregular in having thirteen syllables. This matters not at all since the important effect is to enlarge a sense of possible disgust should these **lords** *and* **own**ers turn out less virtuous than their **faces** show. The disgust ignites finally as the last line flares with sudden rage:

> For **sweet**est **things** turn **sour**est **by** their **deeds;**
> Lilies that **fest**er **smell** far **worse** than **weeds**.

The even tone of the regular 'For sweetest ...' is transformed by the force of the **trochaic** inversion which makes the beat on the first syllable of *Lilies* so emphatic. Of course, the **image** of rotting lilies is immensely powerful, but it is the variation of the measure which embodies the emotion in the poem.

Most often rhythmic effects work across several lines. Here is **Christopher Marlowe**'s (1564–93) Doctor Faustus lamenting that he has the individual immortal soul that is about to be claimed by hell-fire.

> Or **why** is **this** immortal **that** thou **hast**?
> **Ah**, Pythagoras' metempsychosis, were **that** true
> This **soul** should **fly** from **me** and be **changed**
> Unto some **brut**ish **beast**: all **beasts** are **happy**
> For when **they** die
> Their **souls** are **soon** dissolved in **el**ements;

Now the basic measure of *Doctor Faustus* is **iambic pentameter**, but looking for regularity in these lines I am soon at sea. Line 1 is regular, line 5 nearly so but for the 'weak' ending on an unstressed syllable, **happ**y. The real interest lies elsewhere, in the effects created by the long second line, which seems to me impossible to scan conventionally but is a gift to an actor in its rolling out the syllables of that impressive-sounding phrase *'Py-thag-or-as' met-em-psy-cho-sis'*. From there the lines run on in a sequence that, allowing for the mid-line pause at *brutish beast*, joins together (**enjambment**) five lines, including the simple change of key enabled by the very short *For when they die*. In the final line the drawn out sound of *dissolved* means that the line does not need the normal fifth stress as it mimes Faustus's longing. Marlowe, like very many poets, is using a measure and playing changes upon it to create the particular rhythms he needs for his meaning.

Sometimes such variations will be yet more obvious, as in the remarkable inventiveness of **John Donne's** *Songs and Sonets*. There he devises stanza forms which combine pentameter, tetrameter, trimeter and complete irregularity. Consider his love poem 'The Sun Rising':

> She'is all states, and all princes, I,
>   Nothing else is.
> Princes do but play us; compared to this,
> All honour's mimic; all wealth alchemy.

Any attempt to scan this metrically will very likely confuse us, but the lover's defiant exultation is marvellously carried in the rhythmic boldness of the staccato and surging effects. Later, eighteenth-century critics could not stomach Donne's free-handed way with formal metres, and it is significant that the revival of interest in his work and that of his Jacobean contemporaries, including the dramatists, did not take place until the early twentieth century and the beginnings of the '*free verse*' movement (see Chapter 5).

T. S. Eliot (1888–1965) was central to that movement, though he claimed emphatically that it did not exist. But what Eliot did insist is 'the very life of verse' is the 'contrast between fixity and flux'. By this he meant that successful verse lines do not continue to repeat an established pattern as exactly as possible, but operate in relation to it. In his essay 'Reflections on "Vers Libre"' he wrote:

> the most interesting verse that has yet been written in our language has been done either by taking a very simple form, like the iambic pentameter, and constantly withdrawing from it, or taking no form at all, and constantly approximating to a very simple one.

It is as though the 'simple form' radiates a force-field within which the poet works, sometimes closer to the exact pattern, sometimes further away.

Attending to set measures therefore, for the reader as for the poet, is not a matter of slavish notation but of sensing variation, the tension between 'fixity and flux'. It was Eliot too who coined the phrase 'the auditory imagination' by which he meant 'the feeling for syllable and rhythm, penetrating far below conscious levels of thought and feeling, invigorating every word'. It is this 'auditory imagination' that we exercise and develop as we read poetry. Knowledge of the technicalities of **scansion** should increase our awareness as we read and give us the means to understand quite how a rhythmic effect is being produced. But if such knowledge is elevated into an exclusive mystery tasselled with Greek terminology it is serving neither the reader nor poetry. The knowledge should enrich, not replace reading, that is *hearing*. In the noises of words and the rhythms of their combinations we hear all the complex accumulations of meaning with which in all areas we try to make ourselves understood.

MEASURE: AN HISTORICAL OVERVIEW

The history of measure in poetry in English is not only a matter of technical description. The changes in the style of the line are also a matter of culture and relate first to the history of the British Isles and later to the English-speaking world at large. In this part of the chapter I want to re-visit the measures outlined above but in a chronological way that connects them to linguistic and cultural change.

English medieval poetry moved through a varied series of metrical forms, especially in the latter part of the period. In Old English, the Anglo-Saxon verse line was heavily accented, usually, as in the eighth-century epic *Beowulf*, a pure stress line of four accents, but also alliterated and with a mid-line **caesura**. This pattern helped both reciter and audience as the poem was performed aloud (see Chapter 1). The lines tended also to be self-contained and sequential rather than carrying the sense fluidly through a long run of lines.

In the Middle English period, roughly the time between the Norman Conquest in 1066 and the rise of printing at the end of the fifteenth century, the language underwent considerable change: to look at a page of Old English is to see in effect a foreign language, whereas Chaucer, though unfamiliar, is at the least recognizable to the modern reader.

One important change was in syntax. Old English, like Latin and like German, is an *inflected* language, that is the forms of certain words change according to their exact role in the meaning of a sentence. Because a verb-ending, for instance, includes the information of what its subject and tense are, the verb can be variously placed within a sentence and still make the necessary connections. In Old English verbs commonly come at the end of a clause or sentence, as in this line from the poem 'The Seafarer':

bitre breostceare gebiden hæbbe

The most literal translation of this might be:

bitter the cares in my breast [I] have abided

with the 'I' inserted because it had occurred in the original several lines earlier. Using modern English syntax we might have:

I have endured the most bitter anxiety.

Such a change in syntax clearly altered how a poet will order the space of the line. But the change has another significance for poetry. When a poet reverts to the *inversion* of subject and verb in the manner of Old English, this is especially noticeable and becomes foregrounded as a *gestural* feature in itself: '*When Suzi was a baby / A baby Suzi was.*' When **Ezra Pound** wrote his twentieth-century version of 'The Seafarer' he wanted to keep a sense of the original, and give a shock to what he felt had become stale modern usage. Therefore, using a version of Anglo-Saxon strong-stress metre, he compresses it and inverts:

Bitter breast-cares have I abided.

Both in this chapter and Chapter 5, 'Free verse', we shall explore further this tendency to make something new through going back to the old.

A second new influence, felt particularly in vocabulary, was the absorption of the Romance languages, Latin, Italian and especially French following the Norman Conquest of 1066. Metrics in the later Middle Ages, roughly 1300–1500, became more various. The four-stressed Old English line continues in works like *Sir Gawayne and the Grene Knight* and, as we see here, in **William Langland**'s (?1331–?1400) *Piers Plowman*, although the caesura becomes less evident:

In a **som**er **seas**on whan **soft** was the **sonne**
I **shope** me in **shrouds** as **I** a **shepe** were.

But in the same period the ballad, or common metre, described above, flourished with its four/three stress pattern, a measure also employed in the *carol*, originally a dance song with a refrain:

> As I lay upon a night
> Alone in my longing,
> Me thought I saw a wonder sight,
> A maiden child rocking.

This example, from the fourteenth century, is also counting syllables to form an alternating pattern, as does this other verse from the same period where we can hear a virtually regular stress-syllable tetrameter:

> Jesus Christ my lemmon swete,   [beloved]
> That diyedst on the Rode Tree,   [rood/cross]
> With all my might I thee beseche,
> For thy woundes two and three.

Chaucer was certainly writing ***stress-syllable tetrameter*** in his early 'Romance of the Rose':

> Ful gay was al the ground, and quaint,
> And powdred, as men had it peint,
> With many a fresh and sundry flowr,
> That casten up ful good savour.

Savour, one of the many French words that came into English at this time, will have been stressed as *savour*. Chaucer was well acquainted with French poetry and introduced new forms like the ***rondeau*** into English (see Chapter 7), and it was probably with influence from this source that he came to lengthen his line to ten syllables (***decasyllabic***) and to stress five of them, thus working it into ***iambic pentameter*** as in these lines from 'The Pardoner's Prologue' in his *The Canterbury Tales*:

> For I wol preche and begge in sondry landes,
> I wol not do no labour with myne handes.

Chaucer's iambic pentameter did not, however, immediately dominate English metres. Between 1400 and the mid-sixteenth century poets were if anything yet more eclectic in their choice of line forms. But there were several influences at work that eventually consolidated English measures, and other aspects of poetic form, in the sixteenth century.

The first of these had to do with the development of the language and the attitude towards it among the literate classes. The fourteenth and fifteenth centuries saw an accelerated standardization of the language around the version of English that included the dominant city of London. Vocabulary, usage and spelling became much closer to what we can recognize as modern English, and the development of printing in the fifteenth and early sixteenth centuries carried this more and more widely. At the same time, the official dominance and prestige of Latin and French, dating back to the Roman Empire and the Norman Conquest, began to decline among the English educated classes. Increasingly they became interested and committed to their own language as a valued medium of literate as opposed to 'common', or vernacular, communication. Yet, like Chaucer, this elite was multi-lingual and open to continental influence to a greater degree than at any time since. So, for instance, **Sir Thomas Wyatt** (1503–42) and **Henry Howard, Earl of Surrey** (1517–47) in the early and mid-sixteenth century imitated the forms, including the sonnet, they came to know mainly from the Italian poet **Petrarch** (1304–74), and fashioned the long Italian line into English decasyllables. They were educated in the Greek and Latin classics as they had been recovered during the *Renaissance*, and this classicism provided them with a self-conscious interest in *prosody* and metrics which could act as a theoretical framework for their experiments in poetry. One result of this was the development of scansion using the classical terminology I described earlier.

In some ways what was happening was that English poetry was being 'civilized' out of its earlier rough and rude habits by the exercise of sophisticated continental models that appealed to the courtier poets. **Thomas Campion** (1567–1620), for instance, urged *quantitative* metrics on English poetry so that it might have the honour 'to be the first that after so many years of barbarism could second the perfection of the industrious Greekes and Romaines'. Practice, including Campion's own, was happily different, and despite the self-consciousness of poetic theory English verse in the Tudor 1500s continued to be pragmatically irregular, indeed all but 'barbarous', as the later, more determinedly classical, imitators and critics of the eighteenth century found it to be.

THE SPECIAL CASE: IAMBIC PENTAMETER

But we should pause at the **iambic pentameter** because it did become the single most prominent and influential of English measures. It is the line of the Tudor sonnet tradition and, most significantly, of the Elizabethan and Jacobean verse drama of the late sixteenth and early seventeenth centuries. Later, **Milton** employs it in *Paradise Lost*, and, returning to a rhyming form, it is the basis of the heroic couplet of the eighteenth century. **Wordsworth** and other **Romantic** poets used it extensively, and it has continued to figure in verse in English up to our own day.

The reasons why the iambic pentameter became and remained so dominant are far from simple to determine. For a long time it was claimed, without much scientific support, that it is the metre most 'natural' to English speech rhythms. We might occasionally catch someone saying, 'he works at Mister Minit down the street', or 'you'll never see a team as good as Stoke', or even 'another losing season for the Jets', but this argument must be difficult to prove linguistically. It also begs the question of what was 'natural' to English speech rhythms when strong-stress patterns were dominant in English verse. The argument would have to depend on establishing a distinct change in English speech patterns in the later Middle Ages as the Old English inheritance shifted into early modern English.

Alternative theories emphasize the artifice, rather than the naturalness, of the verse line and relate the development of the iambic pentameter to social and cultural change. Thus its channelling through the work of a cosmopolitan, courtly elite might be seen as the process of 'smoothing' the crude stresses of the traditional English verse line into an equable alternation of tone more acceptable to an urbane ruling class seeking to ease itself away culturally from the populace and their 'vulgar' **verses**. The role of poetry in the sophisticated play of the 'courtly love' tradition of **Renaissance** courts, and the emphasis given to this in literary history, certainly aids this impression. An instance of such socio-cultural division might be the scene at the end of Shakespeare's *A Midsummer Night's Dream* where the thumping versifying of the 'rude mechanicals" play is mocked by the aristocratic audience.

But such a theory also needs to consider the Renaissance enthusiasm for the vernacular among a ruling group who, by their privileged access to Latin and French, had been able to set themselves culturally apart from the people for centuries. Moreover, if the literate elites were so keen to maintain their cultural superiority, they might have been expected to foster a more regular and classically inspired metric than the iambic pentameter proved to be. In practice the iambic pentameter became an extremely flexible instrument, especially in the dramatic speech of the growingly popular Elizabethan stage. It can meet the demands of sounding plausibly like real people speaking – a demand that led to the abandonment of rhyme in favour of **blank verse** and of endless alliteration – whilst retaining the capacity for high-flown rhetoric. It can be fluent and continuous or set and reiterative as required. It possesses the capacity to move between 'fixity and flux' that Eliot noted. So to better understand its success we need to look further into its technicalities. We might ask two questions: why is it *iambic* and why is it a *pentameter*?

In musical terms the sequence of unstressed syllable followed by stressed syllable, ˘ \, can be described as a rising rhythm. What are the implications of basing a measure on this impetus? The linguist Otto Jespersen in his 1933 essay 'Notes on Metre' made this suggestion:

> As a stressed syllable tends, other things being equal, to be pronounced with higher pitch than weak syllables, a purely 'iambic' line will tend towards a higher tone at the end, but according to general phonetic laws this is a sign that something more is to be expected. Consequently it is in iambic verses easy to knit line to line in natural continuation.

Following Jespersen's proposal, iambic metre aids verse which seeks to emphasize continuity. The argument of a **sonnet** with its frequently elaborate syntax and the 'real speech' of characters in a drama both need this kind of continuity. (A sense of closure, as Jespersen remarks, can be brought about by consecutive rhyming, as in the *couplet* that closes the sonnet or scenes in Elizabethan drama.) This kind of continuity, where the reader's expectancy is carried along by a developing syntax that crosses the verse line, is well exemplified in **Shakespeare**'s Sonnet 140:

> Be wise as thou art cruel; do not press
> My tongue-tied patience with too much disdain,
> Lest sorrow lend me words, and words express
> The manner of my pity-wanting pain.

The lines where the sense runs on without pause across the line-break (*enjambment*) – 'do not press / My tongue-tied patience' – show in particular how the stress on the last syllable of each line carries us forward to the next stage of the statement. We can also see that while the first two lines can be readily measured out as conventional iambics, it is more likely that we will read them aloud – for the sense – with a very different sequence of stresses:

> Be **wise** as thou art **cru**el; do not **press**
> My **tongue-tied pati**ence with too **much** dis**dain**,

Again we see the basis of the regular measure acting as underlying pattern to enable particular variation.

The opening of Sonnet 91 presents a quite different effect:

> Some glory in their birth, some in their skill,
> Some in their wealth, some in their body's force,
> Some in their garments, though new-fangled ill,
> Some in their hawks and hounds, some in their horse;

The first words might be read conventionally, 'Some **glor**y …', or, for greater attack, by stressing both opening syllables, '**Some glor**y …' The following three certainly demand a stress on the opening syllable, '**Some** in their **wealth** …' The movement here is not based on *syntactical* continuity in the manner of Sonnet 140, but on parallel repetitions. For the required emphasis the stress pattern is reversed from the iambic ˘ \ to give a *trochaic foot*, \ ˘. As we have seen with 'Simple Simon' above, a strong trochaic presence seems to fit the manner of emphatic repetition. An entirely trochaic metre tends towards the closure of every line and to require constant repetition to move the poem along. Longfellow's *Hiawatha*, written in trochaic tetrameter, is a famous example:

> By the shores of Gitche Gumee,
> By the shining Big-Sea-Water
> Stood the wigwam of Nokomis,
> Daughter of the Moon, Nokomis.

The trochaic metre lays each line down in a fixed fashion, often inverting subject and verb – 'Stood the wigwam ... Rose the firs ...' – drawing attention to the verse's artifice. There is, though, a limited call for the manner, and so a technical answer as to why verse since the sixteenth century has been mainly iambic may be because the rising rhythm assists the continuity of speech in dramatic poetry, and the fluency of thought and argument in much lyric poetry.

Now what of the '**pentameter**' half of the term, the line's length – why are *ten* syllables divided iambically? We have looked above at different lengths of line and at the difficulty of sustaining long lines without the tendency to fall into two. We can hear this happening in these 'fourteeners' from **Arthur Golding**'s sixteenth-century translation of the Roman poet Ovid:

> Then sprang up first the golden age, which of itself maintained
> The truth and right of everything, unforced and uncon-
> strained.

But a longish line does clearly assist fluency and continuity, so how did the ten-syllable line achieve such popularity?

One theory points not to a decisive break between the pure, four-stressed line of medieval verse and the succeeding pentameter, but to the things they have in common. The older line tended to spring its four stresses across ten and frequently more syllables. But as Middle English modulated into early modern English there was a great reduction of inflections and consequently many syllables ceased to be sounded, for example that terminal *e* as in the legendary 'ye olde Englishe tea shoppe' now resurrected for tourists. Basil Cottle in his book *The Triumph of English 1350–1400* compares the sentence 'The goode laddes wenten faste to the blake hill', which in Chaucer's day would have sounded each *e* and amounted to fifteen syllables, with its modern equivalent: 'The good lads went fast to the black hill', a total of only nine syllables. Following this suggestion we might recognize a tendency for the medieval line of twelve or fourteen syllables to become a line nearer to ten as the spoken language changed.

There is another possible point of continuity. The implication seems to be that the pentameter line contracts its number of sylla-bles but adds a stress over the earlier four-stress norm. But does it? The formally described measure usually gives us five stresses, but do we always *hear*, and more importantly *speak*, five stresses? As we have seen, in performance there is continual and often considerable variation. Being sure to articulate five stresses in many ten-syllable lines will sound forced and ridiculous. Many critics, as well as performers, maintain that the iambic pentameter often contains one stress so weak, or carries such force in the word-order, that the norm is *four* stresses, not the regulation five. Yet the fifth, 'extra' stress can provide another option for modula-tion and variation. Poetry in early modern English shed older devices such as alliteration and the regular caesura so as to achieve a verse that could mime actual speech whilst still keeping a measure for rhetorical power. In evolving the four/five stress decasyllabic line it also extended the range of its stress base and thus its flexibility.

In an intriguing essay called 'The Dimension of the Present Moment' the Czech poet and scientist **Miroslav Holub** (1923–98) writes that while he can imagine eternity he finds great difficulty in figuring the present moment. In a series of swift, light-footed spec-ulations he draws upon experimental psychology and musical data to propose that 'In our consciousness, the present moment lasts about three seconds, with small individual differences.' This disarm-ingly simple notion he then allies with studies of the poetic line in a number of different languages, with the broad conclusion that the outer edge of momentary attention at three to four seconds corre-sponds to the normal limit of the poetic line. We do not have to attach ourselves strongly to the science of these ideas to feel their possibilities. The neural effects of how we hear the rhythm and length of a line seem close to how we apprehend.

As we have seen, the analysis of measured verse can be quite complex. Much more difficult would be understanding how, within the processes of evolution, the first human speech for record and reiteration depended upon the recurrences of what we now call the poetic line. Nonetheless it is in the simple clap of the hand, the tap of the foot, that these metrical patterns have their foundation.

Summary
In this chapter on formal verse we have examined:

- the poetic line and the importance of rhythm, beat and cadence;
- the distinction between rhythm and metre;
- the four main classifications of formal metres: syllabics, quantitative, accentual and accentual-syllabic;
- the terminology and method of scansion;
- the importance of variation, 'breaking the rules' of metre;
- the historical development of metre in English poetry with special reference to the iambic pentameter.

FURTHER READING

Attridge, Derek (1995) *Poetic Rhythm: An Introduction*, Cambridge: Cambridge University Press.

Fussell, Paul (1979) *Poetic Metre and Poetic Form*, New York: Random House.

Holub, M. (1990) *The Dimension of the Present Moment*, tr. D. Young, London: Faber & Faber.

Leech, Geoffrey N. (1969) *A Linguistic Guide to English Poetry*, London: Longman; Chapter 7.

Stallworthy, Jon (1996) 'Versification' in M. Ferguson, M. J. Salter and J. Stallworthy (eds) *The Norton Anthology of Poetry*, fourth edition, New York: Norton.

'FREE VERSE'

When this Verse was first dictated to me, I consider'd a
Monotonous Cadence, like that used by Milton & Shakespeare
& all writers of English Blank Verse, derived from the modern
bondage of Rhyming, to be a necessary and indispensible part
of Verse. But I soon found that in the mouth of a true Orator
such monotony was not only awkward, but as much a bondage
as rhyme itself. I therefore have produced a variety in every
line, both of cadences & number of syllables. Every word and
every letter is studied and put into its fit place; the terrific
numbers are reserved for the terrific parts, the mild & gentle
for the mild & gentle parts, and the prosaic for inferior parts;
all are necessary to each other. Poetry Fetter'd Fetters the
Human Race.

(William Blake, 'To the Public', *Jerusalem*)

So much of the theory, and the spirit, that gave rise to the notion of
'free verse' in the twentieth century can be seen in this address by
William Blake (1757–1827) writing at the outset of the nineteenth.
This chapter will explore the origins of what came to be known in
the twentieth century as 'free verse', and look at the many directions this approach to the **poetic line** has taken.

As Blake acknowledges, poets have frequently chafed at the formal demands they inherit, which is why Shakespeare and Milton 'derived' their verse from **rhyme** and wrote **blank verse**. We have seen too in Chapter 4 how measured verse regularized the numbers of **cadences** and **syllables**, but that this regularity was not always strict in practice. But Blake finds their measures 'monotonous'. He wants 'variety in every line' and it is the *regulation* of **beat** that becomes the later liberators' complaint against measured verse. When **Ezra Pound** (1885–1972) joined the argument he urged poets 'to compose in the sequence of the musical phrase, not in sequence of a metronome'. **Rhythm** then should not be timed by a pre-set mechanism, but suit the demands of the individual poem according to whether it is, as Blake writes, 'terrific' or 'mild & gentle'.

'FREE VERSE' AND LIBERATION

But the argument is not only technical but part of a wider claim to liberation. Blake's view of poetry is visionary, and for him its true voice is the original voice of humankind. He states that his verse is 'dictated' to him, not composed within the schemes of tradition. This pristine utterance of 'a true Orator' comes from divine inspiration and cannot be so confined. Indeed those schemes are but another part of the chains of culture that bind the natural freedom that is our original state: 'Poetry Fetter'd Fetters the Human Race.' Although two of its major influences in the twentieth century, Ezra Pound and **T. S. Eliot** (1888–1965), shared nothing of this **Romantic** temperament, poetic allegiance or philosophy of Blake's, there is a powerful part of the **poetics** of 'free verse' which appeals to broader hopes of liberation. Let us therefore look at the modern evolution of 'free verse' as both a formal and a cultural development.

THE BIBLICAL LINE

In respect of 'numbers', the biblical line might be described as 'free'. These lines are from Psalm 136:

> O give thanks unto the Lord; for he is good; for his mercy endureth forever.

> O give thanks unto the God of gods: for his mercy endureth
> forever.
> O give thanks to the Lord of lords: for his mercy endureth
> forever.
> To him who alone doeth great wonders: for his mercy endureth
> forever.
> To him that by wisdom made the heavens: for his mercy
> endureth forever.

The lines here are defined by what is to be said. Each has a variety of 'cadences & number of syllables' and is **end-stopped**, that is the end of the line coincides with the end of the sentence. But they do feature other gestures to make the lines memorable, notably the recurrence at their opening – 'O give thanks ...', 'To him who ...' – and their closing – 'endureth forever' – that rhetorical device of **anaphora**. The variations are built around these common elements. Blake's line in *Jerusalem*, and elsewhere, is inspired by the biblical model and sometimes employs similar recurrences:

> The land of darkness flamed, but no light and no repose:
> The land of snows of trembling & of iron hail incessant:
> The land of earthquakes, and the land of woven labyrinths:
> The land of snares & traps & wheels & pit-falls & dire mills:

Another, earlier poet, **Christopher Smart** (1722–71), in his *Jubilate Agno*, uses the style for his own distinctive devotional purposes:

> For I will consider my Cat Jeoffrey.
> For he is the servant of the Living God, duly and daily serving
> him.
> For at the first glance of the glory of God in the East he
> worships him in his way.
> For is this done by wreathing his body seven times round with
> elegant quickness.
> For then he leaps up to catch the musk, which is the blessing of
> God upon his prayer.

Smart is drawing upon the general manner of the verses in the Bible as set out in the **Authorized Version** of 1611, especially that of the Psalms, the Song of Solomon and the Magnificat of the Virgin Mary in Chapter I of St Luke's Gospel, from which he takes the reiterative use of 'For ...'

Both Blake and Smart were working in effect outside the mainstream of verse style. The model of the biblical verse gave them the amplitude their imaginations demanded but also a form closer to popular knowledge. After all, in the eighteenth century the Bible was heard and read in churches, chapels and homes daily and weekly and would be far more widely familiar than the couplets of famous London poets of the period such as **Alexander Pope**. We can see a similar turn towards the biblical line in another poet seeking to mark a poetic space distinct from the tradition, the American **Walt Whitman** (1819–92).

'Not a whisper comes out of him of the old stock talk and rhyme of poetry ... No breath of Europe.' This is Whitman in an (anonymous) review of his own first book in 1855. The main marker of what Whitman called his 'language experiment' is his shattering of measure. Again we can see the influence of the biblical line, augmented apparently by translations he knew of the Hindu sacred texts of the Bhagavad Gita. He makes great use of *anaphoric* structures, series and enumerations, exclamations and declamations ('Endless unfolding word of ages!'), but what he called his 'barbaric yawp' is also capable of the most delicate detailing of image and rhythm in moments of quietude such as this from Chant 6 of his huge poem 'Song of Myself':

> A child said *What is the grass?* fetching it to me with full
>   hands;
> How could I answer the child? I do not know what it is any
>   more than he.
>
> I guess it must be the flag of my disposition, out of hopeful
>   green stuff woven.
>
> Or I guess it is the handkerchief of the Lord,
> A scented gift and remembrancer designedly dropt,
> Bearing the owner's name someway in the corners, that we
>   may see and remark, and say *Whose?*

'Guessing' is characteristic of Whitman's style of intuition by which a flower on his window-sill 'satisfies me more than the metaphysics of books', so that the American colloquialism is a way of talking to the reader and suggesting a whole way of knowing. The use of the commonplace image of the dropped handkerchief in this mystical context is also part of Whitman's democratic voice. The placing of 'someway' too is a matchless touch as it particularizes the handkerchief that bit more and lengthens the line in a way that suggests the slightly longer time taken to look at it. For all his use of formulae and the round oratory of his public manner, Whitman does flex this style of free line to the most various and often subtle and intimate uses. This passage from one of his war poems, 'A Sight in the Daybreak Gray and Dim', begins, and nearly ends, with a tetrameter, but in between he stretches the lines with the utmost tact:

> Curious I halt and silent stand,
> Then with light fingers I from the face of the nearest the first
>   just lift the blanket;
> Who are you elderly man so gaunt and grim, with well-gray'd
>   hair, and flesh all sunken about the eyes?
> Who are you my dear comrade?

The inversion – 'silent stand' – in the first line establishes some formality, though he has admitted to simple curiosity. But then the tremulousness of his lifting the blanket is carried with those seemingly unnecessary, but in fact vital, extra syllables, 'the first', followed by the 'just' of 'just lift'. Generally Whitman makes his line coincide with the unit of sense and hardly ever uses run-on lines. Often the longer lines do have some kind of *caesura*, but here he manages to articulate the whole slow length in order to mime the action described.

The studied informality of Whitman's line is of a piece with his drawing his subject-matter from his contemporary scene of ferries, streets and locomotives, and a *diction* happy to employ words like 'higgled', 'draggled', 'soggy' alongside the more high-toned 'esculent', 'obstetric', 'gneiss' and 'sextillions'. All of this constitutes Whitman's radical – and deliberately American – revision of poetic norms, and anticipates not only the work of his most obvious imitators like **Allen Ginsberg** (1926–98) but all the significant changes to poetry in the twentieth century.

The line of **Gerard Manley Hopkins** (1844–89) is somewhat different but bears comparison with Whitman's. We noted in Chapter 4 how the *iambic pentameter*, with its equable alternation of unstressed and stressed syllables, can be seen as an imposition of educated sophistication. Hopkins wanted to return to the 'roughness' and energy of medieval strong-stress metres. Accordingly he developed a measure he called *sprung rhythm*, named for the way the line 'springs' across a varying number of unstressed syllables from one strong stress to the next. He also disturbs the alternating decorum of standard lines by jamming stresses together, jagging the lines with stutters of punctuation as his matter demands. This is from one of his religious sonnets, 'Carrion Comfort'.

> Not, I'll not, carrion comfort, Despair, not feast on thee;
> Not untwist – slack they may be – these last strands of man
> In me ór, most weary, cry *I can no more*. I can;
> Can something, hope, wish day come, not choose not to be.

Since the poem rhymes and uses the set form of the *sonnet*, it can hardly be called 'free verse', but the lines themselves entirely disrupt traditional measure with their staccato, gasping urgency which Hopkins makes yet more emphatic by his extra marking of stress. In his *Journals* Hopkins writes: 'You must not slovenly read it with your eyes but with your ears as if the paper were declaiming it at you ... Stress is the life of it.'

MODERNISM AND 'FREE VERSE'

Hopkins' poetry was not published until 1918 by which time the *modernist* movement towards 'free verse' was well under way. We have already glanced in Chapter 4 at **Ezra Pound**'s 1913 poem 'The Seafarer', a poem described as 'from the Anglo-Saxon'. Like Hopkins he has sought out earlier, less equable poetic manners to help him escape the inherited voice. Thus the poem is *syntactically* jagged, consonantal, often *alliterative* and set in irregular lines containing strong stresses.

Hung with **hard ice-flakes**, where **hail scur flew**,
There I heard **naught** save the **harsh sea**
And **ice-cold wave**, at **whiles** the **swan cries**,
Did for my **games** the **gann**et's **clam**or,
Sea-fowls' **loud**ness was for **me laugh**ter,
The **mews' sing**ing **all** my **mead-drink**.

'Some knowledge of the Anglo-Saxon fragments ... would prevent a man's sinking into contentment with a lot of wish-wash that passes for classic or "standard" poetry', wrote Pound in a typically pugnacious essay. The sentiment that verse must be dragged from 'contentment', a lolling posture wafted by the zephyrs of familiar rhythms, liquid consonants and mild assonance, pervades modernist *poetics* at this time. It goes along with disgust with the complacent, platform eloquence, 'the conventional oompa oompa', as one critic called it, of Edwardian public poetry in the years before the First World War.

Pound, who wrote bitterly about the waste and misery of the war, wanted to 'make it new', to avoid being gathered into the 'standard' voice of the time. So he sought models not only in Anglo-Saxon but in Provencal, Italian, Chinese and Japanese poetry. He put himself through a programme of defamiliarization aimed at making changes not just to his poetic line, but to subject-matter and, as we have seen in Chapter 3 describing the use of *persona*, to his tone of voice, 'casting off as it were complete masks of the self in each poem'.

T. S. Eliot was engaged in the same process at the same time and he found a decisive influence when, in 1906, he read the poems of **Jules Laforgue** (1860–87). The phrase 'free verse' translates the earlier French phrase *'vers libre'*, and French poetry, most obviously in the prose poems of **Charles Baudelaire** (1821–67) and **Arthur Rimbaud** (1854–91), had been seeking its own departures from formal verse, especially from the dominant *alexandrine*. **Stéphane Mallarmé** (1842–98) wrote in 1891:

We are now witnessing a spectacle which is truly extraordinary, unique in the history of poetry: every poet is going off by himself with his own flute, and playing the songs he pleases. For the first time since the beginning of poetry, poets have stopped singing bass. Hitherto ... if they wished to be accompanied, they had to be content with the great organ of official metre.

In 1886 Laforgue described the direction his verse was taking in a letter:

> I forget to rhyme, I forget the number of syllables, I forget to set it in stanzas – the lines themselves begin in the margin just like prose. The old regular stanza only turns up when a popular quatrain is needed. I'll have a book like this ready when I come to Paris. I'm working on nothing else. This place is a dump: eating, smoking, twenty minutes in the bath to digest – and the rest of the time: what else can you do but write poetry?

I quote the circumstantial details here since they suggest something of the mood between exasperation and lassitude that Eliot evidently responded to in Laforgue and which colours the voices of decisive early poems such as 'Portrait of a Lady' and 'The Love Song of J. Alfred Prufrock'. In contrast to the anguish which wrings Hopkins' line, or the bracing stridency of 'The Seafarer', Laforgue cultivates the notion, though very knowingly, that he has fallen into *vers libre* as an idle accident. The speaker in Eliot's 'Portrait of a Lady' doodles with the line as with the emotions. Look here at the varying lengths and listen for where you think the stresses fall:

> Doubtful, for a while
> Not knowing what to feel or if I understand
> Or whether wise or foolish, tardy or too soon …
> Would she not have the advantage, after all?
> This music is successful with a 'dying fall'
> Now that we talk of dying –
> And should I have the right to smile?

In fact Laforgue did not forget to **rhyme**, though he did so irregularly, and Eliot follows suit. The paragraph in which the above lines come rhymes *a b c d e c a f f g e* (see Chapter 6, 'Rhyme'). As we have already seen in Chapter 3, Eliot's idea of the line focuses on 'the contrast between fixity and flux'. The flux, though, is crucial, for it embodies the casual and colloquial quality of voice that – as we saw in Chapter 3 on 'Tones of voice' – shook early twentieth-century poetry. In 1917 Eliot wrote:

> One of the ways by which contemporary verse has tried to escape the rhetorical, the abstract, the moralising, to recover (for that is its purpose) the accents of direct speech, is to concentrate its attention on trivial or accidental or commonplace objects.

In Eliot's early work the line registers personalities too unsure of themselves to be either certain or anguished. Idleness however has always been the accusation of traditionalists decrying 'free verse'. If this could be associated with the supposed debility of foreigners so much the better.

THE MINIMAL LINE

The general movement of 'free verse' then is towards a democratic informality that has a more flexible rhythm and a wider, more colloquial, range of words. Once freed of measure the line has gone in two main different directions. One has been towards minimalism, reduction, and the other towards expansiveness, spread.

Poets who heard the full metrical line as 'oompa oompa' have wanted to purge verse of elevation or pretension, to strip it down to the barest elements, highlighting words with *semantic* content and minimizing *syntactic* connection. The line in consequence tends to be short. We have already seen in Chapter 3 how **William Carlos Williams** finds a register for his 'nervous' sensibility – 'I didn't go in for long lines' – and, 'very much an American kid', his own cultural situation. The lines he uses in poems like 'An Early Martyr' and 'Pastoral' are self-effacingly brief, and could be briefer, as in this conversation with his barber:

> Of death
> the barber
> the barber
> talked to me
>
> cutting my
> life with

sleep to trim
my hair –

It's just
a moment
he said, we die
every night –

This poem eventually ends warmly, but this matter-of-fact opening is the more discomfiting for its terseness. This is not Williams at his most easy-going, as the opening inversion shows, 'Of death / the barber / the barber / talked to me', though the repetition is a recognizable exclamation of surprise. The line-breaks not only quietly disrupt expectations of the metred line and the grammatical unit, but also enable effects like the small shock of 'cutting my / life'. Overall, what these broken lines do is make the ordinary chat of someone passing the time of day with his barber extraordinary. They wield shears, but hairdressers don't often draw our attention to death. Like many of Williams's poems its casual, jotted manner belies the penetration of his glance at the mundane.

In poems like these the visual element is important too. Isolated on a full page, the poem has a concentrated look, as though distilled on to the whiteness. Here is the strong visual dimension of the poem's 'deliberate space'. Williams has testified to the part played by the introduction of the typewriter into the compositional process. Its brisk mechanical motion and the even impress of the ink promotes a sense of simple fixity and of the poem as a physical object.

We sometimes speak of 'setting down' words. This has the sense of writing as something simple and fixed. 'Putting it down in black and white' carries an associated idea of clarity. This kind of insistence can be seen in a poet who takes the Williams manner into a different scene, **Gary Snyder** (1930–). Their common aesthetic goal is to approximate the physical world as directly as possible on to the page, and to attend to sense and impression rather than thought. Snyder, a pastoral poet of the North American wilderness, likens the poem to a 'riprap', a mountain path laid in single stones: 'Lay down these words / Before your mind like rocks. / placed solid, by hands.' The lines have a set, bitten-off quality. This is the opening of Snyder's 'Mid-August at Soughdough Mountain Lookout':

> Down valley a smoke haze
> Three days heat, after five days rain
> Pitch glows on the fir-cones
> Across rocks and meadows
> Swarms of new flies.

Writing of Williams, Kenneth Burke wrote: 'The process is simply this: There is the eye, and there is the thing upon which that eye alights; while the relationship existing between the two is a poem.'

This aesthetic normally finds its voice, or image, in the spare, minimalist short line. It is also an aesthetic which, in the modernist manner, aims for a depersonalized voice, one that allows only the minimum intervention of conscious thought or emotional response. Writing of Williams, another great American modern poet, **Wallace Stevens** (1879–1955), puts it like this: 'What Williams gives, on the whole, is not sentiment but the reaction from sentiment, or, rather, a little sentiment, very little, together with acute reaction.'

Williams, says Stevens, has 'a sentimental side', but continually reacts against it. This is of a piece with his 'passion for the anti-poetic', a passion shared by many of the poets who, like Snyder, embraced Williams's *poetics*, and especially the guarded austerity of the short, free line. It is a guardedness with respect to the 'I' as original poet, composer or artist, sole origin of the art-work, that found its extreme manifestation in 'found' or 'ready-made' works of art such as Marcel Duchamp's urinal (signed 'R. Mutt') offered for exhibition in New York in 1917 (and rejected).

However, such a line is not bound to such a relaxed tone, nor does it necessarily muffle its speaking voice. The Welsh poet **R. S. Thomas** (1913–2000) often uses a similar line, sometimes in an equally impersonal way, and with something of Snyder's stone-laying quality, if with a different, more imposing tone. But in 'Welsh', Thomas uses the same line for a very distinctly character-ized voice:

> Why must I write so?
> I'm Welsh, see:

> A real Cymro,
> Peat in my veins.
> I was born late:
> She claimed me,
> Brought me up nice,
> No hardship;
> Only the one loss,
> I can't speak my own
> Language –

Of course this sounds much different from American colloquialism, but the spare line fits this grim, resentful shortness with words.

EXPANSIVENESS AND 'FIELD COMPOSITION'

The other stream of writing encouraged by the freeing of the line is expansive. Here the poem can be deliberately, often ostentatiously casual, keen to amble through the everyday bombardment of impressions without feeling any pressure to arrive at significance. The line is simply and suitably longer. Here is **Frank O'Hara** (1926–66) afoot with his own kind of vision in the beginnings of some nearly randomly chosen poems:

- Ah nuts! It's boring reading French newspapers
  in New York as if I were a Colonial waiting for my gin

- The spent purpose of a perfectly marvellous
  life suddenly glimmers and leaps into flame

- I'm getting tired of not wearing underwear
  and then again I like it
                              strolling along
  feeling the wind blow softly on my genitals

- Totally abashed and smiling
                              I walk in
                              sit down and
                              face the frigidaire

> • Light   clarity   avocado salad in the morning
>   after all the terrible things I do how amazing it is
>   to find forgiveness and love

References to literature, art, music and the writing of poems abound in O'Hara's work, but the sub-text of these lines, like many of his openings, is 'this is not a poem'. He shares the modernist fear of the pretentiously 'poetic', but rather than trying to pare it away he seeks to bury it by being talkative and exuberant. Strong feeling, often in gusts, frequently lies at the heart of his poems, but it comes unsuspected, surrounded by the trivial, so that its true surprise is maintained. 'Light   clarity   avocado salad in the morning … the terrible things I do …', camouflage 'big' words like 'forgiveness' and 'love', and that abrupt line-turn, 'how amazing it is / to find', arrests us before we have seen them coming.

Because it denies itself a formal stance, the democratically collo-quial register of modern 'free verse' is always wrestling with the issue of how to phrase something really serious. It wants both to be part of the daily flow of everyday language and to be marked off from it. The poem above beginning 'Ah nuts!', 'Les Luths', is a love poem and, at heart, feels the traditional pains of love:

> everybody here is running around after dull pleasantries and
> wondering if *The Hotel Wentley Poems* is as great as I say it is
> and I am feeling particularly testy at being separated from
> the one I love by the most dreary of practical exigencies money
> when I want only to lean on my elbow and stare into space feeling
> the one warm beautiful thing in the world breathing upon my
>   right rib.

The run-on lines here go to an extreme of cutting into the **syntax**, but in a way that does not disorient the reader as much as empha-size the speaker's impatience. The rhythm slows, though, over that hint at self-mockery in 'particularly testy' and 'the most dreary of practical exigencies'. Both are cleverly cumbersome prosy phrases between which, almost by the way, comes the crucial emotional phrase, itself a consciously recognized cliché: 'the one I love'. The lover's sweet pain is then drawn out in that last line with its series of firm stresses and a breath-catching **caesura**: 'the **one warm**

beautiful **thing** in the **world** / **breath**ing upon my **right rib**'. Free verse of course does not do away with *rhythm*. What it does do is bring in the opportunity for very particular, intuitive variation, Blake's 'variety in every line'.

For all the apparent self-effacement that goes with modernist free verse, whether by clipping the voice, or by having it speak from within a spinning cloud of scenes, impressions and knowingness, idiosyncrasy is at the heart of it. The essential point of the unmeasured line is that it is *particular* to its occasion, bespoke, not already patterned. **Charles Olson** (1910–70), in his influential essay 'PROJECTIVE VERSE' (1950), insists that when a poet departs from 'closed form' 'he [*sic*] puts himself in the open – he can go by no track other than the one the poem under hand declares for itself'. In effect Olson is developing the Romantic tradition in which **Coleridge** formulated the idea of 'organic form'. This means, as Olson's countryman **Ralph Waldo Emerson** put it in the nineteenth century, that a poem is not an artefact, but 'like the spirit of a plant or an animal, it has an architecture of its own'. These ideas will be considered more fully in Chapter 8, 'Image – imagination – inspiration'.

Olson's aim is the entire abandonment of what he derided as the 'honey-head', 'the sweetness of meter and rime', in favour of a line based in the *breathing* rhythms of the poet. As for Williams, the typewriter becomes important since its precise calibrations enable the poet to 'score' the poem not only with conventional punctuation but by exact spacings, multiple margins and symbols such as the /. This 'field composition' can thus configure a poem across the full dimensions of the page. This is from Olson's poem 'The Distances':

> Death is a loving matter, then, a horror
>
>   we cannot bide, and avoid
> by greedy life
>   we think all living things are precious
>   – Pygmalions
>
>
>   a German inventor in Key West
> who had a Cuban girl, and kept her, after her death
> in his bed
>   after her family retrieved her
> he stole the body again from the vault

There is a concrete effect here in the use of the space of the page, but Olson's aim is more radical yet. He is trying to break, or at least stretch, the conventions of grammar and *syntax* in, as he puts it himself, 'the attack, I suppose, on the "completed thought," or, the Idea, yes?' Sentences may have deferred, or no, full-stops, parentheses may be opened but not necessarily closed in order to simulate the processes of uncompleted thought – perhaps the peculiar individuality of consciousness itself. In 'PROJECTIVE VERSE' Olson understandably shies away from 'an analysis of how far a new poet can stretch the very conventions on which communication by language rests'.

In the quest for liberation, 'free verse', as we have seen, has always sought connection with the 'naturalness' of speech. The entry on 'Free Verse' in *The Princeton Encyclopedia of Poetry and Poetics* describes it like this:

> All poetry restructures direct experience by means of devices of equivalence; all poetry has attributes of a naturalizing and an artificializing rhetoric. However, more explicitly than metrical poetry ... free verse claims and thematizes a proximity to lived experience. It does this by trying to replicate, project, or represent perceptual, cognitive, emotional, and imaginative processes. Lived experience and replicated process are unreachable goals, but nevertheless this ethos is what continues to draw writers and readers to free verse.

But 'Why imitate "speech"?' asked the American poet **Robert Grenier** (1941–) in 1971. 'I HATE SPEECH', he continues. He does so because he sees the injunction to follow the spoken word as but another constraint upon the poet. Moreover, since we have less awareness of the impositions of speech patterns, he argues that they form a constraint less obvious and so more confining than, say, an *iambic pentameter*. Grenier asks another, yet more radical question: 'where are the words most themselves?' I suspect it's impossible to answer. To ask it implies a utopian longing. Classical *rhetoric* and *poetics* asserted that words are 'most themselves' when fashioned to best effect. The *Romantic* reaction, which remains dominant to our own day, privileges 'natural speech' as the pristine source for poetry. For Grenier – in a cadence reminiscent of Whitman – 'It isn't the spoken any more than the written': the words of the poem are 'words occurring' whether they hail from the written, the spoken, the dreamt, or wherever else.

Following Grenier we might well ask: do we need more 'ordinary speech', or touched-up 'ordinary speech' as poetry can often be? Poetry that is, in **James Fenton**'s (1949–) witty characterization, 'strictly free', can be as clichéd as a leaden sonnet. But as a poet seeks rhythms that are truly surprising, she or he will sense when the sequence of words is falling into the easy arms of the reader's expectation, whether that be a dull metric or a worn colloquialism. But the background noise of rhythmical as of all other linguistic cliché that surrounds the poet's ear is now more prolific than ever. It inhabits what we call 'information'. To be 'free' in verse is to be heard beyond that blurry, familiar noise – to be distinct. But 'the words most themselves' are not waiting somewhere else for us to find them, but are in the midst, part of the storm. The poetic line has to be tuned from the clamour. Here is **Geoffrey Hill** (1932–), forcing his way, head-down, through the blizzard of contemporary cliché in part 21 of his *Speech! Speech!*:

> SURREAL is natural | só you can discount
> ethics and suchlike. Try perpetuity
> *in vitro*, find out how far is HOW FAR.
> I'd call that self | inflicted. Pitch it
> to the CHORUS like admonition. Stoics
> have answers, but nót one I go for.

Summary
In this chapter on 'free verse' we have considered:

- the origins of 'free verse';
- its associations to ideas of liberation;
- the variations of the 'biblical' verse line;
- modernism and 'free verse';
- the use of 'free verse' as a feature of a democratic, informal style;
- minimal and expansionist styles;
- the opening of the page towards 'field composition';
- whether poetry should be close to speech.

FURTHER READING

Eliot, T. S. (1999) *Selected Essays*, London: Faber & Faber; see 'Reflections on "Vers Libre" '.

Hartman, Charles O. (1980) *Free Verse: An Essay on Prosody*, Princeton: Princeton University Press.

Kennedy, X. J. and Gioa, Dana (1998) *An Introduction to Poetry*, ninth edition, New York: Longman; see Chapter 11, 'Open Form'.

Koch, K. (1998) *Making Your Own Days: The Pleasures of Reading and Writing Poetry*, New York: Simon & Schuster; see Chapter 2, 'Music'.

Mayes, Frances (1987) *The Discovery of Poetry*, Orlando: Harcourt, Brace, Jovanovich; see Chapter 7, 'Free Verse'.

Olson, Charles, 'Projective Verse' in D. Allen (ed.) (1960) *The New American Poetry 1945–1960*, New York: Grove Press; or W. N. Herbert and M. Hollis (eds) (2000) *Strong Words, Modern Poets on Modern Poetry*, Tarset: Bloodaxe.

Preminger, A. and Brogan, T. V. F. (eds) (1993) *The New Princeton Encyclopedia of Poetry and Poetics*, Princeton: Princeton University Press; see entry 'Free Verse'.

Williams, W. C. ([1954] 2000) 'On Measure – Statement for Cid Corman' in W. N. Herbert and M. Hollis (eds) *Strong Words, Modern Poets on Modern Poetry*, Tarset: Bloodaxe.

RHYME AND
OTHER NOISES

Mister Harris, plutocrat,
Wants to give my cheek a pat,
If a Harris pat means a Paris hat
Bé bé!
Mais je suis toujours fidèle, darlin', in my
fashion
Oui, je suis toujours fidèle, darlin', in my
way!

I think these lines from a song by **Cole Porter** (1891–1964) include my favourite rhyme. They are blindingly simple but utterly ingenious in the way they manipulate *Harris, plutocrat/pat* and then carry the sounds into the brilliant inversion of '*Harris pat*' and '*Paris hat*'. The song, 'I'm Always True to You in My Fashion' from Porter's musical *Kiss Me Kate* (1948), plays a series of such rhymes: *vet/pet, Tex/checks/sex* – and it would not be impossible to believe that a less decorous version might have made the Harris word-play yet better by substituting 'ass' for 'cheek'.

   Rhyme is a play with words and its first effect is pleasure. It comes from delighted surprise as words, remote from each other in meaning but which happen to sound alike, are made to coincide. One aspect of this delight can sport with meaning:

Moses supposes
His toe-ses
Are roses
But Moses supposes
Erroneously.

Rhyme can make language disorderly because following its nose can entirely subvert normal sense, especially when words are corrupted to fit. But in other ways rhyme might be said to organize language into tidy shapes. There is a kind of 'click' as this happens, like the neat fastening of a catch or two pieces of a jigsaw. With Porter's conjuring, this occurs with such bewildering speed that we are still figuring out what has happened as the song celebrates not only the character's but the composer's coup. Rhyme is often – as here – a matter of surface, and gains the kind of admiration we give to a magic or acrobatic trick.

But good rhymes can also embody meaning and we can see this here. The cheerfully and endearingly cynical character who sings Porter's song has a formula that works: the (euphemistic) 'Harris pat' = a 'Paris hat'. Algebra could not provide a neater equation.

AGAINST RHYMING

It must be said at the outset that many poets have done without the various qualities of rhyme. **Shakespeare** and the other Elizabethan dramatists employed *blank verse* as part of their simulation of natural speech, and when **Milton** came to write *Paradise Lost* he dismissed 'the troublesome and modern bondage of Rhyming' which had vexed many previous poets by its 'constraint to express many things otherwise, and for the most part worse than else they would have expressed them'. He described it in his preface to *Paradise Lost* as

A thing of it self, to all judicious ears, trivial and of no true musical delight; which consists only in apt Numbers, fit quantity of Syllables, and the sense variously drawn out from one Verse into another.

Milton's objections, that searching for 'the jingling sound of like endings' inhibits free expression, have been shared by many poets ever since, especially in the twentieth century. **Judith Wright** (1915–2000) writes in her manifesto poem 'Brevity':

> Rhyme, my old cymbal,
> I don't clash you as often,
> or trust your old promises
> of music and unison.

Nevertheless many people still like and expect poems to rhyme, and despite hundreds of years of blank verse and a hundred years of *'free verse'*, rhyme is far from dead. In considering rhyme in a poem, therefore, we will always want to decide whether it is 'a thing of it self' or a deeply integrated part of the expression.

DEFINITIONS

In the rest of this chapter I want to describe and explore the definitions, and the different kinds, patterns and purposes, of rhyme. This will include considering its aesthetic effects; how it works to provide closure to poems and parts of poems – and how sometimes it does not; how it helps to structure poems; and how rhyme confirms meanings and helps us to discover them. Whilst most attention will be given to *end-rhymes*, I shall also consider related sound effects such as *alliteration* and *assonance* and *consonance*.

According to *The New Princeton Encyclopedia of Poetry and Poetics*, although every language contains the capacity for rhyme there are some four thousand poetries which make no use of it. In English, however, poems have often been known as 'rhymes', so closely has this feature been associated with the art. Rhyme is another characteristic that comes through the oral tradition and an additional *mnemonic* feature in early verse. It has always persisted in popular forms such as Cockney *rhyming slang* and is intrinsic to rapping and much contemporary *performance*, or *'slam'*, poetry.

Again *recurrence* is at the heart of the pleasures of rhyme. Essentially it exploits an aspect of the coincidence of language. As I've already suggested in Chapter 1, language works not only along the axis of reference (meaning), but also among its own incidental associations. The *chance* that two words sound similar is one of these, and one with a great range of available subtlety. Thus a poem in rhyme is working along two axes: one travelling 'horizontally' along the line of its syntactically organized meaning, and another travelling 'vertically' down the line that connects the rhyming words. It is like an echo that still reverberates as the words move on. The relationship between these two axes is what determines how successful a poem in rhyme is. It should possess and connect reason and rhyme.

The *definition* of rhyme in English has to do with the arrangement of consonants and vowels. The family of rhyming effects can be described in the following seven types, where C = the consonant and V = the vowel. The recurring sound is highlighted.

1   **C** V C   **C** V C e.g. **b**at **b**it = alliteration.
2   C **V** C   C **V** C e.g. c**oo**l f**oo**d = assonance.
3   C V **C**   C V **C** e.g. kna**ck** / so**ck** = consonance (could just be used as an end-rhyme – see 5 below).
4   C **V C**   C **V C** e.g. s**ock** / r**ock** = full or strict rhyme.
5   C V **C**   C V **C** e.g. **cri**ck / **cra**ck = half, or slant, or pararhyme.
6   **C V** C   **C V** C e.g. **kna**ck / **gna**t = reverse rhyme.
7   **C V C**   **C V C** e.g. **wood** / **would** = identical rhyme or *rime riche*.

Purists would argue that the only 'proper' rhyming is **full rhyme** (4), where *the last two or more sounds are in accord and the difference occurs early in the line.* Two polysyllables that illustrate this are d**emonstrate** and r**emonstrate**. The kinship of these two words is, however, disappointingly close to make a good rhyme. We would certainly prefer a greater difference, something like **Tony Harrison**'s (1937–) matching **haemorrhoid** and **unemployed** in his 'Divisions II'. Notice here that the success of the rhyme also depends on the rhythmic combination of syllables: *employed* would fill the rhyme but would not synchronize as well as **un***employed*.

PATTERNS OF RHYME

Poets employ **end-rhymes** in a variety of patterns. The most apparent is the rhyming of successive lines into **couplets**, as here in **Mary Barber**'s (?1690–1757) 'The Conclusion of a Letter to the Rev. Mr C–':

> Her Husband has surely a terrible **Life**;
> There's nothing I dread, like a verse-writing **Wife**:
> Defend me, ye Powers, from that fatal **Curse**;
> Which must heighten the Plagues of 'for better for **worse**'!

In analysing verse we give each rhyming sound a letter, beginning with *a*, so these lines can be seen to rhyme *a a b b*. ***Triplets***, as in 'To Sapho' by **Robert Herrick**, therefore go:

Sapho, I will choose to go	*a*
Where the northern winds do blow	*a*
Endless ice, and endless snow:	*a*
Rather than I once would see,	*b*
But a winter's face in thee,	*b*
To benumb my hopes and me.	*b*

Or, in **Gertrude Schackenberg**'s (1953–) 'Supernatural Love':

My father at the dictionary-stand	*a*
Touches the page to fully understand	*a*
The lamplit answer, tilting in his hand …	*a*

Whereas couplets are commonly employed in extended ***verse paragraphs***, triplets usually form separate ***stanzas***, although they may, like Schackenberg's poem, run the syntax from one stanza to the next. Stanzas are often built around alternating rhyme schemes such as this *a b a b* scheme, also by Herrick:

> Gather ye rose-buds while ye may,
>> Old Time is still a flying:
> And this same flower that smiles to day,
>> To morrow will be dying.

Alternatively a ***quatrain*** might rhyme its outside lines and its inside lines in the pattern of *a b b a* as does **Ben Jonson** (1572–1637) in 'An Elegy':

> Though beauty be the mark of praise,
>> And yours of whom I sing be such

> As not the world can praise too much,
> Yet is't your virtue now I raise.

These basic patterns – *aa bb, abab, abba* – highlight the obvious binary qualities of rhyme with all its implications of balance, symmetry and the division and completeness of even number. The pairings of rhyme imply that nothing is odd, loose or stands apart. It is an aesthetic of harmony and completion. As we shall see, longer stanza forms can use more elaborate rhyme schemes as part of the structure of the poem. As they do so the rhymes might appear much further apart, some seemingly abandoned until they are given their more remote echo. It is in the identification of such structures and echoes that the notation of rhyme is so helpful.

THE BEAUTIES OF RHYME

The aesthetic attraction – the beauties of rhyme – are very different. Sometimes we might gorge on a wonderful excess as in a poem like **Carol Ann Duffy**'s (1955–) 'Mrs Sisyphus' (1999) which plays exultantly on the words 'jerk', 'kirk', 'irk', 'berk', 'dirk', 'perk', 'shriek', 'cork', 'park', 'dork', 'gawk', 'quirk', 'lark' and 'mark'. There's plenitude too in these quatrains from 'Long After Heine' by **Gwen Harwood** (1920–95) which combine *internal rhyme* (rhyme *within* the line) with ingenious end-rhyme:

> The baby **screamed** with colic
>   the windows **streamed** with **rain**.
> She **dreamed** of a demon lover
>   like Richard **Chamberlain**.
>
> He **towered**, **austerely** perfect
>   in **samurai brocade**,
> and **hushed** the **howling** baby
>   with one **swish** of his **blade**.

I've highlighted here not only the full end-and internal rhymes but also the ***assonance*** of vowel sounds (towered/austerely/samurai), the *alliteration* of hushed/howling as well as the *half-rhyme* hushed/swished.

These are comic examples, but the religious poetry of **George Herbert** (1593–1633) also works with rhyme to shape his affirmations. This is the first stanza of 'Virtue':

> Sweet day, so cool, so calm, so bright,
> > The bridal of the earth and sky:
> The dew shall weep thy fall tonight;
> > For thou must die.

The next two of the four *ab ab* quatrains employ different rhymes for the key word 'die': 'eye' and 'lie'. None of these is exotic, for simplicity is everything in this poem. In the last stanza 'die' is replaced by its opposite in meaning, 'lives':

> Only a sweet and virtuous soul,
> > Like seasoned timber, never gives;
> But though the whole world turn to coal,
> > Then chiefly lives.

The attraction of this poem is in its carefully judged composure. Its metre is simple and steady and the process of its imagery and ideas is paralleled in each of the first three stanzas with that recurrent crucial rhyme of the -ie sound. Then as the rhyme changes so does the sense. The poem has the pleasures of calmness, compact shape and balance, and the straightforward rhymes are an important part of this. They show us that whilst we may think of the effects of rhyme as things in themselves – especially when the juxtapositions have real surprise – rhyme is really set into the whole character and meaning of the poem.

RHYMING AND MEANING

Because of its binary character, rhyme is often used for *closure* of various kinds: the two parts come together like the shutting of a lid. A familiar example is the way that Shakespearean verse drama usually indicates the end of a scene by a rhyming *couplet*. The exhaustion of *King Lear* is tolled in a sequence of four rhyming couplets, including the last words of the faithful servant Kent:

> I have a journey, sir, shortly to go;
> My master calls me; I must not say no.

Couplets like this are called **closed** because they contain a whole sentiment or idea within their clearly defined boundaries. It is a **stichic** verse, which is to say that the poem proceeds mainly through distinctly punctuated lines as opposed to verse that flows through several lines. To eighteenth-century **neo-classical** poets who saw verse as primarily **rhetorical**, that is a way of clarifying and persuading an audience of long-established truths, the couplet is attractive because of its capacity for memorable summary. It is its associations with the effort of speech-making and grand persuasion that have gained this kind of couplet the title '**heroic**'.

For important summations or conclusions the couplet can have an **epigrammatic** quality, sometimes using humour and the surprise of the rhyme to achieve a deflating effect. This is well shown in these lines by **John Gay** (1685–1732). The poem is called 'The Man and the Flea' and features a series of creatures – a hawk, a crab, a snail, and of course a Man – discoursing upon how all Creation has been made for their particular benefit. Man boasts:

> 'I cannot raise my worth too high;
> Of what vast consequence am I?'

and gets an unexpected riposte:

> 'Not of th' importance you suppose,'
> Replies a flea upon his nose.

The pride of Man is instantly deflated by this unlooked-for intervention which opens the pretensions of *high/I* to the ridicule of a flea, a flea educated enough, moreover, to employ words like *suppose*. Man cannot see what's on his nose, never mind what's under it. Thus *suppose/nose* is a rhyme that is not merely incidental but essential to the meaning and tone of the poem.

In lines longer than Gay's four-beat **tetrameter**, this summative quality of the rhyming couplet can be complemented by related devices, as in these lines by the style's foremost exponent **Alexander Pope** (1688–1744). Here, in 'An Essay on Criticism', he wants to square the paradox that what we take to be natural, and thus think of as free and unconstrained, is in fact governed by laws of its own:

> Those RULES of old discovered, not devised,
> Are Nature still, but Nature methodized;
> Nature, like liberty, is but restrained
> By the same laws which first herself ordained.

In this instance the pairs of rhyming words, *devised/methodized* and *restrained/ordained*, are associated in their meaning as well as in the coincidence of their sound. The second couplet, moreover, encapsulates a move in understanding. This goes from the notion of restraint alone, which we might think of as a troublesome leash, to the acknowledgement that the natural order is fixed. Pope is aiming to build an argument and so tries to carry us along the line of his thought in a series of clear, separate steps.

To this end the balanced self-containment of the rhyme is often paralleled by a balancing effect within the **iambic pentameter**. Usually this will involve a tiny break, or **caesura**, halfway through the line. The lines above illustrate this, although, as so often in verse, the variation from it is equally important:

> Those RULES of old discovered, | not devised,
> Are Nature still, | but Nature methodized;
> Nature, like liberty, | is but restrained
> By the same laws which first herself ordained.

The caesura in the first line here is delayed so as to emphasize the assertion and push the opposite idea back before it. The alliteration, or **head-rhyme**, helps drive home the essential idea – 'discovered, not devised'. The second line is evenly balanced with Nature on both sides of the equation pivoting on 'but'. The third line allows an extra pause after 'Nature' to allow us to dwell on the relationship between Nature and liberty. But then the line runs on into the next, sweeping us along to its firm conclusion without any further break. That everything is *ordained*, with the strong implication of divine ordinance, is for Pope the last word. Although the closed couplet can be used for more intimate purposes, as here in Pope's 'Epistle to Dr. Arbuthnot',

> The Muse but served to ease some friend, not wife,
> To help me through this long disease, my life –

its primary mode is public and oratorical.

This is less true of *open couplets* where the sense can run on through lines with much less punctuation without necessarily matching components of meaning to the rhyming pairs. In these lines from **Christopher Marlowe**'s (1564–93) poem 'Hero and Leander', the narrator is striving to describe the beauties of the handsome youth Leander:

> Even as delicious meat is to the taste,
> So was his neck in touching, and surpassed
> The white of Pelops' shoulder. I could tell ye
> How smooth his breast was, and how white his belly,
> And whose immortal fingers did imprint
> That heavenly path, with many a curious dint,
> That runs along his back; but my rude pen
> Can hardly blazon forth the loves of men,
> Much less of powerful gods;

These lines are rapturous rather than studied and the sensual excitement does not accept the measure of the couplet but flows enthusiastically across the endings. In running, or *strophic*, lines like these the rhyme is much less important, serving the overall organization of the verse rather than its specific meanings. As Marlowe's own dramatic verse shows, when the pentameter is used like this, rhyme becomes redundant since the necessary pressure in the verse is felt in other ways.

But not all open couplets tend towards blank verse. In this poem, 'The Not-Returning', **Ivor Gurney** (1890–1937) thinks of home from the trenches of the Western Front:

Never more delight comes of the roof dark lit
With under-candle-flicker nor rich gloom on it,
The limned faces and moving hands shuffling the cards,
The clear conscience, the free mind moving towards
Poetry, friends, the old earthly rewards.
No more they come. No more.
Only the restless searching, the bitter labour,
The going out to watch stars, stumbling blind through the
   difficult door.

As in this excerpt, Gurney's poem uses a mixture of couplets and triplets – although the poem's first line stays unrhymed. Unusually too for rhyming verse, the measure is irregular. So what purpose does the rhyme serve here? I think that in this case the customary tidying qualities of rhyme are an ironic counterpoint to the roughed-out quality of the verse. The last phrase, 'stumbling blind through the difficult door', is the poem's keynote. In its awkward word-order – 'Never more delight comes' – and its varying verse lines and punctuation, the poem embodies a sense of stumbling. It seems written under stress, almost improvised. The rhymes then represent some object to stumble towards, something to help keep coherence as the speaker feels everything dissolving. That the rhyme *lit/on it* is crude, and is succeeded by two **triplets** which include **half-rhymes**, shows how ironically distant these agonized, nearly broken lines are from the 'finish' we usually associate with rhyme. That it bothers with such artifice seems an irrelevance, but that it does so turns out to sound poignant and defiant.

BUILDING POEMS WITH RHYMES

I want to turn now to see how rhyme schemes work as part of the architecture of whole poems or the larger sections of poems. The couplet sequence of *aa bb cc* is **plain rhyme**, but as soon as rhyming words are separated further, even only to *ab ab*, then the lines connecting the rhymes criss-cross and the rhyme becomes **interlaced**. *Terza rima* is a good basic example of this. It works in three line units, *aba bcb cdc*. This is the opening of **Thomas Kinsella**'s (1928–) 'Downstream':

> Again in the mirrored dusk the paddles **sank**. *a*
>> We thrust forward, swaying both as **one**. *b*
>> The ripples widened to the ghostly **bank** *a*
>
> Where willows, with their shadows half un**done**, *b*
>> Hung to the water, mowing like the **blind**. *c*
>> The current seized our skill. We let it **run**. *b*
>
> Grazing the reeds, and let the land un**wind** *c*
>> In stealth on either hand. Dark woods: a **door** *d*
>> Opened and shut. The clear sky fell be**hind**, *c*

Terza rima was devised by the Italian poet **Dante** (1265–1321) for his huge three-part poem *The Divine Comedy* and has found counterparts in the poetries of several European languages since. His choice of these **tercets** within his three-part scheme was meant to allude to the Holy Trinity, but more generally the mode suggests forward movement and continuity. The closed couplet can be seen to be continually starting afresh, but with *terza rima* there is a sense of perpetual motion and of everything being connected. The closure occurs with a single line, rhyming with the previous middle rhyme, as Kinsella does in his own very Dantean poem:

> The slow, downstreaming dead, it seemed, were blended
>> One with those silver hordes, and briefly shared
>> Their order, glittering. And then impended
>
> A barrier of rock that turned and bared
>> A varied barrenness as towards its base
>> We glided – blotting heaven as it towered –
>
> Searching the darkness for a landing place.

On the face of it, it seems strange that rhyme, that depends upon evenness, should work to bind three-line systems so strongly. The *villanelle*, originally a French form (see Chapter 7, 'Stanza'), is a more compact example. This is a form comprising five three-line stanzas (**tercets**), each rhyming *aba*, and a closing **quatrain** rhyming *abaa*. The first and third lines of stanza 1 are also repeated alternately at the end of each succeeding stanza, culminating in the reappearance of both as the last two lines of the poem. Strictly, the rhymes should also be the same full sounds throughout. Here are the first two stanzas of **Dylan Thomas**'s (1914–53) *elegy* for his father, 'Do Not Go Gentle into That Good Night':

> Do not go gentle into that good night,
> Old age should burn and rave at close of day;
> Rage, rage against the dying of the light.
>
> Though wise men at their end know dark is right,
> Because their words had forked no lightning they
> Do not go gentle into that good night.

Thomas confines himself throughout to end-words which rhyme with 'night' or 'day', two words of course which represent the poles within the poem's subject. To fashion such a feverishly emotional poem – 'Rage, rage against the dying of the light' – within this extreme discipline produces a special tension. The strict limitations on the poem's means, especially in the tight permissions of the rhyme scheme, construct a vessel to compress, and withstand, the pressure of the poem's feeling.

In 'Lycidas', in memory of his drowned friend Edward King, **Milton** shapes a highly individual version of the *elegy* which varies its line length, and, instead of stanzas, employs *strophic* verse paragraphs of different duration. Rhyme is also used in unexpected ways that move between *plain* and *interlaced* patterns. In this fifth paragraph the poet is asking one of the commonest angry questions of the grief-stricken: how could the divine powers allow this to happen? Associating King (Lycidas) with poetry, he concludes with the anguished recollection that in the ancient Greek myth the Muse could not intervene to save even her own son, the greatest of poets, Orpheus.

> Where were ye nymphs, when the remorseless deep   *a*
> Closed o'er the head of your loved Lycidas?   *b*
> For neither were ye playing on the steep,   *a*
> Where your old Bards, the famous Druids lie,   *c*
> Nor on the shaggy top of Mona high,   *c*

Nor yet where Deva spreads her wizard stream: *d*
Ay me! I fondly dream –   *d*
Had ye been there – for what could that have done?   *e*
What could the Muse herself that Orpheus bore,   *f*
The Muse herself for her enchanting son   *e*
Whom universal Nature did lament,   *g*
When by the rout that made the hideous roar,   *f*
His gory visage down the stream was sent,   *g*
Down the swift Hebrus to the Lesbian shore?   *f*

Earlier in this paragraph the rhymes in the first seven lines have stayed plain, but then as the poet becomes more fraught the pace accelerates and the pattern changes. As we see the vision of Orpheus' severed head tumbling downriver to the sea, the rhymes rush over one another, a tumult caught in the only triple rhyme – *bore, roar, shore* – words we feel here as deep, harsh sounds. The first pairing is separated by two lines, the second by only one. These words beat heavily through the lines. We will see as well that the paragraph contains one, solitary unrhymed line, line 2: the abandoned *Lycidas*. When Milton brings his poem to its resolved conclusion 130 lines later, he gives his final paragraph the conventional harmony of **ottava rima** – a stanza form rhyming *ababbcc*. The fifth paragraph has no such orderly convention, but the way the poet patterns both sections, irregular and regular, shows how important rhyme is to the structure of the poem.

All these instances of the different ways rhyme is used show the connection between rhyme and meaning. But rhyme does always depend upon coincidence, and its use might just be a celebration of the happy anarchy within the language that enables us to bring words together that would otherwise never keep company, like **Tony Harrison** rhyming *lah-di-dah/Panama*. We might exult for instance in discovering the rhyme in loan-words into English such as *crouton/futon*, and make the most of it by working up a poem about bed-crumbs. Some poems might be generated by pre-set rhymes, throwing their intentions entirely upon the mercy of rhyming accidents, or, as some have done, upon another's isolated rhymes which the poet then writes 'towards', as does **John Ashbery** (1927–) in 'The Plural of "Jack-In-The-Box" '.

SO, IS RHYME 'A THING IN ITSELF'?

Thinking about rhyme in this way takes us once more into the great poetic conundrum: does the poem find words to refine its intended meanings, the emotion or idea which is there, as the philosopher Collingwood says, 'before we express it', or are its meanings generated out of the energies of language? **Dylan Thomas**'s rhymes in 'Do Not Go Gentle ...' evidently belong in the first category, for his choices, *night, light, bright, sight*, belong predominantly in the same area of meaning, or **semantic** field as linguists call it, and others, *right, flight, height* and *day, way, pray*, can all be said to 'belong' in the arena where we might think of death and dying. They form that *vertical* axis of association as the poem proceeds *horizontally* along and through its lines. (Since *they* in line 5 is a pronoun, and less substantive than any other of the rhyme words, it might be said to be out of key with the rest of the poem, though this would be a hard judgement.)

Another poem in which we can readily recognize the associations, if not the separate meanings, of the rhyme words is **Ben Jonson**'s (1573–1637) 'On My First Son'.

> Farewell, thou child of my right hand, and joy;
> My sin was too much hope of thee, loved boy:
> Seven years thou wert lent to me, and I thee pay,
> Exacted by thy fate, on the just day.
> O could I lose all father now! For why
> Will man lament the state he should envy,
> To have so soon 'scaped world's and flesh's rage,
> And, if no other misery, yet age?
> Rest in soft peace, and asked, say, 'Here doth lie
> Ben Jonson his best piece of poetry.'
> For whose sake henceforth all his vows be such
> As what he loves may never like too much.

The hard, self-reproachful thought here is that the poet invested too much in the joy of having a child, and prepared himself too little for what fate might bring. Still hard, and raging with exclamation, is the question of why we do not envy the release death brings, most of all from the trials of age. If there is one surprising rhyme it is *lie/poetry*. This is the idea most particular to the poet, and it might be objected that this drawing of attention to his craft has no necessary place in the lament. We might think that even as he is saying that all his poetry is as naught compared with his son, he is reminding us that he is a poet. The defence would be that this discounting of his art is part of throwing off delusion and vanity. We might even consider whether Jonson intends – or subconsciously produces – a pun on *lies* in the sense of deceit, thus associating, as writing of the period often did, poetry and untruth. The closing couplet, *such/much*, has a roughness to it befitting the baleful resolution to make this awful distinction between *loving* and *liking*. Throughout this poem there is a heavy-minded restraint in the way the rhyme words are fixed together.

This is very different from the rhyming of this next poem which, in subject and style, is a modern imitation of sixteenth- and seventeenth-century modes. This is the *abba abba octave* of 'Sonnet 23' by **John Berryman** (1914–72).

> They may suppose, because I would not cloy your ear –
> If ever these songs by other ears are heard –
> With 'love', suppose I loved you not, but blurred
> Lust with strange images, warm, not quite sincere,
> To switch a bedroom black. O mutineer
> With me against these empty captains! gird
> Your scorn again above all at *this* word
> Pompous and vague on the stump of his career.

As a set, *heard, blurred, word* can be seen to have some affinity but it is not shared by *gird*, while *ear, sincere, mutineer, career* appear to have none at all. But, as the central exclamation exhorts, the poem is raising a mutiny against the conventions of the love sonnet with its familiar circuit of 'love', 'heart' and 'beauty'. 'I want a verse fresh as a bubble breaks', he writes in the sonnet's **sestet**, and this will involve unexpected rhymes more promiscuous than chaste in their associations. Berryman and Jonson have quite different approaches to rhyming. We might imagine Jonson looking down his classical nose at Berryman's extrovert style, and indeed think ourselves that he makes rhyme, as Milton says, 'a thing in itself', rather than something that serves the poem's sentiment and ideas.

But whichever our preference, we can see how both poets employ rhyme as part of their total meaning, not just as a bolted-on device.

As we have seen there are several different kinds of correspondences in the company of rhyme from the *'head-rhymes'* of *alliteration*, the chimes of *assonance*, through *half-rhymes*, to the full, prominent *end-rhymes* of the *couplet*. Rhymes can also echo from the middle of lines, or diagonally from end to middle or back. **Emily Dickinson** (1830–86) is one of the subtlest and most determined technicians of rhyme. One reason she is so is that her poems seem at first so artless, even clumsy, and so conventional as to resemble nursery rhymes or the most mundane of hymns. But we soon see that here is an exceptional verbal intelligence which undermines the conventions she works with to produce through her wry styling the most astonishing sentiments and ideas. She makes much use of half-rhyme, both in endings and across the bodies of lines. A recluse herself, it is perhaps not surprising to see how often the word *room* appears in her work and in only a handful of poems we can find it rhymed with *tomb*, *name*, *storm* and *firm*. This obliqueness is entirely characteristic and part of her philosophy, as when she writes,

> Tell all the Truth but tell it slant –
> Success in Circuit lies.

This first stanza of a burial poem – 216 in the standard edition, for she gave none of her poems titles – encapsulates the variety of her rhyming. I have highlighted all the rhyming effects.

> **S**afe in their Alabaster Ch**ambers** –
> **Untouched** by **M**orning –
> And **untouched** by **N**oon –
> Lie the **m**eek **members** of the Resurrection –
> **Raf**ter of **Satin** – and **Roof** of **Stone**!

The criss-crossing here is very intricate. We can see, for instance, how *noon* and *stone* are half-rhymes, and how they slant to bring in the *s* and *t* sounds of *satin*. This echo in *satin* and *stone* is especially effective because of the opposite nature of the substances associated here in the material of the coffin and the tomb, both so far from the light of *noon*. Similarly the assonance of *rafter* and *satin* – the one word reminding us of hardness, the other of softness – combine, as do the consonants of *rafter* and *roof*. Moreover we might see in *rafter* an **eye-rhyme** – that is a combination of letters that look as though they might rhyme although they do not – with the poem's first word *safe*. There are other delicate and eerie effects which help create the unnerving sense of this stanza such as the steady and then varying pace and beat of the rhythm. Then there is that astonishingly rich word *alabaster* whose *a* sounds are different from the others in that line and which carries such **connotations** of deathly, clay-like whiteness. But the web of rhyming effects ensure complex associations between different words and lead to more and more implications. It is a brief, enigmatic poem but one that shows so much of what the poet has available in rhyme and other sounds.

> Summary
> In this chapter on rhyme we have considered:
>
> - rhyme and word-play;
> - the arguments against using rhyme; blank verse;
> - definitions of different kinds of rhyme;
> - the character of different rhyme schemes;
> - the aesthetic purposes of rhyme and how rhyme can enhance meaning;
> - how rhyme schemes can shape a whole poem;
> - a summary of the arguments for and against the use of rhyme.

FURTHER READING

Dickinson, Emily (1951) *The Complete Poems of Emily Dickinson*, Boston, London: Little, Brown.

Harrison, Tony (1984) *Selected Poems*, Harmondsworth: Penguin Books.

Hollander, J. (1989) *Rhyme's Reason: A Guide to English Verse*, new edition, London: Yale University Press.

Koch, Kenneth (1998) *Making Your Own Days: The Pleasures of Reading and Writing Poetry*, New York: Simon & Schuster; Chapter 2, 'Music'.

Thomas, Dylan (1952) *Collected Poems 1934–52*, London: J. M. Dent.

Wesling, Donald (1980) *The Chances of Rhyme, Device and Modernity*, Berkeley: University of California Press.

STANZA

Let me hear a staff, a stanze, a verse.

(Shakespeare, *Love's Labour's Lost,*

Act IV, Scene ii)

When the comic character Holofernes makes this demand he is either showing off by using three words where one will do, or he is uncertain which word to use. He wants Nathaniel to read him some poetry and in the 1590s the word **stanza**, to refer to a grouping of lines, was quite new in English. But, with the sixteenth century's attraction to Italian models, it was coming to displace the Old English word *staff*. The French word **verse**, then as now, could refer to a group of lines or a single line, or simply mean poetry in the generic sense.

After Nathaniel has read a dozen or so lines Holofernes interrupts him, complaining, 'You find not the apostrophus, and so miss the accent': in other words he is missing the correct places to pause. As we have seen, timing is essential to all aspects of the rhythm of poetry both for its sense and effects, so the 'apostrophus' – the pause – is vital.

DEFINITIONS

The original sense of *stanza* in Italian is 'stopping-place', a place to take a stand, and more particularly 'room'. These associated senses are exactly appropriate to the established sense of **stanza** in poetry. A poem in stanzas is one comprising a series of groups of lines shaped in the same way, and usually, although not always, of the same length. As each group ends, the poem has a momentary stopping-place. The structure of each stanza itself provides a space for the words to work, for what, in his overblown way, Holofernes calls 'the elegancy, facility and golden cadence of poetry'.

For the American poet **Kenneth Koch** (1925–2002) a stanza is 'nothing more than organizing other forms of poetic music – rhythm and rhyme'. It is true that the organization of stanzas has traditionally been based on metrical patterns and on rhyme schemes. As we have looked at the variety of individual poetic lines and of their connections through rhyme, so in considering the stanza we are examining larger combinations. But I think the purposes of the stanza go beyond the gathering of rhythm and rhyme. The stanza provides its own aesthetic experience for both the poet and the reader. It also serves necessary functions for several different kinds of poems. In this chapter I want to explain some of those functions and suggest the nature of their aesthetic attraction. In doing so I shall widen the topic by including a description of free-standing forms, such as the **sonnet**. I consider those here because their shapes are basically stanzaic, and I shall present them in the context of the kinds of stanza they most resemble.

With the stanza, once again we can look for origins in the **mnemonics** of the **oral tradition**. The stanza of the oral tradition, as we saw in the discussion of **ballad** form in Chapter 2 ('Deliberate space'), draws together the measures of the line, the repetitions of rhyme, and sometimes refrain, into comprehensible and memorable shapes. These normally coincide with sections of the ballad's narrative. The listener therefore is receiving the progress of the poem in distinct sections, like milestones along the way, and the performer has the same benefits of this segmentation, as well as the chance to recapitulate before going on. The stanza, even in the simple four-line ballad, is therefore eminently practical.

ALTERNATING VOICES

Such division need not only serve long *narrative* poems. Any poem that requires a balance or sequencing of voice or topic can use stanza-form. Here is an excerpt from a sardonic poem from the thirteenth century given the title 'How Death Comes':

Wanne mine eyhnen misten,	[eyes mist over]
And mine heren sissen,	[hearing hisses]
And my nose coldet,	
And my tunge foldet,	
And my rude slaket,	[face goes slack]
And mine lippes blaken,	
[...]	

Thanne I schel flutte	[shall pass]
From bedde to flore,	
From flore to here,	[shroud]
From here to bere,	[bier]
From bere to putte,	[grave]
And te putt fordut.	[closed up]
Thanne lyd mine hus uppe mine nose	[lies my house upon my nose]
Of al this world ne give I it a pese!	[jot]

The poem has a very simple two-part structure: *Wanne* and *Thanne*. The simple **anaphoric** structure – *And / And / And // From / From / From* – simply devises instances of the two conditions, and the stanza-break marks the movement from one to the other. Many early poems use stanzas in this balancing way, or, in similar fashion, to itemize various things on the way to their main argument. Thus another medieval poem, 'The Five Joys of Mary', recounts each of those joys in a centrepiece of five stanzas preceded by an introduction and closed by a prayer. In a more worldly mood, 'Bring Us in Good Ale' is repetitious in a way we know all too well as each boozy stanza implores 'Bring us in no browne bread ... no beefe ... no mutton ... no egges ...' etc., but 'Bring us in good ale'. Sequences which mark time as they elaborate variations on the theme usually use stanzas for each piece of their working.

The obvious artifice of stanza-form has meant that it finds little place in verse-drama where a greater impression of naturalness is needed. **Blank verse** generally is non-stanzaic, although this is much less true in the twentieth century. However, there are poems which make use of dialogue, usually in the form of an argument, and stanzas offer an obvious way of marking and balancing the speakers. The debate in which students and schoolmen exercised their powers of **rhetoric** was a staple of medieval and early modern education, and the argument between Body and Soul was a regular topic which also featured largely in **Renaissance** poetry. In poems like **Andrew Marvell**'s (1621–78) 'A Dialogue Between Soul and Body', the exchange is set out formally in ten-line stanzas of rhyming couplets. The poem as we have it is thought to be incomplete, but here are four lines of each of the complaints towards the other:

Soul

Oh, who shall from this dungeon raise
A soul enslaved so many ways?
With bolts of bones that fettered stands
In feet, and manacled in hands;
[…]

Body

But physic yet could never reach
The maladies thou me dost teach:
Whom first the cramp of hope does tear,
And then the palsy shakes of fear:
[…]

Another poem which uses this dialogue form is **William Wordsworth**'s (1770–1850) encounter with the child in 'We Are Seven', a poem we have already encountered in Chapter 3, 'Tones of voice'. The debate is between the worldly poet and 'the cottage Girl', and though the poem uses the simple ballad stanza, the dialogue is not always divided between them. At the end, for instance, the adult's exasperation and the child's insistence cut across each other:

'How many are you, then,' said I,
'If they two are in heaven?'
Quick was the little Maid's reply,
'O Master! we are seven.'

'But they are dead; those two are dead!
Their spirits are in heaven!'
'Twas throwing words away; for still
The little Maid would have her will,
And said, 'Nay we are seven!'

From these instances we can see how there are kinds of poems – narrative, sequenced and in dialogue – which virtually demand stanzaic form. But there are many stanzaic poems which do not fall even partly into these categories. I want now to consider a series of different types of shorter and then longer stanza forms and explore the effects they achieve in relation to their subject.

ONE-LINE FORMS

A one-line stanza must, on the definition given above, be a contradiction in terms. There are indeed few instances to be found, and some that might be considered are single-line sections of much longer poems. For example **Geoffrey Hill** (1932–) begins his sequence *The Triumph of Love* with the one-line poem,

Sun-blazed, over Romsley, a livid rain-scarp

and concludes it with the 150th poem,

Sun-blazed, over Romsley, the livid rain-scarp.

Between are poems of widely varying lengths, but making a deliberate stopping-place after but one line, and then recalling it at the end of the volume with that one change from *a* to *the*, is bound to make us dwell on the image evoked.

This gesture draws upon the spareness, the isolation of a few words taken out of the torrent of verse that so attracted *modernist* poets like **Ezra Pound** and **H. D. (Hilda Doolittle)** (1886–1961).

Stanzas are meant to combine lines and then present them for attention in the space marked by the boundaries for the eye or ear. Isolating single lines makes this more intense. It is a technique **Jorie Graham** (1951–) employs extensively, as here, in 'Self-Portrait as Hurry and Delay [Penelope at her Loom]'. The poem does not consist wholly of one-line stanzas but here is its conclusion:

17

the shapely and mournful delay she keeps alive for him the
    breathing

18

as the long body of the beach grows emptier awaiting him

19

gathering the holocaust in close to its heart growing more
    beautiful

20

under the meaning of the soft hands of its undoing

21

saying Goodnight goodnight for now going upstairs

22

under the kissing of the minutes under the wanting to go on
    living

23

beginning always beginning the ending as they go to sleep
    beneath her.

Actually section 16, which has four lines, ends 'it is', thus flowing straight into 17, 'the shapely and mournful delay ...' But, as with each of the succeeding line-stanzas, 17 can be read as a beginning. Graham clearly wants this ambiguity of connection and separation besides creating a slow sensual effect by the pauses between her long lines.

TWO-LINE FORMS

We have seen in the chapter on *rhyme* (Chapter 6) how the *couplet* works not as a stanzaic form but within longer poems. But even without rhyme these small units have had an enduring attraction for poets right up to our own day. Proportion, symmetry, counterpart, felt as intuitively satisfying, seem basic to this. **Matthew Welton** (1969–) has a pair of poems, 'The Wonderment of Fundament' and the 'The Fundament of Wonderment'. Each has four sections consisting of two couplets. Usually the rhymes are full, but occasionally, as in this section, he uses *half-rhymes*:

> She makes her music, loosening her hands:
> the moment holds. But if the evening ends
>
> the coffee place will crowd, and trains will leave,
> and fields absorb what light the moon might give.

The gentle, seeming randomness of incident and imagery in these poems and their sportive *word-play* might seem at odds with the clarifying briskness of the eighteenth-century couplet, but each seems to me to act upon the other: the poem's wandering is given shape by the couplet while the normally firm outlines of the form are softened.

A few poets in English have experimented with a verse-form consisting of couplets adapted from Persian, Arabic and other poetries called the *ghazal*. Classically the form rhymed *aa ba ca da ea fa* ..., and in subject tends towards melancholy and a limited range of topic and imagery. **Judith Wright** (1915–2000) has adapted the form, not attempting the rhyme scheme but usually closing each pair of lines. In her sequence *The Shadow of Fire* she maintains a meditation upon the passing of time and age especially by evoking the seasons and the world of nature. This is one of the shorter poems, 'Dust', after the Japanese poet **Bashō** (1644-94):

> In my sixty-eighth year drought stopped the song of the rivers,
> Sent ghosts of wheatfields blowing over the sky.
>
> In the swimming-hole the water's dropped so low
> I bruise my knees on rocks which are new acquaintances.
>
> The daybreak moon is blurred in a gauze of dust.
> Long ago my mother's face looked through a grey motor-veil.
>
> Fallen leaves on the current scarcely move.
> But the azure kingfisher flashes upriver still.
>
> Poems written in age confuse the years.
> We all live, said Bashō, in a phantom dwelling.

In her poem 'Brevity' (see Chapter 6, 'Rhyme') Wright speaks of her attraction to 'honed brevities' and 'inclusive silences', and the limitations imposed by her version of the ghazal ensure terseness and a stoical self-containment.

THREE-LINE FORMS

As stanzas stretch to three lines, so that emphasis on brevity can give way to greater expansiveness. In Chapter 6 ('Rhyme') we have seen how *terza rima* separates stanzas whilst spinning a thread to bind them together. There are two kinds of three-line stanzas, the *triplet* and the *tercet*. The *triplet* is the more traditional form in that it rhymes all three lines in a *monorhyme*, *aaa bbb ccc*, etc. Prolific rhyme usually tends to the comic, and the triplet is the form **John Donne** (?1571–1631) uses in his verse letters where he wants a comparatively informal, jocular tone. This is one of those 'are you still alive, why haven't you written' openings, 'To Mr T. W.':

> Pregnant again with th' old twins hope and fear,
> Oft have I asked for thee, both how and where
> Thou wert, and what my hopes of letters were[.]

But before we think the triplet an essentially cheery form we should look at **Thomas Hardy**'s (1840–1928) adaptation of it in his 'The Convergence of the Twain (Lines on the Loss of the *Titanic*)'. Here, in the third stanza, he evokes the sunken liner on the ocean floor:

> Over the mirrors meant
> To glass the opulent
> The sea-worm crawls – grotesque, slimed, dumb, indifferent.

There is no skip to these lines. There's symmetry in the **monorhyme** and in the double length of line 3, but there is a sombre awkwardness to the rhythm. That long third line especially just seems to stare at us unblinkingly.

Tercet is a more general term for the three-line stanza which might include other rhyme-patterns such as *terza rima*, but, particularly in the twentieth century, the grouping need not be rhymed. **Wallace Stevens** (1879–1955) came to use the form extensively. As this quotation shows, however, his tercets are often not self-contained units. The passages, like this from 'Notes Toward a Supreme Fiction', 1, V, frequently stretch themselves across the stanza divisions:

> The elephant
> Breaches the darkness of Ceylon with blares,
>
> The glitter-goes on surfaces of tanks,
> Shattering velvetest far-away. The bear,
> The ponderous cinnamon, snarls in his mountain
>
> At summer thunder and sleeps through winter snow.

In such a case we might wonder what the point of the stanza is. As these lines show, Stevens is often exotic in his imagery, but he can also be quite prosy, especially when his ideas are to the fore. In both moods his sentences often enlarge themselves, stretching that bit further and creating their own rhythmic period. Stevens was always interested in ideas of order set against the flux of the world – what he called elsewhere 'the meaningless plungings of water and the wind' – and the seemingly arbitrary tercet imposes an orderliness upon the ranging of his thought and imagination. He writes of a 'blessed rage for order', and he has an obvious rage for symmetry since his tercets are often formed into larger sub-sections and those into yet larger ones. This is true of 'Notes Toward a Supreme Fiction' where they are gathered into sevens, the sevens into tens, and the tens into three large sections. He varies this slightly at the very end of this long poem, but the intuitive desire for shapeliness is always apparent, even if it is contending against the varied character of his sentences.

William Carlos Williams (1883–1963) was a freer versifier than his near-contemporary Stevens, but also, especially in his late career, developed his own version of three-part form. He saw this as part of a new **prosody** too elaborate to detail here, but one obvious feature is its visual element as he steps this poem, 'Asphodel, That Greeny Flower', down and across the page:

> I cannot say
> > that I have gone to hell
> > > for your love
> but often
> > found myself there
> > > in your pursuit.

FOUR-LINE FORMS

Four-line forms are usually known as **quatrains** and reckoned to be the most common verse form in European poetry. Before the twentieth century quatrains would normally be rhymed either *abab*, *abba* – sometimes known as **envelope rhyme** – or *aabb*. As we have seen, it is the usual structure for the **ballad**, but also for far too many tones and styles to itemize here. As a whole poem, the compact and balanced quality of the quatrain lends itself to the witty compression of the **epigram**. **Tony Harrison**'s (1937–) mordant quip on secret police listening-devices, 'The Bedbug', is a good modern example:

> Comrade, with your finger on the playback switch,
> Listen carefully to each love-moan,

And enter in the file which cry is real, and which
A mere performance for your microphone.

By contrast the shorter lines of **Alfred Lord Tennyson**'s (1809–92) long poem of grief, 'In Memoriam', use the form for an utterly different emotional state:

> I sometimes hold it half a sin
>   To put in words the grief I feel;
>   For words, like Nature, half reveal
> And half conceal the Soul within.

There is a very delicate modulation in the third and fourth lines here as that parenthesis, 'like Nature', and the extra, internal rhyme *half conceal*, cause a catch in the voice of the stanza's regular progress.

Having looked at three- and four-line stanzas this is a good point to consider a pattern which combines them to produce a form in itself.

Originally a simple Italian and French 'rustic' song, the **villanelle** has been formalized, especially in the use English-language poets have made of it. The modern villanelle has a nineteen-line pattern that uses *five tercets* and a *final quatrain*. Strictly, these rhyme *aba* throughout, and the first and third lines recur at fixed points later in the poem. These reiterations and refrains seem to lend themselves to slow, mournful subjects, such as **Dylan Thomas**'s 'Do Not Go Gentle into That Good Night' that we looked at in Chapter 6. Certainly these first lines from three of the twentieth century's most notable villanelles suggest as much:

> Time will say nothing but I told you so
>
>   (**W. H. Auden**, 'If I could tell you')

> It is the pain, it is the pain, endures.
>
>   (**William Empson**, 'Villanelle')

> I wake to sleep, and take my waking slow.
>
>   (**Theodore Roethke**, 'The Waking')

Here, to demonstrate the whole form, is the whole of Thomas's villanelle with its remarkably tight structure. I have marked the recurring lines.

> **Do not go gentle into that good night**,
> Old age should burn and rave at close of day;
> *Rage, rage against the dying of the light.*
>
> Though wise men at their end know dark is right,
> Because their words had forked no lightning they
> **Do not go gentle into that good night**.
>
> Good men, the last wave by, crying how bright
> Their frail deeds might have danced in a green bay,
> *Rage, rage against the dying of the light.*
>
> Wild men who caught and sang the sun in flight,
> And learn, too late, they grieved it on its way,
> **Do not go gentle into that good night**.
>
> Grave men, near death, who see the blinding sight
> Blind eyes could blaze like meteors and be gay,
> *Rage, rage against the dying of the light.*
>
> And you, my father, there on the sad height,
> Curse, bless, me now with your fierce tears, I pray.
> **Do not go gentle into that good night**.
> *Rage, rage against the dying of the light.*

FIVE-, SIX-, SEVEN-LINE STANZAS

Of course, there is no reason why a stanza might not consist of any number of lines. Thus we can have five-line **quintets**, six-line **sestets** and seven-line **septets**, and in many respects their effects will be similar to those of the **quatrain**. The obvious variation is between odd and even numbers. In his 'Songs of Experience' **William Blake** (1757–1827) can use the quintet to disrupt the expectations of evenness, the comforts of balance that the quatrain gives. This is from 'A Little Girl Lost':

> To her father white
> Came the maiden bright:
> But his loving look,
> Like the holy book
> All her tender limbs with terror shook.

We do not expect a father's 'loving look' to bring terror, especially as the couplets seem to have a child-like simplicity. The disruption we then experience is mimed in that fifth, clashing longer line.

The **sestet** is again a stanza form that can offer closure, most often by developing the subject through the first four lines, perhaps by running them on, and using a rhyming couplet to cap it. **Wordsworth**'s famous poem 'I Wandered Lonely as a Cloud' closes in this way as he recalls the sudden sight of lakeside daffodils:

> For oft, when on my couch I lie
> In vacant or in pensive mood,
> They flash upon that inward eye
> Which is the bliss of solitude;
> And then my heart with pleasure fills,
> And dances with the daffodils.

Keith Douglas (1920–44) uses the sestet differently. In his 'How to Kill' his four stanzas are self-enclosed except for a bridge in the middle of the poem between stanzas 2 and 3. Here, as a tank-commander in the North African battlefield, he gives the order to fire, and

> Death, like a familiar, hears
>
> and look, has made a man of dust
> of a man of flesh. This sorcery
> I do. Being damned, I am amused
> to see the centre of love diffused
> and the waves of love travel into vacancy.
> How easy it is to make a ghost.

Douglas rhymes *abccba*, an '*envelope*' scheme which encloses the speaker's chilling confession of amusement at the instant evaporation of the human target at its centre. The outside half-rhymes, *dust / ghost*, also associate to convey the dissolution into death.

Six lines also form the basis for one of the most interesting of poetic forms, the **sestina**. This began with the Provencal *troubadour* poets of the Middle Ages, notably **Arnaut Daniel** who was at work in the late 1100s. The sestina consists of six six-line stanzas, and concludes with an *envoi* of three lines. In its English versions it usually uses a ten-syllable line. However, instead of a rhyme scheme, the sestina repeats a series of six end-words in each stanza, but in a fixed pattern of variation in which the sixth moves up to first in the next stanza and the others take up other corresponding positions. The three-line envoi then contains all the six repeated words. So, the words at the end of the lines of 'Paysage Moralisé' by **W. H. Auden** (1907–73) are arranged in this pattern:

St 1	valleys mountains water islands cities sorrow
St 2	sorrow valleys cities mountains islands water
St 3	water sorrow islands valleys mountains cities
St 4	cities water mountains sorrow valleys islands
St 5	islands cities valleys water sorrow mountains
St 6	mountains islands sorrow cities water valleys
Envoi	

> It is our sorrow. Shall it melt? Then water
> Would gush, flush, green these mountains and these
>   valleys,
> And we rebuild our cities, not dream of islands.

Normally too the last word of the poem is the same as the last word of its first line, though not in 'Paysage Moralisé'. In this poem, however, five of the six words belong easily in the same field of meaning, and the addition of the sixth, *sorrow*, adds a potential emotional charge that pulses through the poem.

There is a relentless, incantatory quality to the sestina, one that is obviously sustained in the longer version of the double sestina. In her book *The Discovery of Poetry* Frances Mayes points out that the numerology of sixes probably had specific significance to medieval writers. She also shows how each word of the six appears in adjacent lines to every other word twice. Thus, if we construct a hexagon with the points ABCDEF, and draw diagonal lines indicating these pairings, we have a graphically perfect hexagon with a symmetrical pattern of interior triangles. This net or cat's-cradle structure presents the poet with a tensile form to hold subjects that reverse and reflect upon themselves without necessary resolution.

The seven-line stanza or **septet** can vary metre and rhyme scheme, or indeed have none. In English the form became established by **Geoffrey Chaucer** (?1343–1400) in his long poem on the Trojan war, 'Troilus and Criseyde'. He derived the stanza from French models in which the form was traditionally used for formal celebration and came to have the name of *rhyme royal*. As Chaucer uses it, it has a ten-syllable line and rhymes *ababbcc* and has the alternative name of 'Troilus stanza'. The stanza is large and flexible enough to serve many purposes and was widely used in English poetry, including by Shakespeare in his narrative poem 'The Rape of Lucrece', until the early seventeenth century. Some practitioners introduced a longer seventh line, an **alexandrine** of six beats instead of the usual five. We might sense this to be a kind of pediment, the three/three evenness of the beat giving a base to the stanza. **John Donne** does this in 'The Good Morrow', which with typical eccentricity he rhymes *ababccc*. **Wordsworth**, in 'Resolution and Independence', also employs the longer last line but with Chaucer's rhyme scheme:

> All things that love the sun are out of doors;
> The sky rejoices in the morning's birth;
> The grass is bright with rain-drops; – on the moors
> The hare is running races in her mirth;
> And with her feet she from the plashy earth
> Raises a mist; that, glittering in the sun,
> Runs with her all the way, wherever she doth run.

Later in this poem the stanza carries far different moods, though none more sober than the tone of **W. H. Auden** in his great poem 'The Shield of Achilles' in which he uses both seven- and eight-line stanzas:

> A ragged urchin, aimless and alone,
>   Loitered about that vacancy, a bird
> Flew up to safety from his well-aimed stone:
>   That girls are raped, that two boys knife a third,
>   Were axioms to him, who'd never heard
>   Of any world where promises were kept,
> Or one could weep because another wept.

Each stanza in this poem is self-contained, and in this instance we can see the poet's very deliberate space containing a remarkable summarizing range. Each line contains its own clear *image* or idea, but the stanza is not only a sequence but a coordinated, sorrowing thought about dehumanization. The stanza's fulcrum lies in its *syntax*, specifically that colon at the end of line three. It is from that point that the observation of the particular child moves on to generalization. The purpose of the enclosure of the stanza's room could not be better illustrated.

EIGHT-LINE STANZAS

Eight-line stanzas are dominated by a particular form of Italian origin still known as **ottava rima** or, more rarely, **ottava toscana**. *Ottava rima* uses a ten-syllable line which rhymes *abababcc*. It is a form to be found in several European poetries and came into English with the enthusiasm for Italian literature and culture of the sixteenth-century Tudor poets such as **Sir Thomas Wyatt** (1503–42). It was used subsequently by many poets including **Milton**, who chooses *ottava rima* as a stabilizing orthodox stanza after the turbulent series of different shapes he has used throughout 'Lycidas' (see Chapter 6, 'Rhyme'). There the form is used with the greatest gravity, though the final couplet has a definite upbeat effect:

> At last he rose and twitched his mantle blue:
> Tomorrow to fresh woods and pastures new.

George Gordon, Lord Byron (1788–1824), however, finds a quite different tone possible. He called the form 'the half-serious rhyme', and from this fragment, written on the back of the manuscript of his great serio-comic poem 'Don Juan', we can see what he means.

> I would to Heaven that I were so much clay,
>   As I am blood, bone, marrow, passion, feeling –
> Because at least the past were passed away,
>   And for the future – (but I write this reeling,
> Having got drunk exceedingly to-day,
>   So that I seem to stand upon the ceiling)
> I say – the future is a serious matter –
> And so – for God's sake – hock and soda-water!

With the parenthesis, the verse seems to be running out of control in the queasiness of the speaker's hangover, but the metre and rhyme scheme hold it together and the variation of tone shows what can be encompassed in this space. The 6–2 pattern of the *ottava rima* enables the development of an idea, or mood, in the six lines, and then, in the couplet, the chance of a decisive conclusion. But it has great flexibility too in that the couplet can swivel at the very last to take the tone and subject in a different direction as Byron does above. Alternatively the divisions can be muted over a larger span of stanzas to produce more continuity. Among modern poets to use the form is **W. B. Yeats** (1865–1939) in such poems as 'The Circus Animals Desertion' and 'Among School Children'.

An eight-line pattern that constitutes a form in itself is the *triolet*. A French form, but pronounced in English to rhyme with 'get' and 'debt' – as **W. E. Henley** (1849–1903) does in his 'Easy is the Triolet' – the form uses just two rhymes and repeats some lines. The rhyme scheme is *ABaAabAB*, with the capital letters indicating the repeated lines. Edmund Gosse calls the form 'a tiny trill of epigrammatic melody' and it is given to the quick-footed lightness that repeated rhymes always bring. **Hardy** uses it for bird-talk. This is his 'Birds at Winter Nightfall':

> **Around the house the flakes fly faster,**
> **And all the berries now are gone**
> From holly and cotonea-aster
> **Around the house. The flakes fly! – faster**
> Shutting indoors that crumb-outcaster

> We used to see upon the lawn
> **Around the house. The flakes fly faster,**
> *And all the berries now are gone*!

Hardy's punctuation and syntactic play with 'Around the house the flakes fly faster' is so playful it amounts almost to a **parody** of the form.

THE SONNET

Having looked at how **couplets**, **quatrains**, **sestets** and **ottava rima** work, we can move now to one of the most prominent and important of forms, and one which combines some or all of these elements: the **sonnet**. The sonnet has been, and continues to be, successful not only in English but in a wide variety of European languages. Again its name comes from Italian, *sonetto* meaning a little sound or song, and its origins lie in medieval Italian poetry. **Dante** (1265–1321) and **Petrarca (Petrarch)** (1304–74) established the form and it was popularized in English during the sixteenth century. Normally the sonnet in English has fourteen lines of **iambic pentameter**. In Italian the line is **hendecasyllabic** (eleven syllables), and the French sonnet uses the **alexandrine** (twelve syllables).

There are three principal forms of sonnet.

1 The **Italian**, or **Petrarchan**, style: fourteen lines in divisions of eight and six, the **octave** and the **sestet**. The octave rhymes *abbaaba*, and the sestet either *cdecde*, *cdccdc*, or patterns that avoid closing the poem with a couplet such as *cdcdcd*.

2 The **Spenserian**: fourteen lines in three quatrains and a couplet and rhyming *abab bcbc cdcd ee*.

3 The **Shakespearean**, or **English**: also foregrounds the quatrain/couplet pattern and rhymes *abab cdcd efef gg*.

From this we can see that the Petrarchan sonnet can require as few as four rhyme sounds, *abcd*, whereas the Spenserian requires five, *abcde*, and the Shakespearean seven, *abcdefg*. In part this reflects the greater difficulty of rhyming in English as opposed to Italian which has a great predominance of -*o* and -*a* word-endings.

But the more significant distinction is in the *thought structure* of the different styles. The **Petrarchan** is essentially a two-part structure: an idea or subject is expounded in the octave, and then, with the change of rhyme, there is a **volta**, or 'turn', after which the sestet responds to or resolves the opening proposition. The limitation to only two sounds makes the octave very compact as the rhymes overlap.

The **Spenserian** and **Shakespearean** by contrast might be said to be more volatile. Here, the changes in rhyme-sound from quatrain to quatrain encourage new turns of thought and a step-by-step movement towards the definite closure provided by the couplet. Since the main distinction is between the Petrarchan 8/6 pattern and the 4/4/4/2 of the Spenserian and Shakespearean styles, I shall concentrate on comparing just the two types.

The sonnet tradition, especially when closest to the Italian models, is associated with the sixteenth and seventeenth centuries and with a certain manner of (usually anguished) love poetry. For my example of the Petrarchan style, however, I have chosen a twentieth-century sonnet, a love poem but with a different tone. This is by **Edna St Vincent Millay** (1892–1950):

I, being born a woman and distressed	*a*
By all the needs and notions of my kind,	*b*
Am urged by your propinquity to find	*b*
Your person fair, and feel a certain zest	*a*
To bear your body's weight upon my breast:	*a*
So subtly is the fume of life designed,	*b*
To clarify the pulse and cloud the mind,	*b*
And leave me once again undone, possessed.	*a*

Think not for this, however, the poor treason	*c*
Of my stout blood against my staggering brain,	*d*
I shall remember you with love, or season	*c*
My scorn with pity, – let me make it plain:	*d*
I find this frenzy insufficient reason	*c*
For conversation when we meet again.	*d*

Millay's sonnet demonstrates the function of the turn after line 8 to perfection. The *octave* bears witness to the erotic force still exerted by the speaker's lover. Then in the sestet she musters her 'staggering brain' to resist the conquering sexual attraction to someone she evidently feels is bringing her nothing but distress. This balance in the poem also represents the see-saw between heart and head, body and mind, sexual urge and good sense, that suffuses the whole poem. On the one hand she can choose very controlled, distant words like 'propinquity' to refer to being physically close, and then confess to being 'once again undone, possessed'. For the conclusion, which, though it is not a couplet, is clinched in the last two lines, she manages a put-down so haughty she might be returning a visiting-card. However, we always feel that the control mimed by this carefully controlled sonnet is hard-won and precarious. We feel that as soon as the last full-stop goes down she might collapse.

The sonnet seems particularly suited to walking this fine line between self-control and tumultuous emotion. We see it often in **Shakespeare**'s sonnets where we can frequently read a counter-implication beneath the ostensible argument. Here, employing of course the *quatrain* and *couplet* pattern of the **English** sonnet, is his Sonnet 138:

When my love swears that she is made of truth	*a*
I do believe her, though I know she lies,	*b*
That she might think me some untutored youth,	*a*
Unlearnèd in the world's false subtleties.	*b*
Thus vainly thinking that she thinks me young,	*c*
Although she knows my days are past the best,	*d*
Simply I credit her false-speaking tongue;	*c*
On both sides thus is simple truth suppressed.	*d*
But wherefore says she not she is unjust?	*e*
And wherefore say not I that I am old?	*f*
O, love's best habit is in seeming trust,	*e*
And age in love loves not to have years told.	*f*
Therefore I lie with her, and she with me,	*g*
And in our faults by lies we flattered be.	*g*

In this poem the lovers are exchanging duplicities. He, the speaker, is pretending to be younger than he is whilst she is pretending that she is faithful to him. But he knows she is unfaithful and she knows he is not so youthful. Moreover he knows that she knows this, just as she knows that he knows of her unfaithfulness, and this 'he knows that she knows that he knows ...' goes on and on. The final couplet suggests that they get by with this tacit understanding, *lying* together in both senses of the term. The sonnet structure is good for argument and here its phases enable this mutual deceit to be revealed layer by layer. But is the speaker as worldly, even cynical, as he maintains? Do we also sense an emotional discomfort: anxiety about the fragility of the relationship, the pain of betrayal, the deep embarrassment of dishonesty?

Love has not been the sole subject in the sonnet tradition, however. **Milton** and **Wordsworth** used the Petrarchan form to write polemical political sonnets. Like **Donne**, **Gerard Manley Hopkins** (1884–1899) found the argumentative capacity of the sonnet fit for his tussles with his conscience and God, and the intense emotions of his 'terrible sonnets' batter and strain the form to its utmost.

Another important feature of the sonnet in all its styles is the prevalence of *sequences*. Following Petrarch's example, Spenser, Sidney, Shakespeare and many other Elizabethan poets composed their sonnets into extensive sequences, usually exploring different aspects of one theme. This has been continued in such series as **Elizabeth Barrett Browning**'s (1806–61) 'Sonnets from the Portuguese', **George Meredith**'s (1828–1909) 'Modern Love', which varies the structure by adding two lines, and by many twentieth-century poets besides Millay including **Allen Tate** (1899–1979), **John Berryman** (1914–72), **Robert Lowell** (1917–77), **Geoffrey Hill** (1932–) and **Marilyn Hacker** (1942–). **Tony Harrison**'s sequence 'The School of Eloquence', ongoing over many years, uses a sixteen-line iambic pentameter pattern, strictly rhymed but stretching the form extensively in regard to vocabulary and line-endings, and covering a remarkable range of personal and social themes. **James K. Baxter** (1926–72) divides his thirty-nine 'Jerusalem Sonnets' into pairs of lines, sometimes rhyming, and other modern sonneteers such as **Ted Berrigan** (1934–83) ruffle the form yet more radically. 'The Sand Coast Sonnets' of **Les Murray**

(1938–) include several styles, including one with fifteen lines, and indeed in his exuberant hymn to extravagance, 'The Quality of Sprawl', he cites as an example 'The fifteenth to twenty-first / lines in a sonnet, for example.'

But the attraction of the sequence persists, not merely because of tradition, but perhaps because of the opportunity to write extensively to related themes, even over many years, and to do so in a way which presents lots of new beginnings. It does not need the thread of *narrative*, even if each sonnet contains a little 'story' in its set progress. Rather the sequence enables fresh angles, different tones from the intimate and meditative to the comic and polemical. When written over time, for both writer and reader there is the nice juxtaposition of continuity of form against the other likely changes in subject, mood or style. This large-scale attraction of the sequence counterparts the enduring appeal of its component parts: the flux that can be contained – sometimes only just – within the single sonnet's walls.

NINE-LINE STANZAS

Returning to our numerical progress, nine lines would at first seem an arbitrary choice for a stanza. Just as we can see how the sonnet might plausibly be extended to sixteen lines but, *pace* Les Murray, not fifteen, so a nine-line stanza seems to lack the satisfying symmetry of even numbers. In fact the form devised by **Edmund Spenser** (?1552–99) for his long, fantastical narrative poem *The Faerie Queene* proved successful not only for his poem but for many subsequent poets.

The *Spenserian stanza* comprises nine iambic lines, eight *iambic pentameters* and one, closing, *hexameter*. The rhyme scheme is *ababbcbcc*. In common with its nearest relative, *ottava rima*, the stanza is short enough to be pointed and precise, and ample enough for description and dilation. The ninth, longer, line has the effect of securing the footing of the stanza against its seeming imbalance. Here 'the gentle knight' confronts a monster:

> Therewith she spewd out of her filthie maw   [mouth]
> A floud of poison horrible and blacke,
> Full of great lumps of flesh and gobbets raw,

Which stunck so vildly, that it forst him slacke   [vilely]
His grasping hold, and from her turne him backe.
Her vomit full of bookes and papers was,
With loathly frogs and toades, which eyes did lacke,
And creeping sought way in the weedy gras:
Her filthie parbreake all the place defiled has.   [vomit]

Spenser's verse is notably sensuous in its descriptions and it was perhaps that association which drew some of the Romantic poets to imitate his stanza. **John Keats** (1795–1821) did so in 'The Eve of St Agnes', and **Percy Bysshe Shelley** (1792–1822) chose the form for his elegy to Keats, 'Adonais':

He will awake no more, oh, never more! –
Within the twilight chamber spreads apace
The shadow of white Death, and at the door
Invisible Corruption waits to trace
His extreme way to her dwelling-place;
There eternal Hunger sits, but pity and awe
Soothe her pale rage, nor dares she to deface
So fair a prey, till darkness, and the law
Of change, shall o'er his sleep the mortal curtain draw.

Both Spenser and Shelley use the length of the stanza to its utmost to unfurl long sentences whose controlled syntax is enhanced by the discipline of the few permitted rhymes.

Modern poets introduce more variation into stanzas clearly inspired by these examples. **Robert Lowell** (1917–77) employs only four rhymes in this poem, based on the hellfire sermons of the eighteenth-century Massachusetts preacher Jonathan Edwards, 'Mr Edwards and the Spider'. He places them mainly in the pattern of *ababcccdd*, and varies his line length whilst retaining Spenser's final hexameter:

What are we in the hands of the great God?
It was in vain you set up thorn and briar
  In battle array against the fire
  And treason crackling in your blood;
    For the wild thorns grow tame

> And will do nothing to oppose the flame;
> Your lacerations tell the losing game
> You play against the sickness past your cure.
> How will the hands be strong? How will the heart endure?

Lowell also follows his predecessors in his grand tone, but just how different a register can be struck in this type of stanza can be heard in these lines from **Philip Larkin**'s (1922–85) 'Church Going'. Out on an excursion, the speaker has paused to venture into an empty church:

> Yet stop I did: in fact I often do,
> And always end much at a loss like this,
> Wondering what to look for; wondering, too,
> When churches fall completely out of use
> What we shall turn them into, if we shall keep
> A few cathedrals chronically on show,
> Their parchment, plate and pyx in locked cases,
> And let the rest rent-free to rain and sheep.
> Shall we avoid them as unlucky places?

Larkin uses the stanza in a deliberately understated way. Indeed, it's easy to read the familiar manner of these lines, with their commonplace phrases like 'in fact', 'at a loss' and 'out of use', without noticing how carefully they are crafted. For instance, at first sight – or hearing – the stanza appears to have an odd unrhymed sixth line, *show*. The rest of the poem reveals the same pattern until we notice that *show* can be heard as a half-rhyme with the *a* rhymes, *do* and *too*, and that this pattern obtains throughout the poem, mainly with half-rhymes like *on/stone/organ* and *font/don't/meant*. It is as though Larkin's structure is half-hidden from eye and ear.

Why? Maybe it is a game he is playing with himself and the reader in which he strikes a bluff, common-man pose, as unpretentious and unpoetical as can be, but quietly belies this by exercising such subtle but recognizable skill. Maybe too he needs the demands of the form as a discipline in composition. He is wedded to 'ordinary speech' in his choice of words and phrase, but in order to guard against an ease that might become just sloppy, he imposes these unobtrusive restraints upon himself to reach further levels of concentration. As with the demands of all set forms, Larkin's acceptance of these limits might be pushing him towards articulating more interesting things than he would otherwise say.

We could continue our progress through successive stanza lengths to examine **ten-line** stanzas such as those **A. D. Hope** (1907–2000) employs in his 'On an Engraving by Casserius', the ingenious, **eleven-line** stanza of **John Donne**'s 'The Relic', and on to the enormous elaboration of Spenser's **eighteen-line** stanzas of his marriage hymn 'Epithalamium' with its variations of line-length and mixture of *plain* and *interlaced* rhyming. However, the main features and resources of the longer stanza forms are by now established, so I will turn to some other line sequences which, like the sonnet, are forms in themselves.

RONDEAU AND RONDEL

The *rondeau* and the *rondel* are often associated with the *triolet* (see above) and also have their origin in medieval French poetry. Like much poetry their ancestry is in song, and especially the dance-songs of *rondes* or rounds.

Formalized into a literary convention, the **rondeau** became a fifteen-line form divided into a *quintet*, a *quatrain* and a *sestet*, and employing just two rhymes. A further distinguishing feature is that the first line is half-repeated at lines 9 and 15. Experiment with these French forms was popular among English poets of the late nineteenth century, and again we can find **Hardy** using the form in 'The Roman Road':

> The Roman road runs straight and bare
> As the pale parting-line in hair
> Across the heath. And thoughtful men
> Contrast its days of Now and Then,
> And delve, and measure, and compare;
>
> Visioning on the vacant air
> Helmed legionaries, who proudly rear
> The Eagle, as they pace again
>   The Roman Road.

> But no tall brass-helmed legionnaire
> Haunts it for me. Uprises there
> A mother's form upon my ken,
> Guiding my infant steps, as when
> We walked that ancient thoroughfare,
> > The Roman Road.

A Roman road is a fine ancient subject weighted with significance. But in his typically contrary way Hardy upsets this expectation by saying that its meaning for him lies wholly in a childhood memory of his mother. The shift comes with that angular, almost awkward, 'Uprises there ...' By the time 'The Roman Road' comes round for the third time its associations have become quite different.

The vestiges of these forms' beginnings in dance are surely visible in their continual return to where they begin. The first two of the *rondel*'s fourteen lines recur in the last two lines of the second quatrain, and in the last two of the third and last section which is a sestet. **Austin Dobson**'s (1840–1921) 'Too hard it is to sing' is an example. Confusingly, his near-contemporary **Algernon Charles Swinburne** (1837–1909) introduced a variant he called a *roundel* and wrote a 'century' of them, some lamenting the death of the composer Wagner. By this time, though, we are surely exhausting knowledge that might truly be called basic, though readers who want to see how a present-day poet employs the form might seek out **Sophie Hannah**'s (1971–) 'The End of Love' which works enjoyably round the refrain 'The end of love should be a big event. / It should involve the hiring of a hall.'

BORROWING FORMS

Scanning these different types of stanza and free-standing poetic forms, we can see how often their origins are in other languages and how much interchange there is between these different poetries. Most are European, but, as we have seen with the *haiku* (see Chapter 2, 'Deliberate space'), and the *ghazal* earlier in this chapter, other traditions have also been influential.

The *pantoum* is the Europeanization of the Malay form *pantun*. This is based on a four-line form which rhymes *ab ab*, but it also includes internal rhymes and various kinds of correspondences between images and ideas. This translation is from Ruth Finnegan's excellent anthology *The Penguin Book of Oral Poetry*.

> Broken the pot, there's still the jar,
>> Where folk can come and wash their feet.
> And when the mynah's flown afar,
>> For comfort there's the parakeet.

In addition to the end-rhymes, there is *jar/mynah*, and *come/comfort*, and these correspondences match the overall idea of compensation: the pot is broken, there is the jar; the mynah bird goes, there is still the parakeet.

In its European versions the **pantoum** consists of quatrains in which the second and fourth lines of each stanza become the first and third lines of the next, and so on. Eventually the very first line will reappear as the poem's last line and the third line of the poem as the third last. The form is too long to illustrate here but over its length it contrives a criss-crossing, mesmeric quality.

INVENTING STANZAS AND THE VERSE PARAGRAPH

The tension between containment and expansion is present in all these stanzas and their related forms. The history of poetry shows a nearly regular alternation between these competing demands. Here I want to compare a poem that uses most features of strict stanzaic shaping, but does so over a whole poem that has no given definition, with a passage that is not stanzaic at all but would be better called a **verse paragraph**.

The first is **Andrew Marvell**'s (1621–78) poem 'The Coronet'.

> When for the thorns with which I long, too long,
>> With many a piercing wound,
>> My saviour's head have crowned,
> I seek with garlands to redress that wrong:
>> Through every garden, every mead, [field]
> I gather flowers (my fruits are only flowers),
>> Dismantling all the fragrant towers [head-dresses]
> That once adorned my shepherdess's head.
> And now when I have summed up all my store,

Thinking (so I myself deceive)
So rich a chaplet thence to weave   [coronet]
As never yet the king of glory wore:
Alas, I find the serpent old
That, twining in his speckled breast,
About the flowers disguised does fold,
With wreaths of fame and interest.
Ah, foolish man, that wouldst debase with them,
And mortal glory, heaven's diadem!
But thou who only couldst the serpent tame,
Either his slippery knots at once untie;
And disentangle all his winding snare;
Or shatter too with him my curious frame,
And let these wither, so that he may die,
Though set with skill and chosen out with care:
That they, while thou on both their spoils dost tread,
May crown thy feet, that could not crown thy head.

There is no point in pretending that this poem is easy. It is usually thought that poetry is betrayed by paraphrase, but this is the kind of arguing poem that benefits from our trying to put it into our own words.

The argument, I believe, goes like this. The poet, conscious that his sins have long served to add to the pain from the crown of thorns of his saviour Christ, resolves to make amends by turning from writing light love-verses to poems glorifying Christ. Thus, in the system of *images* the poem employs, he will no longer make 'garlands' for his 'shepherdess's head', but a coronet of flowers to replace the crown of thorns. But, as he does so, he realizes that this new ambition is in truth driven by selfish 'fame and interest', that Satan, 'the serpent old', is subverting him. Finally he is reminded that only Christ can tame, 'disentangle', Satan's 'winding snare', and that in crushing Satan underfoot Christ will also tread on the poet's vain verses. Thus, paradoxically, the poem that was meant as a 'coronet' will crown Christ's feet rather than his head.

It is a testing, complex argument and this is engrossed in the structure of the verse and of the *syntax*. First the verse moves between *iambic pentameters* and the sharper, more emphatic three- and four-beat lines. Second the rhyme scheme alters from the *enve-*

lope pattern of *abba* to an **interlaced**, entangled pattern from where 'the serpent old' appears in line 13. This resolves briefly into plainness again with the exclamatory **couplet** 'Ah, foolish man, that wouldst debase with them, / And mortal glory heaven's diadem!' only to overlap again in the lines about Satan's 'slippery knots' and 'winding snare'. Finally, as the poem reaches its closing assertion, simplicity is restored with the couplet:

> That they, while thou on both their spoils dost tread,
> May crown thy feet, that could not crown thy head.

The poem's sentence structure is equally complex. The first sixteen lines consist of just two real sentences, each eight lines long. The first has the main verb 'I seek' and the second 'I find'. As he struggles to establish his task, and then to overcome the unsuspected pride that is undermining his good intentions, the poet wrestles through elaborate sentences beset with parentheses and subordinate clauses. Even the final sentence, the last eight lines, is not structurally straightforward but fights through more elaborate obstruction before arriving at the paradox that his flowers – that is his verses – must be withered and trodden underfoot to fulfil their proper function. Of course the whole poem can then be seen as a paradox since it cannot help but be the poem that the poet *says* he despises. Both the stanza and sentence structure of what the poet calls 'my curious frame' are tortuous, even tortured, as the work struggles to make sense of its contradictions. I'm tempted to say there was no ready-made stanza pattern that could accommodate Marvell's effort here, but that he needed this complex patterning to embody the difficulty of ideas and feeling. A 'freer' form would not do.

We can see such a 'freer' form, however, in the **verse paragraph** that displaces stanza form in much Romantic poetry. 'Frost at Midnight' is one of the poems **Samuel Taylor Coleridge** (1772–1834) called 'a conversation poem'. In some ways the relaxation into blank verse and the abandonment of stanza patterns by Wordsworth, Coleridge and, before them, such poets as **James Thomson** (1700–48), is a return to the resources of Shakespearean blank verse, and even more so to the example of Milton. But whilst Wordsworth especially often sought Milton's grander notes, part of his and Coleridge's revolutionary poetic enterprise in *The Lyrical Ballads* of 1798 was, as we saw in Chapter 3 ('Tones of voice'), to find an easier, more commonplace **register** for both description and meditation. Here is the last part of Coleridge's 'Frost at Midnight' in which the poet is speaking over his sleeping child:

> Therefore all seasons shall be sweet to thee,
> Whether the summer clothe the general earth
> With greenness, or the redbreast sit and sing
> Betwixt the tufts of snow on the bare branch
> Of mossy apple-tree, while the nigh thatch
> Smokes in the sun-thaw; whether the eave-drops fall
> Heard only in the trances of the blast,
> Or if the secret ministry of frost
> Shall hang them up in silent icicles,
> Quietly shining to the quiet Moon.

This paragraph is the shortest of the four that constitute the poem's seventy-four lines. Clearly the poet has not been tempted into symmetry of organization. It is unrhymed and uses a **decasyllabic** line, but without regular **stress**. Looking to the future of his child, the poet is in a reverie of hopefulness. The ten lines are but one sentence in which the assertion 'all seasons shall be sweet to thee' is the heart of the main clause. The seasons are then illustrated through the slow series of descriptive clauses in which details like the robin 'betwixt the tufts of snow', and the steam of 'the sun-thaw', seem to emerge as perfect **images** before his dreaming eye. But the slow stream of the sentence does not dribble away. The gentle action of the final image, 'the secret ministry of frost' hanging the icicles from the house-eaves, is strong enough to balance the sentence. The very last line,

> Quietly shining to the quiet Moon

has just those four firm stresses, and this, combined with the simple effect of repeating 'quiet', produces a wonderful sense of peace. The verse paragraph, by dispensing with so many of the staples of stanza-form, can risk becoming flatly 'prosy'. But in 'Frost at Midnight' we can see how the comparative looseness of the structure suits the movement of the poet's mind, his conversation with himself. Yet the line is still vital. That last line in particular shows us how the separate definition of the line gains an effect not available in prose.

In the twentieth century *'free verse'* has largely jettisoned stanza-form as part of its liberation, although it is striking how often in non-metrical and unrhymed verse line-groupings are routinely employed. For my final example, however, I have chosen a twentieth-century poem which patterns line and rhyme into a highly individual stanza shape, and one as subtle and demanding as any in the history of poetry. The poem is 'What Are Years?' by **Marianne Moore** (1887–1972).

> What is our innocence,
> what is our guilt? All are
>   naked, none is safe. And whence
> is courage: the unanswered question,
> the resolute doubt –
> dumbly calling, deafly listening – that
> in misfortune, even death,
>   encourages others
>   and in its defeat, stirs
>
>   the soul to be strong? He
> sees deep and is glad, who
>   accedes to mortality
> and in his imprisonment rises
> upon himself as
> the sea in a chasm, struggling to be
> free and unable to be,
>   in its surrendering
>   finds its continuing.
>
>   So he who strongly feels,
> behaves. The very bird,
>   grown taller as he sings, steels
> his form straight up. Though he is captive,
> his mighty singing
> says, satisfaction is a lowly

thing, how pure a thing is joy.
This is mortality,
This is eternity.

Because of the way these enigmatic sentences run across the strangely uneven lines, we might not much notice the rhymes, though some, *to be / to be*, are **monorhymes** and others, *others/stirs*, are **half-rhymes**. We will quickly see that the poem comprises three nine-line stanzas, but until we look closely we might not see that these employ a **syllabic** pattern to define the lines, and that this is symmetrical. The architectonic pattern of the poem is shown in Table 1.

If we look at the syllabic pattern we can see that each stanza is organized in exactly the same way. The first two and the last two lines of each stanza have the same number of syllables (six) as do the third and seventh (seven) and the fourth and sixth (nine). Only the fifth – middle – line stands alone with five syllables. Since the

Table 1: The architectonic pattern of 'What Are Years?'

#	stanza 1		stanza 2		stanza 3	
	no. sylls.	rhyme	no. sylls.	rhyme	no. sylls	rhyme
1	6	a	6	a	6	a
2	6	b	6	b	6	b
3	7	a	7	a	7	a
4	9	c	9	c	9	c
5	5	d	5	d	5	d
6	9	d	9	a	9	e
7	7	e	7	a	7	f
8	6	f	6	e	6	e
9	6	f	6	e	6	e

poem consists of three nine-line stanzas, and each stanza contains three different syllable groups – aside from the solitary fifth – it looks likely that the sense of the ratio 3:9 is important to the poem's building even without trying to attach any significance to these numbers themselves. Especially if it is turned through ninety degrees, this gives each stanza an arching shape centring on the solitary line:

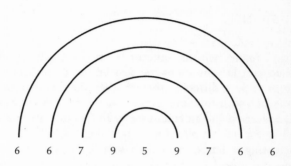

Those three centre lines are: *the resolute doubt // upon himself as // his mighty singing*, but lacking a verb we can't make syntactic sense of this. However, if we see the poem as an effortful, determined acceptance, and therefore defiance, of death, realized in stanza 3 through the image of the captive bird rising and steeling himself to sing, then we can see that the *mighty singing* the bird takes *upon himself* constitutes the powerful paradox of *resolute doubt*. Looking at the last words of the unrhymed lines in each stanza we see: *are / question / death // who / rises / as // bird / captive / singing / joy*. Again we might read the heart of the poem's statement through this vertical axis. (If the third stanza had only three unrhymed lines like the first two, then the 3:9 ratio would have another dimension and the puzzling analyst's cup would run over. Perhaps it is significant that it is the word *joy* that surpasses symmetry.)

There is I believe a strong element of the puzzle in Moore's poem. It has a problem-setting mischievous quality perhaps meant to belie the seriousness of its subject. This subject – mortality and how to face up to it – is difficult to approach without platitude and involves its own tensions. Under these pressures the poem is both awkward and elegant: awkward at first reading with those odd lines with their abrupt, anti-syntactical breaks, but eventually elegant as we come to see its underlying architecture. The poem shows there is no contradiction between the force of emotion and idea and the shaping of expression: 'So he who strongly feels, / behaves.'

CONCLUSION

Set forms can at first seem daunting to the beginning poet. Nonetheless *practice* within some of them can teach writers and readers important lessons about the possibilities of verbal pressure. Also they provide a different sort of liberation in that the page before us is not so intimidatingly blank. We have a shape to fill, lines of demarcation and an end in sight. To feel this it is not necessary to plunge into complex forms like the villanelle or sestina. Simply accepting a determinant such as a fixed number of words, perhaps on the model of another poem, can suffice to sense the concentration of poetic expression. Like all these poets before us we will then gain the experience of surprising ourselves with the words we find when we must think twice, or three, or more times before we can make the fit.

Summary
In this chapter on the stanza we have looked at:

- definitions of the stanza as space and as pause, and its mnemonic qualities;
- its use in dialogue forms;
- a series of stanzas ranging from one-line forms to longer, complex forms;
- several forms related to stanza-form but distinct in themselves, notably the sonnet;
- examples of forms borrowed from other poetries;
- invented stanza-forms and the use of the verse paragraph.

FURTHER READING

Furniss, Tom and Bath, Michael (1996) *Reading Poetry, An Introduction*, Hemel Hempstead: Prentice Hall/Harvester Wheatsheaf; Chapter 12, 'The Sonnet'.

Hopkins, G. M. (1970) *The Poems of Gerard Manley Hopkins*, ed. W. H. Gardner and N. H. Mackenzie, Oxford: Oxford University Press; see the religious 'terrible sonnets'.

Mayes, Frances (1987) *The Discovery of Poetry*, Orlando: Harcourt, Brace, Jovanovich; see Chapter 8, 'Traditional and Open Forms'.

Millay, Edna St Vincent (1992) *Selected Poems*, Manchester: Carcanet Press.

Shakespeare, William (1986) *The Sonnets and A Lover's Complaint*, ed. J. Kerrigan, Harmondsworth: Penguin Books.

IMAGE –
IMAGINATION –
INSPIRATION

What is man's body? It is a spark from the fire
It meets water and it is put out.
What is man's body? It is a bit of straw
It meets fire and it is burnt.
What is man's body? It is a bubble of water
Broken by the wind.

> (a Gond poem, from central India)

IMAGE AND METAPHOR

We might call someone 'daft as a brush', or say of something that doesn't work very well that it's a 'lemon', or, faced with an important decision, that we are 'at the crossroads', or that 'things are looking up'. All these phrases are making use of the figure of speech known as **metaphor**. There are many different types of metaphor but they share the characteristic of *saying one thing in terms of another*. At the heart of metaphor, the *vehicle* which connects the subject of the utterance with the quality being evoked, is an **image**: the brush, the lemon, the crossroads, the act of lifting the eyes. In the poem at the head of this chapter our body is first a spark, then a piece of straw, then a bubble.

Metaphor is often seen to be the essence of the 'poetic', and prose that is coloured in such ways is often labelled 'poetic'. But poetry does not have a monopoly on metaphor, or upon the vivid evocations of descriptive imagery. The historian **Bede** (?672–735) likened human life to the flight of a bird that happens to swoop through a window and crosses a lighted hall before disappearing once more into the dark. Daily speech and prose of many kinds will often try to 'paint pictures' and to use metaphor, and not only to decorate a passage. Metaphor and its infinite variety of images is intrinsic to human language, and because consciousness of language is inseparable from poetry, we are bound to look at how it works, and at some of the interesting questions that follow.

At its simplest, in the *simile*, the metaphoric connection is explicitly made by using 'like' or 'as'. Here **Anne Askew** (1521–46) begins her poem 'The Ballad Which Anne Askew Made and Sang When She Was in Newgate' with a simile:

> Like as the armèd knight
> Appointed to the field,
> With this world will I fight
> And faith shall be my shield.

In other works the metaphorical intention will be more implicit and sometimes more sustained, as in this anonymous lyric, 'The Silver Swan', dated at 1612:

> The silver swan, who living had no note,
> When death approached, unlocked her silent throat;
> Leaning her breast against the reedy shore,
> Thus sung her first and last, and sung no more:
> 'Farewell, all joys; Oh death, come close mine eyes;
> More geese than swans now live, more fools than wise.'

Whatever sad event or melancholy mood inspired the poet, we will read this as a human lament. The swan is a *figure* through which the human voice complains not only of death, but of the condition of the world she leaves behind. The notion that the swan is mute until its last moments presents the neat poignancy, especially for a poet, of first words being last words and vice versa. Perhaps too the swan's muteness is a figure of restraint which makes her final condemnation of the world that much more powerful.

'The Silver Swan' works as an extended metaphor, or **conceit**, that is a figure of speech that is carried on beyond one moment's likening to develop a substantial idea.

'*Conceit*' is an interesting word in this context. The word has two other tendencies within its meanings. First it can mean an idea, or conception, but we are most likely to think of a second sense associated with self-regard and vanity. Our literary term has something of both implications for it carries an idea, but the working through of the metaphor is likely to display an ingenuity which some readers will find affected and self-admiring, as though the poet is keener to show off than to say something significant. Seventeenth-century poetry has often been disparaged for being 'lost in conceits', and the *Oxford English Dictionary* quotes one commentator who praises the classical Greek poets for ignoring such fancy elaboration: 'they did not call the waves "nodding hearse-plumes" … or laburnums "dropping wells of fire"'.

WORDS AND 'THINGS AS THEY ARE'

This takes us deep into one of the most ancient controversies about poetry, and indeed language. The essence of human language is that it puts one 'thing' in place of another 'thing', that is that it makes unrelated sounds – and later their accompanying visual symbols – stand in for objects in the world and for what occurs in our minds. Metaphor – this further 'standing in place of' – extends this remove. Words are already images of the things they represent – 'signifiers' of what they signify – but metaphors are images of images. Writing on the theory of art in Book 10 of *The Republic*, the Greek philosopher **Plato** (327–247 BC) saw 'poetic' language as *refracting* reality. He likened this crucial difference between how a thing is and how words can represent it to the way refraction makes a stick look bent when it is put in water. This difference made Plato and other philosophers and other thinkers uneasy. They worry that the capacity of language to invent and elaborate, for its sounds and images to work upon the emotions, can carry us away from reality and truth.

The developing scientific culture of the seventeenth century also put a high premium on simple clarity in speaking and writing and set its face against 'the easie vanitie of fine speaking'. Sir Thomas Sprat in *The History of the Royal Society* (1667) complained of the 'many mists and uncertainties these specious Tropes and Figures have brought on our Knowledge'. Rather than 'pretty conceits' (it is interesting that the feminizing adjective 'pretty' is often applied here) we should aim, argued Sprat, for a plain, direct language which has 'as many words as there be things', and no more.

RHETORIC

Despite these anxieties, the Greeks and every part of western culture since have acknowledged and sought to deploy the power of words. So much of social life, and especially public life, involves persuasion. As soon as a parent encounters the question 'why?' at least some part of her response is going to involve persuasion. Teachers must find ways of engaging their pupils in their subject, lawyers seek to convince juries and of course politicians aim to sway us towards voting for them. All are using ***rhetoric***: that is, language shaped to persuade.

The full effect, however, has never been confined simply to the words themselves. The great Athenian orator **Demosthenes** (383–22 BC) is said to have had to overcome a stutter, and legend says he did so by practising his delivery with a pebble in his mouth and pitching his voice against the sound of the sea. Nowadays all manner of public figures take advice on how to present themselves on television and elsewhere. The ***pitch*** of the voice and the choice of a jacket are part of rhetorical effect, part indeed of another, newer sense of 'image'.

But our primary interest here is with the verbal dimension of rhetoric. We have already explored how different tones such as anger, pathos, humour, can be represented in poetic forms (Chapter 3, 'Tones of voice'). Within these registers orators deploy particular constructions of phrase for effect. They might for instance begin a series of sentences in exactly the same way. **Martin Luther King**'s celebrated civil rights speech which reiterates the phrase 'I have a dream that ... I have a dream that ... I have a dream today' at the opening of each paragraph is an example of this. Such devices – or rhetorical figures – are quite deliberate, and many generations of students and orators, following the models of antiquity like Demosthenes and the Roman **Cicero** (106–43 BC), practised them assiduously and knew all their specific names. One of them, *anaphora* – the figure used by King in his speech – we have encountered in a variety of contexts already.

We do not need to explore the particulars of classical rhetoric here. But it is important to recognize that there is a long tradition in which its deliberate strategies are also part of poetry. The Roman orator Cicero declared that 'the poet is very near kinsman of the orator', by which he meant that both aim to use all the resources of language, both intellectual and sensuous, to persuade the listener or reader.

Of course, if we become too aware of the design the poet or orator has upon us then we might recoil, thinking that we are being got at, even conned. This is why we often use the word 'rhetoric' pejoratively, as in 'that's just rhetoric', meaning that the words sound good but lack substance. For a poet like **Alexander Pope**, who worked with the grain of this classical rhetorical tradition, devices must appear to arise in the argument simply to clarify what we can readily recognize as true. He puts it neatly in these lines from 'An Essay on Criticism':

> True wit is Nature to advantage dressed,
> What oft was thought, but ne'er so well expressed;
> Something whose truth convinced at sight we find,
> That gives us back the image of our mind.

But the eighteenth-century *neo-classical* tradition Pope represents sees poetry as at heart rhetorical: that is, that it consciously deploys the armoury of *metre*, *rhythm*, *syntax*, *metaphor*, *image*, *tone*, word-choice and word-sound in the interests of clarity and argument. And this implies an actual or potential consensus of poet and audience. The audience will recognize in the poet's words what they in fact already knew to be true. Moreover it is the character of poetry, through its own techniques, to be especially impressive and memorable. At the end of his 'Anatomy of the World' **John Donne** evokes God speaking to the prophet Moses:

>   ...   He spake
> To Moses, to deliver unto all,
> That song: because he knew they would let fall
> The Law, the prophets, and the history,
> But keep the song still in their memory.

The 'song', the poem, will stay fixed in the people's mind after commandments or prophets' words are forgotten.

So we have a long-rumbling argument. Because it is a highly gestured, emphatic utterance, and often makes greater use of metaphoric and descriptive imagery, along with other verbal devices, poetry can have great rhetorical power. But these very qualities can also be seen to give it a dangerous potency. After all, we can be persuaded towards bad as well as good. Thus poetry has often been seen as the distracting enemy of truth and right reason whether religious or scientific. The possibility that poetry might signify nothing has been a horror to many for centuries – as well as a source of joy to others.

IMAGE AND MEANING

The clash of substance and fantasy in poetry is brilliantly evoked in **John Ashbery**'s (1927–) 'And *Ut Pictura Poesis* Is Her Name'. After a few lines the poem needs, says the poet, 'a few important words ... low-keyed, / Dull-sounding ones.' But no sooner are these provided by the comically mundane information that 'She approached me / About buying her desk' than the poem is in a street of 'bananas and the clangor of Japanese instruments'. The poet ponders this 'seesaw':

>   Something
> Ought to be written about how this affects
> You when you write poetry:
> The extreme austerity of an almost empty mind
> Colliding with the lush, Rousseau-like foliage of its desire to
>   communicate
> Something between breaths, if only for the sake
> Of others and of their desire to understand you and desert you
> For other centers of communication, so that understanding
> May begin, and in doing so be undone.

Ashbery's idea is that disciplined, highly focussed verbal communication will be so 'austere' it could come only from 'an almost empty mind'. But 'the desire to communicate' is never so austere, but resembles – and inevitably a metaphor now appears – a luxuriant, multi-coloured forest full of monkeys and tigers of the kind painted by the French artist 'Douanier' Rousseau. For this poet, imagery seems to be the condition of thought.

IMAGINATION

Now, if we put together *image* and the related noun *imagination*, we bring to mind a powerful concept of the nature of poetry. Our 'rhetorical' poets formulating their careful arguments, or the 'makars' of the traditional ballad who did not seek for novelty and surprise in their metaphors but just to tell a communal story, might be seen as constructors of poetry. There might even be said to be something *mechanical* about their method.

By contrast, imagination suggests a quite different definition of poetry. Poetry is seen not as conscious process but as natural surprise, an utterance arising from the *non-rational* processes of the mind. Unbidden associations and illuminations spring into the poet's mind and thence to the page. 'Imagination' has often been regarded suspiciously. The words 'vain' and 'false' have often been attached to it. For **Samuel Johnson** (1709–84) 'imagination' was a negative state of mind he associated with depression: 'Imagination never takes such firm possession of the mind as when it is found empty and unoccupied.'

But for the **Romantic** poets following hard upon Johnson at the end of the eighteenth century, the mental processes lying in the 'empty and unoccupied mind' could be welcomed as absorbing and in some profound sense truer than the processes of the conscious mind. In **William Wordsworth**'s (1770–1850) early poem 'Expostulation and Reply', the poet's 'good friend Matthew' accuses him of idleness. The poet replies:

> 'The – eye it cannot choose but see;
> We cannot bid the ear be still;
> Our bodies feel, wher'er they be,
> Against or with our will.

'Nor less I deem that there are Powers
Which of themselves our minds impress;
That we can feed this mind of ours
In a wise passiveness.'

In a companion poem, 'The Tables Turned', he exhorts his friend to quit his books:

Enough of Science and of Art;
Close up those barren leaves;
Come forth, and bring with you a heart
That watches and receives.

Our minds can absorb knowledge passively, intuitively. What better justification could you have for closing these particular 'barren leaves' right now?

These poems of Wordsworth's are of course themselves rhetorical in that they seek to persuade 'Matthew', and by extension us as readers. But we see here a commitment to a different kind of knowing, a different kind of mental action. It is imagination in the sense of the mind's capacity to form concepts *beyond* those devised from external objects. The articulation of such apprehensions becomes one of the major tasks of Wordsworth's poetry, and again and again he recalls moments, often from childhood, and usually alone in the natural world, when such strange, profound sensations overtake him.

Here in 'There Was a Boy' he is beside 'the glimmering lake' at nightfall, imitating owls and listening to the echoes of his voice come back over the water from the surrounding hills:

Then, sometimes, in that silence, while he hung
Listening, a gentle shock of mild surprise
Has carried far into his heart the voice
Of mountain torrents; or the visible scene
Would enter unawares into his mind
With all its solemn imagery, its rocks,
Its woods, and that uncertain heaven received
Into the bosom of the steady lake.

What is 'carried far into his heart', enters 'unawares into his mind', is impossible to paraphrase. Wordsworth is not trying to convey a shared idea that his rhetorical skill will make us recognize, but to *express* something from his own deep experience that is at or beyond the limits of words. It is not an idea that 'oft was thought', but an experience unique to him. Poetry is the means by which he can seek to recall and understand such 'mild surprise', and aim to put it before others. Poetry can do this because of its elasticity as a medium and because its sensuousness, especially the movements of its rhythms, can embody the charged excitement he remembers. Poetry now seems to be coming as though from elsewhere: the poet is *inspired*.

INSPIRATION

'Inspiration' is one of the great clichés associated with poetry. Until confronted by the cold realities of the creative writing class, would-be poets loiter by guttering candles impatient for the moon-flash of the poem's arrival. As all my emphasis so far upon craft and artifice suggests, poetry is not always and only the product of such miraculous moments. Nonetheless, and however easy it is to satirize the poet's earnest expectancy, the experience of inspiration, and what it means for poetry, needs to be taken very seriously.

Inspiration in this sense is a metaphorical term. Its ground is the literal act of breathing, the taking in of air – respiration. **Walt Whitman** (1819–92), at the outset of 'Song of Myself', gathers his poetic forces in part through the act of breathing. Indeed his poetic being is intensely physical as it carries through to the voice:

> My respiration and inspiration, the beating of my heart, the
> passing of blood and air through my lungs,
> The sniff of green leaves and dry leaves, and of the shore and
> dark-color'd sea rocks, and the hay in the barn,
> The sound of the belch'd words of my voice loos'd to the eddies
> of the wind[.]

For Whitman poetic inspiration is part of the untutored natural world. The lusty character of his poetic sensibility is marked here in the literalness of phrases like 'the passing of blood and air through

my lungs', and the wonderful, arrestingly unpoetic 'belch'd words'.
Inspiration for Whitman is no delicate zephyr.

This natural association has always been present in the
metaphorical sense of inspiration. First it described religious experience where it is 'a breathing or infusion into the mind or soul'
(*Oxford English Dictionary*): the great translator **William Tyndale**
(?1494–1536) wrote of 'scripture geven by inspiracioun of god'.
Later it acquired the more general meaning: 'the suggestion, awakening, or creation of some feeling or impulse especially of an
exalted kind' (*OED*) – even if, occasionally, it might come courtesy
of less exalted means. Thomas Hogg writes in his life of Shelley of
'the soft inspiration of strong sound ale'.

There is always something unbidden about the coming of such
'suggestion ... awakening ... feeling ... impulse'. **Ralph Waldo
Emerson** (1803–82) in his essay 'The Poet' (1844) writes that 'the
poet knows that he speaks adequately ... only when he speaks
somewhat wildly'. He describes the process like this:

> As the traveller who has lost his way, throws his reins on his horse's
> neck, and trusts to the animal to find his road, so must we do with the
> divine animal who carries us through the world. For if in any manner
> we can stimulate this instinct, new passages are opened for us into
> nature, the mind flows into and through things hardest and highest,
> and the metamorphosis is possible.

He goes on to say that the quest to 'stimulate this instinct' is
'why bards love wine, mead, narcotics, coffee, tea, opium, the
fumes of sandal-wood and tobacco'. All these too have their place
among the familiar accessories of poetic inspiration. Emerson,
however, regards them as '*quasi*-mechanical': 'that which we
owe to narcotics is not an inspiration, but some counterfeit
excitement and fury'. Emerson's retreat from this implication of
his ideas is hurried and anxious as he insists that 'sublime vision
comes to the pure and simple soul in a clean and chaste body'.
But he is grand and emphatic on the character of the mental
state of true poetry. It is a condition, he writes, that helps the
poet 'escape the custody of that body in which he is pent up,
and of that jail-yard of individual relations in which he is
enclosed'.

Emerson describes well the Romantics' impulse towards transcendent freedom, a release from the dull drudgery of daily life. Images like these of imprisonment and release abound in his work, as they do in that of his contemporary **Henry David Thoreau** (1817–62). We meet them too in the images contrasting the earthbound and free flight that recur in the poems of **John Keats** (1795–1821) envying his nightingale, and **Percy Bysshe Shelley** (1792–1822) saluting the skylark: 'Higher still and higher / From the earth thou springest / Like a cloud of fire.'

A poem that, as Emerson puts it, 'speaks somewhat wildly' is **Samuel Taylor Coleridge**'s (1772–1834) 'Kubla Khan'. It begins:

> In Xanadu did Kubla Khan
> A stately pleasure-dome decree:
> Where Alph, the sacred river, ran
> Through caverns measureless to man
>   Down to a sunless sea.

Nothing in the poem tells us who this Kubla Khan is and it has no narrative movement. But it is instantly effective with its exotic names and setting – Abyssinia, Mount Abora and Paradise come later – and the rhythmic excitement which tumbles the lines headlong towards that strongly stressed 'Down', and the sudden flat expanse of the 'sunless sea'. We are engaged but we probably don't know why.

As the poem continues, other equally mysterious but fascinating images follow:

> But oh! that deep romantic chasm which slanted
> Down the green hill athwart a cedarn cover!
> A savage place! as holy and enchanted
> As e'er beneath a waning moon was haunted
> By woman wailing for her demon-lover!

The 'woman wailing for her demon-lover' 'beneath a waning moon' is a tremendously compelling image, but is actually only there as part of a simile to describe the 'savage place'. In effect the image of comparison does not serve to elaborate the 'deep romantic chasm'. Instead it disrupts the expectations of metaphor by taking over the interest of the lines. Then no sooner has it usurped the description and caught our fascination than it vanishes. We know no more of the woman or her demon-lover. We do, however, return to the chasm in a series of images of pell-mell violence with the hectic eruption of the river from a 'mighty fountain' before, 'meandering with a mazy motion', it sinks 'in tumult to a lifeless ocean'. Strangely, ominously, at this moment Kubla can hear 'Ancestral voices prophesying war!'

So far the poem seems to be alternating between these images and those of peace and repose in the picture of the 'stately pleasure-dome' set in its grounds with gardens, an 'incense-bearing tree' and walls and towers. This recurs immediately after the voices as 'the shadow of the dome' reappears 'floating midway on the waves'. Yet more strangely, it is now 'A sunny pleasure-dome with caves of ice!' All these images suggest a wondrous culture and civilization, but the alternating images are of the 'savage place', of the turbulent river out of whose roar can be heard those prophetic voices.

Now it changes once more with another memorable image:

> A damsel with a dulcimer
> In a vision once I saw:
> It was an Abyssinian maid,
> And on her dulcimer she play'd,
> Singing of Mount Abora.

With the *dulcet* **connotations** of the instrument the lines speak of a wondrous softness and ease. Now too we have another new presence: 'I'. Moved by this vision, the newly revealed poet rises to a transport of creative enthusiasm:

> Could I revive within me
> Her symphony and song,
> To such a deep delight 'twould win me,
> That with music loud and long,
> I would build that dome in air,
> That sunny dome! those caves of ice!
> And all who heard should see them there,
> And all should cry, Beware! Beware!
> His flashing eyes, his floating hair!

> Weave a circle round him thrice,
> And close your eyes with holy dread,
> For he on honey-dew hath fed,
> And drunk the milk of Paradise.

The exclamations which are such a feature of the poem become more numerous and intense in these lines and they pour breathlessly onward in one sustained rapture.

Yet what does it mean? The poet says 'could I *revive* within me ... I would build ...' Perhaps he is writing of his anguished eagerness to recover his own creative strength, or, with less emphasis on 'revive', that he is inspired by this vision to emulate the wondrous creative labour of Kubla Khan and the Abyssinian maid. As his fever mounts, the point of view changes yet again so that the poet is seen by 'all' as he desires to be seen, an exciting, even dangerously daemonic figure. The pace of the lines includes remarkable rhythmic changes as between

> His **flash**ing eyes, his **float**ing hair!
> **Weave** a circle round him thrice,
> And **close** your **eyes** with **holy dread**

where the first line vaults across just two strong stresses, to the second which begins with the strong stress on 'Weave', to the regular four beats of the third. This exhilarates the reader, catching us up into this strange thrill of the exotic and the possession of glamorous fantasy.

Famously, this is where the poem ends. It is subtitled 'a Vision in a Dream. A Fragment', for Coleridge said it was incomplete. Nor could he complete it, for he claimed that the poem was indeed the recollection of a dream, and that as he was 'instantly and eagerly' writing it down, he was 'unfortunately called out by a person on business from Porlock'. By the time he returned to his vision and the paper, he found that 'the rest had passed away like the images on the surface of a stream into which a stone has been cast, but alas! without the after restoration of the latter!' This is a great story of the excitement and the fragility of the imagination. It suggests the utter mystery of the process, as Emerson has it, 'the mind flowing into and through things hardest and highest'.

The apparent mystery, however, has not prevented scholars from trying to uncover its more material sources. Coleridge wrote the poem after lodging overnight at a farmhouse at Culbone near the Somerset–Devon coast in south-west England. His biographer Richard Holmes convincingly describes how the steep, enclosed combe leading down from Culbone to the sea might have encouraged Coleridge's imaginings of the 'sacred river'. More securely, Coleridge himself tells us that he was reading *Purchas' Pilgrimage*, a travel history which includes this passage:

> In Xanada did Cublai Can build a stately Pallace, encompassing sixteene miles of plaine ground with a wall, wherein are fertile Meddows, pleasant Springs, delightful Streames, and all sorts of beasts of chase and game, and in the midst thereof a sumptuous house of pleasure, which may be removed from place to place.

Put this together with the note on the earliest known manuscript of the poem in which Coleridge describes it as 'a sort of Reverie brought on by two grains of opium taken to check a dysentery', and a picture emerges of a poem resulting from some observations of the surrounding countryside, a bit of near-copying from a book, and the hallucination caused by medicine he had taken for an attack of diarrhoea. So much, we might think, for the inscrutable mysteries of the imagination.

But surely we should be wrong to think so. Even if we allow all these mundane contributions, the incomprehensible brilliance of the poem remains. And it is a brilliance that exceeds the subtlest interpretations, always finally elusive. The response of **William Hazlitt** in 1816 catches its quality when he writes: 'It is not a poem but a musical composition. We could repeat the opening lines to ourselves not the less often for not knowing the meaning of them.' Other readers have worried about not knowing the meaning. Coleridge's friend, the essayist **Charles Lamb**, wrote to Wordsworth on 26 April 1816 about hearing Coleridge read 'Kubla Khan' aloud

> so enchantingly that it irradiates & brings heaven and Elysian bowers into my parlour while he sings or says it, but there is an observation Never tell thy dreams, and I am almost afraid that Kubla Khan is an

owl that wont bear day light, I fear lest it should be discovered by the lantern & clear reducting [reducing] to letters, no better than nonsense or no sense.

Lamb seems undecided whether to place the greater weight in his mind upon the irradiation the poem brings him, or on his anxiety that the cold light of examination will show it to be meaningless. He has not been alone. The poem retains its great ambivalence. Whatever it is 'about', and whatever the true nature of its inspiration, the poem – and the legend that surrounds it – is a great model of one kind of poetic imagination. And a great poem.

POETRY AND LIBERATION

In Chapter 5 on 'Free verse' we saw how **William Blake** saw formal verse as a symptom of imprisonment: 'Poetry Fetter'd Fetters the Human Race.' His fellow *Romantics*, **Wordsworth**, **Coleridge**, **Keats**, **Shelley** and, in America, **Emerson**, **Thoreau** and **Whitman**, are all absorbed by the aspiration towards freedom. Their view of the range of the mind, conscious and unconscious, what they call 'imagination', is one part of this. For most, at least in their younger days, this striving for poetic freedom was part of a longing for political freedom, for the unfettering of the human race from poverty and oppressive government. The first generation of Romantic poets – notably Blake, Wordsworth, Coleridge – spent their youth amid the ferment of change that brought about the American Revolution of 1776 and the French Revolution of 1789. The second generation of Byron, Shelley and Keats were also politically radical. In a very large degree Romantic poetry is a poetry of liberation. Sometimes, as in poems such as Shelley's 'England in 1819' or his 'The Mask of Anarchy', protest and the hope of freedom are *rhetorically* expressed. But for poetry the idea of liberation must amount to more than serving as propaganda.

The Romantic account of poetry, which sees it as the inspired space of imagination, dramatically marks it off from other kinds of language. Poetry does not only look and sound different because of its peculiar formal characteristics, but its relation to language is quite other. 'Other', that is, to the way language normally works, which is to convey consecutive narratives or arguments in sentences governed by the efficient rules of **syntax**. Necessarily the speaker and hearer or writer and reader share the norms of communication and would prefer not to mystify one another. Nonetheless the Romantic ideas and processes of imagination and inspiration present a radical challenge to poetry as **rhetoric**. As Charles Lamb's unease demonstrates, with a poem like 'Kubla Khan' the nature of the communication is different. It does not fall readily within the dominant circle of making sense. It may be , as Lamb fears, 'no sense'.

It is with these ideas in mind that I now want to move to consider poetry in the context and practice of one of the most profound liberation movements of our own day: the liberation of women.

FEMINIST POETICS

In the last forty or so years, the profile of poetry written by women, both past and present, has risen considerably, and as a result at least three topics have presented themselves. The first might be called historical and has to do with the revaluation of women poets of the past whose work has been passed over by literary history. Second there is the question of subject-matter: have women poets written about different topics by virtue of their interests and experience as women? Third there is a more complex and controversial discussion about the language and style of poetry by women: is it fundamentally distinct from that of men at the level of language and form? More controversially, *should* women poets strive to write in ways that are clearly female and owe less and less to the predominantly male poetic tradition? If poetry is itself 'other' with respect to conventional language use, is women's poetry 'other' in a yet further respect? These are the issues I want to discuss in this section.

In an essay called 'The Woman Poet: Her Dilemma', the contemporary Irish poet **Eavan Boland** (1944–) writes that the beginning woman poet very quickly becomes conscious of the 'silences' that have preceded her and which still surround her: 'Women are a minority within the expressive poetic tradition. Much of their experience lacks even the most rudimentary poetic precedent.'

These 'silences' are those of all the women who might have written but for the assumptions of masculine dominance. They will become an indefinable part of her purpose as a poet. Boland's argument is that the overwhelming masculine presence in the poetic tradition inhibits and excludes the woman poet. 'Poetry' is the 'One' and she is the 'Other'. As **Elizabeth Barrett Browning** (1806–61) wrote, 'I look everywhere for grandmothers and see none.'

Many poets and scholars in recent years have sought to discover, or re-discover, these 'grandmothers'. Aside from the anthologies of women's poetry, the newer historical anthologies of poetry in English published for the academic market now feature work entirely absent from their predecessors of even twenty years ago.

Two examples are Roger Lonsdale's *Eighteenth-Century Women Poets* (1990) and *British Literature 1640–1789* (1996) edited by Robert DeMaria Jr. One of DeMaria's new inclusions is this poem by **Anna Laetitia Barbauld** (1743–1825). It is called 'Washing Day' and begins, in *neo-classical* manner, with the *Muses*:

> The Muses are turned gossips; they have lost
> Their buskined step, and clear high-sounding phrase,
> Language of gods. Come, then, domestic Muse,
> In slip-shod measure loosely prattling on
> Of farm and orchard, pleasant curds and cream,
> Or drowning flies, or shoe lost in the mire
> By little whimpering boy, with rueful face;
> Come, Muse, and sing the dreaded *Washing Day*.

Like many other poems in this collection, Barbauld's poem is not to be found in many previous anthologies of eighteenth-century poetry precisely because such a subject would not have been deemed fit for poetry, even when treated in the wry and ironic manner in which Barbauld approaches it. Here we arrive at the second topic concerning women's poetry: subject-matter. Where is the justifying precedent for writing about women's 'actual experience', subjects that have always been a central part of women's lives such as domestic tasks and childcare, what Boland calls the 'snips and threads of an ordinary day'? It might be argued that 'the dreaded Washing Day' is a slight matter, and that attending to it only reinforces women's domestic role. But that would not be a reason to exclude a poem like **Mary Jones**'s (?–1778) 'After the

Small Pox', which is a mordant criticism of the masculine perceptions that oblige women to rely on a pretty face. **Pope**'s 'The Rape
of the Lock' (1714) is fascinated with female beauty, and smallpox is
a real if shadowy presence in his poem. But it is poets like Jones and
Lady Mary Wortley Montagu (1689–1762), in her poem 'The
Small-Pox', who can tell us of the woman's pain at hiding 'this lost
inglorious face':

> How false and trifling is that art you boast;
> No art can give me back my beauty lost!
> In tears, surrounded by my friends I lay,
> Mask'ed o'er, and trembling at the light of day.

The work of recovery of barely noticed women poets is changing
the perception of past poetry and helping fill the 'silences' that
Boland laments.

This re-examination of what poetry might be written about has
also drawn twentieth-century women poets to reconsider some of
the celebrated biblical and classical myths featuring women, such as
the stories of Eve, Medusa, Persephone and Eurydice that feature
in the tradition. **H. D. (Hilda Doolittle)** (1886–1961) rewrote the
story of Helen of Troy in her long poem 'Helen in Egypt', and more
recently **Carol Ann Duffy** (1955–) has fashioned a satirical re-
telling of mythical stories from the woman's point of view in her
volume *The World's Wife*. In her re-telling, the mythical Cretan
aviator, whose pride in his artificial wings led him to fly too close to
the sun, is about to take off from a hillock. Watching him, the long-
suffering 'Mrs Icarus' ruefully concludes that her husband is yet
another 'total, utter, absolute, Grade A pillock'.

So these are two of the issues regarding women and poetry: the
historical matter concerning equality of attention to poetry by
women, both past and present, and the revision and extension of
subject-matter. Both bear upon the confidence and opportunity
available to the beginning poet. But besides these is that third topic:
is poetry by women *marked* as such by its different use of language
and form? How is it, or how might it be, different?

One response to this question has come in the poetry and prose
of the American poet **Adrienne Rich** (1929–). In an influential
essay, 'When We Dead Awaken: Writing as Re-Vision' (1971), Rich

makes her argument by describing the evolution of her own poetic career. She characterizes her early work as formal and distantly impersonal in that it uses metre, rhyme and stanza-form and strives for an 'objective, observant tone'. She quotes her poem 'Aunt Jennifer's Tigers' (1951) in which a woman is embroidering a screen. This is the middle stanza:

> Aunt Jennifer's fingers fluttering through her wool
> Find even the ivory needle hard to pull.
> The massive weight of Uncle's wedding band
> Sits heavily upon Aunt Jennifer's hand.

The voice of these early poems was either third-person or cast in a male persona. Formal style, she writes, acted 'like asbestos gloves, it allowed me to handle materials I couldn't pick up bare-handed'.

Gradually the character of her work changed. Her new work

> was jotted in fragments during children's naps, brief hours in a library, or at 3:00am after rising with a wakeful child. I despaired of doing any continuous work at this time. Yet I began to feel that my fragments and scraps had a common consciousness and a common theme, one which I would have been very unwilling to put on paper at an earlier time because I had been taught that poetry should be 'universal', which meant, of course, nonfemale. Until then I had tried very much *not* to identify myself as a female poet.

These 'fragments and scraps' develop into a freer, less formal style, and she eventually finds the confidence to use the pronoun 'I'. She cites her poem 'Planetarium', about 'Caroline Herschel, 1750–1848, astronomer, sister of William; and others', as one where at last 'the woman in the poem and the woman writing the poem become the same person':

> I am a galactic cloud so deep   so invo-
> luted that a light wave could take 15
> years to travel through me   And has
> taken   I am an instrument in the shape
> of a woman trying to translate pulsations

> into images   for the relief of the body
> and the reconstruction of the mind.

Unlike the voice of 'Aunt Jennifer's Tigers', the truly female poet will turn away from the inherited forms of the male-dominated tradition and be marked by a freer verse style in the manner of the 'field composition' we saw in Chapter 5, 'Free verse', and by confidently assuming the pronoun 'I'. For Rich, feminist poetics is liberationist and therefore Romantic. It challenges previous forms, is iconoclastic towards the tradition – indeed, rejects the viability and desirability of a single tradition – and it values subjective experience and its expression.

But none of this is uncontroversial. Eavan Boland, whilst admiring Rich, is sceptical towards what she senses as 'separatist thinking', and the identification of the poet, male or female, as a Romantic outsider. She writes in her essay, 'In Search of a Language':

> The poet's vocation – or, more precisely, the historical construction put upon it – is one of the single most problematic areas for any woman who comes to the craft. Not only has it been defined by a tradition which could never foresee her, but it is constructed by men about men, in ways which are poignant, compelling and exclusive.

She argues against that exclusion, that 'women have a birthright in poetry', and must have 'the fullest possible dialogue' with the poetry of the past, most of which is male. She warns too against what she calls 'the Romantic heresy'. The turn towards the subjective in poetry since Wordsworth has often declined, she believes, into 'self-consciousness and self-invention'. She is worried that women poets might replicate Romantic ideas which have asserted the realm of imagination as *so* separate from, *so* different in kind, as to be incompatible with the process and experience of daily life. Her own poetry is not averse to the matter-of-fact in either setting or diction, as the opening to this poem 'Contingencies' suggests:

> Waiting in the kitchen for power cuts,
> on this wet night, sorting candles,

feeling the tallow,
brings back to me
the way women spoke in my childhood –

with a sweet mildness in front of company,
or with a private hunger in whispered kisses,
or with the crisis-bright words
which meant

you and you alone were their object –

Boland's style is one which aims to maintain connections with ideas and language as they are current outside the spaces of poetry. If for Rich the female poet must rise up from the kitchen table, for Boland it is a place to work. Her own subjectivity is part of her poetry as it is of her essays, but she is also a poet of argument not at all drawn towards 'no sense'.

Yet many writers do envision a female writing whose difference from male practice is deeply inscribed in how the language is used. The contemporary French writer **Hélène Cixous** writes: 'Nearly the entire history of writing is confounded with the history of reason of which it is at once the effect, the support, and one of the privileged alibis.'

For Cixous, then, presumably real female writing will exist outside what I called earlier 'the dominant circle of making sense'. Thus it might seem possible to recognize an alliance of 'others'. Women stand outside these masculine norms, and so does poetic imagination which, from the point of view of reason, makes 'no sense'. In some poets – including men – she glimpses opposition to this dominance. In a rhetoric as strikingly urgent in its images of imprisonment and liberation as that of Emerson, she writes:

There have been poets who would go to any lengths to slip something by at odds with the tradition – men capable of loving love and hence capable of loving others and of wanting them, of imagining the woman who would hold out against oppression and constitute herself as a superb, equal and hence 'impossible' subject, untenable in a real social framework.

But can this vision be seen in specific poems? Does 'Kubla Khan' 'slip something by at odds with the tradition'? Does the poetic 'other' have a further female dimension?

Some critics have seen the composure of 'masculine' techniques such as the *iambic pentameter*, and indeed grammar itself, as targets for subversion by female poets. Biographically it is easy to represent the reclusive **Emily Dickinson** (1830–86) as so at odds with her circumstances as to be 'an "impossible" subject' unable to be at ease in the 'real social framework' of a small New England town. But her deeper, transgressive difference at the level of language and style is often now seen in her rough broken surfaces and distorted hymn stanzas. Here are the last two stanzas of a poem which begins 'Why – do they shut Me out of Heaven?' (248):

> Wouldn't the Angels try me –
> Just – once – more –
> Just – see – if I troubled them –
> But don't – shut the door!
>
> Oh, if I – were the Gentleman
> In the 'White Robe' –
> And they – were the little Hand – that knocked –
> Could – I – forbid?

Dickinson violates conventional syntax, ignores expectations to be more transparent, and is simply irregular even in her punctuation. Perhaps here we can see the kind of rejection of reasonable 'male writing' that Cixous and others seek.

Women's poetry, and ideas about it, is a rapidly changing and disparate field. To recur to my three topics, poets, scholars and readers – mainly women – continue to revalue the historical legacy. The resulting discoveries, and the writings of our contemporaries, are expanding ideas of what subjects poetry can take on. The third topic is more complex. All theories about poetry have prescriptions to claim that poems should be written this way and not that. Because in these times writing by women is urgently connected to social and political imperatives, the searches for definition and prescription are understandably forceful. To be free of the inherited forms of the male-dominated tradition was obviously going to be a desire of women writing. I think that its radical edge is bound up with the alternatives to the consecutive norms of rational process embodied in the notion of the imagination and the notion of inspiration.

But if we look at women poets since the seventeenth century or, closer to our own time, at poets considered in this book like **Millay** and **Moore**, highly formal poets writing in the current of the tradition, can we say with confidence that in the styles and subjects they adopted they were wrong from the start? Is it true, as Cixous has claimed, that 'there has not yet been any writing that inscribes femininity', and how secure a definition of 'femininity' can there be? How *different* female use of language really is still needs, and is receiving, subtler analysis. A great deal of life in the writing and reading of contemporary poetry is energized by these questions, and out of it will come new poems to surprise us beyond our current imaginings.

Summary
In this final chapter we have considered:

- the working of simple metaphor and image in speech and in poetry and some more complex instances;
- the historical debates around the relation of words, and metaphor in particular, to the representation of reality;
- the purpose and character of rhetoric in relation to poetry;
- the concepts of imagination and inspiration and the impact they have had on poetry and ideas about poetry and the poet;
- how these concepts are allied to liberationist ideas;
- the ways in which poetry by women might be different from that of men and some different ideas about feminist poetic theory and practice.

FURTHER READING

Boland, E. (1995) *Object Lessons: The Life of the Woman and the Poet in Our Time*, Manchester: Carcanet Press.

Cixous, Hélène ([1976] 1981) 'The Laugh of the Medusa', tr. Keith Cohen and Paula Cohen, in Elaine Marks and Isabelle de Courtivon (eds) *New French Feminisms*, Brighton: Harvester Press. Reprinted in D. Walder (ed.) (1990) *Literature in the Modern World: Critical Essays and Documents*, Oxford: The Open University/Oxford University Press.

Kermode, Frank (1957) *The Romantic Image*, London: Routledge.

Lawler, Justus George (1994) *Celestial Pantomime, Poetic Structures of Transcendence*, New York: Continuum.

Rich, Adrienne (1971) 'When We Dead Awaken: Writing as Re-Vision' in (1980) *On Lies, Secrets, and Silence: Selected Prose 1966–1978*, London: Virago.

Stevens, Wallace (1960) 'Imagination as Value' in *The Necessary Angel*, London: Faber.

Vickers, Brian (1988) *In Defence of Rhetoric*, Oxford: Clarendon Press.

Yeats, W. B. (1900) 'The Symbolism of Poetry' in N. Jeffares (ed.) (1964) *Selected Criticism*, London: Macmillan.

CONCLUSION

I began the last chapter with a simple metaphor, 'daft as a brush'. This odd, yet ordinary phrase is a minute instance of the ceaseless creativity in human language, just one of its 'endless associations'. Such images are at the heart of our language-use. They enhance our rhetorical powers and suffuse that strangely occurring mental state we call the imagination, something made as much by the *unconscious* as the conscious mind.

Compared with what Ashbery calls 'other centers of communication', the poem is a free space for rhetoric and imagination. It might be tightly argumentative or loosely associative. It can combine so many aspects of experience, knowledge and ways of speaking. Some philosophy, a child's bedtime memories, geology and evolution, a bit of slang or verbal playfulness, can coexist here as in no other form. Because its sounds are so important – what we inaccurately but inevitably call the 'music' of its rhythms and metres – so much part of the power of this mix is, as the critic Justus George Lawler says, 'nonconceptual'. Poetry works with ideas, but also within the subjective mystery of our consciousness, with the *qualia* of the mind when it is operating other than in its consecutive series. Its kinds are sometimes playful, and sometimes deal with the words that engage the most serious topics we know. In all its presences, the fascination, pleasure and impact of poetry inhere in the patterns of the language where form and content are truly indivisible:

> O body swayed to music, O brightening glance,
> How can we know the dancer from the dance?

GLOSSARY

This glossary provides short definitions to all the specialist terms used in this book. These terms appear in the text set in *bold as well as italic*. The definitions here, however, are necessarily short. Most are discussed at some length in the appropriate chapters and these are indicated. Much fuller accounts, with many examples and histories, can be found in the exhaustive and excellent 1,383 pages of the *Princeton Encyclopedia*:

Preminger, A. and Brogan, T. V. F. (eds) (1993) *The New Princeton Encyclopedia of Poetry and Poetics*, Princeton: Princeton University Press.

Accent In poetry, emphasis upon one particular syllable in speech, e.g. **mass**-ive. See also *stress*, the term most often used in this book for accent, and *beat*.

Accentual-syllabic A measured line of verse which forms a pattern of *accented* and non-accented syllables. Also called *stress-syllable* lines. (See Chapter 4.)

Alexandrine A twelve-syllable line of verse, usually *accentual-syllabic*, with six *stresses* and a *caesura*. The standard metre in French poetry, sometimes adapted into English. (See Chapter 4.)

Allegory　A literary work in which characters, settings and actions are all devised to represent, or *symbolize*, abstractions such as 'Good', 'Evil', 'Wisdom', etc. What is being represented does not emerge from the *image* or *symbol*, but is set in advance with the *images* chosen to fit.

Alliteration　The repetition of *consonants* close enough together to be noticed by the ear. Usually appears on stressed syllables. *Alliterative* verse, mostly in Old and Middle English, normally forms the line by including three stressed syllables beginning with the same consonant. Occasionally alliteration is also referred to as *head-rhyme* when the repetition is on the first syllables of words.

Anadiplosis　A *rhetorical figure* in which the last word of one phrase or sentence is repeated as the first word of the next. E.g. 'If you say he's mad, mad he is.'

Anapest　A *metrical foot* of three *syllables* in which the first two are unstressed and the third stressed: ˇ ˇ \. The term comes from classical *prosody* where it referred to a foot of two short followed by one long syllable. (See Chapter 4.)

Anaphora　A *rhetorical figure* which repeats the same word or word at the beginning of a series of phrases, sentences or lines: e.g. 'Mad she thought him. Mad he seemed to be. Mad he surely was.'

Assonance　The reiteration of the same *vowel* sounds close enough together to be noticed by the ear. It is more aural than alliteration because it is not so visible on the page: e.g. 'the proud cow by the plough'.

Ballad　A narrative poem in simple form that is derived from the *oral* rather than literate tradition. The origins of the 'traditional', or 'popular', or 'folk' ballad are therefore usually unknown, and it will have been altered through its transmission as it has been sung or recited down the years. When more modern poets have wanted to recall simplicity they have re-invented the form, most famously Wordsworth and Coleridge in their *Lyrical Ballads* (1798). (See Chapter 2.)

Ballad metre The *metre* most commonly used for the ballad. Normally it alternates lines of four and three *stresses*. Also known sometimes as *common metre*.

Bard A traditional word for poet. It derives from the Celtic cultures of Wales and Ireland whose tribes had bards who recorded and recited verses, often in elaborate forms, both to entertain and to keep the history of their people. In modern use it is often at best mock-serious, as when Shakespeare is referred to simply as 'the Bard'.

Beat The effect in all sorts of poetry in English, whether it has *measure* or is '*free verse*', relies upon the effective placing of accent or stress. A poem therefore will have *beats* (i.e. stresses, accents) but also an overall *beat*, or *rhythm*. It is helpful to think of beat in poetry in a similar way as we do in music, though the relationship between poetry and music is eventually complex. (See also *metre* and Chapters 4 and 5.)

Blank verse Lines of poetry that do not use *rhyme*. In English it began in the sixteenth century and became the staple form for the dramatic verse of Shakespeare and his contemporaries because it enabled greater continuity, and sounded closer to natural speech than rhymed verse.

Burden Occasionally referred to as *burthen*, it has several meanings. The most prominent is that of *refrain*, or repeated chorus, usually of a song or song-like poem. It can also refer to the main theme or sentiment of a poem (or other utterance). Closer to music, a burden is the underlying aural effect, like the bass-line. More rarely, especially in biblical contexts, it can refer to a deliberate raising of the voice to recite.

Cadence The word derives from Latin and Italian words for 'fall', hence we sometimes also speak of the *fall* of a poem. This especially refers to the *rhythm* as a poem, sentence or line reaches its close. More generally cadence refers to the overall rhythmic movement of one or more lines with regard to the placing of *stress* and other aural features.

Caesura A term from Greek and Latin *prosody* denoting the break in a metrical verse line, usually at halfway. It is in effect a pause, but a distinct one that falls at virtually the same point in the series of formal lines.

Carol Originally a medieval dance-song, usually for ring-dancing. Commonly it had three-/four-*stress* lines, all rhyming the same (i.e. *monorhyme*), and a shorter line that rhymed with its *burden*. As this dance tradition died out with the religious Reformation of the sixteenth century, the term became transferred to the modern sense of popular Christmas songs.

Classics, classical 'The Classics' is generally taken to refer to the literatures of Ancient Greece and Rome. For western poets, especially during and since the *Renaissance*, this literature has been held to set the standard of quality and permanence. Many also wrote in Latin and/or Greek and transposed classical forms into their own languages. When enthusiasm for classical models was most pronounced – for instance in English poetry of the late seventeenth and early eighteenth centuries – poets have been dubbed *neo-classical*. The terminology of versification (*prosody*) has been derived from classical examples even though they do not strictly apply to poetry in English. (See Chapter 4.)

Closed couplet A couplet which matches sentence structure, i.e. closes with a full-stop or other strong punctuation. (See *couplet*, *open couplet* and Chapter 6.)

Common metre Four-line *stanzas* alternating four- and three-beat lines. Another name for *ballad metre*.

Conceit A particularly striking *metaphor*. Usually it is part of a larger pattern of *images*, sometimes continued through the whole poem – see *trope*. The more prominent and ingenious it is, the more marked and self-conscious its artificiality. Its English heyday was in the seventeenth century in the work of Donne and other poets.

Concrete poetry Visual poetry: strictly the words, or just letters, are offered as images which are as far as possible abstract, that is they are detached from their usual *semantic* function. Work of this kind, at the borders of poetry and visual art, was most prominent in the 1950s and 1960s. Less strictly, other poems such as the *emblem* poems of the seventeenth century might be said to have 'concrete' elements but their words are used semantically. All literate poetry, simply by its appearance in the space of the page, can be said to have a visual, if not 'concrete' dimension. (See also *sound poem* and Chapter 2.)

Connotative The secondary or additional meaning of a word besides its primary meaning, i.e. its *connotation* as opposed to its *denotation*.

Consonance Broadly this can mean the overall harmony or concord of sounds. More specifically it includes the correspondence of certain sounds, as for instance in *assonance* or *alliteration*.

Couplet A successive pair of lines that rhyme, notated: *aa bb*, etc. In *closed couplets* the second rhyme coincides with a full-stop or other firm punctuation. In *open couplets* the sense is free to run on through to the next line and the sentence might end mid-line. *Heroic couplets* are closed couplets in *iambic pentameter* that employ a *caesura*. Adapted from *classical* models, it was seen as a *metre* suited to large, public subjects. In English it was used most prominently in the later seventeenth and eighteenth centuries.

Dactyl A metrical foot of three *syllables* in which the first syllable is *stressed* and the second two are unstressed: \ ˘ ˘. The term comes from classical *prosody* where it was used to refer to long and short syllables. (See Chapter 4.)

Decasyllabic A line consisting of ten *syllables*.

Dialect Localized language-use that has vocabulary, pronunciation and idiom particular to itself.

Diction The choice or selection of words or phrases used. (See *lexis*.)

Dimeter   A poetic line of *two feet*, usually containing two *stresses*.

Dramatic monologue   A poem in which an identified character, or *persona*, is the sole speaker: that is, the voice in the poem is 'playing' a role as in drama.

Elegy   A poem occasioned by the death of someone. It will normally expand to become a more general meditation on mortality. Usually quite formal in style and manner.

Emblem   An 'emblem poem' is one that incorporates a visual image into the poem on the page. In English it was most practised in the seventeenth century.

End-rhyme   A rhyme that occurs at the end of a line

End-stopped   A line in which the end of the line and the punctuated end of the sentence or clause coincide, as in the *closed couplet*. The opposite of *enjambment*.

Enjambment   A term from French, meaning to step or stride across, it is the continuation, or run-on, of one line of poetry into the next; that is, the *syntax* flows through the line-break. The opposite of *end-stopped*.

Envelope rhyme   A rhyme, usually a *couplet*, that is enclosed within another pair of rhyming lines: e.g. *a bb a*.

Envoi   Originally from French medieval poetry: a short poem or section which acts as the conclusion, summary or 'send-off' of the whole work. It might repeat a *refrain* from earlier in the poem. (See also *sestina*.)

Epic   A long narrative poem, usually telling the tale of a single hero or group involved in a great historical event. The stories are likely to be legendary and involve divine as well as human characters. There is often a national or communal dimension to the epic in that it tells a story taken as vital to the collective history. Originally epics were *oral*, remembered and recounted over time. Epic was long held to be the ultimately significant poetic form. The major epics of western culture include Homer's *Iliad* and *Odyssey* (Greek), Virgil's *Aeneid* (Latin), Dante's *Divine Comedy* (Italian) and Milton's *Paradise Lost* (English). *Mock epic* is the imitation of epic features but with a comic dimension that pokes fun at the pretensions and pettiness of the characters involved.

Epigram Not always in poetic form, it is a piece of writing that compresses an observation or saying into a very short space. It is often *satiric* and must be witty. In verse it will normally take the form of a *couplet* or *quatrain*.

Eye-rhyme Two words that look similar enough to rhyme but when voiced do not, e.g. *cough* and *rough*. It must be remembered that pronunciation changes over time and that what seems to modern ears to be only an eye-rhyme might once have been an aural rhyme. It might also be used very deliberately as part of the visual aspect of poetry (see also *half-rhyme*).

Figure, figurative In the context of poetry, the word is used in the same sense as in the phrase 'figure of speech', that is an expression used to lend colour or force to speech or writing. Most often this will be in the form of an *image* or *metaphor*. In *rhetoric*, figures are of many particular, defined types such as *anaphora*, *anadiplosis* and hyperbole (deliberate over-statement or exaggeration for effect).

Foot A segment of a poetic line in *metre*. Normally this will be a combination of *stressed* and unstressed *syllables*. For example, an *iambic* foot consists of two syllables, the first unstressed and the second stressed: *ti tum*. Thus, an *iambic pentameter* consists of *five* such feet. (See also *trochee*, *anapest*, *dactyl* and Chapter 4.)

Fourteener A poetic line containing *fourteen syllables*; less often used to describe a poem of fourteen lines (see *sonnet*).

Free verse / vers libre Most often taken to refer to poetry that has no recurring *metrical* pattern to its lines and does not use *rhyme*. More strictly it might avoid all kinds of *recurrence* such as *stanza* pattern and repetition of words or phrases as in any kind of *refrain*. The French phrase *vers libre* was often used in the early twentieth century because of the influence of French poets. (See Chapter 5.)

Full rhyme   A rhyme in which the words involved have the last two or more sounds in accord and thus the only difference is earlier in the word, or line. The typical pattern is therefore CVC as in *knock/mock, insulate/regulate*. Sometimes known as strict rhyme. (See *rhyme, half-rhyme, head-rhyme, eye-rhyme* and Chapter 6.)

Genre, sub-genre   In literature genre refers to the classification into 'types' or 'forms' or 'kinds', e.g. poetry, novel, drama. Within these, distinct genres or 'sub-genres' might be defined, e.g. lyric, dramatic and epic poetry, and within lyric such forms as the *sonnet*. Some controversy arises as to whether a genre is defined by form or subject: e.g. is the *elegy* defined by its subject – commemoration of the dead – or by traditional formal characteristics? How tightly or prescriptively genre can be defined has been a long-standing argument in literary studies as theorists propose new criteria and different classifications.

Ghazal   A lyric poem in which a single rhyme predominates: *aa ba ca da ea*. Its origins are in Arabic, Persian and Turkish poetry. There has been considerable modern interest in imitating the form in English by poets including Adrienne Rich.

Haiku   A short form derived from Japanese poetry. Strictly it consists of just seventeen *syllables*, disposed across three lines in the pattern 5–7–5. Its subjects are normally resonant, momentary observations, often of the natural world. Translation into English that transposes its strict count is very difficult. However, imitation of the form in English, sometimes strict, sometimes less so, has been very popular since the early twentieth century. (See Chapter 2.)

Half-rhyme   A kind of *rhyme* in which the consonants of the two words sound the same but the vowels differ, e.g. *buck / back*. Sometimes known as *pararhyme*. (See also *rhyme, head-rhyme, eye-rhyme, full rhyme* and Chapter 6.)

Head-rhyme (see *alliteration*) The rhyming of the same consonants at the beginning of successive words, e.g. *big bucks*. As in *alliteration* except that it will apply to only two words, not several.

Hendecasyllabic A poetic line of *eleven syllables*. (See *decasyllabic*.)

Heroic couplet Closed couplets in *iambic pentameter* that employ a *caesura*. Adapted from *classical* models, it was seen as a *metre* suited to large, public subjects. In English it was used most prominently in the later seventeenth and eighteenth centuries.

Hexameter A verse line of *twelve syllables* in *six feet*, normally with a *caesura*. The line is uncommon in English poetry but in *classical* Greek and Latin poetry it was the line for *epic* and other major poetry. Because of its prestige the *vernacular* languages of western Europe sought to find an equivalent. In Italian this was the *hendecasyllabic*, in French the standard *alexandrine*. In English the equivalent standard settled as the *iambic pentameter*. (See Chapter 4.)

Iamb, iambic An iamb is a term derived from classical *prosody*. In Latin it denoted a *metrical foot* consisting of a short *syllable* followed by a long. In English *accentual-syllabic* verse it became the most common foot in the form of an *unstressed* syllable followed by a stressed syllable, notated: ˘ \ e.g. reverse, denounce. The *iambic pentameter* is a line of five iambic feet, and is the most familiar metre in the history of poetry in English. It is for instance the basic metre of the verse of Shakespeare's plays. (See Chapter 4.)

Image, imagery A general term, not confined to poetry. Essentially, in *rhetoric* it is a *metaphorical* device whereby one thing is described in terms of another but with an emphasis upon a mental picture. In *Romantic* poetry it becomes more expressive in itself, coming to stand in place of something not directly described. Thus, for example, when Wordsworth in 'Tintern Abbey' seeks to describe the 'presence', the 'sense sublime' he feels to be in Nature, he writes that 'its dwelling is the light of setting suns', and this acts as an image to try to specify what he apprehends. In this sense an image is close to the more fixed notion of the *symbol*. (See also *allegory*.) Modernist poetry set great store by representing the concrete and sensuous, and thus by the importance of the image. Hence the brief flowering of *imagism*.

Imagism   A tendency in *modernist* poetry, and briefly a movement with an 'Imagist Manifesto', which urged the necessity of concrete, sensuous images as the basis of poetic practice. Its proponents included Ezra Pound and H. D. (Hilda Doolittle). For Pound the image should constitute 'direct treatment of the thing', but he also broadened his definition to 'an image is that which presents an intellectual and emotional complex in an instant of time'.

Inspiration   Literally, from the Latin, 'breathing in'. This source points to the notion that poetic inspiration is natural and unconscious. It refers to the arrival in the poet's mind of ideas, words, images, figures that have no conscious source and result from no discernible craft or effort. In many poetries it has had divine associations – the breath of God or the will of the *Muse* flows through the poet. It is very important to *Romantic poetics*. It can also be associated with the processes of the unconscious mind as understood by modern psychology. (See Chapter 8.)

Interlaced rhyme   *Rhyme* where the correspondences are not adjacent as they are in the *couplet*, but two or more lines apart, therefore producing a criss-cross effect with other rhymes. In the sixteenth and seventeenth centuries there was a vogue for drawing lines, or brackets, to show the pattern of rhymes, hence visualizing the interlacing effect. (See Chapter 6.)

Internal rhyme   A correspondence of word-sounds *within* the line rather than, as in conventional *rhyme*, at the end of lines. (See Chapter 6.)

Irony   A *figure* in which what is *said* is the opposite of what is actually *meant*. The reader either realizes what is true from the beginning, or (hopefully) comes to understand it. *Dramatic irony* is when a speaking character thinks the opposite of what the audience, from its superior vantage point, knows to be true.

Lexis, lexical Vocabulary, but for linguists especially the words that carry substantive meaning, as opposed to those which serve grammatical functions like *the, at, to,* etc. Words can thus be divided into lexical items and grammatical items. (See *diction, semantic, syntax.*)

Limerick A highly popular form of comic verse that features in written and oral traditions. It is often nonsensical and frequently bawdy. Its form is very strict: *five* lines rhyming *aabba*; lines 1, 2 and 5 have *three stresses* and lines 3 and 4 *two*. (See Chapter 2.)

Lyric The type of poetry most readily associated with the chanted or sung origins of poetry, traditionally to the harp-like stringed instrument known as the *lyre*. We still refer to the words of songs of all kinds as lyrics, and poetry closest in style and span to songs, as opposed to poems that tell substantial stories or are the medium for drama, is defined as lyric. The form lends itself to expressions of personal feeling such as love. It is the most prominent of all the poetic genres.

Measure The term for the organization of the poetic line into a recurring set pattern, such as *metre*. Whether the line counts *syllables* or *stresses*, it is regular. More broadly, measure can be understood as the sense of order in other equivalences such as the length and form of *stanzas*. (See Chapter 4.)

Metaphor A broad and complex area. Most simply it can be described as a *figure* which expresses one thing in terms of another by suggesting a likeness between them. There are many different kinds of metaphor, some so embedded that it can be hard to remember that they are figures. For example, if someone is called 'bright as a button' their 'brightness' is being emphasized by the association with a shiny button. But 'bright', meaning clever, is itself a metaphor in which light is taken to stand for intelligence. Another type lies in a phrase like 'give me a hand'. Here 'help' is represented by the limb to represent what may or may not be physical assistance. Metaphors are structured into *tenor, vehicle* and *ground*. In the phrase 'happy as a clam', the tenor is what is being said, i.e. 'X is happy'; the vehicle which carries this meaning is the 'clam'; the ground is what happiness and clams have in common, whatever that is. (See *allegory, conceit, image, symbol, trope*.)

Metre A specific recurring pattern of poetic *rhythm*. Typically in English a metred line will have a set number of *syllables*, or *stresses*, or a combination of stressed and unstressed syllables, i.e. *accentual/stress-syllabic* metre. (See *foot* and Chapter 4.)

Mnemonic That which aids the memory. Many features of poetry, such as *metre*, *rhyme* and repetition of words and phrases, help memorization and are thus mnemonic. This is especially important in the *oral tradition*. (See also *Muse*.)

Modernism In literature and the other arts, a loose, experimental movement in the twentieth century which sought to break with preceding styles. In poetry, modernism made formal challenges to such long-standing features as the verse-line, *rhyme* and *stanza*, initiating '*free verse*'. Also conventional *narrative* coherence was sometimes replaced by jump-cut juxtaposition of incidents without clear time-sequence or conclusion. Further, the assumed single speaking 'I' dissolves into a more elusive voice or voices, sometimes seemingly speaking from the unconscious as well as conscious mind, or from a characterized voice or *persona*.

Monorhyme A poem, or part of a poem, which employs the same *end-rhyme*, throughout, i.e. aaa …

Monosyllable A word comprised of just a single *syllable*. (A *polysyllable* is thus a word made up of two or more syllables.)

Muse The idea of the Muse, or Muses, comes from Greek antiquity. Originally they were daughters of the principal god Zeus and Mnemosyne (goddess of memory; see *mnemonic*) and provided inspiration to artists in different *genres* and to different *sub-genres* within the arts: for instance Calliope was the Muse of *epic* poetry, Erato of love poetry and Euterpe of tragedy and *lyric* poetry. Poets since inspired by *classicism*, like Milton, used the convention of calling upon the Muses for inspiration (see Chapter 3). Much more loosely, and often comically, the Muse is invoked as a *metaphor* for poetic *inspiration* (see Chapter 8). Feminist poets and critics have questioned the mythology of male poets recognizing the feminine in only this way, and reconsidered what 'the Muse' might mean for women poets.

Narrative   A narrative poem is most simply a poem that tells a story, that is a relation of events or facts placed in time so as to suggest causal connection and a pointed conclusion. (See *modernism* and Chapter 3.)

Naturalization   A term sometimes used in the context of whether readers feel they 'know where they are' in a poem, that is with respect to who is 'speaking', what the setting is, what the process of the poem is. For example, naturalization might be more readily achieved in a *narrative* poem like Chaucer's *The Canterbury Tales* than in a *modernist* poem by John Ashbery (see Chapter 3).

Neo-classical   Style that deliberately tries to emulate that of the *classics*, in English poetry most often used of work of the late seventeenth and early eighteenth centuries.

Octave   In poetry, an *eight*-line *stanza* (see *ottava rima*). Also the first *eight*-line section of a *Petrarchan sonnet*.

Ode   A form of lyrical poetry, usually of considerable length, that treats significant subjects such as mortality, and often public events. Its tone is serious and the line and *stanza* forms often elaborate. Its origins are in ancient Greek poetry where its name denoted chanting or singing.

Onomatopoeia   Traditionally this refers to the phenomenon of words sounding like what they mean, e.g. *swoosh, tick-tock*. Linguistically this has been widely discredited, but onomatopoeic effects can be made once the context of the meaning has been established. *The New Princeton Encyclopedia of Poetry and Poetics*: 'Sounds can never precede meaning: they can only operate on meanings already *lexically* created.'

Open couplets   Couplets in which the sentence or clause does not close with the completion of the *rhyme* but runs on into the next line or lines. Thus, unlike in the *closed couplet*, the *syntax* and *couplet* do not coincide; indeed, sentences might end mid-line. (See *couplet*, *end-stopped*, *rhyme* and Chapter 6.)

Oral, oral tradition   Poetry composed to be recited or chanted for a listening audience rather than the printed page. The roots of its tradition, as in the *ballad*, lie in pre-print, or even pre-literate, cultures. Closer to our own day it continues to exist in such fields as children's rhymes and *performance poetry*. (See *ballad* and Chapter 2.)

Ottava rima (ottava Toscana)   An *eight-line stanza* rhyming *ababacc*. As its name and occasional alternative name suggest, it is Italian in origin. (See Chapter 7.)

Pantoum   The Europeanization of the Malay form *pantun*. This is based on a four-line form which rhymes *ab ab*, but it also includes internal rhymes and various kinds of correspondences between images and ideas. (See Chapter 7.)

Parody   An imitation of the style of a work, an author or a type of poem (or other writing) for mocking comic effect. (See *pastiche*.)

Pastiche   Like *parody* an imitation, but the result is meant to be a work in itself, not merely a mockery of its model.

Pastoral   Originally a type of poetry which pretends to imitate the simple songs of shepherds and extol the unaffected virtues of rural life as against metropolitan sophistication and corruption. It can therefore have a satirical or political edge. Under *classical* influence, it was seen as the most basic of forms, thus suited to a beginning poet before attempting the more demanding tasks of writing *lyric* and *epic*. More loosely it has come to refer to any poetry about rural life. (See Chapter 3.)

Performance poetry   Generally any poetry presented to an audience in performance, as opposed to on the page, or indeed *from the*

page. In recent years it has come to refer to poetry of large, enter-
taining verbal effect designed to impress a listening audience, the
poets sometimes in competition with each other in what has
become called a *slam*. (See *oral tradition*.)

Persona   *Classical poetics* made a distinction between poetry in
which the poet speaks in her or his own voice, and poems where it
is understood that the poet has fashioned a character, or mask –
i.e. a persona. For twentieth-century poets like Yeats, Pound and
Eliot, the use of a persona became a way of freeing themselves
from *Romantic* assumptions that identified the speaker in the
poem with the poet. A strand of twentieth-century criticism regularly insists on seeing the 'I' in the poem as a persona rather than
the writer him- or herself. (See *dramatic monologue, modernism*
and Chapter 3.)

Petrarchan   After the Italian poet Francesco Petrarca [English:
Petrarch] (1304–74). The adjective has two main references: (1) to
the type of 8/6 *sonnet* form devised by Petrarch, and (2) to a tradition of love poetry characterized by longing and a virtually
religious devotion to the beloved. This last sense by no means
represents the full character of Petrarch's work.

Phoneme   The most basic items in the sounds of a language.
Phonemes are the sounds actually used by a particular language
that will make a difference to the meaning of a word, e.g. if the
phoneme that is the *k* sound in *cat* is replaced by the *m* phoneme
we have the entirely different meaning of *mat*. (See *syllable* and
Chapter 4.)

Pitch   In speech the contour of intonation, for example the rising
movement in which questions are asked or denials delivered. In
English poetry the patterning of *accent / stress* crucially affects the
pitch and therefore sense of lines. (See also *tone*.)

Plain rhyme   Adjacent *rhyme* such as the *couplet*, or patterns such
as *abbacddc*, i.e. schemes that do not overlap, are not *interlaced*.

Poetics   A very broad term now frequently used to refer to many things other than literature. Regarding poetry, its simplest sense has to do with the theory of poetry, especially assuming that verse can be basically distinguished from prose. More locally it is used to refer to the explicit theory, or practical principles, of some particular poet, or movement, or period in poetry.

Polysyllable   A word consisting of two or more *syllables* (*see monosyllable*).

Praise-song   A song or poem composed to express admiration, usually of a person or deity. Most often used in relation to traditional, *oral* poetries.

Prosody   The study and analysis of verse form, mainly the sound-patterning of *rhyme* and *stanza*, and especially of lines in *metre*. (See *scansion*.)

Quantity / quantitative   English, and therefore poetry in English, is *stress* based. However, other languages, including Greek and Latin, measure the lengths, or quantities, of *syllables*, thus producing lines comprising patterns of these lengths. The terminology of this quantitative *prosody* has been carried over into English usage. (See Chapter 4.)

Quatrain   A *stanza* of *four* lines, normally rhymed. (See Chapter 7.)

Quintet   A term sometimes used to denote a *stanza* of *five* lines.

Recurrence   Quite simply that which recurs in the formal properties of a poem. Thus *metre, rhyme, stanza, alliteration* and all kinds of regular repetitions are instances of recurrence, as is the poetic line itself as it *recurs* to the left-hand margin. Arguably, truly *'free verse'* would have no recurrent features.

Refrain   One or more lines repeated at intervals like a chorus. (See *burden*.)

Register   The choice of words and forms and tones appropriate to speaking to various audiences. (See also *pitch*, *tone*.)

Reiteration   Saying something over again – see *recurrence*.

Renaissance   Literally 're-birth' (from the French; in Italian *rinascimento*). It refers to the 'rebirth' of knowledge of and emulation of the arts, architecture and culture of *classical* Greece and Rome, first in Italy and then more widely in western Europe. The dates ought not to be stipulated too closely, but should stretch at least from Dante (1265–1321) to Shakespeare (1564–1616) and Milton (1608–74). Despite this admiration for the *Classics*, the period sees the development of valued literature in Italian, French, German, English and the other spoken vernacular languages of Europe.

Rhetoric   The art of persuasion in language, first in speech-making but extending into writing. This was seen as a crucial public skill in Ancient Greece and Rome and through the *Renaissance. Classical* rhetoric employed formal *devices*, or *tropes*, with which to enhance argument and hold an audience, and these practices were maintained by Renaissance and *neo-classical* poets. More broadly, rhetoric can still be used to denote any deliberate strategies by which a poet forms the argument of his or her poem. In *Romantic poetics*, however, this idea of poetry as rhetoric – depending for its eloquence upon given forms – tends to be replaced by the notion of expression, personal in both content and form to the author. (See Chapters 3 and 8.)

Rhyme   The positioning of words of similar sound for effect, normally at the ends of lines. There are many different varieties and patterns of rhyme. (See Chapter 6.)

Rhyme royal   A *stanza* of *seven decasyllabic* lines, rhyming *ababbcc*. Its first use in English was by Chaucer in his *Troilus and Criseyde* (?1482) and thus the scheme is sometimes called the Troilus stanza. (See Chapter 7.)

Rhyming slang   A Cockney (London dialect) feature in which things are referred to by a code phrase which bears no relation in meaning but does rhyme, e.g. 'apples and pears' for 'stairs'.

Rhythm The Ancient Greek philosopher Plato called rhythm 'order in movement', and it is generally understood to be the 'flow' in the sounding of the line and the succession of lines. It will therefore include the effects of the sounds of individual words and *beat* or *stress*. There may be a recurring *measure* as in *metrical* verse, but *'free verse'* will also have rhythm.

Romantic A various term with very many shades of meanings and applications which make brief generalization very insecure. Historically the Romantic period in poetry is generally held to begin in English in the late eighteenth century, most notably in the poetry of Blake, Wordsworth and Coleridge, and remain the dominant influence throughout the nineteenth century. This has sometimes been called the 'Romantic revival' on account of its being a rediscovery of the styles of Shakespeare and his contemporaries which had been occluded by the *neo-classical poetics* of the late seventeenth and eighteenth centuries. It is also seen to have rediscovered the forms and stories of the native, vernacular cultures of Europe, for example in the *ballad*, as against the dominant inheritance from Ancient Greece and Rome. This points to the degree to which Romanticism has been defined against *classicism*. Against the formality, respect for ancient models and *rhetorical* styles of neo-classicism, Romanticism has been held to promote formal freedom, independence and personal expressivity: nature as against culture, inspiration as against learning. (See Chapter 8.)

Rondeau Originally a French dance-song for *rondes*, or rounds. Formalized into a literary convention, the *rondeau* became a fifteen-line form divided into a *quintet*, a *quatrain* and a *sestet*, and employing just two rhymes. (See Chapter 7.)

Rondel Another stanza form based on dance-song in the manner of the *rondeau*. It has fourteen lines shaped in two *quatrains* and a *sestet*. The first two lines of the poem recur in the last two lines of the second *quatrain* and in the last two lines of the *sestet*. There is also a little known variant known as a *roundel*. (See *rondeau* and Chapter 7.)

Satire   Not confined to poetry, it is a mode which mocks some prevailing aspect of society or fashion. Its aim is always claimed to be didactic: that is to punish vice, corruption and pride through ridicule in order to improve conduct and society. (See Chapter 3.)

Scansion   The practice of analysing, or *scanning*, lines of verse in order to determine their *rhythmical*, or more usually *metrical*, features. (See *prosody* and Chapter 4.)

Semantic   Those aspects of language which pertain to meaning (see *lexis*). In poetry the semantic function of a word can be especially important as against, or in relation to, the use of words for their sound or sensuous qualities where meaning is secondary or even non-existent. A *semantic field* is the appearance of words in some proximity whose meanings can be associated.

Septet   A *seven-line stanza*, rhyming or unrhymed. (See Chapter 6.)

Sestet   Refers to a *six-line stanza*, but more often to the second *six-line* section of the 8/6 Petrarchan sonnet. (See Chapter 7.)

Sestina   A form of *six six-line stanzas* concluding with *an envoi of three lines*. In its English versions it usually uses a ten-syllable line. However, instead of a rhyme scheme, the sestina repeats a series of six *end-words* in each stanza, but in a fixed pattern of variation in which the sixth moves up to first in the next stanza and the others take up other corresponding positions. The three-line envoi then contains all the six repeated words. (See Chapter 7.)

Simile   A basic form of *metaphor* in which the comparison is directly conjoined, usually with 'like' or 'as', e.g. 'black as coal'.

Slam   A competition in which *performance poets* compete, with the audience voting for the winner.

Sonnet   A major, long-lived lyrical form consisting of *fourteen* lines. Strictly, and most often, these are configured in one of several different rhyming patterns. The major ones in English are the *Petrarchan* model divided into sections of 8/6 lines and the

Shakespearean in 4/4/4/2. The standard line is *iambic pentameter*. More recent sonnets, or 'sonnets', have dispensed with rhyme and pentameter, and sometimes with fourteen lines. (See Chapter 7.)

Sound poem A poem which makes no attempt to use words *semantically* but attends only to the sounds of the words, or sometimes does not use recognizable words at all. Its effects might be *onomatopoeic*. (See also *concrete poetry* and Chapter 2.)

Spenserian stanza After Edmund Spenser (?1552–99), it comprises nine *iambic* lines, eight iambic *pentameters* and one closing *hexameter*. The rhyme scheme is *ababbcbcc*. (See Chapter 7.)

Sprung rhythm A term coined by Gerard Manley Hopkins (1844–89) to describe his variation on *strong-stress metre*. His long lines are built around a few strong *accents* with a varying number of unstressed *syllables* between them. The *rhythm* thus 'springs' across from one strong *beat* to the next. (See Chapter 4.)

Stanza A group of lines shaped in the same way, with the lines usually, although not always, of the same length. Traditionally they would be rhymed, but by no means always, especially in the twentieth century. Stanzas can vary greatly in length and structure. They serve the function of segmenting the poem and providing pauses in its progression. (See Chapter 7.)

Stichic The name given to a series of lines in which the grammatical sentence and the line coincide, i.e. there is no run-on or *enjambment* (see *strophic*). This produces an abrupt, staccato effect.

Stress The effect in all sorts of poetry in English, whether it has measure or is 'free', relies upon the effective placing of *accent* or *beat*, e.g. '**mass**-ive'. A poem therefore will have beats (i.e. stresses, accents) but also an overall beat, or rhythm.

Stress-syllable Also *accentual-syllabic*, a metre in which *accented* and non-accented (stressed/unstressed) *syllables* alternate. They can be either *unstressed/stressed* (*iambic*) or stressed/unstressed (*trochaic*), or in other patterns (see *anapest, dactyl,* Chapter 4).

Strong-stress metre   A metre which depends upon a fixed series of strong *beats* without counting the number of unstressed *syllables* in between. (See *sprung rhythm* and Chapter 4.)

Strophic   The common, though with respect to *classical* models, loose meaning is of a series of lines in which the grammatical sentence flows on across the line-endings (see *stichic*). This is often a feature of the *verse paragraph*.

Syllabics   Measured lines which count *syllables*, not *stresses*.

Syllable   The segment of a word uttered with a single effort of articulation, e.g. *seg-ment* (two syllables), *ar-tic-u-la-tion* (five syllables). It is syllables that bear *stress*. (See *phoneme, syllabics* and Chapter 4.)

Symbol   This is part of the same family as *allegory, figure, image* and *metaphor*: that is a symbol is something that stands in for, or represents something else. Thus the colour red conventionally symbolizes danger. Symbols evolve, or are invented, because the 'something else' is usually a complex idea or emotion, an abstraction not easily expressible. The distinctions between symbol and *image* etc. are not wholly clear, but the tendency is for symbols to have more fixed significations.

Syntax   The arrangement of words into sentences by the given rules of the language in order to create meaning. (See *lexis, semantics*).

Tenor   (1) The tone, sometimes style of writing or speaking, often related to the level of formality (see *pitch, register* and Chapter 3).
    (2) In *metaphor*, tenor refers to *what* is being signified, *vehicle* to the *image* being used to signify, and *ground* to that which *tenor* and *vehicle* are seen to have in common. (See *metaphor*.)

Tercet   A *three-line* verse form, usually rhymed and usually employed as a separate *stanza*.

Terza rima   A verse form which links *tercets* together by *interlaced rhyme: aba bcb cdc ded,* etc. A form introduced by Dante for his *Divine Comedy*. (See Chapter 7.)

Tetrameter   An *accentual-syllabic* line of *four beats*, hence four *feet*. (See Chapter 4.)

Tone   A much-used if imprecise term to indicate the 'mood', 'colour', 'atmosphere' of a piece of writing as conveyed by its word-choice, rhythms, etc. An analogy can be made with the tones of a speaking voice, which is what is done in Chapter 3 of this book. (See also *pitch, tenor.*)

Transmission   The process by which poetry is conveyed to audiences through time. The term is most often used in the context of *oral* transmission of poetry and song, like the *ballad*, which is not written down.

Trimeter   An *accentual-syllabic* line of *three beats*, hence three *feet*. (See Chapter 4.)

Triolet   A fixed *stanza* form derived from French. It has eight lines of which the first is repeated three times, the second twice, and there are just two rhymes i.e. *AbaAabAB* (the capital letter denoting the repeated lines). (See Chapter 7.)

Triplet   A sequence or *stanza* of *three* rhyming lines.

Trochee, trochaic   A *metrical foot* consisting of *two syllables*, the first *stressed* the second unstressed, notated: \ ˇ. (In contrast see *iamb* and Chapter 4.)

Trope   A *figure* or *metaphorical* device, usually as part of deliberate *rhetoric*, where an *image* is standing for something else. The term is often used to refer to an extended metaphor where the scheme of likeness is carried on through several related images. (See also *allegory*, *symbol*.)

Troubadour   Poet-songwriters of eleventh- and twelfth-century southern France. Their *lyrics* were mostly on the themes of love, especially unrequited love for a superior or unattainable beloved. Although the men are best known, there were women troubadours. Their themes and styles had great influence on later *Renaissance* love poetry, and reappeared in the interest of some modern poets, notably Ezra Pound. (See also *Petrarchan*.)

Verse / poetry   (1) The terms 'verse' and 'poetry' are often used interchangeably, although this book is using 'poetry' almost exclusively. The use of verse often implies poetry that is formal, usually *metrical* and *rhymed*. It is sometimes seen as an old-fashioned term, or one to be used for simpler, less serious poetry, as in 'children's verse' or 'comic verse'.
(2) A verse can also refer to a single line (the French *vers*), or a number of lines (*verses*), or as another word for *stanza*, especially if it is one that is short and comparatively straightforward like a *quatrain*.

Verse paragraph   A *stanza* form, but one that does not have a recurrent length or other set shape. It is thus a looser form, and although it does occur in longer *couplet* poems, it is mostly used in *blank verse*. (See *strophic* and Chapter 7.)

Villanelle   Originally a simple Italian and French 'rustic' song. The modern villanelle has a nineteen-line pattern that uses *five tercets* and a *final quatrain*. Strictly, these rhyme *aba* throughout, and the first and third lines recur at fixed points later in the poem. (See Chapter 7.)

Volta   The Italian word for *turn*, used for the moment when the 8/6 *Petrarchan sonnet* changes over from the *octet* to the *sestet*. It is a 'turn' in the rhyme scheme, almost always corresponding with a sentence pause and a 'turn of thought'. (See Chapter 7.)

Word-play A very general term to suggest the manipulation of words, especially when used for the sounds in themselves, or their various *connotations*. As 'play' suggests, there is often something sportive or *ludic* about this.

BIBLIOGRAPHY

ANTHOLOGIES

Several new anthologies of poetry in English have been published in recent years which have been re-edited from manuscripts, widened the range of their selection and provide excellent supporting information. Among these are:

Ferguson, M., Salter, M. J. and Stallworthy, J. (eds) (1996) *The Norton Anthology of Poetry*, New York: Norton. See essay 'Versification' by Jon Stallworthy.

Pearsall, D. (ed.) (1998) *Chaucer to Spenser: An Anthology*, Oxford: Blackwell.

Payne, M. and Hunter, J. (eds) (2003) *Renaissance Literature: An Anthology*, Oxford, Blackwell.

DeMaria, R. (ed.) (1996) *British Literature 1640–1789: An Anthology*, Oxford: Blackwell.

Fairer, D. and Gerrard, C. (eds) (2003) *Eighteenth-Century English Poetry: An Annotated Anthology*, second edition, Oxford: Blackwell.

Lonsdale, R. (ed.) (1990) *Eighteenth-Century Women Poets*, Oxford: Oxford University Press.

Wu, D. (ed.) (1998) *Romanticism: An Anthology*, second edition, Oxford: Blackwell.

—— (ed.) (2001) *Romantic Women Poets: An Anthology*, Oxford: Blackwell.

Schmidt, M. (ed.) (1999) *The Harvill Book of Twentieth-Century Poetry in English*, London: Harvill Press.

Leonard, J. (ed.) (1998) *Australian Verse: An Oxford Anthology*, Melbourne: Oxford University Press.

Finnegan, R. (ed.) (1982) *The Penguin Book of Oral Poetry*, Harmondsworth: Penguin Books.

Heaney, S. and Hughes, T. (eds) (1982) *The Rattle Bag*, London: Faber and Faber.

The premier reference work in English for world poetry is:

Preminger, A. and Brogan, T. V. F. (eds) (1993) *The New Princeton Encyclopedia of Poetry and Poetics*, Princeton: Princeton University Press.

Alexander, M. (ed.) (1995) *Beowulf*, Harmondsworth: Penguin Classics.

Allen, D. (ed.) (1960) *The New American Poetry 1945–1960*, New York: Grove Press.

Arkell, D. (1979) *Looking for Laforgue: An Informal Biography*, Manchester: Carcanet Press.

Ashbery, J. (1985) *Selected Poems*, Manchester: Carcanet Press.

Attridge, D. (1982) *The Rhythms of English Poetry*, London and New York: Longman.

—— (1995) *Poetic Rhythm: An Introduction*, Cambridge: Cambridge University Press.

Auden, W. H. (1966) *Collected Shorter Poems 1927–57*, London: Faber and Faber.

Barrell, J. (1988) *Poetry, Language and Politics*, Manchester: Manchester University Press.

Bate, W. J. (1971) *The Burden of the Past and the English Poet*, London: Chatto and Windus.

Bernstein, C. (1986) 'Writing and Method' in R. Silliman (ed.) *In the American Tree*, Orono, Maine: National Poetry Foundation.

Berryman, J. (1973) *The Freedom of the Poet*, New York: Farrar, Strauss & Giroux.

—— (1990) *Collected Poems 1937–71*, London: Faber and Faber.

Blake, W. (1961) *Poetry and Prose of William Blake*, fourth edition, ed. G. Keynes, London: Nonesuch Press.

Bloom, V. (1983) *Touch Mi, Tell Mi*, London: Bogle-L'Ouverture.

Boland, E. (1990) *Outside History*, Manchester: Carcanet Press.

—— (1995) *Object Lessons: The Life of the Woman and the Poet in Our Time*, Manchester: Carcanet Press.

Boland, E. and Strand, M. (eds) (2001) *The Making of a Poem*, London and New York: Norton.

Brathwaite, E. K. (1973) *The Arrivants: A New World Trilogy*, Oxford: Oxford University Press.

—— (1995) *History of the Voice: The Development of Nation Language in Anglophone Caribbean Literature*, New York: New Beacon Books.

Brogan, T. V. F. (ed.) (1994) *The New Princeton Handbook of Poetic Terms*, Princeton: Princeton University Press.

Browning, R. (1974) *Selected Poems*, ed. I. Armstrong, London: Bell.

Cave, N. (1997) *The Boatman's Call*, London: Mute Records.

Chaucer, G. (1957) *The Works of Geoffrey Chaucer*, ed. F. N. Robinson, London: Oxford University Press.

Cixous, H. ([1976] 1981) 'The Laugh of the Medusa', tr. Keith Cohen and Paula Cohen, in Elaine Marks and Isabelle de Courtivon (eds) *New French Feminisms*, Brighton: Harvester Press. Reprinted in D. Walder (ed.) (1990) *Literature in the Modern World: Critical Essays and Documents*, Oxford: Open University/Oxford University Press.

Cottle, B. (1969) *The Triumph of English 1350–1400*, London: Blandford Press.

Davies, R. T. (1963) *Mediaeval English Lyrics: A Critical Anthology*, London: Faber and Faber.

Dennett, D. C. (1984) *Elbow Room: The Varieties of Free Will Worth Wanting*, Oxford: Clarendon Press.

Dickinson, E. (1951) *The Complete Poems of Emily Dickinson*, Boston, London: Little, Brown.

Douglas, K. (1966) *Collected Poems*, London: Faber and Faber.

Duffy, C. A. (1999) *The World's Wife*, London: Picador.

Easthope, A. (1983) *Poetry as Discourse*, London: Methuen.

Eliot, T. S. (1974) *Collected Poems 1909–62*, London: Faber and Faber.

—— (1999) *Selected Essays*, London: Faber and Faber.

Emerson, R. W. (1987) *The Essays of Ralph Waldo Emerson*, Boston: Harvard University Press.

Empson, W. (1961) *Seven Types of Ambiguity*, London: Penguin Books.

Ewart, G. (ed.) (1980) *The Penguin Book of Comic Verse*, Harmondsworth: Penguin Books.

Fenton, J. (1983) 'Letter to John Fuller' in *The Memory of War and Children in Exile, Poems 1968–1983*, Harmondsworth: Penguin Books.

—— (2002) *An Introduction to English Poetry*, London: Viking.

Finley, M. I. (1979) *The World of Odysseus*, revised edition, Harmondsworth: Pelican.

Furniss, T. and Bath, M. (1996) *Reading Poetry: An Introduction*, Hemel Hempstead: Prentice Hall/Harvester Wheatsheaf.

Fussell, P. (1979) *Poetic Metre and Poetic Form*, New York: Random House.

Graham, J. (1987) *The End of Beauty*, New York: The Ecco Press.

Graham, W. S. (1979) *Collected Poems 1942–1977*, London: Faber and Faber.

Greenfield, S. (2000) *The Private Life of the Brain*, London: Allen Lane, Penguin Books.

Grenier, R. (1986) 'On Speech' in R. Silliman (ed.) *In the American Tree*, Orono, Maine: National Poetry Foundation.

Grigson, G. (1975) *The Penguin Book of Ballads*, Harmondsworth: Penguin Books.

Gross, H. (ed.) (1979) *The Structure of Verse, Modern Essays on Prosody*, New York: The Ecco Press.

Gunn, T. (1961) *My Sad Captains*, London: Faber and Faber.

—— (1993) *Shelf Life: Essays, Memoirs and an Interview*, Ann Arbor: University of Michigan Press.

Gurney, I. (1982) *Collected Poems of Ivor Gurney*, ed. P. J. Kavanagh, Oxford: Oxford University Press.

Hall, D. (1991) *To Read a Poem*, Orlando: Harcourt, Brace, Jovanovich.

Hannah, S. (1995) *The Hero and the Girl Next Door*, Manchester: Carcanet Press.

Hardy, T. (1976) *The Complete Poems of Thomas Hardy*, ed. James Gibson, London: Macmillan.

Harrison, T. (1984) *Selected Poems,* Harmondsworth: Penguin Books.

Hartman, C. O. (1980) *Free Verse, An Essay on Prosody*, Princeton: Princeton University Press.

Harwood, G. (1991) *Collected Poems*, Oxford: Oxford University Press.

Herbert, W. N. and Hollis, M. (eds) (2000) *Strong Words, Modern Poets on Modern Poetry*, Tarset: Bloodaxe.

Hill, G. (1998) *The Triumph of Love*, Harmondsworth: Penguin Books.

—— (2000) *Speech! Speech!*, Harmondsworth: Penguin Books.

Hollander, J. (1989) *Rhyme's Reason: A Guide to English Verse*, new edition, London: Yale University Press.

Holmes, R. (1989) *Coleridge: Early Visions*, London: Hodder & Stoughton.

Holub, M. (1990) *The Dimensions of the Present Moment*, tr. D. Young, London: Faber and Faber.

Hope, A. D. (1986) *Selected Poems*, Manchester: Carcanet Press.

Hopkins, G. M. (1970) *The Poems of Gerard Manley Hopkins*, ed. W. H. Gardner and N. H. Mackenzie, Oxford: Oxford University Press.

—— (1980) *Selected Prose*, ed. G. Roberts, Oxford: Oxford University Press.

Jakobson, R. (1987) *Language in Literature*, ed. K. Pomorska and S. Rudy, Cambridge, Mass.: Harvard University Press.

Jespersen, O. (1979) 'Notes on Metre' in H. Gross (ed.) *The Structure of Verse, Modern Essays on Prosody*, New York: The Ecco Press.

Keats, J. (1975) *Letters of John Keats, A Selection*, ed. R. Gittings, Oxford: Oxford University Press.

Kennedy, X. J. and Gioa, D. (1998) *An Introduction to Poetry*, ninth edition, New York: Longman.

Kinsella, T. (1996) *Collected Poems*, Oxford: Oxford University Press.

Koch, K. (1998) *Making Your Own Days: The Pleasures of Reading and Writing Poetry*, New York: Simon & Schuster.

Kostelanetz, R. (ed.) (1980) *Text-Sound Texts*, New York: Morrow.

Larkin, Philip (1988, 2003) *Collected Poems*, London: Faber and Faber.

Lawrence, D. H. (1957) *The Complete Poems of D. H. Lawrence*, London: Heinemann.

Leech, G. N. (1969) *A Linguistic Guide to English Poetry*, Harlow: Longman.

Lennard, J. (1996) *The Poetry Handbook: A Guide to Reading Poetry for Pleasure and Practical Criticism*, Oxford: Oxford University Press.

Lowell, R. (1950) *Poems 1938–1949*, London: Faber and Faber.

—— (1973) *The Dolphin*, London: Faber and Faber.

Mallarmé, S. (1956) *Selected Prose Poems, Essays and Letters*, tr. Bradford Cook, Baltimore: Johns Hopkins University Press.

Mayes, F. (1987) *The Discovery of Poetry*, Orlando: Harcourt, Brace, Jovanovich.

Millay, E. St Vincent (1992) *Selected Poems*, Manchester: Carcanet Press.

Miller, J. H. (ed.) (1966) *William Carlos Williams: A Collection of Critical Essays*, Englewood Cliffs: Prentice Hall.

Milton, J. (1971) *Paradise Lost*, ed. A. Fowler, Harlow: Longman.

Montefiore, J. (1987) *Feminism and Poetry, Language, Experience, Identity in Women's Writing*, London and New York: Pandora.

Moore, M. (1984) *Complete Poems*, London: Faber and Faber.

Morgan, E. (1985) *Collected Poems*, Manchester: Carcanet Press.

Murphy, F. (ed.) (1969) *Walt Whitman: A Critical Anthology*, Harmondsworth: Penguin Books.

O'Hara, F. (1991) *Selected Poems*, ed. D. Allen, Manchester: Carcanet Press.

Olson, C. (1993) *Selected Poems*, University of California Press. See also D. Allen (ed.) (1960) *The New American Poetry 1945–1960*, New York: Grove Press.

Opie, I. and P. (1959) *The Lore and Language of Schoolchildren*, Oxford: Oxford University Press.

Ostriker, A. S. (1986) *Stealing the Language: The Emergence of Women's Poetry in America*, London: The Women's Press.

Pinsky, R. (1988) *Poetry and the World*, New York: The Ecco Press.

—— (2002) *Democracy, Culture and the Voice of Poetry*, Princeton: Princeton University Press.

Porter, Cole (1977) *The Best of Cole Porter*, London: Chappell.

Pound, E. (1954) *Literary Essays*, London: Faber and Faber.

—— (1970) *The Translations of Ezra Pound*, London: Faber and Faber.

Raworth, T. (2003) *Collected Poems*, Manchester: Carcanet Press.

Rees-Jones, D. (2004) *Consorting with Angels: Modern Women Poets*, Tarset: Bloodaxe.

Rich, A. (1980) *On Lies, Secrets, and Silence: Selected Prose 1966–1978*, London: Virago.

Ricks, C. (1984) *The Force of Poetry*, Oxford: Oxford University Press.

Roethke, Theodore (1965) *The Collected Verse of Theodore Roethke: Words for the Wind*, Bloomington: Indiana University Press.

Rothenberg, J. (ed.) (1968) *Technicians of the Sacred: A Range of Poetries from Africa, America, Asia and Oceania*, New York: Vintage Books.

Rothenberg, J. and Quasha, G. (eds) (1974) *America: A Prophecy. A New Reading of American Poetry from Pre-Columbian Times to the Present*, New York: Vintage Books.

Schackenberg, G. (1986) *The Lamplit Answer*, London: Hutchinson.

Schmidt, M. (1998) *Lives of the Poets*, London: Weidenfeld & Nicolson.

Sexton, A. (1964) *Selected Poems*, London: Oxford University Press.

Shakespeare, W. (1986) *The Sonnets and A Lover's Complaint*, ed. J. Kerrigan, Harmondsworth: Penguin Books.

Silliman, R. (ed) (1986) *In the American Tree*, Orono, Maine: National Poetry Foundation.

Smart, C. (1980) *Jubilate Agno (Fragment B)*, ed. K. Williamson, Oxford: Clarendon Press.

Snyder, G. (1966) *Collected Poems*, London: Fulcrum Press.

Stead, C. K. (1967) *The New Poetic: Yeats to Eliot*, Harmondsworth: Pelican.

Stein, G. (1971) *Look At Me Now and Here I Am: Writings and Lectures 1909–45*, Harmondsworth: Penguin Books.

Stevens, W. (1997) *Collected Poetry and Prose*, New York: The Library of America.

Tedlock, J. (1972) *Finding the Center: Narrative Poetry of the Zuni Indians*, New York: The Dial Press.

Thomas, D. (1952) *Collected Poems 1934–52*, London: J. M. Dent.

Thomas, R. S. (1963) *The Bread of Truth*, London: Hart Davis.

—— (2000) *Collected Poems 1945–90*, London: Phoenix.

Vickers, B. (1988) *In Defence of Rhetoric*, Oxford: The Clarendon Press.

Walder, D. (ed.) (1990) *Literature in the Modern World: Critical Essays and Documents*, Oxford: The Open University/Oxford University Press.

Welton, M. (1999) in Michael Schmidt (ed.) *New Poetries II: An Anthology*, Manchester: Carcanet Press.

Wesling, D. (1980) *The Chances of Rhyme, Device and Modernity*, Berkeley: University of California Press.

Whitman, W. (1969) *Walt Whitman: A Critical Anthology*, ed. F. Murphy, Harmondsworth: Penguin Books.

—— (1975) *The Complete Poems*, ed. F. Murphy, Harmondsworth: Penguin Books.

Williams, W. C. (1958, 1969) *I Wanted to Write a Poem*, Boston: Beacon Press (1958 edition); London: Cape (1969 edition).

—— (1962) *Pictures from Breughel and Other Poems*, New York: New Directions.

—— (1966) *William Carlos Williams: A Collection of Critical Essays*, ed. J. H. Miller, Engelwood Cliffs: Prentice Hall.

—— (1972) *William Carlos Williams: Penguin Critical Anthology*, ed. C. Tomlinson, Harmondsworth: Penguin Books.

—— (1976) *William Carlos Williams: Selected Poems*, ed. C. Tomlinson, Harmondsworth: Penguin Books.

—— (1988) *Collected Poems*, Manchester: Carcanet Press.

Woof, P. and Harley, M. (2002) *Wordsworth and the Daffodils*, Grasmere: The Wordsworth Trust.

Wordsworth, W. (1977) *Poems Volume I*, ed. J. O. Hayden, Harmondsworth: Penguin Books.

Wright, J. (1992) *Collected Poems*, Manchester: Carcanet Press.

Yeats, W. B. (1966) *Collected Poems*, London: Macmillan.

INDEX

Page references for the glossary are in **bold**